Trends in Logic

Volume 70

TRENDS IN LOGIC
Studia Logica Library

VOLUME 70

Series Editor

Heinrich Wansing, *Department of Philosophy, Ruhr University Bochum, Bochum, Germany*

Editorial Board Members

Arnon Avron, *Department of Computer Science, University of Tel Aviv, Tel Aviv, Israel*
Katalin Bimbó, *Department of Philosophy, University of Alberta, Edmonton, Canada*
Giovanna Corsi, *Department of Philosophy, University of Bologna, Bologna, Italy*
Janusz Czelakowski, *Institute of Mathematics and Informatics, University of Opole, Opole, Poland*
Roberto Giuntini, *Department of Philosophy, University of Cagliari, Cagliari, Italy*
Rajeev Goré, *Australian National University, Canberra, Australia*
Andreas Herzig, *IRIT, University of Toulouse, Toulouse, France*
Wesley Holliday, *UC Berkeley, Lafayette, USA*
Andrzej Indrzejczak, *Department of Logic, University of Lódz, Lódz, Poland*
Daniele Mundici, *Mathematics and Computer Science, University of Florence, Firenze, Italy*
Sergei Odintsov, *Sobolev Institute of Mathematics, Novosibirsk, Russia*
Ewa Orlowska, *Institute of Telecommunications, Warsaw, Poland*
Peter Schroeder-Heister, *Wilhelm-Schickard-Institut, Universität Tübingen, Tübingen, Germany*
Yde Venema, *ILLC, Universiteit van Amsterdam, Amsterdam, The Netherlands*
Andreas Weiermann, *Vakgroep Zuivere Wiskunde en Computeralgebra, University of Ghent, Ghent, Belgium*
Frank Wolter, *Department of Computing, University of Liverpool, Liverpool, UK*
Ming Xu, *Department of Philosophy, Wuhan University, Wuhan, China*
Jacek Malinowski, *Institute of Philosophy and Sociology, Polish Academy of Sciences, Warszawa, Poland*

Assistant Editor

Daniel Skurt, *Ruhr-University Bochum, Bochum, Germany*

Founding Editor

Ryszard Wojcicki, *Institute of Philosophy and Sociology, Polish Academy of Sciences, Warsaw, Poland*

The book series Trends in Logic covers essentially the same areas as the journal Studia Logica, that is, contemporary formal logic and its applications and relations to other disciplines. The series aims at publishing monographs and thematically coherent volumes dealing with important developments in logic and presenting significant contributions to logical research.

Volumes of Trends in Logic may range from highly focused studies to presentations that make a subject accessible to a broader scientific community or offer new perspectives for research. The series is open to contributions devoted to topics ranging from algebraic logic, model theory, proof theory, philosophical logic, non-classical logic, and logic in computer science to mathematical linguistics and formal epistemology. This thematic spectrum is also reflected in the editorial board of Trends in Logic. Volumes may be devoted to specific logical systems, particular methods and techniques, fundamental concepts, challenging open problems, different approaches to logical consequence, combinations of logics, classes of algebras or other structures, or interconnections between various logic-related domains.

This book series is indexed in SCOPUS.

Authors interested in proposing a completed book or a manuscript in progress or in conception can contact either chris.wilby@springer.com or one of the Editors of the Series.

Davide Rizza

Model Theory: The Algebraic Basics

Springer

Davide Rizza
Arts and Humanities Building
University of East Anglia
Norwich, UK

ISSN 1572-6126   ISSN 2212-7313   (electronic)
Trends in Logic
ISBN 978-3-032-03795-4   ISBN 978-3-032-03796-1   (eBook)
https://doi.org/10.1007/978-3-032-03796-1

© The Editor(s) (if applicable) and The Author(s), under exclusive license to Springer Nature Switzerland AG 2025

This work is subject to copyright. All rights are solely and exclusively licensed by the Publisher, whether the whole or part of the material is concerned, specifically the rights of translation, reprinting, reuse of illustrations, recitation, broadcasting, reproduction on microfilms or in any other physical way, and transmission or information storage and retrieval, electronic adaptation, computer software, or by similar or dissimilar methodology now known or hereafter developed.
The use of general descriptive names, registered names, trademarks, service marks, etc. in this publication does not imply, even in the absence of a specific statement, that such names are exempt from the relevant protective laws and regulations and therefore free for general use.
The publisher, the authors and the editors are safe to assume that the advice and information in this book are believed to be true and accurate at the date of publication. Neither the publisher nor the authors or the editors give a warranty, expressed or implied, with respect to the material contained herein or for any errors or omissions that may have been made. The publisher remains neutral with regard to jurisdictional claims in published maps and institutional affiliations.

This Springer imprint is published by the registered company Springer Nature Switzerland AG
The registered company address is: Gewerbestrasse 11, 6330 Cham, Switzerland

If disposing of this product, please recycle the paper.

Preface

This book offers a gentle introduction to model theory that stresses basic applications to algebra. Although the book might be used by any mathematically minded reader as a first point of contact with this area of mathematical logic, its main motivation is to provide philosophers with an entry point into model theory that should enable them to achieve the following main goals:

(1) To understand Tarski's classic results concerning algebraically closed and real closed fields
(2) To appreciate the deep interplay between logic and other areas of mathematics or mathematical enquiry that model theory has made possible (in this book connections with field theory, basic algebraic geometry, measurement theory and voting theory are discussed).

It is perhaps unnecessary to explain why the first goal is worthwhile. Tarski's results are not only fundamental to the later development of model theory but they are also logical milestones, comparable to Gödel's incompleteness theorems. The main reason why the latter, but not the former, are firmly set within the logic curriculum is that Gödel's results do not rest on any sophisticated number theory, whereas Tarski's call for algebraic background knowledge that no standard logic course will cover. Existing introductions to model theory conform to this omission, while more advanced texts can take the needed background for granted, or briskly summarise it. Either approach makes important model-theoretic content impervious to the beginner with a philosophical background.

This book is an attempt to overcome the difficulty by offering an integrated discussion of algebraic and model-theoretic topics, where the needed algebraic ideas and results are developed as the discussion of model-theoretic topics progresses. Moreover, algebraic concepts are viewed, whenever possible, from a logical standpoint (e.g. polynomials are defined as terms in a particular formal language, quotient structures are introduced by means of interpretations).

This approach has one distinctive advantage beyond bringing within reach contributions to mathematical logic often neglected by philosophers: it provides several examples of what may be called a logical analysis of mathematical content. Many model-theoretic results show certain standard mathematical notions or properties to be specialisations of more abstract, logical concepts. For instance, the notion of an existentially closed structure specialises to that of an algebraically closed fields when restricted to the class of fields; similarly, the notions of linear independence in vector spaces and algebraic independence in fields are concrete instances of an abstract, logical notion of independence, and so on.

The interplay between model-theory and algebra is mathematically significant because it opens the possibility of using logical techniques to tackle algebraic problems, while also promoting new ways of studying algebraic objects and leading to the formulation of new questions about them.

These phenomena are of special philosophical interest because they illustrate in a concrete manner the process of mathematical concept formation, as well as its effects on the reconstruction and expansion of a field of enquiry. Such themes play an important role in the understanding of mathematical practice, a philosophical concern that has been growing over the past two decades, and has found a first important connection with model theory in the recent monograph by John Baldwin (see [6]).

What has been said speaks to the relevance of the second goal aforementioned. Any philosophical study of mathematical practice must be alert to the fact that mathematical material does not merely grow by accumulation but also, and perhaps more importantly, through the intensification of connections between distinct areas of investigation. To learn model theory along the lines proposed in this book is not only to follow its interrelationship with algebraic ideas but also to see its ramifications towards geometric content (especially algebraic and real algebraic geometry) and beyond pure mathematics.

The material thus presented, however basic, is hoped to help expand philosophical study in a direction that includes model theory as a fruitful starting point for reflection on the growth, articulation and refinement of mathematical ideas.

A few remarks on the book's content and some ways of using it may now be apposite. One special feature of the approach followed is that it seeks to minimise assumed knowledge. To this end, a short, preliminary chapter summarises the amount of basic set theory presupposed throughout the text. The more demanding theory of ordinal and cardinal numbers, is thoroughly covered in the appendix, which, however, it is necessary to have mastered only from Chap. 7.

The chapter series proper opens with a very gentle introduction to the key algebraic objects discussed in the remainder of the book, namely, groups, vector spaces and fields. This introduction is not given in the form of an abstract presentation, but is structured around a concrete decision-making problem, whose solution leads to the employment of some concrete structures. The unusual perspective adopted is hoped to help the reader unacquainted with algebra make a first, explicitly motivated contact with it. The decision-making problem from Chap. 1 is briefly returned to in the first section of Chap. 5.

Chapter 2 marks the beginning of logical development. The key topics are first-order syntax and semantics, induction on the complexity of terms and formulae, definability and interpretations.

Chapter 3 provides a brief overview of the general interplay between theories and models, followed by the axiomatic formulations of most of the formal theories studied in the following chapters.

Chapter 4 discusses morphisms between structures and diagrams. A philosophical appendix presents Robinson's model-theoretic account of potential truth.

Chapter 5 is devoted to ultraproducts and their applications. After a preliminary section that motivates the abstract notion of ultrafilter in terms of the electoral idea of a 'decisive coalition', the definition of ultraproduct is given and Łos' theorem proved. Ultraproducts are then used to construct the real field, to develop a small amount of Nonstandard Analysis and finally to prove the compactness theorem and characterise elementary classes. An appendix discusses a little known but very interesting application of filters and ultrafilters to philosophy, due to C.I. Lewis.

Chapter 6 covers essential algebraic background concerning fields, which plays a central role in chapters 7 to 10. The axiomatisation of the class of algebraically closed fields in a fixed characteristic, central to the next two chapters, is given here.

Chapter 7 opens with proofs of the Löwenheim-Skolem theorems, followed by Vaught's test for the completeness of a theory. The test is then applied to theories of dense linear orders, the theory of infinite vector spaces over a fixed field, the theory of torsion-free, divisible abelian groups and, finally, the theories of algebraically closed fields in a fixed characteristic. A first discussion of the model-theoretic notion of algebraic closure concludes the chapter.

Chapter 8 proves quantifier elimination, and thus model completeness, for the theories discussed in Chap. 7. The proofs are applications of several quantifier elimination tests. The notions of strongly minimal and o-minimal theory are introduced through familiar applications of quantifier elimination. The chapter closes with a discussion of the abstract dimension theory supported by a minimal set.

Chapter 9 revolves around the concept of type and discusses the key properties of atomic and saturated structures, as well as including fundamental results like the Ryll-Nardzewski's characterisation of $\aleph_0$-categorical theories and the Omitting Types theorem. As an aside, Sect. 9.3 is devoted to an application of types to measurement theory (a deeper connection between model theory and measurement theory is studied in Chap. 12).

Chapter 10 builds on the previous model-theoretic study of algebraically closed fields. It includes a characterisation of algebraically closed fields as the infinite fields with quantifier elimination and develops the rudiments of algebraic geometry in affine space, including a model-theoretic proof of Hilbert's *Nullstellensatz*. The last two sections of Chap. 10 discuss respectively Morley rank and the equivalence, in the context of algebraic geometry, between this model-theoretic notion and that of the dimension of an algebraic variety.

Chapter 11 is a self-contained introduction to the theories of real closed fields and ordered real closed fields. Their key model-theoretic properties are proved and used to obtain classic results of real algebraic geometry like the Artin-Lang theorem and Milnor's Curve Selection theorem.

Chapter 12 is about Fraïssé limits. The first half of the chapter proves their existence and uniqueness, as well as quantifier elimination and saturation in the case of finite relational languages. The second half of this chapter contains an important application of Fraïssé limits to the classification of measurement scales.

A standard introduction to basic model theory would cover Chaps. 2 to 9, and include Chap. 1 only if students have to gain familiarity with thinking about structures. Chapter 2 may be skimmed by a reader who has taken a first course in logic, except possibly for its final section on definability and interpretations. Chapters 10 to 12 are wholly independent of each other and may be selected or omitted at will, but all rely on material covered in the first nine chapters.

Norwich, UK   Davide Rizza

Acknowledgments

I would like to record my gratitude to Silvio Bozzi, who read a very early draft of this book and offered extremely helpful suggestions for improvement.

I am very grateful to Anna Dmitrieva, Lorna Gregory and Jonathan Kirby for helping me understand model theory.

Finally, I wish to express my thanks to Paolo Mancosu and Marco Panza for their encouragement.

Any of the book's shortcomings remain of course my sole responsibility.

D.R.

Preliminaries

In the following chapters we shall need various ideas and results from set theory. Only a very modest amount of it is required to follow the first six chapters (with the exception of the appendix to Chap. 2 and the final theorem of Sect. 4.2). On the other hand, an understanding of ordinal and cardinal numbers is required in Chaps. 7 to 9 (set-theoretic ideas are then marginal in the remaining three chapters). The necessary concepts and results are contained in the Appendix, which includes exercises, with full solutions given in Appendix B.

The rudiments of set theory required in the first half of the book are quickly summarised here, with references to textbooks that provide more details and exercises (the first two chapters of [31] are especially recommended). We also review certain simple proof strategies (proof by contradiction, by contraposition) and techniques (proofs by arithmetical induction) that we frequently resort to. The reader familiar with basic set theory and with doing proofs may skip this preliminary chapter.

Inclusion, Union, Intersection of Sets. The symbol $\mathbb{N}$ refers to the set of **natural numbers**, which we also describe as a list within curly brackets, thus: $\mathbb{N} = \{0, 1, 2, \ldots\}$. The symbol $\mathbb{Z}$ refers to the set of **integers** $\mathbb{Z} = \{\ldots, -2, -1, 0, 1, 2, \ldots\}$ and $\mathbb{Q}$ to the set of **rational numbers**, whose elements are of the form a/b, with a, b integers and b different from zero. We take the rules to add and multiply integers and rational numbers to be known. We adopt the general notation $X = \{x, y, z, \ldots\}$ to describe a set X whose elements are $x, y, z, \ldots$.

If x is an **element** of a given set X, we write $x \in X$ ('x belongs to X'), and if x is not an element of X, then we write $x \notin X$. In this notation, $0 \in \mathbb{N}$ and $1/2 \notin \mathbb{Z}$. We use the symbol $\emptyset$ for the empty set, which has no elements. Note that the set $\{\emptyset\}$ has exactly one element, unlike its element.

If each element of a set X is also an element of the set Y, we write $X \subseteq Y$ ('X is included in Y') or, equivalently, $Y \supseteq X$ ('Y includes X'). If $X \subseteq Y$, we call X a **subset** of Y and Y a **superset** or an **extension** of X. It is clear that $\mathbb{N} \subseteq \mathbb{Z}$ and $\mathbb{Z} \subseteq \mathbb{Q}$. Note that X is a subset of Y exactly when $x \notin Y$ implies $x \notin X$. It follows that, for any set X, $\emptyset \subseteq X$, since $\emptyset$ has no elements.

If X is not a subset of Y, we write $X \not\subseteq Y$: thus $\mathbb{Q} \not\subseteq \mathbb{Z}$. If $X \subseteq Y$ and at least one element of Y is not an element of X, we say that X is strictly included in Y and we write $X \subset Y$ or $Y \supset X$. When both $X \subseteq Y$ and $Y \subseteq X$ hold, we conclude $X = Y$ (by the principle of extensionality, an axiom of set theory that enables us to identify sets with the same elements).

Exercise 0.1 Show that:

(i) For every set X, $\emptyset \subseteq X$ and $X \subseteq X$.
(ii) $X \subseteq Y$ and $Y \subseteq Z$ imply $X \subseteq Z$.
(iii) There are sets X, Y with common elements and such that $X \not\subseteq Y$ and $Y \not\subseteq X$.

Given a set X, its subsets are the elements of the set $\mathcal{P}(X)$, known as **powerset** of X. If $X = \mathbb{Q}, \mathbb{N}, \mathbb{Z} \in \mathcal{P}(\mathbb{Q})$. By Exercise 0.1-(i), $\emptyset, X \in \mathcal{P}(X)$.

Exercise 0.2 List the elements of $\mathcal{P}(X)$ when $X = \emptyset$, when $X = \{\emptyset\}$ and when $X = \{\emptyset, \{\emptyset\}\}$.

We are often interested in isolating a subset Y of some given set X or, equivalently, in singling out $Y \in \mathcal{P}(X)$ by means of a linguistic condition C that identifies the elements of Y. In this case, we write $Y = \{y \in X : C(y)\}$ and we say that Y is the set of those elements in X that satisfy the condition C. For instance, we may isolate $Y = \{y \in \mathbb{N} : 0 < y\} \subseteq \mathbb{N}$ (where $C(y)$ is the inequality $0 < y$), containing the positive numbers. The set $Y \subseteq \mathbb{N}$ is also isolated by the condition $y \neq 0$, when viewed as a subset of $\mathbb{N}$, but not when viewed as a subset of $\mathbb{Z}$. Relative to $\mathbb{Z}$, the conditions $y \neq 0$ and $0 < y$ isolate different subsets.

We operate upon sets by taking their **union**, **intersection** and **relative complement**. If X, Y are sets, then $X \cap Y$, the intersection of X and Y, is the set of elements common to X and Y. For instance, $\mathbb{N} \cap \mathbb{Z} = \mathbb{N}$, whilst the intersection between $\mathbb{N}$ and the non-positive integers is $\{0\}$. The union $X \cup Y$ is the set whose elements belong to X, to Y or to both. For instance $\mathbb{N} \cup \mathbb{Z} = \mathbb{Z}$. The complement of X relative to Y, referred to as the set $X - Y$, is obtained by deleting from X the elements of $X \cap Y$. Thus, for instance, $\mathbb{Z} - \mathbb{N} = \{\ldots, -2, -1\}$, $\mathbb{N} - \mathbb{Z} = \emptyset$ and $\{0, 1\} - \{2\} = \{0, 1\}$.

Exercise 0.3 The standard way of proving that two sets are equal consists in showing separately that each of them is included in the other. Use this approach to verify the following set-theoretic equalities under the hypothesis $X, Y \subseteq Z$:

(i) $X - (X \cap Y) = X - Y$
(ii) $X \cap (X \cup Y) = X$
(iii) $Z - (X \cup Y) = (Z - X) \cap (Z - Y)$

We also take unions and intersections of entire families of sets. For instance, given the finite sequence $X_1, \ldots, X_n$, its union is $X_1 \cup \ldots \cup X_n$, also denoted by the more concise equivalent $\bigcup_{i=1}^{n} X_i$. In a similar fashion, the intersection of the last family of sets is $X_1 \cap \ldots \cap X_n$ or $\bigcap_{i=1}^{n} X_i$. If $X_1, X_2, X_3, \ldots, X_n, \ldots$ is an infinite sequence of sets, its union and intersection are $\bigcup_{i \in \mathbb{N}} X_i$ and $\bigcap_{i \in \mathbb{N}} X_i$. At times, we only require that the sets in a given family be labelled by a set of indices I (which may be finite or infinite and, if infinite, may be different from $\mathbb{N}$). The notation for a family of sets indexed over I is $\{X_i : i \in I\}$. The union and intersection of such a family are $\bigcup_{i \in I} X_i$ and $\bigcap_{i \in I} X_i$.

Relations and Functions. Given two sets X, Y, the set $X \times Y$, known as the **Cartesian product** of X and Y, is the set of all ordered pairs (x, y), where $x \in X$ and $y \in Y$. If, e.g. $X = \mathbb{N}$ and $Y = \mathbb{Z}$, $(1, -1) \in \mathbb{N} \times \mathbb{Z}$ and $(-1, 1) \notin \mathbb{N} \times \mathbb{Z}$. Two ordered pairs $(x, y), (u, v) \in X \times Y$ are identical if, and only if, $x = y$ and $u = v$.

If $X = Y$, $X \times X$ and X^2 are equivalent notations: they both refer to the set of ordered pairs whose components belong to X. In an analogous manner, $X \times X \times X = X^3$, $X \times X \times X \times X = X^4$, and so on. The elements of $X^3, X^4, \ldots, X^n, \ldots$ are, respectively, ordered triples, ordered quadruples, ..., ordered n-**tuples** (we shall often refer simply to pairs, triples, n-tuples, assuming that they are ordered). The n-tuples $(x_1, \ldots, x_n), (y_1, \ldots, y_n) \in X^n$ are equal if, and only if, for every $i = 1, \ldots, n$, $x_i = y_i$.

A relation like 'greater than' over the set $\mathbb{N}$, usually symbolised by $>$, can be conceived as a set of ordered pairs. It suffices to think of the ordered pair (x, y) as a way of codifying the information conveyed by the inequality $x > y$. If we identify $>$ with a subset of $\mathbb{N}^2$, then it makes sense to write $> \subseteq \mathbb{N}^2$, $(2, 1) \in >$ and $(1, 2) \notin >$. We shall, as a rule, prefer to the last membership relations the more familiar alternatives $2 > 1$ and $1 \not> 2$ (the latter will in fact be formalised as a negated inequality, as soon as we have at our disposal, in Chap. 2, the linguistic resources of first-order logic).

What we noted with respect to $>$ applies to relations involving n items, for any $n \leq 1$: we call such relations n-**ary** (1-ary and 2-ary relations are also known as 'unary' and 'binary', respectively). A n-ary relation R over the set X is a subset of X^n. In symbols:

$$R \subseteq \underbrace{X \times \ldots \times X}_{n\ times} = X^n.$$

A unary relation over X is an element of $\mathcal{P}(X)$, while a binary relation over X is, as we noted, a subset of X^2. For $n \geq 2$, a n-ary relation is an element of $\mathcal{P}(X^n)$. Perhaps the simplest example of a n-ary relation (with $n \geq 2$) is the diagonal over X^n, i.e. the set containing exactly the n-tuples from X^n whose components are all identical.

Among the binary relations over X, we single out the **unary functions** over X. A unary function f over X is a subset of X^2 such that, if $(x, y) \in f$ and $(x, z) \in f$, then $y = z$. We usually write $f(x) = y$ in place of $(x, y) \in f$ and, in this case, we refer to x as an **argument** of f and to y as a **value** of f. A unary function is therefore a single-valued, binary relation: $f(x) = y$ and $f(x) = z$ cannot simultaneously hold if y, z are distinct. The binary relation $\{(x, y) \in \mathbb{N}^2 : y = 2x\}$ is a function because, given x, $f(x) = 2x$ is uniquely determined.

If X, Y are distinct sets, then any single-valued, binary relation f included in $X \times Y$ is a function from X to Y. If $\{x \in X : (x, y) \in f\} \neq X$, we say that f is a **partial function** from X to Y. Otherwise, we say that f is a **total function** from X

to Y. The function from $\mathbb{N}$ to $\mathbb{N}$ defined by the condition $f(n) = n - 1$ is a partial function because it contains no pair of the form $(0, f(0))$. By contrast, the function defined by the condition $f(n) = n + 1$ is a total function because, for any $n \in \mathbb{N}$, it contains the pair $(n, n + 1)$.

Whenever f is a unary function from X to Y, we write:

$$f : X \longrightarrow Y.$$

The set X is called the **domain** of f, and the set Y is called the **codomain** of f. The set $f[X] = \{y \in Y : \text{for some } x \in X, y = f(x)\}$ is called the **image** of f. It is easy to generalise the set-theoretic account of unary functions to n-**ary functions**. A n-ary function g over a set X is a single-valued subset of X^{n+1}. In other words, if $(x_1, \ldots, x_n, y), (x_1, \ldots, x_n, z) \in g$, then $y = z$. If X, Y are distinct sets, a n-ary function with values in Y is a single-valued subset of $X^n \times Y$.

The function $f : \mathbb{N} \longrightarrow \mathbb{Z}$ determined by the condition $f(x) = -x$ has domain $\mathbb{N}$, codomain $\mathbb{Z}$ and the set of non-positive integers as its image. The binary function $g : \mathbb{N}^2 \longrightarrow \mathbb{Z}$, determined by the condition $g(x, y) = x - y$, has codomain $\mathbb{Z}$ and its codomain coincides with its image. Whenever the codomain and image of a function are the same set, we say that the function is **surjective** or **onto** (its codomain). A surjective function is also called a **surjection**. The function $f(x) = -x$ discussed earlier is not a surjection. Nonetheless, it does assign distinct values to distinct arguments (if $x \neq y$, then $-x \neq -y$). Any function with this feature is **injective** or an **injection** or, lastly, a **one-to-one** function. It is easy to see that the function $g(x, y) = x - y$ described above is not one-to-one.

In general, a n-ary function f is injective if $(x_1, \ldots, x_n) \neq (y_1, \ldots, y_n)$ implies $f(x_1, \ldots, x_n) \neq f(y_1, \ldots, y_n)$ or, equivalently, if $f(x_1, \ldots, x_n) = f(y_1, \ldots, y_n)$ implies $(x_1, \ldots, x_n) \neq (y_1, \ldots, y_n)$, where $(x_1, \ldots, x_n), (y_1, \ldots, y_n)$ are arguments of f.

If a function is at once injective and surjective, it is called a **bijection** or a **one-to-one correspondence**. If $\mathbb{E} \subseteq \mathbb{N}$ is the set of even numbers, the function $f : \mathbb{N} \longrightarrow \mathbb{E}$ determined by the condition $f(x) = 2x$ is a bijection from $\mathbb{N}$ to $\mathbb{E}$.

Given a function $f : X \longrightarrow Y$ if $A \subseteq X$, we use the notation $f \upharpoonright A$ to indicate the **restriction** of f to the domain A: if we view f as a set of ordered pairs, $f \upharpoonright A$ is the subset of f whose elements are of the form (a, y), with $a \in A$.

Exercise 0.4

(i) Given the function $f : \mathbb{N} \longrightarrow \mathbb{Q}$, determined by the equality $f(x) = x/3$, find $A \subseteq \mathbb{N}$ such that the image of $f \upharpoonright A$ is $\mathbb{N}$.
(ii) If $f \subseteq X \times Y$ is an injection, show that the set $f^{-1} = \{(y, x) : (x, y) \in f\}$ is a function. Describe a function f such that the set f^{-1} is not a function.
(iii) If $f : X \longrightarrow Y$ is surjective, is there an injective restriction of f? What assumption concerning the selection of elements from a family of sets do we rely upon to show that this is always the case?

Preliminaries

If f, g are functions and the range of f is included in the domain of g, it makes sense to apply f, g consecutively, in this order, to an argument of f, say x. The result is $g(f(x))$, a uniquely determined value. Since a value exists for any x in the domain of f, $g(f(x))$ is a function with the same domain as f and the same codomain as g. We also write $(gf)(x)$ and call gf the **composition** of g after f. Composition may be itself viewed as a binary function, whose arguments are functions. We shall mainly denote it by juxtaposition, but may at times adopt a special symbol for it, namely ∘: in such cases the composition of g after f is the function $g \circ f$. An important property of composition is that it is **associative**, i.e., if f, g, h are functions such that the range of f is included in the domain of g and the range of g is included in the domain of h, then we have:

for any x in the domain of f, $(h(gf))(x) = ((hg)f)(x))$

or, equivalently:

$$h \circ (f \circ g) = ((h \circ g) \circ f).$$

This is to say that an application of h to the composition gf is the same function as an application of the composition hg to f (the reader wishing for an illustration of associativity may want to set $x = 4$ and let f be multiplication times 2, g addition of -3 and h division by 5, with f, g, h functions from the domain $\mathbb{Q}$ to the codomain $\mathbb{Q}$).

Using functions, we can generalise Cartesian products to arbitrary families of sets, finite or infinite. If I is a set of indices, the family of sets $\{X_i : i \in I\}$ determines the Cartesian product $\Pi_{i \in I} X_i$, whose elements are the functions with domain I and codomain $\bigcup_{i \in I} X_i$ such that $f(i) \in X_i$. It is as if each one of these functions described one 'I-tuple'.

Exercise 0.5 If I is the finite set $\{1, 2, \ldots, n\}$, find a bijection between $X_1 \times \ldots \times X_n$ and $\Pi_{i \in I} X_i$.

Countability. A set X is said to be **countable** if there is an injection $f : X \longrightarrow \mathbb{N}$ or, equivalently, if there is a surjection $g : \mathbb{N} \longrightarrow X$. When the image of an injection f is included in the set $\{0, 1, \ldots, k\} \subseteq \mathbb{N}$, we say that X is **finite**. If a countable set is not finite, it is **countably infinite** or **denumerable**.

The set $\mathbb{N}$ is obviously countably infinite and so are $\mathbb{Z}$ and $\mathbb{Q}$. Moreover, the Cartesian product and the union of finitely many countable sets is countable. More generally, the union of a countable family of countable sets is countable.[1]

If X_1, X_2 are countably infinite sets, then there are injections $f_1 : X_1 \longrightarrow \mathbb{N}$ and $f_2 : X_2 \longrightarrow \mathbb{N}$. It follows that $f_2^{-1} f_1$ and $f_1^{-1} f_2$ are injections from X_1 to X_2 and

---

[1] These results are nicely proved in [22], pp. 73–77, but sections 2, 3 from chapter 4 of [31] are also worth consulting.

from X_2 to X_1, respectively (we are not excluding that one of X_1, X_2, or both, may be $\mathbb{N}$).

By an important set-theoretic result, known as the **Cantor-Bernstein theorem**, whenever we have a pair of injections between two sets, we can conclude that there is a bijection between them. The converse statement is clear: a bijection f yields itself and f^{-1} as the desired injections. An especially clear proof of the Cantor-Bernstein theorem may be found in [22], p. 75, but see also the elegant discussion in [31], pp. 65–66.

The Cantor-Bernstein theorem proves especially helpful when we use injections to compare set sizes. Given two sets X, Y, we stipulate that the size of X is not greater than the size of Y—in symbols $|X| \leq |Y|$—if, and only if, there is an injection $f : X \longrightarrow Y$. The sets X, Y have the same size—in symbols $|X| = |Y|$—if, and only if, there is a bijection between them. By the Cantor-Bernstein theorem, our two definitions of size comparison imply the antisymmetry of this relation: in other words, $|X| \leq |Y|$ and $|Y| \leq |X|$ jointly imply $|X| = |Y|$.

It is meaningful to ask whether there are enough functions to compare any two sets relative to size. An axiom of set-theory known as the Axiom of Choice (see Appendix) ensures that there are.

Although we have only mentioned countable sets so far, there are infinite sets that fail to be countable: we refer to them as **uncountable** sets. The Appendix refines the rough partition of sets into countable and uncountable ones by the introduction of **cardinal numbers**, which are special sets employed to assign a canonical 'size' to each set.

Arithmetical Induction. Repeatedly in the chapters to come we will want to prove an infinite sequences of statements of the same form, each referring to one particular natural number.[2] In symbols, given the sequence of statements $C(0), C(1), C(2), \ldots, C(n), \ldots$, our aim will be to prove that $C(n)$ holds, for each $n \in \mathbb{N}$. There is a uniform way of tackling this situation, which consists in carrying out a **proof by arithmetical induction**.

A proof by arithmetical induction (shortly to be described) relies on the fact that the set $\mathbb{N} = \{0, 1, 2, 3, \ldots\}$ may be regarded as a sequence generated by 0 under the repeated application of the **successor function** s, determined by the condition $s(n) = n + 1$ (the successor function is a bijection between $\mathbb{N}$ and $\mathbb{N} - \{0\}$). In order to prove that a condition $C(n)$ is satisfied by every natural number, it thus suffices to prove that C is satisfied by 0 and that, if C is satisfied by n, then it is satisfied by $s(n) = n + 1$, i.e. that $C(n)$ implies $C(n + 1)$. If so much can be established, then it is possible to conclude that the generation of $\mathbb{N}$ from 0 by means of the successor function parallels the preservation of C from 0 through the successive elements of $\mathbb{N}$. Each $n \in \mathbb{N}$ other than 0 inherits C from the immediately preceding number after finitely many stages. Thus C is satisfied by every natural number. The argument just described provides a template for all proofs by arithmetical induction.

---

[2] The reader unfamiliar with the discussion to follow may find it helpful to consult [78], especially section 6.1.

Preliminaries

More concisely, to prove by arithmetical induction that every natural number satisfies C, it suffices to prove that: (i) $C(0)$ holds and that (ii) if $C(n)$ holds, so does $C(n+1)$. Part (i) is called the **inductive base** of the proof and part (ii) is called the **inductive step** of the proof. The assumption, within the inductive step, that $C(n)$ holds is the **inductive hypothesis**.

At times, we may only want to prove that C holds from some number k on. In any such case, it suffices to replace $C(0)$ by $C(k)$ in the inductive base.

Suppose that C is the condition:

$$1 + \ldots + n = \frac{n(n+1)}{2}.$$

If $n \in \mathbb{N}$ is a positive number, to show that $C(n)$ holds is to show that the sum of the first n positive natural numbers is $n(n+1)/2$. We prove it by induction. The inductive base requires that we establish $C(1)$, which is easily done by substituting 1 for n in the equality above and verifying that the equality holds under the substitution.

The inductive hypothesis is that $C(n)$ holds. We use it to show that $C(n+1)$ holds as follows:

$$\frac{(n+1)(n+2)}{2} = \frac{(n+1)n + (n+1)2}{2}$$
$$= \frac{n(n+1)}{2} + (n+1)$$
$$= (1 + \ldots + n) + (n+1),$$

where the equality between the first and last term in the above chain of equations is $C(n+1)$ and the inductive hypothesis is used in the transition from the second to the third equality.

Exercise 0.6 If $n \in \mathbb{N}$ is odd, it can be represented as $2k - 1$, for some positive $k \in \mathbb{N}$. Prove by induction that, for each positive $n \in \mathbb{N}$, the sum of the first n odd numbers is n^2.

Exercise 0.7 View the set $\mathbb{N}$ as an ordered set and suppose that $X \subseteq \mathbb{N}$ has no least element. Prove by induction that $X = \emptyset$.

When we view the natural numbers as an ordered set, and not only as a sequence generated from 0 by iterating the rule 'add one', the statement of Exercise 0.7 is equivalent to the statement that, if $X \subseteq \mathbb{N}$ is not the empty set, then it has a least element, a result known as the **least number principle**. The equivalence between the least number principle and the statement from the last exercise can be established using a proof by contraposition, to be described in the next section.

Exercise 0.8 Suppose that, given an arithmetical condition C, $C(0)$ holds and that, for each n, if $C(n)$ holds, then $C(n+1)$ holds. Use the least number principle to show that $C(n)$ holds for every $n \in \mathbb{N}$ (*Hint*: consider what happens when C stops holding at some point).

Proof by Contradiction and Contraposition. It is often convenient to deduce a statement B from a statement A using a **proof by contradiction**. The goal is to rule out the possibility that A and the negation of B can simultaneously hold. This may be done by assuming both and then deriving some statement S and its negation from the assumptions.[3]

The technique of proof by contradiction is used, e.g., to prove that there is no rational number a/b such that $(a/b)^2 = 2$. We suppose that a, b have no common factors (which could be cancelled anyway). Our assumption A is therefore the statement that $a/b \in \mathbb{Q}$ (i.e. $a, b \in \mathbb{Z}$ and $b \neq 0$) and a, b have no common factors. We may also assume $a, b > 0$. The statement B we want to prove is $(a/b)^2 \neq 2$. We proceed by contradiction and assume both A and the negation of B, i.e. $(a/b)^2 = 2$.

The equality $(a/b)^2 = 2$ is equivalent to $a^2/b^2 = 2$ or $a^2 = 2b^2$. The reader can check that, if a^2 is a multiple of 2, i.e. an even number, so is a. Thus, since every even number is of the form $2k$, with $k \in \mathbb{N}$, there is k such that $a = 2k$. Then $4k^2 = a^2 = 2b^2$. Since $b^2 = 2k^2$, we conclude that b is even. We have now deduced the statement S, to the effect that a, b are even, and we can deduce the negation of this statement from A alone, since we supposed that a, b did not have any common factors (if they had both been even, 2 would have been a common factor).

From this contradiction we conclude that the negation of B is incompatible with A, i.e. that A implies B.

Another basic technique to prove that A implies B is called **proof by contraposition**. Instead of proving B from A, it is often convenient to prove the negation of A from the negation of B. To see why this strategy suffices to prove B from A, suppose that the negation of B implies the negation of A: if A holds, the negation of B cannot hold because it implies the negation of A and A, together with its negation, determines a contradiction. It follows from this contradiction that B holds. In other words, A implies B.

As an illustration, suppose we wish to prove that, if a^2 is even, then a is even. We may obtain this result by proving that, if a is not even, then a^2 is not even. We leave the proof to the reader.

---

[3] A useful reference for the proof strategies discussed here is chapter 3 of [78].

Contents

1	**Some Concrete Structures**		1
	1.1	Purpose of This Chapter	1
	1.2	Choices	2
	1.3	Permutations	6
	1.4	Decompositions	10
	1.5	Towards Model Theory	17
	1.6	Groups, Fields, Vector Spaces	18
2	**Languages and Structures**		23
	2.1	Terms	23
	2.2	Formulae	28
		2.2.1 Open and Closed Formulae	31
	2.3	Structures	33
		2.3.1 Satisfaction	34
		2.3.2 Valuations	38
	2.4	Consequence	42
	2.5	Induction on the Complexity of Formulae	46
	2.6	Definability and Interpretability	51
		2.6.1 Interpretations	54
	2.7	The Size of a Language	61
3	**Theories and Models**		63
	3.1	The Connection Between Theories and Models	63
	3.2	Groups and Vector Spaces	65
		3.2.1 Abelian Groups	66
		3.2.2 Vector Spaces	67
	3.3	Rings and Fields	70
		3.3.1 Polynomial Rings	72
	3.4	Addendum: Quotient Rings and Noetherian Rings	73

4	**Morphisms, Substructures and Extensions**	77
4.1	Maps That Preserve Formulae	77
4.2	Elementary Maps, Substructures, Extensions	79
	4.2.1 Substructures and Extensions	82
	4.2.2 Downward Löwenheim-Skolem Theorem	86
4.3	Diagrams	87
4.4	Addendum: Potential Truth	92
5	**Ultraproducts**	97
5.1	Product Structures and Voting	97
	5.1.1 Pairwise Aggregations	99
5.2	Ultraproducts	103
	5.2.1 Cofinite Subsets of I	104
	5.2.2 The Ultraproduct Construction	105
5.3	The Real Field R	110
	5.3.1 Sequential Completeness of R	114
5.4	A Touch of Nonstandard Analysis	116
5.5	The Compactness Theorem	122
5.6	Elementary Classes	124
5.7	Addendum: The Unity of the World	125
6	**Essential Field Theory**	131
6.1	Simple Field Extensions	131
6.2	Algebraic and Finite Extensions	134
	6.2.1 Finite Extensions	136
6.3	Algebraically Closed Fields	138
6.4	The Characteristic of a Field	140
6.5	Field Embeddings	141
6.6	Addendum: The Primitive Element Theorem	144
7	**Complete Theories**	147
7.1	The Löwenheim-Skolem Theorems	147
7.2	Completeness	151
7.3	Infinite Vector Spaces and Divisible Groups	153
7.4	Dense Linear Orders	156
7.5	Transcendence Bases	160
7.6	Algebraic Closure	165
7.7	Completeness of ACF_0 and ACF_p	167
7.8	Model-Theoretic Algebraic Closure	170
8	**Model-Completeness and Quantifier Elimination**	177
8.1	Model-Completeness	178
	8.1.1 Applications to Dense Linear Orders	181
	8.1.2 Applications to Discrete Linear Orders	183
	8.1.3 Model-Completeness and Completeness	184
8.2	Preservation and Axiomatisability	185

	8.3	Chains and Lindström's Test	189
		8.3.1 Universal-Existential Axiomatisability	193
	8.4	Quantifier Elimination	195
		8.4.1 Morleysation	197
		8.4.2 Skolemisation	197
	8.5	Finite Relational Languages	199
	8.6	T-Closures	201
	8.7	Algebraically Prime Models	204
	8.8	Strongly Minimal Theories	208
		8.8.1 Minimality and Dimension	210
	8.9	o-Minimal Theories	215
9	**Types**		**219**
	9.1	Key Definitions	219
		9.1.1 The Space $S_n(T)$	222
	9.2	Isolated and Algebraic Types	222
	9.3	An Application to Measurement Theory	225
	9.4	Types and Elementary Maps	229
	9.5	Homogeneity	232
	9.6	Atomic Structures	236
		9.6.1 The Omitting Types Theorem	239
		9.6.2 The Ryll-Nardzewski Theorem	242
	9.7	Saturated Structures	244
		9.7.1 Small Theories	247
		9.7.2 Monster Models	249
10	**Algebraically Closed Fields and Algebraic Geometry**		**253**
	10.1	Specialisations	254
		10.1.1 Infinite Fields with Quantifier Elimination	255
	10.2	Elements of Algebraic Geometry	261
		10.2.1 The Fundamental Correspondence	262
		10.2.2 Radical and Primary Ideals	263
		10.2.3 Hilbert's *Nullstellensatz*	266
	10.3	Morley Rank	269
		10.3.1 Basic Properties of Morley Rank	270
		10.3.2 Morley Degree	272
		10.3.3 Morley Rank in Strongly Minimal Structures	274
	10.4	Dimension of an Algebraic Variety	279
		10.4.1 A Note on ω-Stable Theories	285
11	**Real Closed Fields**		**287**
	11.1	Ordered Fields	287
	11.2	Real Closed Fields	293
	11.3	Quantifier Elimination of $\mathsf{RCF}_<$ and Its First Consequences	298
		11.3.1 o-Minimal Fields	305
		11.3.2 Decidability	308

11.4	Applications to Real Geometry and Algebra	311
	11.4.1 Extensions	312
	11.4.2 The Artin-Lang Theorem and Hilbert's 17th Problem	315
	11.4.3 Milnor's Curve Selection	317

12 Fraïssé Limits and Measurement Scales ... 321
- 12.1 Classes of Finitely Generated Substructures ... 321
- 12.2 Fraïssé Limits ... 324
- 12.3 Two Concrete Limits ... 328
  - 12.3.1 Finite Linear Orders ... 328
  - 12.3.2 Finite Vector Spaces over $\mathbb{F}_q$ ... 330
- 12.4 Saturation and Quantifier Elimination ... 331
- 12.5 Measurement Scales over $\mathbb{Q}$ ... 333
- 12.6 Model-Theoretic Preliminaries ... 335
- 12.7 Group-Theoretic Preliminaries ... 337
- 12.8 A Theorem on Rational Scale Types ... 340

Further Reading ... 345
- Intermediate Model Theory ... 345
- Historical and Philosophical Themes ... 346

A Set-Theoretic Background ... 347
- A.1 Well-Ordered Sets ... 347
- A.2 Classes and Sets ... 352
- A.3 Comparability and Towers ... 354
  - A.3.1 Order Properties of Towers ... 356
- A.4 Transfinite Recursion ... 358
  - A.4.1 An Application of Transfinite Recursion ... 362
- A.5 Cardinal Numbers ... 364
  - A.5.1 Cardinal Arithmetic ... 367
  - A.5.2 Regular Cardinals ... 372
- A.6 The Axiom of Choice ... 372
  - A.6.1 Zorn's Lemma ... 373

B Solutions to Exercises ... 377
- B.1 Chapter 1 ... 377
- B.2 Chapter 2 ... 384
- B.3 Chapter 3 ... 400
- B.4 Chapter 4 ... 403
- B.5 Chapter 5 ... 411
- B.6 Chapter 6 ... 422
- B.7 Chapter 7 ... 429
- B.8 Chapter 8 ... 444
- B.9 Chapter 9 ... 467
- B.10 Chapter 10 ... 482

B.11	Chapter 11	489
B.12	Chapter 12	500
B.13	Appendix A	504

Bibliography 517

Index 521

Chapter 1
Some Concrete Structures

Abstract Our main aim in this book is to develop the basic ideas of model theory as well as a thorough account of their application to certain specific classes of mathematical objects like linear orders, groups, vector spaces and fields. This chapter is mainly written for the reader who may not have encountered any of these mathematical objects before, or can only recollect a passing acquaintance with them.

1.1 Purpose of This Chapter

In the following chapters we shall study various classes of structures. This chapter introduces some concrete representatives from each of these classes. We thus naturally progress from a direct study of concrete objects in this chapter to the more abstract standpoint of the subsequent chapters.

In order to make even the introduction of concrete structures less daunting to the beginner, we do not discuss them primarily as mathematical objects but as solutions to a decision problem studied by social scientists.[1]

It is, arguably, a more effective pedagogic strategy—and likely one more agreeable to the non-expert—to provide an easily grasped motivation for engaging with specific mathematical content, than it may be to present such content without any motivation beyond its mathematical interest. Moreover, the focus on a decision problem distinctive of this chapter is not unique to it: in the course of the book, we highlight several connections between model theory and social science (in Chaps. 5, 9 and 12 respectively). The material covered in this chapter is in fact closely connected to the opening of Chap. 5.

The reader who is only interested in mathematical logic may skip all applications to social science without loss of continuity. Our hope, however, is that every reader

---

[1] Our intention is evidently not to provide an historical account of the way certain structures have acquired prominence in mathematical investigations. To learn about the problems that motivated their study, the reader may choose to consult histories of modern algebra like [77] or, more recently, [13] and [21].

© The Author(s), under exclusive license to Springer Nature Switzerland AG 2025
D. Rizza, *Model Theory: The Algebraic Basics*, Trends in Logic 70,
https://doi.org/10.1007/978-3-032-03796-1_1

will find it of interest to see how model-theoretic ideas can be naturally related to fields of enquiry that do not constitute their typical reference.

1.2 Choices

Suppose that a municipal committee is to choose between three contractors A, B, C, who have made bids to carry out the construction of a public hospital. The committee wishes to rank the contractors and assign the construction works to the top-ranked one. To this end, the committee may take into account three distinct ranking criteria, namely (i) value for money, (ii) technical qualifications and (iii) anticipated date of completion (the faster contractor, if qualified to carry out a high-quality job, may thus be preferred).

One way to make use of the criteria is to rank A, B, C relative to each of them and use the three rankings thus generated to run pairwise majority elections. To see how this procedure works, let the ranking determined by criterion (i) be $A < B < C$ (where $A < C$), the ranking determined by criterion (ii) be $B < C < A$ (where $B < A$) and the ranking determined by criterion (iii) be $C < A < B$ (where $C < B$). The symbol $<$ indicates that the alternative to its right is strictly better than the alternative to its left.

Pairwise majority selections are now easy to make. In the pair $\{A, B\}$, because B is better than A with respect to two out of three criteria, B is the preferred option, i.e. the committee accepts the ranking $A < B$. In a similar fashion, the pairs $\{B, C\}$, $\{C, A\}$ give rise to the rankings $B < C$, $C < A$. The resulting list of rankings, namely:

$$A < B, B < C, C < A, \tag{1.1}$$

does not support the choice of a contractor. Clearly, A cannot be chosen, since it is inferior to B. The same is true of B, relative C. Finally, C can't be chosen, being inferior to A. A choice could have been made if the overall rankings had satisfied one particular formal property. The property in question can be stated as follows:

for any $x, y, z \in \{A, B, C\}$, $x < y$ and $y < z$ imply $x < z$.

The above statement expresses the **transitivity** of the binary relation $<$. A transitive relation $<$ is incompatible with the pairwise rankings displayed in 1.1. To see this, note that, by transitivity, $A < B$ and $B < C$ imply $A < C$, while $A < C$ and $C < A$, imply the contradiction $A < A$ (no alternative is inferior to itself). The decision-making problem we have briefly illustrated thus depends on the fact that, even if (i), (ii), (iii) determine transitive rankings of A, B, C, the application of pairwise majority to these rankings does not guarantee a transitive outcome.

We may look at this situation slightly more formally. Each of the criteria (i)–(iii) ranks A, B, C transitively but also in such a way that, for any two distinct alternatives, one is inferior to the other. In symbols, if $x, y \in \{A, B, C\}$ and $x \neq y$, then $x < y$ or $y < x$. This last property is known as **trichotomy**. Moreover, no alternative $x \in \{A, B, C\}$ is inferior to itself, i.e. $x \not< x$. The last property makes $<$ an **irreflexive** relation.

Definition 1.1 Given a nonempty set X, a binary relation $<$ over X is a **linear order** of X if, and only if, it is irreflexive, transitive and satisfies trichotomy.

Exercise 1.2

(i) Prove that, if $<$ is a linear order of X and $x, y \in X$, then $x < y$ and $y < x$ cannot simultaneously hold.
(ii) If X is a finite set linearly ordered by $<$, the element $x \in X$ such that, for every $y \in X$, $y \neq x$ implies $y < x$, is called the *maximum* of X (relative to $<$). Define the *minimum* of X in an analogous manner and prove that, if X is linearly ordered by $<$, then it has a maximum and a minimum. In this case, describe a procedure to find the maximum and minimum of X. Finally, prove that the maximum and the minimum of X are unique.
(iii) List the possible linear orders of $X = \{A, B, C\}$.

When a decision procedure determines a linear order of $\{A, B, C\}$, it is possible to rely upon it to select the 'best' alternative: this is nothing but the maximum relative to the given linear order. Pairwise procedures, however, may fail to determine a linear order of $\{A, B, C\}$. We have seen that this is the case with pairwise majority. Its failure to deliver a transitive outcome is often referred to as **Condorcet's paradox**.[2]

Condorcet's paradox is more than a formal puzzle, on account of the fact that pairwise procedures are relatively inexpensive to implement and, for this reason, they are often adopted by decision-makers (see e.g. the remarks in [66]: p.3). In any such context, the decision-making goal is the construction of a finite linear order, from which a unique choice can be made. The information supplied to achieve the goal is a set of independent pairwise evaluations or rankings. As we have seen, the structure of the available information may prove irreconcilable with the set goal.

The problem persists when ties are allowed.[3] By trichotomy, a linear order over X rules out the possibility that $x \neq y$ and that none of x, y is inferior to the other. If we allow the latter possibility and require that any two alternatives be comparable, we no longer have a linear order but a closely related type of ranking.

---

[2] Arrow's theorem, which we shall state and prove in Chap. 5, shows that the same issue affects every other nontrivial pairwise procedure.

[3] The example discussed above rules out ties because an odd number of criteria is involved. If, however, four criteria had been adopted, ties might have occurred.

Definition 1.3 If $X \neq \emptyset$ and $\succcurlyeq$ is a binary relation over X, we say that $\succcurlyeq$ is a **complete preorder** if, and only if, $\succcurlyeq$ is transitive and, for any $x, y \in X$, either $x \succcurlyeq y$ or $y \succcurlyeq x$ holds.

Intuitively, the expression $x \succcurlyeq y$ says that x is not inferior to y. Thus, x may either be superior to y or equivalent to y.[4] We say that x, y are equivalent, or tied, if, and only if, both $x \succcurlyeq y$ and $y \succcurlyeq x$ hold. When this is the case, we simply write $x \sim y$. If, on the other hand, $x \succcurlyeq y$ holds and x, y are not equivalent, we write $x \succ y$.

Exercise 1.4 Suppose that $X \neq \emptyset$ and $\succcurlyeq$ is a complete preorder over X.

(i) Show that, for every x, $x \succcurlyeq x$.
(ii) Using the definition of $\sim$, verify that, for every $x \in X$, $x \sim x$. This is to say that $\sim$ is a reflexive relation.
(iii) Verify that, if $x, y \in X$ and $x \sim y$, then $y \sim x$. This is to say that $\sim$ is a symmetric relation.
(iv) Verify that $\sim$ is transitive.

Exercise 1.4 shows that $\sim$ is a reflexive, symmetric and transitive relation, i.e. an **equivalence relation**. We can use $\sim$ to establish a systematic link between complete preorders and linear orders. Let X be a nonempty set completely preordered by $\succcurlyeq$, and let $[x]$ the *equivalence class* of x, i.e. the set $\{y \in X : x \sim y\}$. We order $\sim$-equivalence classes as follows: $[x] > [y]$ if, and only if, $x \succ y$ (i.e. $x \succcurlyeq y$ but not $y \succcurlyeq x$). Because $x \succcurlyeq x$ certainly holds, the relation $\succ$ is irreflexive. As a consequence, $>$ is irreflexive.

Exercise 1.5 Show that, if $[x] > [y]$ holds, the same relation holds between $[u], [v]$, whenever $u \sim x$ and $v \sim y$. (*Hint*: show that $u \sim x$ implies $u \succ y$ and use this result to deduce $u \succ v$).

The last exercise ensures that equivalence classes are ordered in a way that does not depend on the choice of representatives. If this did not happen, we could have $[x] > [y]$ and $[v] > [u]$, with $x \sim u$ and $y \sim v$: the same equivalence classes would be simultaneously ordered in two different ways. Because we have ruled this undesirable possibility out, we say that the relation $>$ is *well defined*.

Exercise 1.6

(i) Verify that the relation $>$ is transitive.
(ii) Show that, if $[x], [y]$ are distinct equivalence classes (thus $x \not\sim y$), then $[x] > [y]$ or $[y] > [x]$.
(iii) Given a complete preorder $\succcurlyeq$ on $X = \{A, B, C\}$, find a variant of Condorcet's paradox in which pairwise majority comparisons give rise to the overall rankings $A \succ B$, $B \succ C$ and $C \sim A$.

---

[4] As in the case of two distinct candidates who receive the same number of votes.

1.2 Choices

The three linear orders of A, B, C used above to illustrate Condorcet's paradox have one notable feature: each alternative is top-ranked, middle-ranked and bottom-ranked by exactly one linear order. Because of this, it is plausible to think that, all things considered, no alternative is superior to any other. An analogous suggestion is pertinent to the variant of Condorcet's paradox from Exercise 1.6-(iii). If we adopt this standpoint, we can think of any configuration of linear orders that produces Condorcet's paradox as a null contribution to a ranking procedure. From this standpoint, what might have initially looked like a puzzle can now be regarded as a clearly defined problem. We would like to find a way of detecting the null contributions to a decision outcome, given a pairwise procedure.

Such contributions do not occur only in the format we have described. If, for instance, we modify the example at the beginning of this section to include four contractors A, B, C, D and eleven ranking criteria, we may consider a situation in which the criteria produce the following linear orders (the number to the left of a linear order counts the criteria that generate the same ranking):

$$3 \; D > A > B > C$$
$$6 \; B > D > C > A$$
$$2 \; D > C > A > B.$$

If, for each one of the three linear orders listed above, only two of the criteria that generate it are considered, the six criteria so isolated determine a version of Condorcet's paradox over $\{A, B, C\}$. Nonetheless, pairwise majority on the full set of linear orders yields the comparisons:

$$B > A, C > A, B > C, B > D, D > A, D > C,$$

which determine the linear order $B > D > C > A$, with maximum B. If we choose to treat Condorcet's paradox as a null contribution to decision-making, then we need a way of identifying and 'cancelling' this contribution out even when it is integrated into more extensive rankings that produce a transitive outcome, as in the example just described (more complicated ones, in which several Condorcet paradoxes are interlaced within a list of linear orders, are not too difficult to devise). A successful treatment of the kind suggested may still prove insufficient to solve the type of problem illustrated by Condorcet's paradox: we do not know whether pairwise majority gives rise to outcomes that are not transitive under configurations of preferences different from those singled out by the paradox.

In Sect. 1.4 we shall show that this is not the case and that Condorcet's paradox and its variants are the single source of all problems that affect decision-making by pairwise comparisons. We shall also provide a systematic way of treating the paradox as a null contribution under whichever guise it manifests itself.

Before turning to these tasks, we devote Sect. 1.3 to examining Condorcet's paradox in terms of some basic algebraic and geometric ideas. The algebraic notions we make use of provide a very accessible entry point into topics that later chapters will return to in greater detail.

Exercise 1.7 In the example at the beginning of the present section, the ranking criterion (i) determines a linear order that can be broken down into the pairwise comparisons $A < B, B < C, A < C$. Describe the linear order determined by criteria (ii) and (iii) in a similar fashion. Represent the pairwise comparisons produced by each criterion as an array of three rows and three columns. Verify that it suffices to switch pairwise comparisons across columns twice to obtain from the original array a new array in which no row determines a linear order. What issue with pairwise comparisons does this result highlight?

Exercise 1.8 Given the alternatives A, B, C, D, describe four linear orders that pairwise majority turns into a non-transitive ranking.

1.3 Permutations

Given three candidates A, B, C and three voters who linearly order the candidates according to preference, pairwise majority procedures based on voters' preferences may fail to express a winner. In particular, pairwise majority can generate non-transitive outcomes supported by transitive preferences. This is just a restatement of Condorcet's paradox in the context of voting.

Whether we refer the paradox to an election or another decision-making process, as we have done in Sect. 1.2, its occurrence depends on the combinatorial structure of preference rankings. In the simple case of three alternatives A, B, C, the structure we are interested in admits of an easy description that can be obtained by looking at the possible 'anagrams' of the word ABC.

Omitting the symbol $>$, we suppose that ABC stands for $A < B < C$. Any other linear order of the set $\{A, B, C\}$ can then be represented as a suitable rearrangement of the letters A, B, C that make up the word ABC. If, for instance, we wish to transform $A < B < C$ into $B < A < C$, we only have to interchange A, B. We introduce a notation that allows us to keep track of the permutations we carry out on ABC and to study how they interact. Let us call a the permutation that interchanges A, B. The action of a on $\{A, B, C\}$ is completely specified by the equalities:

$$a(A) = B \; a(B) = A \; a(C) = C. \tag{1.2}$$

Any permutation other than a is described in a similar manner, by specifying what it swaps each element of $\{A, B, C\}$ with. There are as many distinct permutations as there are anagrams of ABC. An anagram is obtained by two consecutive choices: the choice of one letter from the set $\{A, B, C\}$ followed by the choice of one letter from the remaining pair (the third letter in the anagram is uniquely determined by these choices). There are $3 \times 2 = 6$ possible choices and, thus, 6 anagrams. We wish to describe the permutations corresponding to each anagram. Besides a, we have e, the permutation that keeps each letter fixed, and b, specified as follows:

$$b(A) = B \; b(B) = C \; b(C) = A. \tag{1.3}$$

1.3 Permutations

Instead of listing the remaining permutations, we are going to compute them using the ones we already know. To this end, we must make a couple of preliminary observations. First, the consecutive application of two permutations is a permutation. We call xy the permutation obtained by an application of y followed by an application of x. If $x = a$ and $y = b$, it is easy to verify that $ab(A) = A$, $ab(B) = C$ and $ab(C) = B$. Moreover, if x is a permutation, its action on $\{A, B, C\}$ is unaffected by any applications of e, before or after it. In symbols, we have the equality:

$$ex = xe. \tag{1.4}$$

Exercise 1.9

(i) Apply b to b and describe the permutation b^2.
(ii) Show that $b^3 = e$.
(iii) Show that $a(b(ab)) = e$.
(iv) Show that a, ab e ab^2 are distinct permutations.
(v) Show that $ba = ab^2$ and deduce that ab and ba must be distinct permutations.
(vi) If $b^2 = a$, then $b^2 b = ab$. Since $b^3 = e$, we obtain $e = ab$, which cannot hold, because ab does not fix all of A, B, C. We deduce that $b^2 \neq a$ (we have just carried out a quick proof by contradiction). Adapt this argument to prove that none of the equalities $b^2 = b$, $b^2 = ab$, $b^2 = ab^2$ can hold.

Exercise 1.9 shows that e, b, b^2, a, ab, ab^2 are all distinct. Consequently, the set

$$S_3 = \{e, b, b^2, a, ab, ab^2\}.$$

contains every permutation of ABC. It follows that $x, y \in S_3$ implies $xy \in S_3$. We call the set S_3, regarded as a set of permutations that can be consecutively applied, the **symmetric group on three letters**. We refer to the consecutive application of two permutations as their **composition**. Composition in S_3 has some noteworthy properties, which we briefly examine. First, if $x, y, z \in S_3$, then:

$$(xy)z = x(yz). \tag{1.5}$$

The last equality states that permutations satisfy the **associative law**. Since permutations are functions, the associativity of their composition follows from the associativity of functional composition (see the Preliminaries).[5]

Exercise 1.10 Using the associative law, prove that, if x, y, z, w are permutations, then $x(yzw) = (xyz)w = (xy)(zw)$.

---

[5] Even though we are focussing on the permutations of a three-element set, the same is true of any finite or infinite set of permutations. The associative law is also satisfied by the familiar arithmetical operations.

Invoking the associative law and the equalities $a^2 = e = b^3$ we can use Exercise 1.9-(v) and Exercise 1.10 to deduce:

$$ab^2ab^2 = (ab^2)(ab^2) = (ba)(ab^2) = b(a(ab^2)) = b((aa)b^2) = b(eb^2) = bb^2 = e.$$

Exercise 1.11

(i) Which elements of S_3 are b^2abab^2a and $abab^2abab^2$?
(ii) Find, for each $x \in S_3$, a permutation $y \in S_3$ such that $xy = e = yx$. The permutation y is called the **multiplicative inverse** of x.

Multiplicative inverses are uniquely determined. If y, z are such that $xy = e = xz$, the associative law implies:

$$z = ze = z(xy) = (zx)y = ey = y.$$

Thus z and y cannot be distinct.

Exercise 1.12 Show that **cancellation** holds, i.e. $xz = yz$ implies $x = y$.

Note that the uniqueness of inverses and cancellation only depend on the fact that a binary operation is associative and invertible: thus, they are not properties peculiar to S_3. We shall soon work with them and other formal properties in a general setting. At present, though, we wish to restrict attention to S_3 and finally relate it to Condorcet's paradox.

The connection is provided by the fact that, if we focus on the permutations e, b, b^2 and apply them to any, initially given linear order, say $A < B < C$, we obtain a version of Condorcet's paradox. We call $C_3 = \{e, b, b^2\}$ the **cyclic group of order three**. The qualification 'cyclic' depends on the fact that, if we keep iterating b, we go back to $b = b^4$ at the fourth iteration, after generating C_3: equivalently, $b^3 = e$.

Condorcet's paradox arises from the interaction between linear orders generated by the action of C_3 on a fixed arrangement of A, B, C and pairwise majority comparisons. An illuminating representation of their interaction, which throws into relief its problematic character, is provided by viewing both S_3 and the space of election outcomes over $\{A, B, C\}$ geometrically.

The symmetric group S_3 has a geometric representation obtained by conceiving A, B, C as labels attached to the vertices of an equilateral triangle (see Fig. 1.1 below). Suppose that the triangle ABC is rotated around its barycentre (in this case, the intersection of its heights). A 120° rotation around the barycentre moves A into the position originally occupied by B, B into the position originally occupied by C and C into the initial position of A. Such a rotation corresponds to the action of b on ABC. The cyclic group C_3 algebraically describes the counterclockwise rotations of ABC around its barycentre. The permutations a, ab, ab^2, by contrast, describe the reflections of ABC relative to one of its three heights. In short, S_3 represents the symmetries of an equilateral triangle.

1.3 Permutations

Fig. 1.1 Geometric representation of Condorcet's paradox

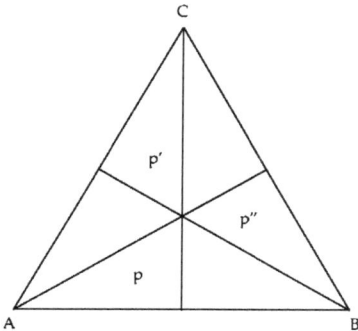

Among the elements of S_3, a, ab, ab^2 represent symmetries of ABC that only affect one dimension (one side of the triangle ABC), whereas the elements of C_3 represent two-dimensional symmetries. This difference in dimensionality sheds light on Condorcet's paradox as soon as we represent the outcomes of a decision or an election as points lying on the triangle ABC. Any such point may be determined by three non-negative rational coordinates q, r, s. The number q records the proportion of voters who rank A at the top; the number r, the proportion of voters who rank B and the top and the number s records the proportion of voters who rank C at the top. Thus $q + r + s = 1$.

The set of triples (q, r, s) such that $0 \le q, r, s \in \mathbb{Q}$ and $q + r + s = 1$ is a rational coordinate system over the triangle ABC. Each triple uniquely identifies a point on ABC, namely its barycentre when the vertices A, B, C are assigned the respective weights q, r, s. The vertices A, B, C have coordinates $(1, 0, 0)$, $(0, 1, 0)$ and $(0, 0, 1)$ respectively, and the intersection of the triangle's heights has coordinates $(1/3, 1/3, 1/3)$.

Without computing any further coordinate values, we see that the interior of ABC is partitioned by its heights into six triangular regions (see Fig. 1.1), each of which corresponds to a permutation of ABC or, equivalently, a linear ordering of A, B, C. The region delimited by A, the midpoint of AB and the point $(1/3, 1/3, 1/3)$ codifies the linear order $A > B > C$ (its two edges lying in the interior of ABC are disregarded). The adjacent region determined by the midpoint of AC codifies $A > C > B$, and so on.

The six triangular regions within ABC correspond to the permutations of the word ABC: thus, the action of a permutation in S_3 takes an election outcome from one triangular region to another. The permutations in C_3 move an outcome across regions in alternating fashion. Figure 1.1 illustrates Condorcet's paradox as the pairwise outcome of three linear orders p, p', p'' such that p', p'' are obtained from p under the application of b, b^2 respectively.

The configuration determined by p, p', p'' is a two-dimensional one (it determines three points that do not line on the same line), which is projected on each side of the triangle ABC when a pairwise majority ranking is computed. For instance, the configuration is projected on AB when A, B are compared.

The projection thus effected deletes the information concerning the position of C relative to A, B. As Donald Saari has pointed out (see [65], pp.148–149), the cyclic outcome of Condorcet's paradox is produced by the projection of an essentially two-dimensional configuration onto a one-dimensional configuration.

If we take Condorcet's paradox as a null contribution to a decision-making process, than it is reasonable to look for a way of filtering the Condorcet contribution out and only retaining information that does not get 'projected' and distorted when decomposed into pairwise comparisons. We achieve this goal for decisions involving four alternatives in the next section: the strategy we describe, due to Donald Saari, readily generalises to any finite number of alternatives.

1.4 Decompositions

A set of four alternatives $\{A_1, A_2, A_3, A_4\}$ determines six pairwise comparisons, involving $\{A_1, A_2\}$, $\{A_1, A_3\}$, $\{A_1, A_4\}$, $\{A_2, A_3\}$, $\{A_2, A_4\}$ and $\{A_3, A_4\}$ respectively.

For concreteness, it may be helpful to think of the four alternatives as candidates and to call 'preference' any linear order of the candidates (or some of them). An election involving n voters thus determines a list n preferences, which we shall call a **profile**. Given a profile, we can extract from it information concerning pairwise comparisons. If, for instance, we focus on the pair $\{A_i, A_j\}$, we can find, from the given profile, the number of voters expressing the preference $A_i > A_j$ and the number of voters expressing the preference $A_j > A_i$: we denote the difference between the last two numbers by $d_{i,j}$. If A_i beats A_j in a pairwise majority election, then $d_{i,j} > 0$. A tie yields $d_{i,j} = 0$.

Since there are six pairwise comparisons of interest, information concerning them is recorded by the six rational numbers $d_{1,2}, d_{1,3}, d_{1,4}, d_{2,3}, d_{2,4}, d_{3,4}$ (the last numbers are, in particular, integers. It is, however, convenient to work with the set $\mathbb{Q}$ instead of just $\mathbb{Z}$, as will soon become clear). Since we wish to consider these numbers as a single, finite sequence, we write their list as:

$$(d_{1,2}, d_{1,3}, d_{1,4}, d_{2,3}, d_{2,4}, d_{3,4}). \tag{1.6}$$

Any list like the one above is called a **vector**. We shall use the notation in 1.6 to represent an arbitrary vector whose six components are rational numbers. Since vectors are, in the present context, ordered 6-tuples of rational numbers, the set of vectors is the Cartesian product $\mathbb{Q}^6$. Not all vectors in $\mathbb{Q}^6$ are equally relevant to our goals. We isolate a subset of special interest by:

Definition 1.13 A vector in $\mathbb{Q}^6$ is **strongly transitive** if, and only if, for every triple $\{i, j, k\}$, where i, j, k are distinct elements of $\{1, 2, 3, 4\}$:

$$d_{i,j} + d_{j,k} = d_{i,k}.$$

1.4 Decompositions

The vector $(0, 0, 0, 0, 0, 0)$ is certainly strongly transitive and so is the vector $(-1, 0, 0, 1, 1, 0)$. By contrast, the vector $(1, 0, -1, 1, 1, 1)$ is not strongly transitive because $d_{1,3} + d_{3,4} = 0 + 1 = 1$ but $d_{1,4} = -1$.

The purpose of Definition 1.13 is to identify the vectors encoding pairwise information that gives rise to a complete preorder or a linear order of A_1, A_2, A_3, A_4. The strongly transitive vectors are, in particular, free from Condorcet-type paradoxes.

To see why, suppose that n voters express n linear orders of four candidates, which in turn determine a strongly transitive vector d. Let us focus on the pairs $\{A_i, A_j\}$ and $\{A_j, A_k\}$: if it turns out that a majority of the voters prefers A_i and A_j and that a majority of the voters (non necessarily the same) prefers A_j to A_k, then $d_{i,j}$ and $d_{j,k}$ are positive quantities that codify the relations $A_i > A_j$ and $A_j > A_k$ respectively. By strong transitivity, $d_{i,k} = d_{i,j} + d_{j,k} > 0$, i.e. $d_{i,k}$ codifies the preference $A_i > A_k$. Pairwise comparisons determine in this case an overall transitive ranking.

The next exercise identifies some key features of strongly transitive vectors.

Exercise 1.14

(i) Given the vector $(d_{1,2}, d_{1,3}, d_{1,4}, d_{2,3}, d_{2,4}, d_{3,4}) \in \mathbb{Q}^6$ and $r \in \mathbb{Q}$, we obtain from them the vector:

$$r(d_{1,2}, d_{1,3}, d_{1,4}, d_{2,3}, d_{2,4}, d_{3,4}) = (rd_{1,2}, rd_{1,3}, rd_{1,4}, rd_{2,3}, rd_{2,4}, rd_{3,4}).$$

Show that, if $(d_{1,2}, d_{1,3}, d_{1,4}, d_{2,3}, d_{2,4}, d_{3,4})$ is strongly transitive, so is $r(d_{1,2}, d_{1,3}, d_{1,4}, d_{2,3}, d_{2,4}, d_{3,4})$.

(ii) Given two vectors

$$(d_{1,2}, d_{1,3}, d_{1,4}, d_{2,3}, d_{2,4}, d_{3,4}), (f_{1,2}, f_{1,3}, f_{1,4}, f_{2,3}, f_{2,4}, f_{3,4}) \in \mathbb{Q}^6,$$

their sum is the vector

$$(d_{1,2} + f_{1,2}, d_{1,3} + f_{1,3}, d_{1,4} + f_{1,4}, d_{2,3} + f_{2,3}, d_{2,4} + f_{2,4}, d_{3,4} + f_{3,4}).$$

Show that the sum of two strongly transitive vectors is strongly transitive.

(iii) If $d_1, d_2 \in \mathbb{Q}^6$ and $r_1, r_2 \in \mathbb{Q}$ the vector $r_1 d_1 + r_2 d_2$ is a **linear combination** of d_1, d_2. Show that a linear combination of two strongly transitive vectors is strongly transitive. If $d_1, d_2, d_3 \in \mathbb{Q}^6$ are strongly transitive and $r_1, r_2, r_3 \in \mathbb{Q}$, show that the linear combination $r_1 d_1 + r_2 d_2 + r_3 d_3$ is strongly transitive. Generalise this result to n strongly transitive vectors (use arithmetical induction on the number of vectors).

Let $\mathcal{T}$ be the set of all strongly transitive vectors in $\mathbb{Q}^6$. This set contains $(0, 0, 0, 0, 0, 0)$ and every linear combination of finitely many among its elements. The last two properties are also true of $\mathbb{Q}^6$. We refer to $\mathbb{Q}^6$ as a **vector space** of which $\mathcal{T}$ is a **subspace**.

Exercise 1.15 Show that every element of $\mathbb{Q}^6$ can be represented as a linear combination with coefficients in $\mathbb{Q}$ of any six vectors r_i, $i = 1, 2, 3, 4, 5, 6$ whose only non-zero component is the i-th. Moreover, show that no selection of five elements from your choice of six suffices to generate $\mathbb{Q}^6$.

By Exercise 1.15 and results that we shall establish in Chap. 3, the least number of vectors required to generate $\mathbb{Q}^6$ is six. We call this number the **dimension** of $\mathbb{Q}^6$. Because the dimension of $\mathbb{Q}^6$ is finite and $\mathcal{T}$ is a strict subset of $\mathbb{Q}^6$, it is plausible to conjecture that $\mathcal{T}$ has a finite dimension ≤ 6 and that it should be possible to generate it as the set of all linear combinations determined by finitely many, strongly transitive vectors.

Lemma 1.16 $\mathcal{T}$ *is a three-dimensional subspace of* $\mathbb{Q}^6$. *Every vector in* $\mathcal{T}$ *can be represented as a linear combination with coefficients in* $\mathbb{Q}$ *of the three vectors* $(-1, 0, 0, 1, 1, 0)$, $(0, -1, 0, -1, 0, 1)$ *and* $(0, 0, -1, 0, -1, -1)$.

Proof If $(d_{1,2}, d_{1,3}, d_{1,4}, d_{2,3}, d_{2,4}, d_{3,4}) \in \mathcal{T}$, then, for every distinct $i, j, k \in \{1, 2, 3, 4\}$:

$$d_{i,j} + d_{j,k} = d_{i,k}.$$

For each index i ($i \in \{1, 2, 3, 4\}$), we define the vector b_i, whose only non-zero components are $d_{i,j} = 1$ e $d_{j,i} = -1$. Thus, for instance, the only non-zero components of b_1 are $d_{1,2}, d_{1,3}, d_{1,4}$: in particular, we have $b_1 = (1, 1, 1, 0, 0, 0)$. The vectors listed in the statement of the lemma are $b_2 = (-1, 0, 0, 1, 1, 0)$, $b_3 = (0, -1, 0, -1, 0, 1)$ and $b_4 = (0, 0, -1, 0, -1, -1)$. We check that the following conditions hold:

(1) the vectors b_1, b_2, b_3, b_4 are strongly transitive;
(2) every strongly transitive vectors can be represented as a linear combination with coefficients in $\mathbb{Q}$ of b_2, b_3, b_4;
(3) part (2) does not hold for any two of the vectors b_2, b_3, b_4.

Toward proving (1), we note that, by the definition of b_i, $d_{i,j} + d_{j,k} = d_{i,j} = 1 = d_{i,k}$, since i is different from j, k (i.e. $d_{j,k} = 0$). Similar considerations lead to $0 = d_{j,k} = d_{j,i} + d_{i,k} = -d_{i,j} + d_{i,k} = -1 + 1$. It now follows that, for each $i \in \{1, 2, 3, 4\}$, the vector b_i is strongly transitive. We turn to part (2). If $d = (d_{1,2}, d_{1,3}, d_{1,4}, d_{2,3}, d_{2,4}, d_{3,4})$ is strongly transitive, we can rewrite its components in terms of $d_{1,2}, d_{1,3}$ and $d_{1,4}$, thus:

$$d = (d_{1,2}, d_{1,3}, d_{1,4}, -d_{1,2} + d_{1,3}, -d_{1,2} + d_{1,4}, -d_{1,3} + d_{1,4})$$

It is easy to check that $d = -d_{1,2}b_2 - d_{1,3}b_3 - d_{1,4}b_4$. In other words, d is a linear combination with coefficients in $\mathbb{Q}$ of the vectors b_2, b_3, b_4. We are left with part (3). Consider b_2: this vector cannot be represented as a linear combination with coefficients in $\mathbb{Q}$ of the vectors b_3, b_4 because the first component of b_2 is 1, while the first component of both b_3 and b_4 is 0. Analogous observations show that b_3

1.4 Decompositions

cannot be represented as a linear combination with rational coefficients of b_2, b_4 and that the same is true of b_4 relative to b_2, b_3.

We conclude that $\mathcal{T}$ is a three-dimensional subspace of $\mathbb{Q}^6$ because it is generated by three vectors but it cannot be generated by any two of them.[6] □

The elements of $\mathcal{T}$ identify the circumstances under which it is safe to use pairwise majority comparisons. We already know that some profiles, which cannot be strongly transitive, generate pairwise majority outcomes that are cyclic. Our goal is to identify vectors in $\mathbb{Q}^6$ that correspond to such profiles. To this end, we begin by looking at a familiar version of Condorcet's paradox, given by the four-voter profile:

$$A_1 > A_2 > A_3 > A_4$$
$$A_2 > A_3 > A_4 > A_1$$
$$A_3 > A_4 > A_1 > A_2$$
$$A_4 > A_1 > A_2 > A_3,$$

where each voter expresses exactly one linear order of the candidates. The profile determines the vector $v = 2(0, 0, -1, 1, 0, 1)$, which is not, as expected, strongly transitive. The pairwise majority outcomes associated with our profile are:

$$A_1 > A_2, A_2 > A_3, A_3 > A_4, A_4 > A_1.$$

They determine a cyclic preference, starting at A_1 and ending with A_1, which in turn may be described by a cyclic reading of the sequence $(1, 2, 3, 4)$ (in which 1 precedes 2 and follows 4). Two consecutive indices i, j from $(1, 2, 3, 4)$ record the pairwise ranking of A_i, A_j. The sequence $(1, 2, 3, 4)$ can be used to determine a unique vector: it suffices to set $d_{i,j} = 1$ if i, j are consecutive in $(1, 2, 3, 4)$ and $d_{i,j} = 0$ otherwise (e.g. $d_{4,1} = 1$ implies that, in v, $d_{1,4} = -1$). Even though we do not reconstruct v in this way, we obtain the vector $v' = (0, 0, -1, 1, 0, 1)$ such that $v = 2v'$. In other words, v is a linear combination of v'.

The last remarks show that it is convenient to describe the vectors corresponding to cyclic election outcomes as linear combinations of simpler vectors, which can be determined by sequences of numerical indices (taken from the set $\{1, 2, 3, 4\}$) and whose components belong to the set $\{-1, 0, 1\}$. Given an index-sequence $a = (h, i, j, k)$ (with $h, i, j, k \in \{1, 2, 3, 4\}$), we define the **cyclic vector** c_a associated with a to be the vector whose nonzero components are exactly $d_{h,i} = d_{i,j} = d_{j,k} = d_{k,h} = 1$.

Exercise 1.17

(i) Given the sequences $\alpha = (1, 2, 3)$, $\beta = (1, 2, 4)$ and $\gamma = (1, 3, 4)$, determine c_α, c_β and c_γ respectively.

---

[6] Alternative choices of generating vectors could not reduce their number, as we shall clarify in Chap. 3. This is why we can talk about *the* dimension of a vector space or subspace.

(ii) Show that $(1, 3, 2), (1, 4, 2)$ and $(1, 4, 3)$ determine the cyclic vectors $-c_\alpha, -c_\beta$ and $-c_\gamma$ respectively.
(iii) Show that none of the vectors c_α, c_β and c_γ can be expressed as a linear combination of the remaining vectors.
(iv) Represent the cyclic vector associated with $(1, 2, 4, 3)$ as a linear combination of c_α, c_β and c_γ (*Hint*: decompose $(1, 2, 3, 4)$ into two sequences of three elements each).
(v) Show that no linear combination of c_α, c_β and c_γ other than $(0, 0, 0, 0, 0, 0)$ is strongly transitive.

Let us now consider the set S of all linear combinations with coefficients in $\mathbb{Q}$ of the vectors $b_2, b_3, b_4, c_\alpha, c_\beta$ and c_γ (recall that b_2, b_3, b_4 generate $\mathcal{T}$). A generic element of S is of the form:

$$q_1 b_2 + q_2 b_3 + q_3 b_4 + q_4 c_\alpha + q_5 c_\beta + q_6 c_\gamma. \tag{1.7}$$

Clearly, $S \subseteq \mathbb{Q}^6$. We wish to show that the converse inclusion also holds, i.e. $S = \mathbb{Q}^6$. To this end, we only have to check that S is a six-dimensional subspace of $\mathbb{Q}^6$ (any n-dimensional subspace of a n-dimensional vector space coincides with the whole space). We do so indirectly, by showing that any five elements from the set $B = \{b_2, b_3, b_4, c_\alpha, c_\beta, c_\gamma\}$ fail to generate the whole of S.

If this were not the case, a linear combination of five elements from B would represent the sixth. Toward a contradiction, let us assume that b_4 could be represented as a linear combination of $b_2, b_3, c_\alpha, c_\beta, c_\gamma$. There would then exist $r_1, \ldots, r_5 \in \mathbb{Q}$ such that:

$$r_1 b_2 + r_2 b_3 + r_3 c_\alpha + r_4 c_\beta + r_5 c_\gamma = b_4,$$

and we would be able to deduce from the last equality the following:

$$r_1 b_2 + r_2 b_3 - b_4 = -r_3 c_\alpha - r_4 c_\beta - r_5 c_\gamma,$$

where the left-hand side is strongly transitive but the right-hand side is cyclic (by Exercise 1.17-(v)), which is impossible. The same contradiction arises for every choice of five elements from B. It follows that S is generated by B but it cannot be generated by any strict subset of B. We conclude that S is a six-dimensional subspace of $\mathbb{Q}^6$, i.e. $S = \mathbb{Q}^6$. *Every element of $\mathbb{Q}^6$ can be represented as the sum of a strongly transitive vector and a linear combination of cyclic vectors.*

With this information at our disposal, we can easily introduce a decision procedure P such that, given an arbitrary profile p and its associated vector $v_p \in \mathbb{Q}^6$: (i) cancels the cyclic part of v_p; (ii) makes use of the strongly transitive part of v_p to produce a complete preorder of $\{A_1, A_2, A_3, A_4\}$ based on pairwise majority comparisons. For each $i \in \{1, 2, 3, 4\}$, we set:

$$P(A_i) = \frac{1}{4} \left(d_{i,1} + d_{i,2} + d_{i,3} + d_{i,4} \right) \ (i = 1, 2, 3, 4) \tag{1.8}$$

1.4 Decompositions

To illustrate how the procedure P works, let us consider the profile p below, where each preference is preceded by the number of voters expressing it:

$$3 \; A_1 > A_2 > A_3 > A_4$$
$$6 \; A_2 > A_3 > A_1 > A_4$$
$$6 \; A_4 > A_3 > A_1 > A_2$$
$$3 \; A_4 > A_1 > A_2 > A_3.$$

Here $v_p = (6, -6, 0, 6, 0, 0) = 6(1, -1, 0, 1, 0, 0) = 6c_\alpha$. Consequently, $P(A_1) = 1/4(d_{1,1} + d_{1,2} + d_{1,3} + d_{1,4}) = 1/4(0 + 6 - 6 + 0) = 0$. It is easy to see that $P(A_2) = P(A_3) = P(A_4) = 0$. In general, if $r \in \mathbb{Q}$, any profile that determines the vector rc_α gives rise to a complete tie under P.

Exercise 1.18 Show that, given $r, s, t \in \mathbb{Q}$ and a profile p whose associated vector is $rc_\alpha + sc_\beta + tc_\gamma$, the procedure P assigns 0 to each alternative.

Moreover, if p is a profile and $v_p = u + w$, with u strongly transitive and w a linear combination of cyclic vectors, then the procedure P separates the contributions u, w in the score it assigns each A_i ($i \in \{1, 2, 3, 4\}$). This can be easily seen by setting $u = (d_{1,2}, d_{1,3}, d_{1,4}, d_{2,3}, d_{2,4}, d_{3,4})$, $w = (d'_{1,2}, d'_{1,3}, d'_{1,4}, d'_{2,3}, d'_{2,4}, d'_{3,4})$ and using P on the vector:

$$u + w = (d_{1,2}, d_{1,3}, d_{1,4}, d_{2,3}, d_{2,4}, d_{3,4}) + (d'_{1,2}, d'_{1,3}, d'_{1,4}, d'_{2,3}, d'_{2,4}, d'_{3,4}).$$

Because P sets to 0 the contribution of w to each pairwise majority comparison, it also allows us to compute the components of u. To see this, let us set:

$$u + w = (a_{1,2}, a_{1,3}, a_{1,4}, a_{2,3}, a_{2,4}, a_{3,4}).$$

Then:

$$P(A_i) = \frac{1}{4}\left(a_{i,j} + \sum_{k \neq j} a_{i,k}\right),$$

where $1/4$ multiplies $a_{i,j}$ and the remaining components of $u + w$ whose first index is i and whose second index differs from j. Because cyclic contributions are separated and cancelled out under P, for any $i \neq j$ we obtain:

$$P(A_i) - P(A_j) = \frac{1}{4}\left(d_{i,j} + \sum_{k \neq j} d_{i,k} - d_{j,i} - \sum_{k \neq i} d_{j,k}\right)$$

$$= \frac{1}{4}\left(2d_{i,j} + \sum_{k \neq j} d_{i,k} + \sum_{k \neq i} d_{k,j}\right)$$

$$= \frac{1}{4}\left(2d_{i,j} + \sum_{k \neq j,i}(d_{i,k} + d_{k,j})\right)$$

$$= \frac{1}{4}(2d_{i,j} + 2d_{i,j})$$

$$= d_{i,j},$$

where the fourth equality depends on strong transitivity and the fact that, if k differs from i, j, then k can take only one of two distinct values. Each difference $P(A_i) - P(A_j)$ thus determines the (i, j)-th component of u, the strongly transitive part of $u + w$.

Exercise 1.19

(i) Given the profile p:

$$A_1 > A_2 > A_3 > A_4$$
$$A_2 > A_3 > A_4 > A_1$$
$$A_3 > A_4 > A_1 > A_2$$
$$A_4 > A_1 > A_2 > A_3$$
$$A_3 > A_1 > A_2 > A_4$$
$$A_4 > A_2 > A_1 > A_3$$
$$A_3 > A_1 > A_4 > A_2$$
$$A_3 > A_4 > A_1 > A_2.$$

where each preference is expressed by one voter, determine the vector $v_p \in \mathbb{Q}^6$ and express v_p as the sum of a strongly transitive vector and a linear combination of cyclic vectors.

(ii) Repeat the previous exercise on the 33 voter profile below, in which every preference is preceded by the number of voters expressing it:

$$6 \quad A_1 > A_2 > A_3$$
$$5 \quad A_3 > A_2 > A_1$$
$$9 \quad A_1 > A_3 > A_2$$
$$2 \quad A_2 > A_3 > A_1$$
$$1 \quad A_3 > A_1 > A_2$$
$$10 \quad A_2 > A_1 > A_3.$$

The ideas discussed in this section are a fragment of a broader and deeper analysis. The interested reader may consult [66, 67, 69], the papers in which Donald Saari devised the techniques we have briefly studied.

1.5 Towards Model Theory

In a 1954 paper, Alfred Tarski, one of the pioneers of model theory, described this subject in the following terms:

'problems studied in the theory of models concern mutual relations between sentences of formalized theories and mathematical systems in which these sentences hold'.

In the preceding sections we have encountered certain 'mathematical systems', e.g. linear orders of three alternatives, the symmetric group S_3 or the vector space $\mathbb{Q}^6$. Although a connection between the study of such objects and formalised sentences was not explicit, it begins to emerge as we review some ideas left implicit in the preceding sections.

For example, in Sect. 1.2 we isolated the class of linear orders by means of a finite list of sentences, namely:

(a) for every x, $x \not< x$;
(b) for every x and y, if $x \neq y$, either $x < y$ or $y < x$;
(c) for every x, y and z, $x < y$ and $y < z$ imply $x < z$.

Although we have not yet declared by what means the above sentences might be formalised, it is not difficult to see that, once formalised, they could be used as a formal theory of the class of linear orders, which, in turn, would be **axiomatised** by these sentences. By definition, a linear order has to be a mathematical system in which the given sentences hold.

A subtler use of language was implicit in our treatment of the vector space $\mathbb{Q}^6$. In that context we were interested in isolating the subspace $\mathcal{T}$. This was in effect done by linguistic means. If, for any $i, j \in \{1, 2, 3, 4\}$, we treat $d_{i,j}$ from Sect. 1.4 as a variable, the elements of $\mathcal{T}$ are exactly the vectors in $\mathbb{Q}^6$ that satisfy the twenty-four simultaneous conditions:

$$(d_{1,2} + d_{2,3} = d_{1,3}) \text{ and } (d_{1,2} + d_{2,4} = d_{1,4}) \text{ and } (d_{1,3} + d_{3,4} = d_{1,4}) \text{ and } \ldots$$
$$\text{and } (d_{3,4} = d_{3,2} + d_{2,4}).$$

Each of the above conditions may be seen as an equation in three unknowns. The simultaneous solutions to the full list of equations are precisely the components of strongly transitive vectors. Because equations are linguistic objects and simultaneous equations may be conjoined into a single statement like the one above, we have **defined** the subspace $\mathcal{T}$ of $\mathbb{Q}$ by means of a statement, which, if not fully formalised, comes very close to a formalisation (we would need to remove 'and' and use in its stead a symbol whose semantic behaviour is explicitly specified).

There is at least a third way in which the connection between mathematical systems and linguistic statements has entered our investigations, which differs from the previous two, namely axiomatisation and definition. While examining the group S_3, we effectively showed that some of its properties, e.g. the existence of a unique multiplicative inverse for any one of its elements, could be stated and proved without making any special reference to S_3. In order to prove uniqueness we only required the following statements to hold:

(i) for every x, y, z, $x(yz) = (xy)z$;
(ii) there is z such that, for every x, $xz = x = zx$.

The statements just listed hold in S_3 but, wherever they hold (in whichever mathematical system they are satisfied), we can deduce that the object z mentioned in (ii) is unique. If two distinct objects c, d satisfied condition (ii), we could link them by the chain of equalities $c = dc = (dc)d = d(dc) = dc = d$. The uniqueness of z is therefore not only true in S_3 but, more generally, in any mathematical system in which (i) and (ii) hold. We say that the uniqueness of z is a **consequence** of statements (i) and (ii).

The remarks just made indicate that even very simple engagement with mathematical content involves, more or less overtly, a linguistic element. This realisation would be an inconsequential truism if the development of model theory had not shown that paying sharper attention to this linguistic element, suitably prepared for mathematical use (via formalisation) discloses a new outlook on existing mathematical objects and theories,[7] as well as a new field of investigation with its distinctive problems and techniques.

In the next chapter we make our entry into this new environment by describing several formal languages and introduce the notion of a structure for a fixed formal language. With these notions at our disposal we shall be able to define key concepts like 'theory', 'model', 'consequence', 'definability', 'interpretation' and to carry out, in particular, a model-theoretic study of several algebraic theories.

1.6 Groups, Fields, Vector Spaces

The problem discussed in the first four sections of this chapter has not been chosen at random. Its treatment requires the introduction of certain algebraic structures, which belong to classes we shall axiomatise and study from a model-theoretic standpoint. These are the classes of groups, fields and vector spaces (we shall in fact only work with certain special subclasses of each class). In due course we shall also consider certain subclasses of the class of linear orders but for the moment we focus on algebraic structures alone.

Because we are not too far away from being able to axiomatise these classes, it is worth singling out their defining properties. As far as groups are concerned, we may again turn to the symmetric group S_3: we have already observed that it has a (unique)

---

[7] As Daniel Lascar insightfully and elegantly put it: 'je dirai plutôt que le théoricien des modèles est un mathématicien qui s'écoute parler. Il s'intéresse aux mêmes objets que le mathématicien classique [...]. Mail, il est conscient, par exemple, que le langage qui est nécessaire à la définition d'un corps valueé est beaucoup plus simple que celui requis pour définir un anneau noethérien. Et ces faits, il sait les exploiter.' [37, p.238].

neutral element e and that every element x of S_3 has a unique multiplicative inverse x^{-1}, i.e. $xx^{-1} = e = x^{-1}x$. Since the composition of permutations is associative, the following statements hold in the mathematical system S_3:

(1) for every $x, y, z, x(yz) = (xy)z$;
(2) for every $x, ex = x = xe$;
(3) for every $x, xx^{-1} = e = x^{-1}x$.

Clearly, (1)–(3) do not make any distinctive reference to S_3. We call any mathematical system in which they hold a **group** (this definition will be linguistically sharpened when we describe a formal language for groups in the next chapter).

Further examples of groups are now easy to produce: if we consider the set $\mathbb{Z}$ and the binary operation of addition between two integers, we easily see that this operation is associative and that that $\mathbb{Z}$ contains the additive neutral element 0 (thus, in $\mathbb{Z}$, the expressions ex, xe correspond to the more familiar expressions $0+x, x+0$) as well as the additive inverse $-n$ for each integer n. In short, $\mathbb{Z}$ is a group relative to arithmetical addition.

Similar considerations apply to $\mathbb{Q} - \{0\}$ equipped with multiplication. The neutral element is 1, and the inverse of a/b is b/a. It is equally straightforward to check that the set $\mathbb{Q}^2$ is a group relative to component-wise addition. So are $\mathbb{Q}^3, \mathbb{Q}^4, \ldots, \mathbb{Q}^n, \ldots$.

Exercise 1.20 Any structure that satisfies statement (1) above is a **semigroup**. A structure that satisfies statements (1) and (2) is a **monoid**. Every group is therefore both a semigroup and a monoid.

(i) Describe a monoid that is not a group.
(ii) Describe a semigroup that is not a monoid.
(iii) Explain why statement (3) cannot follow from statements (1) and (2).

Groups are not singled out by a unique set of statements. Multiple choices may work.

Exercise 1.21 If a nonemptyset G is equipped with a binary operation, denoted by juxtaposition, that satisfies the following statements:

(i) for every $x, y, z, x(yz) = (xy)z$;
(ii) for every x, y, there are u, z such that $xu = y$ and $zx = y$,

show that, relative to the operation of juxtaposition, G is a group.

Note that S_3 contains elements that do not satisfy the equality $xy = yx$, while this is not true of $\mathbb{Z}$ or $\mathbb{Q} - \{0\}$, relative to addition and multiplication respectively. We call the latter groups **abelian**. The class of abelian groups is singled out by conditions (1) to (3) above together with the following statement:

(4) for every $x, y, xy = yx$.

We shall mainly be interested in certain classes of abelian groups, to be described in Chap. 3. In Chap. 12, however, we deal with groups that are not abelian.

Exercise 1.22 The symmetric group S_3 is the smallest non-abelian group, since every group with less than six elements is abelian. Show that a three-element group must be abelian.

Exercise 1.23 Let G be a group. If $x, y \in G$, we call $xy \in G$ the *product* of x, y. We say that G is **finitely generated** if there is a finite subset F of G such that every element of G can be expressed as a product of elements of F and their inverses. Every finite group is clearly finitely generated.

(i) Show that the group $\mathbb{Z}$ of the integers equipped with addition is finitely generated.
(ii) Show that $\mathbb{Q}$ equipped with addition is not finitely generated.
(iii) What about $\mathbb{Q} - \{0\}$ equipped with multiplication?

Our discussion of groups easily allows us to realise that the set $\mathbb{Q}$ may be viewed as a group in two distinct, related ways. On the one hand, $\mathbb{Q}$ is an abelian group relative to addition, with neutral element 0 and inverse $-r$, for any $r \in \mathbb{Q}$. On the other hand, $\mathbb{Q} - \{0\}$ is an abelian group relative to multiplication, with neutral element 1 and inverse $1/r$, for any $r \in \mathbb{Q} - \{0\}$. The two, distinct group operations are related by the **distributive law**:

(5) for every x, y, z, $(x + y)z = xz + yz$.

Any set K with distinct elements 0, 1 and endowed with two group operations behaving like addition and multiplication in $\mathbb{Q}$ is a **field**. This means, in particular, that K is an additive group with a neutral element called 0, $K - \{0\}$ is a multiplicative group with a neutral element called 1 and $0 \neq 1$ (we shall see an explicit, axiomatic presentation of fields in Chap. 3).

Because a field may be regarded as an abelian group in an additive and in a multiplicative way, it is useful to have two distinct formulations of (1)–(4). So far we have formulated them in what is known as multiplicative notation (the symbol 1 is at times chosen over e). The additive notation yields:

(1′) for every x, y, z, $x + (y + z) = (x + y) + z$;
(2′) for every x, $0 + x = x = x + 0$;
(3′) for every x, $x + (-x) = 0 = (-x) + x$;
(4′) for every x, y, $x + y = y + x$.

Using the definitions of group and field now available, we may in the first instance define a vector space as a field acting on a group. It is easy to illustrate this initial definition with reference to the additive, abelian group $\mathbb{Q}^6$, on which the field $\mathbb{Q}$ acts 'multiplicatively', in the sense that, if $s \in \mathbb{Q}$, then it acts on the vector $(q_1, q_2, q_3, q_4, q_5, q_6) \in \mathbb{Q}^6$ by multiplying each of its components, thus:

$$s(q_1, q_2, q_3, q_4, q_5, q_6) = (sq_1, sq_2, sq_3, sq_4, sq_5, sq_6).$$

To have a vector space, we require a field acting on an abelian group in such a way that it satisfied the conditions listed in the next exercise.

1.6 Groups, Fields, Vector Spaces

Exercise 1.24 If $u, v \in \mathbb{Q}^6$ and $r, s \in \mathbb{Q}$, show that the following equalities hold:
(5) $s(u+v) = su + sv$;
(6) $(r+s)v = rv + sv$;
(7) $(rs)v = r(sv)$;
(8) $1v = v$.

An abelian group G that satisfies the equalities (5)–(8) relative to the action of a fixed field K is a **vector space over the field** K. In particular, $K^2, K^3, \ldots, K^n, \ldots$ are vector spaces over K, much like the finite Cartesian products of $\mathbb{Q}$ were vector spaces over $\mathbb{Q}$.

Exercise 1.25 Can $\mathbb{Q}$ be regarded as a vector space over $\mathbb{Q}$?

Chapter 2
Languages and Structures

Abstract The quote from Alfred Tarski in Sect. 1.5 described model theory as a study of the relation between mathematical systems and the sentences that hold in them. Tarski's description sets the key tasks of this chapter: we shall define mathematical systems (or structures, as we shall call them), the linguistic resources we are to employ, and a relation of 'satisfaction', to spell out, in mathematically useful terms, what we mean when we say that a particular statement 'holds' in a particular system or structure.

In Sects. 2.1 and 2.2 we specify our linguistic resources. In particular, we explicitly declare which symbol sets or alphabets we use and describe how we concatenate symbols to form expressions belonging to our syntactic categories, i.e. terms and formulae.

The next task is to establish a systematic correspondence between formulae and certain set-theoretic objects, which we refer to as structures. This task is carried out in Sect. 2.3, where the notion of a structure, which is relative to a fixed language, is introduced and its connection with a formal language is given by defining the satisfaction relation. Satisfaction holds between a formula φ and a structure $\mathcal{M}$ when φ is true (in a sense to be specified) in $\mathcal{M}$, possibly after the variables occurring in φ have been assigned fixed values.

In Sect. 2.4 we use satisfaction to define the fundamental relation of consequence between two formulae or, more generally, between a set of formulae and a formula.

Section 2.5 is devoted to proofs by induction on the complexity of formulae, which yield some abstract theorems about formulae and provide an opportunity to practice a proof technique that will often be appealed to in later chapters.

The last section, Sect. 2.6, contains a succinct discussion of definability and interpretations.

2.1 Terms

Let us briefly consider the sentences (1') to (4') given in Sect. 1.6 as an axiomatisation of the class of abelian groups.

Although the sentences we have employed were not formulated in a specified formal language, some of their constituents already functioned in the way we want our syntax to behave. Let us explicitly identify these constituents: some of them are variables like x, y, z, which occur e.g. in (1'). We next have the constant symbol 0, employed to designate the (unique) neutral element of a group. Finally, we introduced two function symbols, namely $+$ and $-$, in order to describe, respectively, the group operation and the operation associating with any given group element its inverse.

Variables, constant symbols and function symbols are exactly the linguistic items we shall use to define our first syntactic category, that of terms. As a matter of fact, we have also employed brackets to write expressions like $x + (y + z)$. We might, however, dispense with them by a slight change of notation (which we rely upon only in this chapter). Instead of the *infix* notation $x + y$, we might make use of the *prefix* notation $+ xy$. We are then able to write e.g. $+ + xyz$ instead of $(x + y) + z$ and $+ x + yz$ instead of $x + (y + z)$. We can also write $+ x - z$ in place of $x + (-z)$.

Restricting attention for the moment to the language of groups only, we focus on expressions in prefix notation that may be used to name individual group elements. Consider for instance the additive group based on the set of integers $\mathbb{Z}$ and the string $+ + xyz$: if we assign $1, 2, 3 \in \mathbb{Z}$ to the variables x, y, z respectively, then the string $+ + xyz$ names the number $(1 + 2) + 3 = 6$. Similarly, when 0 refers to the neutral element of $\mathbb{Z}$, the string $+ 0 + + x - y + y - x$ names the integer $0 + ((1 + (-2)) + (2 - 1)) = 0$.

It is clear that the same strings could be used to name elements of a different group, say S_3. In this latter case we could have assigned $a, b, b^2 \in S_3$ to the variables x, y, z respectively, thus naming the group element $(ab)b^2 = a$ by the string $++xyz$ and the group element $e = e((ab^{-1})(ba^{-1}))$ by the string $+ 0 + + x - y + y - x$.

We may think of strings naming group elements as constant symbols (0 directly names the neutral element of a group) or 'operation schemas', telling us which operations, and how nested, to apply in order to determine a specific group element. For instance $+ + xyz$ tells us that we have to add whichever group elements are associated with x, y and then add the group element associated with z to their sum.

Not all finite sequences of variables and symbols from $\{0, +, -\}$ are 'operation schemas' in the sense just described. For instance, the sequence of symbols $+ + -$ does not provide information about the arguments to which the function symbols $+, +, -$ should be applied. In a similar manner, $0xy$ exhibits possible arguments but no operation that might be applied to them. Finally, $+ xyz$ contains arguments and a function symbol, but the arguments are too many, since we require $+$ to take only two.

In view of these difficulties, we see the importance of declaring the number of arguments that each function symbol takes. This number is known as the function symbol's **arity**. In our running example, the arity of $+$ is 2 and the arity of $-$ is 1. It is customary to think of arity as a function **a** defined on a set of constant and function symbols. The arity of a constant symbol is always 0, while the arity of a function symbol is a positive natural number. In the context of group theory, a

2.1 Terms

is a function with domain $\{0, +, -\}$ and codomain $\mathbb{N}$, defined by the equalities: $a(0) = 0, a(-) = 1, a(+) = 2$.

For the sake of later generalisation, it is convenient to split $\{0, +, -\}$ into a constant symbol set $\mathbf{C} = \{0\}$ and a function symbol set $\mathbf{F} = \{+, -\}$. The linguistic resources we use to build strings that name group elements are uniquely determined by the triple:

$$\tau = (\mathbf{F}, \mathbf{C}, \mathbf{a}). \tag{2.1}$$

A τ-string is a *finite* sequence of variables[1] and symbols in $\mathbf{C} \cup \mathbf{F}$. A set S of τ-strings is *closed* under $+$ and $-$ if, and only if, the conditions $r, s \in S$ imply $+rs \in S$ and $-r \in S$. Let V be the set of variables. We are interested in a special set of τ-strings, which we call the set of τ-terms.

Definition 2.1 The set T of τ-**terms** is the smallest set of τ-strings such that $\mathbf{C} \cup V \subseteq T$ and T is closed under $+$ and $-$.

In the context of Definition 2.1 'the smallest set' means that any set U of τ-strings that contains $\mathbf{C} \cup V$ and is closed under $+, -$ includes T.

The definition of T does not merely allow us to select certain linguistic strings of interest from a larger set of strings. It also yields a proof technique. To show that a statement A holds for every τ-term, it suffices to consider the set S of τ-terms for which A holds and then prove that: (i) $\mathbf{C} \cup V \subseteq S$ and (ii) S is closed under $+, -$. If (i) and (ii) can be proved, the fact that T is the smallest set to satisfy the last conditions implies $T \subseteq S$. By definition of S, $S \subseteq T$. We conclude $T = S$, i.e. A holds for every τ-term.

The proof technique just outlined is known as **induction on the complexity of terms** and it consists of two parts. Part (i) above is the **inductive base** and part (ii) is the **inductive step**.

We are now in a position to carry out an actual proof by induction on the complexity of terms. We wish to show that any τ-term t can be transformed into a τ-term t' by substituting every occurrence of the variable x in t with any fixed τ-term.[2] Thus, for instance, we may substitute x with 0 in the τ-term $+0x$ and obtain the τ-term $+00$, or we may substitute x with $+xy$ in $+0x$ and obtain $+0+xy$. For any τ-terms $t(x), s$ (we use the notation $t(x)$ to emphasise the fact that x may occur, possibly more than once, in t), let $t(x/s)$ be the τ-term obtained by replacing each occurrence of x in t with the term s.

Instead of restricting attention to $t(x)$, we consider τ-terms of the form $t(x, y)$, i.e. whose variables are elements of $\{x, y\}$ (the τ-terms $0, y, +0x, +x + xy$ are of the prescribed type). What we are about to prove applies to the special case of $t(x)$, which is a τ-term of the form $t(x, y)$ containing no occurrences of y.

---

[1] We assume that we are always in possession of an infinite sequence of variables $x_0, x_1, x_2, \ldots,$ which we may also write in an index-free notation, e.g. as $x, y, z, \ldots$.

[2] If x does not occur in t, the substitution is vacuous and t' is t.

Lemma 2.2 *If $t(x, y), r, s$ are τ-terms, then $t(x/r, y/s)$ is a τ-term.*

Proof We proceed by induction on the complexity of terms. Let S be the set of τ-terms for which the lemma holds. If $t(x, y)$ is a constant symbol, then $t(x, y) = 0$ and the substitution of r, s for x, y respectively is vacuous. As a consequence $t(x/r, y/r) = 0$. If $t(x, y)$ is a variable, then $t(x, y) = x$, $t(x, y) = y$ or $t(x, y) \neq x, y$. In the first case, $t(x/r, y/s) = r$. In the second case $t(x/r, y/s) = s$. In the remaining cases, $t(x/r, y/s) = t(x, y)$. In any case we end up with a τ-term.

So far we have only used the fact that variables and the symbol 0 are τ-terms. We are now going to use the fact that τ-terms are closed under $+$ and $-$. We suppose that the lemma holds for $t_1(x, y), t_2(x, y)$: this is our **inductive hypothesis**. We make use of the inductive hypothesis to show that the lemma holds.

Consider the τ-term $t(x, y) = +t_1(x, y)t_2(x, y)$. By the inductive hypothesis, $t_1(x/r, y/s), t_2(x/r, y/s)$ are τ-terms.

By closure under $+$, the string $+t_1(x/r, y/s)t_2(x/r, y/s)$ is a τ-term, namely $t(x/r, y/s)$. We leave the proof that S is closed under $-$ as an exercise. The induction is complete. $\square$

Exercise 2.3 Complete the proof of Lemma 2.2.

Exercise 2.4

(i) Identify the inductive base and the inductive step in the proof of Lemma 2.2.
(ii) Let s be a τ-string. Any τ-string s^* obtained from s by deleting the symbols occurring in s consecutively from left to right and from some point on is called an *initial segment* of t. Prove by induction on the complexity of terms that no initial segment of a τ-term is a τ-term.

So far we have given a full treatment of terms only relative to a specific symbol set, appropriate for groups. Other structures are associated with different symbol sets, which gives rise to different sets of terms. When talking about fields, we may for instance extend the symbol sets used for groups to $\mathbf{C}' = \{0, 1\}$ and $\mathbf{F}' = \{+, -, \cdot\}$, where the relevant arity function a' is determined by the equalities: $\mathsf{a}'(0) = 0 = \mathsf{a}'(1), \mathsf{a}'(-) = 1$ and $\mathsf{a}'(+) = 2 = \mathsf{a}'(\cdot)$. Setting $\tau' = (\mathbf{F}', \mathbf{C}', \mathsf{a}')$, we are in a position to define the set of τ'-terms and carry out proofs by induction on their complexity.

Exercise 2.5 Define the set of τ'-terms along the lines of Definition 2.1 and prove Lemma 2.2 for τ'-terms.

The τ'-strings $\cdot xy, +\cdot xz + \cdot xx \cdot xy, +1 + z + \cdot zz \cdot z \cdot zz$ are τ'-terms. If we abbreviate $\underbrace{x \cdots \cdots x}_{n \text{ times}}$ by x^n and return to the familiar, infix notation,[3] the τ'-terms just listed are $x \cdot y$, $x \cdot z + x \cdot y + x^2$ and $1 + z + z^2 + z^3$. In the next chapter, we

---

[3] We should, in principle, stipulate how brackets are to be inserted in expressions like an iterated product of x times itself. Since we are always going to work with binary operations that are associative, the positioning of brackets does not in fact matter.

2.1 Terms

shall refer to the τ'-terms as the set of *polynomials with integer coefficients*. If **C'** is extended to include a constant symbol for each $q \in \mathbb{Q}$, the corresponding set of terms is the set of *polynomials with rational coefficients*.

An alternative way of expanding the linguistic resources used for groups arises when we consider vector spaces over a fixed field, say $\mathbb{Q}$. In this context we continue to have exactly one constant symbol: $\mathbf{C}'' = \{0\} = \mathbf{C}$. By contrast, the set of function symbols is no longer finite, because it supplements $+$ and $-$ with an infinite set of unary function symbols. More precisely, the new function symbol set is $\mathbf{F}'' = \mathbf{F} \cup \{f_q\}_{q \in \mathbb{Q}}$, where the arity a'' of the old symbols stays the same and $\mathsf{a}''(f_q) = 1$ for every $q \in \mathbb{Q}$.

Definition 2.6 The set T'' of τ''-terms is the smallest set of τ''-strings including $\mathbf{C}' \cup V$ and closed under the function symbols in $\mathbf{F}''$.

A generic τ''-term is of the form:

$$f_{q_1} x_1 + \ldots + f_{q_n} x_n$$

and names a linear combination of the vectors assigned to the variables $x_1, \ldots, x_n$. Suitable modifications of $\mathbf{F}''$ allow us to identify sets of terms for vector spaces over any given field K.

The examples we have gone through make it plain to see what an abstract definition of τ-term, relative an arbitrary symbol set $\mathbf{C} \cup \mathbf{F}$ and the corresponding arity function a, must be like.

Definition 2.7 The set of τ-terms is the smallest set of τ-strings that contains $\mathbf{C} \cup V$ and is closed under the function symbols in $\mathbf{F}$.

From now on we shall simply talk about terms, as opposed to τ-terms. It will generally be clear from the context of our discussion whether we are working with a specific symbol set or with an arbitrary one. Under certain circumstances, it is helpful to have a name for terms in which no variables occur.

Definition 2.8 A **closed term** is a variable-free term.

It is also useful to single out those terms in which the application of function symbols is not iterated.

Definition 2.9 Let $\mathbf{C} \cup \mathbf{F}$ be a set of constant and function symbols, with associated arity function a. An **unnested term** is either a constant symbol from $\mathbf{C}$, a variable or a term of the form $f(x_1, \ldots, x_n)$, where $f \in \mathbf{F}$ and $\mathsf{a}(f) = n$.

Exercise 2.10

(i) Give examples of two closed τ''-terms and of two τ''-terms that are not closed.
(ii) Describe the set of τ-terms when $\mathbf{C} = \emptyset = \mathbf{F}$.

2.2 Formulae

Since we are not only interested in naming objects, but also in formalising mathematical statements, we have to integrate terms into more complex strings that involve new symbols. These strings we call formulae.

We begin with an illustrative example. In Sect. 1.2 we gave axioms that single out the class of linear orders: the only terms used in that context were variables like x, y, z, which we integrated into inequalities like $x < y$ or $y < z$, involving the symbol $<$.

Our discussion of preorders required the introduction of further symbols like $\prec, \succcurlyeq, \sim$. With their help we could transpose simple statements (e.g. 'x is smaller than y', 'x is equivalent to y') into a symbolic notation. These symbols are not function symbols because they do not contribute to naming a uniquely determined object. What they do is state that their arguments stand in a particular relation: we unsurprisingly refer to them as **relation symbols**.

We can now view the alphabet of a formal language as consisting not only of the symbols in $\mathbf{F} \cup \mathbf{C}$, which we use to form terms, but also of the relation symbols in a further set $\mathbf{R}$, which may or may not be empty. Notationally, we refer to the elements of $\mathbf{R}$ by the letters $P, Q, R, \ldots$ or possibly indexed variants like $P_0, P_1, P_2, \ldots,$ $R_0, R_1, R_2, \ldots$ et caetera.

The arity function a, defined on $\mathbf{F} \cup \mathbf{C}$, extends to $\mathbf{R} \cup \mathbf{F} \cup \mathbf{C}$. For instance, $\mathsf{a}(<) = 2 = \mathsf{a}(\succcurlyeq)$. Even though we shall mainly focus on binary relations, the elements of $\mathbf{R}$ may be of any finite arity. If, for instance, we wished to introduce a symbol $B \in \mathbf{R}$ to describe the relation that holds between three points on a line when one lies between the other two, we should have $\mathsf{a}(B) = 3$. A symbol $D \in \mathbf{R}$ used to describe the relation that holds between four points A, B, C, D when the segment AB is congruent to the segment CD should have arity equal to 4. If we adopted a symbol $Q \in \mathbf{R}$ to describe the relation that holds among six rational numbers when, in the given order, they determine a strongly transitive vector in $\mathbb{Q}^6$, we must set $\mathsf{a}(Q) = 6$.

Remark 2.11 We do not regard the identity symbol $=$ as a relation symbol in $\mathbf{R}$. Whereas the symbols in $\mathbf{R}$ denote distinct relations across distinct contexts of enquiry, the identity symbol shall always denote the relation of identity or equality.

Given the sets $\mathbf{R}, \mathbf{F}, \mathbf{C}$, we call $\tau = (\mathbf{R}, \mathbf{F}, \mathbf{C}, \mathsf{a})$ a **signature**. The signature τ determines, as we know, the set of τ-terms. It also determines the set of **atomic τ-formulae**:

Definition 2.12 An atomic τ-formula is either an equality of the form $t = s$, where t, s are τ-terms, or a string of the form $Rt_1 t_2 \ldots t_n$, where $R \in \mathbf{R}$, $\mathsf{a}(R) = n$ and $t_1, t_2, \ldots, t_n$ are τ-terms. Henceforth we shall employ the notation $Rt_1 \ldots t_n$ and the notation $R(t_1 \ldots t_n)$ interchangeably.

When the signature we are working it is unambiguous, we may simply talk about atomic formulae, as opposed to atomic τ-formulae. It follows from Definition 2.14

2.2 Formulae

that, if $\mathbf{R} = \emptyset$, the only atomic formulae are equations. In Sect. 2.6 it will be helpful to work with a particular subset of the set of atomic formulae, defined below.

Definition 2.13 Given a signature τ, a τ-string is said to be an **unnested atomic formula** iff it is either of the form $R(x_1, \ldots, x_n)$, where $R \in \mathbf{R}$ and $\mathsf{a}(R) = n$, or of the form $t = s$, where t, s are unnested τ-terms.

Given a signature τ, the atomic τ-formulae alone are insufficient to convey the mathematical content we might wish to formalise. For instance, they do not allow us to negate an atomic formula, to state an alternative or to say that two atomic formulae hold at once.

These issues already arise when we single out the class of linear orders using the signature $\tau = (\{<\}, \mathsf{a})$. Irreflexivity requires that we negate the atomic formula $x < x$; transitivity, that we assert $x < y$ and $y < z$ simultaneously, and, moreover, an implication from their conjunction to $x < z$; trichotomy, that we state the alternative $x = y$ or $x < y$ or $y < x$. We use special symbols known as **connectives** to deal with all the issues raised. The connectives are called:

- **negation**, symbolised by $\neg$ ('not');
- **conjunction**, symbolised by $\wedge$ ('and');
- **disjunction**, symbolised by $\vee$ ('or');
- **implication**, symbolised by $\rightarrow$ ('if ... then').

Using the connectives, we arrive at the following semiformal axiomatisation of linear orders:

- for all x, $\neg(x < x)$;
- for all x, y, z$((x < y \wedge y < z) \rightarrow (x < z))$;
- for all x, y$(\neg(x = y) \rightarrow (x < y \vee y < x))$.

For a complete formalisation we only have to add the **universal quantifier** $\forall$ as an abbreviation of the locution 'for all'. We thus arrive at a list of formal axioms for linear orders, namely:

(a) $\forall x \neg (x < x)$;
(b) $\forall x \forall y \forall z ((x < y \wedge y < z) \rightarrow (x < z))$;
(c) $\forall x \forall y (\neg(x = y) \rightarrow (x < y \vee y < x))$.

A linear order $\mathcal{O}$ with domain O is **dense** if, for every $x, z \in O$, $x < z$ implies that there is $y \in O$ such that $x < y$ and $y < z$. In order to express the last property, we introduce the **existential quantifier** $\exists$ ('there is'), which allows us to define the dense linear orders as the linear orders satisfying the following formula:

(d) $\forall x \forall z \forall z (x < z \rightarrow \exists y (x < y \wedge y < z))$.

Definition 2.14 Let $\mathbf{L}$ be the set $\{\neg, \wedge, \vee, \rightarrow, \exists, \forall, =\ (,)\} \cup \{x_i\}_{i \in \mathbb{N}}$. Given a signature $\tau = (\mathbf{R}, \mathbf{F}, \mathbf{C}, a)$, we call $\mathcal{L}(\tau) = \mathbf{L} \cup \mathbf{R} \cup \mathbf{F} \cup \mathbf{C}$ the **first-order language of signature** τ. Unless it should prove necessary to specify the signature, we shall refer to a first-order language simply as $\mathcal{L}$, with a possible addition of subscripts, instead of $\mathcal{L}(\tau)$.

For every signature, we have an associated first-order language: distinct languages correspond to distinct contexts of enquiry and, in particular, distinct classes of structures, as we shall see in the next section. On the other hand, all first-order languages share the symbol set **L**, whose elements are often called *logical symbols*, as opposed to the *extra-logical* symbols in the signature τ. In the course of this book, we shall often be interested in certain specific languages. They are listed below for future reference:

- the language of sets $\mathcal{L}_= = \mathbf{L}$;
- the language of linear orders $\mathcal{L}_< = \mathbf{L} \cup \{<\}$;
- the language of groups $\mathcal{L}_g = \mathbf{L} \cup \{0, +, -\}$;
- the language of rings $\mathcal{L}_r = \mathcal{L}_g \cup \{1, \cdot\}$;
- the language of vector spaces over the field K $\mathcal{L}_v = \mathcal{L}_g \cup \{f_s\}_{s \in K}$;
- the language of ordered groups $\mathcal{L}_{g<} = \mathcal{L}_g \cup \mathcal{L}_<$;
- the language of ordered rings $\mathcal{L}_{r<} = \mathcal{L}_r \cup \mathcal{L}_<$.

A first-order language $\mathcal{L}$ determines a set of finite sequences of symbols, the $\mathcal{L}$-strings. Since we are not interested in every possible string, we single out the subset of $\mathcal{L}$-strings that we shall call $\mathcal{L}$-formulae. We do so, as in the case of terms, by specifying what syntactic operations we admit as legitimate to produce $\mathcal{L}$-formulae from atomic $\mathcal{L}$-formulae.

Definition 2.15 The set $\mathcal{F}$ of $\mathcal{L}$-formulae is the smallest set of $\mathcal{L}$-strings such that:

(1) $\mathcal{F}$ includes the atomic $\mathcal{L}$-formulae;
(2) if $\mathcal{F}$ contains the $\mathcal{L}$-strings ϕ, ψ, then it also contains the $\mathcal{L}$-strings $\neg(\phi)$, $(\phi \vee \psi)$, $(\phi \wedge \psi)$, $(\phi \rightarrow \psi)$;
(3) if $\mathcal{F}$ contains the $\mathcal{L}$-string ϕ and x is a variable, then $\mathcal{F}$ also contains the $\mathcal{L}$-strings $\forall x \phi$ and $\exists x \phi$.

Parts (2) and (3) of Definition 2.15 spell out what are known as the *closure properties* of $\mathcal{F}$. They amount to the fact applying connectives and quantifiers (with variables) to a formula produces a formula. Exactly as the definition of terms allowed us to introduce proofs by induction on their complexity, Definition 2.15 allows us to introduce proofs by **induction on the complexity of formulae**. The proof strategy should sound familiar: let S be the set of $\mathcal{L}$-formulae for which a certain result A holds. If every atomic $\mathcal{L}$-formula is in S and S is closed under the application of connectives and quantifiers, then $S = \mathcal{F}$ and A is true of every formula.

We postpone a detailed discussion of proofs by induction on the complexity of formulae until Sect. 2.5.

Exercise 2.16 Express the formal properties stated in Exercise 11.22 and in Exercise 1.24 by means of $\mathcal{L}(\tau)$-formulae, for suitable choices of the signature τ.

Exercise 2.17

(i) Write $\mathcal{L}_=$-formulae saying, respectively, that there are two distinct elements and that there are three distinct elements;

2.2 Formulae

(ii) A linear order is said to be without endpoints if it lacks a least and a greatest element. State this property as a $\mathcal{L}_<$-formula;

(iii) Suppose $a, b \in G$ where G is a nonempty set that determines a group $\mathcal{G}$ relative to the operation of juxtaposition. We say that a, b commute if $ab = ba$. The **centre** Z_G of $\mathcal{G}$ is the set containing the elements of G that commute with every element of G. Use a $\mathcal{L}_g$-formula to state that, if the centre of a group coincides with the whole group, then the group in question is abelian;

(iv) Given a group $\mathcal{G}$, written additively, suppose that there is $x \in G$ such that $nx = \underbrace{x + \ldots + x}_{n\,times} = 0$. The set $\{x, x + x, x + x + x, \ldots, nx\}$ is called a **cycle** of order n in G. Write a $\mathcal{L}_g$-formula saying that there is a cycle of order four.

Some substrings of a formula are themselves formulae: we call them **subformulae**. Given a specific formula ϕ, the following list of rules provides a mechanical way of identifying the subformulae of φ:

(1) if φ is atomic, then φ is its only subformula;
(2) if φ is obtained from ϕ by an application of $\neg$, the subformulae of φ are $\neg\phi$ and the subformulae of ϕ;
(3) if φ is obtained from ϕ, ψ by an application of $\vee$, the subformulae of φ are: $(\phi \vee \psi)$, the subformulae of ϕ, and the subformulae of ψ. Analogous rules hold for $\wedge$ and $\rightarrow$;
(4) if φ is obtained from ϕ by prefixing $\exists x$ to it, the subformulae of φ are $\exists x \phi$ and the subformulae of ϕ. An almost identical rule holds for the quantifier $\forall$.

Exercise 2.18 Use rules (1)–(4) above to list the subformulae of each formula used to define a linear order.

2.2.1 Open and Closed Formulae

Let us consider the $\mathcal{L}_<$-formulae $x < y$ and $\forall x \neg (x < x)$: no variable in the first formula is subject to the action of a quantifier, whereas both occurrences of the variable x in the second formula are. In the first case, we say that x and y **occur free** in $x < y$, whereas in the second case we say that x does not occur free in $\forall x \neg (x < x)$ or that it is a **bound variable** of $\forall x \neg (x < x)$. A variable x is free in a formula ϕ if ϕ contains any free occurrences of x and it is bound in ϕ otherwise. For instance, x is free in $\forall x (x < y) \vee x < x$, since two occurrences of x in the last formula are free.

If ϕ is a formula, let $\ell(\phi)$ be the set of variables that occur free in ϕ. The rules listed below allow us to compute the set of free variables associated with a formula.

(1) if φ is atomic, $\ell(\varphi)$ is the set of all variables occurring in φ;
(2) if φ is of obtained from ϕ by an application of $\neg$, then $\ell(\varphi) = \ell(\phi)$;

(3) if φ is obtained from ϕ, ψ by an application of $\wedge$, $\vee$ or $\to$, then $\ell(\varphi) = \ell(\phi) \cup \ell(\psi)$;
(4) if φ is obtained from ϕ by prefixing to it $\exists x$ or $\forall x$, then $\ell(\varphi) = \ell(\phi) - \{x\}$.

Example 2.19 If φ is the $\mathcal{L}_<$-formula $x < y$, rule (1) immediately yields $\ell(\varphi) = \{x, y\}$.

Example 2.20 If φ is the $\mathcal{L}_<$-formula:

$$\forall x_1 (x_1 < y \to \exists x_2 (y < x_2 \wedge x_2 < x)),$$

we note that φ is obtained by prefixing $\forall x_1$ to a specific $\mathcal{L}_<$-formula. Rule (4) applies and it yields:

$$\ell(\varphi) = \ell((x_1 < y \to \exists x_2 (y < x_2 \wedge x_2 < x))) - \{x_1\}. \tag{2.2}$$

Next, we apply rule (3) to $(x_1 < y \to \exists x_2(y < x_2 \wedge x_2 < x))$. It follows that $\ell((x_1 < y \to \exists x_2(y < x_2 \wedge x_2 < x))) = \ell(x_1 < y) \cup \ell(\exists x_2(y < x_2 \wedge x_2 < x))$. By rule (1), $\ell(x_1 < y) = \{x_1, y\}$. By rules (4), (3) and (1), applied in this order, we obtain:

$$\begin{aligned}
\ell(\exists x_2(y < x_2 \wedge x_2 < x)) &= \ell(y < x_2 \wedge x_2 < x) - \{x_2\} \\
&= \ell(y < x_2) \cup \ell(x_2, x)) - \{x_2\} \\
&= \{y, x_2\} \cup \{x_2, x\}) - \{x_2\} \\
&= \{y, x_2, x\} - \{x_2\} \\
&= \{y, x\}.
\end{aligned}$$

The above chain of equalities allows us to deduce:

$$\ell((x_1 < y \to \exists x_2(y < x_2 \wedge x_2 < x))) = \{x_1, y\} \cup \{y, x\} = \{x_1, y, x\}.$$

By (2.2), we conclude $\ell(\varphi) = \{y, x\}$.

Exercise 2.21 Use rules (1)–(4) to compute $\ell(\varphi)$ when:

(i) φ is the $\mathcal{L}_<$-formula $(\forall x(x < y \to \exists x_2(y < x_2 \wedge x_2 < x)) \vee x_2 < y)$;
(ii) $\mathcal{L}_P$ is the first-order language obtained by adding the unary relation symbol P to $\mathcal{L}_=$ and φ is the $\mathcal{L}_P$-formula $(P(y) \to \exists y(P(y) \to (x = z)))$;
(iii) φ is the $\mathcal{L}_{g<}$-formula $(\forall x_1 \exists x_2 \forall x_3 (x_1 + x_2 = x_2) \to \exists y(y + 0 < 0))$.

We conclude this section with the important:

Definition 2.22 Let $\mathcal{L}$ be a first-order language. A $\mathcal{L}$-formula φ is **closed** if $\ell(\varphi) = \emptyset$ and **open** otherwise. Closed $\mathcal{L}$-formulae are also called $\mathcal{L}$-**sentences**

In the next chapter, we shall define a first-order *theory* as a particular set of sentences.

2.3 Structures

We now know what a formula is. Formulae are of interest because they may be true or false, hold or fail in a given context. This context of reference is what we call a structure. To be precise, we work with the notion of a $\mathcal{L}$-structure, where $\mathcal{L}$ is a fixed language.

We want $\mathcal{L}$-structures to be objects whose distinctive features correspond to the extra-logical symbols in the first-order language $\mathcal{L}$. Once more we postpone an abstract definition and look at some examples for motivation first.

Example 2.23 Consider the set $\mathbb{Z}$ and the language $\mathcal{L}_<$. When we treat $\mathbb{Z}$ as a linear order, we associate with the symbol $<$ a specific subset $<^{\mathbb{Z}}$ of $\mathbb{Z}^2$, i.e. the set-theoretic relation 'smaller than' on the integers. We may think of this association as a way of using (the signature of) $\mathcal{L}_<$ to equip $\mathbb{Z}$ with a binary relation, thus obtaining the pair:

$$(\mathbb{Z}, <^{\mathbb{Z}}).$$

Example 2.24 Consider the set $\mathbb{Z}$ and the language $\mathcal{L}_{g<}$. When we treat $\mathbb{Z}$ as an ordered group, we effectively expand the linear order:

$$(\mathbb{Z}, <^{\mathbb{Z}})$$

to a richer structure. We do so by associating a distinguished integer $0^{\mathbb{Z}}$ with the constant symbol $0 \in \mathbf{C}$ (the superscript $\mathbb{Z}$ enables us to tell apart the number and the constant symbol naming it). Moreover, to the function symbol $+ \in \mathbf{F}$, of arity 2, we associate the binary operation of addition $+^{\mathbb{Z}} \subseteq \mathbb{Z}^3$, whose elements are of the form $(x, y, x+y)$. Finally, to the function symbol $- \in \mathbf{F}$, of arity 1, we associate the unary operation $-^{\mathbb{Z}} \subseteq \mathbb{Z}^2$, whose elements are of the form $(n, -n)$. The resulting structured set is the ordered quadruple:

$$(\mathbb{Z}, \{\{<^{\mathbb{Z}}\}, \{+^{\mathbb{Z}}\} \cup \{-^{\mathbb{Z}}\}, \{0^{\mathbb{Z}}\}),$$

in which we collect together the set-theoretic objects associated with each set in the signature of $\mathcal{L}_{g<}$.

Example 2.25 Consider the set $\mathbb{Q}^6$ and the language $\mathcal{L}_v$. When treating $\mathbb{Q}^6$ as a vector space over the field $\mathbb{Q}$, we associate the constant symbol $0 \in \mathbf{C}$ with a distinguished element of $\mathbb{Q}^6$, namely $(0, 0, 0, 0, 0, 0)$ (for notational convenience we avoid using $0^{\mathbb{Q}}$ in this context). To the function symbol $+ \in \mathbf{F}$ we associate the binary operation $+^{\mathbb{Q}^6}$ of component-wise addition between elements of $\mathbb{Q}^6$, which is a subset of $\mathbb{Q}^6 \times \mathbb{Q}^6 \times \mathbb{Q}^6$. We may represent $+^{\mathbb{Q}^6}$ as the set:

$$\{((a_1, \ldots, a_6), (b_1, \ldots, b_6), (c_1, \ldots, c_6)) \in Q^6 : c_i = a_i +^{\mathbb{Q}} b_i, i \in \{1, \ldots, 6\}\}$$

where $+^{\mathbb{Q}}$ is ordinary addition on $\mathbb{Q}$. In a similar manner, to the symbol $- \in \mathbf{F}$ we associate a unary operation on $\mathbb{Q}^6$, namely:

$$-^{\mathbb{Q}^6} = \{((a_1, \ldots, a_6), (b_1, \ldots, b_6)) \in \mathbb{Q}^6 : b_i = -^{\mathbb{Q}} a_i, i \in \{1, \ldots, 6\}\}.$$

Finally, for each $r \in \mathbb{Q}$, we assign to the unary function symbol $f_r \in \mathbf{F}$ the set:

$$f_r^{\mathbb{Q}^6} = \{((a_1, \ldots, a_6), (b_1, \ldots, b_6)) \in \mathbb{Q}^6 : b_i = r a_i, i \in \{1, \ldots, 6\}\}.$$

where juxtaposition is ordinary multiplication in $\mathbb{Q}$.

We have used the signature $\tau(\mathbf{C}, \mathbf{F}, \mathsf{a})$ of $\mathcal{L}_v$ to equip the set $\mathbb{Q}^6$ with the features of a vector space. The resulting structured set is:

$$(\mathbb{Q}^6, \{+^{\mathbb{Q}^6}, -^{\mathbb{Q}^6}\} \cup \{f_r^{\mathbb{Q}^6}\}, \{0^{\mathbb{Q}^6}\}).$$

An abstract generalisation of what we have seen in the preceding examples leads to:

Definition 2.26 Let $\mathcal{L}$ be a first-order language of signature $\tau = (\mathbf{R}, \mathbf{F}, \mathbf{C}, \mathsf{a})$. A $\mathcal{L}$-**structure** $\mathcal{M}$ is an ordered quadruple $(M, \mathbf{R}^M, \mathbf{F}^M, \mathbf{C}^M)$ such that:

(1) $M \neq \emptyset$;
(2) $\mathbf{R}^M$ contains a j-ary relation $R^M \subseteq M^j$ for each $R \in \mathbf{R}$ such that $\mathsf{a}(R) = j$;
(3) $\mathbf{F}^M$ contains a k-ary function $f^M \subseteq M^{k+1}$ for each $f \in \mathbf{F}$ such that $\mathsf{a}(f) = k$;
(4) $\mathbf{C}^M$ contains an object $a \in M$ for each $c_a \in \mathbf{C}$.

The set M from the last definition is the **domain** of the $\mathcal{L}$-structure $\mathcal{M}$. When we deal with a finite signature, we may not view a structure as an ordered quadruple: if, for instance, there are n_1 relation symbols, n_2 function symbols and n_3 constant symbols, then $\mathcal{M}$ will typically be conceived as an ordered $(n_1 + n_2 + n_3)$-tuple.

2.3.1 Satisfaction

Structures are the domains of discourse of languages. Formulae, if they are to be understood as meaningful statements, say something that may be true or false in a structure. We now want to give explicit conditions under which, given a first-order language $\mathcal{L}$, a $\mathcal{L}$-formula ϕ is satisfied by a $\mathcal{L}$-structure $\mathcal{M}$. This satisfaction relation will allow us to distinguish true and false statements about a particular context of reference. Before providing the definition of satisfaction, let us restrict attention to a few simple examples involving atomic formulae.

Example 2.27 Up to renaming variables, there only are two atomic $\mathcal{L}_=$-formulae, that is $x = x$ and $x = y$. The second formula is satisfied by a $\mathcal{L}_=$-structure $\mathcal{M}$ exactly when x, y name the same element of M. The first formula is satisfied by

2.3 Structures

$\mathcal{M}$ irrespective of which element of M is named by x. Since variables name objects assigned to them, their values are determined only after an assignment σ has been fixed. Formally speaking, an assignment σ in $\mathcal{M}$ is a function with domain V and codomain M. We call the respective values of x, y $\sigma(x), \sigma(y) \in M$. When $\sigma(x)$ and $\sigma(y)$ are the same object $a \in M$, we say that the formula $x = y$ is satisfied by $\mathcal{M}$ with σ and we write:

$$\mathcal{M} \models x = y[\sigma].$$

Example 2.28 Consider the $\mathcal{L}_<$-structure $(\mathbb{Z}, <^\mathbb{Z})$. Given σ, we can determine whether $x < y$ is satisfied by $(\mathbb{Z}, <^\mathbb{Z})$ with σ by comparing $\sigma(x), \sigma(y)$. If e.g. $\sigma(x) = -2$ and $\sigma(y) = 3$, we conclude:

$$(\mathbb{Z}, <^\mathbb{Z}) \models x < y[\sigma].$$

Example 2.29 Consider the group of additive integers. This is a $\mathcal{L}_g$-structure with domain $\mathbb{Z}$. To determine whether $x_1 + x_2 = x_3 - x_4$ (where the right-hand side abbreviates the term $x_3 + (-x_4)$) is satisfied by the group $(\mathbb{Z}, +^\mathbb{Z}, -^\mathbb{Z}, 0^\mathbb{Z})$ with an assignment σ, we need to compute the σ-values of the $\mathcal{L}_g$-terms $x_1 + x_2$ and $x_3 - x_4$. If, for instance, $\sigma(x_1) = -1 = \sigma(x_4)$ and $\sigma(x_2) = 5 = \sigma(x_3)$, then $\sigma(x_1 + x_2) = \sigma(x_1) +^\mathbb{Z} \sigma(x_2) = 4$ and $\sigma(x_3 - x_4) = \sigma(x_3) +^\mathbb{Z} (-^\mathbb{Z}\sigma(x_4)) = 6$. We conclude:

$$(\mathbb{Z}, +^\mathbb{Z}, -^\mathbb{Z}, 0^\mathbb{Z}) \not\models x_1 + x_2 = x_3 - x_4[\sigma].$$

The last example implicitly relied on rules to compute the values of terms that contain function symbols. The rules are natural enough and can be easily stated. Let $\mathcal{G}$ be a $\mathcal{L}_g$-structure (which need not be a group) and σ an assignment with domain V and codomain G. Then σ can be extended to the full set T of terms (which includes V) by the following stipulations: (i) $\sigma(t + s) = \sigma(t) +^G \sigma(s)$; (ii) $\sigma(-t) = -^G \sigma(t)$, where t, s are terms. The reader may practice applying (i) and (ii) by computing $\sigma((x_1 + (-x_2 + x_3)) - x_4)$ in $(\mathbb{Z}, +^\mathbb{Z}, -^\mathbb{Z}, 0^\mathbb{Z})$ when $\sigma(x_i) = i$ and when $\sigma(x_i) = 1 - 2i$, with $i = 1, 2, 3, 4$.

Strictly speaking, rules (i) and (ii) apply to an extension of σ with domain T but, for the sake of notational simplicity, we identify this extension with σ itself.

Rules (i)–(ii) can be generalised to an arbitrary signature $\tau = (\mathbf{R}, \mathbf{F}, \mathbf{C}, a)$. For each k-ary function symbol $f \in \mathbf{F}$, we need the condition:

★ $\sigma(f t_1 \ldots t_n) = f^\mathcal{M}(t_1^{\mathcal{M},\sigma}, \ldots, t_k^{\mathcal{M},\sigma}) \in M$, where $t_1, \ldots, t_k$ are τ-terms, $\mathcal{M}$ a $\mathcal{L}$-structure such that the signature of $\mathcal{L}$ is τ. The notation $t_i^{\mathcal{M},\sigma}$ indicates that we evaluate the term t_i in $\mathcal{M}$ with the assignment σ.

Exercise 2.30

(i) Given the signature of $\mathcal{L}_g$, describe three distinct assignments under which the term $x + y$ names the same nonzero object in $(\mathbb{Z}, +^{\mathbb{Z}}, -^{\mathbb{Z}}, 0^{\mathbb{Z}})$.
(ii) Let $\mathcal{L}_s$ be a first-order language of signature $\tau = (\mathbf{F}, \mathbf{C}, \mathsf{a})$, where $\mathbf{C} = \{0\}$, $\mathbf{F} = \{s\}$ and $\mathsf{a}(s) = 1$. Determine a $\mathcal{L}_s$-structure with domain $\mathbb{N}$ and such that every element of $\mathbb{N}$ is named by a distinct τ-term.
(iii) Prove by induction on the complexity of terms that, if σ', σ'' are extensions of the same assignment σ to T and if they satisfy the rule $\star$ for each function symbol in $\mathbf{F}$, then $\sigma'(t) = \sigma''(t)$ for each $t \in T$.

Exercise 2.31 Let $\mathcal{L}$ be a first-order language of signature $\tau = (\mathbf{F}, \mathsf{a})$, with $\mathbf{F} = \{f\}$ and $\mathsf{a}(f) = k$ and let $t(x_1, \ldots, x_n)$ be a τ-term whose variables figure among $\{x_1, \ldots, x_n\}$. Fix a $\mathcal{L}$-structure $\mathcal{M}$ and an assignment σ such that $\sigma(x_i) = a_i \in M$, $i = 1, \ldots, n$. If $s_1, \ldots, s_n$ are τ-terms such that $s_i^{\mathcal{M},\sigma} = a_i$, $i = 1, \ldots, n$, prove by induction on the complexity of terms that $t^{\mathcal{M},\sigma}$ and $t(x_1/s_1, \ldots, x_n/s_n)^{\mathcal{M},\sigma}$ name the same element of $\mathcal{M}$. Provide two concrete illustrations of the result just stated. By this exercise, the restriction to unnested terms does not produce any loss of generality from a semantic standpoint.

We are now ready to tackle the definition of satisfaction. We start with atomic formulae. Let $\mathcal{M}$ be a $\mathcal{L}$-structure and $\sigma : T \longrightarrow M$ an assignment. Suppose that $R \in \mathbf{R}$ (the relation symbol set in the signature of $\mathcal{L}$) and $\mathsf{a}(R) = n$. Consider the terms $t_1, \ldots, t_n$ and the atomic formula $Rt_1 \ldots t_n$. We know that $R^{\mathcal{M}} \subseteq M^n$ and that, given σ, $t_1^{\mathcal{M},\sigma}, \ldots, t_n^{\mathcal{M},\sigma} \in M$: either $(t_1^{\mathcal{M},\sigma}, \ldots, t_n^{\mathcal{M},\sigma}) \in R^{\mathcal{M}}$ or $(t_1^{\mathcal{M},\sigma}, \ldots, t_n^{\mathcal{M},\sigma}) \notin R^{\mathcal{M}}$. In the former case we say that $\mathcal{M}$ **satisfies** $Rt_1 \ldots t_n$ **with** σ and we write:

$$\mathcal{M} \models Rt_1 \ldots t_k[\sigma].$$

In the latter case we say that $\mathcal{M}$ does not satisfy $Rt_1 \ldots t_n$ with σ and we write $\mathcal{M} \not\models Rt_1 \ldots t_k[\sigma]$. The last remarks are summarised below:

$$\mathcal{M} \models Rt_1 \ldots t_k[\sigma] \text{ if, and only if, } (t_1^{\mathcal{M},\sigma}, \ldots, t_k^{\mathcal{M},\sigma}) \in R^{\mathcal{M}}.$$

We can now extend the definition of satisfaction from the set of atomic formulae to the set of all formulae. To this end we only have to observe that, if φ is a formula in some fixed first-order language, either φ is atomic or φ is obtained by the application of connectives and quantifiers to less complex formulae (i.e. formulae with fewer connectives or quantifiers).

Our definition will be, in essence, a device to ensure that, for any formula φ, structure $\mathcal{M}$ and assignment σ, the satisfaction relation between φ, $\mathcal{M}$ and σ can be iteratively reduced to the satisfaction relation between various subformulae of φ, $\mathcal{M}$ and σ. After finitely many iterations, we reach the atomic subformulae of φ, for which satisfaction is already defined.

2.3 Structures

In the definition to follow, we adopt the notation $\sigma(x/a)$ to designate any variant of the assignment σ obtained from it by fixing $a \in M$ as the value of x (it is not excluded that $\sigma(x/a)$ may just be σ).

Definition 2.32 (Satisfaction) Let ϕ, ψ be $\mathcal{L}$-formulae, $\mathcal{M}$ a $\mathcal{L}$-structure and σ an assignment.

(1) $\mathcal{M} \models Rt_1 \ldots t_k[\sigma]$ if, and only if, $(t_1^{M,\sigma}, \ldots, t_k^{M,\sigma}) \in R^M$;
(2) $\mathcal{M} \models \neg\phi[\sigma]$ if, and only if, $\mathcal{M} \models \phi[\sigma]$ does not hold;
(3) $\mathcal{M} \models (\phi \vee \psi)[\sigma]$ if, and only if, $\mathcal{M} \models \phi[\sigma]$ or $\mathcal{M} \models \phi[\sigma]$;
(4) $\mathcal{M} \models (\phi \wedge \psi)[\sigma]$ if, and only if, $\mathcal{M} \models \phi[\sigma]$ and $\mathcal{M} \models \phi[\sigma]$;
(5) $\mathcal{M} \models (\phi \to \psi)[\sigma]$ if, and only if, $\mathcal{M} \models \phi[\sigma]$ implies $\mathcal{M} \models \psi[\sigma]$;
(6) $\mathcal{M} \models \forall x \phi[\sigma]$ if, and only if, for every $a \in M$, $\mathcal{M} \models \phi[\sigma(x/a)]$;
(7) $\mathcal{M} \models \exists x \phi[\sigma]$ if, and only if, for some $a \in M$, $\mathcal{M} \models \phi[\sigma(x/a)]$.

From now on we shall systematically abbreviate the locution 'if, and only if' by the standard shorthand 'iff'. Some remarks are in order to clarify the rationale behind Definition 2.32 and to illustrate the way it is applied.

Remark 2.33 Definition 2.32 is not meant to provide any deep insight. Its role is to link together the notions of formula, assignment and structure in a serviceable, and perhaps predictable, way. For instance, part (1) of the definition says that an atomic formula is satisfied by a structure precisely when the relation it names actually holds in the given structure, among the named arguments. Part (2) says that the negation of a formula is satisfied by a structure with a fixed assignment exactly when the original formula is not so satisfied.

Remark 2.34 It is worth clarifying how part (2) of Definition 2.32 works by looking at an example. Let us consider the $\mathcal{L}_<$-formula $x < y$, the $\mathcal{L}_<$-structure $\mathcal{N} = (\mathbb{N}, <^{\mathbb{N}})$ and an assignment σ such that $\sigma(x) = \mathbf{1}, \sigma(y) = \mathbf{0}$. To show $\mathcal{N} \models \neg(x < y)[\sigma]$, we must show that $\mathcal{N} \models x < y[\sigma]$ does not hold. By (1), the last satisfaction relation holds exactly when the ordered pair $(1^{\mathbb{N}}, 0^{\mathbb{N}}) = (\sigma(x), \sigma(y))$ belongs to $<^{\mathbb{N}}$. Because $(1^{\mathbb{N}}, 0^{\mathbb{N}}) \notin <^{\mathbb{N}}$, we conclude that the satisfaction relation $\mathcal{N} \models x < y[\sigma]$ fails. Equivalently, $\mathcal{N} \not\models x < y[\sigma]$. By (2), this is to say that $\mathcal{N} \models \neg(x < y)[\sigma]$.

Remark 2.35 In part (3) of Definition 2.32, the disjunction 'or' is inclusive. Both alternatives may hold.

Remark 2.36 In part (5) of Definition 2.32, we do not require that $\mathcal{M} \models \phi[\sigma]$ should hold. We only require that, under the hypothesis that the relation $\mathcal{M} \models \phi[\sigma]$ holds, the relation $\mathcal{M} \models \phi[\sigma]$ hold.

Remark 2.37 By part (5) of Definition 2.32, $\mathcal{M} \not\models (\phi \to \psi)[\sigma]$ implies $\mathcal{M} \models \phi[\sigma]$ and $\mathcal{M} \models \neg\psi[\sigma]$.

Example 2.38 Consider the $\mathcal{L}_<$-structure $(\mathbb{Z}, <^{\mathbb{Z}})$, the $\mathcal{L}_<$-formula $\forall x \neg(x < x)$, and a fixed assignment σ. In order to determine whether $(\mathbb{Z}, <^{\mathbb{Z}}) \models \forall x \neg(x < x)[\sigma]$, we apply part (6) of Definition 2.32. To this end, we must take each variant

$\sigma(x/n)$ of σ, with $n \in \mathbb{Z}$, and determine whether $N \models \neg(x < x)[\sigma(x/n)]$. It is clear that, for every $n \in \mathbb{Z}$, $(n,n) \notin <^{\mathbb{Z}}$. It follows that $(\mathbb{Z}, <^{\mathbb{Z}}) \not\models \neg(x < x)[\sigma(x/n)]$, for every $n \in \mathbb{Z}$. We conclude $(\mathbb{Z}, <^{\mathbb{Z}}) \models \forall x \neg(x < x)[\sigma]$.

If ρ is any assignment other than σ, the argument just given for σ works for ρ, too, and allows us to conclude $\mathbb{Z} \models \forall x \neg(x < x)[\rho]$. The last satisfaction relation holds irrespective of the choice of assignment. For this reason we simply write $(\mathbb{Z}, <^{\mathbb{Z}}) \models \forall x \neg(x < x)$ and we say that $(\mathbb{Z}, <^{\mathbb{Z}})$ is a model of the sentence $\forall x \neg(x < x)$.

Example 2.38 illustrates a general situation. Given a $\mathcal{L}$-structure $\mathcal{M}$ and a $\mathcal{L}$-sentence ϕ, $\mathcal{M} \models \phi[\sigma]$ for every assignment σ, since any assignment has the same (vacuous) effect on ϕ, in which no variables occur free. In this case we simply write $\mathcal{M} \models \phi$ and say that $\mathcal{M}$ **models the sentence** ϕ or that $\mathcal{M}$ **is a model of** ϕ.

Example 2.39 Consider again the linear order $(\mathbb{Z}, <^{\mathbb{Z}})$, the $\mathcal{L}_<$-formula $\neg \forall x (x < y)$ and an assignment σ such that $\sigma(y) = 3$. In order to determine whether the last formula is satisfied by $(\mathbb{Z}, <^{\mathbb{Z}})$ with σ, we first apply part (2) of Definition 2.32. We must verify $(\mathbb{Z}, <^{\mathbb{Z}}) \not\models \forall x (x < y)[\sigma]$.

To do so, we rely on part (6) of Definition 2.32. In particular, we look for $n \in \mathbb{Z}$ such that $(\mathbb{Z}, <^{\mathbb{Z}}) \models \neg(x < y)[\sigma(x/n)]$: the value $n = 4$ does what we want (infinitely many alternative values work).

Part (6) allows us to conclude $(\mathbb{Z}, <^{\mathbb{Z}}) \not\models \forall x (x < y)[\sigma]$. Moreover, because we found a variant $\sigma(x/n)$ of σ such that $(\mathbb{Z}, <^{\mathbb{Z}}) \models \neg(x < y)[\sigma(x/n)]$, we can use part (7) of Definition 2.32 to conclude $(\mathbb{Z}, <^{\mathbb{Z}}) \models \exists x \neg(x < y)[\sigma]$. The last argument is reversible: from the assumption $(\mathbb{Z}, <^{\mathbb{Z}}) \models \exists x \neg(x < y)[\sigma]$ we can deduce $(\mathbb{Z}, <^{\mathbb{Z}}) \models \neg \forall x (x < y)$.

The last example is a particular verification of a general result. For any $\mathcal{L}$-structure and any $\mathcal{L}$-formula ϕ:

$$\mathcal{M} \models \neg \forall x \phi[\sigma] \text{ iff } \mathcal{M} \models \exists x \neg \phi[\sigma].$$

The above satisfaction relations are always simultaneous. If one fails, the other cannot hold.

Exercise 2.40

(i) Imitating the argument from Example 2.39, prove that $\mathcal{M} \models \forall x \neg \phi[\sigma]$ iff $\mathcal{M} \models \neg \exists x \phi[\sigma]$.
(ii) Prove that $\mathcal{M} \models \neg \exists x \neg \phi[\sigma]$ iff $\mathcal{M} \models \forall x \phi[\sigma]$.
(iii) Prove that $\mathcal{M} \models \neg \forall x \neg \phi[\sigma]$ iff $\mathcal{M} \models \exists x \phi[\sigma]$.

2.3.2 Valuations

If $\mathcal{M}$ is a $\mathcal{L}$-structure and ϕ is any $\mathcal{L}$-sentence, either $\mathcal{M} \models \phi$ or $\mathcal{M} \models \neg \phi$ (assignments do not matter, as we have observed earlier). In the first case we also say that ϕ is true in $\mathcal{M}$; in the second, that ϕ is false in $\mathcal{M}$.

2.3 Structures

Once we know whether or not a fixed $\mathcal{L}$-structure $\mathcal{M}$ is a model of certain $\mathcal{L}$-sentences, say $\phi_1, \ldots, \phi_n$, we can easily infer whether or not $\mathcal{M}$ is a model of any $\mathcal{L}$-sentence built from $\phi_1, \ldots, \phi_n$ by means of the connectives. E.g. $\mathcal{M} \models \phi$ and $\mathcal{M} \models \neg\psi$ imply $\mathcal{M} \models \phi \wedge \neg\psi$ and $\mathcal{M} \models \phi \vee \psi$ (by parts (3) and (4) of Definition 2.32 respectively). It is useful to introduce a mechanical method to deal with multiple applications of connectives.

Given a first-order language $\mathcal{L}$, we call *prime* any $\mathcal{L}$-sentence that is is either atomic, of the form $\forall x \phi$ or of the form $\exists x \phi$. Let S be the set of prime $\mathcal{L}$-sentences: a *valuation* v on S is a function with domain S and codomain $\{0, 1\}$ such that, for each $\phi \in S$:

$$v(\phi) = \begin{cases} 0 \text{ if } \mathcal{M} \models \neg\phi \\ 1 \text{ if } \mathcal{M} \models \phi \end{cases}$$

The equality $v(\phi) = 0$ indicates that a sentence ϕ is false in $\mathcal{M}$, while $v(\phi) = 1$ signals that ϕ is true in $\mathcal{M}$. We extend v^4 to the connectives by the following stipulations, in which $max\{v(\phi), v(\psi)\}$ is the maximum of $v(\phi), v(\psi)$ and $min\{v(\phi), v(\psi)\}$ is their minimum:[5]

(1) $v(\neg(\phi)) = 1 - v(\phi)$;
(2) $v((\phi \vee \psi)) = max\{v(\phi), v(\psi)\}$;
(3) $v((\phi \wedge \psi)) = min\{v(\phi), v(\psi)\}$;
(4) $v((\phi \to \psi)) = max\{(1 - v(\phi)), v(\psi)\}$.

Using (1), we easily see that ϕ is true in $\mathcal{M}$ exactly when $\neg\phi$ is false in $\mathcal{M}$ and vice versa: if $v(\phi) = 1$, we compute $v(\neg(\phi)) = 1 - 1 = 0$ and if $v(\phi) = 0$, then $v(\neg(\phi)) = 1 - 0 = 1$. Using (2), we see that $\mathcal{M} \models (\phi \vee \psi)$ holds exactly when $\mathcal{M}$ models at least one of ϕ, ψ, in which case the maximum of $v(\phi), v(\psi)$ is certainly 1. It follows from (3) that $v((\phi \wedge \psi)) = 1$ exactly when $v(\phi) = 1 = v(\psi)$, i.e. when both ϕ, ψ are true in $\mathcal{M}$.

Let us finally consider a sentence of the form $(\phi \to \psi)$. By Definition 2.32, $\mathcal{M} \models \neg(\phi \to \psi)$ when $\mathcal{M} \models \phi$ and $\mathcal{M} \models \neg\psi$. In this event we must have $\mathcal{M} \models (\phi \wedge \neg\psi)$. It follows that $\mathcal{M} \models (\phi \to \psi)$ iff $\mathcal{M} \models \neg(\phi \wedge \neg\psi)$ iff $\mathcal{M} \not\models (\phi \wedge \neg\psi)$. The last condition holds exactly when $\mathcal{M}$ does not model at least one of $\phi, \neg\psi$, i.e. when $\mathcal{M} \models \neg\phi$ or $\mathcal{M} \models \psi$. Put more compactly, $\mathcal{M} \models (\neg(\phi) \vee \psi)$, which in turn yields:

$$v((\phi \to \psi)) = v(\neg(\phi) \vee \psi) = max\{(1 - v(\phi)), v(\psi)\}.$$

Exercise 2.41 Suppose $\mathcal{M} \models \phi$, $\mathcal{M} \models \neg\psi$ and $\mathcal{M} \models \neg\chi$. Compute: (i) $v((\psi \vee (\chi \to \neg\psi)))$; (ii) $v((\chi \wedge (\psi \vee \chi)))$; (iii) $v((\chi \to (\psi \vee (\neg\phi \wedge \chi))))$.

---

[4] And identify the extension with v itself for notational convenience.
[5] Note that $max\{1, 1\} = 1 = min\{1, 1\}$ and similar equalities hold for 0.

We shall henceforth omit the brackets whose absence does not compromise the intelligibility of a formula. In particular, we systematically write $\neg \phi$ instead of $\neg(\phi)$ and $\phi \vee \psi$ instead of $(\phi \vee \psi)$, following a similar policy for the remaining connectives. In the absence of parentheses, negation takes precedence over all binary connectives.

So far we have focussed on sentences evaluated in a fixed structure, but it is simple enough to consider their possible evaluations across all structures. If, for instance, ϕ, ψ are $\mathcal{L}$-sentences, their evaluations in any $\mathcal{L}$-structure $\mathcal{M}$ can only give rise to one of four possible combinations. If $\mathcal{M} \models \phi, \psi$, then $v(\phi) = 1 = v(\psi)$; if $\mathcal{M} \models \phi, \neg\psi$, then $v(\phi) = 1$ and $v(\psi) = 0$; if $\mathcal{M} \models \neg\phi, \psi$, then $v(\phi) = 0$ and $v(\psi) = 1$; if, finally, $\mathcal{M} \models \neg\phi, \neg\psi$, then $v(\phi) = 0 = v(\psi)$. We have thus listed four distinct valuations, all of which we called v to avoid notational clutter. The reader will not struggle to see (or prove by arithmetical induction) that, when n distinct formulae are involved, 2^n distinct valuations arise.

Example 2.42 The valuation $v(\phi \vee \psi)$ takes a uniquely determined value that only depends on $v(\phi), v(\psi)$. The latter values determine four possibilities, presented across four rows in the table below. When the corresponding values of $v(\phi \vee \psi)$ are also specified in each row, we obtain the **truth table** of $\phi \vee \psi$:

ϕ	ψ	$\phi \vee \psi$
1	1	1
0	1	1
1	0	1
0	0	0

Given a sentence in which several connectives occur, the computation of its possible values in a truth-table is carried out by first considering the connective nested inside the largest number of parentheses. The values obtained for it are then employed to determine values for the next connective, nested inside the second largest number of parentheses. This procedure must stop when there are no connectives left to consider. Here is a simple illustration of the procedure that only involves two connectives:

ϕ	ψ		$\neg(\phi \vee \psi)$
1	1	**0**	1
0	1	**0**	1
1	0	**0**	1
0	0	**1**	0

The column highlighted in bold is computed last.

Exercise 2.43 Compute the truth-tables of: (i) $\neg\phi$; (ii) $\phi \wedge \psi$; (iii) $\phi \rightarrow \psi$; (iv) $\neg(\phi \rightarrow \neg\psi)$; (v) $(\phi \rightarrow \psi) \wedge (\psi \rightarrow \phi)$.

2.3 Structures

We say that two $\mathcal{L}$-sentences ϕ, ψ are equivalent iff $v(\phi) = v(\psi)$ in every $\mathcal{L}$-structure. When this is the case, we write $\phi \equiv \psi$. Note that $\equiv$ is not a connective, nor is $\phi \equiv \psi$ a $\mathcal{L}$-formula.

Exercise 2.44 Verify the following equivalences:

a $\neg(\neg\psi \wedge \neg\phi) \equiv \psi \vee \phi$;
b $\neg(\neg\psi \vee \neg\phi) \equiv \psi \wedge \phi$;
c $\psi \rightarrow \phi \equiv \neg\psi \vee \phi$;
d $\phi \rightarrow (\psi \rightarrow \chi) \equiv \psi \rightarrow (\phi \rightarrow \chi)$;
e $\phi \wedge (\psi \vee \chi) \equiv (\phi \wedge \psi) \vee (\phi \wedge \chi)$;
f $\phi \vee (\psi \wedge \chi) \equiv (\phi \vee \psi) \wedge (\phi \vee \chi)$;
g $\phi \equiv \phi \wedge (\phi \vee \psi)$;
h $\phi \equiv \phi \vee (\phi \wedge \psi)$.

Exercise 2.45 The connective $\leftrightarrow$, called **biconditional**, is sometimes introduced under the stipulation that $\phi \leftrightarrow \psi$ abbreviates $(\phi \rightarrow \psi) \wedge (\psi \rightarrow \phi)$. Show that the $\mathcal{L}$-sentence $\phi \leftrightarrow \psi$ is true in every $\mathcal{L}$-structure iff $\phi \equiv \psi$.

Exercise 2.46 Explain how the above discussion of valuations can be adapted to arbitrary formulae.

Let us consider again the truth-table:

ϕ	ψ	$(\phi \vee \psi)$
1	1	1
0	1	1
1	0	1
0	0	0

If we restrict attention to the rows in which $(\phi \vee \psi)$ is true, we realise that this is the case when ϕ and ψ are true, or when ϕ is false and ψ is true, or, finally, when ϕ is true and ψ is false. We can effectively transcribe the last remark as a formula, i.e. the following disjunction of conjunctions:

$$(\phi \wedge \psi) \vee (\neg\phi \wedge \psi) \vee (\phi \wedge \neg\psi),$$

which has the same truth-table as $\phi \vee \psi$ (the reader is invited to check this). The two formulae are therefore equivalent. The approach just illustrated is completely general and enables us produce, for any sentence obtained by finitely many applications of the connectives to prime sentences, and equivalent disjunction of conjunctions. We call disjunction of conjunctions of prime sentences and their negations **disjunctive normal forms**. By Exercise 2.46, the notion of disjunctive normal form can be extended to include formulae. We shall find disjunctive normal forms of atomic formulae very useful in Chaps. 8 and 11.

Exercise 2.47 Determine disjunctive normal forms for the formulae (i) $\phi \to \neg\psi$ and (ii) $(\phi \vee \neg\psi) \wedge (\neg\phi \vee \psi)$. Show that the same sentence can have distinct disjunctive normal forms.

2.4 Consequence

Using the satisfiability relation, we can look at some of the arguments given in Chap. 1 from a logical standpoint. Consider, for instance, the proof from Sect. 1.3, to the effect that each element of S_3 has a unique multiplicative inverse. Although a direct verification of this result is certainly possible, we have obtained it by a general train of thought, which it is instructive to review.

Example 2.48 Let us consider the following $\mathcal{L}_g$-sentences:
1. $\forall x \forall y \forall z (x + (y + z) = (x + y) + z)$;
2. $\forall x (x + 0 = x \wedge 0 + x = x)$;
3. $\forall x \exists y (x + y = 0 \wedge y + x = 0)$.

Let φ be the conjunction of (1), (2), (3). To prove that a group (and, in particular S_3) has uniquely determined multiplicative inverses is, in model-theoretic terms, to prove that, for any $\mathcal{L}_g$-structure $\mathcal{G}$:

$$\mathcal{G} \models \varphi \implies \mathcal{G} \models \forall x \forall y \forall z ((x + y = 0 \wedge x + z = 0) \to y = z). \tag{2.3}$$

We shall now establish the implication just stated after a more formal fashion than previously adopted. Using Definition 2.32-(6), we first fix an assignment σ such that $\mathcal{G} \models x + y = 0 \wedge x + z = 0[\sigma]$. Set $\sigma(x) = a, \sigma(y) = b$ and $\sigma(z) = c$, with $a, b, c \in G$.

Clearly, $a +^G b \in G$ and $c +^G (a +^G b) \in G$.[6] By hypothesis $\mathcal{G} \models (1)$. Since (1) is a sentence, it holds with every assignment. Thus $\mathcal{G} \models (1)[\sigma]$ and, by part (6) of Definition 2.32, $\mathcal{G} \models x + (y + z) = (x + y) + z[\sigma(x/a, y/b, z/c)]$. Less formally, $c +^G (a +^G b) = (c +^G a) +^G b$.

Since $\mathcal{G} \models x + y = 0 \wedge x + z = 0[\sigma]$, Definition 2.32-(4) implies that $\mathcal{G} \models x + y = 0[\sigma]$, i.e. $a +^G b = 0$. If $G \models x + y = 0[\sigma]$, the terms $z + (x + y)$ and $z + 0$ name the same element of G relative to σ. We can therefore conclude $c +^G (a +^G b) = c +^G 0$. By the same clue, $c +^G a = 0$ implies $(c +^G a +^G b = 0 +^G b$. Because, as we already know, $(c +^G a) +^G b = c +^G (a +^G b)$, we deduce $c +^G 0 = 0 +^G b$.

We now use the condition $\mathcal{G} \models (2)$ to obtain $\mathcal{G} \models z + 0 = z[\sigma]$: thus $c +^G 0 = c$. For essentially the same reason $0 +^G b = b$. Consequently, the equality $c +^G 0 = 0 +^G b$ reduces to $c = b$, i.e. $\sigma(z) = \sigma(x)$.

---

[6] Every $\mathcal{L}_g$-term names an element of G.

2.4 Consequence

We have shown that, under the hypothesis $\mathcal{G} \models \varphi$, the satisfaction relation $\mathcal{G} \models x + y = 0 \wedge x + z = 0[\sigma]$ implies $\mathcal{G} \models x = z[\sigma]$. Definition 2.32-(5) now yields:

$$\mathcal{G} \models (x + y = 0 \wedge x + z = 0) \to y = z[\sigma],$$

where σ is an arbitrarily chosen assignment. This is to say that the same conclusion holds for $\sigma(x/d, y/e, z/f)$, however $d, e, f \in G$ are chosen. By part (6) of Definition 2.32:

$$\mathcal{G} \models \forall x \forall y \forall x((x + y = 0 \wedge x + z = 0) \to y = z).$$

The argument just given does not in fact rely on (3): we do not need to know that an element exists to prove that, *if* it exists, then it is unique.

By Example 2.48, any model of the $\mathcal{L}_g$-sentence φ must also be a model of the $\mathcal{L}_g$-sentence $\forall x \forall y \forall z((xy = e \wedge xz = e) \to y = z)$. We say that $\forall x \forall y \forall z((xy = e \wedge xz = e) \to y = z)$ is a *consequence* of φ (in fact, of (1) $\wedge$ (2)) and, with an abuse of notation,[7] we write:

$$\varphi \models \forall x \forall y \forall x((xy = e \wedge xz = e) \to y = z).$$

Example 2.48 provides an instance of the following, abstract definition:

Definition 2.49 Let ϕ, ψ be $\mathcal{L}$-sentences. We say that ψ is a **consequence** of ϕ, in symbols $\phi \models \psi$, iff for every $\mathcal{L}$-structure $\mathcal{M}$, $\mathcal{M} \models \phi$ implies $\mathcal{M} \models \psi$.

By Definitions 2.49 and 2.32-(5), if ϕ, ψ are $\mathcal{L}$-sentences, $\phi \models \psi$ iff for every $\mathcal{L}$-structure $\mathcal{M}$, $\mathcal{M} \models \phi \to \psi$. The consequence relation $\phi \models \psi$ fails iff there is a $\mathcal{L}$-structure $\mathcal{M}$ such that $\mathcal{M} \models \phi \wedge \neg \psi$. Whenever this is the case, we refer to $\mathcal{M}$ as a **countermodel**, i.e. a model of the sentences $\psi, \neg\psi$ that provides a counterexample to the hypothesis $\phi \models \psi$. When a countermodel exists, we write $\phi \not\models \psi$.

We have already verified that, if φ is as in Example 2.48, $\varphi \not\models \forall x \forall y(x + y = y + x)$. The symmetric group S_3 provides a countermodel.

Exercise 2.50 Extend the notions of consequence and equivalence to arbitrary formulae.

Certain consequence relations can be established by considering only the semantics of connectives and quantifiers (i.e. the abstract definition of satisfaction from Sect. 2.3).

Example 2.51 Consider the $\mathcal{L}$-sentence $\forall x(\phi(x) \wedge \psi(x))$, where we highlight the fact that the variable x may occur in ϕ or ψ,[8] and suppose $\mathcal{M} \models \forall x(\phi(x) \wedge \psi(x))$.

---

[7] That is, assigning different meanings to the same symbol.
[8] On grounds of convenience and clarity, which shall mention or omit to mention variables that may occur in certain formulae. For this reason, we may treat $\forall x \phi$ and $\forall x \phi(x)$ as two distinct ways of describing the same formula.

By Definition 2.32-(6), $\mathcal{M} \models \phi(x) \wedge \psi(x)[\sigma(x/a)]$, for every $a \in M$. We invoke Definition 2.32-(4) to deduce:

$$\mathcal{M} \models \phi(x)[\sigma(x/a)] \text{ e } \mathcal{M} \models \psi(x)[\sigma(x/a)], \text{ for every } a \in M.$$

Two applications of Definition 2.32-(6) yield:

$$\mathcal{M} \models \forall x \phi(x) \text{ e } \mathcal{M} \models \forall x \psi(x),$$

i.e. $\mathcal{M} \models \forall x \phi(x) \wedge \forall x \psi(x)$. We have proved:

$$\forall x(\phi(x) \wedge \psi(x)) \models \forall x \phi(x) \wedge \forall x \psi(x).$$

Exercise 2.52 Verify the following consequence relations, where ϕ, ψ are arbitrary formulae, not necessarily sentences:

(i) $\exists x(\phi(x) \vee \psi(x)) \models \exists x \phi(x) \vee \exists x \psi(x)$;
(ii) $\forall x(\phi(x) \rightarrow \psi(x)) \models \forall x \phi(x) \rightarrow \forall x \psi(x)$;
(iii) $\forall x \phi(x) \vee \forall x \psi(x) \models \forall x(\phi(x) \vee \psi(x))$;
(iv) $\neg \forall x \forall y \forall z \phi(x, y, z) \models \exists x \exists y \exists z \neg \phi(x, y, z)$;
(v) $\exists x \exists y \exists z \neg \phi(x, y, z) \models \neg \forall x \forall y \forall z \phi(x, y, z)$;
(vi) If $x \notin \ell(\phi)$, $\phi \rightarrow \forall x \psi(x) \models \forall x(\phi \rightarrow \psi(x))$;
(vii) If $x \notin \ell(\psi)$, $\forall x \phi(x) \rightarrow \psi \models \exists x(\phi(x) \rightarrow \psi)$ (*Hint*: proceed by contradiction).

Definition 2.53 Two formulae ϕ, ψ are **equivalent** iff $\phi \models \psi$ and $\psi \models \phi$. The formulae ϕ, ψ are **independent** iff $\phi \not\models \psi$ and $\psi \not\models \phi$.

Exercise 2.54 Show that two sentences are equivalent in the sense of the previous section iff they are equivalent in the sense of Definition 2.53. In view of this result, we may simply write $\phi \equiv \psi$ whenever each of ϕ, ψ is a consequence of the other.

Exercise 2.55

(i) Find a $\mathcal{L}_=$-formula $\phi(x)$ such that $\phi(x) \not\equiv \phi(u)$, where x, u are variables; Prove that:
(ii) if $u, v \notin \ell(\phi(x, y))$, then $\phi(x, y)[\sigma] \equiv \phi(u, v)[\sigma(u/\sigma(x), v/\sigma(y))]$;
(iii) if $u, v \notin \ell(\phi)$, then $\forall x \exists y \phi(x, y) \equiv \forall u \exists v \phi(u, v)$;
(iv) if $\phi \equiv \psi$, then $\forall x \phi \equiv \forall x \psi$ and $\exists x \phi \equiv \exists x \psi$;
(v) if $x, y \notin \ell(\psi)$ and $u, v \notin \ell(\phi)$, then $\forall x \exists y \phi \vee \exists u \forall v \psi \equiv \forall x \exists y \exists u \forall v(\phi \vee \psi)$.

To prove that two sentences are equivalent, we have to establish two consequence relations. To show that they are independent, we must produce two countermodels. To prove that n sentences are independent it is enough to find n countermodels, each of which satisfies $n - 1$ sentences and the negation of the remaining sentence (the reader is encouraged to show why n countermodels suffice).

2.4 Consequence

Example 2.56 Let $\mathcal{L}_\sim$ be $\mathbf{L} \cup \{\sim\}$, where $\sim$ is a binary relation. We shall verify that the three $\mathcal{L}_\sim$-sentences:

(i) $\forall x(x \sim x)$;
(ii) $\forall x \forall y(x \sim y \to y \sim x)$;
(iii) $\forall x \forall y \forall x((x \sim y \wedge y \sim x) \to x \sim z)$

are independent. First, we show that $(i) \wedge (ii) \not\models (iii)$. Our countermodel is the $\mathcal{L}_\sim$-structure $\mathcal{Z}_1 = (\mathbb{Z}, \sim^{\mathbb{Z}})$, where:

$$n \sim m \text{ iff } -1 \leq m - n \leq 1.$$

Note that $\mathcal{Z}_1 \models (i) \wedge (ii)$. Since, for every $n \in \mathbb{Z}$, $-1 \leq 0 = n - n$ and $n - n = 0 \leq 1$, $\mathcal{Z}_1 \models \forall x(x \sim x)$. Moreover, if $\mathcal{Z}_1 \models x \sim y[\sigma]$ with $\sigma(x) = m$ and $\sigma(y) = n$, for any $m, n \in \mathbb{Z}$, then $-1 \leq m - n \leq 1$. The difference $m - n$ can only take values in $\{-1, 0, 1\}$. The same must be true of the difference $n - m$. As a consequence, $\mathbb{Z} \models y \sim x[\sigma]$. We conclude:

$$\mathbb{Z} \models x \sim y \to y \sim x[\sigma(x/m, y/n)], \text{ for every } m, n \in \mathbb{Z},$$

In short, $\mathcal{Z}_1 \models (ii)$. We can easily verify that $\sim$ is not transitive: $4 \sim^{\mathbb{Z}} 3$ and $3 \sim^{\mathbb{Z}} 2$ do not imply $4 \sim^{\mathbb{Z}} 2$. More formally, if $\sigma(x) = 4$, $\sigma(y) = 3$ e $\sigma(z) = 2$, we have:

$$\mathcal{Z}_1 \models x \sim y \wedge y \sim z \wedge \neg(x \sim z)[\sigma(x/4, y/3, z/2)].$$

For at least three distinct values of x, y, z, $\mathcal{Z}_1$ models the formula $x \sim y \wedge y \sim z \wedge \neg(x \sim z)$, which is equivalent to $\neg((x \sim y \wedge y \sim z) \to x \sim z)$. Using Definition 2.32-(7), we arrive at:

$$\mathcal{Z}_1 \models \exists x \exists y \exists z \neg((x \sim y \wedge y \sim z) \to x \sim z).$$

By Exercise 2.40:

$$\mathcal{Z}_1 \models \neg \forall x \forall y \forall z((x \sim y \wedge y \sim z) \to x \sim z).$$

We conclude that $\mathcal{Z}_1 \models (i) \wedge (ii)$ and $\mathbb{Z} \models \neg(iii)$. To complete our independence proof, we require two more countermodels: finding them is left as an exercise.

Exercise 2.57 Find two $\mathcal{L}_\sim$-structures $\mathcal{M}, \mathcal{N}$ such that $\mathcal{M} \models (i) \wedge (iii)$ and $\mathcal{M} \not\models (ii)$, whilst $\mathcal{N} \models (ii) \wedge (iii)$ and $\mathcal{M} \not\models (i)$. Also show that $(ii) \wedge (iii) \wedge \forall x \exists y(x \sim y) \models (i)$.

By Example 2.56 and Exercise 2.57, the sentences we used to define an equivalence relation in Chap. 1 cannot be further reduced.

Exercise 2.58 Given the $\mathcal{L}_g$-sentences (1), (2) and (3) from Example 2.48:

(i) verify that $Z' \models (2) \wedge (3)$ and $Z' \not\models (1)$, where $Z' = (\mathbb{Z}, \oplus, -, 0)$ and $m \oplus n = m + n$ if $m \neq n$ and $m \oplus n = 0$ otherwise;
(ii) find a $\mathcal{L}_g$-structure $\mathcal{M}$ such that $\mathcal{M} \models (1) \wedge (3)$ and $\mathcal{M} \not\models (2)$;
(iii) find a $\mathcal{L}_g$-structure $\mathcal{N}$ such that $\mathcal{N} \models (1) \wedge (2)$ and $\mathcal{M} \not\models (3)$.

If $\phi_1, \ldots, \phi_n, \psi$ are formulae and $\mathcal{M} \models \phi_1, \ldots, \mathcal{M} \models \phi_n$ imply $\mathcal{M} \models \psi$, we say that ϕ is a consequence of the set $\{\phi_1, \ldots, \phi_n\}$ and we write:

$$\{\phi_1, \ldots, \phi_n\} \models \psi.$$

We thus obtain no more than a notational variant of $\phi_1 \wedge \ldots \wedge \phi_n \models \psi$. This variant is however helpful to suggest a way of defining the consequence relation between a, possibly infinite, set of sentences T and a sentence ψ. If T is infinite, only the conjunctions of finitely many elements of T are sentences, but we can still refer to a structure $\mathcal{M}$ that satisfies every sentence $\phi \in T$. Thus, we say that ψ is a consequence of the set T iff, whenever $\mathcal{M}$ is a model of T (i.e. a model of every sentence in T), $\mathcal{M} \models \psi$. In this event we write:

$$T \models \psi.$$

If $T = \{\phi\}$, $T \models \psi$ coincides with $\phi \models \psi$. Given two $\mathcal{L}$-sentences ϕ, ψ, if $T \models \phi$ implies $T \models \psi$, we say that ψ is a T-consequence of ϕ. We know that a group with three elements is abelian. If T contains the $\mathcal{L}_g$-sentences from Example 2.48 and an additional sentence stating the existence of exactly three distinct objects, then the $\mathcal{L}_g$-sentence $\forall x \forall y (x + y = y + x)$ is a T-consequence of ϕ (with ϕ as in Example 2.48) but not a consequence of ϕ.

When ψ is a T-consequence of ϕ and ϕ is T-consequence of ψ, we say that ϕ, ψ are T-equivalent.

Exercise 2.59 Prove that, for any set T of $\mathcal{L}$-sentences, if any two $\mathcal{L}$-formulae are equivalent, then they are T-equivalent.

2.5 Induction on the Complexity of Formulae

The notions of consequence and equivalence with which we have just made our acquaintance provide a useful background to state and prove a few important results.

We collect these results here because they are all proved by induction on the complexity of formulae. Although we have outlined this proof technique in Sect. 2.2, we have not given any detailed account of it or shown how it is used. This section will provide extensive practice with it.

2.5 Induction on the Complexity of Formulae

A proof by induction on the complexity of formulae rests on the fact that $\mathcal{F}$, the set of formulae in a first-order language $\mathcal{L}$, contains the atomic $\mathcal{L}$-formulae and is closed under the application of both connectives and quantifiers. Closure may be stated thus:

if $\phi, \psi \in \mathcal{F}$ and x is a variable, then $\{\neg\phi, \phi \vee \psi, \phi \wedge \psi, \phi \rightarrow \psi, \exists x \phi, \forall x \phi\} \subseteq \mathcal{F}$.

As already remarked, inductive proofs on the complexity of formulae are used to show that a certain result A holds for every element of $\mathcal{F}$. To achieve this goal we consider the set $S \subseteq \mathcal{F}$, consisting of the formulae for which A holds. In principle, S may be empty. The inductive base proves that S contains all of the atomic $\mathcal{L}$-formulae and is therefore nonempty.

The second and final part of an inductive proof on the complexity of formulae, known as the inductive step, consists in showing that S and $\mathcal{F}$ have the same closure properties. Since $\mathcal{F}$ is, by definition, the smallest extension of the set of atomic formulae to satisfy these closure properties, the equality $S = \mathcal{F}$ follows. In other words, A holds for every element of $\mathcal{F}$.

Before turning to examples, it is helpful to examine the inductive base and the inductive step in greater detail. In the inductive base we take into account terms $t, s, t_1, \ldots, t_n$ and we want to show that A holds for equalities of the form $t = s$ and formulae of the form $Rt_1 \ldots t_n$, where R is a n-ary relation symbol in $\mathcal{L}$.

In the inductive step, we suppose that A holds for ϕ, ψ and use this assumption, known as inductive hypothesis, to prove that A also holds for $\neg\phi, \phi \vee \psi, \phi \wedge \psi, \phi \rightarrow \psi$. This is not enough, because we must consider quantifiers as well. Using the inductive hypothesis that A holds for $\psi(x)$, we prove that A also holds for $\exists x \psi(x), \forall x \psi(x)$.

Once all of the above tasks have been carried out, the induction is complete.

The theorems we are now going to prove by induction are similar to the result about disjunctive normal form discussed at the end of Sect. 2.3. They show that it is always possible to replace a formula with an equivalent one in a canonical form. In order to state the results to follow, we need to fix some terminology.

We say that a formula is in **prenex form** if it consists of a string of quantifiers, its **prefix**, followed by a quantifier-free formula, its **matrix**. For instance, the formula $x < y$ is in prenex form, while the formula $\forall x(x < y) \rightarrow \exists z(z < x)$ is not. A formula in prenex form whose prefix is empty or only contains universal quantifiers is called a **universal formula**. A formula in prenex form whose prefix is empty or only contains existential quantifiers is called an **existential formula**. The sentence $\forall x \exists y(x + y = 0)$ is neither universal nor existential, but it is in prenex form.

Under certain conditions, shortly to be stated, arbitrary formulae are T-equivalent to a universal or to an existential formula.

Lemma 2.60 *Let $\mathcal{L}$ be a first-order language and T a set of $\mathcal{L}$-sentences. If every universal $\mathcal{L}$-formula is T-equivalent to an existential $\mathcal{L}$-formula, then every $\mathcal{L}$-formula is T-equivalent to a universal $\mathcal{L}$-formula.*

Proof We consider the set $S \subseteq \mathcal{F}$ of $\mathcal{L}$-formulae that are T-equivalent to a universal $\mathcal{L}$-formula.

(1) **Inductive Base**. An atomic $\mathcal{L}$-formula is of the form $t = s$ or $Rt_1 \ldots t_n$, where R is a n-ary relation symbol in $\mathcal{L}$ and $s, t, t_1, \ldots, t_n$ are terms. Since atomic $\mathcal{L}$-formulae are both universal and existential and certainly equivalent to themselves, they are all in S.

(2) **Inductive Step**.

Connectives. By the inductive hypothesis, $\phi, \psi \in S$: in other words, ϕ, ψ are T-equivalent to universal $\mathcal{L}$-formulae, say $\forall x_1 \ldots \forall x_n \alpha$ and $\forall z_1 \ldots \forall z_m \beta$ respectively, where α, β are quantifier-free. By a simple generalisation of Exercise 2.55-(ii), we may assume that the variables $x_1, \ldots, x_n$ do not occur in β and that the variables $z_1, \ldots, z_m$ do not occur in α.

We now show $\neg \phi \in S$. Note that, by the inductive hypothesis, ϕ is equivalent to $\forall x_1 \ldots \forall x_n \alpha$ and, by the lemma's hypothesis, the last formula is T-equivalent to an existential $\mathcal{L}$-formula, which we may call $\exists y_1 \ldots \exists y_p \gamma$ (we do not assume that equivalence preserves the number of quantifiers or the names of variables). It follows that $\neg \phi$, equivalent to $\neg \forall x_1 \ldots \forall x_n \alpha$, must also be T-equivalent to $\neg \exists y_1 \ldots \exists y_p \gamma$. By Exercise 2.54-(iv),(v), the last formula is equivalent, and thus T-equivalent, to the universal formula $\forall y_1 \ldots \forall y_p \neg \gamma$. We can conclude from this chain of T-equivalences that $\neg \phi$ is T-equivalent to a universal formula, i.e. $\neg \phi \in S$.

Next, we show $\phi \vee \psi \in S$. By the inductive hypothesis, $\phi \vee \psi$ is T-equivalent to $\forall x_1 \ldots \forall x_n \alpha \vee \forall z_1 \ldots \forall z_m \beta$. Because $z_1, \ldots, z_m$ do not occur in α, $z_1, \ldots, z_m \notin \ell(\alpha)$. For a similar reason, $x_1, \ldots, x_n \notin \ell(\beta)$.

We can therefore use a generalisation of Exercise 2.52-(iii) to deduce:

$$\forall x_1 \ldots \forall x_n \alpha \vee \forall z_1 \ldots \forall z_m \beta \equiv \forall x_1 \ldots \forall x_n \forall z_1 \ldots \forall z_m (\alpha \vee \beta).$$

Noting that equivalence implies T-equivalence and that the last formula is universal, we conclude $\phi \vee \psi \in S$.

What we have established makes it easy to infer $\phi \wedge \psi \in S$ and $\phi \rightarrow \psi \in S$. We know from Exercise 2.50 that $\phi \wedge \psi \equiv \neg(\neg \phi \vee \neg \psi)$ and $\phi \rightarrow \psi \equiv \neg \phi \vee \psi$. We have shown that S is closed under $\neg$ and $\vee$. By closure under $\neg$, $\neg \phi, \neg \psi \in S$. By closure under $\vee$, $\neg \phi \vee \neg \psi \in S$. Another appeal to closure under $\neg$ yields $\neg(\neg \phi \vee \neg \psi) \in S$. Since $\phi \wedge \psi$ is equivalent to $\neg(\neg \phi \vee \neg \psi)$ and every formula in S is T-equivalent to a universal formula, $\phi \wedge \psi$ must be T-equivalent to such a formula, i.e. $\phi \wedge \psi \in S$. A similar argument guarantees $\phi \rightarrow \psi \in S$.

Quantifiers. The inductive hypothesis is that $\psi(x)$ is T-equivalent to a universal $\mathcal{L}$-formula $\forall y_1 \ldots \forall y_m \alpha$. We may assume $x \in \ell(\psi)$.[9] By Exercise 2.55-(iv), $\forall x \forall y_1 \ldots \forall y_m \alpha$ is equivalent to $\forall x \psi(x)$ and the first $\mathcal{L}$-formula is universal.

---

[9] If not, $\forall x \psi$ and $\exists x \psi$ are equivalent to ψ and the inductive hypothesis alone guarantees their T-equivalence to a universal $\mathcal{L}$-formula.

2.5 Induction on the Complexity of Formulae

We deal with $\exists x \psi(x)$ by means of the closure properties already established for S. Because S is closed under negation and $\psi(x) \in S$ by inductive hypothesis, we obtain $\neg\psi(x) \in S$. By what we have just established, $\forall x \neg \psi(x) \in S$. Closure under negation now implies $\neg\forall x \neg \psi(x)$, which is equivalent to $\exists x \psi(x)$ by Exercise 2.40-(iii). It follows that $\exists x \psi(x) \in S$.

The induction is complete. □

Exercise 2.61 Prove, with reference to Lemma 2.60, that $\phi(x) \in S$ implies $\exists x \phi(x) \in S$ without making use of the corresponding implication for $\forall$ (*Hint*: set up a proof by contradiction).

Exercise 2.62 Prove that, if every $\mathcal{L}$-formula is T-equivalent to a universal $\mathcal{L}$-formula, then every $\mathcal{L}$-formula is T-equivalent to an existential $\mathcal{L}$-formula.

In the following chapters, we shall use induction on the complexity of formulae to obtain results that invariably involve semantic properties. Any such result is preserved by equivalence. In other words, if it holds for ϕ and $\psi \equiv \phi$, it continues to hold for ψ. We saw this in the proof of Lemma 2.60, where, in effect, we could reduce closure under $\wedge, \rightarrow, \exists$ to closure under the connectives $\neg, \vee, \forall$.

Because of this, we can systematically abridge inductive proofs. Instead of reducing closure under certain connectives and quantifiers to closure under a fixed set of connectives and quantifiers, we simply omit the reduction and restrict the inductive step to $\neg, \vee$ and $\exists$. This is to say that we run an induction only on those formulae that can be obtained from the atomic formulae by finitely many applications of two connectives and one quantifier. We can prove that this restricted induction is not less general than an induction involving the other connectives and quantifiers, because every formula is equivalent to a formula in which only the connectives $\neg, \vee$ and only the quantifier $\exists$ occur.

The last statement is a special case of the **substitution theorem**, which we shall now prove, unsurprisingly by induction.

Theorem 2.63 (Substitution) *Let ϕ, ψ, χ $\mathcal{L}$-formulae, ψ a subformula of ϕ. If $\psi \equiv \chi$ and ϕ' is obtained by substituting χ to ψ in ϕ, then $\phi \equiv \phi'$.*

Proof Let S be the set of formulae for which the substitution theorem holds.

Inductive base: if ϕ is atomic, its only subformula is ϕ. We deduce that $\phi = \psi$ and $\phi' = \chi$. By hypothesis, $\psi \equiv \chi$: thus, $\phi \equiv \phi'$. The set S must therefore include every atomic formula.

Inductive step: *Connectives* The inductive hypothesis is $\varphi_1, \varphi_2 \in S$ for any assignment. Our goal is to show $\neg\varphi_1, \varphi_1 \vee \varphi_2, \varphi_1 \wedge \varphi_2, \varphi_1 \rightarrow \varphi_2 \in S$.

We focus on $\neg\varphi_1$ and suppose that ψ is a subformula $\neg\varphi_1$. It follows that ψ, if it does not coincide with $\neg\varphi_1$ (a case easy to deal with), is a subformula φ_1. By inductive hypothesis $\varphi_1 \equiv \varphi_1'$. The reader may verify that the definition of equivalence implies $\neg\varphi_1 \equiv \neg\varphi_1'$. We conclude $\neg\varphi_1 \in S$.

We turn to $\varphi_1 \vee \varphi_2$. A subformula ψ of the latter formula, other than the formula itself, is a subformula of φ_1 or a subformula of φ_2. It may be a subformula of both: we suppose it is. The substitution of χ to ψ in $\varphi_1 \vee \varphi_2$ leads to the formula $\varphi_1' \vee \varphi_2'$, where, by inductive hypothesis, $\varphi_1 \equiv \varphi_1'$ and $\varphi_2 \equiv \varphi_2'$. The reader may readily check that $\varphi_1 \vee \varphi_2 \equiv \varphi_1' \vee \varphi_2'$, i.e. $\varphi_1 \vee \varphi_2 \in S$. This argument applies with minor modifications when ψ is a subformula of φ_1 only or of φ_2 only.

Consider $\varphi_1 \wedge \varphi_2$. We may reason as in the case of $\vee$ and rely on the fact that, if $\varphi_1 \equiv \varphi_1'$ and $\varphi_2 \equiv \varphi_2'$, then $\varphi_1 \wedge \varphi_2 \equiv \varphi_1' \wedge \varphi_2'$. We conclude $\varphi_1 \vee \varphi_2 \in S$. An analogous train of thought allows us to conclude $\varphi_1 \to \varphi_2 \in S$.

Quantifiers. We finally suppose that the substitution theorem holds for $\varphi(x)$ and for any assignment. If ψ is a subformula of $\exists x \varphi(x)$ other than the formula itself, then ψ is a subformula of $\varphi(x)$. Once χ is substituted to ψ in $\varphi(x)$ the resulting formula $\varphi'(x)$ is equivalent to $\varphi(x)$ by inductive hypothesis. This is to say that, given a structure $\mathcal{M}$ and an assignment σ, $\mathcal{M} \models \varphi(x)[\sigma]$ iff $\mathcal{M} \models \varphi'(x)[\sigma]$.

Assume $\mathcal{M} \models \exists x \varphi(x)[\sigma]$. By Definition 2.32, $\mathcal{M} \models \varphi(x)[\sigma(x/a)]$, for some $a \in M$. We use the inductive hypothesis to obtain $\mathcal{M} \models \varphi'(x)[\sigma(x/a)]$, which implies $\mathcal{M} \models \exists x \varphi'(x)[\sigma]$. This argument can be reversed. It follows that $\exists x \varphi(x) \equiv \exists x \varphi'(x)$, where $\exists x \varphi'(x)$ is the result of the substitution of χ to ψ in $\exists x \varphi(x)$.

The final part of the inductive step, involving $\forall$, is left as an exercise. □

Exercise 2.64 Complete the proof of the substitution theorem.

An application of the substitution theorem shows that every $\mathcal{L}$-formula φ is equivalent to a $\mathcal{L}$-formula in which only the connectives $\neg$, $\wedge$ and the quantifier $\exists$ occur. Given a formula φ, we substitute: (i) any subformula of φ of the form $\forall x \psi$ with the equivalent formula $\neg \exists \neg \psi$; (ii) any subformula of φ of the form $\psi \wedge \chi$ with the equivalent formula $\neg(\neg \psi \vee \neg \chi)$ and (iii) any subformula of φ of the form $\psi \to \chi$ with the equivalent formula $\neg \psi \vee \chi$. Substitutions (i)-(iii) can be carried out systematically, working on the subformulae included within the largest number of parentheses at each step of the procedure.

Exercise 2.65 For each of the following $\mathcal{L}_=$-formulae, find an equivalent $\mathcal{L}_=$-formula in which only the connectives $\neg$, $\wedge$ and only the quantifier $\exists$ occur.

(i) $\forall x \forall y (x = y \to \forall z (x = z \wedge x = x))$;
(ii) $\forall x \neg (x = z) \wedge \forall y (x = y \to \exists z (x = x))$.

We conclude this section with a proof by induction of the *prenex form theorem*. The theorem asserts that, up to equivalence, there is no difference between the set of $\mathcal{L}$-formulae and the set of $\mathcal{L}$-formulae in prenex form.

Theorem 2.66 (Prenex Form) *Every $\mathcal{L}$-formula is equivalent to a $\mathcal{L}$-formula in prenex form.*

Note that, if $\phi \equiv \psi$ and the prenex form theorem holds for ψ, it also holds for ϕ. The theorem is, in other words, invariant under equivalence. In view of the substitution theorem, we can establish it by an abridged induction on the complexity of formulae.

Proof The inductive base is straightforward. An atomic $\mathcal{L}$-formula is in prenex form and it is certainly equivalent to itself.

The inductive hypothesis is that ϕ and ψ are equivalent to formulae in prenex form. Because, in the prefix of a prenex form, universal and existential quantifiers can alternate in any pattern, we adopt the symbols Q and q, possibly equipped with numerical indices, to indicate generic quantifiers, without specifying whether they are $\exists$ or $\forall$. Moreover if Q or q is a fixed quantifier, we indicate the other quantifier by Q' or q' respectively (i.e., if Q were $\exists$, Q' would be $\forall$ and vice versa).

Let $Q_1x_1 \ldots Q_nx_n\alpha$ and $q_1y_1 \ldots q_my_m\beta$ be the formulae in prenex form equivalent to ϕ, ψ respectively. By Exercise 2.55-(iii), we may assume that $y_1, \ldots, y_m \notin \ell(Q_1x_1 \ldots Q_nx_n\alpha)$ and $x_1, \ldots, x_n \notin \ell(q_1y_1 \ldots q_my_m\beta)$.

It is easy to see that $\neg \phi$ is equivalent to $Q'_1x_1 \ldots Q'_nx_n\neg\alpha$. Moreover, $\phi \vee \psi$ is equivalent to:

$$Q_1x_1 \ldots Q_nx_n\alpha \vee q_1y_1 \ldots q_my_m\beta,$$

and a straightforward generalisation of 2.55-(v) yields:

$$Q_1x_1 \ldots Q_nx_nq_1y_1 \ldots q_my_m(\alpha \vee \beta).$$

We turn to quantifiers. The inductive hypothesis is that $\phi(x)$ is equivalent to a prenex form $Q_1x_1 \ldots Q_nx_n\alpha$. In this case, $Qx\phi(x) \equiv QxQ_1x_1 \ldots Q_nx_n\alpha$, where $QxQ_1x_1 \ldots Q_nx_n\alpha$ is in prenex form. □

Exercise 2.67 Find an equivalent prenex form for each of the following $\mathcal{L}_=$ formulae.

(i) $\forall x(x = y \to \exists z \forall x(z = x))$;
(ii) $\forall x \exists y(x = y) \to \exists y \forall x(x = y)$.

Is is true or false that a formula is equivalent to a unique formula in prenex form?

Exercise 2.68 Prove by induction on the complexity of formulae that, whenever two assignments σ, σ' coincide on $x_1, \ldots, x_n$ (i.e. $\sigma(x_i) = \sigma'(x_i)$ con $i = 1, \ldots, n$), they satisfy exactly the same $\mathcal{L}$-formula of the form $\phi(x_1, \ldots, x_n)$, whose variables occur among $\{x_1, \ldots, x_n\}$. In other words, prove that, for every $\mathcal{L}$-structure $\mathcal{M}$, $\mathcal{M} \models \phi(x_1, \ldots, x_n)[\sigma]$ iff $\mathcal{M} \models \phi(x_1, \ldots, x_n)[\sigma']$.

2.6 Definability and Interpretability

In the previous section we have always associated the free occurrences of variables within a formula with values uniquely determined by a fixed assignment. If we drop the focus on one fixed assignment of values to all variables, we can use formulae to *define* subsets of a structure's domain, or subsets of its Cartesian products.

To clarify the preceding remarks, let us fix a $\mathcal{L}$-structure $\mathcal{M}$ and a $\mathcal{L}$-formula $\phi(x, y_1, y_2)$. Suppose we fix the values assigned to y_1, y_2 and let them be, respectively, $a_1, a_2 \in M$. We do not, by contrast, fix the value of x, but run a search over M for those values under which $\phi(x, y_1, y_2)$ is satisfied in $\mathcal{M}$ while y_1, y_2 are assigned the fixed values a_1, a_2. At the end of our search we have isolated a subset X of M (perhaps the empty set). Because we have arrived at this result by keeping the values of y_1, y_2 fixed, it is convenient to adopt the notation:

$$\phi(x, a_1, a_2),$$

and treat the last expression as a formula (even though, strictly speaking, it does not correspond to our syntactic definition of a formula and we have not made use of partial assignments either).

We say that X is *defined* by the formula $\phi(x, a_1, a_2)$. If we had worked with a formula of the form $\psi(x_1, x_2, a_1, a_2)$, we would have defined a set of ordered pairs, i.e. a subset of M^2, instead of a subset of M. The general situation can now be made the content of a formal definition.

Definition 2.69 Let $\mathcal{M}$ be a $\mathcal{L}$-structure, $X \subseteq M^n$ and $A \subseteq M$. We say that X is **definable over** A iff there are a $\mathcal{L}$-formula $\phi(x_1, \ldots, x_n, y_1, \ldots, y_k)$ and $a_1, \ldots, a_k \in A$ such that $X = \{(b_1, \ldots, b_n) \in M^n : \mathcal{M} \models \phi(b_1, \ldots, b_n, a_1, \ldots, a_k)\}$. If $k > 0$, we say that X is **definable with parameters** or parametrically definable. If $k = 0$, we say that X is **definable without parameters**.

Example 2.70 Consider the structure $(\mathbb{Z}, +^{\mathbb{Z}})$ for the language with a single binary function symbol $+$ in its signature. The binary function $+^{\mathbb{Z}}$ is ordinary addition. It is easy to see that the formula $\exists y (y + y = x)$ is satisfied with any assignment that associates x with an even integer. We say that $\exists y (y + y = x)$ defines (without parameters) the subset of $\mathbb{Z}$ containing all even integers.

Example 2.71 Let $\mathbb{N}$ be a $\mathcal{L}_=$-structure. Consider the formula $x_1 = 1 \lor x_1 = 2$. It is obvious that it defines the finite set $\{1, 2\} \subseteq \mathbb{N}$. By suitably increasing the number of disjunct and of parameters, it is easy to see that we can parametrically define any finite subset of $\mathbb{N}$. For instance, the set $\{a_1, \ldots, a_n\} \subseteq \mathbb{N}$ is defined by the disjunction:

$$x_1 = a_1 \lor \ldots \lor x_1 = a_n.$$

The negation of the latter formula, by contrast, defines an infinite subset of $\mathbb{N}$, which contains all but finitely many natural numbers. We call such a set **cofinite** because its complement in $\mathbb{N}$ is a finite set.

We shall prove that the finite and cofinite subsets of $\mathbb{N}$ are all the parametrically $\mathcal{L}_=$-definable subsets of this structure. In particular, the set of even numbers is not definable in $\mathbb{N}$.

An analogous situation presents itself when we consider an arbitrary infinite $\mathcal{L}_=$-structure, i.e. an infinite set.

A $\mathcal{L}$-structure $\mathcal{M}$ whose (parametrically) definable subsets are exactly the finite and cofinite subsets is called **minimal**. We shall explore some of the momentous consequences of minimality in Chap. 8, once we have more interesting examples of minimal structures at hand.

2.6 Definability and Interpretability

Example 2.72 Let $\mathcal{Z}$ be the $\mathcal{L}_r$-structure $(\mathbb{Z}, +^\mathbb{Z}, -^\mathbb{Z}, \cdot^\mathbb{Z}, 0, 1)$. By a number-theoretic result known as Lagrange's theorem,[10] every natural number can be expressed as the sum of four squares. It follows that the $\mathcal{L}_r$-formula:

$$\exists y_1 \exists y_2 \exists y_3 \exists y_4 (x = y_1 \cdot y_1 + y_2 \cdot y_2 + y_3 \cdot y_3 + y_4 \cdot y_4)$$

defines $\mathbb{N}$ in $\mathbb{Z}$ (without parameters). Rename this last formula $\psi(x)$. Then the $\mathcal{L}_r$-formula:

$$\exists x (\psi(x) \wedge y + x = z)$$

defines the set of pairs of integers (m, n) such that $m \leq n$. In short, order is definable in $\mathcal{Z}$.

Remark 2.73 Frequently in the next subsection, and later in the book, it will be convenient to assign shorthand names to certain formulae, exactly as we have done in the above example by calling an explicitly given $\mathcal{L}_r$-formula $\psi(x)$. In order to indicate that we label certain formulae by a shorthand name, we shall use the notation ':='. Thus, for instance, we may write:

$$\psi(x) := \exists y_1 \exists y_2 \exists y_3 \exists y_4 (x = y_1 \cdot y_1 + y_2 \cdot y_2 + y_3 \cdot y_3 + y_4 \cdot y_4),$$

where the whole string of symbols is not a $\mathcal{L}_r$-formula but a declaration that one particular formula will be referred to simply as $\psi(x)$.

Exercise 2.74 Show that order is definable in the field of rational numbers, viewed as a $\mathcal{L}_r$-structure.

Exercise 2.75 Let $\mathcal{M}$ be a $\mathcal{L}$-structure and $X, Y \subseteq M^n$ be sets defined by the formulae ϕ_X, ψ_Y respectively (with or without parameters). Show that the sets $X \cap Y, X \cup Y, M^n - X, (X - Y) \cup (Y - X)$ are definable (with or without parameters).

Exercise 2.76 Let $\mathcal{M}$ be a $\mathcal{L}$-structure and $X \subseteq M^n$ be a set defined by the formula ϕ_X (with or without parameters). Show that the set $M \times X$ is definable (with or without parameters).

Exercise 2.77 Let $\mathcal{M}$ be a $\mathcal{L}$-structure and $X \subseteq M^{n+1}$ be a set defined by the formula ϕ_X (with or without parameters). Let $\pi : X \longrightarrow M^n$ be the projection function whose action is given by $\pi((x_1, \ldots, x_{n+1})) = (x_1, \ldots, x_n)$. Show that $\pi[X]$ is definable.

Remark 2.78 From now on we adopt, for any positive integer n, the notation $\phi(M^n)$ to indicate the subset of a given set M^n that is defined by the formula $\phi(x_1, \ldots, x_n)$.

---

[10] For proof of this result, see e.g. [23], pp.375–377.

2.6.1 Interpretations

Let $\mathcal{L}, \mathcal{L}^*$ two languages, which may have distinct signatures. If $\mathcal{M}$ is a $\mathcal{L}$-structure and $\mathcal{N}$ a $\mathcal{L}^*$-structure, it may in certain cases be possible to define or 'interpret' $\mathcal{N}$ inside $\mathcal{M}$.

In rough terms, which we shall refine, it may be possible to isolate a 'copy' of $\mathcal{N}$ whose elements are tuples in M^k, for some $k \geq 1$. The constants, relations and functions of $\mathcal{N}$ are then definable in M^k by means of suitable $\mathcal{L}$-formulas.

Here is an example that may be familiar:

Example 2.79 We let $\mathcal{L} = \mathcal{L}_r$ and let $\mathcal{L}^*$ be a relational version of the language of rings, in which the binary operation symbols $+, \cdot$ are replaced by the respective ternary relation symbols $R_+, R_\times$.

Consider the $\mathcal{L}_r$-structure $\mathcal{Z}_r = (\mathbb{Z}, +^{\mathbb{Z}}, -^{\mathbb{Z}}, \cdot^{\mathbb{Z}}, 0^{\mathbb{Z}}, 1^{\mathbb{Z}})$ and the $\mathcal{L}^*$-structure $\mathcal{Q}^* = (\mathbb{Q}, R_+^{\mathbb{Q}}, R_\times^{\mathbb{Q}}, 0^{\mathbb{Q}}, 1^{\mathbb{Q}})$.

Note that $\mathcal{Q}^*$ is just the rational field presented in a relational signature (e.g. $\mathcal{Q}^* \models R_\times(1/2, 3/5, 3/10)$ just because $(1/2)(3/5) = 3/10$).

The set $\mathbb{Z}^2$ is the domain from which we are going to define a copy of the rational field. The idea is to use ordered pairs of integers in N as codes of 'numerator-denominator' pairs to define the constants and relations of $\mathcal{Q}^*$. Because denominators must be non-zero, we must restrict attention to the subset N of $\mathbb{Z}^2$ whose elements satisfy the formula $\phi(x) := \neg(x_2 = 0)$.

The constants $0^N, 1^N \in N$ are defined respectively by the formulae $x_1 = 0 \wedge x_2 = 1$ and $x_1 = 1 \wedge x_2 = 1$.

Because we identify two fractions if one is obtained from the other by multiplying both its numerator and denominator by a common non-zero integer, we carry out the same identification in N, by means of the formula:

$$\phi_=(x_1, x_2, y_1, y_2) := x_1 \cdot y_2 = y_1 \cdot x_2.$$

It is easy to verify that the formula $\phi_=$ determines an equivalence relation on N. What remains to be done is to define addition and multiplication between pairs of integers.

The relation R_+^N, which codifies the addition of rational numbers, is defined by the formula:

$$\phi_+(x_1, x_2, y_1, y_2, z_1, z_2) := z_2(x_1 y_2 + y_1 x_2) = z_1 \cdot x_2 \cdot y_2,$$

which stands for the familiar equality:

$$\frac{x_1}{x_2} + \frac{y_1}{y_2} = \frac{z_1}{z_2}.$$

Finally, the relation $R_\times^N$, which codifies multiplication, is defined by the formula:

2.6 Definability and Interpretability

$$\phi_\times(x_1, x_2, y_1, y_2, z_1, z_2) := z_2(x_1 y_1) = z_1 \cdot x_2 \cdot y_2.$$

It is easy to see that the $\mathcal{L}^*$-structure $\mathcal{N} = (N, R_+^N, R_\times^N, 0^N, 1^N)$ is a field, relationally presented, and that each rational number is represented by a pair (in fact, several identifiable pairs) in N. Moreover, we may now view the set of integers $\mathbb{Z}$ as the subset of N defined by the formula $x_2 = 1$.

The construction we have described is quite general. We shall encounter it again in Chap. 4, when we look at certain $\mathcal{L}_r$-structures called *integral domains*, which share many key properties with $\mathcal{Z}_r$. We shall find it helpful to take advantage of the fact that, if $\mathcal{A}$ is an integral domain, then a field known as its *field of fractions* $Q(\mathcal{A})$ can be defined on A^2, the domain of $\mathcal{A}$, exactly as $\mathbb{Q}$ was defined on $\mathbb{Z}^2$. Thus, in particular, an integral domain may also be viewed as a subset of its associated field of fractions.

Keeping Example 2.79 as a concrete reference in mind, we now define the notion of interpretation, which provides the concept we need to think in general terms about defining structures inside other structures.

Definition 2.80 Let $\mathcal{L}, \mathcal{L}^*$ be first-order languages, $\mathcal{M}$ a $\mathcal{L}$-structure and $\mathcal{N}$ a $\mathcal{L}^*$-structure. A n-dimensional **interpretation** Γ **of** $\mathcal{N}$ **in** $\mathcal{M}$ consists of:

(1) a $\mathcal{L}$-formula $\partial_\Gamma(x_1, \ldots, x_n)$;
(2) for each unnested atomic $\mathcal{L}^*$-formula $\phi(x_1, \ldots, x_m)$, a $\mathcal{L}$-formula $\phi_\Gamma(x_{11}, \ldots, x_{1n}, \ldots, x_{m1}, \ldots, x_{mn})$, in which a n-tuple of distinct variables occurs for each x_i in the original formula;
(3) a surjective function $f_\Gamma : \partial_\Gamma(M^n) \longrightarrow N$ such that, for any unnested, atomic $\mathcal{L}^*$-atomic formula $\phi(x_1, \ldots, x_m)$ and any m-tuple of sequences $(a_{j1}, \ldots, a_{jn})$, with $j = 1, \ldots, m$:

$$\mathcal{M} \models \phi_\Gamma(a_{11}, \ldots, a_{1n}, \ldots, a_{m1}, \ldots, a_{mn}) \iff$$
$$\mathcal{N} \models \phi(f_\Gamma(a_{11}, \ldots, a_{1n}), \ldots, f_\Gamma(a_{m1}, \ldots, a_{mn})).$$

Remark 2.81 Example 2.79 describes a two-dimensional interpretation Γ, where $\partial_\Gamma(x_1, x_2)$ is the formula $\neg(x_2 = 0)$, f_Γ is the function determined by the condition $f_\Gamma(m, n) \mapsto m/n \in \mathbb{Q}$, and the required $\mathcal{L}$-atomic formulae are given in the example. Note that we have associated with the atomic $\mathcal{L}^*$-formula $x_1 = x_2$ the $\mathcal{L}$-formula $x_1 y_2 = x_2 y_1$, which determines an *equivalence relation* between pairs of integers.

Exercise 2.82 Whenever Γ is an n-dimensional interpretation of a $\mathcal{L}^*$-structure $\mathcal{N}$ in a $\mathcal{L}$-structure $\mathcal{M}$, $\mathcal{M}$ must model certain $\mathcal{L}$-sentences, called **admissibility conditions** for Γ. These sentences say that:

(i) the $\mathcal{L}$-formula $\phi_=(x_1, \ldots, x_n, y_1, \ldots, y_n)$, associated with the $\mathcal{L}^*$-formula $x = y$, defines an equivalence relation on $\partial_\Gamma(M^n)$.

(ii) For notational convenience, suppose the signature of $\mathcal{L}^*$ contains a unary relation symbol P. The $\mathcal{L}$-formula $\phi_P(x_1, \ldots, x_n)$ associated with $P(x_1)$ by the interpretation Γ, satisfies the following condition, for any $(a_1, \ldots, a_n), (b_1, \ldots, b_n) \in \partial_\Gamma(M^n)$:

$$\mathcal{M} \models \phi_=(a_1, \ldots, a_n, b_1, \ldots, b_n) \text{ and } \mathcal{M} \models \phi_P(a_1, \ldots, a_n)$$
$$\implies \mathcal{M} \models \phi_P(b_1, \ldots, b_n).$$

An entirely similar (but notationally more cumbersome) condition must hold for each m-ary relation symbol R in the signature of $\mathcal{L}^*$ and for any unnested atomic $\mathcal{L}^*$-formula $R(x_1, \ldots, x_m)$.

(iii) For any atomic $\mathcal{L}^*$-formula of the form $c = x_1$, if $\phi_c(x_1, \ldots, x_n)$ is the $\mathcal{L}$-formula associated with it by the interpretation Γ, there is a n-tuple $(a_1, \ldots, a_n) \in M^n$ such that, for any $(b_1, \ldots, b_n) \in M^n$:

$$\mathcal{M} \models \phi_c(b_1, \ldots, b_n) \iff \mathcal{M} \models \phi_=(a_1, \ldots, a_n, b_1, \ldots, b_n).$$

(iv) For any atomic $\mathcal{L}^*$-formula of the form $f(x_1, \ldots, x_k) = y$, if

$$\phi_f(x_{11}, x_{1n}, \ldots, x_{k1}, \ldots, x_{kn}, y_1, \ldots, y_n)$$

is the $\mathcal{L}$-formula associated with it by Γ, there is $(b_1, \ldots, b_n) \in M^n$ such that, for any $a_{ij} \in M$ ($1 \leq i \leq k, 1 \leq j \leq n$) and any $(c_1, \ldots, c_m) \in M^n$:

$$\mathcal{M} \models \phi_f(a_{11}, a_{1n}, \ldots, a_{k1}, \ldots, a_{kn}, c_1, \ldots, c_n) \iff$$
$$\mathcal{M} \models \phi_=(b_1, \ldots, b_n, c_1, \ldots, c_n).$$

When the admissibility conditions are satisfied we say that $\phi_=$ defines a **congruence** over M^n (in essence, a congruence is a structure-preserving equivalence relation). State each of the admissibility conditions as a $\mathcal{L}$-sentence and show that they hold in Example 2.79.

Exercise 2.83 Let $\mathcal{L}$ be a language. A $\mathcal{L}$-formula is said to be an **unnested formula** iff the atomic subformulae occurring in it are unnested.

Show that every atomic $\mathcal{L}$-formula ϕ is equivalent both to a universal unnested formula $\varphi_\forall$ and an existential unnested formula $\varphi_\exists$.

(*Hint*: fix a simple signature and work out the result for your example, then generalise it.)

Exercise 2.84 Let $\mathcal{L}$ be a language. Prove by induction on the complexity of formulae that every $\mathcal{L}$-formula is equivalent to an unnested $\mathcal{L}$-formula.

Exercise 2.85 Let $\mathcal{M}$ be a $\mathcal{L}$-structure, $\mathcal{N}$ a $\mathcal{L}^*$-structure and Γ a n-dimensional interpretation of $\mathcal{N}$ into $\mathcal{M}$. Let $\bar{x}$ be a shorthand notation to designate a n-tuple of variables, e.g. $(x_1, \ldots, x_n)$. Prove by induction on the complexity of formulae that,

2.6 Definability and Interpretability

for every $\mathcal{L}^*$-formula $\phi(x_1, \ldots, x_k)$, there is a $\mathcal{L}$-formula $\phi_\Gamma(\overline{x}_1, \ldots, \overline{x}_k)$, such that, for any $\overline{a}_1, \ldots \overline{a}_k \in M^n$:

$$\mathcal{N} \models \phi(f_\Gamma(\overline{a}_1), \ldots, f_\Gamma(\overline{a}_k)) \iff \mathcal{M} \models \phi_\Gamma(\overline{a}_1, \ldots, \overline{a}_k).$$

(*Hint*: recursively define rules to convert a $\mathcal{L}^*$-formula ϕ into a $\mathcal{L}$-formula ϕ_Γ. Since Γ is given, this rule is available for atomic unnested formulae. By Exercise 2.84, we may restrict attention to unnested $\mathcal{L}$-formulae and run a proof by induction on them. The rules relative to the propositional connectives are easy to formulate: we need one for each quantifier because we are describing a syntactic transformation. The rule for the existential quantifier is:

$$[\exists x \varphi]_\Gamma = \exists x_1 \ldots \exists x_n (\partial_\Gamma(x_1, \ldots, x_n) \wedge \phi_\Gamma).$$

The rule for the universal quantifier is:

$$[\forall x \varphi]_\Gamma = \forall x_1 \ldots \forall x_n (\partial_\Gamma(x_1, \ldots, x_n) \to \phi_\Gamma).$$

Note that, if we replace $\forall x \varphi$ with the equivalent formula $\neg \exists x \neg \varphi$ and apply the previous rules to it, we do not obtain $[\forall x \varphi]_\Gamma$).

2.6.1.1 Quotient Structures and Quotient Groups

In Sect. 1.2 of the previous chapter we have constructed a linear order from a complete preorder by means of an equivalence relation. In the terminology of this section, we have in fact verified that the given preorder satisfied the admissibility conditions for a one-dimensional interpretation of the linear order inside it. It is an instructive exercise to try and reformulate the example from Sect. 1.2 in model-theoretic terms.

This example is in fact a special case of a general type of interpretation, which we shall refer to as the interpretation of a **quotient structure** $\mathcal{N}$ into a structure $\mathcal{M}$. The expression 'quotient structure' depends on the fact that we use an equivalence relation that is also a congruence to 'factor out' the objects identified by it.

A simple but common and important example of quotient structure occurs in group theory. We first look at an example that will matter in the next chapter.

Example 2.86 Consider the additive group of integers, i.e. the $\mathcal{L}_g$-structure $\mathcal{Z} = (\mathbb{Z}, +^{\mathcal{Z}}, -^{\mathcal{Z}}, 0^{\mathcal{Z}})$. We designate by $5\mathbb{Z}$ the set of all multiples of 5 (obtained by multiplying each element of $\mathbb{Z}$ by 5). It is easy to verify that this subset of $\mathbb{Z}$ is itself a group relative to the restriction of the group operations on $\mathbb{Z}$. We refer to the group with domain $5\mathbb{Z}$ as a *subgroup* of $\mathcal{Z}$.

Let us identify two integers when they differ by a multiple of 5. Thus, for instance, we identify the distinct integers $5, -10$ and 25, but we also identify the distinct integers $2, 7, -8$. Since every integer can be written in the form $i + 5k$, where $i \in \{0, 1, 2, 3, 4\}$ and $k \in \mathbb{Z}$, it is clear that we are partitioning $\mathbb{Z}$ into five disjoint sets, i.e. five equivalence classes. The equivalence relation associated with these equivalence classes is defined by the $\mathcal{L}_g$-formula $\exists z(x - y = 5z)$ (here and below, we take $5z$ to be an abbreviation of the term $z + z + z + z + z$).

Let us now call the resulting equivalence classes $5\mathbb{Z} = [0], [1], [2], [3], [4]$ respectively. The set $\{[0], [1], [2], [3], [4]\}$ can be turned into a $\mathcal{L}_g$-structure $\mathcal{Z}_5$ that is a group. To this end we simply define the group operations $+_5$ and $-_5$ by the stipulations:

$$[a] +_5 [b] = [a + b] \text{ and } -_5 [b] = [-b].$$

Using the conditions just stated we see e.g. that $[3] +_5 [4] = [7] = [2]$, since 7 and 2 are identified. In general, the sum of two equivalence classes is always one of $[0], [1], [2], [3], [4]$. For an example involving additive inverses, consider $-_5 [2] = [-2] = [3]$.

We can now easily see that there is a one-dimensional interpretation Γ of $\mathcal{Z}_5$ in $\mathcal{Z}$. First of all $\partial(x) := (x = x)$.

The unnested atomic $\mathcal{L}_g$-formulae relative to $\mathcal{Z}_5$ are interpreted in $\mathcal{Z}$ as follows:

- $(x = y)_\Gamma := \exists z(x - y = 5z)$.
- $(x = 0)_\Gamma := \exists y(x = 5y)$;
- $(x + y = z)_\Gamma := \exists w((x + y) - z = 5w)$;
- $(x = -y)_\Gamma := \exists z(x + y = 5z)$,

The surjection f_Γ is given by the condition $f(n) = [n]$. We refer to $\mathcal{Z}_5$ as the quotient group of $\mathbb{Z}$ relative to $5\mathbb{Z}$ and also refer to it, as a $\mathcal{L}_g$-structure, by the notation $\mathbb{Z}/5\mathbb{Z}$.

Remark 2.87 We can obviously trivially adapt the construction given in the above example to build, for any fixed, positive integer, the group $\mathbb{Z}/n\mathbb{Z}$.

It is possible to look at Example 2.86 from a slightly different standpoint. Suppose we take as our starting point the formula $\partial(x)$ and the list of $\mathcal{L}_g$-formulae used to interpret the unnested atomic formulae of $\mathcal{L}_g$. Let us call this linguistic information Γ. It is quickly verified that $\mathcal{Z}$ models the admissibility conditions of Γ. What we have shown in the example is that there are a $\mathcal{L}_g$-structure $\mathcal{Z}_5$ and a surjection f from $\mathbb{Z}$ into $\{[0], [1], [2], [3], [4]\}$ such that Γ and f interpret $\mathcal{Z}_5$ in $\mathcal{Z}$.

This result is a special case of a general theorem, which also implies that $\mathcal{Z}_5$ is uniquely determined, in a sense to be specified. In order to state the relevant theorem, we introduce an important definition.

Definition 2.88 Let $\mathcal{L}, \mathcal{L}^*$ be languages. A n-dimensional **interpretation** Γ of $\mathcal{L}^*$ in $\mathcal{L}$ consists of:

(1) a $\mathcal{L}$-formula $\partial_\Gamma(x_1, \ldots, x_n)$;
(2) for each unnested atomic $\mathcal{L}^*$-formula $\phi(x_1, \ldots, x_m)$, a $\mathcal{L}$-formula

2.6 Definability and Interpretability

$\phi_\Gamma(x_{11}, \ldots, x_{1n}, \ldots, x_{m1}, \ldots, x_{mn})$, in which a n-tuple of distinct variables occurs for each x_i in the original formula.

Note that, in the above definition, we have not made any reference to structures: instead, we have defined an interpretation only in terms to signatures. With this notion of interpretation in hand we can state:

Theorem 2.89 *Let $\mathcal{L}, \mathcal{L}^*$ be languages and Γ be a n-dimensional interpretation of $\mathcal{L}^*$ in $\mathcal{L}$. If $\mathcal{M}$ is a $\mathcal{L}$-structure that models the admissibility conditions of Γ, then:*

(1) there are a $\mathcal{L}^*$-structure $\mathcal{N}$ and a surjection $f : \partial(M) \longrightarrow N$ such that Γ with f is an interpretation of $\mathcal{N}$ in $\mathcal{M}$;

(2) for any $\mathcal{L}^*$-structure $\mathcal{A}$ and surjection $g : \partial(M) \longrightarrow A$ such that Γ with g is an interpretation of $\mathcal{A}$ in $\mathcal{M}$, $\mathcal{A}$ and $\mathcal{N}$ are isomorphic (we shall explain what this means in the course of the proof).

Proof In order to prove (1), i.e. the existence of a suitable $\mathcal{N}$, we consider the definable equivalence relation given by Γ. Suppose that its defining formula is $\phi_= (x_1, \ldots, x_n, y_1, \ldots, y_n)$. We abbreviate n-tuples as in Exercise 2.85 and write $\phi_= (\overline{x}, \overline{y})$.

We let the domain of $\mathcal{N}$ be the set of equivalence classes determined by $\phi_=$ in $\partial(M)$. Because $\mathcal{M}$ models the admissibility conditions of Γ, if R is k-ary relation symbol in the signature of $\mathcal{L}^*$ and $b_1, \ldots, b_k$ are equivalence classes of N, we define the k-ary relation R^N as follows:

$$(b_1, \ldots, b_n) \in R^N \iff \mathcal{M} \models \phi_R(\overline{a}_1, \ldots, \overline{a}_k),$$

where $\overline{a}_i$ is a n-tuple from the equivalence class b_i. The admissibility condition for R ensures that the definition of R^N is sound, i.e. that this relation does not switch from holding to not holding (or vice versa) as we switch from certain equivalence class representatives to others.

In an entirely similar manner, if h is a m-ary function symbol in the signature of $\mathcal{L}^*$ nd $b_1, \ldots, b_m, b_{m+1}$ are equivalence classes of N, we define h^N by the stipulation:

$$h^N(b_1, \ldots, b_m) = b_{m+1} \iff \mathcal{M} \models \phi_h(\overline{a}_1, \ldots, \overline{a}_m) = \overline{a}_{m+1},$$

where, as before, $\overline{a}_i$ is a n-tuple from the equivalence class b_i. Again, the admissibility conditions ensure that h^N is a (single-valued) function, i.e. that its value does not change as we change representatives of the equivalence classes b_i in $\partial(M)$. Constants are treated in a similar manner.

We have built a $\mathcal{L}^*$-structure $\mathcal{N}$ from equivalence classes. The function f that sends each element of $\partial(M)$ into the corresponding equivalence class is clearly surjective. Thus $\mathcal{N}$ is interpreted by Γ and f in $\mathcal{M}$.

To prove (2), we suppose $\mathcal{A}$ and g are given. We define a function $i : N \longrightarrow A$ by the stipulation that, for any $\overline{a} \in \partial(M)$:

$$i(f(\overline{a})) = g(\overline{a}).$$

Since f is surjective, each element of N is of the form $f(\overline{a})$. Next we show that i is well-defined, i.e. that $i(f(\overline{a}))$ does not change its value if we choose $\overline{a}'$ to be an element of the equivalence class of $\overline{a}$ distinct from $\overline{a}'$.

Consider the $\mathcal{L}^*$-formula $x = y$. Because f, g determine interpretations, we have:

$$\mathcal{N} \models f(\overline{a}) = f(\overline{a}') \text{ implies } \mathcal{M} \models \phi_=(\overline{a}, \overline{a}'), \text{ which in turn implies } \mathcal{A} \models g(\overline{a}) = g(\overline{a}').$$

The above chain of implications can be reversed, thus showing the injectivity of i. Because f, g are surjective, we conclude that i is a bijection.

Finally, for any unnested atomic $\mathcal{L}^*$-formula $\alpha(x_1, \ldots, x_k)$ we have:

$$\mathcal{N} \models \alpha(f(\overline{a}_1), \ldots, f(\overline{a}_k)) \text{ iff } \mathcal{M} \models \phi_\alpha(\overline{a}_1, \ldots, \overline{a}_k) \text{ iff } \mathcal{A} \models \alpha(i(f(\overline{a}_1)), \ldots, i(f(\overline{a}_k))).$$

This is to say that unnested atomic $\mathcal{L}^*$-formulae are preserved back and forth between $\mathcal{N}$ and $\mathcal{M}$. The function i is therefore a bijection that preserves all the structural information (concerning constants, functions and relations) of $\mathcal{N}$. In technical terms it is an isomorphism (a general definition of isomorphism will be given in Chap. 4). Because $\mathcal{N}, \mathcal{A}$ are structurally identical, we may think of the quotient structure $\mathcal{N}$ as a canonical representative of a $\mathcal{L}^*$-structure interpretable in $\mathcal{M}$ with Γ and f.

Since any other interpretable $\mathcal{L}^*$-structure is formally identical to $\mathcal{N}$, we think of $\mathcal{N}$ as uniquely determined. □

Example 2.90 We restate the previous example in abstract, group-theoretic terms. Let $\mathcal{G}$ be an abelian group and $\phi_H(x)$ a $\mathcal{L}_g$-formula that defines a subgroup $\mathcal{H}$ of $\mathcal{G}$. The nested formula $\psi(x_1, x_2) := \phi_H(x_2 - x_1)$ defines an equivalence relation that partitions G. By Exercise 2.83, we may alternatively take $\psi(x_1, x_2)$ to be the unnested equivalent $\exists u \exists v (u = -x_2 \wedge v = x_1 + u \wedge \phi_H(v))$ of the original nested formula.

Using $\partial(x) := x = x$, we obtain a one-dimensional interpretation Γ of $\mathcal{L}_g$ in $\mathcal{L}_g$ if we set:

- $(x = y)_\Gamma := \psi(x, y)$.
- $(x = 0)_\Gamma := x = 0$;
- $(x + y = z)_\Gamma := \psi(x + y, z)$;
- $(x = -y)_\Gamma := \psi(x, -y)$.

By Theorem 2.89, the canonical $\mathcal{L}_g$-structure determined by Γ is the quotient group G/H, whose elements are the equivalence classes determined by ψ. We call these equivalence classes **cosets**.

Exercise 2.91 Verify that $\psi(x_1, x_2)$ determines an equivalence relation and that $\mathcal{G}$ models the admissibility conditions for Γ. Moreover, show that each coset in G/H can be represented as $a +^G H = \{b \in G : b = a +^G h \text{ for some } h \in H\}$, with $a \in G$.

2.7 The Size of a Language   61

Remark 2.92 Example 2.90 has an immediate generalisation in case the subgroup $\mathcal{H}$ is not $\mathcal{L}_g$-definable. It suffices to add a unary relation symbol P to the signature of $\mathcal{L}_g$ to and expand $\mathcal{G}$ to a structure in which $P^G = H$. The quotient structure G/H is then obtained via a one-dimensional interpretation of $\mathcal{L}_g$ in the expanded language with $\partial(x) := P(x)$.

Remark 2.93 If $\mathcal{G}$ is an abelian group, any subgroup $\mathcal{H}$ determines a quotient group G/H. If, however, $\mathcal{G}$ is not abelian, only certain specific subgroups, known as normal subgroups, will work. For a detailed discussion, see e.g. [23], pp.49–53. We shall have to bear this in mind only in Chap. 12, since we do not encounter non-abelian groups before then.

2.7 The Size of a Language

In the following chapters it will often be important to know how large the set of $\mathcal{L}$-formulae is, for any given language $\mathcal{L}$ in use. At this stage the reader only needs to accept and bear in mind the last line of this section. For the sake of completeness, we supply the general argument leading to it, which, however, rests on set-theoretic results proved in the Appendix.

Section A.5 from the Appendix shows how we can assign a cardinal number to the symbol set of $\mathcal{L}$ (a symbol set here consists of variables, connectives, quantifiers and the symbols in the signature of $\mathcal{L}$). By the notation used in the Appendix, if $\mathcal{L}$ has a countable symbol set, then its denumerable size is indicated by $\aleph_0$ ('aleph zero'), otherwise its size is indicated by some uncountable cardinal κ.

In order to determine the size of the set of $\mathcal{L}$-formulae, which we denote by $|\mathcal{L}|$, we provide an argument that is not sensitive to the difference between a countable and uncountable language, and thus is given below only for uncountable languages. Because, by Theorem A.72 from the Appendix, every cardinal number satisfies the equality $\kappa \times \kappa = \kappa$, we have by iteration, $\kappa^n = \kappa$. The symbols in $\mathcal{L}$ enable us to produce exactly κ strings of symbols of length n, for any $n \in \mathbb{N}$. Let S_n be the set of $\mathcal{L}$-strings of symbols of length n.

The union $S = \bigcup_{n \in \mathbb{N}} S_n$ can be injected into the Cartesian product $\mathbb{N} \times K$, where $|K| = \kappa$. Since the size of this Cartesian product is κ by Corollary A.75, the last set has size κ. It follows from the Cantor-Bernstein theorem that S has size κ. The set of $\mathcal{L}$-formulae is a subset of S and thus its size is at most κ. Because there are only countably many logical symbols in $\mathcal{L}$, its signature must contain κ symbols. For each distinct symbol we can easily write an atomic formula in which it occurs. Thus, we can distinguish at least κ $\mathcal{L}$-atomic formulae. We conclude that $|\mathcal{L}| = \kappa$.

Consequently, if $\mathcal{L}$ is countable, $|\mathcal{L}|$ is countably infinite.

Chapter 3
Theories and Models

Abstract The last chapter provides the background knowledge needed to begin our study of first-order theories and their models. In Sect. 3.1 we discuss them from an abstract perspective, in order to describe a fundamental correspondence between sets of sentences in a first-order language and the corresponding classes of first-order structures (this correspondence essentially amounts to what is known as a Galois connection).

Sections 3.2 and 3.3 introduces some of the algebraic theories that will feature prominently in the rest of the book. More precisely, in Sect. 3.2 we axiomatise some special classes of abelian groups as well as vector spaces, whose dimension theory we outline for future reference. In Sect. 3.3 we axiomatise rings and fields but focus essentially on rings, and especially rings of polynomials, which will play a very important role in the second half of the book.

Special properties of polynomial rings, which will be used throughout Chap. 10, are proved in the Addendum.

3.1 The Connection Between Theories and Models

If Σ is a set of sentences, we shall refer to the set of its consequences as $\Sigma^\vDash$. More formally: $\Sigma^\vDash = \{\phi : \Sigma \vDash \phi\}$. If $\Sigma = \Sigma^\vDash$, we say that Σ is a **deductively closed** set of sentences.

Definition 3.1 A **theory** is a deductively closed set of sentences with at least one model.

The set of $\mathcal{L}_<$-sentences $\{\forall x(\neg(x < x)), \exists x(x < x)\}$ is not a theory because it has no models. The set L of $\mathcal{L}_<$-sentences used to isolate the class of linear orders among the $\mathcal{L}_<$-structures is not a theory because it is not deductively closed. For instance, $\forall x \forall y (x < y \to \neg(y < z)) \notin \mathsf{L}$. On the other hand, $\mathsf{L}^\vDash = \mathsf{LO}$ is a theory: it is deductively closed by construction and e.g. the linear order $\mathbb{Z}$ is a model. We call LO the **theory of linear orders**.

Clearly, L and LO have the same consequences. We say that L axiomatises LO. In general, if Σ is a theory, we say that Γ **axiomatises** Σ iff $\Gamma^\vDash = \Sigma$. In this event, we call Γ an axiomatisation of Σ. In the remainder of this book we shall identify the axiomatisations we study with their deductive closures.

Our definition of axiomatisability ensures that a theory is always axiomatised, possibly by itself. It is of interest to determine whether certain convenient axiomatisations are available, e.g. involving only a finite number of sentences. If Σ is a theory that can be axiomatised by a finite set of sentences (and, thus, by a single sentence), then Σ is said to be **finitely axiomatisable**. The best alternative to a finite axiomatisation is a quasi-finite axiomatisation, which involves infinite sequences of sentences of the same form. Starting in this chapter, we shall introduce and examine several finitely axiomatised and quasi-finitely axiomatised theories (LO is one of them and it is clearly finitely axiomatised).

Exercise 3.2 A set of sentences is **inconsistent** if the set of its consequences contains both a sentence ϕ and its negation. A set of sentences is **consistent** if it is not inconsistent. Show that a set of sentences is consistent iff it has a model (*Hint*: any sentence follows from a theory that has no models).

Theories may be extracted from classes of structures.

Definition 3.3 If **K** is a class of structures, the **theory of K**, in symbols $Th\mathbf{K}$, is the set of sentences modelled by each structure in **K**.

The reader is encouraged to verify that $Th\mathbf{K}$ is a theory in the sense of Definition 3.1. A class of structures **K** is **elementary** if there is a set of sentences Σ such that $Th\mathbf{K} = \Sigma$. When **K** has exactly one element M, ThM is the *theory of M*, i.e. the set of sentences modelled by M.

Whilst classes of structures determine theories, sets of sentences (and thus, in particular, theories) determine certain classes of structures in turn. If Σ is a set of sentences, we call $Mod\,\Sigma$ the class of its models.

Exercise 3.4 Show by induction on the complexity of formulae that, given a $\mathcal{L}$-structure M, for every $\mathcal{L}$-sentence ϕ, either $\phi \in ThM$ or $\neg\phi \in ThM$ (but not both). Check that ThM is a theory.

A $\mathcal{L}$-theory T is **complete** iff, for every $\mathcal{L}$-sentence ϕ, either $\phi \in T$ or $\neg\phi \in T$. A theory may not be complete but every theory can be extended to (i.e. is included in) at least one complete theory.

Exercise 3.5 Prove that every theory T has a complete extension. This is to say that there is a complete theory S such that $T \subseteq S$ (*Hint*: use Exercise 3.2).

There is an important correspondence between $\mathcal{L}$-sentences and $\mathcal{L}$-structures. We describe it by means of two operations s, t: s assigns to any set of sentences Σ the class $Mod\,\Sigma$ and t associates with a class of structures **K** the set of sentences $Th\mathbf{K}$. The operations s and t interact in a noteworthy manner:

Lemma 3.6

(a) $\Sigma \subseteq ts(\Sigma)$;
(b) $\mathbf{K} \subseteq st(\mathbf{K})$;

(c) $\Sigma = \Sigma^\models$ iff $\Sigma = ts(\Sigma)$;
(d) **K** is elementary iff $\mathbf{K} = st(\mathbf{K})$.

Proof

(a) If $\phi \in \Sigma$, every model of Σ is a model of ϕ. This is to say that ϕ is a sentence modelled by every structure in $Mod\Sigma$. By definition, $\phi \in ThMod\Sigma$;
(b) exercise;
(c) if $\Sigma = \Sigma^\models$ and $\phi \in ThMod\Sigma$, for every $M \in ModS$ we have $M \models \phi$. This is to say that ϕ is a consequence of Σ, i.e. that $\phi \in \Sigma$, since Σ is deductively closed. We have thus established $ThMod\Sigma \subseteq \Sigma$. The converse inclusion follows from (a). Suppose $\Sigma = ThMod\Sigma$. If $\Sigma \models \phi$, then $M \models \phi$ for every M in $Mod\Sigma$, i.e. $\phi \in ThMod\Sigma = \Sigma$. As a consequence, $\Sigma = \Sigma^\models$;
(d) exercise.

□

Exercise 3.7 Show that:

(i) if S, T are sets of sentences, $S \subseteq T$ implies $s(T) \subseteq s(S)$.
(ii) if **H**, **K** are classes of structures, $\mathbf{H} \subseteq \mathbf{K}$ implies $t(\mathbf{K}) \subseteq t(\mathbf{H})$.
(iii) if S, T are theories, $ModS = ModT$.

Our treatment of the basic ideas and results of model theory will keep moving back and forth between abstract developments and applications. In keeping with this style of presentation, we turn to a few examples of theories, whose model-theoretic properties will constitute the subject of later investigations.

3.2 Groups and Vector Spaces

The main theories we shall be interested in axiomatise classes of algebraic structures. We devote separate subsections to them, in which we single out some of their models, which will be familiar in view of the previous chapters. Before turning to these theories, we describe a $\mathcal{L}_=$-axiomatisation of the class of infinite sets.

This axiomatisation, which we shall refer to as IS, simply consists of an infinite sequence of sentences $\{\varphi_n\}_{n \in \mathbb{N}}$ such that φ_n states the existence of at least n distinct objects. We have seen in Chap. 2 how we might express the existence of two or three distinct objects by means of $\mathcal{L}_=$-formula: the generalisation to an arbitrary, finite number of objects is straightforward and left to the reader.

The theory IS is sufficiently simple for us to expect that it should be easy enough to establish its model-theoretic properties. Even though this is certainly true, we postpone their discussion until the next chapter.

3.2.1 Abelian Groups

The **theory of groups** G is the (deductive closure of the) following set of $\mathcal{L}_g$-sentences:

(1) $\forall x \forall y \forall z (x + (y + z) = (x + y) + z)$;
(2) $\forall x (x + 0 = x)$;
(3) $\forall x (x + (-x) = 0)$;

The **theory of abelian groups** AG is obtained from the above, finite axiomatisation by adding the $\mathcal{L}_g$-sentence:

(4) $\forall x \forall y (x + y = y + x)$.

Our model-theoretic study of abelian groups is restricted to a special subclass of them, which we can isolate by means of an addition of infinitely many $\mathcal{L}_g$-sentences to AG.

The quotient group $\mathbb{Z}/n\mathbb{Z}$ (see Sect. 2.6) is an abelian group that models the sentence $\exists x (\underbrace{x + \ldots + x}_{n\ times} = 0)$. For convenience, we shall henceforth abbreviate the term $\underbrace{x + \ldots + x}_{n\ times}$ as nx. Note that, for no positive $n \in \mathbb{N}$ is the last sentence modelled by the additive group of integers $\mathcal{Z}$. We say that $\mathbb{Z}/n\mathbb{Z}$ is a torsion group, while $\mathcal{Z}$ is **torsion-free**.

Definition 3.8 The theory of **torsion-free abelian groups** is obtained from AG by adding to it the set of sentences $\{\forall x (x \neq 0 \rightarrow nx \neq 0) : n > 1\}$.

Exercise 3.9 Show that no finite model of AG is torsion-free.

Among torsion-free abelian groups, we restrict attention to those that look like the additive group of rational numbers. Unlike $\mathcal{Z}$, the latter group models the sentence $\forall y \exists x (nx = y)$, for each $n > 1$. No such sentence is modelled by $\mathcal{Z}$.

Definition 3.10 The theory of **torsion-free, divisible abelian groups** DAG is obtained by adding to the theory of torsion-free abelian groups the infinite set of sentences $\{\forall x \exists y (ny = x) : n \geq 1\}$.

By Exercise 3.9, finite abelian groups are automatically torsion groups.[1] It is not the case, however, that infinite groups are automatically torsion-free. An example of infinite torsion group, which is also divisible, is the quotient group $\mathbb{Q}/\mathbb{Z}$, whose cosets are of the form:

$$m/q + \mathbb{Z}, \text{ where } m/q \in \mathbb{Q}, m, q \in \mathbb{Z} \text{ and } q \neq 0.$$

We can easily check that $\mathbb{Q}/\mathbb{Z}$ is an infinite group. Two cosets $m/q + \mathbb{Z}, m'/q' + \mathbb{Z}$ are the same iff $m/q - m'/q' \in \mathbb{Z}$. It thus suffices to produce an infinite set S of

---
[1] The same is true for finite groups that are not abelian.

rational numbers such that the difference between any two of its elements is not an integer. It will follow that the elements of S are representatives of infinitely many distinct equivalence classes, or cosets, of $\mathbb{Q}/\mathbb{Z}$. We may set $S = \{1/n : n \geq 1\}$. For any $1/n, 1/m \in S$ with $n < m$, we easily see that $0 < (m-n)/mn < 1$, i.e. $1/n - 1/m \notin \mathbb{Z}$. The set S is clearly infinite: consequently, $\mathbb{Q}/\mathbb{Z}$ is an infinite group.

Exercise 3.11 Show that $\mathbb{Q}/\mathbb{Z}$ is divisible but not torsion-free.

Although not every divisible group is also torsion-free, every ordered, divisible abelian group is. An **ordered abelian group** is a model of LO∪AG and, in addition, of the $\mathcal{L}_{g<}$-sentence:

$$\forall x \forall y \forall z (x < y \to x + z < y + z).$$

By adding the divisibility sentences from Definition 3.10, to the set:

$$\text{LO} \cup \text{AG} \cup \{\forall x \forall y \forall z (x < y \to x + z < y + z)\}$$

we obtain DAG$_<$, the theory of **ordered, divisible abelian groups**.

Exercise 3.12 Recall that a linear order is dense if it models the sentence $\forall x \forall y (x \neq y \to \exists z (x < z \land z < y))$. A linear order is said to be without endpoints if it models the sentences $\forall x \exists y (x < y)$ and $\forall y \exists x (x < y)$. Prove that, if the $\mathcal{L}_{g<}$-structure $\mathcal{G}$ is a divisible, ordered abelian group containing $x \neq 0$, then $\mathcal{G}$ is also a dense, linear order without endpoints. Deduce that $\mathcal{G}$ must be infinite and show that $\mathcal{G}$ is torsion-free.

In Chaps. 8 and 9 we shall establish the basic model-theoretic properties of DAG$_<$.

3.2.2 Vector Spaces

Let $\mathcal{K}$ be a field, viewed as a $\mathcal{L}_r$-structure. Let $\{f_q\}_{q \in K}$ be a set of unary function symbols, one for each $q \in K$. Suppose

$$\mathcal{V} = (V, +^V, -^V, 0^V, \{f_q\}_{q \in K})$$

is a $\mathcal{L}_v$-structure that models the abelian the group axioms (1)-(4) as well as the following sentences (in which, for any field element q, $f_q x$ is more naturally rewritten as qx):

(5) $\forall x \forall y (q(x+y) = qx + qy)$ for each $q \in K$;
(6) $\forall x ((q+r)x = qx + rx)$, for each $q, r \in K$;
(7) $\forall x ((qr)x = q(rx))$, for each $q, r \in K$;
(8) $\forall x (1x = x)$,

We call $\mathcal{V}$ a **vector space over the scalar field** $\mathcal{K}$ and refer to the elements of V as **vectors**. Note that each one of the items (5), (6), (7) above describes infinitely many $\mathcal{L}_v$-sentences of the same form. These sentences, together with the abelian group axioms, determine the **theory** $\mathsf{VS}_\mathcal{K}$ **of vector spaces over the field** $\mathcal{K}$.

Remark 3.13 For any choice of field $\mathcal{K}$, the unary function f_0 (where 0 is the additive neutral element of K) is usually designated by 0_K. If $\mathcal{V} \models \mathsf{VS}_\mathcal{K}$, then $\mathcal{V} \models 0_K(v) = 0^V$.[2] Symmetrically, for any $q \in K$ we have $q(0^V) = 0^V$. It is customary to adopt the notation 0^V for both the neutral element of the abelian group $(V, +^V, -^V, 0^V)$ and for 0_K.

An exhaustive and remarkably clear treatment of the basic properties of vector spaces may be found in the first three chapters of [5]. Here we shall briefly focus only on a few properties that are of special model-theoretic interest. Their proofs will follow from abstract model-theoretic results developed in Chap. 8.

Suppose $\mathcal{V} \models \mathsf{VS}_\mathcal{K}$ and $A \subseteq V$. We introduce the notation $acl_\mathcal{V}(A)$ to indicate the subset of V that is obtained by interpreting all of the $\mathcal{L}_v$-terms on vectors in A in every possible way.

Exercise 3.14 Show that $acl_\mathcal{V}(A)$ is the domain of a vector space over the scalar field $\mathcal{K}$.

We refer to $acl_\mathcal{V}(A)$, viewed as a $\mathcal{L}_v$-structure, as the *subspace of $\mathcal{V}$ generated by A* or as the **linear closure** of A. Note that, when A is a finite set of vectors, e.g. $\{v_1, \ldots, v_n\}$, the linear closure $acl_\mathcal{V}(A)$ is the set obtained by interpreting all of the $\mathcal{L}_v$-terms in at most the variables $x_1, \ldots, x_n$ on $v_1, \ldots, v_n$ respectively. The notion of linear closure allows us to introduce a number of key notions, which we state in several definitions.

Definition 3.15 A set of vectors A is said to be **linearly independent** iff, for any $v \in A$, $v \notin acl_\mathcal{V}(A - \{v\})$. In other words, no v in A is contained in the subspace generated by $A - \{v\}$.

Example 3.16 Consider $\mathbb{Q}^2$ as the domain of a vector space over $\mathbb{Q}$. If $A = \{(0, 1), (1, 0)\}$, it is easy to see that A is a linearly independent set. Both $acl_\mathcal{V}(\{(1, 0)\})$ and $acl_\mathcal{V}(\{(0, 1)\})$ are obtained by interpreting the set of $\mathcal{L}_v$-terms in at most one variable x on a fixed vector, that is $(1, 0)$ or $(0, 1)$. The set of terms in question contains 0 and, by the properties of vector spaces, qx for each $q \in \mathbb{Q}$. This means that $acl_\mathcal{V}(\{(1, 0)\}) = \{(q, 0) : q \in \mathbb{Q}\}$ and $acl_\mathcal{V}(\{(0, 1)\}) = \{(0, q) : q \in \mathbb{Q}\}$. Clearly $(0, 1) \notin acl_\mathcal{V}(\{(1, 0)\})$ and $(1, 0) \notin acl_\mathcal{V}(\{(0, 1)\})$. It follows that A is linearly independent.

Definition 3.17 A set of vectors A is said to be a **basis** for $\mathcal{V}$ iff $V = acl_\mathcal{V}(A)$. In other words, if A is a basis for $\mathcal{V}$, the whole of $\mathcal{V}$ is the linear closure of A.

---

[2] Since $0_K(v) + 0_K(v) = (0_K + 0_K)(v) = 0_K(v)$, adding the group inverse of $0_K v$ to the first and last term in this chain of equalities yields the desired conclusion.

3.2 Groups and Vector Spaces

Example 3.18 Consider again $\mathbb{Q}^2$ as the domain of a vector space over $\mathbb{Q}$. A generic element of $\mathbb{Q}^2$ is an ordered pair (q, r), with $q, r \in \mathbb{Q}$. We may rewrite the ordered pair as $q(1, 0) +^{\mathbb{Q}^2} r(0, 1)$, which is the interpretation of the $\mathcal{L}_v$-term $qx_1 + rx_2$ on A. It now follows that $acl_\mathcal{V}(A) = \mathbb{Q}^2$, i.e. that the whole vector space is generated by A. It follows that A is a basis. A trivial adaptation of this argument yields finite bases for every vector space over $\mathbb{Q}$ with domain $\mathbb{Q}^n$.

Definition 3.19 A vector space $\mathcal{V}$ is said to be **finite-dimensional** iff there is a finite set of vectors $A \subseteq V$ such that $acl_\mathcal{V}(A) = V$.

Example 3.20 Every $\mathbb{Q}^n$, or indeed K^n, where K is the domain of some fixed field, is a finite-dimensional vector space.

Exercise 3.21 In each of the following exercises, we take $\mathbb{Q}^2$ and $\mathbb{Q}^3$ to be the domains of the vector spaces $\mathcal{V}_2, \mathcal{V}_3$ over the scalar field $\mathbb{Q}$, respectively.

(i) Given $v_1 = (4, 6)$, $v_2 = (2, 3)$, $v_3 = (5, 1) \in \mathbb{Q}^2$, show that $acl_{\mathcal{V}_2}(\{v_1, v_2\}) = acl_{\mathcal{V}_2}(\{v_1\})$ and that $acl_{\mathcal{V}_2}(\{v_2, v_3\}) = \mathbb{Q}^2 = acl_{\mathcal{V}_2}(\{v_1, v_3\})$.
(ii) Setting $v_1 = (1, 0, 1)$, $v_2 = (0, 5, 2) \in \mathbb{Q}^3$, show that $acl_{\mathcal{V}_3}(\{v_1, v_2\}) \neq \mathbb{Q}^3$ and find v_3 such that $acl_{\mathcal{V}_2}(\{v_1, v_2, v_3\}) = \mathbb{Q}^3$.
(iii) Show that, in $\mathbb{Q}^3$, $q(1, 0, 1) + r(0, 5, 2) = 0$ iff $q = r = 0$.
(iv) Show that v_3 from (ii) does not belong to $acl_{\mathcal{V}_2}(\{v_1, v_2\})$.

We have defined 'finite-dimensional' without introducing any notion of dimension. We can easily do so for finite-dimensional vector spaces.

Definition 3.22 Let $\mathcal{V}$ be a **finite-dimensional** vector space. The **dimension** of V is the number of elements of a basis A for $\mathcal{V}$.

By Exercise 3.21-(i), there may be more than one basis for the same vector space. Thus, the above definition is sound if we can guarantee that all bases have the same size. Although we could prove this directly for vector spaces using linear algebra, we shall postpone a proof until Chap. 8, where we obtain this result as a corollary of a more abstract model-theoretic characterisation of dimension. This characterisation specialises to the notion of dimension we have just defined in the context of models of VS_K.

The model-theoretic characterisation of dimension essentially rests on the logical notion of *algebraic closure*, which specialises to linear closure when referred to vector spaces. The notation we have employed for linear closure so far, namely $acl_\mathcal{V}(A)$, is used for the abstract model-theoretic notation for algebraic closure.

In our further discussion of vector spaces we shall not restrict attention to finite-dimensional spaces alone, but consider infinite-dimensional spaces as well.

Definition 3.23 A model $\mathcal{V}$ of VS_K is said to be infinite-dimensional iff it is not finite-dimensional.

Exercise 3.24 Prove that $\mathcal{V}$ is **infinite-dimensional** iff for each positive $n \in \mathbb{N}$, V contains n linearly independent vectors.

3.3 Rings and Fields

A $\mathcal{L}_r$-structure $\mathcal{R}$ that models the sentences (1) to (4) and

(6) $\forall x \forall x \forall z \cdot ((x \cdot y) = x \cdot (y \cdot z))$;
(7) $\forall x \forall y \forall z (x \cdot (y + x) = (x \cdot y) + (x \cdot z))$;
(8) $\forall x \forall y (x \cdot y = y \cdot x)$;
(9) $\forall x (1 \cdot x = x)$,

is called a **commutative ring with unity**. If, in addition, $\mathcal{R}$ models the sentence:

$$\forall x \forall y (x \cdot y = 0 \to (x = 0 \lor y = 0)),$$

then $\mathcal{R}$ is said to be an **integral domain**. Finally, if $\mathcal{R}$ models the sentences (1) to (9) and, in addition:

(10) $\neg(1 = 0)$;
(11) $\forall x \exists y (\neg(x = 0) \to x \cdot y = 1)$,

then $\mathcal{R}$ is said to be a **field**. Note that a field is automatically an integral domain[3] but not vice versa. For instance the integers, viewed as a $\mathcal{L}$-structure, are a commutative ring with unity but not a field, since they lack multiplicative inverses, i.e. they fail to model (11). We refer to the conjunction of (1) to (11) as the **theory of fields**.

Although we shall focus our model-theoretic study only on fields (in fact, only on specific classes of fields), and not on rings, it will be essential to have a basic knowledge of quotient rings. Our work on interpretations from Chap. 2 puts us in an easy position to understand them.

3.3.0.1 Quotient Rings

Let $\mathcal{R} = (R, +^\mathbb{R}, -^R, \cdot^R, 0^R, 1^R)$ be a commutative ring with unity[4] If we restrict our attention to the signature of $\mathcal{L}_g \subseteq \mathcal{L}_r$, we may view $\mathcal{R}$ as an abelian group. This restriction amounts to focussing on what is known as the $\mathcal{L}_g$-**reduct** of $\mathcal{R}$, namely the $\mathcal{L}_g$-structure $(R, +^\mathbb{R}, -^R, 0^R)$. Let us add one unary relation symbol P to $\mathcal{L}_g$: the $(\mathcal{L}_g \cup \{P\})$- **expansion** of $(R, +^\mathbb{R}, -^R, 0^R)$ to a group $(R, +^\mathbb{R}, -^R, 0^R, P^R)$, in which $P^R = I$ is a particular kind of subgroup (shortly to be described), is the object that interests us.

---

[3] Fields enforce ordinary arithmetic so if the product of two field elements equals zero, one of the two elements must be zero.

[4] Since these are the only rings we are going to work with, in the context of this book 'ring' always means 'commutative ring with unity'.

3.3 Rings and Fields

We already know how to view R/I as a quotient group, relative to the language $\mathcal{L}_g$. To view R/I as a ring, we have to ensure that I has one important feature, spelled out in the following definition.

Definition 3.25 Let $\mathcal{R}$ be a ring, $\mathcal{J}$ a subgroup of the $\mathcal{L}_g$-reduct of $\mathcal{R}$. We say that the expansion of $\mathcal{J}$ to the language $\mathcal{L}_r$ is an **ideal** of $\mathcal{R}$ iff for any $a \in R$ and $b \in J$, $a \cdot^R b \in J$.

Trivially, any ring is also an ideal of itself (in fact, an ideal $\mathcal{I}$ of $\mathcal{R}$ is the whole ring iff $1^R \in I$). For any positive integer n, the set $n\mathbb{Z}$ of integer multiples of n is an ideal of the ring of integers. Not every additive subgroup of a ring is automatically an ideal. Consider for example the field of rational numbers $\mathcal{Q}$: the integers are a subgroup but not an ideal of $\mathcal{Q}$.

From now on we shall refer to ideals by capital letters I, J, K and treat them notationally as subsets of ring domains, even though they carry ring structure.

We obtain the **quotient ring** R/I by taking the one-dimensional interpretation Γ of $\mathcal{L}_r$ in $\mathcal{L}_r \cup \{P\}$ given by $\partial(x) := x = x$, the nested formula $\psi(x_1, x_2) := P(x_2 - x_1)$ and the exact same conditions listed in Example 2.90, with an obvious addition for multiplication, namely:

$$(x \cdot y = z)_\Gamma := \psi(x \cdot y, z).$$

By Exercise 2.91, the cosets of R/I (i.e. the equivalence classes determined by $\psi(x_1, x_2)$) are of the form $a +^R I$, with $a \in R$. We add and multiply cosets in the obvious way. More precisely, we view R/I as a $\mathcal{L}_r$-structure by endowing it with the ring operations $+^*, \cdot^*$, defined as follows:

$$(a +^R I) +^* (b +^R I) = (a +^R b) +^R I \text{ and } (a +^R I) \cdot^* (b +^R I) = (a \cdot^R b) +^R I.$$

Because $\mathcal{I}$ is an ideal, the **distributive law** (7) is modelled by R/I. Note that I itself plays the role of neutral element relative to $+^*$, i.e. it is the 'zero' of R/I, while $1 +^R I$ is the quotient ring's unit. Ideals with special features ensure that quotients are rings with additional properties.

Definition 3.26 Let $\mathcal{R}$ be a ring, I an ideal of R. We say that I is a **maximal ideal** iff $I \neq R$ (or, equivalently, $1^R \notin I$) and there is no ideal strictly between I and R.

Definition 3.27 Let $\mathcal{R}$ be a ring, I an ideal of R. We say that I is a **prime ideal** iff $I \neq R$, for any $a, b \in R$, $ab \in I$ implies $a \in I$ or $b \in I$.

Theorem 3.28 *Let $\mathcal{R}$ be a ring, I an ideal of R.*

(1) The quotient ring R/I is a field iff I is a maximal ideal.
(2) The quotient ring R/I is an integral domain iff I is a prime ideal.

The last theorem is proved in this chapter's appendix.

3.3.1 Polynomial Rings

Recall from Sect. 2.1 that, supplementing $\mathcal{L}_r$ with one distinct constant symbol q for each element of the set $\mathbb{Q}$, we obtain an expanded language whose terms are the **polynomials in one indeterminate** x with coefficients in $\mathbb{Q}$. We call $\mathbb{Q}[x]$ both the set just described and the ring based on it. In general, if $\mathcal{K}$ is a field, a suitable enrichment of $\mathcal{L}_r$ gives rise to the set of polynomials in one indeterminate with coefficients in K, denoted by $K[x]$.

Example 3.29 As soon as we agree to abbreviate $x \cdot x, x \cdot x \cdot x, \ldots$ by $x^2, x^3, \ldots$ and use juxtaposition instead of the symbol $\cdot$, expressions like $1 + x + 2x^3 - 3x^4$ and $4 - x + x^7$ are polynomials in $\mathbb{Q}[x]$, as are $0x^2 + 2$ and $x^2 + x^2 + x^2$. We take the reader to be familiar with polynomial addition (add the coefficients of the same powers of x together) and multiplication (add the exponents of distinct powers of x and multiply their coefficients). Relative to these operations, $\mathbb{Q}[x]$ is a ring.

Remark 3.30 In the sequel, we refer to generic polynomials in a polynomial ring $K[x]$ by expressions such as $f(x), g(x), h(x), p(x), q(x)$.

Remark 3.31 An element of $K[x]$ is both a ring element and a term in a first-order language. In the second capacity, we are able to evaluate it on K, simply by assigning $a \in K$ to the variable x. Thus, for instance, given the term $1 + x + x^3$ in $\mathbb{Q}$ and an assignment in σ in $\mathbb{Q}$ such that $\sigma(x) = 2$, we obtain $\sigma(1 + x + x^3) = 11$. We are especially interested in assignments yielding the value 0 (with $\sigma(x) = 2$ again, we have e.g. $\sigma(x^2 - 4) = 0$). These assignments determine field elements that are called **roots** of polynomials.

Each polynomial $f(x) \in K[x]$ has a **degree** ∂f. If $f(x)$ is a constant symbol, its degree is zero. Otherwise $f(x)$ can be written in the form $a_0 + a_1 x + a_2 x^2 + \ldots + a_n x^n$. The highest n such that $a_n \neq 0$ is the $\partial f(x)$ (thus e.g. $1 + x$ is of degree 1 and $x^4 + 3x^5 + 0x^7$ is of degree 5).

Division with remainder (or long division, as it is sometimes called) can be carried out on polynomials in the ring $K[x]$, where $\mathcal{K}$ is a field. For any $f(x), g(x) \in K[x]$ there are unique polynomials $q(x), r(x)$ such that:

$$g(x) = q(x)f(x) + r(x), \text{ where } 0 \leq \partial r(x) < \partial f(x).$$

Polynomial division with remainder makes it easy to deduce certain key properties of polynomial rings.

Exercise 3.32 Show that, if $\mathcal{K}$ is a field and $a \in K$ a root of $f(x) \in K[x]$, there is a polynomial $g(x) \in K[x]$ such that $f(x) = (x - a)g(x)$. Use this result to prove that, if $\partial f(x) = n$, $f(x)$ cannot have more than n distinct roots in any field $L \supseteq K$.

Using division with remainder, it is possible to show that every ideal in $K[x]$ is of the form $f(x)K[x]$, i.e. it is the set of all polynomial multiples of a single, fixed polynomial $f(x)$. We call this the **ideal generated by** $f(x)$ and denote it by $\langle f(x) \rangle$.

3.4 Addendum: Quotient Rings and Noetherian Rings

When this ideal is maximal, $K[x]/\langle f(x)\rangle$ is a field that extends $\mathcal{K}$, since any field element $a \in \mathcal{K}$ may be identified with the coset $a + \langle f(x)\rangle$.

Example 3.33 The term $x^2 - 2$ is a polynomial in $\mathbb{Q}[x]$. Because the square root of 2 is not a rational number, this polynomial has no roots in $\mathbb{Q}$. It follows that the ideal $I = \langle x^2 - 2\rangle$ is a maximal ideal of $\mathbb{Q}[x]$. To see this, suppose there is an ideal J strictly included between I and $\mathbb{Q}[x]$. We arrive at a contradiction by showing that $1 \in J$. Since $I \subset J$ and $I \neq J$, there is $g(x) \in J - I$. Using polynomial division we deduce the existence of $q(x), r(x)$ such that $g(x) = q(x)(x^2 - 2) + r(x)$, with $\partial r(x) < \partial f(x)$. Because $q(x)(x^2 - 2) \in J$ and $g(x) \in J$, we must have $r(x) \in J$ because J is an additive group. Now the degree of $r(x)$ is either 0 or 1. If it is zero, then $r(x)$ is a (constant symbol designating) a rational number r and it is easy to see that $J \supseteq \langle r\rangle = \mathbb{Q}[x]$. If $\partial r(x) = 1$, then there are $a(x), b(x) \in \mathbb{Q}[x]$ such that $x^2 - 2 = a(x)r(x) + b(x)$, where the degree of $b(x)$ is zero, i.e. $b(x) = s$ for some $s \in \mathbb{Q}$. Since $x^2 - 2$ and $a(x)r(x)$ are in J, $b(x)$ must be in J and $J = \mathbb{Q}[x]$. It follows that the quotient ring $\mathbb{Q}[x]/\langle x^2 - 2\rangle$ is in fact a field extending $\mathbb{Q}$. In this field extension of $\mathbb{Q}$, the polynomial $Y^2 - 2$ with coefficients in the field $\mathbb{Q}[x]/\langle x^2 - 2\rangle$ (where 2 stands for $2 + \langle x^2 - 2\rangle$) has a root, namely the coset $x + \langle x^2 - 2\rangle$.

Evaluating $Y^2 - 2$ on this coset and using the quotient ring operations $+^*, \cdot^*$, we obtain:

$$(x + \langle x^2 - 2\rangle) \cdot^* (x + \langle x^2 - 2\rangle) +^* (2 + \langle x^2 - 2\rangle) = (x^2 + \langle x^2 - 2\rangle) +^* (2 + \langle x^2 - 2\rangle)$$
$$= (x^2 - 2) + \langle x^2 - 2\rangle = \langle x^2 - 2\rangle.$$

Our claim follows from the fact that $\langle x^2 - 2\rangle$ is the zero element of the quotient ring $\mathbb{Q}[x]/\langle x^2 - 2\rangle$.

In Chap. 10 we study polynomial rings in a finite number of indeterminates with coefficients over the field K. The notation for n indeterminates is simply $K[x_1, \ldots, x_n]$. These polynomial rings are defined exactly as the polynomial rings in one indeterminate. Their elements are again terms in an expansion of $\mathcal{L}_r$ with constant symbols naming the elements of a fixed field K. What changes is that more variables than one can be involved in the formation of terms. For instance, the ring $K[x_1, x_2]$ includes $K[x_1]$ but also contains additional elements like $x_1^2 x_2^3 + 3x_1 x_2$.

Polynomial rings have a special property: they are **Noetherian** rings. This is to say that they do not have infinitely long, increasing chains of ideals. A proper definition and a proof of this fact are given in the Addendum to follow. The fact that polynomial rings are Noetherian will play an important role when we study the elements of algebraic geometry in Chap. 10.

3.4 Addendum: Quotient Rings and Noetherian Rings

In this addendum we drop superscripts on ring operations and constants.

Definition 3.34 Let $\mathcal{R}$ be a ring, I an ideal of R. We say that I is a **prime** ideal iff $ab \in I$ implies $a \in I$ or $b \in I$.

Theorem 3.35 *Let $\mathcal{R}$ be a ring, I an ideal of R.*

(1) The quotient ring R/I is a field iff I is a maximal ideal.
(2) The quotient ring R/I is an integral domain iff I is a prime ideal.

Proof We prove (1) by means of two contrapositions. If I is a maximal ideal and R/I is not a field, then there is a coset $a + I$ different from I (i.e. with $a \notin I$) and lacking a multiplicative inverse. In this case we can show that $I + aR$ is a proper ideal (i.e. $\neq R$) that strictly extends I, against maximality. The only part of this verification that requires some work is the proof that $J = I + aR$ is not the whole of R. If it were, we would have $1 = m + ar$, for some $m \in I, r \in R$. In this case $(r + I)(a + 1) = ar + I = ar + m + I = 1 + I$ ($m \in I$ implies $m + I = I$): this contradicts the assumption that $a + I$ did not have a multiplicative inverse.

Next we suppose that R/I is a field but I is not a maximal ideal. In this case there is an ideal J strictly included between I and R. We pick $a \in J - I$ and show that $a + I$ cannot have a multiplicative inverse in R/I. If there was one, say $r + I$, we could deduce $ar + I = 1 + I$, which in turn implies the existence of $m \in I$ such that $ar + 1 = m$ or $1 = m - ar$. Since $m - ar \in J$, we contradict the assumption that J is not the whole of R.

The proof of (2) is immediate. If R/I is an integral domain and $ab \in I$, then $(a + I)(b + I) = ab + I = I$. Integrality now implies that $a + I = I$ or $b + I = I$. The first condition is equivalent to $a \in I$ and the second to $b \in I$. The converse implication is proved in a similar manner. □

Definition 3.36 A ring $\mathcal{R}$ is said to be **Noetherian** iff every ideal $I \subseteq R$ is finitely generated, i.e. of the form $\langle a_1, \ldots, a_k \rangle$, with $a_1, \ldots, a_k \in R$, k a positive integer.

Example 3.37 Let $\mathcal{Z}$ be the ring of integers. Every ideal $I \subseteq \mathbb{Z}$ is of the form $n\mathbb{Z}$ for some positive integer n. To see why, it suffices to consider the least positive element of I, which exists by the least number principle if I is not the trivial ideal $\{0\}$. If $m \in I$, we can divide m by n with remainder and the expression $m = qn + r$ implies that $r \in I$. In this case r must be 0 or n would not be least. So every element of I is a multiple of n and, since I is a subgroup of the additive integers, all multiples of n are in it: we conclude $I = n\mathbb{Z}$. The exact same argument goes through with polynomial division. Thus, in a polynomial ring $K[x]$, with $\mathcal{K}$ a field, every ideal is of the form $f(x)K[x]$, i.e. it is the set of polynomial multiples of $f(x)$. Rings with this property are known as **principal ideal domains**. Clearly, principal ideal domains are Noetherian.

Unlike the sketchy definition of Noetherian rings stated at the end of the previous section, Definition 3.36 does not mention chains of ideals. It is nonetheless equivalent our original definition by the following, useful result:

3.4 Addendum: Quotient Rings and Noetherian Rings

Theorem 3.38 *Let $\mathcal{R}$ be a ring. The following statements are equivalent:*

(1) $\mathcal{R}$ is Noetherian;
(2) Ascending Chain Condition (ACC): if $\{I_n\}_n \in \mathbb{N}$ is an ascending (or, equivalently, increasing) chain of ideals, i.e. $I_1 \subseteq I_2 \subseteq I_3 \subseteq \ldots$, then it stabilises after finitely many steps. In other words, $I_k = I_{k+1}$ for sufficiently large k;
(3) Every nonempty set S of ideals has a maximal element J relative to inclusion. This is to say that $J \in S$ is not strictly included in any ideal of S.

Proof To prove that (1) implies (2), we take the union I of the chain $I_1 \subseteq I_2 \subseteq I_3 \subseteq \ldots$ and note that it is an ideal. By (1), I is finitely generated. Its generators, however, must occur in some I_m along the chain. It follows that $I \subseteq I_m$. We must have equality because I_m is, by construction, a subset of I.

To prove that (2) implies (3), we proceed by contraposition. If there were a nonempty set of ideals S without a maximal element, we could select I_1 in S and find, since I_1 cannot be maximal, I_2 that strictly extends it. This argument can be recursively iterated to determine an increasing chain of ideals that does not stabilise.

To prove that (3) implies (1), we consider and ideal I and the set S of finitely generated ideals included in I. Because this is nonempty, (3) guarantees the existence of a maximal, finitely generated ideal $J \subseteq I$. If J differed from I, we could find $a \in I - J$ and produce the finitely generated ideal $J + aR$ (a and the finitely many generators of J determine a finite set of generators), which is in S but strictly extends J, against its maximality. It follows that $I = J$ and I is finitely generated. □

The following important theorem, known as **Hilbert's basis theorem** will be used in Chap. 10.

Theorem 3.39 *Let $\mathcal{R}$ be a ring. If $\mathcal{R}$ is Noetherian, then the polynomial ring $R[x]$ is Noetherian.*

Proof We proceed by contraposition, suppose that $R[x]$ is not Noetherian and find an increasing chain of ideals in R that does not stabilise. Because $R[x]$ is not Noetherian, there is at least one ideal $I \subseteq R[x]$ that is not finitely generated. Let $f_1(x)$ be a polynomial of least degree in I. Since $\langle f_1(x) \rangle$ does not generate I, $I - \langle f(x) \rangle$ is a nonempty set. We let $f_2(x)$ be a polynomial of least degree in the last set. Because $\langle f_1(x), f_2(x) \rangle$ does not generate I, we can make one further choice and find f_3 of least degree. After n iterations we obtain a set of polynomials $\{f_1(x), \ldots, f_n(x)\}$ that cannot generate I. It follows again that there is a polynomial $f_{n+1}(x)$ of least degree in $I - \langle f_1(x), \ldots, f_n(x) \rangle$.

The result of the iterative process described is a sequence of polynomials $\{f_i(x)\}_{i \in \mathbb{N}}$ such that $\partial f_i(x) \leq \partial f_j(x)$ whenever $i \leq j$. Let n_i be $\partial f_i(x)$ and a_i the coefficient (in R) of the subterm of degree n_i in $f_i(x)$.

We claim that the chain of ideals:

$$a_1 R \subset a_1 + a_2 R \subset \ldots \subset a_1 R + \ldots + a_n R \subset \ldots$$

does not stabilise, i.e. ACC fails (we leave it to the reader to check that each set in the above chain is actually an ideal). To prove our claim, we proceed by contradiction and suppose that the chain stabilises from $k \in \mathbb{N}$. It follows in particular that $a_{k+1} \in a_1 R + \ldots + a_k R$, i.e. that there are $r_1, \ldots, r_k$ such that:

$$a_1 r_1 + \ldots + a_k r_k = a_{k+1}.$$

Consider the polynomial

$$f_{k+1}(x) - (a_1 r_1 x n_{k+1} - n_1 f_1(x) + \ldots + a_k r_k x n_{k+1} - n_k f_k(x)).$$

This polynomial is of degree smaller than $n_{k+1} = \partial f_{k+1}(x)$ by construction. It certainly does not lie in the ideal $\langle f_1, \ldots, f_n \rangle$ but it must lie in I. It follows that, when choosing $f_{k+1}(x)$, we have not chosen a polynomial of least degree in $I - \langle f_1(x), \ldots, f_n(x) \rangle$. This contradiction proves that ACC fails in R. □

Remark 3.40 Given a fixed field $\mathcal{K}$, we may view the ring $K[x_1, x_2]$ as a ring $R[x_2]$, where the ring of coefficients is $K[x_1]$. By Hilbert's basis theorem, $K[x_1, x_2]$ is therefore Noetherian. Iterating this argument, we conclude that, for any $n \in \mathbb{N}$, the polynomial ring $K[x_1, \ldots, x_n]$ is Noetherian.

Remark 3.41 If $\mathcal{R}$ is a Noetherian ring and I is an ideal of R, then R/I is Noetherian. This is because there is a bijective correspondence between the ideals of R/I and the ideals of R that include I (see e.g. [14], p.246). By means of this correspondence we can use the generating set of an ideal of R, which we know to be finite, to determine a finite generating set for an ideal of R/I.

Chapter 4
Morphisms, Substructures and Extensions

Abstract The theories we are interested in have many different models: a natural way to relate them is to consider the maps between them that preserve first-order information. This chapter is entirely devoted to such maps.

In Sect. 4.1 we focus on maps that preserve atomic formulae. In Sect. 4.2 we restrict attention to the elementary maps, which transfer all first-order information back and forth between structures, and define the notion of substructure and elementary substructure in terms of maps. In Sect. 4.3 we study an important connection between the existence of maps relating structures and the satisfiability of sentences, using the fundamental notions of a diagram and an elementary diagram. In Sect. 4.4 we offer a philosophical application of some model-theoretic ideas discussed in this chapter.

4.1 Maps That Preserve Formulae

We are interested in maps primarily because they interact with both languages and structures. The latter interaction is perhaps more obvious, because maps are semantic objects with a domain and codomain, which, for our purposes, will always be the domains of two structures. The interaction between languages and maps is given by:

Definition 4.1 Let $\mathcal{M}, \mathcal{N}$ be $\mathcal{L}$-structures, $f : M \longrightarrow N$ a function, and $\phi(x_1, \ldots, x_n)$ a $\mathcal{L}$-formula. We say that f **preserves** $\phi(x_1, \ldots, x_n)$ iff, for any $a_1, \ldots, a_n \in M$:

$$\mathcal{M} \models \phi(a_1, \ldots, a_n) \implies \mathcal{N} \models \phi(f(a_1), \ldots, f(a_n)).$$

We say that f **strongly preserves** $\phi(x_1, \ldots, x_n)$ iff it preserves both $\phi(x_1, \ldots, x_n)$ and its negation.

In the first instance, we focus on maps that preserve atomic information.

Definition 4.2 Let $\mathcal{L}$ be a first-order language of signature τ and $\mathcal{M}, \mathcal{N}$ be $\mathcal{L}$-structures. A function $f : M \longrightarrow N$ is a **homomorphism** iff:

(a) for any constant symbol $c \in \tau$, $f(c^M) = c^N$;
(b) for any n-ary function symbol $g \in \tau$ and $a_1, \ldots, a_n \in M$:

$$f(g^M(a_1, \ldots, a_n)) = g^N(f(a_1), \ldots, f(a_n));$$

(c) for any m-ary relation symbol $R \in \tau$ and $a_1, \ldots, a_m \in M$:

$$R^M a_1, \ldots, a_m \implies R^N f(a_1), \ldots, f(a_m).$$

If, in (c), $R^M a_1, \ldots, a_m \iff R^N f(a_1), \ldots, f(a_m)$, then h is said to be a **strong homomorphism**. If $\mathcal{L}$ has no relation symbol (as is the case of $\mathcal{L}_g, \mathcal{L}_r$ or $\mathcal{L}_v$), every homomorphism is a strong homomorphism.

Exercise 4.3 Let M, N be $\mathcal{L}$-structures, $h : M \longrightarrow N$ a homomorphism, σ an assignment on M and $h\sigma$ an assignment on N. Prove by induction on the complexity of terms that, if $t(x_1, \ldots, x_n)$ is a $\mathcal{L}$-term whose variables occur among $x_1, \ldots, x_n$, then, relative to the assignments $\sigma, h\sigma$:

$$h(t^M(a_1 \ldots a_n)) = t^N(h(a_1) \ldots h(a_n))$$

In view of the last exercise we can easily verify that homomorphisms preserve all atomic formulae.

Lemma 4.4 Let $\mathcal{M}, \mathcal{N}$ be $\mathcal{L}$-structures and $h : M \longrightarrow N$ a homomorphism. If R is a m-ary relation symbol in the signature of $\mathcal{L}$, $t_1(x_1, \ldots, x_n), \ldots t_m(x_1, \ldots, x_n)$ are $\mathcal{L}$-terms and $a_1, \ldots, a_n \in M$, then:

$$\mathcal{M} \models Rt_1(a_1, \ldots, a_n), \ldots, t_m(a_1, \ldots, a_n)$$

implies

$$\mathcal{N} \models Rt_1(h(a_1), \ldots, h(a_n)), \ldots, t_m(h(a_1), \ldots, h(a_n)).$$

Proof Suppose $\mathcal{M} \models Rt_1(a_1, \ldots, a_n), \ldots, t_m(a_1, \ldots, a_n)$. This is the case iff $R^M t_1^M(a_1, \ldots, a_n), \ldots, t_m^M(a_1, \ldots, a_n)$. Because h is a homomorphism, the last condition is equivalent to:

$$R^N h(t_1^M(a_1, \ldots, a_n)), \ldots, h(t_m^M(a_1, \ldots, a_n))$$

In view of Exercise 4.3, we obtain:

$$R^N t_1^N(h(a_1), \ldots, h(a_n)), \ldots, t_m^N(h(a_1), \ldots, h(a_n)).$$

This is to say that $\mathcal{N} \models Rt_1(h(a_1), \ldots, h(a_n)), \ldots, t_m(h(a_1), \ldots, h(a_n))$. □

4.2 Elementary Maps, Substructures, Extensions

Our definition of a strong homomorphism does not require the strong preservation of equalities, i.e. strong homomorphisms (and thus homomorphisms) may not be injective, even though they strongly preserve the atomic formulae that are not equalities.

Example 4.5 Let $\mathcal{L}$ be a language that contains only the two symbols $\prec, \sim$, denoting two distinct binary relations. Suppose that $\mathcal{M}, \mathcal{N}$ are $\mathcal{L}$-structures and that, in particular, $\mathcal{M}$ is the complete preorder (discussed in Sect. 1.2) $a \sim b \prec c$ and $\mathcal{N}$ is the linear order $c' \prec a' \prec b'$. The function $h : \mathcal{M} \longrightarrow \mathcal{N}$, defined by the conditions $h(a) = c' = h(b), h(c) = b'$ is a strong homomorphism, but it is not injective because $a \neq b$ and $h(a) = h(b)$.

If a homomorphism strongly preserves every atomic formula, including equalities, then it is injective. We have a special name for such a map, recorded below.

Definition 4.6 Let $\mathcal{M}$ $\mathcal{N}$ be $\mathcal{L}$-structures and $f : \mathcal{M} \longrightarrow \mathcal{N}$ a function. Then f is said to be an $\mathcal{L}$**-embedding** iff it strongly preserves all atomic formulae.

Exercise 4.7 Prove that a homomorphism preserves every quantifier-free formula iff it is an embedding.

Exercise 4.8 Consider two distinct squares Ξ and Φ, one of which, say Φ, has its diagonal traced out, while the other does not. Let $\mathcal{L}$ be a first-order language whose signature only contains a binary relation symbol R. We view Ξ and Φ as $\mathcal{L}$-structures. Let a, b, c be the vertices of Ξ and d, e, f those of Φ. We interpret R in Ξ and Φ respectively by stipulating that it holds between two points if there is a line joining them.

(i) describe a map from Ξ to Φ that preserves the atomic $\mathcal{L}$-formulae;
(ii) is your map an embedding? Can there be an embedding of Ξ into Φ?

Exercise 4.9 Show that, if $\mathcal{M}, \mathcal{N} \models \mathsf{LO}$, then any homomorphism $h : \mathcal{M} \longrightarrow \mathcal{N}$ is already an embedding.

Exercise 4.10 Let $\mathcal{L}$ be the language whose signature contains the single binary operation symbol $+$. Show that there is no elementary embedding of the $\mathcal{L}$-structure $(\mathbb{N} - \{0\}, +^\mathbb{N})$ into the $\mathcal{L}$-structure $(\mathbb{N}, +^\mathbb{N})$.

Although homomorphisms and embeddings are important, the natural maps between first-order structures are the ones strongly preserving the totality of first-order information. These maps allow us to transfer sentences or formulae with parameters back and forth across structures. We discuss them in the next section.

4.2 Elementary Maps, Substructures, Extensions

If we drop the restriction to atomic formulae used in the definition of embeddings, we arrive at elementary maps.

Definition 4.11 If $\mathcal{M}, \mathcal{N}$ are $\mathcal{L}$-structures and $f : M \longrightarrow N$ strongly preserves every $\mathcal{L}$-formula, then f is an **elementary embedding** or an **elementary map** between $\mathcal{M}$ and $\mathcal{N}$.

Remark 4.12 If $\mathcal{M}, \mathcal{N}$ are $\mathcal{L}$-structures and $f : M \longrightarrow N$ is an elementary map, then for every $\mathcal{L}$-formula $\psi(x_1, \ldots, x_n)$ and every $a_1, \ldots, a_n \in M$, we have:

$$\mathcal{M} \models \psi(a_1, \ldots, a_n) \iff \mathcal{N} \models \psi(f(a_1), \ldots, f(a_n)).$$

If the domain of f is $A \subseteq M$ instead of the whole of M, and the last condition holds only relative to a choice of $a_1, \ldots, a_n$ within A, not M, f is said to be **partial elementary**.

Exercise 4.13 Let $\mathcal{M}, \mathcal{N}$ be $\mathcal{L}$-structures, $f : M \longrightarrow N$ a surjective function. Show that f is an embedding iff it is an elementary embedding.

Exercise 4.14 Let M, N be finite sets and $f : M \longrightarrow N$ an injection that is not surjective. Show that f is not a $\mathcal{L}_=$-elementary map.

Exercise 4.15 A **category** $\mathscr{C}$ consists of:
(a) a collection $ob(\mathscr{C})$ of **objects**;
(b) for each A, B in $ob(\mathscr{C})$, a collection $\mathscr{A}(A, B)$ of **arrows** from A to B;
(c) for each A, B, C in $ob(\mathscr{C})$, a binary function called **composition**, sending a pair (g, f) of arrows from $\mathscr{A}(B, C), \mathscr{A}(A, B)$ respectively into $g * f$ in $\mathscr{A}(A, C)$;
(d) for each $A \in ob(\mathscr{A})$, an arrow $1_A \in \mathscr{A}(A, A)$ called the **identity** on A;

satisfying two conditions, namely:

(1) **Associativity**: for each f in $\mathscr{A}(A, B)$, g in $\mathscr{A}(B, C)$ and h in $\mathscr{A}(C, D)$:

$$(h * g) * f = h * (g * f),$$

where equality means that both compositions describe the same arrow.
(2) **Identity Laws**: for each arrow f in $\mathscr{A}(A, B)$, $f * 1_A = f = 1_B * f$.

Show that the collection **K** of $\mathcal{L}$-structures, together with the elementary maps as arrows, determines a category. What about the collection of $\mathcal{L}$-structures that model a first-order theory T?

By Exercise 4.13, a surjective embedding is elementary. We have a special name for such a map, namely **isomorphism**. We say that two $\mathcal{L}$-structure $\mathcal{M}, \mathcal{N}$ are **isomorphic** when there is an isomorphism $h : M \longrightarrow N$ between them, in which case we write $\mathcal{M} \simeq \mathcal{N}$.

Isomorphisms strongly preserve all first-order information, as a straightforward induction on the complexity of formulae, left to the reader, shows.

Theorem 4.16 Let $\mathcal{L}$ be a first-order language, $\mathcal{M}, \mathcal{N}$ two $\mathcal{L}$-structures and $f : M \longrightarrow N$ an isomorphism. Then f strongly preserves every $\mathcal{L}$-formula.

4.2 Elementary Maps, Substructures, Extensions

Remark 4.17 If $\mathcal{M}, \mathcal{N}$ are $\mathcal{L}$-structures and $\mathcal{M} \simeq \mathcal{N}$, then $\mathcal{M}, \mathcal{N}$ model exactly the same $\mathcal{L}$-sentences. The converse statement does not hold. For instance (see Exercise 4.36), any two models of IS satisfy the same $\mathcal{L}_=$-sentences. If M, N are infinite sets such that $|M| < |N|$, it is clear that there cannot be any isomorphism between them, because there is no bijection between them.

Due to its importance, we have a special name for the property of modelling the same $\mathcal{L}$-sentences, which is weaker than the property of being isomorphic.

Definition 4.18 Let $\mathcal{M}, \mathcal{N}$ be $\mathcal{L}$-structures. We say that $\mathcal{M}, \mathcal{N}$ are **elementarily equivalent**, and write $\mathcal{M} \equiv \mathcal{N}$, iff $\mathcal{M}, \mathcal{N}$ are models of the same $\mathcal{L}$-sentences.

Example 4.19 By Remark 4.17, any two infinite sets are elementarily equivalent relative to $\mathcal{L}_=$. We shall see in Chap. 8 that the $\mathcal{L}$-structure $(\mathbb{Q}, <^{\mathbb{Q}})$ is elementarily equivalent (but not isomorphic) to the real line $(\mathbb{R}, <^{\mathbb{R}})$ (the latter is obtained from the real field introduced in the next chapter by dropping from its signature every symbol but $<$). We shall prove in Chaps. 7 and 8 that e.g. the models of $\mathsf{DAG}_<$ or of $\mathsf{VS}_\mathbb{Q}$ are all elementarily equivalent.

Exercise 4.20

(i) Define a (strong) homomorphism from the additive group of integers $\mathbb{Z}$ onto the quotient group $\mathbb{Z}/5\mathbb{Z}$. More generally, define a (strong) homomorphism from $\mathbb{Z}$ onto the quotient group $\mathbb{Z}/n\mathbb{Z}$.
(ii) Given the vector spaces $\mathbb{Q}^2, \mathbb{Q}^3$, show that the function $f : \mathbb{Q}^2 \longrightarrow \mathbb{Q}^3$ defined by the condition $f((p, q)) = (p, q, 0)$ is a $\mathcal{L}_v$-embedding.
(iii) Let $\mathsf{P}_n(\mathbb{Q})$ be the subset of $\mathbb{Q}[x]$ consisting of the polynomials of degree at most 2. Show that the vector space with domain $\mathbb{Q}^3$ and $\mathsf{P}_2(\mathbb{Q})$ are isomorphic as $\mathcal{L}_v$-structures. Generalise this result to $\mathsf{P}_n(\mathbb{Q})$.
(iv) Show that a k-dimensional vector space over the field of rational numbers is isomorphic to the vector space with domain $\mathbb{Q}^k$.
(v) Define a binary relation $\sim$ on $\mathbb{N} \times \mathbb{N}$ by the following condition:

$$(m, n) \sim (p, q) \text{ iff } m + q = p + n.$$

(a) Show that $\sim$ is an equivalence relation.
(b) Call $[m, n]$ the equivalence class containing (m, n), and $\mathbb{N} \times \mathbb{N}/\sim$ the set of equivalence classes determined by $\sim$ over $\mathbb{N} \times \mathbb{N}$. Define the operations $+', -'$ on $\mathbb{N} \times \mathbb{N}/\sim$ as follows:

$$[m, n] +' [p, q] = [m + p, n + q], \quad -'[m, n] = [n, m].$$

Show that, if $(m, n) \sim (a, b)$ and $(p, q) \sim (c, d)$ then $[m + p, n + q] = [a + c, b + d]$ and $[n, m] = [b, a]$.
(c) Verify that the $\mathcal{L}_g$-structure $(\mathbb{N} \times \mathbb{N}/\sim, +', -', [0, 0])$ is an abelian group.
(d) Define an $\mathcal{L}_g$-isomorphism from $\mathbb{N} \times \mathbb{N}/\sim$ onto $\mathbb{Z}$.

Isomorphisms whose domain and codomain coincide are of special interest and for this reason they receive a special name.

Definition 4.21 Let $\mathcal{M}$ be a $\mathcal{L}$-structure. An isomorphism $\alpha : \mathcal{M} \longrightarrow \mathcal{M}$ is called an **automorphism** of $\mathcal{M}$.

Remark 4.22 Let $\mathcal{M}$ be a $\mathcal{L}$-structure. We call the set of its automorphisms $Aut(\mathcal{M})$. This set is nonempty, because it contains at least the identity automorphism of $\mathcal{M}$, i.e. the function $\iota : \mathcal{M} \longrightarrow \mathcal{M}$ such that $\iota(a) = a$, for each $a \in \mathcal{M}$. Moreover, if $f, g \in Aut(\mathcal{M})$, the functional composition fg is also an automorphism of $Aut(\mathcal{M})$, as the reader may easily verify. Finally, if $f \in Aut(\mathcal{M})$, the function f^{-1}, i.e. the set of ordered pairs (y, x) such that $(x, y) \in f$, is a bijection, because f is, and a strong homomorphism, because f is. Clearly, we may view $Aut(\mathcal{M})$ as the domain of a $\mathcal{L}_g$-structure, once we interpret 0 on ι, $+$ on functional composition and $-$ on the operation of taking the inverse function. The resulting $\mathcal{L}_g$-structure is a group.

We next turn our attention to maps that link structures whose domains are related by inclusion: these maps will lead us to the fundamental notions of substructure and extension, as well as their elementary counterparts.

4.2.1 Substructures and Extensions

Let $\mathcal{M}, \mathcal{N}$ be $\mathcal{L}$-structures, with $M \subseteq N$. The **inclusion map** $\iota_M : M \longrightarrow N$ is the restriction of the identity map on N to the subset M. By definition, for every $a \in M$, $\iota_M(a) = a$. In this context we do not only regard $\mathcal{M}$ a structure in its own right, but also as a structured part of $\mathcal{N}$. More precisely, we have:

Definition 4.23 Let $\mathcal{M}, \mathcal{N}$ be $\mathcal{L}$-structures, with $M \subseteq N$. We say that $\mathcal{M}$ is a **substructure** of $\mathcal{N}$ (or, equivalently, that $\mathcal{N}$ is an **extension** of $\mathcal{M}$) iff the inclusion map $\iota_M : M \longrightarrow N$ is an embedding. If ι_M is an elementary embedding, then $\mathcal{M}$ is an **elementary substructure** of $\mathcal{N}$ (or, equivalently, $\mathcal{N}$ is an **elementary extension** of $\mathcal{M}$).

We adopt the notation $\mathcal{M} \sqsubseteq \mathcal{N}$ to indicate that $\mathcal{M}$ is a substructure of in symbols $\mathcal{N}$, and the notation $\mathcal{M} \preceq \mathcal{N}$ to indicate that $\mathcal{M}$ is an elementary substructure of $\mathcal{N}$.

Exercise 4.24 Show that, if $\mathcal{M}, \mathcal{N}$ are $\mathcal{L}$-structures and ι_M is an embedding, then $\mathcal{M}, \mathcal{N}$ have the same constants, each n-ary relation R^M of $\mathcal{M}$ is the restriction of R^N to M^n and, for any k-ary function symbol in the signature of $\mathcal{L}$ and any k-tuple $(a_1, \ldots, a_k) \in M^k$, $f^M(a_1, \ldots, a_k) = f^N(a_1, \ldots, a_k)$.

Remark 4.25 An explicit way of saying that the inclusion map $\iota_M : M \longrightarrow N$ is an embedding is to say that, for every atomic formula $\varphi(x_1, \ldots, x_n)$ and any $a_1, \ldots, a_m \in M$:

4.2 Elementary Maps, Substructures, Extensions

$$\mathcal{M} \models \varphi(a_1, \ldots, a_n) \iff \mathcal{N} \models \varphi(a_1, \ldots, a_n).$$

This is the same as saying that, for every *quantifier-free* formula $\varphi(x_1, \ldots, x_n)$, the above equivalence holds (by an induction on the complexity of formulae that stops at the connectives). If φ is an arbitrary formula whose free variables occur among $x_1, \ldots x_n$, the same notation says that ι_M is an elementary embedding.

Example 4.26 Let the $\mathcal{L}_v$-structures $\mathbf{Q}_j, \mathbf{Q}_k$, with $j < k$, be the vector spaces over $\mathbb{Q}$ with respective domains $\mathbb{Q}^j, \mathbb{Q}^k$. Clearly $\mathbf{Q}_j$ is not a substructure of $\mathbf{Q}_k$ because $\mathbb{Q}^j$ is not a subset of $\mathbb{Q}^k$. If, however, we consider the map $\alpha : \mathbb{Q}^j \longrightarrow \mathbb{Q}^k$ such that $\alpha((q_1, \ldots, q_j)) = (q_1, \ldots, q_j, 0, \ldots, 0)$, we see that $\alpha[\mathbb{Q}^j]$ is the domain of a substructure of $\mathbf{Q}_k$, obtained simply by restricting the vector space operations to $\alpha[\mathbb{Q}^j]$. Moreover, this substructure is isomorphic to $\mathbf{Q}_j$. As the next exercise shows, this example illustrates a general situation.

Exercise 4.27 Let $\mathcal{M}, \mathcal{N}$ be $\mathcal{L}$-structures and $f : M \longrightarrow N$ an embedding. Show that, once we restrict the relations and operations of $\mathcal{N}$ to $f[M]$ (which already contains the constants of N), we obtain a substructure $\mathcal{M}_f$ of $\mathcal{N}$ such that $\mathcal{M} \simeq \mathcal{M}_f$.

Exercise 4.28 Consider the $\mathcal{L}_<$-structures $(\mathbb{N}, <^\mathbb{N}), (\mathbb{Z}, <^\mathbb{Z})$. Certainly $(\mathbb{N}, <^\mathbb{N}) \sqsubseteq (\mathbb{Z}, <^\mathbb{Z})$ but it is not the case that $(\mathbb{N}, <^\mathbb{N}) \preceq (\mathbb{Z}, <^\mathbb{Z})$. Explain why and decide whether the linear order $(\mathbb{Q}, <^\mathbb{Q})$ is an elementary extension of the linear order $\mathbb{Z}$.

Example 4.29 Consider the $\mathcal{L}_<$-structures $\mathcal{M} = (\mathbb{Q}, <)$ and $\mathcal{N} = (\mathbb{R}, <)$, where $\mathcal{N}$ is the ordered set of real numbers (see Chap. 5). By results to be proved in Chap. 8, $\mathcal{M} \preceq \mathcal{N}$. If, however, we expand the language to $\mathcal{L}_{r<}$ and, consequently, expand $\mathcal{M}$ to the ordered field of rational numbers and $\mathcal{N}$ to the ordered field of real numbers, we lose the relation of elementary substructure because the field of rational numbers is a model of the sentence $\neg(\exists x (x^2 - (1+1)) = 0)$ while the field of real numbers models the sentence $\exists x (x^2 - (1+1)) = 0$ because $\sqrt{2}$ is a real number.

Exercise 4.30 Let $Tm(x_1, \ldots, x_n)$ be the set of all $\mathcal{L}$-terms whose free variables occur among $x_1, \ldots, x_m$ and M be a $\mathcal{L}$-structure. If $a_1, \ldots, a_n \in M$, show that $Tm(a_1, \ldots, a_n)$—the subset of M whose elements are named by the interpretations of the terms in $Tm(x_1, \ldots, x_n)$ on the a_i ($i = 1, \ldots, n$)—determines a substructure of M.

By Remark 4.25, if $\mathcal{M} \sqsubseteq \mathcal{N}$, then for any $a_1, \ldots, a_n \in M$, $\mathcal{M} \models \varphi(a_1, \ldots, a_n)$ implies $\mathcal{N} \models \varphi(a_1, \ldots, a_n)$, for any quantifier-free formula $\varphi(x_1, \ldots, x_n)$. We restate this fact by saying that quantifier-free formulae are **preserved by substructures** or **hereditary**. Because, moreover, $\mathcal{N} \models \varphi(a_1, \ldots, a_n)$ implies $\mathcal{M} \models \varphi(a_1, \ldots, a_n)$, we say that quantifier-free formulae are **preserved by extensions**. Formulae in which quantifiers occur are not in general preserved by substructures or extensions. If, however, we restrict attention to universal or existential formulae, we obtain:

Lemma 4.31 *Let $\mathcal{M}, \mathcal{N}$ be $\mathcal{L}$-structures such that $\mathcal{M} \sqsubseteq \mathcal{N}$. If $a \in M$ and $\varphi(x_1, \ldots, x_n, y)$ is a quantifier-free $\mathcal{L}$-formula:*
(a) $\mathcal{N} \models \forall x_1 \ldots \forall x_n \phi(x_1, \ldots, x_n, a)$ implies $\mathcal{M} \models \forall x_1 \ldots \forall x_n \phi(x_1, \ldots, x_n, a)$;
(b) $\mathcal{M} \models \exists x_1 \ldots \exists x_n \phi(x_1, \ldots, x_n, a)$ implies $\mathcal{N} \models \exists x_1 \ldots \exists x_n \phi(x_1, \ldots, x_n, a)$.

Proof To prove (a), consider the formula $\phi(x_1, \ldots, x_n, x_{n+1})$. For any assignment of $a_1, \ldots, a_n \in M$ to $x_1, \ldots, x_n$ respectively, which also assigns a to x_{n+1}:

$$\text{if } \mathcal{N} \models \phi(a_1, \ldots, a_n, a) \text{ then } \mathcal{M} \models \phi(a_1, \ldots, a_n, a).$$

If $\mathcal{N} \models \forall x_1 \ldots \forall x_n \phi(x_1, \ldots, x_n, a)$, then $\mathcal{N} \models \phi(a_1, \ldots, a_n, a)$ for any $a_1, \ldots, a_n, a \in M$. By the above implication, the same is true of $\mathcal{M}$. The definition of satisfiability now yields:

$$\mathcal{M} \models \forall x_1 \ldots \forall x_n \phi(x_1, \ldots, x_n, a).$$

Part (b) is very simple. Note that the proof goes through even if we consider a quantifier-free formula $\varphi(x_1, \ldots, x_n, y_1, \ldots, y_m)$ and work with an assignment of $b_1, \ldots, b_m \in M$ to $y_1, \ldots, y_m$ respectively. □

A more concise way of stating the last lemma is to say that universal formulae are preserved by substructures whilst existential formulae are preserved by extensions. We shall prove in Chap. 8 that Lemma 4.31 has a converse: any formula that is preserved by substructures is equivalent to a universal formula and any formula that is preserved by extensions is equivalent to an existential formula.

Example 4.32 Let $\mathcal{G}$ be a $\mathcal{L}_g$-structure. If $\mathcal{G}$ is an abelian group and $\mathcal{H} \sqsubseteq \mathcal{G}$, then, because **GA** is a set of universal sentences, the previous lemma implies that $\mathcal{H}$ must be an abelian group. The same argument applies if $\mathcal{G}$ is a group: in that case any substructure must be a group.

Let us now drop the unary function symbol '$-$' from the signature of $\mathcal{L}_g$. In the reduced signature $\mathcal{L}_g^-$, an arbitrary substructure $\mathcal{H}$ of a group $\mathcal{G}$ is no longer required to be closed under inverses. In other words, not every $\mathcal{L}_g^-$-substructure of $\mathcal{G}$ is a subgroup of $\mathcal{G}$. On the other hand, if $\mathcal{H} \sqsubseteq \mathcal{G}$, then $\mathcal{H}$ certainly has an associative binary operation interpreting $+$ and a neutral element 0 relative to it. This is because the last conditions are expressed by universal $\mathcal{L}_g^-$-sentences true in $\mathcal{G}$. A model of these sentences is called a **monoid**. Conversely, if $\mathcal{H}$ is a monoid obtained by restricting $+^G$ to H (which will then contain 0^G, the neutral element of $+^G$), it is easy to check that its inclusion in $\mathcal{G}$ is a $\mathcal{L}_g^-$-embedding.

A concrete example of what we have remarked is provided by the $\mathcal{L}_g^-$-structure $(\mathbb{N}, +^\mathbb{N}, 0^\mathbb{N})$, which is a $\mathcal{L}_g^-$-substructure of $(\mathbb{Z}, +^\mathbb{Z}, 0^\mathbb{Z})$. Clearly, $\mathbb{N}$ cannot be the domain of a $\mathcal{L}_g$-substructure of the additive group of integers because the function $-^\mathbb{Z}$ cannot be restricted to $\mathbb{N}$.

If we further drop the constant symbol from $\mathcal{L}_g^-$ and call the resulting language $\mathcal{L}_g^{--}$, then the set of positive multiples of 3, which is not a monoid and thus not a

4.2 Elementary Maps, Substructures, Extensions

$\mathcal{L}_g^{--}$-substructure of the additive group of integers, is nonetheless a substructure of $(\mathbb{Z}, +^{\mathbb{Z}})$. Sets endowed with an associative, binary operation are called **semigroups**. Thus, the $\mathcal{L}_g^{--}$-substructures of $(\mathbb{Z}, +^{\mathbb{Z}})$ are the additive semigroups included in the set of integers.

Example 4.33 If $\mathcal{K}$ is a field in the language $\mathcal{L}_r$, then the $\mathcal{L}_r$-substructures of $\mathcal{K}$ are the integral domains included in K. The reader is invited to verify this statement.

If $\mathcal{M}, \mathcal{N}$ are $\mathcal{L}$-structures and $\mathcal{M} \sqsubseteq \mathcal{N}$, it does not in general follow that $\mathcal{M} \preceq \mathcal{N}$. It is interesting to look closely at what prevents a general argument from establishing the implication. Since embeddings strongly preserve all quantifier-free formulae, an attempt to prove $\mathcal{M} \sqsubseteq \mathcal{N} \implies \mathcal{M} \preceq \mathcal{N}$ by induction on the complexity of formulae can only break down at the inductive step for quantifiers. In fact, this step can almost be carried out: if $\mathcal{M} \models \exists x \phi(x, a_1, \ldots, a_n)$, there is $a \in M$ such that $\mathcal{M} \models \phi(a, a_1, \ldots, a_n)$. By the inductive hypothesis, $\mathcal{N} \models \phi(a, a_1, \ldots, a_n)$ and, consequently $\mathcal{N} \models \exists x \phi(x, a_1, \ldots, a_n)$.

There is, however, insufficient data to establish the converse implication. If $\mathcal{N} \models \exists x \phi(x, a_1, \ldots, a_n)$, then there is $b \in N$ such that $\mathcal{N} \models \phi(b, a_1, \ldots, a_n)$, but we do not know whether $b \in M$ or b can be exchanged with a suitable element a of M. If we postulate that such an element always exists, we can complete the induction and obtain a proof the **Tarski-Vaught test**:

Lemma 4.34 *Let $\mathcal{M} \sqsubseteq \mathcal{N}$. If, for every $a_1, \ldots, a_n \in M$, $\mathcal{N} \models \phi(b, a_1, \ldots, a_n)$ implies the existence of $a \in M$ such that $\mathcal{N} \models \phi(a, a_1, \ldots, a_n)$, then $\mathcal{M} \preceq \mathcal{N}$.*

We may slightly refine Lemma 4.34 by noting that, if the lemma's hypotheses hold relative to a subset A of N, then A is the domain of an elementary substructure of $\mathcal{N}$.

Exercise 4.35 Let $\mathcal{N}$ be a $\mathcal{L}$-structure, $A \subseteq N$. Show that A is the domain of a substructure of N iff it contains all the constants in N and, for any k-ary operation f^N on N^k and k-tuple $a_1, \ldots, a_k$, $f^N(a_1, \ldots, a_k) \in A$.

Exercise 4.36

(i) Show that, if M, N are infinite sets and $f : M \longrightarrow N$ a bijection, then f is a $\mathcal{L}_=$-elementary embedding.
(ii) Let M, N be infinite sets and $f : M \longrightarrow N$ an injection. By part (i), every bijection of N onto N is a $\mathcal{L}_=$-elementary embedding. Use this fact and the Tarski-Vaught test to show that f is a $\mathcal{L}_=$-elementary embedding.

If a structure $\mathcal{N}$ has enough automorphisms, it is easy to apply the Tarski-Vaught test to show that certain substructures of $\mathcal{N}$ are elementary. This is because, if $\mathcal{N} \models \phi(b, a_1, \ldots, a_n)$ and there is $\alpha \in Aut(N)$ that keeps $a_1, \ldots, a_n$ fixed while permuting b with an element of M, the fact that isomorphisms (and thus automorphisms) are elementary maps immediately implies $\mathcal{N} \models \phi(\alpha(b), a_1, \ldots, a_n)$. The test applies and $\mathcal{M} \preceq \mathcal{N}$.

Exercise 4.37 Use the Tarski-Vaught test to prove that $\mathcal{M}$ is an elementary substructure of the linear order $(\mathbb{N}, <)$ iff $\mathcal{M}$ is already the whole structure $(\mathbb{N}, <)$.

Besides allowing us to obtain specific results about elementary substructures, the Tarski-Vaught test allows us to prove general theorems about the existence of elementary substructures. We devote the next subsection to an important one, for the moment only orived for countable languages, i.e. first-order languages whose signature is a countable set of symbols. The theorem also holds for languages of arbitrarily large signature, but we postpone its general proof until Chap. 7.

4.2.2 Downward Löwenheim-Skolem Theorem

Theorem 4.38 *Let $\mathcal{L}$ be a countable, first-order language and $\mathcal{N}$ be an infinite $\mathcal{L}$-structure. For any countable subset $X \subseteq N$, there is a countable, elementary substructure $\mathcal{M} \preceq \mathcal{N}$ such that $X \subseteq M$ and $|M| = |X|$.*

Note that, in general, the set X is not the domain of M: what we are going to prove is that we do not need to add to X any more than a countable set of objects (possibly none) in order to obtain the domain of an elementary substructure of N. Moreover, because we may set $X = \emptyset$, Theorem 4.38 implies that any infinite $\mathcal{L}$-structure has a countable, elementary substructure.

Proof We are going to build an infinite, increasing sequence of sets $X = M_0 \subseteq M_1 \subseteq M_2, \ldots$ such that $X = M_0$ and, for each $i \in \mathbb{N}$, $M_i \subseteq N$ and $|M_i| = |X|$.

Since M_0 is given, we describe a method to construct M_1 from M_0 that will work in general, i.e. to construct M_{n+1} from M_n. Consider the set E of all $\mathcal{L}$-formulae of the form $\exists y \varphi(y, x_1, \ldots, x_n)$ (for some $n \in \mathbb{N}$) such that $N \models \exists y \varphi(y, a_1, \ldots, a_n)$ for some $a_1, \ldots, a_n \in X$.

Because the set of all $\mathcal{L}$-formulae is countable (see Sect. 2.7), its subset E is also countable. Moreover, for each formula in E, N models $\exists y \varphi(y, a_1, \ldots, a_n)$ for at most[1]

$$|X^n| = |X| \text{ choices of } a_1, \ldots, a_n.$$

Since X is countable, so is $|X^n|$. For each list $a_1, \ldots, a_n$, we select one $a \in N$ such that $\mathcal{N} \models \varphi(a, a_1, \ldots, a_n)$. These selections determine a countable set V_1. We set:

$$M_1 = X \cup V_1. \tag{4.1}$$

---

[1] The last equality follows from Theorem A.72 from the Appendix.

Clearly, M_1, the union of two countable sets, is countable. The procedure just described can be iterated to build M_2 from M_1, M_3 from M_2, and so on. In each case we consider the set E and every finite sequence $a_1, \ldots, a_n$ from M_k^n, in order to obtain M_{k+1}. We eventually obtain the infinite sequence $M_0 \subseteq M_1 \subseteq M_2, \ldots$.

Let M be $\bigcup_{i \in \mathbb{N}} M_i$. As a union of countably many, countable sets, M is countable. Moreover, M may be regarded as the domain of a substructure of $\mathcal{N}$. We leave it as an exercise to check that M must contain the constants of N and that it is closed under the functions on N (i.e, if $a_1, \ldots, a_m \in M$ and f is a m-ary function on N, then $f(a_1, \ldots, a_n) \in M$). The relations on N can be straightforwardly restricted to relations on M: if R^N is a k-ary relation on N, we obtain a corresponding relation on M by taking the set of k-tuples in R^N whose components are elements of M. It now follows that the inclusion map from M to N is an embedding.

To show that it is elementary, we apply the Tarski-Vaught test. Suppose $\mathcal{N} \models \varphi(b, a_1, \ldots, a_n)$, with $a_1, \ldots, a_n \in M$ and $b \in N$. Each a_i must belong to some M_i. In particular, there must be a largest $k \in \mathbb{N}$ such that $a_1, \ldots, a_n \in M_k$. Because $\exists x \varphi(x, a_1, \ldots, a_n) \in E$, there is $a \in M_{k+1}$ such that $\mathcal{N} \models \varphi(a, a_1, \ldots, a_n)$. We conclude that M is the underlying set of an elementary substructure of N. □

By Exercise 4.24, the existence of an embedding $f : M \longrightarrow N$ is equivalent to the existence of a substructure of $\mathcal{N}$ isomorphic to $\mathcal{M}$. An easy generalisation of that exercise implies that the same result holds relative to an elementary substructure of $\mathcal{N}$ when f is elementary. In such cases $\mathcal{N}$ carries, as it were, all the first-order information required to ensure that $\mathcal{M}$ is embeddable or, respectively, elementarily embeddable, into it. In the next section we describe this information and use it reduce the existence of embeddings or elementary embeddings to the satisfiability of certain sets of sentences.

4.3 Diagrams

Embeddings guarantee that the atomic information carried by a $\mathcal{L}$-structure $\mathcal{M}$ be realised in some $\mathcal{L}$-structure $\mathcal{N}$. If $\varphi(x_1, \ldots, x_m)$ is an atomic formula or a negated atomic formula, $f : M \longrightarrow N$ an embedding, and $\mathcal{M} \models \varphi(a_1, \ldots, a_m)$, then $\mathcal{N}$ realises the atomic information carried by $\mathcal{M}$ via f, since $\mathcal{N} \models \varphi(f(a_1), \ldots, f(a_m))$.

Remark 4.39 Because, in the sequel, we will often have to deal with atomic formulae and their negations simultaneously, it is convenient to adopt a single term that designates both: we simply call them **literals**.

What we have observed show that, if an embedding of $\mathcal{M}$ exists, atomic information is going to be modelled by the embedding structure $\mathcal{N}$. We now reverse our perspective and describe the atomic information whose models embed a given structure $\mathcal{M}$.

To this end, we expand **C** in the signature of $\mathcal{L}$ by adding to it one distinct constant symbol c_a for each $a \in M$. The newly added constant symbols enable us to view the set of literals modelled by $\mathcal{M}$ as a set of sentences $\mathcal{D}(\mathcal{M})$, known as the *diagram* of $\mathcal{M}$. Slightly more formally, we have:

Definition 4.40 Let $\mathcal{M}$ be a $\mathcal{L}$-structure and $\mathcal{L}_M$ an expansion of $\mathcal{L}$ obtained by adding to the signature of $\mathcal{L}$ a distinct constant symbol for each element of M. Let $\mathcal{M}^+$ be the corresponding expansion of $\mathcal{M}$. The **diagram** of $\mathcal{M}$ is the set $\mathcal{D}(\mathcal{M})$ of all $\mathcal{L}_M$-literals modelled by $\mathcal{M}^+$.

As we shall shortly prove, a model of $\mathcal{D}(\mathcal{M})$ embeds $\mathcal{M}$ or, equivalently, contains an isomorphic copy of $\mathcal{M}$.

Example 4.41 Consider the $\mathcal{L}_r$-structure $\mathcal{Z}_r = (\mathbb{Z}, +^\mathbb{Z}, -^\mathbb{Z}, \cdot^\mathbb{Z}, 0^\mathbb{Z}, 1^\mathbb{Z})$. The language $\mathcal{L}_r$ contains enough closed terms to name each integer. For the positive integers, we use $1, 1+1, 1+1+1, \ldots$ and, for the negative integers, $-1, -(1+1), -(1+1+1), \ldots$. The atomic information carried by $\mathcal{Z}_r$ only involves equalities between sums and products. The literals $(1+1)+(-1) = 1$, $(1+1)(1+1+1) = 1+1+1+1+1+1$ and $\neq (1+1 = 1+1+1)$ carry some of this information. Let us call $\mathcal{D}(\mathcal{Z}_r)$ the set of all literals modelled by $\mathcal{Z}_r$. Strictly speaking, $\mathcal{D}(\mathcal{Z}_r)$ is not the diagram of $\mathcal{Z}_r$ as we have defined it, because the only elements of $\mathbb{Z}$ named by a constant symbol are $0^\mathbb{Z}$ and $1^\mathbb{Z}$. It is clear, however, that we can trivially inter-translate the sentences in $\mathcal{D}(\mathcal{Z}_r)$ with sentences in an expanded signature containing a distinct constant symbol for each integer.

If $\mathcal{N} \models \mathcal{D}(\mathcal{Z}_r)$, then $\mathcal{N}$ is a $\mathcal{L}_r$-structure containing a countably infinite subset whose elements are named by the closed terms $0, 1, -1, 1+1, -(1+1), \ldots$. Whenever an equality between sums and products of these terms holds in $\mathcal{Z}_r$, it must hold in $\mathcal{N}$, and vice versa (since, if it holds in $\mathcal{N}$, it is in $\mathcal{D}(\mathcal{Z}_r)$). It follows that the function $f : \mathbb{Z} \longrightarrow N$, sending each element $a \in \mathbb{Z}$ named by the closed term t_a, or a suitable constant symbol c_a, into the element b of N named by the same term, or constant symbol, is an embedding.

To see this in some detail, we show that f is a homomorphism. Suppose $a_1, a_2, a_3 \in \mathbb{Z}$ and $a_1 + a_2 = a_3$. There are closed $\mathcal{L}_r$-terms t_1, t_2, t_3 naming $a_1, a_2, a_3 \in \mathbb{Z}$ and $b_1, b_2, b_3 \in N$ respectively. By the definition of f, $b_i = f(a_i)$, with $i = 1, 2, 3$. Since $\mathcal{N} \models t_1 + t_2 = t_3$, we have $f(a_1) + f(a_2) = f(a_3) = f(a_1 + a_2)$. By a similar argument, f preserves $-$ and $\cdot$. It is therefore a homomorphism of $\mathbb{Z}$ into N.

Example 4.42 Consider the $\mathcal{L}_r$-structure $\mathcal{Q}_r = (\mathbb{Q}, +^\mathbb{Q}, -^\mathbb{Q}, \cdot^\mathbb{Q}, 0^\mathbb{Q}, 1^\mathbb{Q})$. The language $\mathcal{L}_r$ does not contain enough constant symbols to allow us to name each element of $\mathbb{Q}$ by a closed term. We may, however, expand $\mathcal{L}_r$ to a language that contains one distinct constant symbol for each rational number: let the expanded language be called $\mathcal{L}_\mathbb{Q}$.

The expansion of $\mathcal{L}_r$ to $\mathcal{L}_\mathbb{Q}$ corresponds to an expansion of the $\mathcal{L}_r$-structure $\mathcal{Q}_r$ to the $\mathcal{L}_\mathbb{Q}$-structure $(\mathbb{Q}, +^\mathbb{Q}, -^\mathbb{Q}, \cdot^\mathbb{Q}, \{q\}_{q \in \mathbb{Q}})$, where each rational number is singled out as a constant q, named by a corresponding constant symbol $c_q \in \mathcal{L}_\mathbb{Q}$.

4.3 Diagrams

With these linguistic resources at our disposal, we can use sentences to describe, in a case by case fashion, equalities between sums and products of rationals. We effectively portray the field structure of $\mathbb{Q}$ by means of literals in $\mathcal{L}_\mathbb{Q}$. Calling the set of these literals $\mathcal{D}(\mathcal{Q}_r)$, we have e.g. $c_{1/2} + c_{1/2} = c_1, c_{-3/5} + c_{2/5} = c_{1/5}, \neg(c_{2/3}c_{3/3} = c_{6/3}) \in \mathcal{D}(\mathcal{Q}_r)$.

As in Example 4.41, we see that, if $\mathcal{N} \models \mathcal{D}(\mathcal{Q}_r)$, where $\mathcal{N}$ is a $\mathcal{L}_\mathbb{Q}$-structure, i.e. an expansion of a $\mathcal{L}_r$-structure, $\mathcal{N}$ contains a countably infinite subset that realises field arithmetic on $\mathbb{Q}$. The function that links the denotations of each c_q in $\mathbb{Q}$ and $\mathcal{N}$ respectively determines an embedding of $\mathcal{Q}_r$ into $\mathcal{N}$.

Exercise 4.43 How many sentences does $\mathcal{D}(\mathbb{Z}/3\mathbb{Z})$ contain, relative to a suitable expansion of $\mathcal{L}_r$? Show that, if $\mathcal{N}$ is a $\mathcal{L}_r$-structure with an expansion $\mathcal{N}^+$ such that $\mathcal{N}^+ \models \mathcal{D}(\mathbb{Z}/3\mathbb{Z})$, then $\mathbb{Z}/3\mathbb{Z}$ can be embedded in $\mathcal{N}$.

Exercise 4.44 Let $\mathcal{L}$ a first-order language whose only extra-logical symbol is the relation symbol R, of arity 2. Let $\mathcal{M}$ be a $\mathcal{L}$-structure with six elements and such that any two distinct elements are R-related. Count the sentences in $\mathcal{D}(\mathcal{M})$. Let $\exists^{=6} x \varphi$ be a $\mathcal{L}$-sentence stating the existence of exactly six distinct elements. Show that, if $\mathcal{N} \models \mathcal{D}(\mathcal{M}) \cup \exists_{=6} x \varphi$, then $\mathcal{N}$ is isomorphic to $\mathcal{M}$. Deduce that there is a single $\mathcal{L}$-sentence φ such that any model of φ is isomorphic to $\mathcal{M}$.

Exercise 4.45 Prove that, if $\mathcal{M}$ is a finite $\mathcal{L}$-structure, there is a first-order sentence ψ that describes it up to isomorphism (this is to say that $\mathcal{M} \models \psi$ iff $\mathcal{N} \simeq \mathcal{M}$).

The immediately preceding examples and exercises are consequences of the following theorem.

Theorem 4.46 Let $\mathcal{M}, \mathcal{N}$ be $\mathcal{L}$-structures, $\mathcal{L}_\mathcal{M}$ an expansion of $\mathcal{L}$ obtained by adding a new constant symbol for each element of $\mathcal{M}$. The following statements are equivalent:

(a) there is an embedding $f : \mathcal{M} \longrightarrow \mathcal{N}$;
(b) the $\mathcal{L}_\mathcal{M}$-structure $\mathcal{N}_\mathcal{M}$ (an expansion of $\mathcal{N}$) models $\mathcal{D}(\mathcal{M})$;
(c) $\mathcal{M}$ has an extension $\mathcal{M}'$ that is isomorphic to the structure $\mathcal{N}$, obtained from $\mathcal{N}_\mathcal{M}$ by reducing the language to $\mathcal{L}$.

Exercise 4.47 Prove Theorem 4.46. (*Hint*: use the proof of Theorem 4.49 as a template).

Elementary embeddings do not only preserve atomic information so their existence must depend on suitable extensions of diagrams.

Definition 4.48 Let $\mathcal{M}$ be a $\mathcal{L}$-structure. The **elementary diagram** of $\mathcal{M}$ is the set $\mathcal{ED}(\mathcal{M})$ of all $\mathcal{L}_\mathcal{M}$-sentences modelled by the expansion $\mathcal{M}^+$ of $\mathcal{M}$ to the language $\mathcal{L}_\mathcal{M}$.

An elementary analogue of Theorem 4.46 holds.

Theorem 4.49 Let $\mathcal{M}, \mathcal{N}$ be $\mathcal{L}$-structures, $\mathcal{L}_M$ an expansion of $\mathcal{L}$ obtained by adding a new constant symbol for each element of M. The following statements are equivalent:

(a) there is an elementary embedding $f : \mathcal{M} \longrightarrow \mathcal{N}$;
(b) the expanded $\mathcal{L}_M$-structure $\mathcal{N}_M$ models $\mathcal{ED}(\mathcal{M})$;
(c) $\mathcal{M}$ has an elementary extension $\mathcal{M}'$ that is isomorphic to $\mathcal{N}$, the $\mathcal{L}$-reduct of $\mathcal{N}_M$.

Proof First, we show that (a) implies (b). We are given an elementary embedding $f : \mathcal{M} \longrightarrow \mathcal{N}$. We expand $\mathcal{N}$ to the $\mathcal{L}_M$-structure $\mathcal{N}_M$, where c_a names $a \in M$ and $f(a) \in N$.

To each $\mathcal{L}_M$-sentence $\varphi(c_{a_1}, \ldots, c_{a_n}) \in \mathcal{ED}(\mathcal{M})$, we associate the $\mathcal{L}$-formula $\varphi(x_1, \ldots, x_n)$. We next note that $\mathcal{M}^+$, the expansion of $\mathcal{M}$ to the language $\mathcal{L}_M$, models $\varphi(c_{a_1}, \ldots, c_{a_n})$ iff $\mathcal{M}$ models $\varphi(a_1, \ldots, a_n)$. Because the elementary embedding f preserves the formula $\varphi(x_1, \ldots, x_n)$, we deduce:

$$\mathcal{N} \models \varphi(f(a_1), \ldots, f(a_n)).$$

By construction:

$$\mathcal{N} \models \varphi(f(a_1), \ldots, f(a_n)) \text{ iff } \mathcal{N}_M \models \varphi(c_{a_1}, \ldots, c_{a_n}).$$

We conclude that $\mathcal{N}_M \models \varphi(c_{a_1}, \ldots, c_{a_n})$. Since φ was an arbitrary element of $\mathcal{ED}(\mathcal{M})$, $\mathcal{N}_M \models \mathcal{ED}(\mathcal{M})$.

Next, we show that (b) implies (c). We use the set N to build $\mathcal{M}'$. Because $\mathcal{N}_M \models \mathcal{ED}(\mathcal{M})$, there is a subset $M^+ \subseteq N$, which contains the elements of $\mathcal{N}_M$ named by the constant symbols in $\mathcal{L}_M$. We set $M' = (N - M^+) \cup M$ and a function $h : M' \longrightarrow N$ by the condition:

$$h(x) = \begin{cases} x & \text{if } x \in N - M^+ \\ c_a^{\mathcal{N}_M} & \text{if } x = a \in M \end{cases}$$

where $c_a^{\mathcal{N}_M}$ stands for the denotation in $\mathcal{N}_M$ of the the constant symbol c_a, which names a in M. The function h is a bijection between M' and N: it obviously is an injection between $N - M^+$ and $M' - M$ (these two sets are equal) and an injection between M and M^+, since, for any $a, a' \in M$, $a \neq a'$ implies $c_a \neq c_{a'} \in \mathcal{ED}(\mathcal{M})$, which yields $c_a^{\mathcal{N}_M} \neq c_{a'}^{\mathcal{N}_M}$ (by (b)), i.e. $h(a) \neq h(a')$. Moreover, h is surjective because any element of N either lies in $N - M^+$ or it is is named by some constant symbol c_a, i.e. it is $h(a)$.

We next equip the set M' with structure in such a way that h is a strong homomorphism. For any k-ary relation $R^N \subseteq N^k$, we induce the k-ary relation $R^{M'} \subseteq M'^k$ by stipulating:

$$R^{M'} = \{\langle a'_1, \ldots, a'_k \rangle \in M'^k : \langle h(a'_1) \ldots h(a'_k) \rangle \in R^N\}.$$

4.3 Diagrams

If f^N is a n-ary function, we induce the n-ary function $f^{M'}$ on M' by stipulating:

$$f^{M'}(a'_1, \ldots, a'_n) = h^{-1}(f^N(a_1, \ldots, a_n)),$$

which works on account of the fact that h^{-1} is a bijection like h. Any constants in N are mapped into their h^{-1}-values in M'. The reader may check that, relative to the above definitions, h is actually a $\mathcal{L}$-isomorphism.

To conclude the proof of (c), we only have to show that $\mathcal{M} \preceq \mathcal{M}'$. To this end, we consider an arbitrary formula $\varphi(x_1, \ldots, x_n)$ and suppose $\mathcal{M} \models \varphi(a_1, \ldots, a_n)$. This is to say that $\varphi(c_{a_1}, \ldots, c_{a_n}) \in \mathcal{ED}(\mathcal{M})$. By hypothesis, $\mathcal{N}_{\mathcal{M}} \models \varphi(c_{a_1}, \ldots, c_{a_n})$. Using the fact that $c_{a_1}^{N_M}, \ldots, c_{a_n}^{N_M} \in N$, we obtain:

$$\mathcal{N}_{\mathcal{M}} \models \varphi(c_{a_1}, \ldots, c_{a_n}) \iff \mathcal{N} \models \varphi(c_{a_1}^{N_M}, \ldots, c_{a_n}^{N_M})$$
$$\iff \mathcal{N} \models \varphi(h(a_1), \ldots, h(a_n))$$
$$\iff \mathcal{M}' \models \varphi(a_1, \ldots, a_n)$$

where the last equivalence follows from the fact that h is an isomorphism. If, on the other hand, $\mathcal{M}' \models \varphi(a_1, \ldots, a_n)$, the last argument immediately implies $\mathcal{N}_{\mathcal{M}} \models \varphi(c_{a_1}, \ldots, c_{a_n})$. The condition $\mathcal{N}_{\mathcal{M}} \models \mathcal{ED}(\mathcal{M})$ now implies $\varphi(c_{a_1}, \ldots, c_{a_n}) \in \mathcal{ED}(\mathcal{M})$, i.e. $\mathcal{M} \models \varphi(a_1, \ldots, a_n)$. We have shown that, for any $\mathcal{L}$-formula $\varphi(x_1, \ldots, x_n)$:

$$\mathcal{M} \models \varphi(a_1, \ldots, a_n) \text{ iff } \mathcal{M}' \models \varphi(a_1, \ldots, a_n).$$

This is to say that the inclusion of $\mathcal{M}$ into $\mathcal{M}'$ is elementary, i.e. $\mathcal{M} \preceq \mathcal{M}'$.

Finally, we show that (c) implies (a). Let f be an isomorphism from $\mathcal{M}'$ to $\mathcal{N}$. The inclusion map $\iota_\mathcal{M}$ is an elementary embedding of $\mathcal{M}$ into $\mathcal{M}'$, by definition of elementary extension. The map $f \circ \iota$ is the composition of two elementary embeddings. It is therefore an elementary embedding of $\mathcal{M}$ into $\mathcal{N}$. □

Exercise 4.50 In the proof of Theorem 4.49, verify that h is a strong homomorphism.

Let $\mathcal{M}$ be a $\mathcal{L}$-structure that models the theory T. The last two theorems specify what we have to prove in order to show the existence of models of T that are extensions or, respectively, elementary extensions of $\mathcal{M}$. For extensions, we need to show that $T \cup \mathcal{D}(\mathcal{M})$ has a model and, for elementary extensions, that $T \cup \mathcal{ED}(\mathcal{M})$ does. In typical cases both sets of sentences are infinitely large. Any results that should simplify the problem of determining whether these sets have any models are therefore very useful. A key theorem offering the desired simplification is the **compactness theorem**, which reduces the existence of global models (for a whole set of sentences) to the existence of local models (for each finite subset).

4.4 Addendum: Potential Truth

What we have learned so far about substructures and extensions allows us to examine a philosophical contribution of Abraham Robinson's contained in [61].[2] In this paper, Robinson is concerned with the fact that mathematical work very frequently deals with infinite totalities like the set of natural numbers $\mathbb{N}$, the set of integers $\mathbb{Z}$, or the set of continuous functions defined on a fixed interval, among many possible examples. In fact, we have already encountered theories with infinite models and we easily come across theories that have only infinite models (e.g. the theory of non-zero vector spaces over the field of rational numbers). We have also seen that the logical framework we adopt to talk about structures involves references to infinite totalities. For instance, as noted at the end of Chap. 2, the set of all formulae in a language $\mathcal{L}$ is infinite even when the signature of $\mathcal{L}$ is finite.

Robinson draws a contrast between finite objects that can have a direct reference or a concrete presentation and infinite totalities that lack a direct reference. One may, for instance, concretely describe the group S_3 as a lists of permutations of three letters whose action can be specified, but it is no longer possible to offer a similar 'implementable' description of the symmetric group of an infinite set X, i.e. the group of all permutations acting on X.

Robinson's attitude towards the simultaneous presence, within mathematics, of objects of investigation that admit of a direct reference and objects that do not, consists in taking assertions concerning the latter as 'meaningless' (where 'meaning' is understood as direct reference to a specifiable or presentable object). The result of this attitude is not to dismiss or somewhat exclude infinitary mathematics from mathematical work, but to offer an explication of its function and of its connection with the objects supporting a direct reference.

It is on one aspect of this connection that we shall focus in the remainder of this section. Its significance is evinced by the fact that formal or abstract mathematical results, which involve the notion of infinity, may have practical consequences. This is clear in the domain of applied mathematics, but the same phenomenon occurs within pure mathematics. In the latter context, Robinson offers a number-theoretical example that we now describe. The axiomatic theory known as Peano Arithmetic[3] implies, among its consequences, the following sentence, originally proved by Euler:

---

[2] Closely related and of equal interest is [62].

[3] A list of the axioms may for instance be found in [26], p.33.

4.4 Addendum: Potential Truth

$$\neg \exists x \exists y \exists z (\neg(x=0) \wedge \neg(y=0) \wedge \neg(z=0) \wedge x^3 + y^3 = z^3).$$

Robinson refers to the last sentence as a formal theorem, i.e. a result that depends on T, a theory whose models are infinite, rather than on directly interpretable or more concrete principles. Nonetheless, the sentence implies that in practice no integers satisfying the relation $x^3 + y^3 = z^3$ will be found. In other words, any attempt at concretely exhibiting specific integers will be accompanied by a failure of the last relation. The latter fact is what Robinson understands to be a practical consequence of the formal theorem, which refers to the infinite totality of integers.

From Robinson's standpoint it is thus a noteworthy fact that 'formal' (i.e. not directly or concretely interpretable) statements can have an impact upon practice. This fact makes it especially desirable to find a way of showing how the formal truth of such principles might arise from the concrete basis of mathematics or logic. A partial attempt at realising this project appears in the Appendix of [61], which shows how, by model-theoretic means, it may be possible to reduce the truth of a statement in an infinite structure to the 'potential truth' of the same statement in its finitely generated substructures.

In the remainder of this section we shall explain how this reduction works: that it is only a partial realisation of Robinson's project depends on the fact that the set of finitely generated substructures of an infinite structure may well be infinite. Robinson restricts attentions to $\mathcal{L}$-structures, where $\mathcal{L}$ is a relational language, i.e. a language whose signature only contains relation symbols. If $\mathcal{M}$ is a $\mathcal{L}$-structure and $A \subseteq M$, A is automatically the domain of a $\mathcal{L}$-structure, obtained by intersecting any k-ary relation $R^{\mathcal{M}}$ of $\mathcal{M}$ with A^k. In particular, the finite subsets of M are already domains of finite substructures.

The central notion on which Robinson's model-theoretic reduction of truth to potential truth rests is that of a direct system.

Definition 4.51 A set Δ of $\mathcal{L}$-structures is a **direct system** relative to inclusion[4] iff for any $\mathcal{A}, \mathcal{B} \in \Delta$ there is $\mathcal{C}$ such that $\mathcal{A}, \mathcal{B} \sqsubseteq \mathcal{C}$.

In other words, any two structures in a direct system have a common extension.

Example 4.52 Let $\mathcal{L}$ be a relational language whose signature contains only the ternary relation symbols Add and $Mult$. We consider the $\mathcal{L}$-structure $\mathcal{N} = (\mathbb{N}, Add^{\mathbb{N}}, Mult^{\mathbb{N}})$, whose relations $Add^{\mathbb{N}}, Mult^{\mathbb{N}} \subseteq \mathbb{N}^3$ describe addition and multiplication respectively. The subsets of $\mathbb{N}$ of the form $\{0, \ldots, k-1\}$ $(1 \leq k)$ are domains of substructures of $\mathcal{N}$. Let Δ be the set of all such substructures: then Δ is a direct system relative to inclusion. We may thus view $\mathcal{N}$ as approximated by the direct system of its finite initial segments.

Remark 4.53 For any relational language $\mathcal{L}$, the set of finite substructures of a fixed structure $\mathcal{M}$ is a direct system relative to inclusion.

---

[4] The notion of a direct system can be described in more general terms by means of morphisms that are not necessarily inclusions but in the present context it suffices to focus on inclusions.

A direct system determines a structure that is unique up to isomorphism, known as its **direct limit**. The direct limit of a direct system Δ relative to inclusion is easily described. To this end, we represent Δ as the family of $\mathcal{L}$-structures $\{\mathcal{A}_i\}_{i \in I}$. The direct limit of Δ is a structure based on $M = \bigcup_{i \in I} A_i$. Suppose $\mathcal{L}$ contains the k-ary relation symbol R. Then $R^M a_1 \ldots a_k$ iff there is $\mathcal{A}_i$ such that $\{a_1, \ldots, a_k\} \subseteq A_i$ and $R^{\mathcal{A}_i} a_1 \ldots a_k$. The reader may check that the parametric atomic formula $R a_1 \ldots a_k$ is modelled by some $\mathcal{A}_i$ whose domain includes the given, finite set of parameters iff it is modelled by all such structures in Δ.

As we shall shortly see, direct systems lift the following notion of potential truth to truth to their direct limits.

Definition 4.54 Let $\mathcal{L}$ be a relational language, $\Delta = \{\mathcal{A}_i\}_{i \in I}$ a direct system of $\mathcal{L}$-structures relative to inclusion. If $\overline{a}$ is a tuple of parameters from A_i and $\varphi(\overline{a})$ a parametric $\mathcal{L}$-formula, then:

(a) if $\varphi(\overline{a})$ is atomic, then $\varphi(\overline{a})$ is potentially true in $\mathcal{A}_i$ iff $\mathcal{A}_i \models \varphi(\overline{a})$;
(b) if $\varphi(\overline{a}) := \theta_1(\overline{a}) \vee \theta_2(\overline{a})$ then $\varphi(\overline{a})$ is potentially true in $\mathcal{A}_i$ iff $\theta_1(\overline{a})$ is or $\theta_2(\overline{a})$ is;
(c) of $\varphi(\overline{a}) := \exists x \psi(x, \overline{a})$ then $\varphi(\overline{a})$ is potentially true in $\mathcal{A}_i$ if there is $\mathcal{A}_k \sqsupseteq \mathcal{A}_i$ such that, for some $b \in A_k$, $\psi(b, \overline{a})$ is potentially true in every extension of $\mathcal{A}_k$.

When referred to a finite context, fixed by a parametric atomic formula, potential truth coincides with truth. When existential quantification intervenes, potential truth requires a shift to an extended finite context, a sort of enlargement of the domain of enquiry.

The next theorem shows that potential truth proliferates throughout a direct system Δ.

Lemma 4.55 *Let $\mathcal{A}_i \sqsubseteq \mathcal{A}_k$ be elements of the direct system of $\mathcal{L}$-structures Δ. For any tuple $\overline{a}$ whose elements are included in A_i and any $\mathcal{L}$-formula $\varphi(\overline{a})$, $\varphi(\overline{a})$ is potentially true in $\mathcal{A}_i$ iff it is potentially true in $\mathcal{A}_k$.*

Proof The assertion is proved by an induction on the complexity of formulae. We leave the inductive base and the inductive step for connectives to the reader. As for the existential quantifier, let us first suppose that $\varphi(\overline{a})$ is of the form $\exists x \psi(x, \overline{a})$ and that it is potentially true in $\mathcal{A}_i$. In this case there is an extension $\mathcal{A}_j$ of $\mathcal{A}_i$ such that $b \in A_j$ and $\psi(b, \overline{a})$ is potentially true in every extension of $\mathcal{A}_j$. Because Δ is direct, there is a common extension of $\mathcal{A}_k, \mathcal{A}_j$, say $\mathcal{A}_s$. The formula $\psi(b, \overline{a})$ is potentially true in every extension of $\mathcal{A}_s$ because $\mathcal{A}_s$ extends $\mathcal{A}_j$. We have thus found an extension of $\mathcal{A}_k$ such that $\psi(b, \overline{a})$ is potentially true in all of its extensions. We conclude that $\exists x \psi(x, \overline{a})$ is potentially true in $\mathcal{A}_k$.

If, conversely, $\exists x \psi(x, \overline{a})$ is potentially true in $\mathcal{A}_k$, then $\psi(b, \overline{a})$ is potentially true in all extensions of some $\mathcal{A}_s \sqsupseteq \mathcal{A}_k$ such that $b \in A_s$. But $\mathcal{A}_s$ is already an extension of $\mathcal{A}_i$ so $\exists x \psi(x, \overline{a})$ is potentially true in $\mathcal{A}_i$. □

If $\overline{a}$ is a fixed tuple and $\varphi(\overline{a})$ a parametric formula, then the latter formula holds or fails simultaneously in every structure from Δ whose domain contains the

4.4 Addendum: Potential Truth

parameters in $\bar{a}$. In the light of this observation we proceed to deduce the announced reduction of truth to potential truth.

Theorem 4.56 *Let $\mathcal{M}$ be the direct limit of a direct system Δ of $\mathcal{L}$-structures and $\bar{a}$ be a tuple of elements of M. A formula $\varphi(\bar{a})$ is potentially true in every $\mathcal{L}$-structure from Δ containing $\bar{a}$ iff $\mathcal{M} \models \varphi(\bar{a})$.*

Proof We proceed again by an induction on the complexity of formulae, leaving the inductive base and inductive step for connectives to the reader.

Suppose $\varphi(\bar{a})$ is of the form $\exists y \psi(x, \bar{a})$. If $\mathcal{M} \models \exists x \psi(x, \bar{a})$, there is $b \in M$ such that $\mathcal{M} \models \psi(b, \bar{a})$. We fix $\mathcal{A}_i \in \Delta$ in which the tuple of parameters $\bar{a}$ occurs. Because $\mathcal{M}$ is a direct limit, b occurs in the domain of some $\mathcal{A}_j$. The fact that Δ is a direct system implies the existence of an extension $\mathcal{A}_j$ of $\mathcal{A}_i$ containing b. By the inductive hypothesis, $\psi(B, \bar{a})$ is potentially true in every extension of $\mathcal{A}_j$: we conclude that $\exists x \psi(x, \bar{a})$ is potentially true in $\mathcal{A}_i$. Since $\mathcal{A}_i$ was arbitrarily chosen, $\exists x \psi(x, \bar{a})$ is potentially true in every element of Δ in which the parameters $\bar{a}$ occur.

To complete the proof, we suppose that $\exists x \psi(x, \bar{a})$ is potentially true in some $\mathcal{A}_i$ from the set Δ (and thus, by the remark preceding this theorem, in every structure containing the parameters in $\bar{a}$). then $\psi(b, \bar{a})$ is potentially true in some extension of $\mathcal{A}_i$ containing $b, \bar{a}$. It is therefore potentially true in every element of Δ containing $b, \bar{a}$, by the remark preceding this theorem. The inductive hypothesis yields $\mathcal{M} \models \psi(b, \bar{a})$, which in turn implies $\mathcal{M} \models \exists x \psi(x, \bar{a})$. □

Chapter 5
Ultraproducts

Abstract In this chapter we prove the compactness theorem using a model-theoretic construction known as an ultraproduct.

To motivate the introduction of ultraproducts, we briefly revisit ideas from Chap. 1. An ultraproduct arises from a family of factor structures, which may be conceived as voters choosing which first-order formulae will hold in the product structure.

The reference to elections is not purely metaphorical. There is an interesting connection between pairwise decision methods and ultraproducts, which we explore in Sect. 5.1. The ultraproduct construction is given in Sect. 5.2, where the fundamental theorem governing the first-order properties of ultraproducts (i.e. Łos' theorem) is proved.

Instead of moving immediately to compactness, in Sect. 5.3 we obtain the field of real numbers from an ultraproduct construction. The reader who has never seen a rigorous construction of the reals will find this section helpful for its own sake and as a point of reference for the discussion of real closed fields in Chap. 11. The reader who is familiar with a rigorous construction of the reals is likely not to have encountered the one we describe and may find it of some interest.

Essentially the same ultraproduct construction employed in Sect. 5.3 can be iterated on the real field to obtain an elementary extension containing infinitely small and large numbers. In Sect. 5.4 we provide a brief taster of the model-theoretic approach to real analysis afforded by this expansion of numerical resources.

We finally prove compactness in a few equivalent forms in Sect. 5.5 and proceed to obtain from it a characterisation of elementary classes (introduced in Sect. 3.1) in terms of ultraproducts.

The addendum, i.e. Sect. 5.7, discusses a little known but remarkable application of ultrafilters due to the philosopher Clarence Irving Lewis.

5.1 Product Structures and Voting

Our first goal in this chapter is to describe a model-theoretic construction that allows us to build a single product $\mathcal{L}$-structure out of a family of possibly distinct factor

$\mathcal{L}$-structures in such a way that the first-order information carried by the factor structures determines which $\mathcal{L}$-formulae are satisfied with any given assignment by the product structure.

Given a family $\{\mathcal{A}_i\}_{i \in I}$ of factor $\mathcal{L}$-structures indexed over a set I, we take their product structure to be based on the Cartesian product $\times_{i \in I} A_i$. If u is an element of the last Cartesian product (i.e. a function from I to $\bigcup_{i \in I} A_i$), let u_i be its i-th component or i-the projection.

We aim to equip the Cartesian product $\Pi_{i \in I} A_i$ with $\mathcal{L}$-structure in such a way that its relations, functions and constants can be defined in terms of the relations, functions and constants carried by the family $\{\mathcal{A}_i\}_{i \in I}$. Moreover, we would like to determine whether or not a $\mathcal{L}$-formula holds under an assignment in the product structure by looking at the projections of the assignment and the corresponding satisfaction relation in the factor structures.

This twofold goal can be achieved if, instead of working with $\Pi_{i \in I} A_i$, we work with a suitable quotient of this Cartesian product. In order to determine what equivalence relation might deliver the desired quotient, it is helpful to consider a simple example in the language $\mathcal{L}_<$, which also gives rise to a celebrated theorem of social choice theory.

Let $\{\mathcal{A}_1, \mathcal{A}_2, \mathcal{A}_3\}$ be distinct linear orderings of the same three-element set $\{a, b, c\}$. In particular, let $\mathcal{A}_1$ be the ordering $a <_1 b <_1 c_1$, $\mathcal{A}_2$ be the ordering $b <_2 c <_2 a$, and $\mathcal{A}_3$ be the ordering $b <_3 a <_3 c$.

We would like to determine a $\mathcal{L}_<$-structure with domain $\times_{i \in \{1,2,3\}} A_i$ that is also a model of LO. At least two problems immediately arise. First, we need a rule to guarantee that trichotomy holds. For instance, we wish to compare $u = (a, b, c)$ and $v = (b, c, a)$, viewed as functions from $\{1, 2, 3\}$ to $\{a, b, c\}$. Since $u(1) <_1 v(1)$ and $u(2) <_2 v(2)$ but $v(3) <_3 u(3)$, we cannot read the relative ranking of u, v off the last inequalities. We may however determine it by mans of a majority election. In this case, because two out of three projections rank the first component below the second component, we should have $u < v$ in the product structure.

Majority rule thus helps, but it is not sufficient to resolve the difficulties of the case. When $\mathcal{A}_1, \mathcal{A}_2, \mathcal{A}_3$ describe the Condorcet's paradox discussed in Chap. 1, majority rule prevents the product structure from modelling transitivity, once we consider the constant pairs (a, a, a), (b, b, b) and (c, c, c).

One way to overcome this problem consists again in viewing the index set $\{1, 2, 3\}$ as a set of voters, this time equipped with an associated family of decisive coalitions. If all voters in a decisive coalition agree on a particular ranking, then what the remaining voters say is immaterial. The decisive coalition approach provides an intuitive route into the abstract notion of an *ultrafilter* over a set, which will be defined in the next section, where it plays a key role in constructing a well-behaved product $\mathcal{L}$-structure from a, typically infinite, family of $\mathcal{L}$-structures.

In this section we deduce a set of formal properties that a family of decisive coalitions is to satisfy if it is to determine a linear order on a Cartesian product of linear orders. This set of formal properties determines precisely an ultrafilter of coalitions over the set of voters.

5.1 Product Structures and Voting

We can essentially deduce the relevant properties from a single decomposability assumption on the design of a voting procedure. As we shall see in the subsection to follow, this particular design always trivialises products over a finite set of preferences. Because our natural setting involves products of infinitely many structures, this latter inconvenience will not affect us: it is however of interest to social choice theorists, as we shall shortly see.

5.1.1 Pairwise Aggregations

We now return to the scenario illustrated in Sect. 1.1, in which a family I of voters or decision-makers produces a set of distinct rankings of a fixed set of alternatives A.

More formally, we let L be the set of linear orders over A. The elements of L^I, the set of Cartesian products of I linear orders over A, are usually called *preference profiles*: generic profiles will be denoted by $\Pi A_i, \Pi B_j$.[1]

We suppose that a given function $f : \mathsf{L}^I \longrightarrow \mathsf{L}$ assigns one linear order over A to each preference profile: we can, in other words, *aggregate* the preferences in any profile into a single linear ordering of the elements of A. We thus refer to f as an **aggregation**.

In the spirit of Chap. 1, we restrict attention to aggregations based on pairwise comparisons or **pairwise aggregations**. To this end, we require that an aggregation f be decomposable into a collection of pairwise comparisons between alternatives.

Definition 5.1 Let $A, I, \mathsf{L}, \mathsf{L}^I, f$ be as above. We say that f is a **pairwise aggregation** iff, for any $a, b \in A$ and any profiles $\Pi A_i, \Pi B_i$:

$$\{i \in I : \mathcal{A}_i \models a < b\} = \{i \in I : \mathcal{B}_i \models a < b\}$$

implies $(f(\Pi A_i) \models a < b \iff f(\Pi B_i) \models a < b)$.

When an aggregation f is pairwise, we can determine how it ranks any two alternatives simply by looking at the pairwise procedure obtained from f by restricting it to the subset $\{a, b\} \subseteq A$. In other words, f may be viewed as a collection of pairwise procedures.

We are now ready to define decisive coalitions.

Definition 5.2 A set $C \subseteq I$ is a **decisive coalition** for the aggregation f, the profile ΠA_i and the ordered pair (a, b), with $a, b \in A$ iff $\{i \in I : \mathcal{A}_i \models a < b\} = C$ implies $f(\Pi A_i) \models a < b$.

Exercise 5.3 Show that, if C is a decisive coalition for f, (a, b) and a given profile in L^I, then it is a decisive coalition for f, a, b and every profile in L^I.

---
[1] Our discussion in the present section relies upon [38] and [15].

By Exercise 5.3, we can take decisive coalitions to depend on a procedure f and a pair of alternatives a, b alone. As the next result shows, any fixed pair of alternatives determines a family of decisive coalitions corresponding to that pair.

Lemma 5.4 *Let f be a pairwise aggregation, $a, b \in A$ a pair of alternatives: f determines a family $\mathcal{C}_{a,b} \subseteq \mathcal{P}(I)$ of decisive coalitions relative to (a, b).*

Proof For each profile ΠA_i, we consider the subsets of I of the form $C_{a,b}(\Pi A_i) = \{i \in I : A_i \models a < b\}$. Let C be the family of all sets $C_{a,b}(X)$, as X ranges over the set of profiles. We define a function $g : C \longrightarrow \{0, 1\}$ as follows:

$$g(C_{a,b}(\Pi A_i)) = \begin{cases} 1 \text{ if } f(\Pi A_i) \models a < b \\ 0 \text{ if } f(\Pi A_i) \models \neg(a < b) \end{cases}$$

Obviously, if $i \in C' = C_{a,b}(\Pi A_i)$ and $a < b$ is satisfied by $\mathcal{A}_i$, then $b < a$ is automatically satisfied by A_j for any $j \in I - C_{a,b}(\Pi A_i)$. If we restrict attention to a, b, this situation arises in every profile in which $\mathcal{A}_i \models a < b$ iff $i \in C'$. We conclude that g is independent of the choice of profile, i.e. it is an actual (single-valued) function.

We now define $\mathcal{C}_{a,b} = g^{-1}(1) = \{C_{a,b}(\Pi A_i) : f(\Pi A_i) \models a < b\}$. It remains to check that $\mathcal{C}_{a,b}$ is a family of decisive coalitions for (a, b). If $C \in \mathcal{C}_{a,b}$, then $C = C_{a,b}(\Pi A_i)$. Pick any profile ΠB_i. If, for every $i \in C_{a,b}$, $\mathcal{B}_i \models a < b$, then the pairwise election on $f \{a, b\}$ is fixed and its outcome is identical with $f(\Pi A_i)$ because f is a pairwise aggregation.

If, on the other hand, C is decisive relative to (a, b), there is a profile in L^I such that $C = \{i \in I : A_i \models a < b\}$. Clearly $g(C) = 1$, i.e. $C \in \mathcal{C}_{a,b}$. □

Decisive coalitions exhibit distinctive features in presence one additional, non-triviality assumption on pairwise aggregations:

Definition 5.5 An aggregation $f : \mathsf{L}^I \longrightarrow \mathsf{L}$ is **unanimous** iff I is decisive on every pair of alternatives.

If a pairwise aggregation is unanimous, it can generate every linear order on A. The assumption of unanimity is thus of significance because it ensures that enough distinct outcomes can arise. It excludes from consideration trivial pairwise methods, e.g. those yielding only a fixed outcome. Unanimous, pairwise aggregations have absolute decisive coalitions, i.e. no longer relative to a fixed pair of alternatives.

Lemma 5.6 *Let f be a unanimous, pairwise aggregation and a, b, c, d alternatives in A that are all distinct. Then $\mathcal{C}_{a,b} = \mathcal{C}_{c,d}$, i.e. a coalition is decisive relative to (a, b) iff it is decisive relative to (c, d).*

Proof It suffices to show that $\mathcal{C}_{a,b} \subseteq \mathcal{C}_{c,d}$, since the same argument, with the pairs of alternative interchanged, leads to the reverse inclusion. Suppose $C \in \mathcal{C}_{a,b}$ and that, for every $i \in C$ and a given profile ΠA_i, $\mathcal{A}_i \models c < d$. This is to say that, for any $j \in I - C$, $A_j \models d < c$.

5.1 Product Structures and Voting

We consider the profile ΠB_i such that, for each $i \in C$, $B_i \models c < d \wedge a < b$ and, for each $j \in I - C$, $B_j \models c < a \wedge b < d$. Because f is pairwise, $f(\Pi A_i)$ and $f(\Pi B_i)$ must model the same inequality involving c, d. We can compute this inequality with reference to ΠB_i.

The fact that C is decisive over (a, b) implies $f(\Pi B_i) \models a < b$. The fact that f is unanimous further implies $f(\Pi B_i) \models c < a \wedge b < d$. Because $f(\Pi B_i)$ is a linear order, it follows from $c < a$, $a < b$ and $b < d$ that $c < d$. Thus $f(\Pi A_i) \models c < d$ and C is decisive on (c, d). □

Exercise 5.7 Let f be a pairwise, unanimous aggregation.

(i) Show that, if C is a decisive coalition relative to (a, b) and $c \neq a, b$, then C is decisive relative to (c, b).
(ii) Show that, if C is a decisive coalition relative to (c, b) and $a \neq c, b$, then C is decisive relative to (c, a).
(iii) Show that, if C is a decisive coalition relative to (a, b), then C is decisive relative to (b, a).

In view of Lemma 5.6 and Exercise 5.7, we can now simply talk about the family $\mathcal{C}$ of decisive coalitions determined by a pairwise, unanimous aggregation f, without further relativising the notion. The following three lemmas establish key properties of decisive coalitions.

Lemma 5.8 *If $C \subseteq I$, either C is decisive or $D = I - C$ is.*

Proof Because decisiveness on a pair implies decisiveness on every pair, it is sufficient to fix a pair (a, b) and show that, if C is not decisive for (a, b), then D is. In the former case, there is a profile ΠA_i with $C = \{i \in I : A_i \models a < b\}$ and $f(\Pi A_i) \models b < a$. In this case D is decisive relative to (a, b) and the profile ΠA_i. Consequently, D is decisive. □

Lemma 5.9 *If $C \subseteq I$ is decisive and $C \subseteq D$, then D is decisive.*

Proof We again fix a pair (a, b) and show that, if C is decisive for (a, b), so is D. If $D = I$, there is nothing to prove. Otherwise, we suppose that D is not decisive. In this case, there is a profile ΠA_i such that $D = \{i \in I : A_i \models a < b\}$ and $f(\Pi A_i) \models b < a$. The set $I - D$ is nonempty and, for every $j \in I - D$, $A_j \models b < a$. It follows that $I - D$ is decisive for (b, a). We obtain a contradiction by considering an arbitrary profile ΠB_i in which $C = \{i \in I : B_i \models a < b\}$ and $I - D = \{j \in I : B_j \models b < a\}$. Since C, D are disjoint, such a profile exists but $f(\Pi B_i) \notin L$ because $f(\Pi B_i) \models a < b \wedge b < a$. We conclude that D is decisive. □

Lemma 5.10 *If $C, D \subseteq I$ are decisive, then $C \cap D$ is decisive.*

Proof We rely on the fact that A contains at least three distinct alternatives a, b, c. We consider a profile in which $C = \{i \in I : A_i \models a < b\}$, $D = \{j \in I : A_i \models b < c\}$ and $C \cap D = \{k \in I : A_k \models a < c\}$. The arrangement just described is possible because the relative position of a, b or of b, c does not affect the relative position of a, c. In the profile just described, the decisiveness of C, D and the fact that $f(\Pi A_i) \in L$ imply $f(\Pi A_i) \models a < c$. It follows that $C \cap D$ is decisive for (a, c) and thus decisive. □

We collect the last lemmas and a trivial remark about the empty coalition into a single statement:

Theorem 5.11 *Let f be a unanimous pairwise aggregation and $\mathcal{U} \subseteq \mathcal{P}(I)$ be the set of decisive coalitions determined by f. Then, for any $C, D \in \mathcal{P}(I)$:*

(a) $\emptyset \notin \mathcal{U}$;
(b) *if $C \in \mathcal{U}$ and $C \subseteq D$, then $D \in \mathcal{U}$;*
(c) *if $C, D \in \mathcal{U}$, then $C \cap D \in \mathcal{U}$;*
(d) *if $C \notin \mathcal{U}$, then $I - C \in \mathcal{U}$.*

Given a nonempty set I, a family $\mathcal{U} \subseteq \mathcal{P}(I)$ that satisfies properties (a) to (c) above is called a **proper filter** over I. A family that satisfies properties (a) to (d) is called an **ultrafilter** over I. What Theorem 5.11 states is that any unanimous, pairwise aggregation determines an ultrafilter of decisive coalitions.

Exercise 5.12 Let I be a nonempty set and j be a fixed element of I. Show that the family $\mathcal{U}_j = \{C \subseteq I : j \in C\}$ is an ultrafilter. We call $\mathcal{U}_j$ the **principal** ultrafilter generated by $\{j\}$.

Exercise 5.13

(i) A (proper) strong simple game $\mathcal{G}$ over a nonempty set I is a family of subsets that satisfies the following three conditions: (i) for any $A, B, C \in \mathcal{G}$, $A \cap B \cap C \neq \emptyset$; (ii) if $A \in \mathcal{G}$ and $A \subseteq B$, then $B \in \mathcal{G}$; (iii) $A \in \mathcal{G}$ iff $I - A \notin \mathcal{G}$. Show that $\mathcal{G}$ is an ultrafilter.
(ii) Let $\mathcal{U}$ be an ultrafilter over a nonempty set I and $C \in \mathcal{U}$. Suppose that $C = D \cup E$ and that $D \cap E = \emptyset$. Show that $C \in \mathcal{U}$ or $E \in \mathcal{U}$ (*Hint*: use property (d)).
(iii) Use part (ii) to show that, if $\mathcal{U}$ is an ultrafilter over a finite, nonempty set I, then $\mathcal{U}$ is principal (*Hint*: show that, for some $j \in I$, $\{j\} \in \mathcal{U}$ by successively removing elements from I).

In view of the Exercise 5.13-(iii), we obtain:

Theorem 5.14 (Arrow's Theorem) *Let A, I be finite, nonempty sets such that $|A| \geq 3, |I| \geq 2$. If $f : L^I \longrightarrow L$ is a unanimous, pairwise aggregation, there is a fixed $j \in I$ such that, for every profile ΠA_i, $f(\Pi A_i) = A_j$.*

Exercise 5.15 Let A, I be as in Arrow's theorem. If $\mathcal{U}$ is an ultrafilter over I, define a function f with domain $\mathcal{L}^I$ and codomain the set of $\mathcal{L}_<$-structures based on A as follows:

for any $a, b \in A$, $f(\Pi A_i) \models a < b$ iff there is $V \in \mathcal{U}$ such that $i \in V$ implies $A_i \models a < b$.

Show that f is a unanimous, pairwise aggregation and that its values are linear orders on the domain A.

Together with Theorem 5.11, the last exercise tells us that the decision procedures that determine an ultrafilter of decisive coalitions are exactly the unanimous, pairwise aggregations. When only finitely many voters are involved, Arrow's theorem tells us that a unanimous pairwise aggregation coincides with the choice of a fixed individual.

This result has often been interpreted as a statement to the effect that, with enough alternatives (at least three) and agents (at least two), no voting method is fair. As Donald Saari has pointed out in [68], pp.136–137, the significant issue highlighted by Arrow's theorem is rather that no *pairwise* method is fair. Pairwise methods ignore by design the fact that individual preferences are transitive, since transitivity can only be expressed with reference to two distinct pairs of alternatives, not one. In any concrete context (i.e. when a population is not infinite), pairwise methods can make use of transitivity only in a trivial manner, i.e. by uniformly identifying the aggregate linear order with one individual linear order.

One may therefore take Arrow's theorem as a warning against using pairwise methods or as a recommendation to the effect that such methods should be supplemented with techniques that correct their shortcomings (e.g. the approach we studied in Chap. 1).

From an abstract point of view, i.e. if we allow infinite populations, we do not obtain a generalisation of Arrow's theorem because, as we shall see in the next section, Zorn's lemma[2] guarantees the existence of non-principal ultrafilters over any infinite set.

We can take advantage of them to obtain *ultraproducts* of $\mathcal{L}$-structures, which are constructed by taking quotients of Cartesian products relative to equivalence relations defined in terms of certain ultrafilters. The next section provides full details.

5.2 Ultraproducts

Given an indexed family of $\mathcal{L}$-structures $\{\mathcal{A}_i\}_{i \in I}$, we define an equivalence relation on the Cartesian product $\Pi_{i \in I} A_i$ that makes use of 'decisive coalitions' over I. If u, v are elements of $\Pi_{i \in I} A_i$, we identify them if a decisive coalition included in I says that they are equal. More formally, we identify u, v if the set $\{i \in I : u_i = v_i\}$ is a decisive coalition, i.e. an element of an ultrafilter over I (we take u_i to be the same as $u(i)$, i.e. the value of the function u at i).

So far we only know that, if I is finite, its ultrafilters are bound to contain a singleton or one-element set, say $\{j\}$, for some $j \in I$. Filters containing singletons are called **principal**. In this case the equivalence relation on $\Pi_{i \in I} A_i$ forces this domain to collapse on A_j because any u, v that agree on the j-th projection are going to be identified.

---
[2] The reader unfamiliar with this result should consult Sect. A.6.1 from the Appendix.

With an infinite index set I, we are able to avoid trivialisation only if we can find an ultrafilter over I that is not principal. In the next subsection we show that, when I is infinite, non-principal ultrafilters are always available.

5.2.1 Cofinite Subsets of I

Let I be an infinite set. A subset $V \subseteq I$ is **cofinite** iff $I - V$ is finite. Let $\mathcal{F}$ be the family of cofinite subsets of I. It is left to the reader to verify that:

(a) $\emptyset \notin \mathcal{F}$;
(b) $I \in \mathcal{F}$;
(c) $U, V \in \mathcal{F}$ implies $U \cap V \in \mathcal{F}$;
(d) $U \in \mathcal{F}$ and $U \subseteq V$ imply $V \in \mathcal{F}$.

As we know, a subset of $\mathcal{P}(I)$ with the properties $(a) - (d)$ is a **proper filter** over I. We now restate the definition of ultrafilter.

Definition 5.16 Let I be a nonempty set. An **ultrafilter** $\mathcal{U}$ over I is a proper filter over I that satisfies the following condition:

$$\text{for every } U \in \mathcal{P}(I), \text{ either } U \in \mathcal{U} \text{ or } I - U \in \mathcal{U}.$$

If an ultrafilter $\mathcal{U}$ contains every cofinite subset of I, then it cannot be principal. If it were, it would contain $\{j\}$, for some $j \in I$ and, since $I - \{j\}$ is cofinite, $\{j\}, I - \{j\} \in \mathcal{U}$, which implies $\emptyset \in \mathcal{U}$, against the definition of ultrafilter. Thus, one way to prove the existence of non-principal ultrafilters over an infinite set I consists in showing that, in particular, the filter of cofinite subsets can be extended to an ultrafilter. One way of proving this result relies on Zorn's lemma (see Appendix, Sect. A.6 for a statement and proof). To apply the lemma, we have to identify a suitable partially ordered set and verify that any chain (i.e. linearly ordered subset) has an upper bound, i.e. that the partial order contains an element not smaller than any chain element. In the present context, we consider the set of filters over I, which we shall call $\mathscr{F}$: this set is clearly partially ordered by inclusion. Since, as we are going to check, the chain condition just stated holds in $\mathscr{F}$, Zorn's lemma allows us to deduce the existence of a *maximal* proper filter relative to inclusion. Such a filter is a strict subset of $\mathcal{P}(I)$ not included in any other filter. A maximal proper filter is exactly what we need because the ultrafilters over I are exactly the maximal proper filters in $\mathscr{F}$. It is easy to see that the only filter over I that strictly includes any fixed ultrafilter is $\mathcal{P}(I)$, which is not a proper filter because it contains $\emptyset$. Thus, no proper filter can extend an ultrafilter.

Exercise 5.17 Show that, if $\mathcal{U} \in \mathscr{F}$ is an ultrafilter, then $\mathcal{U}$ is a maximal element in the partial order $(\mathscr{F}, \subseteq)$.

It remains to show that the converse of the last exercise holds, i.e. that maximal proper filters are ultrafilters. To this end consider a maximal element $\mathcal{F}$ of $(\mathscr{F}, \subseteq)$,

5.2 Ultraproducts

select an arbitrary subset $U \subseteq I$ and suppose $U \notin \mathcal{F}$. In this case (see Exercise 5.18) there is $V \in \mathcal{F}$ such that $U \cap V = \emptyset$. The last equality is equivalent to $V \subseteq I - U$: by property (c), $I - U \in \mathcal{F}$ and $\mathcal{F}$ is an ultrafilter.

Exercise 5.18

(i) Let $\mathcal{F}$ be a proper filter and $U \subseteq I$ a set such that for each $V \in \mathcal{F}$, $U \cap V \neq \emptyset$. Show that the set:

$$\mathcal{F}_U = \{Z \subseteq I : \text{for some } V \in \mathcal{F}, U \cap V \subseteq Z\}$$

is a proper filter over I.

(ii) Let $\mathcal{F}$ be a maximal, proper filter over I. If for each $V \in \mathcal{F}$, $U \cap V \neq \emptyset$, prove that $U \in \mathcal{F}$ (use part (i)).

(iii) Prove that, if $\mathcal{F}$ is maximal and $U \notin \mathcal{F}$, then there is $V \in \mathcal{F}$ such that $U \cap V = \emptyset$.

We have established that the maximal filters over I, if any, are exactly the ultrafilters over I. We now use Zorn's lemma to show that any proper filter $\mathcal{F}$ over I can be extended to an ultrafilter.

Lemma 5.19 *Let I be a nonempty set and $\mathcal{F}$ a proper filter over I. There is an ultrafilter $\mathcal{U}$ over I such that $\mathcal{F} \subseteq \mathcal{U}$.*

Proof Let $\mathbb{F}$ be the family of filters over I that extend $\mathcal{F}$. This family is nonempty and it is partially ordered by inclusion. If $\{\mathcal{F}_j\}_{j \in J}$ is a chain of filters over I, each of which includes $\mathcal{F}$, the union $\bigcup_{j \in J} \mathcal{F}_J$ is an upper bound for it in $\mathscr{F}$. This is because $\bigcup_{j \in J} \mathcal{F}_J$ is a filter over I including $\mathcal{F}$, as the reader should verify. Since the hypotheses of Zorn's lemma are verified, $(\mathscr{F}, \subseteq)$ has a maximal element $\mathcal{U}$. By construction $\mathcal{F} \subseteq \mathcal{U}$. Thus, $\mathcal{U}$ is a maximal filter, and thus an ultrafilter, extending $\mathcal{F}$. □

Lemma 5.19 says that, whenever I is infinite, there exists a non-principal ultrafilter $\mathcal{U}$ over I. We next show how to use $\mathcal{U}$ to describe an equivalence relation on $\Pi_{i \in I} \mathcal{A}_i$ and induce relations, functions and constants on the resulting equivalence classes that yield a $\mathcal{L}$-structure $\Pi_\mathcal{U} \mathcal{A}_i$ called the **ultraproduct** of $\{\mathcal{A}_i\}_{i \in I}$.

The connection between an ultraproduct and the associated family of structures is given by Łos' theorem, the fundamental result proved in the next subsection.

5.2.2 The Ultraproduct Construction

We are finally in a position to describe the ultraproduct construction. Our data consists of a family of $\mathcal{L}$-structures $\{\mathcal{A}_i\}_{i \in I}$ and a non-principal ultrafilter $\mathcal{U}$ over the index set I. We first define the following equivalence relation on $\Pi \mathcal{A}_i$:

$$f \sim g \text{ iff } \{i \in I : f(i) = g(i)\} \in \mathcal{U},$$

where f, g are elements of the Cartesian product $\Pi_{i \in I} A_i$ and can therefore be viewed as functions from I to $\bigcup_{i \in I} A_i$ such that $f(i), g(i) \in A_i$. We use the notation $f(i)$ to indicate the i-th component or projection of f, which is its value at the argument i, when we regard f as a function.

We next let $\Pi A_i / \sim$ be the set of $\sim$-equivalence classes and $[f]$ be the $\sim$-equivalence class containing f. Since we want to use $\Pi A_i / \sim$ as the domain of a $\mathcal{L}$-structure, we need to define the relations, functions and constants it is to be equipped with.

Suppose first that $\mathcal{L}$ contains a k-ary relation symbol R. This symbol names the k-ary relation R^i on the structure $\mathcal{A}_i$. We define a k-ary relation $R^\sim$ on $\Pi A_i / \sim$ by requiring:

$$R^\sim([f_1], \ldots, [f_k]) \text{ iff } \{i \in I : \mathcal{A}_i \models R^i(f_1(i), \ldots, f_n(i))\} \in \mathcal{U},$$

Remark 5.20 From now on we adopt $[R^i(f_1(i), \ldots, f_k(i))]$ as shorthand to designate the set $\{i \in I : \mathcal{A}_i \models R^i(f_1(i), \ldots, f_n(i))\}$. We also extend this notation to arbitrary formulae whenever necessary. In other words, we take $[\varphi^i(f_1(i), \ldots, f_k(i))]$ to indicate the set $\{i \in I : \mathcal{A}_i \models \varphi^i(f_1(i), \ldots, f_n(i))\}$.

It is easy to see that $R^\sim$ is well-defined, i.e. that a relation does not hold or fail depending on which $\sim$-equivalence class representatives of its arguments we consider.

We now turn to the function symbols in $\mathcal{L}$. If $\mathcal{L}$ contains a m-ary function symbol F, this symbol names, for each $i \in I$, the m-ary function F^i carried by structure $\mathcal{A}_i$. We define a m-ary function $F^\sim$ on $\Pi A_i / \sim$ by requiring:

$$F^\sim([f_1], \ldots, [f_k]) = [(F^i(f_1(i), \ldots, f_k(i)))],$$

where, in conformity with the last remark, $[(F^i(f_1(i), \ldots, f_k(i)))]$ is the equivalence class containing the function whose i-th projection is $F^i(f_1(i), \ldots, f_k(i))$.

Finally, if $\mathcal{L}$ contains a constant symbol c, then this symbol names the constant c^i in the structure $\mathcal{A}_i$. We define the constant $c^\sim$ on $\Pi A_i / \sim$ by the stipulation:

$$c^\sim = [(c^i)].$$

where $[(c^i)]$ is the equivalence class containing the function whose i-th projection is c^i, the object named by c in $\mathcal{A}_i$, for each $i \in I$.

Exercise 5.21 Let $I, \mathcal{U}, \{A_i\}_{i \in I}$ as above.

(i) Suppose that $f_i \sim g_i$, with $i = 1, \ldots, k$, and let $R \in \mathcal{L}$ be a k-ary relation symbol. Show that $R^\sim([f_1], \ldots [f_k]) \iff R^\sim([g_1], \ldots, [g_k])$.
(ii) Suppose that $f_i \sim g_i$, with $i = 1, \ldots, m$, and let $F \in \mathcal{L}$ be a m-ary function symbol. Show that $F^\sim([f_1], \ldots [f_m]) = F^\sim([g_1], \ldots, [g_m])$.

We have now defined a $\mathcal{L}$-structure on $\Pi A_i / \sim$: we shall refer to it as $\Pi_{\mathcal{U}} A_i$ and call it the **ultraproduct** of the family $\{A_i\}_{i \in I}$ relative to $\mathcal{U}$. Our immediate goal to

5.2 Ultraproducts

prove a fundamental theorem, due to Jerzy Łos, which connects the satisfiability of $\mathcal{L}$-formulae in $\Pi_{\mathcal{U}}\mathcal{A}_i$ with their satisfiability in ultrafilter-many $\mathcal{L}$-structures $\mathcal{A}_i$ (i.e. a subfamily of $\mathcal{L}$-structures whose indices determine an element of $\mathcal{U}$).

The proof of Łos' theorem requires a preliminary lemma on terms. In the course of its proof we adopt, for any $\mathcal{L}$-term t, the notation t^i to designate the interpretation of t in $\mathcal{A}_i$ and $t^\sim$ to designate the interpretation of the same term in $\Pi_{\mathcal{U}}\mathcal{A}_i$.

Lemma 5.22 *For any $\mathcal{L}$-term $t(x_1, \ldots, x_n)$ of $\mathcal{L}$:*

$$t^\sim([f^1], \ldots, [f^n]) = [(t^i(f^1(i), \ldots, f^n(i)))],$$

where $[(t^i(f^1(i), \ldots, f^n(i)))]$ is the equivalence class containing the function whose i-th projection is $t^i(f^1(i), \ldots, f^n(i)) \in A_i$.

Proof If $t = c$ for some constant symbol c, then the result holds by definition of $c^\sim$. If $t = x_j$, suppose that $t^\sim = [f_j]$. The result now follows because $[f_j] = [(f_j(i))]$.

If F is a m-ary function symbol and $t_1(x_1, \ldots, x_n), \ldots, t_m(x_1, \ldots, x_n)$ are terms, the inductive hypothesis is:

$$t_j^\sim([f_1], \ldots, [f_n]) = [(t_j^i(f_1(i), \ldots, f_n(i)))], \, j = 1, \ldots, m.$$

Setting $\overline{f(i)} = (f_1(i), \ldots, f_n(i))$, $\overline{[f]} = ([f_1], \ldots, [f_n])$, we deduce:

$$F^\sim(t_1^\sim(\overline{[f]}), \ldots, t_1^\sim(\overline{[f]})) = F^\sim([(t_1^i(\overline{f(i)}))], \ldots, [(t_m^i(\overline{f(i)}))])$$
$$= [(F^i((t_1^i(\overline{f(i)}), \ldots, t_m^i(\overline{f(i)})))],$$

where the first equality follows from the inductive hypothesis and the second follows from the definition of $F^\sim$. □

We are now ready to prove the fundamental result of this section:

Theorem 5.23 (Łos' Theorem) *Let $\{\mathcal{A}_i\}_{i \in I}$ be a family of $\mathcal{L}$-structures, $\mathcal{U}$ an ultrafilter over I and $\varphi(x_1, \ldots, x_n)$ an arbitrary $\mathcal{L}$-formula. Then:*

$$\Pi_{\mathcal{U}}\mathcal{A}_i \models \varphi([f_1], \ldots, [f_n]) \text{ iff } [\varphi(f_1(i), \ldots, f_n(i))] \in \mathcal{U}.$$

Proof For notational convenience, we set $\overline{f(i)} = (f_1(i), \ldots, f_n(i))$, $\overline{[f]} = [f_1], \ldots, [f_n]$ and $\overline{x} = (x_1, \ldots, x_n)$. Let R be a k-ary relation symbol in $\mathcal{L}$ and $t_1(\overline{x}), \ldots, t_k(\overline{x})$ be $\mathcal{L}$-terms. Suppose:

$$\Pi_{\mathcal{U}}\mathcal{A}_i \models Rt_1(\overline{[f]}) \ldots t_k(\overline{[f]})$$

By definition of $R^\sim$, this is to say that:

$$R^\sim t_1^\sim(\overline{[f]}) \ldots t_k^\sim(\overline{[f]})$$

By Lemma 5.22, $t_j^\sim([\overline{f}]) = [(t^i(\overline{f(i)}))]$. Thus, the last condition can be rewritten as:

$$R^\sim[(t_1^i(\overline{f(i)}))]\ldots[(t_k^i(\overline{f(i)}))].$$

Using the definition of $R^\sim$, we obtain:

$$\Pi_\mathcal{U} A_i \models Rt_1([\overline{f}])\ldots t_k([\overline{f}]) \iff R^\sim t_1^\sim([\overline{f}])\ldots t_k^\sim([\overline{f}])$$
$$\iff R^\sim[(t_1^i(\overline{f(i)}))]\ldots[(t_k^i(\overline{f(i)}))]$$
$$\iff [R^i t_1^i(\overline{f(i)})\ldots t_k^i(\overline{f(i)})] \in \mathcal{U}.$$

This completes the inductive base. We turn to the inductive step. The inductive hypothesis is the following pair of equivalences involving $\mathcal{L}$-formulae φ, ψ:

$$\Pi_\mathcal{U} A_i \models \varphi([f_1],\ldots,[f_n]) \text{ iff } V = [\varphi(f_1(i),\ldots,f_n(i))] \in \mathcal{U};$$

$$\Pi_\mathcal{U} A_i \models \psi([f_1],\ldots,[f_n]) \text{ iff } W = [\psi(f_1(i),\ldots,f_n(i))] \in \mathcal{U}.$$

The last condition is equivalent to:

$$\Pi_\mathcal{U} A_i \not\models \varphi([f_1],\ldots,[f_n]) \text{ iff } [\varphi(f_1(i),\ldots,f_n(i))] \notin \mathcal{U}.$$

By definition of ultrafilter, $V \notin \mathcal{U}$ implies $I - V \in \mathcal{U}$. Consequently, the last condition is equivalent to:

$$\Pi_\mathcal{U} A_i \models \neg\varphi([f_1],\ldots,[f_n]) \text{ iff } I - [\varphi(f_1(i),\ldots,f_n(i))] \in \mathcal{U}.$$

Because $j \in I - [\varphi(f_1(i),\ldots,f_n(i))]$ iff $\mathcal{A}_j \not\models \varphi(f_1(i),\ldots,f_n(i))$ iff $\mathcal{A}_j \models \neg\varphi(f_1(i),\ldots,f_n(i))$, we obtain:

$$\Pi_\mathcal{U} A_i \models \neg\varphi([f_1],\ldots,[f_n]) \text{ iff } [\neg\varphi(f_1(i),\ldots,f_n(i))] \in \mathcal{U}.$$

We next consider the formula $\varphi \wedge \psi$.
If

$$\Pi_\mathcal{U} A_i \models \varphi([f_1],\ldots,[f_n]) \wedge \psi([f_1],\ldots,[f_n])$$

then $\Pi_\mathcal{U} A_i$ models each conjunct. Consider the set:

$$Z = [\varphi(f_1(i),\ldots,f_n(i)) \wedge \psi(f_1(i),\ldots,f_n(i))]$$

Because $V \cap W \subseteq Z$ and $V \cap W \in \mathcal{U}$, $Z \in \mathcal{U}$. We have shown that

5.2 Ultraproducts

$$\Pi_{\mathcal{U}} \models \varphi([f_1], \ldots, [f_n]) \wedge \psi([f_1], \ldots, [f_n]) \text{ implies } Z \in \mathcal{U}.$$

Conversely, if $Z \in \mathcal{U}$, then $Z \subseteq V, W$ implies $V, W \in \mathcal{U}$ and, using the inductive hypothesis), $\Pi_{\mathcal{U}} A_i \models \varphi([f_1], \ldots, [f_n]) \wedge \psi([f_1], \ldots, [f_n])$.

It only remains to prove that, given the $\mathcal{L}$-formula $\phi(y, x_1, \ldots, x_n)$:

$$\Pi_{\mathcal{U}} A_i \models \exists y \phi(y, [f_1], \ldots, [f_n]) \text{ iff } [\exists y \phi(y, f_1(i), \ldots, f_n(i))] \in \mathcal{U}.$$

As a preliminary, we show that, for some $f' \in \Pi_{i \in I} A_i$, $[\phi(f'(i), f_1(i), \ldots, f_n(i))] = [\exists y \phi(y, f_1(i), \ldots, f_n(i))]$ For any choice of f:

$$[\phi(f(i), f_1(i), \ldots, f_n(i))] \subseteq [\exists y \phi(y, f_1(i), \ldots, f_n(i)].$$

We only have to establish the converse inclusion. To this end we note that, for each $j \in [\exists y \phi(y, f_1(i), \ldots, f_n(i))]$, there is c_j such that $\phi(c_j, f_1(j), \ldots, f_n(j))$.

We select f' such that $f'(j) = c_j$ if $j \in [\exists y \phi(y, f_1(i), \ldots, f_n(i))]$ and $f'(j) = a$, for a fixed $a \in A_j$, otherwise. By construction, $[\exists y \phi(y, f_1(i), \ldots, f_n(i)] \subseteq [\phi(f'(i), f_1(i), \ldots, f_n(i)]$ and equality holds.

Now we have:

$$[\exists y \phi(y, f_1(i), \ldots, f_n(i))] \in \mathcal{U} \iff [\phi(f'(i), f_1(i), \ldots, f_n(i))] \in \mathcal{U}$$
$$\iff \Pi_{\mathcal{U}} A_i \models \phi([f'], [f_1], \ldots, [f_n])$$
$$\implies \Pi_{\mathcal{U}} A_i \models \exists y \phi(y, [f_1], \ldots, [f_n])$$

and:

$$\Pi_{\mathcal{U}} A_i \models \exists y \phi(y, [f_1], \ldots, [f_n]) \implies \Pi_{\mathcal{U}} A_i \models \phi([f], [f_1], \ldots, [f_n])$$
$$\iff [\phi(f(i), f_1(i), \ldots, f_n(i))] \in \mathcal{U}$$
$$\implies [\exists y \phi(y, f_1(i), \ldots, f_n(i))] \in \mathcal{U}.$$

Note that, in the last two chains of implications, the second step (which is an equivalence) is due to the inductive hypothesis. The induction is now complete. □

Exercise 5.24 Show that $\Pi_{\mathcal{U}} A_i \models t_1(\overline{[f]}) = t_2(\overline{[f]})$ iff $[t_1(\overline{f(i)}) = t_2(\overline{f(i)})] \in \mathcal{U}$.

Remark 5.25 By Łos' theorem, if T is a theory and each $\mathcal{L}$-structure in $\{A_i\}_{i \in I}$ a model of T, then the ultraproduct $\Pi_{\mathcal{U}} A_i$ is also a model of T.

Remark 5.26 If for, each $i \in I$, we have $A_i = \mathcal{A}$, then $\Pi_{\mathcal{U}} \mathcal{A}$ is called an **ultrapower** of $\mathcal{A}$. By the next exercise, a $\mathcal{L}$-structure $\mathcal{A}$ can be elementarily embedded in any of its ultrapowers. In fact, $\mathcal{A}$ is isomorphic to the elementary substructure of an ultrapower $\Pi_{\mathcal{U}} \mathcal{A}$ whose domain consists of the equivalence classes containing a constant function, i.e. a function $f \in \Pi_{i \in I} A_i$ such that $f(i) = a \in A$, for each $i \in I$. Any ultrapower $\Pi_\mathcal{U} A$ is therefore an elementary extension of $\mathcal{A}$.

Exercise 5.27

(i) Let $\mathcal{U}$ be an ultrafilter over the set I. Show that, if $V \cup W \in \mathcal{U}$, then $V \in \mathcal{U}$ or $W \in \mathcal{U}$.
(ii) Prove the inductive step of Theorem 5.23 for the connective $\vee$.
(iii) Let I, for each $i \in I$, $A_i = A$. In this case we write $\Pi_\mathcal{U} A$ instead of $\Pi_\mathcal{U} A_i$. Define $H : A \longrightarrow \Pi_\mathcal{U} A$ by the equality:

$$H(a) = [f_a], \text{ where } f_a(i) = a, \text{ for each } i \in I.$$

Show that H is an elementary embedding.

The ultraproduct construction is very versatile. In the following sections we discuss three of its basic applications: two of them concern basic analysis, while the third is the announced proof of the compactness theorem.

Our first application is a construction of the ordered field of real numbers from an ultrapower of the ordered field of rational numbers.

5.3 The Real Field R

In this section we consider an ultrapower of the $\mathcal{L}_{r<}$-structure $\mathbf{Q} = (\mathbb{Q}, <, +, -, \cdot, 0, 1)$. The underlying ultrafilter extends the filter of cofinite subsets of $\mathbb{N}$, which we are going to refer to as $\mathcal{U}$. The ultrapower we construct thus involves an infinite sequence of copies of $\mathbf{Q}$, indexed over $\mathbb{N}$.

The rational numbers, considered not only as a field, but also as an ordered field, are a model of LO and of the following $\mathcal{L}_{r<}$-sentences:

(1) $\forall x \forall y \forall z (x < y \rightarrow x + z < y + z)$;
(2) $\forall z (0 < z \rightarrow \forall x \forall y (x \cdot z < y \cdot z))$.

Exercise 5.28 If the $\mathcal{L}_r$-structure $\mathcal{K}$ is a field, it may or may not be ordered. It can be ordered exactly when it admits an expansion $\mathcal{K}_P$ to the language $\mathcal{L}_P = \mathcal{L}_r \cup \{P\}$ that models the following sentences:[3]

(a) $\exists x P x$;
(b) $\neg P 0$;
(c) $\forall x \forall y ((Px \wedge Py) \rightarrow Px + y \wedge Px \cdot y)$;
(d) $\forall x (Px \vee x = 0 \vee P - x)$.

---

[3] As the sentences below show, P is a unary relation symbol. Intuitively, the interpretation P^K of P is meant to single out the set of *positive elements* of K.

5.3 The Real Field R

Show that, if $\mathcal{M}$ is a field endowed with a unary relation that models (a) to (d), then $\mathcal{M}$ can be linearly ordered and, in addition, it models the sentences (1), (2) above (*Hint*: define the binary relation $<^\mathcal{M}$ by the condition $a <^\mathcal{M} b$ iff $b - a \in P^\mathcal{M}$). If, conversely, a $\mathcal{L}_{r<}$-structure $\mathcal{M}$ is an ordered field, then it has a definable unary relation $P^\mathcal{M}$ and a corresponding $\mathcal{L}_P$-expansion that models (a) to (d).

The ordered field Q has a special property: it is Archimedean.

Definition 5.29 An ordered field $\mathcal{K}$ is said to be **Archimedean** if, for every non-zero field element a, there is $n \in \mathbb{N}$ such that $1 <^\mathcal{K} n|a|$.

Here $|a|$ is the **absolute value** of a, defined as follows: if $0 = a$ or $0 <^\mathcal{K} a$, then $|a| = a$, otherwise $|a| = -a$. The integer multiples of $|a|$ are recursively defined by the clauses $1|a| = |a|$ and $(m+1)|a| = m|a| +^\mathcal{K} |a|$.

Exercise 5.30 Let the $\mathcal{L}_{r<}$-structure $\mathcal{K}$ be field ordered by the relation $<$ (we drop the usual superscript for the relations and functions on K). We abbreviate the disjunction $x = y \lor x < y$ by $x \leq y$.

(i) Show that, for any $a, b \in K$, $|a - b| < \epsilon$ iff $b - \epsilon < a$ and $a < b + \epsilon$.
(ii) Use (i) to show that $|a + b| \leq |a| + |b|$.
(iii) Use (ii) to show that $|a| - |b| \leq |a - b|$.

The real field R is, as we shall see, an Archimedean extension of Q. We do not directly construct it from Q but rather 'carve it out' of the (non-Archimedean) ultrapower $\Pi_\mathcal{U} \mathbb{Q} = {}^*\mathbb{Q}$, whose elements are equivalence classes of infinite sequences of the form $a = (a(0), a(1), a(2), \ldots)$, with $a(n) \in \mathbb{Q}$ for each $n \in \mathbb{N}$. We refer to generic elements of $^*\mathbb{Q}$ as $[a], [b], [c], \ldots$.

The structure Q may be viewed as an elementary substructure of $^*\mathbb{Q}$ in the canonical way described at the end of the previous section. In particular, we identify $q \in \mathbb{Q}$ with the equivalence class $[q]$, where q is the constant sequence defined by the condition $q(n) = q$ for each $n \in \mathbb{N}$.

By the definition of ultraproduct, order in $^*\mathbb{Q}$ is given by the condition below:

$$[a] < [b] \text{ iff } \{n \in \mathbb{N} : a(n) < b(n)\} \in \mathcal{U}.$$

Moreover, addition and multiplication satisfy the following conditions:

$$[a] + [b] = [c] \text{ iff } \{n \in \mathbb{N} : a(n) + b(n) = c(n)\} \in \mathcal{U};$$

$$[a][b] = [c] \text{ iff } \{n \in \mathbb{N} : a(n)b(n) = c(n)\} \in \mathcal{U}.$$

Since $\mathbb{N} \in \mathcal{U}$, the sum $[a] + [b]$ contains the component-wise sum $a + b$ and the product $[a][b]$ contains the component-wise product ab. Now it is easy to see that $^*\mathbb{Q}$ is not Archimedean. Consider for instance the equivalence class $[1]$ containing the constant sequence $a = (1, 1, 1, \ldots)$, and the equivalence class ϵ, containing the sequence $b = (1/2, 1/3, 1/4, \ldots)$.

Note that ϵ is positive and, by the next exercise, for no positive integer n is $n\epsilon$ greater than $[1]$. The Archimedean property thus fails. We say that ϵ is a **positive infinitesimal** of the field *Q, whose domain we henceforth designate by *$\mathbb{Q}$.

Exercise 5.31 Let ϵ be the positive infinitesimal just described and $[0]$, $[1/n]$ the elements of *$\mathbb{Q}$ that contain, respectively, the constant sequences $(0, 0, 0, \ldots)$ and $(1/n, 1/n, 1/n)$. Show that:

(i) $[0] < [1/n] < [1]$;
(ii) $\epsilon + \epsilon < [1]$ and $\epsilon + \epsilon + \epsilon < [1]$;
(iii) for any fixed $k \in \mathbb{N}$, $k\epsilon < [1]$;
(iv) $\epsilon < [1/2], [1/3], [1/4], \ldots$, where, for any positive integer k, $[1/k]$ is the element of *$\mathbb{Q}$ containing the constant sequence $(1/k, 1/k, 1/k, \ldots)$.

Because ϵ is a non-zero field element, it has a multiplicative inverse $\epsilon^{-1} \in$ *$\mathbb{Q}$. The inequality $n\epsilon < [1]$ implies $[n] < \epsilon^{-1}$ by the arithmetic of ordered fields. The positive field element ϵ^{-1} is therefore greater than any $[n]$. It is an **infinitely large** element of *$\mathbb{Q}$.

Remark 5.32 Since sentences transfer across elementary extensions, if the Archimedean property of a field could be expressed by a first-order $\mathcal{L}_{r<}$-sentence, it should transfer from $\mathbb{Q}$ to *$\mathbb{Q}$. The fact that it does not shows that the Archimedean property cannot be formulated in the first-order language of ordered fields. The result is not affected by the selection of a richer first-order language: even if we added further structure to *$\mathbb{Q}$, its ordering would remain unaffected.

Because they will be important in this and the next section, we officially define the infinitesimal elements of an arbitrary ordered field.

Definition 5.33 Let the $\mathcal{L}_{r<}$-structure $\mathcal{K}$ be an ordered field. If $a \in K$ and, for any $n \in \mathbb{N}$, $n|a| < 1$, then a is an **infinitesimal** of K. If $0 < a$, then a is a positive infinitesimal. Note that, on our definition, 0 is an infinitesimal.

Exercise 5.34 Show that an ordered field $\mathcal{K}$ is an Archimedean ordered field iff its set of infinitesimals is $\{0\}$.

A key role in our construction of the real field is played by the following:

Definition 5.35 Let $\mathcal{K}$ be an ordered field. An element $a \in K$ is **finite** iff there is $n \in \mathbb{N}$ such that $|a| < n$ (note that in an Archimedean ordered field every element is finite).

Certainly, in any ordered field $\mathcal{K}$, 1 is finite. It is not difficult to see that the sum, difference and product of two finite elements are finite.[4] Moreover, Definition 5.35 implies that infinitesimals count as finite.

---

[4] If $a \neq b \in K$ are finite, there are n, m such that $|a| < m$, $|b| < n$. Then the definition of absolute values implies $|a + b| \leq |a| + |b| < m + n$, $|a + (-b)| \leq |a| + |-b| < m + n$ and $|ab| = |a||b| < mn$.

5.3 The Real Field R

Relative to the signature $\mathcal{L}_{r<}$, the set of finite elements of an arbitrary ordered field determines a substructure of $\mathcal{K}$ that is an ordered ring but not a field (see Exercise 5.36-(i)) and the set of infinitesimals determines an ideal of this ring.

Let us focus again on *Q: we shall call the ordered ring of its finite elements *F, the domain of this ring *$\mathbb{F}$ and the ideal of infinitesimals J. In the next exercise, we show that J is indeed an ideal and, moreover, that it is maximal.

Exercise 5.36

(i) Show that, if $[a] \in J$ is a negative infinitesimal, then its multiplicative inverse is not finite.
(ii) Show that, if $[a] \in {}^*\mathbb{F}$ and $[a]$ is not an infinitesimal, then its multiplicative inverse is finite.
(iii) Show that J is an ideal of the ring *F.
(iv) Show that, if $[b] \in {}^*F$ and $[b]$ is not an infinitesimal, then the only ideal containing $[b]$ is *F.
(v) Deduce from (iv) that J is a maximal ideal of *F.

By Theorem 3.41, the quotient *$\mathbb{F}/J$ is a field. The elements of *$\mathbb{F}/J$ are cosets of the form $[a] + J$.

Exercise 5.37 Let $[a]+J, [b]+J \in {}^*\mathbb{F}/J$. Show that $[a]+J = [b]+J$ iff $[a]-[b]$ is an infinitesimal.

We say that $[a], [b]$ are **infinitely close** iff $[a] - [b] \in J$, in which case we also write $[a] \simeq [b]$. The elements of *$\mathbb{F}/J$ may be thought of as infinitely small segments centred around each finite element of *Q. We now drop the lengthier notation $[a] + J$ to replace it with the simpler (a). Because our eventual goal is to obtain an ordered field, we expand the $\mathcal{L}_r$-structure *$\mathbb{F}/J$ to $\mathcal{L}_{r<}$ by defining the order relation $<^J$ as follows:

$$(a) <^J (b) \text{ iff } [a] \not\simeq [b] \text{ and } [a] < [b].$$

where the superscript-free $<$ is simply the ordering of *Q.

Exercise 5.38 Show that, if the condition $[a] \not\simeq [b]$ from the last definition is dropped, then $<^J$ violates irreflexivity.

The ordered field *$\mathbb{F}/J$ is automatically Archimedean, since it was built from the finite elements of *Q. If $0 <^J < (a)$ then $[a]$ cannot be a positive infinitesimal. Consequently, there is a positive integer n such that $n[a] > [1]$. If $(a) <^J 0$, then $[a]$ is not a negative infinitesimal and we can reason as above on $(-a)$.

We have obtained an Archimedean ordered field that extends Q, which we refer to as the **ordered field of real numbers** R. In the next subsection we prove that R has a property known as sequential completeness, which characterises it, up to isomorphism, as the Archimedean, sequentially complete extension of the ordered rational field Q.[5]

---

[5] We shall not prove that any two ordered, Archimedean fields that are sequentially complete are isomorphic, but see Appendix A of [52] for a discussion of this result.

5.3.1 Sequential Completeness of R

Let the $\mathcal{L}_{r<}$-structure $\mathcal{K}$ be an ordered field. For any infinite sequence $(a_1, a_2, \ldots)$ of elements of K, we say that $(a_1, a_2, \ldots)$ **converges** to $a \in K$ if, for any positive field element ϵ, there is an index $j \in \mathbb{N}$ such that, for every $m \geq j$, $|a - a_m| <^K \epsilon$. Intuitively, a sequence converges to a a if every point of sufficiently large index in the sequence is close to a, and closeness can be fixed at will in K.

If the sequence $(a_1, a_2, \ldots)$ converges to a, we refer to a as the[6] **limit** of $(a_1, a_2, \ldots)$ and to $(a_1, a_2, \ldots)$ as a convergent sequence. Thus, for instance, in any Archimedean ordered field the sequence $(1, 2, 3, 4, \ldots)$ is not convergent, while the sequence $(1/2, 1/3, 1/4, \ldots)$ is convergent and converges to the limit 0.

We now define sequential completeness, also known as Cauchy completeness.

Definition 5.39 Let the $\mathcal{L}_{r<}$-structure $\mathcal{K}$ be an ordered field. A sequence $(a_1, a_2, \ldots)$ of field elements is said to be a **Cauchy sequence** iff, for any positive $\epsilon \in K$, there is $j \in \mathbb{N}$ such that, for every $m, n > j$, $|a_m - a_n| <^K \epsilon$. The field $\mathcal{K}$ is said to be **sequentially complete** iff every Cauchy sequence of field elements has a limit in K.

Exercise 5.40 Show that $(1/2, 1/3, 1/4, \ldots)$ is a Cauchy sequence in $\mathbb{Q}$.

The Archimedean ordered field $\mathbb{Q}$ contains Cauchy sequences that do not have any limit in $\mathbb{Q}$ (e.g. the increasing rational approximations of $\sqrt{2}$ from below). We may thus regard its extension R as a richer field, filling the 'gaps' revealed in $\mathbb{Q}$ by sequences whose elements are more and more densely packed but do not locate a point of convergence.

In order to prove that R is sequentially complete it is convenient to invoke the lemmas proved in the next two exercises (in this respect we follow [50], pp.150–152).

Exercise 5.41 Let the $\mathcal{L}_{r<}$-structure $\mathcal{K}$ be an Archimedean ordered field. We view the ordered field $\mathbb{Q}$ as a subfield of $\mathcal{K}$ (this is because $1 \in K$ and all rational multiples of 1 are elements of K). We are going to prove that, if $a \in K$ and p is a positive rational, there is $q \in \mathbb{Q}$ such that $|a - q| <^K p$. This result shows that, within an Archimedean ordered field, rationals can be used to approximate arbitrary field elements as well as one wishes.

(i) Show that there is $m \in \mathbb{N}$ such that $1/m <^K p$ and that there is a smallest $n \in \mathbb{N}$ such that $ma \geq^K n - 1$.
(ii) Set $q = (n-1)/m$ and evaluate $|a - q| = a - q$.

Exercise 5.42 When dealing with R, we use the notation ' $<$ ' to designate its ordering. Prove that, for any $(a) \in$ R, $|(a)| = (|a|)$ by cases. In particular:

(i) show that the result holds when $[a] \not\approx [0]$ and $[0] < [a]$;

---
[6] Limits in ordered fields are unique.

5.3 The Real Field R

(ii) show that the result holds when $[a] \not\simeq [0]$ and $[a] < [0]$;
(iii) show that the result holds when $[a] \simeq [0]$ (in which case $(a) = (0)$).

Deduce that, if the inequality $|(a) - (q)| < (p)$ holds in R, then the inequality $|[a] - [q]| < [p]$ must hold in $*\mathbb{Q}$.

We are now ready to prove:

Theorem 5.43 *The field R is sequentially complete.*

Proof Whenever convenient, we use $x > y$ as a notational variant of $y < x$.

We are given a Cauchy sequence $((a_1), (a_2), \ldots)$ and we must show that it has a limit (a). We slightly simplify the problem by working with a rational approximation of the given Cauchy sequence. For each positive $n \in \mathbb{N}$, Exercise 5.41 ensures the existence of a rational (q_n) such that:

$$|(a_n) - (q_n)| < (1/n).$$

By Exercise 5.42, $|[a_n] - [q_n]| = [1/n]$, where $[q_n]$ is the element of $*\mathbb{Q}$ that contains the constant sequence $(q_n, q_n, \ldots)$. We now verify that the sequence $((q_1), (q_2), \ldots)$ is a Cauchy sequence. Fix a positive field element (b). We take advantage of the fact $((a_1), (a_2), \ldots)$ is a Cauchy sequence and determine k_1 so large that, for any $m, n > k$:

$$|(a_n) - (a_m)| < (b/3)$$

We can independently choose k_2 to be large enough to satisfy $(1/k) < (b/3)$, due to the fact that R is Archimedean. Setting $k = max\{k_1, k_2\}$, we obtain:

$$|(q_m) - (q_n)| \leq |(q_m) - (a_m)| + |(a_m) - (a_n)| + |(a_n) - (q_n)|$$
$$< (1/m) + (b/3) + (1/n)$$
$$< (b/3) + (b/3) + (b/3)$$
$$= b.$$

Since b is an arbitrary positive element, the sequence $((q_1), (q_2), \ldots)$ is a Cauchy sequence. We make use of this sequence to construct a field element (a) and to show that (a) is the limit of $((a_1), (a_2), \ldots)$.

First, we define $[a] \in *\mathbb{Q}$ to be the equivalence class containing the sequence of rational numbers $a = (q_1, q_2, \ldots)$ (by this definition, we also write $a(i) = q_i$). Next, we show that $[a]$ is finite in $*\mathbb{Q}$, which allows us to ensure that $[a] + J = (a)$ is in fact an element of R. Because $((q_1), (q_2), \ldots)$ is a Cauchy sequence, there is $k \in \mathbb{N}$ such that, whenever $m, n > k$, we obtain:

$$|(q_n) - (q_m)| < (1)$$

In this case, $|[q_n] - [q_m]| < [1]$ and, fixing $m > k$, we obtain:

$$|[q_n]| \leq |[q_n] - [(q_m)]| + |[q_m]| < [1] + |[q_m]|$$

Because all of the equivalence classes involved contain constant sequences, we deduce:

$$|q_n| < 1 + |q_m| \text{ whenever } n > k.$$

Since the set $\{n \in \mathbb{N} : n > k\}$ is cofinite, it is in $\mathcal{U}$. Thus $\{n \in \mathbb{N} : \mathsf{Q} \models |a(n)| < 1+|q_m|\} \in \mathcal{U}$, which is to say that $^*\mathsf{Q} \models |[a]| < 1+|[q_m]|$, i.e. that $[a]$ is finite. We finally show that $(a) \in \mathsf{R}$ is indeed the limit of $((a_1), (a_2), \ldots)$. To this end, we fix a positive field element (b) and consider a rational q such that $(0) < (q) < (b)$. We have to prove that, for sufficiently large $k \in \mathbb{N}$, $m > k$ implies $|(a_m) - (a)| < (b)$.

One way to do so is to pick k so large that the choice of (q_m) allows us to keep the quantity $|(a_m) - (a)| \leq |(a_m) - (q_m)| + |(q_m) - (a)|$ below (b). We can easily keep $|(a_m) - (q_m)|$ below $(b/2)$, by choosing m to be greater than m_1, where m_1 is a positive integer that satisfies the inequality $1/m_1 < b/2$ (the Archimedean property guarantees that this is possible).

Moreover, recalling that $[q_m]$ is the equivalence class containing the constant sequence $(q_m, q_m, \ldots)$, if we designate the n-th element of this sequence by $q_m(n)$ we see that $|q_m(n) - a(n)| = |q_m - q_n|$. The last quantity can be made smaller than $b/2$, as long as $m, n > m_2$, for a sufficiently large integer m_2.

Because $\{n \in \mathbb{N} : n > m_2\} \in \mathcal{U}$, it follows that $|[q_m] - [a]| < [b/2]$. If $|[q_m] - [a]| \not\simeq [b/2]$, we apply Exercise 5.42 to deduce $|(q_m) - (a)| < (b/2)$. Otherwise $|(q_m) - (a)| = (b/2)$. In any case $|(q_m) - (a)| \leq (b/2)$.

We conclude that that, if $k = max\{m_1, m_2\}$, then for any $m > k$:

$$|(a_m) - (a)| \leq |(a_m) - (q_m)| + |(q_m) - (a)|$$
$$< (b/2) + (b/2)$$
$$= (b).$$

□

5.4 A Touch of Nonstandard Analysis

Given the real field R, its ultrapower over $\mathcal{U}$ (the ultrafilter used to build $^*\mathsf{Q}$) is a non-Archimedean elementary extension. It thus contains field elements that are not finite, e.g. the equivalence class determined by the sequence $(1, 2, 3, \ldots, n, \ldots)$, as well as infinitesimals. Abraham Robinson showed that suitable expansions of the real field containing infinitesimals provide a model-theoretic avenue into an alternative approach to real analysis, in which limits and convergence, as well as other standard

5.4 A Touch of Nonstandard Analysis

notions (e.g. continuity, uniform continuity, differentiability etc.), may be replaced by infinitesimal equivalents.

This approach to real analysis is known as Nonstandard Analysis.[7] Although it is not the purpose of this book to offer any comprehensive account of techniques and results employed in Nonstandard Analysis, we can rather swiftly and easily introduce its simplest ideas and use them to develop a nonstandard approach to the study of infinite real sequences.

To this end, we expand $\mathcal{L}_{r<}$ to a first-order language $\mathcal{S}$ for sequences, which contains one constant symbol for each real number, one function symbol for each for each real-valued sequence, one function symbol for the absolute value and a unary relation symbol N to single out $\mathbb{N}$ as a subset of the real field. The symbol $\mathbb{R}$ designates the domain of R, which in this section we view as a $\mathcal{S}$-structure.

The ultrapower of R based on the ultrafilter $\mathcal{U}$ from the preceding section is the $\mathcal{S}$-structure *R, whose domain is *$\mathbb{R}$. Because *$\mathbb{R}$ is an elementary extension of the real field, any first-order property of a sequence that can be stated in $\mathcal{S}$ and only depends on parameters in $\mathbb{R}$ holds in R iff it holds in *$\mathbb{R}$.

This fact goes under the name of the **transfer principle** of nonstandard analysis. It essentially allows us to prove theorems of real analysis in the enriched environment *$\mathbb{R}$, where we may usefully rely on infinitesimals or infinitely large numbers. We shall now see why such additional numbers are useful.

From now on we refer to constants in $\mathbb{R}$ by the letters $a, b, c, \ldots$ and, if they are natural numbers, by the letters $n, m, \ldots$. By contrast, real-valued sequences are designated by the letters s, t, u. When we need to specify the i-th argument of a sequence s, we write $s(i)$.

First of all, since a real-valued sequence is a function from $\mathbb{N}$ to $\mathbb{R}$, we consider the interpretation of the symbol N in *R, which we call *$\mathbb{N}$. By definition of ultrapower, the set *$\mathbb{N}$ is determined by the condition:

$$[s] \in {}^*\mathbb{N} \text{ iff } \{i \in \mathbb{N} : s(i) \in \mathbb{N}\} \in \mathcal{U}.$$

Thus, in particular, the equivalence classes containing a sequence whose values are natural numbers belong to *$\mathbb{N}$. In particular, for any $n \in \mathbb{N}$, $n \in {}^*\mathbb{N}$, i.e. $\mathbb{N} \subseteq {}^*\mathbb{N}$. Moreover, the sequence $(1, 2, 3, 4, \ldots)$ determines an element $[s]$ of *$\mathbb{N}$ that is greater than any $n \in \mathbb{N}$. We thus regard *$\mathbb{N}$ as a linearly ordered enlargement of $\mathbb{N}$ that contains infinitely large integers.

Exercise 5.44

(i) Show that *$\mathbb{N}$ contains infinitely many, infinitely large elements and deduce that *$\mathbb{R}$ contains infinitely many distinct infinitesimals.
(ii) Show that, if $H, K \in {}^*\mathbb{N}$ are infinitely large, then there is $E \in {}^*\mathbb{N}$ such that $H < E$ and $E < K$.

---

[7] This is the title of Robinson's monograph on the subject, see [60].

Since any real-valued sequence $s : \mathbb{N} \longrightarrow \mathbb{R}$ has a name in $\mathcal{S}$ we can associate it with the sequence in $^*\mathbb{R}$ that goes by the same name, which we call $^*s$ and define in the obvious manner.

$$^*s([a]) = [b] \text{ iff } \{i \in \mathbb{N} : s(a(i)) = b(i)\} \in \mathcal{U}.$$

Exercise 5.45 Use the definition of $^*s$ to show that:

(i) if $[a] = n \in \mathbb{N}$, $^*s(n) = s(n)$;
(ii) the domain of $^*s$ is $^*\mathbb{N}$.

By Exercise 5.45, whenever $n \in \mathbb{N}$, $^*s(n) = s(n)$ so we can think of $^*s$ as an infinite sequence that coincides with s at every finite argument and takes further values at infinitely large arguments. The behaviour of $^*s$ at infinitely large arguments codifies its asymptotic behaviour in $\mathbb{R}$, i.e. it provides information as to s converges or not to a real limit.

Theorem 5.46 *Let $s : \mathbb{N} \longrightarrow \mathbb{R}$ be a real-valued sequence. Then s converges to a limit a in $\mathbb{R}$ iff, for any infinitely large integer $H \in {}^*\mathbb{N}$, $^*s(H) \simeq a$.*

Proof The theorem requires that we use $^*\mathbb{R}$ to prove a result about $\mathbb{R}$. Suppose that s converges to a limit a in $\mathbb{R}$. Then R is a model of the following $\mathcal{S}$-sentence (in which we use s as a function symbol):

$$\exists y \forall x_1 (0 < x_1 \rightarrow \exists x_2 (N(x_2) \wedge \forall x_3 ((N(x_3) \wedge x_2 < x_3) \rightarrow |s(x_3) - y| < x_1)))$$

which says that, for any positive x_1, there is $x_2 \in \mathbb{N}$ such that any $x_3 \in \mathbb{N}$ greater than x_2 determines a sequence value $s(x_3)$ closer to y by less than x_1. Using the definition of satisfaction and the fact that $\mathcal{S}$ contains a constant symbol for each real number, we can deduce:

$$\mathsf{R} \models \exists x_2 (N(x_2) \wedge \forall x_3 ((N(x_3) \wedge x_2 < x_3) \rightarrow |s(x_3) - a| < 1/n))$$

and, choosing $k \in \mathbb{N}$ with the properties of x_2 in the above sentence:

$$\mathsf{R} \models \forall x_3 ((N(x_3) \wedge k < x_3) \rightarrow |s(x_3) - a| < 1/n)$$

Note, in particular, that R models the last sentence for each positive $n \in \mathbb{N}$. As a consequence, so does $^*\mathsf{R}$. We now fix an infinitely large integer $H \in {}^*\mathbb{N}$. This integer satisfies the formula $N(x_2) \wedge k < x_3$, since k is a finite integer. Thus, for each positive $n \in \mathbb{N}$:

$$^*\mathsf{R} \models |z(H) - a| < 1/n$$

It follows that $s(H)$ is closer to a than $1/n$ for each positive integer. By definition of infinitesimal, $|s(H) - a|$ is an infinitesimal, i.e. $s(H) \simeq a$. Since H was chosen arbitrarily, the result holds for any infinitely large, positive integer.

5.4 A Touch of Nonstandard Analysis

To conclude the proof, we suppose that, for any infinitely large $H \in {}^*\mathbb{N}$, ${}^*s(H) \simeq a$. Our goal is to deduce that, in $\mathbb{R}$, s converges to a. It suffices to prove that, for any positive integer n, there is $k \in \mathbb{N}$ such that $m > k$ implies $|s(m) - a| < 1/n$. In ${}^*\mathbb{R}$ it is clear that, since ${}^*s(H) \simeq a$, $|{}^*s(H) - a| < 1/n$. As we have seen, the last statement is true of any infinitely large integer in ${}^*\mathbb{N}$, in particular of every integer $K > H$. In other words:

$${}^*\mathbb{R} \models \exists x_2 (N(x_2) \wedge \forall x_3 ((N(x_3) \wedge x_2 < x_3) \to |s(x_3) - a| < 1/n))$$

holds for each $n \in \mathbb{N}$. In other words, each sentence of the above form (there is one for each positive integer) transfers to $\mathbb{R}$. Because $\mathbb{R}$ is Archimedean, every positive quantity x_1 is greater than $1/n$, for n sufficiently large. We conclude:

$$\mathbb{R} \models \forall x_1 (0 < x_1 \to \exists x_2 (N(x_2) \wedge \forall x_3 ((N(x_3) \wedge x_2 < x_3) \to |s(x_3) - a| < x_1))).$$

The last sentence says precisely that s converges to a. □

Exercise 5.47 Use Theorem 5.46 to show that, if s has a limit a in $\mathbb{R}$, this limit must be unique.

Exercise 5.48

(i) Show that $s : \mathbb{N} \longrightarrow \mathbb{R}$ is a Cauchy sequence iff for any infinitely large $H, K \in {}^*\mathbb{N}$, ${}^*s(H) \simeq {}^*s(K)$.
(ii) Show that any convergent sequence is a Cauchy sequence.
(iii) What can we say about s if, for any infinitely large $H \in {}^*\mathbb{N}$, ${}^*s(H)$ is infinitely large?

Exercise 5.49 The real-valued sequence s is said to be **monotonic decreasing** if it satisfies the chain of inequalities $s(0) \geq s(1) \geq s(2) \geq \ldots$. Suppose that this sequence has a lower bound, i.e. that there is $b \in \mathbb{R}$ such that, for every $n \in \mathbb{N}$, $s(n) \geq b$. We shall show that s has a greatest lower bound a, which it converges to.

(i) By transferring a suitable first-order sentence, show that, in ${}^*\mathbb{R}$, b is a lower bound of ${}^*s(H)$ whenever H is infinitely large. Since $s(1) \geq {}^*s(H)$, it follows that ${}^*s(H)$ is a finite element of ${}^*\mathbb{R}$ and, consequently, infinitely close to a real number $a \in \mathbb{R}$ (when $\mathbb{R}$ is viewed as a subset of ${}^*\mathbb{R}$). Because this argument is true for any infinitely large element of ${}^*\mathbb{N}$, $s(n)$ converges to a.
(ii) Show that a is a lower bound of $s(n)$ using the fact that $m \leq n$ implies $s(m) \geq s(n)$ and a suitable first-order sentence.
(iii) Show that, if c is a lower bound of $s(n)$, then $c \leq a$. In other words, a is the greatest lower bound of s.

Exercise 5.50 Let $c \in \mathbb{R}$ be such that $0 < c$ and $c < 1$. Show that the sequence (c^n), whose elements are the positive integer powers of c, converges to 0, following the strategy outlined below:

(i) first, show that $\{c, c^2, c^3, \ldots\}$ is a decreasing sequence. By Exercise 5.49, this sequence has a limit $a \in \mathbb{R}$;
(ii) let H be infinitely large: use the powers c^H, c^{H+1} to show that $a \simeq ac$;
(iii) deduce from $a \simeq ac$ and $c < 1$ that $a = 0$.

We proved in Exercise 5.48-(ii) that every convergent sequence is a Cauchy sequence. We use the next exercise to obtain a nonstandard proof of the converse statement.

Exercise 5.51 Let s be a Cauchy sequence. We shall prove that it is **bounded**, i.e. that there is a positive $m \in \mathbb{N}$ such that, for each $n \in \mathbb{N}$, $|s(n)| \leq m$.

(i) Show that there is $k \in \mathbb{N}$ such that, if $n > k$, then $|s(n)| < |s(k)| + 1$;
(ii) set $m - 1 = max\{|s(0)|, \ldots, |s(k-1)|, |s(k)|\}$ and show that m is an upper bound for each $|s(n)|$.

Lemma 5.52 *If s is a Cauchy sequence, then it converges to a limit $a \in \mathbb{R}$.*

Proof Let s be a Cauchy sequence. By Exercise 5.51, z is bounded. Let m be a bound for s. Then $\mathsf{R} \models \forall x(N(x) \to |s(x)| \leq m)$ and, since the last S-sentence transfers to *$\mathbb{R}$, we can conclude that, for any infinitely large $H \in {}^*\mathbb{N}$, $0 \leq |{}^*s(H)| \leq m$. The last inequalities imply that *$s(H)$ is infinitely close to a real number a.

To see why, note that, if we restrict attention to the set *$\mathbb{F}$ of finite elements of *$\mathbb{R}$ (where 'finite' is defined as in the previous section), then *$\mathbb{F}$ is a ring and the set of infinitesimals J a maximal ideal in this ring. The resulting quotient is the Archimedean ordered field *$\mathbb{F}/J$. The latter field includes an isomorphic copy of $\mathbb{R}$ because, if $b \neq c \in \mathbb{R}$, then $b - c$ is not infinitesimal and $a + J \neq b + J$. Since $\mathbb{R}$ is a maximal Archimedean ordered field (see Appendix A of [52]), it follows that *$\mathbb{F}/J$ is the real field itself, i.e. each distinct coset of *$\mathbb{F}/I$ is of the form $b + I$, with $b \in \mathbb{R}$.

As a consequence, if *$s(H)$ is finite, it is $a + J$, for some $a \in \mathbb{R}$. Since s is Cauchy, by Exercise 5.49-(i), for any infinitely large integers $H, K \in {}^*\mathbb{N}$, *$s(H) \simeq a \simeq {}^*s(K)$. We have shown that, at infinitely large arguments, *s takes values that are infinitely close to $a \in \mathbb{R}$. By Theorem 5.46, $s(n)$ converges to a. □

We conclude this section with a nonstandard characterisation of limit points, leading to a concise and elegant proof of the Bolzano-Weierstrass theorem.

Definition 5.53 Let s be a real-valued sequence. A number $a \in \mathbb{R}$ is a **limit point** of s iff for any positive real r, there is n such that $|s(n) - a| < r$.

Note that the definition of limit point is different from the definition of limit. We do not require $|s(n) - a| < r$ for *every* sufficiently large $n \in \mathbb{N}$, but only that the last equality holds for *some*, sufficiently large $n \in \mathbb{N}$. If $a \in \mathbb{R}$ is a limit point of s, then $|s(n) - a| < r$ for infinitely many values of n. This is because we know that there is at least some m_1 such that $|s(m_1) - a| < r$. Set $|s(m_1) - a| = r_1$: then there is m_2 such that $|s(m_2) - a| < r_1 < r$, and so on. The reader may use this observation

5.4 A Touch of Nonstandard Analysis

to verify that a is a limit point of s iff, for any positive real r and $m \in \mathbb{N}$, there is $n \in \mathbb{N}$ such that $m \leq n$ and $|s(n) - a| < r$.

We can restate the last condition formally as:

$$\mathsf{R} \models \forall x_1(0 < x_1 \rightarrow \forall x_2(N(x_2) \rightarrow \exists x_3(N(x_3) \wedge x_2 < x_3 \wedge |s(x_3) - a| < x_1))).$$

The last $\mathcal{S}$-sentence allows us to prove:

Lemma 5.54 *Let s be a real-valued sequence. The number $a \in \mathbb{R}$ is a limit point of s iff there is some infinitely large $H \in {}^*\mathbb{N}$ such that ${}^*s(H) \simeq a$.*

Proof Suppose that a is a limit point of s. Then:

$$\mathsf{R} \models \forall x_1(0 < x_1 \rightarrow \forall x_2(N(x_2) \rightarrow \exists x_3(N(x_3) \wedge x_2 < x_3 \wedge |s(x_3) - a| < x_1))).$$

The last $\mathcal{S}$-sentence transfers to ${}^*\mathsf{R}$, where we can choose x_1 to take an infinitely small value ϵ and obtain:

$${}^*\mathsf{R} \models \forall x_2(N(x_2) \rightarrow \exists x_3(N(x_3) \wedge x_2 < x_3 \wedge |s(x_3) - a| < \epsilon))).$$

Note that the last formula is not a $\mathcal{S}$-sentence but a $\mathcal{S}$-formula evaluated in ${}^*\mathsf{R}$ under the assignment of ϵ to x_1. If we now choose x_2 to be evaluated on an infinitely large integer $H \in {}^*\mathbb{N}$, we obtain the existence of a larger integer K such that $|{}^*s(K) - a| < \epsilon$, i.e. ${}^*s(K) \simeq a$.

Conversely, if ${}^*s(K) \simeq a$ for an infinitely large integer in ${}^*\mathbb{N}$, then for each positive $n \in \mathbb{N}$, we have $|{}^*s(K) - a| < 1/n$. In fact, for each $m \in \mathbb{N}$:

$${}^*\mathsf{R} \models \exists x_3(N(x_3) \wedge m < x_3 \wedge |s(x_3) - a| < 1/n)).$$

Keeping n fixed, the fact that we can transfer the last $\mathcal{S}$-sentence back to R for each $m \in \mathbb{N}$ implies

$$\mathsf{R} \models \forall x_2(N(x_2) \rightarrow \exists x_3(N(x_3) \wedge x_2 < x_3 \wedge |s(x_3) - a| < 1/n)).$$

Since every positive real is smaller than $1/n$ for some positive integer n, we conclude:

$$\mathsf{R} \models \forall x_1(0 < x_1 \rightarrow \forall x_2(N(x_2) \rightarrow \exists x_3(N(x_3) \wedge x_2 < x_3 \wedge |s(x_3) - a| < x_1))).$$

Thus, a is a limit point for z. □

As a corollary to Lemma 5.54, we finally obtain:

Theorem 5.55 (Bolzano-Weierstrass) *Every bounded sequence has a limit point.*

Proof Let s be a bounded sequence. Then $\mathsf{R} \models \forall x(N(x) \rightarrow |s(x)| < m)$, where m is a fixed constant. The last sentence transfers to ${}^*\mathsf{R}$, where it implies that, for any infinitely large $H \in {}^*\mathbb{N}$, $|{}^*s(H)| \leq m$. It follows that ${}^*s(H)$ is infinitely close to a real a. By Lemma 5.54, a is a limit point of s. □

Even though we have restricted our attention to the study of real-valued sequences, it is not too hard to see that, by expanding the language further to include symbols for real-valued functions, we can use a corresponding expansion of ${}^*\mathbb{R}$ to study continuity and differentiation by means of infinitesimal methods.

With a sufficiently comprehensive framework in hand, we can go far beyond what we have seen. Good introductions to nonstandard analysis are [41] and the early chapters of [19]. A recent refinement of Robinson's approach appears in [32], which is accessible even to the reader who approaches real analysis for the first time.

5.5 The Compactness Theorem

After our analytical detour, we return to the main objective of this chapter, namely proving the compactness theorem. We begin with a simple yet essential definition.

Definition 5.56 A set of first-order sentences Σ is said to be **finitely satisfiable** iff every finite subset Δ of Σ has a model.

Theorem 5.57 (Compactness) *Let Σ be a set of first-order sentences: Σ has a model (or, equivalently, is satisfiable) iff it is finitely satisfiable.*

Clearly, if $\mathcal{M} \models \Sigma$, then $\mathcal{M} \models \Delta$, for any finite $\Delta \subseteq \Sigma$. The non-trivial content of the compactness theorem amounts to the fact that there is a model for a set of simultaneous conditions as soon there are models for each of its distinct, finite subsets.

The reader may guess why ultraproducts are helpful to deliver a proof of compactness: given a model for each finite part of Σ and a suitable ultrafilter, the resulting ultraproduct will model the whole of Σ. The difficult part of the proof consists in finding the required ultrafilter.

To this end, we take Σ to be the indexed set of sentences $\{\varphi_j\}_{j \in J}$ and we separately consider the indexed family $\{\Delta_i\}_{i \in I}$, which consists of the finite subsets of Σ. The ultrafilter we want is based on the set I. We specify it by associating with each $\varphi_j \in \Sigma$ the set $U_j = \{i \in I : \varphi_j \in \Delta_i\}$.

The family $\{U_j\}_{j \in J}$ has the **finite intersection property** (fip): any finite intersection $U_{j_1} \cap \ldots \cap U_{j_n}$, with $U_{j_i} \in \{U_j\}_{j \in J}$, is nonempty. This is because the finite set $\{\varphi_{j_1}, \ldots, \varphi_{j_n}\}$ is included in each U_{j_i}.

Exercise 5.58 Suppose the family $\{U_j\}_{j \in J}$ has the fip. Consider the family $\mathcal{F} \subseteq \mathcal{P}(I)$ determined by the following condition: $V \in \mathcal{F}$ iff there are $U_{j_1}, \ldots, U_{j_n}$ (with $j \geq 1$) such that $U_{j_1} \cap \ldots \cap U_{j_n} \subseteq V$. Show that $\mathcal{F}$ is a proper filter.

The family $\mathcal{F}$ from Exercise 5.58 can be extended to an ultrafilter $\mathcal{U}$ over I. Note that, for each $j \in J$, $U_j \in \mathcal{U}$. We now have a simple and direct proof of compactness.

5.5 The Compactness Theorem

Proof Let the index sets I, J be as above. Suppose that Σ is finitely satisfiable and let $\{\mathcal{M}_i\}_{i \in I}$ be a family of models such that $\mathcal{M}_i \models \Delta_i$. Let $\mathcal{U}$ be the ultrafilter over I we have just described. Consider the ultraproduct $\Pi_{\mathcal{U}} \mathcal{M}_i$. If $\varphi_j \in \Sigma$, then $U_j \subseteq \{i \in I : \mathcal{M}_i \models \varphi_j\}$, which implies $\{i \in I : \mathcal{M}_i \models \varphi_j\} \in \mathcal{U}$. By Łoś' Theorem, $\Pi_{\mathcal{U}} \mathcal{M}_i \models \varphi_j$. Since this is the case for any $j \in J$, $\Pi_{\mathcal{U}} \mathcal{M}_i \models \Sigma$, i.e. Σ is satisfiable. □

In Sect. 3.1, we defined a set of sentences Σ to be consistent iff it has a model. Compactness implies that Σ is consistent iff it is finitely consistent, i.e. iff any finite subset of Σ is consistent. Equivalently, Σ is inconsistent iff it is finitely inconsistent. It also follows from our definition of consistency that, if $\phi \land \neg \phi$ follows from Σ, then $\phi \land \neg \phi$ must follow from a finite subset of Σ. The last remark generalises.

Corollary 5.59 *Let Σ be a set of $\mathcal{L}$-sentences and φ a $\mathcal{L}$-sentence. Then $\Sigma \models \varphi$ implies that, for some finite $\Delta \subseteq \Sigma$, $\Delta \models \varphi$.*

Proof Let $\{\Delta_i : i \in I\}$ be the indexed family of all finite subsets of Σ. Suppose no finite subset $\Delta_i \subseteq \Sigma$ implies φ. This is to say that, for each $i \in I$, $\mathcal{M}_i \models \Delta_i \cup \{\neg \varphi\}$. Consequently, the ultraproduct $\Pi_{\mathcal{U}} \mathcal{M}_i$ is a model Σ and $\neg \varphi$, against the hypothesis $\Sigma \models \varphi$. This contradiction concludes the proof. □

Exercise 5.60 Assume Corollary 5.59 and prove compactness from it.

Exercise 5.61 Let Σ be a set of $\mathcal{L}$-formulae whose free variables occur among $x_1, \ldots, x_n$. Show that there are a $\mathcal{L}$-structure $\mathcal{M}$ and an assignment σ such that $\mathcal{M} \models \Sigma[\sigma]$ iff this is the case for every finite subset of Σ (*Hint*: add n new constant symbols to the signature of $\mathcal{L}$).

Exercise 5.62 Prove that, if a theory T has arbitrarily large, finite models, then it must have an infinite model. Deduce that, if a theory T has only finite models, their size must be bounded.

In some applications, the following equivalent form of compactness proves useful.

Corollary 5.63 *Let Σ be a set of sentences with the property that $\phi, \psi \in \Sigma$ implies $\phi \land \psi \in \Sigma$. Let Ω be a consistent set of sentences. Then $\Omega \cup \Sigma$ is inconsistent iff for some $\varphi \in \Sigma$, $\Omega \models \neg \varphi$.*

Proof If, for some $\varphi \in \Sigma$, $\Omega \models \neg \varphi$, then $\Omega \cup \Sigma$ is inconsistent because $\Omega \cup \Sigma \models \varphi \land \neg \varphi$. Conversely, if $\Omega \cup \Sigma$ is inconsistent, then it is finitely inconsistent, by compactness. Let $\Delta = \{\varphi_1, \ldots, \varphi_n\}$ be a finite, inconsistent subset of $\Omega \cup S$. Because Ω is consistent, Δ must contain elements of Σ: let φ be their conjunction. By hypothesis, Σ contains the conjunctions of its elements, so $\varphi \in \Sigma$. If $\Omega \cup \{\varphi\}$ had a model $\mathcal{M}$, then $\mathcal{M} \models \Omega \cup \Delta$. Since $\Omega \cup \Delta$ has no models, $\Omega \cup \{\varphi\}$ can have none either, i.e. $\Omega \models \neg \varphi$. □

We conclude this section with an Exercise that establishes the equivalence between Compactness and Corollary 5.63.

Exercise 5.64 Suppose that Corollary 5.63 holds and let Σ be a finitely satisfiable set of sentences. We shall prove that Σ is has a model by contradiction. Thus, we assume Σ to be unsatisfiable, i.e. to have no model.

(i) Define a chain of extensions of Σ as follows: set $\Sigma_0 = \Sigma$ and, assuming Σ_n to have been defined, set $\Sigma_{n+1} = \Sigma_n \cup C_n$, where C_n is the set of all conjunctions of two sentences in Σ_n. Show by arithmetical induction that, for any $n \in \mathbb{N}$, Σ_n is finitely satisfiable.
(ii) Set $\Sigma^+ = \bigcup_{n \in \mathbb{N}} \S_n$. Show that Σ^+ is finitely satisfiable and that $\phi, \psi \in \Sigma^+$ implies $\phi \wedge \psi \in \Sigma^+$.
(iii) Since Σ^+ is unsatisfiable, apply Corollary 5.63 to Σ^+ and $\Omega = \emptyset$ to deduce that Σ^+ contains a contradiction and cannot therefore be finitely satisfiable.

5.6 Elementary Classes

Recall from Sect. 3.1 that a class of $\mathcal{L}$-structures **K** is **elementary** iff there is a set of sentences T such that $Th(\mathbf{K}) = T$ or, equivalently, $Mod(T) = \mathbf{K}$. Utraproducts provide a straightforward characterisation of elementary classes, i.e. allows us to give necessary and sufficient conditions for **K** to be elementary.

We say that class **K** is **closed under elementary equivalence** iff $\mathcal{M} \in \mathbf{K}$ and $\mathcal{M} \equiv \mathcal{N}$ imply $\mathcal{N} \in \mathbf{K}$. We say that a class **K** is **closed under ultraproducts** iff, whenever $\{\mathcal{M}_i\}_{i \in I}$ is a family of $\mathcal{L}$-structures in **K** and $\mathcal{U}$ is an ultrafilter over I, then $\Pi_{\mathcal{U}} \mathcal{M}_i \in \mathbf{K}$.

Theorem 5.65 *Let K be a class of $\mathcal{L}$-structures. Then K is elementary iff it is closed under elementary equivalence and ultraproducts.*

Proof If **K** is elementary, then it is axiomatised by a set of sentences T. It is clear that, if $\mathcal{M} \models T$ and $\mathcal{N} \equiv \mathcal{M}$, then $\mathcal{N} \models T$ and $\mathcal{N} \in \mathbf{K}$. Given a family $\{\mathcal{M}_i\}_{i \in I}$ of models of T and an ultrafilter $\mathcal{U}$, Łos' Theorem implies that $\Pi_{\mathcal{U}} \mathcal{M}_i \models T$.

It remains to show that, if **K** is closed under elementary equivalence and ultraproducts, then it is an elementary class. We set $T = Th(\mathbf{K})$ and show that there are no models of T beyond the $\mathcal{L}$-structures in **K**. If $\mathcal{M} \in \mathbf{K}$, obviously $\mathcal{M} \models T$. To prove that, if $\mathcal{M} \models T$, then $\mathcal{M}$ belongs to the class **K**, we proceed indirectly. We show that $\mathcal{M} \equiv \mathcal{N}$ for some $\mathcal{N} \in \mathbf{K}$, where $\mathcal{N}$ is a suitable ultraproduct.

First, we consider $Th\mathcal{M}$, the set of sentences modelled by $\mathcal{M}$. Note that any finite subset Δ_i of $Th\mathcal{M}$ has a model $\mathcal{M}_i$ in **K**. If this was not the case, calling φ the conjunction of the elements of Δ_i, we could deduce $Th\mathbf{K} \models \neg\varphi$. The condition $\mathcal{M} \models T = Th\mathbf{K}$ implies $T \models \varphi \wedge \neg\varphi$, which is impossible because T is consistent.

Let $\{\varphi_j\}_{j \in J}$ be the set of formulae in $Th\mathcal{M}$ and $\{\mathcal{M}_i\}_{i \in I}$ the set of $\mathcal{L}$-structure that model the corresponding finite subsets $\{\Delta_i : i \in I\}$ of $Th\mathcal{M}$. We select an ultrafilter $\mathcal{U}$ over I exactly exactly as we did when proving compactness. By this choice of ultrafilter, the ultraproduct $\mathcal{N} = \Pi_{\mathcal{U}} \mathcal{M}_i$ models $Th\mathcal{M}$. Since $Th\mathcal{M}$ is a complete theory, it has no consistent extensions: the fact that $Th\mathcal{M} \subseteq Th\mathcal{N}$ and the consistency of ThN thus imply $ThM = ThN$. In other words $\mathcal{M} \equiv \mathcal{N}$. But $\mathcal{N}$ was constructed as an ultraproduct of $\mathcal{L}$-structures in **K**, so $\mathcal{N} \in \mathbf{K}$. The closure of **K** under elementary equivalence finally implies $\mathcal{M} \in \mathbf{K}$. □

Some immediate applications of Theorem 5.65 show that certain classes of structures are not elementary.

Exercise 5.66 Show that the class **F** of finite $\mathcal{L}_=$-structures is not elementary (*Hint*: consider the $\mathcal{L}_=$-structure $X_k = \{0, 1, \ldots, k\}$ for each $k \in \mathbb{N}$).

Exercise 5.67

(1) Show that if a class of structures **K** and its complement (i.e. the class of all structures not in **K**) are elementary, then **K** is axiomatised by single sentence (*Hint*: use Compactness on an axiomatisation S_1 of **K** and an axiomatisatoin S_2 of its complement).
(2) Show that the class of infinite $\mathcal{L}_=$-structures is elementary.
(3) Given an alternative argument to the effect that the class of finite $\mathcal{L}_=$-structures is not elementary.
(4) Show that the class of infinite $\mathcal{L}_=$-structures is not finitely axiomatisable or, equivalently, that it is not axiomatised by a single $\mathcal{L}_=$-sentence.

Exercise 5.68 Let $P = \{p_i\}_{i \in \mathbb{N}}$ be the sequence of all prime numbers, listed in order of magnitude. For each $i \in \mathbb{N}$, let the $\mathcal{L}_r$-structure $\mathcal{K}_i$ be a field with the property that $\underbrace{1 + 1 + \ldots + 1}_{p_i \text{ times}} = 0$, where p_i is the i-th prime number. We say that $\mathcal{K}_i$ is a field of characteristic p_i.

Show that, if $\mathcal{U}$ is an ultrafilter over P that extends the filter of cofinite subsets of P, then $\Pi_{\mathcal{U}} \mathcal{K}_i$ is such that, for every $n \in \mathbb{N}$, $\underbrace{1 + 1 + \ldots + 1}_{n \text{ times}} \neq 0$. When a field satisfies this property, we say that it has characteristic 0. This exercise shows that the class of fields of nonzero characteristic is not elementary (more information about the characteristic of a field will be given in the next chapter).

Exercise 5.69 Use compactness directly to provide alternative solutions to Exercises 5.66 and 5.68.

5.7 Addendum: The Unity of the World

In a classic, thought relatively little known, paper [39], Clarence Irving Lewis develops an implicitly model-theoretic approach to a philosophical question concerning the tenability of a metaphysical conception of the world as a necessary system of interdependent facts. Lewis takes 'facts' as information that we may come to acquire in the course of scientific investigation. The established results of enquiry at a point in time determine what Lewis calls a **system of facts**.

> What we truly know of physics, of space, of biological forms, or of the world altogether, will determine a system. But we must be careful here to say "determine" and not "constitute". We are never aware of all the logical consequences of what we know; our knowledge fixates a whole system of facts but can not possibly exhaust it [39, pp.143–144].

According to Lewis, a system of facts resembles an axiomatic system:[8] we may have explicitly given a system's axioms but we may not have generated the totality of its consequences. Furthermore, a system may provide only incomplete information concerning the world in that only further enquiry can test hypotheses yet undecided.

Lewis characterises the logical structure of a system by the following list of conditions (quoted from [39], p.143):

(1) If A is a fact of a given system—call it Σ—and A is inconsistent or incompatible with B, then B is not in Σ.
(2) If C is in Σ and C requires or implies D, then D also is in Σ.
(3) If E and F are facts in Σ, then Σ contains also the joint-fact EF [i.e. $E \wedge F$].

It is clear that conditions (1) and (2) amount to our definition of a theory as a consistent, deductively closed set of sentences. Furthermore, conditions (1) to (3) are also very similar to the defining conditions of a proper filter over a set Σ.

These first impressions can be spelled out in model-theoretic terms, which offer a mathematical articulation of Lewis' standpoint.

We can in fact identify first-order $\mathcal{L}$-theories with filters over a suitable structure, which satisfy exactly conditions (1) to (3). To this end, let us fix a language $\mathcal{L}$ and consider the set S of all $\mathcal{L}$-sentences. We partition S into equivalence classes by identifying equivalent sentences. Let the resulting set of equivalence classes be $\mathscr{S}$ and $[\varphi] \in \mathscr{S}$ the equivalence class containing the sentence φ.

It is easy to show that the binary relation $\leq$, defined on $\mathscr{S}$ by the stipulation:[9]

$$[\varphi] \leq [\psi] \text{ iff } \varphi \models \psi$$

is reflexive and transitive. If, moreover $[\varphi] \leq [\psi]$ and $[\psi] \leq [\varphi]$, then $[\varphi] = [\psi]$. The last property of $\leq$ is called antisymmetry. The structure $(\mathscr{S}, \leq)$ is a set endowed with a reflexive, antisymmetric and transitive relation, i.e. a **partially ordered set**.

The partially ordered set $(\mathscr{S}, \leq)$ has a special property: any two elements $[\varphi], [\psi] \in \mathscr{S}$ have a least upper bound and a greatest lower bound. Their least upper bound is an equivalence class $[\chi]$ such that $[\varphi], [\psi] \leq [\chi]$ and, for every $[\theta]$ simultaneously 'above' (relative to $\leq$) $[\varphi], [\psi]$, we have $[\chi] \leq [\theta]$. Clearly $[\chi] = [\varphi \vee \psi]$. The greatest lower bound of $[\varphi], [\psi]$ is defined by reversing the ordering in the previous definition. Again, it is clear that it is $[\varphi \wedge \psi]$.

---

[8] Some of Lewis' examples are indeed axiomatisations of Euclidean or non-Euclidean geometries.
[9] Note that the relation $\leq$ introduced below is well-defined, i.e. it does not depend on which representatives we choose from the relevant equivalence classes.

5.7 Addendum: The Unity of the World

The existence of least upper bounds and greatest lower bounds for any pair of elements enables us to expand $(\mathscr{S}, \leq)$ by the addition of two binary operations defined as follows:

$$[\varphi] \sqcap [\psi] = [\varphi \wedge \psi]; \; [\varphi] \sqcup [\psi] = [\varphi \vee \psi].$$

Both operations are easily seen to be well-defined, commutative and associative. They also satisfy the following absorption laws:[10]

$$[\varphi] \sqcap ([\varphi] \sqcup [\psi]) = [\varphi]; \; [\varphi] \sqcup ([\varphi] \sqcap [\psi]) = [\varphi].$$

The structure $(\mathscr{S}, \leq, \sqcup, \sqcap)$ is by definition a **lattice**. It is, in fact, a rather special kind of lattice, since it contains a least element, namely the equivalence class **0** of all contradictions and a greatest element, namely the equivalence class **1** of all valid sentences. Moreover, the definitions of $\sqcap, \sqcup$ and two applications of truth-tables readily show that the following equalities, known as distributive laws, hold:

$$[\varphi] \sqcup ([\psi] \sqcap [\chi]) = ([\varphi] \sqcup [\psi]) \sqcap ([\varphi] \sqcup [\chi]);$$

$$[\varphi] \sqcap ([\psi] \sqcup [\chi]) = ([\varphi] \sqcap [\psi]) \sqcup ([\varphi] \sqcap [\chi]).$$

We are thus in presence of a **distributive lattice**. In fact, this lattice can naturally be endowed with a unary operation of **complement** defined as follows:

$$[\varphi]' := [\neg \varphi].$$

For any $[\varphi] \in \mathscr{S}$, complement satisfies the identities:

$$[\varphi] \sqcup [\varphi]' = \mathbf{1}; \; [\varphi] \sqcap [\varphi]' = \mathbf{0}.$$

Distributive, complemented lattices are also known as **Boolean algebras**. By what we have remarked, it is apparent that the class of Boolean algebras is elementary, i.e. it can be axiomatised in a suitable first-order language. The special Boolean algebra $(\mathscr{S}, \leq, \sqcap, \sqcup, ', \mathbf{0}, \mathbf{1})$ is known as a **Lindenbaum algebra**.

The reader may have noted that $\sqcap, \sqcup$, and $'$ behave like the union, intersection and complement of subsets of a fixed set A and that $\mathbf{0}, \mathbf{1}$ behave like $\emptyset$ and A. Although this analogy could be made mathematically more explicit, we shall leave it in suggestive form, since it only serves the purpose of indicating that there is a natural way of defining a proper filter in a Boolean algebra. If we think of $\leq$ as playing the role of set-theoretic inclusion and $\sqcap$ as playing the role of intersection, we arrive at

---

[10] These laws have a trivial verification, since the definitions of $\sqcap$, $sqcup$ allows to reduce them to equivalences between single formulae, which can be verified by means of truth-tables.

Definition 5.70 Let $\mathcal{B} = (B, \leq, \sqcap, \sqcup, ', \mathbf{0}, \mathbf{1})$ be a Boolean algebra. A **proper filter** over B is a subset F of B such that:

(i) $\mathbf{0} \notin F$;
(ii) $a \in F$ and $a \leq b$ imply $b \in F$;
(iii) $a, b \in F$ implies $a \sqcap b \in F$.

Definition 5.70 spells out, in an abstract form, exactly the conditions (1) to (3) identified by Lewis. When we think of his 'facts' as linguistically formulated, Lewis' notion of a system is exemplified, in the context of a first-order language, by a filter over the corresponding Lindenbaum algebra. We shall now prove that, given a first-order language $\mathcal{L}$, the $\mathcal{L}$-theories are exactly the filters over the corresponding Lindenbaum algebra. Thus, for instance, the theories we have discussed in Chap. 3 are systems in Lewis' sense.

Theorem 5.71 Let $\mathcal{L}$ be a first-order language and $\mathcal{S} = (\mathscr{S}, \leq, \sqcap, \sqcup, ', \mathbf{0}, \mathbf{1})$ the associated Lindenbaum algebra. Any $\mathcal{L}$-theory determines a proper filter over $\mathscr{S}$. Conversely, any proper filter over $\mathscr{S}$ determines a $\mathcal{L}$-theory.

Proof Let T be a $\mathcal{L}$-theory. The sentences of T single out a subset F of $\mathscr{S}$. Because T is consistent, $\mathbf{0} \notin F$. Next suppose $[\varphi] \in F$ and $[\varphi] \leq [\psi]$: by definition $\varphi \models \psi$ and, since $\varphi \in T$ and T is deductively closed, we have $\psi \in T$, i.e. $[\psi] \in F$. If, finally, $[\varphi], [\psi] \in F$, then $\varphi, \psi \in T$ and, since T is deductively closed, $\varphi \wedge \psi \in T$, which implies $[\varphi] \sqcap [\psi] \in F$. It follows that F is a proper filter over the Lindenbaum algebra $\mathcal{S}$.

Suppose, on the other hand, that $F \subseteq \mathscr{S}$ is a proper filter. The union of the equivalence classes of elements of F determines a set T of $\mathcal{L}$-sentences. We only have to show that this set is a $\mathcal{L}$-theory. Clearly, T must be consistent. We only have to show that T is deductively closed, i.e. that, if $T \models \varphi$, then $\varphi \in T$. Note that, by compactness, if $T \models \varphi$ then a finite subset of T, say $\theta_1, \ldots, \theta_k$, already implies φ. Because $[\theta_i]$ ($i = 1, \ldots, k$) is in F, and F is a filter, $[\theta_1 \wedge \ldots \wedge \theta_k]$ is also in F. Now we have $[\theta_1 \wedge \ldots \wedge \theta_k] \leq [\psi]$ and the properties of F imply $[\psi] \in F$, i.e. $\psi \in T$, as desired. □

Lewis observes that any system may be included into a complete system, what he calls a 'world', i.e. an exhaustive totality of facts. A world has for Lewis the following features (see [39], p.145):

(4) If A is a fact of a given world—call it Ω—and A is incompatible with B, then B is not in Ω.
(5) If C is not in Ω, then the contradictory of C is in Ω.
(6) If E and F are in Ω, then Ω contains also the joint-fact EF.

Conditions (4) to (6) imply conditions (1) to (3). What they describe is the completion of a system. In model-theoretic terms, Lewis is viewing a world as a complete theory extending a consistent theory. A minor extension of the last theorem enables us to associate completions of theories with ultrafilters in the corresponding Lindenbaum algebra.

5.7 Addendum: The Unity of the World

Theorem 5.72 *Let $\mathcal{L}$ be a first-order language and $\mathcal{S} = (\mathscr{S}, \leq, \sqcap, \sqcup, ', \mathbf{0}, \mathbf{1})$ the associated Lindenbaum algebra. Any complete $\mathcal{L}$-theory determines an ultrafilter over $\mathscr{S}$. Conversely, any ultrafilter over $\mathscr{S}$ determines a complete $\mathcal{L}$-theory.*

Lewis' goal is to discuss the logical structure of systems and their completions in order to reject a conception of the world that has been historically influential in philosophy, and according to which the world is a fixed, necessary system of facts. In his words, the conception of interest seeks to:

> [...] prove for the world that kind of unity which makes every part depend upon every other part or upon the whole. And this in turn has been construed to mean that everything in the world is necessarily as it is, and every fact of it a necessary fact [39, p.146].

We famously encounter the conception outlined in the metaphysical system of Spinoza. Lewis rejects this conception by means of the following argument, whose steps for the moment we only outline in Lewis' terminology:

(i) if any complete system may be conceived, then more than one may be conceived;
(ii) every possible world will contain some facts in common with any other;
(iii) it follows that the actual world contains facts in common with worlds not actual;
(iv) every world, including the actual one, contains mutually independent parts;
(v) reality cannot be such that all its facts are necessary and necessarily related.

We now present Lewis' articulation of his argument with reference to model-theoretic examples whenever they can be invoked to shed light on the structure of his reasoning. Step (i) is made to rest on the possibility of systems that cannot be true of the same world. In formal terms, Lewis is talking about sets of conditions that cannot admit one and the same completion.

That such sets exist may be shown by an easy example, once we understand 'completion' in the sense of the last theorem or, equivalently, in the sense defined at the beginning of Chap. 3. Consider the $\mathcal{L}_=$-sentence φ_2 asserting that there are at least 2 distinct objects and the $\mathcal{L}_=$-sentence φ_3 asserting that there are at least 3 distinct objects. Let ψ_2 be the sentence asserting that there are at most 2 distinct objects. Then $T_2 = \{\psi_2 \cup \varphi_2\}$ is a completion of φ_2 (all models of T_2 are isomorphic and thus elementarily equivalent). At the same time, ψ_2 is incompatible with φ_3 so any completion T of φ_3 will differ from T_2.

If we take 'worlds' to be complete theories, we have shown that step (i) is model-theoretically vindicated. Step (ii) is, from our perspective, an immediate consequence of the fact that complete theories are ultrafilters in a Lindenbaum algebra and thus any two of them necessarily share **1**. Lewis' deduction of (ii) is more elaborate but not too different:

> Second, every world which is logically conceivable will have something in common with every other. For if this were not the case, there could be at most only two such worlds, the world of all the propositions actually true, and the world consisting of the denials of each of these propositions. But the contradictories of true propositions are not all consistent with each other and could not describe a possible world [39, p.148].

Step (iii) is clear given (ii). Complete theories that are mutually incompatible have nonempty intersection. What Lewis goes on to observe is that, in this event, we can identify distinct independent parts of a complete theory by looking at other complete theories. We reconstruct his argument in terms of ultrafilters over a Lindenbaum algebra. If U_1, U_2 are ultrafilters, set $U_1 \cap U_2 = A$. If $[\varphi] \in U_1 - A$, then $[\varphi]' \in U_2$. Taking A as a set of sentences (representing the corresponding equivalence classes), we cannot have $A \models \varphi$ or U_2, being a filter, should contain φ. By a similar argument we cannot have $A \models \neg\varphi$. It follows that φ is independent of A. Because we can repeat this argument for any sentence in $U_1 - A$, the last set is what Lewis calls an 'independent part' of the world U_1. If we had U_3 distinct from U_1, U_2, then $U_1 \cap U_3 = B \neq A$, or $U_3 = U_2$. In this case $U_1 - B \neq U_1 - A$ would be another independent part of U_1.

Lewis argues that there are infinitely many mutually incompatible complete theories,[11] which determine the independent parts of the 'world', i.e. the complete theory meant as a true description of the world. It is worth noting that Lewis's argument does not depend on us ever possessing such a description, but only on the logical structure it must possess.

Step (v) now follows: if it is possible to identify independent parts of the world, these parts cannot be locked together in a necessary order, because they are logically consistent with very different configurations of facts. It does not follow that a 'world' may not be 'axiomatisable', i.e. reduced to a configuration of facts on which every other fact depends. Even in such a close-knit world, however, some of the controlling facts will be mutually independent.

---

[11] In model-theoretic terms, this statement amounts to the claim that a $\mathcal{L}$-theory may have infinitely many distinct completions. The $\mathcal{L}_r$-theory of algebraically closed fields to be introduced in the next chapter is such a theory, as we shall prove in Chap. 8.

Chapter 6
Essential Field Theory

Abstract Many theorems to be discussed in Chaps. 7 to 10 depend on basic field-theoretic results. This chapter reviews them.

Section 6.1 describes the structure of simple fields extensions, obtained from a given field by adjoining a single, new element. Section 6.2 includes the basic results concerning algebraic and finite extensions that will be needed.

Section 6.3 introduces the first-order theory ACF of algebraically closed fields and Sect. 6.4 the extensions of ACF to fields of positive and zero characteristic.

Section 6.5 proves an important theorem about field embeddings, which in turn provides the foundation for our discussion of algebraic closure in the next chapter. The proof makes use of Zorn's lemma, for which the reader is again referred to Sect. A.6 of the Appendix.

Throughout this chapter, we invariably deal with $\mathcal{L}_r$-structures that model the field axioms. For this reason it is helpful to have a symbolic way of designating multiplicative inverses of non-zero field elements. We write $1/x$ to refer to the multiplicative inverse of $x \neq 0$, while y/x designates the product of y and the multiplicative inverse of x. These notational choices should not suggest that we are transgressing the linguistic boundaries imposed upon us by the signature of $\mathcal{L}_r$: they are only special typographical conventions that enable us concisely to identify certain field elements.

6.1 Simple Field Extensions

We have already seen in Chap. 3 that, given a field $\mathcal{K}$, there exist fields including K and obtained by taking quotients of the polynomial ring $K[x]$ *modulo* a maximal ideal.

In this section we consider the extensions of a field obtained by adjoining a new element to it. Our typical setting consists of a pair of fixed fields $\mathcal{K}, \mathcal{M}$ such that $\mathcal{K} \sqsubseteq \mathcal{M}$. We wish to study the extensions of $\mathcal{K}$ generating by adding to K an element of M. If, for instance, $\alpha \in M$, we want to describe the smallest subfield of $\mathcal{M}$ that contains $K \cup \{\alpha\}$. We call this field a **simple extension** of $\mathcal{K}$ and denote it by $\mathcal{K}(\alpha)$. Note that $\mathcal{K}$ is a simple extension of itself.

For reasons to be later clarified, it is illuminating to view field extensions as vector spaces over the field they extend. When $\mathcal{M}$ is a field extension of $\mathcal{K}$, the additive group with domain M can be regarded as the domain of a vector space over K. In model-theoretic terms, the $\mathcal{L}_g$-reduct[1] of $\mathcal{M}$ is an abelian group and we can easily interpret each function symbol f_k ($k \in K$) on a unary function from M to M by setting, for any $a \in M$:

$$f_k a = ka, \text{ where } ka \text{ is the field product of } k, a \text{ relative to } \mathcal{M}.$$

Since viewing M as the domain of a field or of a vector space makes model-theoretic difference, we shall take $\mathcal{M}$ to refer to the field based on M while $\mathcal{M}_K$ is the corresponding vector space over K.

Definition 6.1 Let $\mathcal{M}$ be a field extending $\mathcal{K}$. The **degree** of $\mathcal{M}$ over $\mathcal{K}$, denoted by $[M : K]$, is the dimension of the $\mathcal{L}_v$-structure $\mathcal{M}_K$. The degree of an extension may or may not be finite, as we shall shortly see.

The next exercise looks closely at a simple extension of the rational field $\mathbb{Q}$.

Exercise 6.2 Consider the set

$$\mathbb{Q}(\sqrt{2}) = \{a + b\sqrt{2} : a, b \in \mathbb{Q}\},$$

where $\sqrt{2}$ is a number α such that $\alpha^2 = 2$.

(i) Show that, if $u = a + b\sqrt{2} \in \mathbb{Q}(\sqrt{2})$ is different from 0, there are $c, d \in \mathbb{Q}$ such that $1/v = c + d\sqrt{2}$ satisfies the equality $u/v = 1$.
(ii) Show that, if $a, b \in \mathbb{Q}(\sqrt{2})$, then $a + (-b), a/b \in \mathbb{Q}(\sqrt{2})$. Deduce that $\mathbb{Q}(\sqrt{2})$ is the domain of a $\mathcal{L}_r$-structure that models the field axioms. We call this field $\mathbb{Q}(\sqrt{2})$.
(iii) Let $\sigma : V \longrightarrow \mathbb{Q}(\sqrt{2})$ be an assignment such that $\sigma(x) = \sqrt{2}$ (here V is the set of variables in $\mathcal{L}_r$). Show that $\mathbb{Q}[\sqrt{2}] = \mathbb{Q}(\sqrt{2})$, where $\mathbb{Q}[\sqrt{2}] \subseteq \mathbb{Q}(\sqrt{2})$ is obtained from the polynomial ring $\mathbb{Q}[x]$ by evaluating its elements with σ.
(iv) The set $\mathbb{Q}(\sqrt{2})$ may be viewed as the domain of a $\mathcal{L}_v$-structure that models $VS_\mathbb{Q}$, once we take each f_q to behave exactly like field multiplication by q. Let the resulting $\mathcal{L}_v$-structure be called $\mathbb{Q}(\sqrt{2})_\mathbb{Q}$. Find a basis for this vector space.

The elements of a simple extension have a special form, described in the next lemma.

Lemma 6.3 *Let the $\mathcal{L}_r$-structures $\mathcal{K}, \mathcal{M}$ be fields such that $\mathcal{K} \sqsubseteq \mathcal{M}$. If $\alpha \in M$, then any $\beta \in K(\alpha)$ is of the form $p(\alpha)/q(\alpha)$, where $p(\alpha), q(\alpha) \neq 0$ are linear combinations of powers of α with coefficients in K.*

---

[1] This is the structure we obtain from $\mathcal{M}$ by dropping the 'multiplicative part' of $\mathcal{L}_r$'s signature.

6.1 Simple Field Extensions

Proof Call S the set of all elements of $\mathcal{M}$ of the form $p(\alpha)/q(\alpha)$, with $q(\alpha) \neq 0$. First, we show that S is a subfield of $\mathcal{M}$. Since, by definition $K \cup \{\alpha\} \subseteq S$, it will follow that $\mathcal{K}(\alpha) \subseteq S$. Clearly, S is nonempty, since e.g. $\alpha \in S$ (set $p(\alpha) = \alpha$, $q(\alpha) = 1$).

Next, consider $p(\alpha)/q(\alpha)$, $p'(\alpha)/q'(\alpha)$, with $q(\alpha), q'(\alpha) \neq 0$. Since $\mathcal{M}$ is a field, $q(\alpha)q'(\alpha) \neq 0$ and its multiplicative inverse $1/(q(\alpha)q'(\alpha))$ is in L. Using the existence of additive inverses and the distributive law in $\mathcal{M}$, we obtain:

$$\frac{p(\alpha)}{q(\alpha)} + \left(-\frac{p'(\alpha)}{q'(\alpha)}\right) = \frac{p(\alpha)q'(\alpha) + (-p'(\alpha)q(\alpha))}{q(\alpha)q'(\alpha)} \in S.$$

Moreover, it is not difficult to show that:

$$\frac{p(\alpha)}{q(\alpha)}, \frac{p'(\alpha)}{q'(\alpha)} \in S \text{ implies } \frac{p(\alpha)q'(\alpha)}{q(\alpha)p'(\alpha)} \in S.$$

The details of the above computations are left to the reader (see Exercise 6.4). It follows from them that S is a subfield of $\mathcal{M}$ containing $K \cup \{\alpha\}$. Since, by definition $\mathcal{K}(\alpha)$ is the smallest subfield of $\mathcal{M}$ including $K \cup \{\alpha\}$, we have shown that $\mathcal{K}(\alpha) \subseteq S$.

To prove the converse inclusion, suppose $p(\alpha)/q(\alpha) \in S$. Note that $p(\alpha), q(\alpha)$ are obtained under suitable iterations of the field operations on finitely many elements of K and α. Consequently, $p(\alpha), q(\alpha) \in K(\alpha)$ because the latter set is the domain of a field and, thus, it is closed under the field operations.

If, in addition, $q(\alpha) \neq 0$, the set $K(\alpha)$ contains the multiplicative inverse of $q(\alpha)$, i.e. $1/q(\alpha)$. Since $K(\alpha)$ also contains the products of its elements, we conclude $p(\alpha)/q(\alpha) \in K(\alpha)$. We now have $S \subseteq K(\alpha)$.

Putting together both inclusions, we obtain the equality $S = K(\alpha)$. □

Exercise 6.4 Complete the proof of Lemma 6.3.

Exercise 6.5 Let $\mathcal{K}, \mathcal{M}$ be fields such that $\mathcal{K} \sqsubseteq \mathcal{M}$ and $S = \{\alpha_1, \ldots, \alpha_n\}$ a finite subset of M. Generalise Lemma 6.3 to $\mathcal{K}(\alpha_1, \ldots, \alpha_n) = \mathcal{K}(S)$, the smallest subfield of $\mathcal{M}$ containing $K \cup S$.

Remark 6.6 The last exercise can be further generalised. If $\mathcal{K}, \mathcal{M}$ are as in the Exercise 6.5 and $S \subseteq M$, then the field $\mathcal{K}(S)$ consists of all 'quotients of polynomials' in finitely many elements $\alpha_1, \ldots, \alpha_n$ of S. This observation will be needed in the next chapter.

There are two distinct kinds of field extension. The distinction depends on the fact that, given $\mathcal{K}(\alpha) \sqsupseteq \mathcal{K}$, α may or may not be the root of a non-constant polynomial in $K[x]$. In the former case we say that α is **algebraic over** K while in the latter case we say that α is **transcendental** over K.

Definition 6.7 Let $\mathcal{K}$ be a field and $\mathcal{K}(\alpha)$ be a simple extension of $\mathcal{K}$. We say that $\mathcal{K}(\alpha)$ is a **simple algebraic extension** $\mathcal{K}$ iff there is a non-constant polynomial $p(x) \in K[x]$ such that $p(\alpha) = 0$.

If this is not the case, we say that $\mathcal{K}(\alpha)$ is a **simple transcendental extension** $\mathcal{K}$.

More generally, we have:

Definition 6.8 Let the $\mathcal{L}_r$-structure $\mathcal{K}$ be a model of the field axioms. The field $\mathcal{N}$ is said to be an **algebraic extension** of $\mathcal{K}$ iff every element of $\mathcal{N}$ is algebraic over $\mathcal{K}$. We call an extension of $\mathcal{K}$ **transcendental** iff it is not algebraic.

Lemma 6.3 describes the field extension $\mathcal{K}(\alpha)$ independently of whether α is transcendental or algebraic over K. When, however, $\mathcal{K}(\alpha)$ is an algebraic extension of $\mathcal{K}$, a significant amount of additional information is available. We describe it in the next section.

6.2 Algebraic and Finite Extensions

Simple algebraic extensions are associated with special polynomials that are both monic and irreducible. Here are the relevant definitions:

Definition 6.9 Let $\mathcal{K}$ be a field. A polynomial $p(x) \in K[x]$ is said to be **monic** iff it contains the sub-term x^n and $\partial p(x) = n$. Clearly, if α is the root of a given polynomial $f(x)$, it is also the root of a monic polynomial.

Definition 6.10 Let $\mathcal{K}$ be a field. A polynomial $p(x) \in K[x]$ is said to be **irreducible** iff $p(x) = f(x)g(x)$ implies $\partial f(x) = 0$ or $\partial g(x) = 0$.

Any element α that is algebraic over a given field $\mathcal{K}$ determines a simple extension of $\mathcal{K}$ and a unique, monic and irreducible polynomial $m(x)$ associated with it.

Theorem 6.11 *Let $\mathcal{K}, \mathcal{M}$ be fields such that $\mathcal{K} \sqsubseteq \mathcal{M}$. If $\alpha \in M$ is a root of a non-constant polynomial in $K[x]$, there is an irreducible, monic polynomial $m(x) \in K[x]$ such that:*

(a) $m(\alpha) = 0$;
(b) if $p(x) \in K[x]$ and $p(\alpha) = 0$, then $m(x) \mid p(x)$, i.e. $m(x)$ divides $p(x)$;
(c) $K(\alpha) = K[\alpha]$;
(d) $[K(\alpha) : K] = \partial(m(x))$.

Proof Note that the polynomial $m(x)$ is uniquely determined. Suppose Theorem 6.11 is true of the monic polynomials $m(x)$ and $m'(x)$. By (b), $m(x)$ divides $m'(x)$ and $m'(x)$ divides $m(x)$. Consequently, $\partial m'(x) \leq \partial m(x)$ and $\partial m(x) \leq \partial m'(x)$, i.e. $m(x), m'(x)$ have the same degree. It follows that, for some $a \in K$, $m(x) = am'(x)$. Because $a \neq 1$ implies that $m(x)$ is not monic, we must have $a = 1$ and $m(x) = m'(x)$. If $m(x)$ exists, it is unique.

6.2 Algebraic and Finite Extensions

We now prove its existence. Consider a non-constant polynomial $g(x) \in K[x]$ with least degree such that $g(\alpha) = 0$. If $g(x)$ is monic, we set $g(x) = m(x)$. Otherwise there is a monic polynomial $m(x)$ of the same degree as $g(x)$ and such that $m(\alpha) = 0$. The polynomial $m(x)$ is irreducible. Toward a contradiction, suppose otherwise and let $m(x) = p(x)q(x)$, where $\partial p(x), \partial q(x) \geq 1$. The equality $m(\alpha) = 0$, implies $p(\alpha) = 0$ or $q(\alpha) = 0$. None of the alternatives can hold because each implies the existence of a non-constant polynomial with a root equal to α and of degree smaller than $\partial m(x)$, against the choice of $m(x)$. We have established part (a).

To prove part (b), we suppose that, for some $p(x) \in K[x]$, $p(\alpha) = 0$. Because division with remainder can be carried out in $K[x]$, there are unique $q(x), r(x) \in K[x]$ such that:

$$p(x) = q(x)m(x) + r(x) \text{ and } \partial r(x) < \partial m(x).$$

Because

$$0 = p(\alpha) = q(\alpha)m(\alpha) + r(\alpha) = q(\alpha)0 + r(\alpha),$$

we obtain $r(\alpha) = 0$, which implies $r(x) = 0$ (otherwise $r(x)$ is non-constant, $r(\alpha) = 0$ and $\partial r(x) < \partial m(x)$, against the choice of $m(x)$). We conclude $p(x) = q(x)m(x)$, i.e. $m(x) \mid p(x)$.

To prove part (c), we invoke Lemma 6.3. An arbitrary element of $K(\alpha)$ is of the form:

$$\frac{p(\alpha)}{q(\alpha)} \text{ with } q(\alpha) \neq 0.$$

Clearly, $K[\alpha] \subseteq K(\alpha)$ (set $q(\alpha) = 1$). To obtain the converse inclusion, we must show $p(\alpha)/q(\alpha) \in K[\alpha]$. To this end, we note that $m(x)$ does not divide $q(x)$ (otherwise $q(\alpha) = 0$). The irreducibility of $m(x)$ implies that the only common factors of $m(x), q(x)$ have zero degree. We now make use of the fact that $m(x)K[x] + q(x)K[x]$ is an ideal U of $K[x]$. Since every ideal in $K[x]$ is generated by a single polynomial, $U = \langle f(x) \rangle$, where $f(x)$ divides both $m(x)$ and $q(x)$. It follows that $f(x)$ has zero degree, i.e. $f(x) = a \in K - \{0\}$ (0 is only a factor of the constant polynomial 0). It now follows that $\langle a \rangle = K[x]$. In particular, $1 \in \langle a \rangle$, i.e. there are polynomials $a(x), b(x)$ such that:

$$1 = a(x)q(x) + b(x)m(x) \in K[x].$$

As a consequence, $1 = a(\alpha)q(\alpha) \in K[\alpha]$. Since $q(\alpha) \neq 0$, its multiplicative inverse $1/q(\alpha)$ exists in M and $a(\alpha) = 1/g(\alpha)$. We infer:

$$\frac{p(\alpha)}{q(\alpha)} = p(\alpha)a(\alpha) \in K[\alpha].$$

We have established $K(\alpha) \subseteq K[\alpha]$. Part (c) follows.

We are left with part (d): we set $n = \partial m(x)$ and consider the field elements $1, \alpha, \ldots, \alpha^{n-1} \in K[\alpha]$. For any $p(x) \in K[x]$, there are $q(x), r(x)$ such that $p(x) = q(x)m(x) = r(x)$. As a consequence, for any $p(\alpha) \in K[\alpha]$, $p(\alpha) + r(\alpha)$, with $\partial r(x) < n$. In other words, any element of $K[\alpha]$ is the remainder of the division $p(x) : m(x)$, evaluated at α. It follows that the linear closure[2] of $\{1, \alpha, \ldots, \alpha^{n-1}\}$ is $K[\alpha]$. Moreover, if $\lambda_1 + \lambda_1\alpha + \ldots + \lambda_{n-1}\alpha^{n-1} = 0$ and the coefficients $\lambda_1, \ldots, \lambda_n \in K$ are not all equal to 0, we can deduce the existence of a polynomial of degree at most $n-1 < \partial m(x)$ and with α among its roots. Such a polynomial cannot, however, exist by the choice of $m(x)$. We conclude that the set $\{1, \alpha, \ldots, \alpha^{n-1}\}$ is linearly independent. By construction, it is a basis of $K[\alpha]$. We conclude $[K[\alpha] : K] = n$. Since $K(\alpha) = K[\alpha]$ by part (c), we have proved part (d). □

The polynomial $m(x)$ referred to in the statement of Theorem 6.11 is called the **minimal polynomial** of α. When α is algebraic, the vector space $\mathsf{K}(\alpha)_K$ is finite-dimensional. As we shall shortly see, this continues to be true even if we adjoin finitely elements of M to K.

By contrast, when α is transcendental over K, $\mathsf{K}(\alpha)_K$ is no longer a finite-dimensional vector space.

Lemma 6.12 *Let $\mathcal{K}, \mathcal{M}$ fields such that $K \sqsubseteq \mathcal{M}$ and let $\alpha \in M$ be transcendental over K. The infinite sequence $1, \alpha, \alpha^2, \alpha^3, \ldots$ is linearly independent over K.*

Proof By definition, an infinite sequence is linearly independent iff any finite subsequence is. Toward a contradiction, we suppose that $1, \alpha, \alpha^2, \alpha^3, \ldots$ are not linearly independent, i.e. that some finite subsequence $1, \alpha, \ldots, \alpha^m$ is linearly dependent over K. In other words, there are field elements $\lambda_0, \ldots, \lambda_m \in K$ such that at least one of them is non-zero and the following equality holds in $\mathsf{K}(\alpha)_K$:

$$\lambda_0 + \lambda_1\alpha + \ldots + \lambda_m\alpha^m = 0.$$

In this case, the polynomial $1 + \lambda_2 x + \ldots + \lambda_m x^m \in K[x]$ has α among its roots. The field element α cannot therefore be transcendental over K, against the hypothesis. This contradiction concludes the proof. □

If $\mathcal{K}$ is a field, any field extension $\mathcal{M}$ such that $[M : K]$ is finite is known as a **finite extension** of $\mathcal{K}$. What we know so far is that simple algebraic extensions are finite extensions. In the next subsection we slightly expand our knowledge of finite extensions, especially in view of Sect. 7.4.

6.2.1 Finite Extensions

Although algebraic field extensions may not be finite, finite extensions are algebraic, as the next exercise shows.

---
[2] In the sense of Sect. 3.2.

6.2 Algebraic and Finite Extensions

Exercise 6.13 Let $\mathcal{K}, \mathcal{M}, \mathcal{N}$ be fields such that $\mathcal{K} \sqsubseteq \mathcal{M} \sqsubseteq \mathcal{N}$. Prove that, if $[M : K] = n$, then M is an algebraic extension of K.

Finite iterations of finite extensions continue to be finite, and thus algebraic.

Theorem 6.14 *Let $\mathcal{K}, \mathcal{M}, \mathcal{N}$ be fields such that $\mathcal{K} \sqsubseteq \mathcal{M}$ and $\mathcal{M} \sqsubseteq \mathcal{N}$. If M is a finite extension of K and N is a finite extension of M, then:*

$$[N : K] = [N : M][M : K]$$

Proof Setting $[N : M] = n$ and $[M : K] = m$, let $\{a_1, \ldots, a_m\}$ be a basis of $\mathcal{M}_K$ and $\{b_1, \ldots, b_n\}$ be a basis of $\mathcal{N}_M$. The finite set $\{a_i b_j : 1 \leq i \leq m, 1 \leq j \leq n\}$ is a basis of $\mathcal{N}_K$ with mn elements. □

Exercise 6.15 Show that $\{a_i b_j : 1 \leq i \leq m, 1 \leq j \leq n\}$ is a basis of $\mathcal{N}_K$.

The following lemma is a relatively straightforward consequence of Theorem 6.14.

Lemma 6.16 *Let $\mathcal{K}, \mathcal{M}$ be fields such that $\mathcal{N}_K \sqsubseteq \mathcal{N}_M$ and $\alpha_1, \ldots, \alpha_n \in M$ be algebraic over K. Then $\mathcal{K}(\alpha_1, \ldots, \alpha_n)$ is a finite extension of $\mathcal{K}$.*

Proof We consider the successive, simple extensions:

$$\mathcal{K}(\alpha_1), \mathcal{K}(\alpha_1, \alpha_2), \ldots, \mathcal{K}(\alpha_1, \ldots, \alpha_{n-1}), \mathcal{K}(\alpha_1, \ldots, \alpha_n).$$

Since α_1 is algebraic over K, let $g_1(x) \in K[x]$ be its minimal polynomial. Theorem 6.11 implies that $[K(\alpha_1) : K] = m_1$ with $m_1 = \partial g_1(x)$. Since α_2 is algebraic over K, it is algebraic over $K(\alpha_1)$ (every polynomial in $K[x]$ is a polynomial in $K(\alpha_1)[x]$). Let $g_2(x)$ be the minimal polynomial of α_2 in $K(\alpha_1)[x]$. Clearly, $[K(\alpha_1, \alpha_2) : K(\alpha_1)] = m_2$, where $m_2 = \partial g_2(x)$. Proceeding in this manner we obtain $[K(\alpha_1, \ldots, \alpha_i) : K(\alpha_1, \ldots, \alpha_{i-1})]$ for each $i \in \{2, \ldots, n\}$, so, by Theorem 6.14:

$$[K(\alpha_1, \ldots, \alpha_n) : K] = [K(\alpha_1, \ldots, \alpha_n) : K(\alpha_1, \ldots, \alpha_{n-1})] \cdots$$
$$[K(\alpha_1) : K] = m_1 \cdots m_n,$$

i.e. $\mathcal{K}(\alpha_1, \ldots, \alpha_n)$ is a finite extension of $\mathcal{K}$. □

Exercise 6.17 Prove that, if $\mathcal{K} \sqsubseteq \mathcal{M}$ are fields and $T \subseteq M$ is a set of field elements algebraic over K, then $\mathcal{K}(T)$ is an algebraic extension of $\mathcal{K}$.

Although not every algebraic extension of a field is finite, we can in some cases use the algebraic character of finite extensions to prove results about algebraic extensions. The following theorem is a representative illustration, which will be repeatedly used in the next chapter.

Theorem 6.18 *Let $\mathcal{K} \subseteq \mathcal{M} \subseteq \mathcal{N}$ be fields. If M is an algebraic extension of $\mathcal{K}$ and $\mathcal{N}$ is an algebraic extension of $\mathcal{M}$, then $\mathcal{N}$ is an algebraic extension of $\mathcal{K}$.*

Proof We note that any $\alpha \in N$ has a minimal polynomial with coefficients $b_0, \ldots, b_k \in M$. It follows that α is algebraic over the field $\mathcal{K}(b_0, \ldots, b_k)$, which is a finite extension of $\mathcal{K}$ by Lemma 6.16, due to the fact that $b_0, \ldots, b_k$ are algebraic over $\mathcal{K}$. Since $\mathcal{K}(\alpha, b_0, \ldots, b_k)$ is a simple extension of $K(b_0, \ldots, b_k)$, it is finite (over it) by Theorem 6.11. We can now apply Exercise 6.19 to infer from the equality:

$$[K(\alpha, b_0, \ldots, b_k) : K] = [K(\alpha, b_0, \ldots, b_k) : K(b_0, \ldots, b_k)][K(b_0, \ldots, b_k) : K]$$

the fact that α belongs to a finite extension of K. Such an extension is algebraic, so α is an element of N algebraic over K. Because α was arbitrarily chosen, we conclude that $\mathcal{N}$ is an algebraic extension of $\mathcal{K}$. □

6.3 Algebraically Closed Fields

If $\mathcal{K}, \mathcal{M}$ are fields and $\mathcal{K} \sqsubseteq \mathcal{M}$, the most comprehensive way of obtaining an algebraic extension of $\mathcal{K}$ within $\mathcal{M}$ consists in adjoining every element of M that is algebraic over K.

The resulting set is actually the domain of a substructure, which we call the **relative algebraic closure** of $\mathcal{K}$ in $\mathcal{M}$ and designate by $acl_\mathcal{M}(K)$.

Exercise 6.19 Let $\mathcal{K}, \mathcal{M}$ be fields such that $\mathcal{K} \sqsubseteq \mathcal{M}$ and let $\alpha, \beta \in M$ be algebraic over K. Prove that $\mathcal{K}(\alpha, \beta) = (K[\alpha])[\beta]$, where $(K[\alpha])[\beta]$ is the set of polynomials from the ring $(K[\alpha])[x]$ evaluated at β.

Theorem 6.20 *Let $\mathcal{K}, \mathcal{M}$ fields such that $\mathcal{K} \sqsubseteq \mathcal{M}$. Then $acl_\mathcal{M}(K)$ is the domain of a subfield of $\mathcal{M}$.*

Proof Every element of K is trivially algebraic over K. In particular, $0, 1 \in acl_\mathcal{M}(K)$. Suppose $\alpha, \beta \in acl_\mathcal{M}(K)$: to avoid triviality, we may assume $\alpha, \beta \notin K$ are algebraic over K. Note that $\alpha, \beta \in K(\alpha, \beta) = (K[\alpha])[\beta]$. Because $\beta \neq 0$, α/β is a field element in $(K[\alpha])[\beta]$. We now rely on the fact that β must be algebraic over $K[\alpha]$, since it is already algebraic over K, to infer that $(K[\alpha])[\beta]$ is a finite extension of $K[\alpha]$ (using Theorem 6.11-(d)). Moreover, $K[\alpha]$ is a finite extension of K. It follows that, for some positive integers m, n, $[(K[\alpha])[\beta] : K[\alpha]] = n$ and $[K[\alpha] : K] = n$. Theorem 6.14 yields:

$$[(K[\alpha])[\beta] : K] = mn.$$

We have shown that the field-element α/β belongs to a finite extension of $\mathcal{K}$, i.e. it is an element of M algebraic over K. In short, $\alpha/\beta \in acl_\mathcal{M}(K)$. An almost identical argument shows that $\alpha + (-\beta) \in acl_\mathcal{M}(K)$. This is all that is needed to prove that $acl_\mathcal{M}(K)$ is the domain of a subfield of $\mathcal{M}$. □

6.3 Algebraically Closed Fields

Remark 6.21 The relative algebraic closure $acl_\mathcal{M}(\mathcal{K})$ depends on the ambient field $\mathcal{M}$. A simple illustration of this dependence arises from setting $\mathcal{K} = \mathbb{Q}$ and considering two distinct ambient fields for the rational numbers, namely the real field $\mathbb{R}$ and the complex field $\mathbb{C}$, which may be viewed as the extension $\mathbb{R}[x]/(x^2+1)$ of the real field. Since $\mathbb{C}$ contains a root of $x^2 + 1$, i.e. an element i such that $i^2 = -1$, the field element $i \in \mathbb{C}$ is algebraic over $\mathbb{Q}$, with minimal polynomial $x^2 + 1 \in \mathbb{Q}[x]$. Thus, $i \in acl_\mathbb{C}(\mathbb{Q})$. By contrast, since the square of every real number is nonnegative, $i \notin acl_\mathbb{R}(\mathbb{Q})$. The relative algebraic closure of the rational field in the complex field is therefore different from its relative algebraic closure in the real field.

The relative algebraic closure of a field $\mathcal{K}$ within an extension $\mathcal{M}$ is not only a substructure, but a submodel of $\mathcal{M}$ when $\mathcal{M}$ models the $\mathcal{L}_r$-sentence:

$$\psi_n := \forall y_0 \ldots \forall y_n \exists x (y_0 + y_1 x + \ldots + y_n x^n = 0),$$

for each positive integer $n \geq 1$. Let T be the set of field axioms given in Chap. 3. We set $T \cup \{\psi_n\}_{n \in \mathbb{N}} = \mathsf{ACF}$ and refer to this set of sentences as the **theory of algebraically closed fields** .

Exercise 6.22 Prove that, if $\mathcal{K} \models \mathsf{ACF}$, then $\mathcal{K}$ is infinite. (*Hint*: suppose that K is finite and produce an element of $K[x]$ that has no roots in K).

Remark 6.23 Note that $\mathcal{K} \models \mathsf{ACF}$ iff every non-constant polynomial in $K[x]$ has a root in K.

Remark 6.24 By Remark 6.21, the relative algebraic closure of $\mathcal{K}$ in the extension $\mathcal{M}$ may not contain roots of polynomials with coefficients in $\mathcal{K}$. In other words, the relative algebraic closure of a field does not have to be an algebraically closed field. If, however, $\mathcal{M} \models \mathsf{ACF}$, then the relative algebraic closure of $\mathcal{K}$ in $\mathcal{M}$ is also a submodel of $\mathcal{M}$, i.e. it is an algebraically closed field.

Theorem 6.25 *Let $\mathcal{K}, \mathcal{M}$ be fields such that $\mathcal{K} \sqsubseteq \mathcal{M}$. If $\mathcal{M} \models \mathsf{ACF}$, then $acl_\mathcal{M}(\mathcal{K}) \models \mathsf{ACF}$.*

Proof It suffices to show that every non-constant polynomial $p(x)$ in the polynomial ring $(acl_\mathcal{M}(\mathcal{K}))[x]$ has a root in $acl_\mathcal{M}(\mathcal{K})$. Since $\mathcal{M}$ is algebraically closed, $p(x)$ has a root r in M. Every element of $acl_\mathcal{M}(\mathcal{K})$ is algebraic over K, so it suffices to show that r is one of them.

Let $a_0, \ldots, a_n \in acl_\mathcal{M}(\mathcal{K})$ be the coefficients of $p(x)$. These coefficients are all algebraic over K. Since there are finitely many of them, we know from Lemma 6.16 that $\mathcal{K}(a_0, \ldots, a_n)$ is a finite extension of $\mathcal{K}$. Thus, for some positive $m_1 \in \mathbb{N}$, $[K(a_0, \ldots, a_n) : K] = m_1$. Note that r is algebraic over $K(a_0, \ldots, a_n)$, since it is the root of a polynomial with coefficients in the latter field. If $q(x)$ is the minimal polynomial of r over $K(a_0, \ldots, a_n)$ and $\partial q(x) = m_2$, then $[K(a_0, \ldots, a_n, r) : K(a_0, \ldots, a_n)] = m_2$. By Theorem 6.14, $[K(a_0, \ldots, a_n, r) : K] = m_1 m_2$ and $\mathcal{K}(a_0, \ldots, a_n, r)$ is a finite extension of $\mathcal{K}$. By Exercise 6.13, it is an algebraic extension. It follows that r is algebraic over K. □

Two issues may naturally arise at this point. First, $\mathcal{K}$ may be a subfield of several models of ACF, in which case its relative algebraic closures in each one of these submodels are distinct. A priori, we cannot exclude that the same field may thus have non-isomorphic algebraic closures inside distinct algebraically closed fields. Second, $\mathcal{K}$ may not be a subfield of any algebraically closed field, in which case we may not be able to extend it to a model of ACF.

We shall resolve the first issue in the final section of this chapter. If a field $\mathcal{K}$ is a subfield of several models of ACF, then its relative algebraic closures in these models are not sensitive to the ambient field. They are all isomorphic. For this reason we talk about *the* **algebraic closure** of $\mathcal{K}$. In the next chapter we shall prove that any field can be extended to an algebraically closed field, using the compactness theorem.

Remark 6.26 If $\mathcal{K}$ is a field, algebraists denote its algebraic closure by the symbol $\overline{\mathcal{K}}$. When convenient, we shall conform to the same notation. Our choice to adopt $acl_\mathcal{M}(K)$ in this section depends on the fact that 'algebraic closure' in the field-theoretic sense is a specialisation of a more abstract model-theoretic notion with the same name. This notion will be introduced in the next chapter. As hinted in Chap. 3, the same model-theoretic notion specialises to linear closure in the context of vector spaces.

6.4 The Characteristic of a Field

The set $\{0, 1\}$ can be easily turned into a field. If suffices to define addition and multiplication on $\{0, 1\}$ as the ordinary arithmetical operations, with the single exception that $1 + 1 = 0$.

The lesson to be drawn from this observation is that, in an arbitrary field, the elements named by the closed terms $1 + 1, 1 + 1 + 1, 1 + 1 + 1 + 1, \ldots$ are not necessarily all different from 0. If this turns out to be the case, we say that the field has **characteristic** 0. Otherwise, we say that the field has **positive characteristic**.

Let $1, 2, 3, 4, \ldots$ be abbreviations of the terms $1 + 1, 1 + 1 + 1, 1 + 1 + 1 + 1, \ldots$ respectively. We say that a field has characteristic $p > 0$ if p is the least positive integer such that $\underbrace{1 + \ldots + 1}_{p \, times} = 0$. In this case, p has to be a prime number. For any other positive integer n, there are $a, b > 1$ such that $ab = n$. In this case, since $a, b < n$, a $\mathcal{L}_r$-structure $\mathcal{M}$ of characteristic n must model the sentences:

$$\underbrace{1 + \ldots + 1}_{n \, times} = 0, \neg(\underbrace{1 + \ldots + 1}_{a \, times} = 0) \text{ and } \neg(\underbrace{1 + \ldots + 1}_{b \, times} = 0).$$

It follows that $\mathcal{M}$ is not an integral domain, let alone a field. This problem does not arise in prime characteristic.

Let T be, as in the previous section, a model of the field axioms. We say that $\mathcal{K} \models T$ is a *field of characteristic p* if $\mathcal{K}$ models the $\mathcal{L}_r$-sentence:

(16) $\underbrace{1 + \ldots + 1}_{p \text{ times}} = 0$.

If, on the other hand $\mathcal{K}$ models T and, for each integer $n \geq 1$, the $\mathcal{L}_r$-sentence:

(17) $\chi_n := \neg(\underbrace{1 + \ldots + 1}_{n \text{ times}} = 0)$,

then we say that $\mathcal{K}$ is a field of characteristic 0. It is now clear that we can axiomatise the class of **algebraically closed fields of characteristic** 0, whose elements are models of the theory $\mathsf{ACF}_0 = \mathsf{ACF} \cup \{\chi_n\}_{n \geq 1}$. We can also axiomatise the class of **algebraically closed fields of positive characteristic** p, with p a prime: its elements are models of $\mathsf{ACF}_p = \mathsf{ACF} \cup \{\neg \chi_p\}$.

Exercise 6.27 Show that any field of characteristic 0 contains a copy of the field of rational numbers and that every field of characteristic p contains a copy of the field $\mathbb{Z}/p\mathbb{Z}$.

6.5 Field Embeddings

In Chaps. 7 and 8 (specifically Sects. 7.6 and 8.6) we shall be interested in extending certain embeddings from substructures of fields to models of ACF. In the present section we show that the desired extensions always exist.

The setup we consider involves $\mathcal{A}$, a substructure of some model $\mathcal{K}$ of the field axioms, and $\mathcal{N} \models \mathsf{ACF}$. Our eventual aim is to show that, if $f : A \longrightarrow N$ is an embedding, then we can extend it to $g : M \longrightarrow N$, where $\mathcal{M} \models \mathsf{ACF}$ and $\mathcal{A} \sqsubseteq \mathcal{M}$.

Our first step towards this goal is to note that $\mathcal{A}$ is an integral domain. As noted in Example 2.79, $\mathcal{A}$ may be embedded in a field known as its **field of fractions** $Q(\mathcal{A})$ (with domain $Q(A)$) exactly as the ring of integers is embedded into $\mathbb{Q}$.

The construction of a fraction field for $\mathcal{A}$ is notationally identical to the construction described in Example 2.79.

Exercise 6.28 Recalling that $\mathbb{Q}[x]$ is an integral domain, describe the field of fractions associated with $\mathbb{Q}[x]$. This is known as the field of **rational functions** in the indeterminate x with coefficients in $\mathbb{Q}$.

Given the embedding $f : A \longrightarrow N$ described earlier, we can extend its domain to $Q(A)$.

Exercise 6.29 Let $\mathcal{A}$ be an integral domain, $Q(\mathcal{A})$ its field of fractions and $\mathcal{N}$ a field. Show that, if $f : A \longrightarrow N$ is an embedding, then there is an embedding $g : Q(A) \longrightarrow N$ that extends f (this means that, for any $a \in A$, $f(a) = g(a)$).

Exercise 6.30 Show that a field homomorphism is an embedding iff its kernel is $\{0\}$. Show that the kernel of any field homomorphism is $\{0\}$.

Incidentally, abelian, torsion-free groups support a construction entirely analogous to the one involved in Exercise 6.29.

Exercise 6.31 Let H be an abelian, torsion-free group. For any $a \in H$, let na be an abbreviation of $a + \ldots + a$ iterated n times, with $n \geq 1$. Define the binary relation $\sim$ on $H \times \mathbb{N} - \{0\}$ as follows:

$$(a, n) \sim (b, m) \text{ iff } ma = nb.$$

(i) show that $\sim$ is an equivalence relation;
(ii) show that $D = H \times \mathbb{N} - \{0\}/\sim$, the set of equivalence classes, can be endowed with group structure;
(iii) show that the $\mathcal{L}_g$-structure D is a torsion-free, divisible group. We call D the **divisible hull** of H;
(iv) show that there is an embedding of H into its divisible hull.

So far we know how to extend an embedding $f : A \longrightarrow N$ to $g : Q(A) \longrightarrow N$. Our next step is to suppose that $\mathcal{N} \models \mathsf{ACF}$ and prove the existence of an extension of g whose domain is the algebraic closure of $Q(A)$. This final result is an easy consequence of the following more general theorem (we invite the reader to verify it).

Theorem 6.32 *Let $\mathcal{K}, \mathcal{M}$ be fields such that $\mathcal{K} \sqsubseteq \mathcal{K}$. Suppose that $\mathcal{M}$ is an algebraic extension of $\mathcal{K}$ and that $\mathcal{N}$ is an algebraically closed field. For any embedding $f : K \longrightarrow N$ there is an embedding $g : M \longrightarrow N$ such that, for every $a \in K$, $f(a) = g(a)$.*

Proof Consider the set F, whose elements are pairs of the form (K', h), where $\mathcal{K} \sqsubseteq \mathcal{K}' \sqsubseteq \mathcal{M}$ and $h : K' \longrightarrow N$ extends f. We define $\leq$ on F as follows:

$$(K', h) \leq (K'', h') \text{ iff } K' \sqsubseteq K'' \text{ and } h \subseteq h',$$

for any $(K', h), (K'', h') \in F$. A slight variation of Exercise A.84 from the Appendix shows that $(F, \leq)$ is a partially ordered set. If $\{(K_i, h_i)\}_{i \in I}$ is a chain in F, $(\bigcup_{i \in I} K_i, \bigcup_{i \in I} f_i) = (K', h)$ is an upper bound for it. To show this we must verify that, for each $i \in I$, $(K_i, h_i) \leq (K', h)$, $\mathcal{K}'$ is a field extension of $\mathcal{K}$, as well as a subfield of $\mathcal{M}$. Furthermore, we must verify that h is a homomorphism from K' into N (and thus an embedding by Exercise 6.30).

Because the fields $\{\mathcal{K}_i\}_{i \in I}$ determine a chain, it is easy to see that $\bigcup_{i \in I} K_i$ is the domain of a field that extends each $\mathcal{K}_i$. We refer to this common extension as $\mathcal{K}^+$. The union $\bigcup_{i \in I} h_i$ is a function $h^+ : K^+ \longrightarrow N$, since $\{h_i\}_{i \in I}$ is a chain relative to inclusion (thus, if $h_i(a) = b$ and a is in the domain of h_j, we must have $h_i(a) = h_j(a)$).

6.5 Field Embeddings

We now note that, if $a, b \in K^+$, then there is some $j \in I$ such that $a, b \in K_j$. Then $h^+(a +^M b) = h_j(a +^M b) = h_j(a) +^N h_j(b)$ and $h^+(a \cdot^M b) = h_j(a \cdot^M b) = h_j(a) \cdot^N h_i(b)$. Since $h_j(a) = h^+(a)$ and $h_j(b) = h^+(b)$, the function h^+ is a field homomorphism and thus an embedding.

Zorn's lemma applies and it yields the existence of a maximal element $(K^*, h^*) \in F$. Since h^* embeds K^* into N and extends f, we only have to prove that $K^* = M$. Toward a contradiction, we consider a hypothetical field element $a \in M - K^*$. Since M is an algebraic extension of K, a is algebraic over K and thus over K^*. Let $p(x)$ be the minimal polynomial of a in $K^*[x]$. There is an isomorphism ρ from the field $K^*[x]/\langle p(x) \rangle$ to the field $K^*[a]$, given by the condition:

$$\rho(q(x) + \langle p(x) \rangle) = q(a),$$

where $q(a)$ is the term $q(x)$ evaluated on an assignment σ such that $\sigma(x) = a$. We leave it as an exercise to show that ρ is a $\mathcal{L}_r$-homomorphism and thus an embedding. It is also surjective by construction.

We conclude the proof by deriving a contradiction from the existence of a homomorphism π from $K^*[x]/\langle p(x) \rangle$ to a subfield of N such that that $(K^*(a), \pi \rho^{-1}) \in F$ is greater than (K^*, h^*) in the partial order we have defined earlier. This conclusion will contradict the maximality of (K^*, h^*).

To obtain the required homomorphism π, we note that h^* extends to a ring homomorphism into $N[x]$, which we indicate by the superscript $'$. For any polynomial $p(x) \in K^*[x]$, we obtain $p'(x) \in N[x]$ by replacing the coefficients $a_0, \ldots, a_n$ of $p(x)$ with $h^*(a_0), \ldots, h^*(a_n)$. Because $\mathcal{N} \models \mathsf{ACF}$ and $p'(x)$ is non-constant, there is $r \in N$ such that $p'(r) = 0$.

Let us finally introduce the function $\pi : K^*[x]/\langle p(x) \rangle \longrightarrow N$, defined by the condition:

$$\pi(q(x) + \langle p(x) \rangle) = q'(r).$$

Since:

$$\pi(q(x) + s(x) + \langle p(x) \rangle) = (q(r) + s(r))'$$
$$= q'(r) + s'(r)$$
$$= \pi(q(x) + \langle p(x) \rangle) + \pi(s(x) + \langle p(x) \rangle)$$

and $(q(x)s(x))' = q'(x)s'(x)$, we conclude that π is a field homomorphism and thus an embedding. So is $\pi \rho^{-1} : K^*(a) \longrightarrow N$. Thus $(K^*, h^*) \leq (K^*(a), \pi \rho^{-1})$, while the converse inequality fails because, by hypothesis, $a \notin K^*$. We have reached a contradiction: it follows that $a \in K^*$. Because a was an arbitrary element of M, we can conclude $K^* = M$. □

Exercise 6.33 Show that ρ and $'$ are $\mathcal{L}_r$-homomorphisms.

Exercise 6.34 In this exercise we apply Theorem 6.32 to prove that the algebraic closure of a field is unique up to isomorphism, under the assumptions that algebraic closures exist (as already noted, this further fact will be proved in the next chapter). If $\mathcal{K}$ is a field and $\mathcal{C}_1, \mathcal{C}_2 \models \mathsf{ACF}$ are algebraic closures of $\mathcal{K}$, we think of $\mathcal{K}$ as a subfield common to them. The inclusion maps ι_1, ι_2 from K to C_1, C_2 respectively, are embeddings.

(i) Show that there is an embedding $h : C_1 \longrightarrow C_2$ such that $h\iota_1 = \iota_2$.
(ii) The set $h[C_1]$ is the domain of a subfield of C_2 as well as an algebraic extension of $\mathcal{K}$. Prove that $h[C_1]$ is algebraically closed.
(iii) Show that $h[C_1] = C_2$ and conclude that h is a surjective embedding, i.e. an isomorphism.

6.6 Addendum: The Primitive Element Theorem

In this addendum we prove, for the sake of completeness, a field-theoretic result that will be used only once, in the proof of Lemma 11.23. This result, which we only obtain for fields of characteristic 0, is known as the *primitive element theorem*. It states that, if $\mathcal{K}$ is a field and $\mathcal{M} = \mathcal{K}(\alpha_1, \ldots, \alpha_n)$ an algebraic extension obtained by adjoining finitely many elements (which may be assumed to belong to an extension of $\mathcal{K}$, maybe its algebraic closure), then there is $\gamma \in M$ such that $\mathcal{M} = \mathcal{K}(\gamma)$. Although the primitive element theorem is often proved by means of Galois theory, it can be derived from first principles by an argument due to Bartel Leendert van der Waerden, which we reproduce below.

Theorem 6.35 *Let $\mathcal{K}$ be a field of characteristic 0, $\alpha_1, \ldots, \alpha_n$ field elements algebraic over $\mathcal{K}$. Then $\mathcal{K}(\alpha_1, \ldots, \alpha_n)$ is a simple extension of $\mathcal{K}$.*

Proof An inductive argument allows us to restrict attention to the case given by $n = 2$. It thus suffices to show that $\mathcal{K}(\alpha_1, \alpha_2)$, with α_1, α_2 algebraic over K, is a simple extension of $\mathcal{K}$. To this end we set $\gamma = \alpha_1 + \lambda \alpha_2$, with $\lambda \in K$.

What we are going to show is that, for all but finitely many choices of λ, $\alpha_2 \in K(\gamma)$ and $[K(\gamma) : K(\alpha_1, \alpha_2)] = 1$. Since K is infinite,[3] we can therefore select a field element λ that suits our purpose.

Note that, if $\alpha_2 \in K(\gamma)$, then $\alpha_1 \in K(\gamma)$ and we have the inclusion $K(\gamma) \supseteq K(\alpha_1, \alpha_2)$. It therefore make sense to ask what the degree of $\mathcal{K}(\gamma)$ over $\mathcal{K}(\alpha_1, \alpha_2)$ is.

There is a field extension $\mathcal{M}$ of $\mathcal{K}$ in which the minimal polynomials over K of α_1, α_2 respectively split completely, i.e. can be decomposed as products of distinct linear factors.[4]

---

[3] Being a field of characteristic 0, it contains a copy of $\mathbb{Q}$, which is an infinite field.

6.6 Addendum: The Primitive Element Theorem

Let $m_{\alpha_1}(x)$, $m_{\alpha_2}(x)$ be these minimal polynomials. We next consider the minimal polynomial n_{α_2} of α_2 over $F(\gamma)$. Clearly, $n_{\alpha_2}(x)$ divides $m_{\alpha_2}(x)$.

Since, moreover $\alpha_1 = \gamma - \lambda \alpha_2$, the polynomial $m_{\alpha_1}(\gamma - \lambda x) = h(x)$ in $F(\gamma)[x]$ has α_2 as a root. In other words, $n_{\alpha_2}(x)$ divides both $m_{\alpha_2}(x)$ and $m_{\alpha_1}(\gamma - \lambda x)$.

We claim that the greatest common divisor of $m_{\alpha_2}(x)$ and $m_{\alpha_1}(\gamma - \lambda x)$ does not have degree ≥ 2 for infinitely many choices of λ.

If there was a common divisor of degree ≥ 2, then, apart from the factor $(x - \alpha_2)$, $m_{\alpha_2}(x)$ and $m_{\alpha_1}(\gamma - \lambda x)$ would share at least one other distinct linear factor $(x - \alpha_3)$. Consequently, $m_{\alpha_1}(\gamma - \lambda \alpha_3) = 0$. Setting $\gamma - \lambda \alpha_3 = \alpha_4$, we compute:

$$\alpha_4 = \gamma - \lambda \alpha_3 = \alpha_1 - \lambda \alpha_2 - \lambda \alpha_3 \iff \lambda = \frac{\alpha_1 - \alpha_4}{\alpha_2 - \alpha_3},$$

where α_1, α_4 are roots of $m_{\alpha_1}(\gamma - \lambda x)$ in M and α_2, α_3 are roots of m_{α_2} in M. Since there are only finitely many such roots, we can ensure that there are are values of λ for which $m_{\alpha_1}(\gamma - \lambda x)$ and $m_{\alpha_2}(x)$ are relatively prime.

Since $n_{\alpha_2}(x)$ is a common factor of these latter polynomials, its degree is at most 1, but cannot be 0, or $n_{\alpha_2}(x)$ would not be a minimal polynomial of α_2. We conclude that $n_{\alpha_2}(x)$ must have degree 1, i.e. $\alpha_2 \in F(\gamma)$. The proof is complete. □

---

[4] Distinctness here is guaranteed by a property of polynomials called separability, which is guaranteed to hold in characteristic 0 (see [30], pp.109-111). Compare the remarks on separability in Sect. 10.1.

Chapter 7
Complete Theories

Abstract In Sect. 3.1 we defined $\mathcal{L}$-theory T to be complete iff, for any $\mathcal{L}$-sentence ϕ, exactly one of ϕ, $\neg\phi$ is a consequence of T or, equivalently, an element of $T^{\models}$.

In this chapter we introduce an important test for completeness and apply it to several theories. In particular, we build on the elements of field theory discussed in Chap. 6 to deduce the completeness of ACF_0 and ACF_p, for each prime number p.

The field-theoretic results in this chapter are closely related to the arithmetic of cardinal numbers, covered in the Appendix. We advise the reader unfamiliar with it to study the Appendix, in particular Sect. A.5, before tackling this chapter.

The present chapter is structured as follows: in Sect. 7.1 we prove the Löwenheim-Skolem theorems, which jointly yield a completeness test, discussed in Sect. 7.2. Sections 7.3 to 7.7 contain various applications of the test.

We look at vector spaces and divisible, torsion-free abelian groups in Sect. 7.3, turn to dense linear orders without endpoints in Sect. 7.4, and finally tackle algebraically closed fields in Sects. 7.5 to 7.7.

In Sect. 7.5 we develop a dimension theory for algebraically closed fields. In Sect. 7.6 we prove the existence of algebraic closures and evaluate their size. The knowledge so obtained yields the completeness proofs presented in Sect. 7.7.

Section 7.8 closes the chapter with a discussion of model-theoretic algebraic closure and the dimension theory that can be associated with it under special conditions.

7.1 The Löwenheim-Skolem Theorems

We have proved the Downward Löwenheim-Skolem theorem for countable languages. An appeal to cardinal arithmetic, Theorem A.72 in particular, allows us to generalise it to arbitrary languages.

Theorem 7.1 (Downward Löwenheim-Skolem) *Let $\mathcal{N}$ be an infinite $\mathcal{L}$-structure. For any $X \subseteq N$, there is a $\mathcal{L}$-structure $\mathcal{M} \preceq \mathcal{N}$ such that $X \subseteq M$ and $|M| \leq max\{|X|, |\mathcal{L}|\}$.*

Proof We set $|X| = \kappa$, $|\mathcal{L}| = \lambda$, $\mu = max\{\kappa, \lambda\}$. Our goal, as in the countable version of the theorem, is to enlarge X, if necessary, until it satisfies the Tarski-Vaught test. We still build a countably infinite sequence of extensions of X and take their union as the domain of a substructure of $\mathcal{N}$. Cardinal arithmetic allows us to show that successive extensions do not push the size of X above μ.

Let us set $X = X_0$ and consider the set of $\mathcal{L}$-formulae of the form:

$$\exists x \varphi(x, a_1, \ldots, a_m),$$

where $a_1, \ldots, a_m$ are parameters from X. We add at most one element of N to X for each such formula. In particular, we add one new element of N to X exactly when $\mathcal{N} \models \exists x \varphi(x, a_1, \ldots, a_m)$ and $\varphi(x, a_1, \ldots, a_m)$ is not already satisfied in $\mathcal{N}$ by an element of X.

Consider a formula of the form $\exists x \varphi(x, x_1, \ldots, x_m)$, in which exactly $x_1, \ldots, x_m$ occur free. There are $\kappa^m = \kappa$ (by Theorem A.72) ways of choosing finitely many parameters from X to substitute for $x_1, \ldots, x_m$ respectively. Moreover, there cannot be more than λ distinct formulae of the required form. If we had to add one new object to X for each choice of parameters on each of the λ formulae considered, we would have added $\kappa \cdot \lambda = max\{\kappa, \lambda\} = \mu$ objects overall, by Corollary A.75.

The same calculation is to be repeated for each positive number of parameters $m \in \mathbb{N}$. When we are done, we obtain X_1 from X_0: the former set will contain at most $\aleph_0 \cdot \mu = \mu$ (since μ is an infinite cardinal) elements of N not in X_0. In other words, $|X_1| \leq \mu$.

Repeating exactly the same argument relative to formulae with parameters in X_1, we arrive at X_2, which is again of size at most μ. Proceeding as in the proof of Theorem 4.38, we obtain an increasing sequence of extensions:

$$X_0 \subseteq X_1 \subseteq \ldots \subseteq X_n \subseteq X_{n+1} \subseteq \ldots.$$

If these extensions had all been disjoint, the size of their union would have been bounded above by μ. Since they are not disjoint, the same conclusion holds *a fortiori*. Thus $M = \bigcup_{n \in \mathbb{N}} X_n$ is of size at most μ (since $\aleph_0 \cdot \mu = \mu$). The Tarski-Vaught test now applies (as in the proof of Theorem 4.38) and shows that M is the domain of an elementary substructure $\mathcal{M}$ of $\mathcal{N}$. □

Exercise 7.2

(i) Show that every $\mathcal{L}$-structure $\mathcal{N}$ has an elementary substructure $\mathcal{M}$ of size $\leq |\mathcal{L}|$.
(ii) Let $\mathcal{N}$ be a $\mathcal{L}$-structure of size $\lambda \geq |\mathcal{L}|$. Show that, for any cardinal number κ such that $|\mathcal{L}| \leq \kappa \leq |N|$, $\mathcal{N}$ has an elementary substructure of size κ.

The Downward Löwenheim-Skolem theorem concerns elementary substructures. If we move 'upward', the focus shifts to elementary extensions.

7.1 The Löwenheim-Skolem Theorems

Theorem 7.3 (Upward Löwenheim-Skolem) *Let $\mathcal{M}$ be an infinite $\mathcal{L}$-structure of size κ. For any cardinal number $\lambda > \kappa$, there is $\mathcal{N} \succeq \mathcal{M}$ such that the domain of $\mathcal{N}$ is of size $\geq \lambda$.*

Proof Consider $\mathcal{E}D(\mathcal{M})$ and a set $C = \{c_\alpha\}_{\alpha \in \lambda}$ of constant symbols not occurring in the sentences from $\mathcal{E}D(\mathcal{M})$. By compactness, the set:

$$\mathcal{E}D(\mathcal{M}) \cup \{\neg(c_\alpha = c_\beta) : \alpha \neq \beta, \alpha, \beta \in \lambda\}$$

has a model $\mathcal{N}$, which we may identify with an elementary extension of $\mathcal{M}$. Since the constant symbols in C name λ distinct elements of N, it follows that $|N| \geq \lambda$. □

Exercise 7.4 Explain how compactness is applied in the proof of Theorem 7.3.

In view of the Downward Löwenheim-Skolem theorem, we obtain the following refinement of the Upward Löwenheim-Skolem theorem:

Theorem 7.5 *Let $\mathcal{M}$ be an infinite $\mathcal{L}$-structure. If $\kappa \geq max\{|M|, |\mathcal{L}|\}$, there is an elementary extension of $\mathcal{M}$ of size κ.*

Proof Exercise 7.6. □

Exercise 7.6

(i) Show that if $\mathcal{M}$ is a finite $\mathcal{L}$-structure, then it has exactly one elementary substructure.
(ii) Prove Theorem 7.5. Moreover, show that, if T is a $\mathcal{L}$-theory with an infinite model $\mathcal{M}$, then T has models of arbitrarily large cardinality. Finally, show that, if $|M| \geq |\mathcal{L}|$, then T has models of each cardinality greater than $|M|$.
(iii) Suppose that T is a $\mathcal{L}$-theory with an infinite model. Show that T has a model of cardinality $|\mathcal{L}|$.

Exercise 7.6-(ii) implies that no set of first-order sentences can fix the cardinality of its infinite models. For instance, there is no theory T whose infinite models are all countable. More generally, any first-order theory with infinite models has non-isomorphic models.

A theory whose models are all isomorphic is said to be **categorical**. By what we have pointed out, the notion of categoricity is not relevant to the study of first-order theories with infinite models (e.g. $VS_\mathbb{Q}$ or ACF_0). The following refinement of categoricity is pertinent.

Definition 7.7 Let $\mathcal{T}$ be a $\mathcal{L}$-theory and κ be an infinite cardinal. The theory T is said to be κ-**categorical** iff it has a model of size κ and any two models of T whose domains have size κ are isomorphic.

In the next section, we shall rely on the Löwenheim-Skolem theorems to prove that κ-categorical theories that have only infinite models are complete. We conclude this section with two illustrations of their algebraic usefulness.

Lemma 7.8 *The algebraic closure of the field of rational numbers* $\mathbf{Q}$ *is countably infinite.*

Proof We consider the complex field $\mathbf{C}$, regarded as a $\mathcal{L}_r$-structure. This field is an uncountable model of ACF_0[1] that extends $\mathbf{Q}$. By the Downward Löwenheim-Skolem theorem, $\mathbf{C}$ has a countably infinite, elementary substructure $\mathcal{M}$. Since $\mathcal{M} \equiv \mathbf{C}$, $\mathcal{M}$ is a countably infinite, algebraically closed field, which must extend $\mathbf{Q}$ (any field of zero characteristic contains a copy of the rational field). The relative algebraic closure of $\mathbf{Q}$ in $\mathcal{M}$ is necessarily countable. Because $\mathcal{M} \models \mathsf{ACF}_0$, this relative algebraic closure is in fact an algebraically closed extension of $\mathbf{Q}$. Thus, it is, up to isomorphism, $\overline{\mathbf{Q}}$. It follows that $|\overline{\mathbf{Q}}| = \aleph_0$. □

In Sect. 7.5 we shall look at another, combinatorial way of evaluating the cardinality of $\overline{\mathbf{Q}}$. Our second algebraic application of the Downward Löwenheim-Skolem theorem is the following exercise.

Exercise 7.9

(i) Show that any infinite group, viewed as a $\mathcal{L}_g$-structure, has a countably infinite subgroup.
(ii) A group $\mathcal{G}$ is said to be **simple** when its only normal[2] subgroups are itself and the subgroup consisting of its neutral element only. Let $\mathcal{G}$ be a $\mathcal{L}_g$-structure that is an infinite, simple group and λ an infinite cardinal such that $\lambda \leq |G|$. Prove the existence of a simple subgroup of $\mathcal{G}$ of size λ, following the steps below.

   (a) Show that $\mathcal{G}$ has an elementary subgroup $\mathcal{H}$ of cardinality λ.
   (b) Fix $b \in H$, with $b \neq e$, and show that the set

   $$N_b = \{g_1 b^{\epsilon_1} g_1^{-1} \cdots g_k b^{\epsilon_k} g_k^{-1} : \epsilon_i \in \{-1, 1\}, i = 1, \ldots, k, g_i \in H\}$$

   written multiplicatively, is the domain of $\mathcal{N}_b$, the smallest normal subgroup of $\mathcal{H}$ containing b.
   (c) Show that, for any $a \in H$, $a \in \mathcal{N}_b$. Deduce that $\mathcal{H}$ is simple.

The Löwenheim-Skolem theorems are often qualified as limitative results because they imply that the property 'has size κ' cannot be expressed by a set of first-order sentences when κ is an infinite cardinal. Although this is a point of interest with respect to elementary axiomatisability, it does not quite capture the model-theoretic significance of the Löwenheim-Skolem theorems.

Their essential role lies in the fact that they ensure the existence of elementary substructures or extensions, which continue to model the sentences true in the

---

[1] This is an immediate consequence of Theorem 11.15, see Exercise 11.1.
[2] Let $\mathcal{G}$ be a group. For any $g \in G$, the function α_g, defined by the equality $\alpha_g(h) = g^{-1}hg$, is an automorphism of $\mathcal{G}$. A subgroup $\mathcal{H}$ of $\mathcal{G}$ is normal when, for any $g \in G$, $\alpha_g[H] = H$, i.e. the set H is fixed by each α_g.

7.2 Completeness

structure they extend or restrict. Furthermore, in more concrete settings like group theory or field theory, the existence of elementary substructures or extensions has a direct algebraic significance, as the applications just discussed should have indicated.

7.2 Completeness

Recall that a $\mathcal{L}$-theory T is complete iff $T^\models$ contains exactly one of $\phi, \neg\phi$, for each $\mathcal{L}$-sentence ϕ. In other words, $T \models \phi$ or $T \models \neg\phi$ (but not both) for any $\mathcal{L}$-sentence ϕ.

When proving completeness it is often convenient to adopt the following, equivalent alternative to the definition we have given.

Theorem 7.10 *Let T be a $\mathcal{L}$-theory. Then T is complete iff $T \models \mathcal{M}, \mathcal{N}$ implies $\mathcal{M} \equiv \mathcal{N}$.*

Proof It is clear that if a $\mathcal{L}$-theory is complete, any two of its models must satisfy exactly the same $\mathcal{L}$-sentences. The proof of the converse statement is left to the reader. □

Exercise 7.11 Complete the proof of Theorem 7.10.

Theorem 7.10 provides a useful test for completeness if we have enough information about isomorphic models in some infinite cardinality.

Theorem 7.12 (Vaught's Test) *Let T be a $\mathcal{L}$-theory that has only infinite models. If T is κ-categorical with $|\mathcal{L}| \leq \kappa$, T is complete.*

Proof Suppose $T \models \mathcal{M}, \mathcal{N}$. We show that $\mathcal{M} \equiv \mathcal{N}$. Suppose that $|\mathcal{M}| < \kappa$ and $|\mathcal{N}| > \kappa$ (the other possible cases are treated in an entirely similar way). By the Downward Löwenheim-Skolem theorem, $\mathcal{N}$ has an elementary substructure $\mathcal{N}_0$ of cardinality κ. By the Upward Löwenheim-Skolem theorem, $\mathcal{M}$ has an elementary extension $\mathcal{M}_0$ of cardinality κ. By κ-categoricity, $\mathcal{N}_0 \cong \mathcal{M}_0$. Since $\mathcal{M} \equiv \mathcal{M}_0$ and $\mathcal{N} \equiv \mathcal{M}_0$, the transitivity of $\equiv$ implies $\mathcal{M} \equiv \mathcal{N}$. □

Exercise 7.13 Why does Vaught's test not apply if T is a $\mathcal{L}$-theory with finite models?

Another basic characterisation of completeness, based on maps, is of interest.

Definition 7.14 Let T be a $\mathcal{L}$-theory. We say that T **admits joint elementary embeddings** iff for any two models $\mathcal{M}, \mathcal{N}$ of T there are a $\mathcal{L}$-structure $\mathcal{N}_0$ and maps $f : \mathcal{M} \longrightarrow \mathcal{N}_0$, $g : \mathcal{N} \longrightarrow \mathcal{N}_0$ such that $\mathcal{N}_0 \models T$ and f, g are elementary embeddings.

Theorem 7.15 *Let T be a $\mathcal{L}$-theory. Then T is complete iff it admits joint elementary embeddings.*

Proof If T admits joint elementary embeddings, it is easy to see that any two models of T are elementarily equivalent (the details are left to the reader): Theorem 7.10 then applies. To prove the converse statement, we consider $\mathcal{M}, \mathcal{N} \models T$ and, toward an application of compactness, show that the set $\mathcal{E}D(\mathcal{M}) \cup \mathcal{E}D(\mathcal{N})$ is finitely satisfiable.

We work under the assumption that the constant symbols occurring in the sentences of $\mathcal{E}D(\mathcal{M})$ other than those already occurring in $\mathcal{L}$ do not figure in $\mathcal{E}D(\mathcal{N})$, and vice versa.

Consider a finite subset $A \subseteq \mathcal{E}D(\mathcal{M}) \cup \mathcal{E}D(\mathcal{N})$. We split A into three disjoint subsets A_0, A_1 and A_2. A_0 contains all of the $\mathcal{L}$-sentences in $A \cap (\mathcal{E}D(\mathcal{M}) \cup \mathcal{E}D(\mathcal{N}))$. These sentences are simultaneously modelled by $\mathcal{M}$ and $\mathcal{N}$ (by hypothesis, $\mathcal{M} \equiv \mathcal{N}$) and do not contain any new constant symbols occurring in their respective elementary diagrams. Setting, $A' = A - A_0$ we take A_1 to be the set $A' \cap \mathcal{E}D(\mathcal{M})$ and A_2 the set $A' \cap \mathcal{E}D(\mathcal{N})$.

As we have observed, $\mathcal{M} \models A_0$. Now let $\varphi(a_1, \ldots, a_i)$ be the conjunction of the sentences in A_1, in which the distinct constant symbols $a_1, \ldots, a_i$ occur. Moreover, let $\psi(b_1, \ldots, b_j)$ be the conjunction of the sentences in A_2. Then:

$$\exists x_1 \ldots \exists x_i \varphi(x_1, \ldots, x_i), \exists y_1 \ldots \exists y_j \psi(y_1, \ldots, y_j)$$

are $\mathcal{L}$-sentences. Because $\mathcal{N} \models \exists y_1 \ldots \exists y_j \psi(y_1, \ldots, y_j)$ and T is complete, $T \models \exists y_1 \ldots \exists y_j \psi(y_1, \ldots, y_j)$. Consequently:

$$\mathcal{M} \models \exists x_1 \ldots \exists x_i \varphi(x_1, \ldots, x_i) \land \exists y_1 \ldots \exists y_j \psi(y_1, \ldots, y_j).$$

It follows that an expansion of $\mathcal{M}$ satisfies A. Because A is arbitrary, we have shown that any finite subset of $\mathcal{E}D(\mathcal{M}) \cup \mathcal{E}D(\mathcal{N})$ has a model. By compactness, this whole set of sentences has a model $\mathcal{N}^+$. By Theorem 4.49, $\mathcal{M}$ and $\mathcal{N}$ can be elementarily embedded in $\mathcal{N}^+$. In other words, T admits joint elementary embeddings. □

The next exercise discusses a very simple situation in which any one of Vaught's test, Theorem 7.10 or Theorem 7.15 can with ease be applied to show that the same theory is complete.

Exercise 7.16 Prove that IS is complete in three different ways.

Exercise 7.17 Let $\mathcal{L}_E$ be the language $\mathcal{L}_= \cup \{E\}$, where E is a binary relation symbol.

(i) Axiomatise the $\mathcal{L}_E$-theory EQ^∞ of equivalence relations with infinitely many, infinite equivalence classes and no finite equivalence classes.
(ii) Show that EQ^∞ is complete.

In the next few sections, we use Vaught's test systematically to prove the completeness of several theories.

7.3 Infinite Vector Spaces and Divisible Groups

By Theorem A.54, if the $\mathcal{L}_v$-structure $\mathcal{V}$ is a vector space over a fixed field $\mathcal{K}$, then it has a basis. Any two vector spaces with bases of the same size are isomorphic: this is because any bijection between their bases can be extended to a bijection between the respective linear closures.[3] It follows from these remarks that Vaught's test can be applied to VS_K if we restrict attention to models with sufficiently large bases, as the next theorem clarifies.

Theorem 7.18 *Let* $\mathcal{V} \models \mathsf{VS}_K$ *be a nonzero vector space over the infinite field* $\mathcal{K}$. *If* $|\mathcal{L}_v| < |V| = \kappa$, *then any basis for* V *has size* κ.

Proof Because we can associate each $k \in K$ with the sentence $f_k(0) = 0$, $|K| \leq |\mathcal{L}_v|$. Equality follows from the fact that any finite string of symbols in $\mathcal{L}_v$ consists of a selection from a set of size $|K|$. Our hypothesis now yields $|V| > |K|$. Let $B \subseteq V$ be a basis for V. Since $V = acl_\mathcal{V}(B)$, each non-zero vector $v \in V$ has the following, unique representation:

$$v = k_0 b_0 + \ldots + k_n b_n,$$

where $b_0, \ldots, b_n$ are linearly independent elements of B and $k_0, \ldots, k_n \in K - \{0\}$. An arbitrary representation of v involving n basis vectors requires a finite number of selections (without replacement[4]) from B and the same number of independent selections from $K - \{0\}$ (instead of allowing for coefficients to be 0, we consider sums of more or fewer terms separately). Cardinal arithmetic yields the following chain of equalities:

$$\underbrace{|K| \cdot |B| + \ldots + |K| \cdot |B|}_{n\ times} = n(|K| \cdot |B|) = |K| \cdot |B|.$$

Because V is generated by considering selections of any finite length, the size of V, i.e. κ, equals the size of the disjoint union of the sets of linear combinations of n elements of B, for each positive $n \in \mathbb{N}$. By Corollary A.75, the latter disjoint union has cardinality:[5]

$$|V| = \aleph_0 \cdot (|K| \cdot |B|) = max\{\aleph_0, |K| \cdot |B|\}.$$

---

[3] Any linear combination of vectors $a_1, \ldots, a_n$ from one basis can be uniquely associated with the corresponding linear combination of $f(a_1), \ldots, f(a_n)$, where f is a fixed bijection between bases.

[4] The absence of replacement for selection on an infinite set does not affect the number of available choices because, for any positive $n \in \mathbb{N}$, $|K| = |K| + n$ and $|B| = |B| + n$.

[5] We implicitly rely on the fact that the disjoint union of denumerably many sets, each of cardinality κ is, $\aleph_0 \cdot \kappa$, which the reader may verify as an exercise.

Because B, K are both infinite, $|V| = |K| \cdot |B|$. The inequality $|K| < |V|$ and the equality $|K| \cdot |B| = max\{|K|, |B|\}$ imply that the last product cannot equal $|K|$. We deduce $|V| = |B|$. □

In order to apply Vaught's test to vector spaces, we must restrict attention to theories that have only infinite models. If $\mathcal{K}$ is an infinite field, this is the case of $\mathsf{VS}_K \cup \{\exists x \neg (x = 0)\}$. Alternatively, we may focus on the theories defined below:

Definition 7.19 Let VS_K be the theory of vector spaces over the field $\mathcal{K}$. We call $\mathsf{VS}_K^\infty = \mathsf{VS}_K \cup \mathsf{IS}$ the **theory of infinite vector spaces over** K.

Exercise 7.20 Let VS_K^∞ be the theory of infinite vector spaces over K.

  (i) Show that the theory is κ-categorical for sufficiently large κ.
  (ii) Show that, if K is the domain of a countable field, then VS_K^∞ is κ-categorical for every uncountable cardinal κ.
  (iii) Deduce that, for any choice of K, VS_K^∞ is complete.
  (iv) If $\mathcal{K}$ is a finite field, show that VS_K, the theory of vector spaces over K, is incomplete.
  (v) If $\mathcal{K}$ is a finite field, show that VS_K has infinitely many complete extensions.

Theorem 7.18 has interesting consequences beyond the completeness of VS_K^∞. We record one:

Lemma 7.21 *If $n \geq 1$, $\mathcal{R}^n = (\mathbb{R}^n, +^n, -^n, 0^n)$ and $\mathcal{R} = (\mathbb{R}, +, -, 0)$ are isomorphic models of* DAG *(the superscript n is used to designate operations on n-tuples of real numbers and 0^n is the n-tuple whose components are equal to 0).*

Proof We view $\mathcal{R}^n$, $\mathcal{R}$ as models of $\mathsf{VS}_\mathbb{Q}^\infty$. As such, they have bases of the same cardinality, which, by Exercise 7.22, is $|\mathbb{R}| = |\mathbb{R}^n|$. We deduce that $\mathcal{R}^n$, $\mathcal{R}$ are isomorphic as vector spaces over the rational field. The theorem follows from restricting any given vector space isomorphism to the signature of DAG. □

Exercise 7.22

  (i) Show that the set $\mathbb{R}$ gives rise to a 1-dimensional vector space over the real field but not to a 1-dimensional vector space over the rational field.
  (ii) Prove that the $\mathcal{L}_v$-structure $(\mathbb{R}, +, -, 0, \{q\}_{q \in \mathbb{Q}})$ cannot be finite-dimensional.
  (iii) Show that $|\mathbb{R}| = 2^{\aleph_0}$ and deduce that $dim((\mathbb{R}, +, -, 0, \{q\}_{q \in \mathbb{Q}})) = 2^{\aleph_0}$ (*Hint*: use binary sequences and the Cantor-Bernstein theorem).

Exercise 7.23 Reprove Lemma 7.21 along the following lines:

  (i) The $\mathbb{Q}$-vector space $\mathcal{R}$ has a basis $B = \{x_i\}_{i < 2^{\aleph_0}}$ of size $2^{\aleph_0}$. Show that $\mathcal{R}^n$, a $\mathbb{Q}$-vector space, has a basis B' whose elements are of the form $(0, \ldots, x_i, \ldots, 0)$, where only the i-th component x_i is non-zero and $x_i \in B$.
  (ii) Show that $|B| = |B'|$ and deduce that there is a vector space isomorphism between $\mathcal{R}^n$ and $\mathcal{R}$.

7.3 Infinite Vector Spaces and Divisible Groups

Lemma 7.21 suggests a close connection between torsion-free, divisible abelian groups and vector spaces over $\mathbb{Q}$. On the one hand, any vector space over $\mathbb{Q}$ has a $\mathcal{L}_g$-reduct that is a model of DAG. On the other hand, given a model $\mathcal{M}$ of DAG with non-zero elements, $\mathcal{M}$ is infinite and it can be expanded to a $\mathcal{L}_v$-structure that models $\mathsf{VS}_\mathbb{Q}^\infty$.

We conclude that the non-zero models of DAG are, when expanded to $\mathcal{L}_v$, models of $\mathsf{VS}_\mathbb{Q}^\infty$. The relevant expansion is obtained by interpreting the unary function symbols in $\mathcal{L}_v$ in a rather obvious way.

Lemma 7.24 *Every non-zero model of* DAG *can be expanded to a model of* $\mathsf{VS}_\mathbb{Q}^\infty$.

Proof If $\mathcal{M} \models$ DAG, then $\mathcal{M}$ is torsion-free. Moreover, because $\mathcal{M}$ is non-zero, there is $a \in M$ such that $a \neq 0^\mathcal{M}$. It follows that $a, a+a, a+a+a, \ldots$ are distinct elements of M and, thus, that $\mathcal{M} \models$ IS. In order to expand $\mathcal{M}$ to a $\mathcal{L}_v$-structure, we fix $q \in \mathbb{Q}$. There are integers m, n, with $n > 0$, such that $q = m/n$. The $\mathcal{L}_g$-formula $mx = ny$ defines a set $f_q^\mathcal{M}$ of ordered pairs. If $a, b \in M$ and $(a, b) \in f_q^\mathcal{M}$, then $ma = nb$, where we take ma to be a sum of copies of the additive inverse of a if $m < 0$. The set we have defined is a function: $(a, b) \in f_q^\mathcal{M}$ and $(a, c) \in f_q^\mathcal{M}$ jointly imply $b = c$ (check).

Since $q \in \mathbb{Q}$ was arbitrarily chosen, we have introduced a uniform way of defining a countably infinite family of unary functions on M. Unsurprisingly, when $\mathcal{M}$ is expanded to a $\mathcal{L}_v$-structure $\mathcal{M}_\mathbb{Q}$, equipped with the functions we have defined, $\mathcal{M}_\mathbb{Q} \models \mathsf{VS}_\mathbb{Q}$ (see the next exercise). □

Exercise 7.25 Show that the vector space axioms are indeed satisfied by $\mathcal{M}_\mathbb{Q}$.

Let DAG^+ be $\mathsf{DAG} \cup \{\exists x \neg (x = 0)\}$. Then DAG^+ has only infinite models. Moreover:

Theorem 7.26 *The theory* DAG^+ *is complete.*

Proof Consider two models $\mathcal{M}, \mathcal{N}$ of DAG^+ such that $|N| = |M| = \kappa > \aleph_0$. Models of the prescribed size exist by the Löwenheim-Skolem theorems. By Lemma 7.24, we can expand them to models $\mathcal{M}_\mathbb{Q}, \mathcal{N}_\mathbb{Q}$ of $\mathsf{VS}_\mathbb{Q}^\infty$. By Exercise 7.20-(ii), $\mathsf{VS}_\mathbb{Q}^\infty$ is κ-categorical. The isomorphism between $\mathcal{M}_\mathbb{Q}, \mathcal{N}_\mathbb{Q}$ induces an isomorphism between their $\mathcal{L}_g$-reducts $\mathcal{M}, \mathcal{N}$. This shows that DAG^+ is κ-categorical. By Vaught's test, it is a complete theory. □

By Exercise 7.20-(ii), $\mathsf{VS}_\mathbb{Q}^\infty$ is a complete theory in a countable language that is κ-categorical for any uncountable cardinal κ. We have shown that the same statement holds relative to DAG^+.

We have also encountered theories in a countable language, e.g. IS, that are κ-categorical for any infinite cardinal κ. A natural question is whether there are any countable, complete theories that are categorical in some uncountable cardinalities but not in others.

A central result of model theory, originally proved by Michael Darwin Morley in [51], provides a negative answer to the last question: no complete theory in a countable language can be κ-categorical for some but not all uncountable cardinalities.[6] A first-order theory is either κ-categorical for every uncountable cardinal κ, or for none.

The first case is exemplified by DAG^+. The second case, by the $\mathcal{L}_<$-theories discussed in the next section.

7.4 Dense Linear Orders

The $\mathcal{L}_{r<}$-sentence $\exists x(x^2 - 2 = 0)$ is modelled by the ordered real field R but not by the ordered rational field Q. We are thus able to distinguish between these two $\mathcal{L}_{r<}$-structures using the first-order information they carry. By contrast the $\mathcal{L}_<$-reducts of Q and R are no longer distinguishable. This is because $(\mathbb{Q}, <^{\mathbb{Q}}) \equiv (\mathbb{R}, <^{\mathbb{R}})$.

Although we could prove the last statement directly, it is more interesting to deduce it from the fact that $(\mathbb{Q}, <^{\mathbb{Q}})$, $(\mathbb{R}, <^{\mathbb{R}})$ are models of the same complete $\mathcal{L}_<$-theory. The theory in question is axiomatised by a single sentence, which we now describe.

We first let ψ_1 be the $\mathcal{L}_<$-sentence stating that $<$ is a linear order. Let ψ_2 be the $\mathcal{L}_<$-sentence:

$$\forall x \forall y (x < y \rightarrow \exists z (x < z \land z < y)).$$

A model of $\psi_1 \land \psi_2$ is a *dense* linear order. Now let ψ_3, ψ_4 be respectively the $\mathcal{L}_<$-sentences:

$$\forall x \exists y (y < x) \text{ and } \forall x \exists y (x < y)$$

A model of $\psi_1 \land \psi_2 \land \psi_3 \land \psi_4$ is a dense linear order without a smallest or a largest element relative to $<$. We call the last conjunction of $\mathcal{L}_<$-sentences $\mathsf{DLO}_{-,-}$, the **theory of dense linear orders without endpoints**. Clearly, $(\mathbb{Q}, <^{\mathbb{Q}})$ and $(\mathbb{R}, <^{\mathbb{R}}) \models \mathsf{DLO}_{-,-}$ and $\mathsf{DLO}_{-,-}$ has only infinite models.

Vaught's applies to $\mathsf{DLO}_{-,-}$ because its countable models are all isomorphic to one another. Because one of them is $(\mathbb{Q}, <^{\mathbb{Q}})$, this preliminary result can be established simply by proving that any countable model of $\mathsf{DLO}_{-,-}$ is isomorphic to the ordered rationals.

Before proving the last statement, we look at a closely related result, which is of independent interest.

Lemma 7.27 *The $\mathcal{L}_<$-structure $(\mathbb{Q}, <^{\mathbb{Q}})$ can be embedded into any model of $\mathsf{DLO}_{-,-}$.*

---

[6] Morley's theorem lies beyond the scope of this book. Accessible discussions of its proof may be found e.g. in section 9.5 of [25] or at the end of chapter 5 of [70].

7.4 Dense Linear Orders

Proof Suppose $\mathcal{M} \models \mathsf{DLO}_{-,-}$. Since $\mathbb{Q}$ is countably infinite, its elements can be enumerated as an infinite sequence of the form $\{q_i\}_{i \in \mathbb{N}}$. We recursively define a chain of partial functions from $\mathbb{Q}$ into M. Their union will be the embedding we are looking for.

Our starting point is $f_0 = \{(q_0, a_0)\}$, where $a_0 \in M$ is chosen arbitrarily. We next suppose that we have defined the partial function f_k, with domain $\mathbb{Q}_k = \{q_1, \ldots, q_k\}$, in such a way that $q_i <^{\mathbb{Q}} q_j$ iff $f(q_i) <^M f(q_j)$.

We turn to the next rational number from the given enumeration, namely q_{k+1}, and consider its relative position with respect to $q_1, \ldots, q_k$.

Three possibilities arise: (i) q_{k+1} is smaller than any element of $\mathbb{Q}_k$: in this case we use the fact that $\mathcal{M}$ has no minimum to select $a_{k+1} \in M$ below the elements of $f_k[\mathbb{Q}_k]$; (ii) q_{k+1} is larger than any element of $\mathbb{Q}_k$: in this case we use the fact that $\mathcal{M}$ has no maximum to select $a_{k+1} \in M$ above the elements of $f_k[\mathbb{Q}_k]$; (iii) q_{k+1} lies between q_i and q_j: in this case we use the density of $\mathcal{M}$ to select $a_{k+1} \in M$ within the interval bounded by $f_k(q_i)$ and $f_k(q_j)$. In any case, we set $f_{k+1} = f_k \cup \{(q_{k+1}, a_{k+1})\}$.

The set:

$$f = \bigcup_{i \in \mathbb{N}} f_i$$

is a function because it is the union of the countably infinite chain $f_0 \subseteq f_1 \subseteq \ldots \subseteq f_n \subseteq \ldots$. It is also an embedding from $\mathbb{Q}$ into $\mathcal{M}$. □

Exercise 7.28 Show that f from the proof of the last lemma is an embedding.

Exercise 7.29 Show that the $\mathcal{L}_v$-structure $(\mathbb{Q}, +^{\mathbb{Q}}, -^{\mathbb{Q}}, \{f_q\}_{q \in \mathbb{Q}})$, where $f_q(r) = q \cdot^{\mathbb{Q}} r$, for each $q \in \mathbb{Q}$, is embeddable in any model of $\mathsf{VS}_{\mathbb{Q}}^{\infty}$.

When $\mathcal{M}$ is a countable model of $\mathsf{DLO}_{-,-}$, it is possible to determine a surjective embedding of $(\mathbb{Q}, <^{\mathbb{Q}})$ onto $\mathcal{M}$.

Theorem 7.30 *If $\mathcal{M} \models \mathsf{DLO}_{-,-}$ is countable, then it is isomorphic to $(\mathbb{Q}, <^{\mathbb{Q}})$.*

Proof Let $\{q_i\}_{i \in \mathbb{N}}$ be a fixed enumeration of $\mathbb{Q}$. Since $\mathcal{M} = (M, <^M)$ is countable, let $\{a_i\}_{i \in \mathbb{N}}$ be a fixed enumeration of M.

We build an embedding $f : \mathbb{Q} \longrightarrow M$ in stages, as before, but now taking care that its image should exhaust M. At each stage we generate a partial function including those obtained at earlier stages. The union of these partial functions will be the desired isomorphism f.

We split the construction stages into odd and even, so we can carry out a distinct kind of task depending on the stage's parity.[7]

---

[7] This split approach is helpful for bookkeeping purposes, but not essential. We could have carried out two distinct tasks at each stage of our construction, for instance.

At stage 0, we simply match q_0 and a_0 from the given enumeration and set $f_0 = \{(q_0, a_0)\}$. At each even stage $2n$, with $n \geq 1$, we fix the rational number q_j with least index among the elements of $\mathbb{Q}$ not yet considered.

The number q_j satisfies a finite set of inequalities relative to the rationals already considered, call them $q_{j_1}, \ldots, q_{j_k}$. We match q_j with the element a_k of least index to satisfy a finite system of inequalities obtained from those satisfied by q_j by substituting the parameters $f_{2n-1}(q_{j_s})$ for the parameters q_{j_s} ($s = 1, \ldots, k$).

We can always select a suitable a_k because we can always find solutions to the given inequalities, proceeding as in cases (i) to (iii) from the proof of Lemma 7.27. We thus set $f_{2n} = f_{2n-1} \cup \{(q_j, a_k)\}$.

At each odd stage $2n+1$, we fix a_k with least index among the elements of M not yet considered. This a_k satisfies a finite set of inequalities relative to the elements of M that are already values of f_{2n}. Reasoning as above, we can pair a_k with a suitable $q_j \in \mathbb{Q}$, of least index to satisfy the relevant set of inequalities. We set $f_{2n+1} = f_{2n} \cup \{(q_j, a_k)\}$.

The union $\bigcup_{n \in \mathbb{N}} f_n$ is a function. The domain of f is the whole of $\mathbb{Q}$. To see this, note that q_1 certainly belongs to the domain of f_2 (because we assign it a value in M by the second stage of our construction), q_2 belongs to the domain of f_4 and, in general q_n belongs to the domain of f_{2n}. By a similar argument, relative to the odd stages, we deduce that f is surjective.

Now suppose $q_i <^{\mathbb{Q}} q_j$ and q_i is introduced in the construction of f at an even stage $2n$, while q_j is introduced at a later, odd stage $2m + 1$, with $m \geq n$. Let a_s, a_t be the distinct elements of M matching q_i, q_j respectively.[8]

By trichotomy, we must have $a_s <^M a_t$ or $a_t <^M a_s$. Toward a contradiction suppose $a_t <^M a_s$. Then $f_{2n+2}(q_i) = a_s <^M a_t = f_{2n+2}(q_j)$ iff $q_j <^{\mathbb{Q}} q_i$, because f_{2n+2} is a partial embedding. This result contradicts our assumption. We conclude that $q_i <^{\mathbb{Q}} q_j$ implies $f(q_i) = a_s <^M a_t = f(q_j)$, since $f_{2n+2}(q_i) = f(q_i)$ and $f_{2n+2}(q_j) = f(q_j)$.

Essentially the same argument applies in reverse (from M to $\mathbb{Q}$), yielding:

$$q_j < q_i \text{ iff } f(q_i) < f(q_j).$$

Certainly f is an embedding, since any homomorphism that preserves a strict linear order must be injective. The surjective embedding f is automatically an isomorphism between $\mathcal{M}$ and $\mathbb{Q}$. □

By the last theorem, $\mathsf{DLO}_{-,-}$ is $\aleph_0$-categorical and only has infinite models. Vaught's test applies.

Theorem 7.31 *The $\mathcal{L}_<$-theory $\mathsf{DLO}_{-,-}$ is complete. In particular, it is a complete extension of the incomplete theory of dense linear orders DLO.*

---

[8] The argument to follow continues to work under different hypotheses on the parities of the stages considered.

7.4 Dense Linear Orders

Exercise 7.32

(i) Let $\mathsf{DLO}_{+,+}$ be the theory of dense linear orders with a least element and with a greatest element. Is this theory complete?
(ii) Define $\mathsf{DLO}_{+,-}$ and Let $\mathsf{DLO}_{-,+}$ in the obvious manner and discuss their completeness.
(iii) How many complete extensions of the $\mathcal{L}_<$-theory DLO are there?

Exercise 7.33

(i) There are only two atomic $\mathcal{L}_<$-formulae in which exactly one free variable occurs, namely $x = x$ and $x < x$. Deduce that every quantifier-free $\mathcal{L}_<$-formula $\phi(x)$ in which only x occurs must be satisfied either by every element of $\mathcal{M} \models \mathsf{DLO}_{+,-}$ or by none.
(ii) Show that $\mathsf{DLO}_{+,-}$ cannot imply an equivalence between $\exists y(y < x)$ and a quantifier-free formula $\phi(x)$ in which only x occurs.
(iii) Given a model $\mathcal{N}$ of $\mathsf{DLO}_{+,-}$, find a substructure $\mathcal{M} \sqsubseteq \mathcal{N}$ that is itself a model of $\mathsf{DLO}_{+,-}$ but not an elementary substructure of $\mathcal{N}$.

Although $\mathsf{DLO}_{-,-}$ is countably categorical, it fails to be categorical in any uncountable cardinality.

Theorem 7.34 *Let $\lambda > \aleph_0$ be an infinite cardinal. The theory $\mathsf{DLO}_{-,-}$ is not λ-categorical.*

Proof The next two exercises provide possible proofs. Exercise 7.36 involves a more substantial use of set-theoretic ideas. □

Exercise 7.35 Let $\lambda > \aleph_0$ be an infinite cardinal and L a set of cardinality λ such that $(L, <^L) \models \mathsf{DLO}_{-,-}$.

(i) Take countably many disjoint, isomorphic copies of $(L, <^L)$, index them on $\mathbb{N}$ and order them by their indices. We now define a linear order on the union of disjoint copies. If $a, b \in M$ and a, b belong to the same copy of $(L, <^L)$, they are ordered like their counterparts in L. If, on the other hand, a belongs to the copy of $(L, <^L)$ indexed by m and b to the copy indexed by n then, $a < b$ iff $m < n$. Show that $\mathcal{M} \models \mathsf{DLO}_{-,-}$.
(ii) Consider λ disjoint, isomorphic copies of $\mathbb{Q}$, ordered relative to each other like the elements of $(L, <^L)$. Show that the resulting linear ordering $\mathcal{N}$ is a model of $\mathsf{DLO}_{-,-}$.
(iii) Show that $\mathcal{M}, \mathcal{N}$ have the same cardinality and prove by contradiction that they cannot be isomorphic. Deduce that $\mathsf{DLO}_{-,-}$ is not λ-categorical for any uncountable λ.

Exercise 7.36 Let $\lambda > \aleph_0$ be an infinite cardinal.

(i) For each ordinal $\alpha \in \lambda$ we consider the set $\{\alpha\} \times \mathbb{Q}$, whose elements are of the form (α, p), with $p \in \mathbb{Q}$. We obtain λ sets of the form $\{\alpha\} \times \mathbb{Q}$. Let L be the union of these sets. Show that, if L is ordered by the relation $<_L$, defined by the condition:

$(\alpha, p) <_L (\beta, q)$ iff $\alpha \in \beta$ or $\alpha = \beta$ and $p <^{\mathbb{Q}} q$ (with $\alpha, \beta \in \lambda$ and $p, q \in \mathbb{Q}$),

then $(L, <_L) \models \mathsf{DLO}_{-,-}$
(ii) Show that, if L is ordered by the relation $<^L$, defined by the condition:

$$(\alpha, p) <^L (\beta, q) \text{ iff } \beta \in \alpha \text{ or } \alpha = \beta \text{ and } p <^{\mathbb{Q}} q,$$

then $(L, <^L) \models \mathsf{DLO}_{-,-}$.
(iii) Both $(L, <_L)$ and $(L, <^L)$ are linear orderings of λ copies of $\mathbb{Q}$. Show that $(L, <_L)$ has a first copy of $\mathbb{Q}$ and no last, while $(L, <^L)$ has a last copy of $\mathbb{Q}$, but no first.
(iv) Show that the map $f : \aleph_1 \longrightarrow L$, defined by the condition $f(\beta) = (\beta, p)$, with $p \in \mathbb{Q}$ fixed and $\beta \in \aleph_1$, embeds the ordinal $\aleph_1$ (usually also designated by ω_1) into $(L, <_L)$.
(v) If g was an isomorphism between $(L, <_L)$ and $(L, <^L)$, $g \circ f$ would embed ω_1 into $(L, <^L)$. Let $(g \circ f)(0) = (\beta, p)$. Show that $\beta \neq 0$.
(vi) Show that $g \circ f[\omega_1]$ determines an infinite, descending chain of ordinals below β. Deduce that $\mathsf{DLO}_{-,-}$ is not λ-categorical.

In the next chapter we develop model-theoretic techniques that will allow us to deepen our investigation of linear orders, including those that are not dense.

7.5 Transcendence Bases

In order to use Vaught test to prove the completeness of VS_K^∞ we have taken advantage of the vector space notion of dimension and the fact that the dimension of a sufficiently large vector space coincides with its cardinality.

This approach also works for algebraically closed fields if we adopt suitable notions of independence and basis. Although linear independence will not work, a natural generalisation does. In order to formulate it, we observe that, from a syntactic standpoint, linear independence is defined by considering $\mathcal{L}_v$-terms of the form:

$$\lambda_1 x_1 + \ldots + \lambda_n x_n.$$

If $\mathcal{V} \models \mathsf{VS}_K^\infty$ and $\lambda_1 v_1 + \ldots + \lambda_n v_n = 0$ implies $\lambda_1 = \ldots = \lambda_n = 0$, then the vectors $v_1, \ldots, v_n$ are said to be linearly independent.

Once we view the expression $\lambda_1 x_1 + \ldots + \lambda_n x_n$ as a $\mathcal{L}_r$-term, where $\lambda_1, \ldots, \lambda_n$ are constant symbols designating elements from a fixed field $\mathcal{K}$, linear independence may be understood as a condition on a linear polynomial in n variables (i.e. a polynomial whose terms are all of degree 1). In this case we say that the field elements $a_1, \ldots, a_n$ are linearly independent over K iff the only linear polynomial in $K[x_1, \ldots, x_n]$ that has the n-tuple $(a_1, \ldots, a_n)$ as a root is 0.

7.5 Transcendence Bases

It suffices to drop the restriction 'linear' from the last condition, to obtain the notion of *algebraic independence* we need.

Definition 7.37 Let the $\mathcal{L}_r$-structure $\mathcal{K}$ be a field and $S = \{s_1, \ldots, s_n\} \subseteq K$. We say that the set S is **algebraically independent** over K if the only way for a polynomial $f(x_1, \ldots, x_n) \in K[x_1, \ldots, x_n]$ to equal zero when x_i is evaluated on s_i is for f to be the constant polynomial 0. If S is infinite, we say that it is algebraically independent over K iff every finite subset of S is. By contrast, we say that the field elements $s_1, \ldots, s_n$ are algebraically dependent over K iff there is a non-constant polynomial $f(x_1, \ldots, x_n) \in K[x_1, \ldots, x_n]$ that equals zero when x_i is evaluated on s_i.

Example 7.38 Algebraic independence implies linear independence but the converse implication does not hold. For instance, in the field $\mathbb{R}$, the elements $\sqrt{2}, \sqrt{3}$ are linearly independent over $\mathbb{Q}$ but not algebraically independent over $\mathbb{Q}$.

Exercise 7.39

(i) Verify the statement from Example 7.38.
(ii) Let r be a real number transcendental over $\mathbb{Q}$. Show that there is a real number transcendental over $\mathbb{Q}(r)$ (*Hint*: count the reals that are algebraic over $\mathbb{Q}(r)$).
(iii) Suppose $\mathcal{K} \sqsubseteq \mathcal{M}$ are fields with $S \subseteq M$. Show that, if S is algebraically independent over K, then $S \cap K = \emptyset$ and every element of S must be transcendental over K.
(iv) If $\mathcal{K}$ is a field, $K[x_1, \ldots, x_n]$ is an integral domain. As such, it is possible to extend it to its field of fractions $K(x_1, \ldots, x_n)$. Describe the elements of $K(x_1, \ldots, x_n)$ and the field operations defined on them. Can we extend this construction to arbitrary set of indeterminates X and build the field of fractions $K(X)$?
(v) Suppose $\mathcal{K} \sqsubseteq \mathcal{M}$ are fields with $S = \{s_1, \ldots, s_n\} \subseteq M$. Prove that the following statements are equivalent:

  (a) S is algebraically independent over K;
  (b) s_1 is transcendental over K and, for each $j \in \{2, \ldots, n\}$, s_j is transcendental over $K(s_1, \ldots, s_{j-1})$;
  (c) $K(s_1, \ldots, s_n)$ is isomorphic to $K(x_1, \ldots, x_n)$.

Algebraic independence enables us to introduce a corresponding notion of basis, defined below.

Definition 7.40 Let the $\mathcal{L}_r$-structures $\mathcal{K} \sqsubseteq \mathcal{M}$ be fields. We say that $S \subseteq M$ is a **transcendence base** of $\mathcal{M}$ over $\mathcal{K}$ iff S is a maximal, algebraically independent set over K.

Every transcendence base of $\mathcal{M}$ over $\mathcal{K}$ is a linearly independent set, once we view $\mathcal{M}$ as a vector space over $\mathcal{K}$. Even though transcendent bases can be extended to vector space bases (once we modify the signature), they do not in general coincide with vector space bases, despite their maximality. This is shown in the next exercise, which adapts an instructive example from [34], p.313.

Exercise 7.41

(i) Given the real field R, show the existence of $r \in \mathbb{R}$ such that r is transcendental over $\mathbb{Q}$.
(ii) Given an element $\alpha \in \mathbb{Q}(r)$, verify that, for suitable choices of $f, g \in \mathbb{Q}[x]$, the polynomial $g(x_1)x_2 - f(x_1)$ equals zero when x_1, x_2 are evaluated on r, α respectively.
(iii) Deduce that $\{r\}$ is a transcendence base of R over $\mathbb{Q}$ and verify that $\{r\}$ cannot be a vector space basis for $\mathbb{R}$ over the scalar field $\mathbb{Q}$.

Our main goal in this section is to prove that, if $\mathcal{K} \sqsubseteq \mathcal{M}$ are fields, any two transcendence bases of $\mathcal{M}$ over $\mathcal{K}$ have the same cardinality. Toward a proof of this result, we establish a useful characterisation of algebraic independence.

Theorem 7.42 *Let $\mathcal{K} \sqsubseteq \mathcal{M}$ be fields and $S \subseteq M$ be algebraically independent over K. Then $t \in M$ is transcendental over $\mathcal{K}(S)$ iff $S \cup \{t\}$ is algebraically independent over K.*

Proof Since we have to prove an equivalence, we split it into two implications. We begin with the assumption that $t \in M$ is transcendental over $\mathcal{K}(S)$. We will prove that, if $f(x_1, \ldots, x_n, y) \in K[x_1, \ldots, x_n, y]$ equals zero when we evaluate x_i on $s_i \in S$ and y on t, then f is the constant polynomial 0.

More concisely, we show that $f(s_1, \ldots, s_n, t) = 0$ implies that f is 0. In syntactic terms, we may regard $f(s_1, \ldots, s_n, t) = 0$ as the evaluation on t of a polynomial $f(s_1, \ldots, s_n, y) \in \mathcal{K}(S)[y]$, where the coefficients of f are elements of the field $\mathcal{K}(S)$.

Because t is transcendental over $\mathcal{K}(S)$, t cannot be a root of $f(s_1, \ldots, s_n, y)$ unless the latter polynomial is an algebraic expression involving $s_1, \ldots, s_n$ and y that equals 0 independently of the value assigned to t. What we have to show is that $f(s_1, \ldots, s_n, y)$ is in fact a constant. To this end, we express it in the following form:

$$f(s_1, \ldots, s_n, y) = h_1(s_1, \ldots, s_n)y^k + \ldots + h_k(s_1, \ldots, s_n)y + h_0,$$

where each coefficient h_i is a product of powers of the s_i, i.e. a polynomial from the ring $K[x_1, \ldots, x_n]$ evaluated over $\{s_1, \ldots, s_n\}$. Because the addends of f sum up to zero, each h_i equals zero when evaluated over $\{s_1, \ldots, s_n\}$. By the algebraic independence of the last set, each h_i is the constant polynomial 0. We conclude that f is itself this constant. It follows that $S \cup \{t\}$ is algebraically independent over K.

Conversely, suppose that $S \cup \{t\}$ is algebraically independent over K. In order to verify that t is transcendental over $\mathcal{K}(S)$, we show that, if a polynomial $f \in \mathcal{K}(S)[y]$ satisfies the condition $f(t) = 0$, this polynomial is the constant 0 (thus, t cannot have a minimum polynomial over $\mathcal{K}(S)$, i.e. it is not algebraic over the latter field). The polynomial f has finitely many coefficients, each of which depends only on finitely many elements of S, if any. This is to say that all such coefficients will figure in a finitely generated extension of $\mathcal{K}$, which we may call $\mathcal{K}(s_1, \ldots, s_n)$, where $s_1, \ldots, s_n$ are algebraically independent.

7.5 Transcendence Bases

We know from Exercise 6.5 that the elements of $\mathcal{K}(s_1, \ldots, s_n)$ have the form $f(s_1, \ldots, s_n)/g(s_1, \ldots, s_n)$ with $f, g \in K[x_1, \ldots, x_n]$ and $g \neq 0$. We may therefore represent f as:

$$f(t) = \frac{f_n}{g_n} t^n + \ldots + \frac{f_1}{g_1} t + \frac{f_0}{g_0},$$

where we omit the explicit mention of $s_1, \ldots, s_n$ from f_i/g_i. Note that each g_i is different from 0. Setting $g = g_0 g_1 \cdots g_n$ we are able rewrite $f(t)$ as:

$$f(t) = \frac{1}{g} \left(f_n g_0 g_1 \cdots g_{n-1} t^n + \ldots + f_1 g_0 g_2 \cdots g_n + f_0 g_1 g_2 \cdots g_n \right)$$

Since the first factor of the last product is not zero, the second must be. The algebraic independence of $s_1, \ldots, s_n$ implies that the second factor is the constant 0. It follows that f is this constant. We conclude that t is transcendental over $\mathcal{K}(S)$. □

In view of the last theorem, we obtain a useful way of describing a transcendence base, given in the next exercise.

Exercise 7.43 Let $\mathcal{K} \sqsubseteq \mathcal{M}$ be fields and $S \subseteq M$. Then S is a transcendence base of $\mathcal{M}$ over K iff $\mathcal{M}$ is an algebraic extension of $\mathcal{K}(S)$.

By Zorn's lemma, transcendence bases always exist.

Exercise 7.44 Let $\mathcal{K} \sqsubseteq \mathcal{M}$ be fields. Show that $\emptyset \subseteq M$ is algebraically independent over K. Apply Zorn's lemma to guarantee the existence of a transcendence base of $\mathcal{M}$ over K.

We are finally ready to prove that any two transcendence bases of a given field over a specified subfield have the same size. This result shows that we can use transcendence bases to introduce a well-defined (i.e. uniquely determined) notion of dimension.

Finite transcendence bases are discussed in the next exercise and infinite ones in the proof of the next theorem.

Exercise 7.45 Let $\mathcal{K} \sqsubseteq \mathcal{M}$ be fields and $S = \{s_1, \ldots, s_n\} \subseteq M$ be a transcendence base of $\mathcal{M}$ over K. Following the steps below, prove that, if T is another such transcendence base, then $|T| = n = |S|$.

(i) Consider the set $S_1 = \{s_2, \ldots, s_n\}$. Prove by contradiction that there is $t_1 \in T$ such that t_1 is transcendental over $\mathcal{K}(s_2, \ldots, s_n)$ (Hint: consider the field extension obtained by adding all of T to $\mathcal{K}(S_1)$ and use Theorem 6.18).
(ii) The set $T_1 = \{t_1, s_2, \ldots, s_n\}$ is algebraically independent by part (i) and Theorem 7.42. Prove by contradiction that s_1 is algebraic over $\mathcal{K}(T_1)$.
(iii) Use Exercise 7.41 to show that T_1 is a transcendence base of $\mathcal{M}$ over K (Hint: use Theorem 6.18).

(iv) Consider $S_2 = \{t_1, s_3, \ldots, s_n\}$ and show that there is $t_2 \in T$ such that t_2 is transcendental over $\mathcal{K}(S_2)$. Deduce that we can iterate parts (i) to (iii) and prove $T_2 = \{t_1, t_2, s_3, \ldots, s_n\}$ to be a transcendence base of $\mathcal{M}$ over K.
(v) After n iterations of the argument given in (i)–(iii) we reach a transcendence base $T_n = \{t_1, \ldots, t_n\}$. Show that $T_n = T$.

Theorem 7.46 *Let $K \sqsubseteq \mathcal{M}$ be fields and $S \subseteq \mathcal{M}$ be an infinite transcendence base of $\mathcal{M}$ over K. If T is another such transcendence base, then $|T| = |S|$.*

Proof Because S is infinite, Exercise 7.45 prevents T from being finite. Moreover, since T is a transcendence base, any $s \in S$ is algebraic over $\mathcal{K}(T)$. Because s has a minimum polynomial with coefficients in $\mathcal{K}(T)$, there is a finite set $T_s \subseteq T$ such that s is algebraic over $\mathcal{K}(T_s)$ (check). Thus, we may associate with each $s \in S$ a finite set $T_s \subseteq T$.

First, we show that $\bigcup_{s \in S} T_s \subseteq T$ is a transcendence base of $\mathcal{M}$ over K. To this end, it suffices to show that $\mathcal{M}$ is an algebraic extension of $\mathcal{K}(\bigcup_{s \in S} T_s)$. Because S is a transcendence base, $\mathcal{M}$ is certainly an algebraic extension of $\mathcal{K}(S)$ and, *a fortiori*, of $\mathcal{K}(S)(\bigcup_{s \in S} T_s)$. Moreover, since, by construction, each $s \in S$ is algebraic over $\mathcal{K}(\bigcup_{s \in S} T_s)$, the adjunction of S to $\mathcal{K}(\bigcup_{s \in S} T_s)$ determines the algebraic extension $\mathcal{K}(\bigcup_{s \in S} T_s)(S)$.

By construction, each $s \in S$ is algebraic over $\mathcal{K}(\bigcup_{s \in S} T_s)$. The addition of algebraic elements produces the algebraic field extension $\mathcal{K}(\bigcup_{s \in S} T_s)(S)$. Since $\mathcal{K}(S) \subseteq \mathcal{K}(\bigcup_{s \in S} T_s)(S)$ (by Exercise 7.47) and S is a transcendence base of $\mathcal{M}$ over K, $\mathcal{M}$ is an algebraic extension of $\mathcal{K}(S) \subseteq \mathcal{K}(\bigcup_{s \in S} T_s)(S)$. By Theorem 6.18, $\mathcal{M}$ is an algebraic extension of $\mathcal{K}(\bigcup_{s \in S} T_s)$.

It now follows that from Exercise 7.43 that $\bigcup_{s \in S} T_s$ is a transcendence base of $\mathcal{M}$ over K. We conclude that:

$$\bigcup_{s \in S} T_s = T.$$

Note that, by reversing the roles of S, T in the previous argument, we can prove $S = \bigcup_{t \in T} S_t$, with S_t a suitably chosen finite subset of S. Now we can use each of the representations $\bigcup_{t \in T} S_t$ and $\bigcup_{t \in T} S_t$ to infer $|S| \leq |T|$ and $|T| \leq |S|$ respectively. The Cantor-Bernstein theorem implies $|S| = |T|$.

To show $|T| \leq |S|$, we invoke the Axiom of Choice to well-order S (see Exercise A.60 from the Appendix), which we can enumerate as $S = \{s_\gamma : \gamma \in \alpha\}$ with α a suitable ordinal. Once S is well-ordered, we can index the sets T_s as T_{s_γ}, for each $\gamma \in \alpha$. We now re-describe T as a union of pairwise disjoint, finite sets: to this end, we consider $T'_{s_\gamma} = T_{s_\gamma} - \bigcup_{\delta \in \gamma} T_{s_\delta}$. By definition, $T'_{s_0} = T_{s_0}$. Moreover, for any $T'_{s_\beta}, T'_{s_\gamma}$ we must have $T_{s_\beta} \cap T_{s_\gamma} = \emptyset$ because the set of larger ordinal index does not contain any element of the set of smaller ordinal index. Finally:

$$\bigcup_{\gamma \in \alpha} T_\gamma = T = \bigcup_{\gamma \in \alpha} T'_{s_\gamma}.$$

7.6 Algebraic Closure

This is because, if $t \in T$, then there is T_{s_γ} such that $t \in T_{s_\gamma}$. In particular, there is a least ordinal β such that $t \in T_{s_\beta}$, in which case $t \in T'_{s_\beta}$. Now each set of the union $\bigcup_{\gamma \in \alpha} T'_{s_\gamma}$ has cardinality $\leq \aleph_0$. Because there are $|S|$ such sets, their disjoint union has cardinality $|T| \leq \aleph_0 |S| = |S|$ (verify). We obtain a similar inequality, with the roles of S, T interchanged, by considering an enumeration of T over some ordinal. It follows that $|S| \leq |T|$. The proof is complete. □

Exercise 7.47 In the proof of Theorem 7.46, show that $\mathcal{K}(S) \subseteq \mathcal{K}(\bigcup_{s \in S} T_s)(S)$.

Theorem 7.46, together with its finite version, allows us to introduce a notion of dimension for fields, which we call *transcendence degree*.

Definition 7.48 et $\mathcal{K} \sqsubseteq \mathcal{M}$ be fields. The cardinality of a transcendence base of $\mathcal{M}$ over $\mathcal{K}$ is called the **transcendence degree** of $\mathcal{M}$ over $\mathcal{K}$.

Remark 7.49 Even though Theorem 7.46 is a result about fields, its proof can be transferred almost literally to a model-theoretic setting to develop the abstract dimension theory discussed in Sect. 7.8. In particular, we shall repeat the last proof to obtain Theorem 7.77. The resulting, abstract notion of dimension, which depends on a model-theoretic notion of algebraic closure, specialises to transcendence degree relative to ACF and to linear dimension relative to VS_K^∞, as will become apparent by the end of Chap. 8.

Exercise 7.50 Prove that $\mathcal{M}$ is an algebraic extension of $\mathcal{K}$ iff its transcendence degree over K equals 0.

Exercise 7.51 Prove the following **Exchange lemma** for fields. Let $\mathcal{K} \sqsubseteq \mathcal{M}$ be fields, $S \subseteq L$ and $a \in L$, $b \in S$. If a is algebraic over $\mathcal{K}(S)$ but not algebraic over $\mathcal{K}(S - \{b\})$, then b is algebraic over $\mathcal{K}((S - \{b\}) \cup \{a\})$.

We already know how prove the completeness of ACF_0 by means of Vaught's test: it suffices to show that any two algebraically closed fields of characteristic zero and with the same, uncountable transcendence degree are isomorphic. The same approach works for algebraically closed fields of characteristic p.

A key part of our argument will be the proof that, if $\mathcal{K}$ is an infinite field, its algebraic closure has the same size as the base field. We arrive at this result in Sect. 7.7, after a brief detour in which we use compactness to show that any field can be extended to a model of ACF, i.e. that any field has an algebraic closure (namely, its relative algebraic closure within an algebraically closed extension).

7.6 Algebraic Closure

Let the $\mathcal{L}_r$-structure $\mathcal{K}$ be a field. Appealing to compactness, we shall construct a $\mathcal{L}_r$-structure $\mathcal{M}$ such that $\mathcal{M} \models \mathsf{ACF}$ and $\mathcal{K} \sqsubseteq \mathcal{M}$.

The strategy we adopt to construct $\mathcal{M}$ is simple: setting $\mathcal{K} = \mathcal{K}_0$, we build a chain $\{\mathcal{K}_i\}_{i \in \mathbb{N}}$ of algebraic extensions by successive additions of polynomial roots.

The union of this chain is a model of ACF_0 or ACF_p, depending on the common characteristic of the fields involved. The next lemma shows how to move up the chain.

Lemma 7.52 *Let $\mathcal{K}$ be a field. There is an extension $\mathcal{K}_1$ of $\mathcal{K}$ such that, for every polynomial $f(x) \in K[x]$, f has a root in $\mathcal{K}_1$.*

Proof We work with an expansion of $\mathcal{L}_r$ obtained by adding one new constant symbol c_f for each polynomial $f(x) \in K[x]$. Let $\mathcal{L}_r^C$ be the expanded language. Its introduction allows us to consider atomic sentences of the form $f(c_f) = 0$, which are modelled by any field containing roots of f. We now apply compactness to to the set of $\mathcal{L}_r^C$-sentences:

$$\Sigma = \mathcal{D}(\mathcal{K}) \cup \{f(c_f) = 0 : f(x) \in K[x]\}.$$

Where $\mathcal{D}(\mathcal{K})$ is the diagram of $\mathcal{K}$. Let A be a finite subset of Σ. If $A \subseteq \mathcal{D}(\mathcal{K})$, then $\mathcal{K} \models A$. If A also contains finitely many sentences of the form $f(c_f) = 0$, then the ones not modelled by $\mathcal{K}$ involve polynomials that are irreducible over K.

Suppose, in particular, that $f(x)$ is irreducible over $\mathcal{K}$. In this case $\langle f(x) \rangle$ is a maximal ideal of $K[x]$ and, consequently, $K[x]/\langle f(x) \rangle$ is a field extension of $\mathcal{K}$ in which $f(x)$ has a root. If necessary, we can iterate this process finitely many times to obtain a field extension in which every sentence of the form $f(c_f) = 0$ in A is true. The resulting field extension contains a copy of $\mathcal{K}$, so it models $\mathcal{D}(\mathcal{K})$. We have shown that Σ is finitely satisfiable. By compactness, let $\mathcal{H}$ be a model of Σ.

By part (c) of Theorem 4.46, since $\mathcal{H} \models \mathcal{D}(\mathcal{K})$, there is an extension $\mathcal{K}_1$ of $\mathcal{K}$ such that $\mathcal{H}$, $\mathcal{K}_1$ are isomorphic. Consequently, $\mathcal{K}_1 \models \Sigma$ and a reduct of this structure to the signature of $\mathcal{L}_r$ provides the field extension of $\mathcal{K}$ we want. □

We obtain models of ACF through a recursive application of the last lemma.

Theorem 7.53 *Any field $\mathcal{K}$ is contained in an algebraically closed extension $\mathcal{K}^+$.*

Proof We construct a chain of field extensions recursively. The base of the recursion is given by $\mathcal{K}_0 = \mathcal{K}$. If the field extension $\mathcal{K}_n$ is given, we apply the previous lemma to obtain the field $\mathcal{K}_{n+1} \sqsupseteq \mathcal{K}_n$. It follows that any polynomial $f \in K_n[x]$ has a root in K_{n+1}.

The recursion we have defined yields the infinite sequence of field extensions:

$$\mathcal{K}_0 \sqsubseteq \mathcal{K}_1 \sqsubseteq \mathcal{K}_2 \sqsubseteq \ldots$$

The union

$$K^+ = \bigcup_{n \in \mathbb{N}} K_n.$$

is the domain of a field $\mathcal{K}^+$. Now suppose that $f(x) \in K^+[x]$ is a non-constant polynomial and let $a_0, \ldots, a_n$ be its coefficients. By the definition of K^+, there is a field K_{i_j} such that $a_j \in K_{i_j}$, with $j = 0, \ldots, n$. Let $m = max\{i_0, \ldots, i_n\}$. Because the fields we are considering are linearly ordered by inclusion, $a_0, \ldots, a_n \in K_m$. By construction, $f(x)$ has a root in $K_{m+1} \subseteq K^+$. We conclude that $\mathcal{K}^+ \models$ ACF. The characteristic of $\mathcal{K}^+$ is clearly the same as the characteristic of $\mathcal{K}$. □

Exercise 7.54 Show that, in the proof of the last theorem, K^+ is the domain of a field.

By a simple combinatorial argument, the construction used in the proof of Theorem 7.53 implies:

Lemma 7.55 *If $\mathcal{K}$ is a countable field, then its algebraic closure $\overline{\mathcal{K}}$ is countable.*

We leave the proof as the next exercise.

Exercise 7.56 Prove Lemma 7.55. It is enough to show that, for any $n \geq 1$, $\mathcal{K}_n$ from the proof of Theorem 7.53 can be selected to have a countable domain.

It follows that $\overline{\mathbb{Q}}$ and, for any prime p, $\overline{\mathbb{Z}_p}$, are countably infinite.

Exercise 7.57 Generalise Lemma 7.55. Show that, if $\mathcal{K}$ is a field and $|K| = \lambda$, then $\overline{\mathcal{K}}$ has size λ.

Exercise 7.58 Let $\mathcal{M}$ be an algebraically closed field of characteristic 0 and $A \subseteq M$. We define the algebraic closure of A in $\mathcal{M}$ as follows. First, we consider $\mathcal{M}_A \sqsubseteq \mathcal{M}$, the substructure generated by A, i.e. the smallest substructure of $\mathcal{M}$ to include A in its domain. In general, $\mathcal{M}_A$ is not a field but an integral domain. Let $Q(\mathcal{M}_A)$ be its associated field of fractions.

The algebraic closure $\overline{Q(\mathcal{M}_A)}$ is a model of ACF$_0$. Using a familiar notation, let us refer to this model of ACF$_0$ as $acl_\mathcal{M}(A)$.

Show that, for any $A \subseteq M$:

(i) $A \subseteq acl_\mathcal{M}(A)$;
(ii) If $a \in acl_\mathcal{M}(A)$, then there is a finite set $A_0 \subseteq A$ such that $a \in acl_\mathcal{M}(A_0)$;
(iii) $A \subseteq B$ implies $acl_\mathcal{M}(A) \sqsubseteq acl_\mathcal{M}(B)$;
(iv) If $C = acl_\mathcal{M}(A)$, then $acl_\mathcal{M}(C) = acl_\mathcal{M}(A)$;
(v) $A \subseteq acl_\mathcal{M}(B)$ and $B \subseteq acl_\mathcal{M}(C)$ imply $A \subseteq acl_\mathcal{M}(C)$.

7.7 Completeness of ACF$_0$ and ACF$_p$

We are now ready to prove the completeness of ACF$_0$ and ACF$_p$, for any prime p. Our poof is an immediate consequence of a fundamental result due to Ernst Steinitz, to the effect that any two algebraically closed fields are isomorphic iff they have the same characteristic and transcendence degree.

Before looking at the full proof of Steinitz's theorem, we outline its structure. On the one hand, if we suppose two algebraically closed fields to be isomorphic, it is not too difficult to see that they must have equal characteristic and transcendence bases of the same size.

If, on the other hand, two fields have the same characteristic and transcendence degree, it takes more work to use this information to build an isomorphism between them. The common characteristic of $\mathcal{K}_1, \mathcal{K}_2$ tells us that they must have the same prime subfield $\mathcal{F}$ (which is either a copy of $\mathbf{Q}$ or a copy of $\mathbf{Z}_p$).

The common transcendence degree yields the existence of a bijection between transcendence bases S_1, S_2. It is then relatively easy to arrive at an isomorphism between $\mathcal{F}(S_1)$ and $\mathcal{F}(S_2)$.

To complete the proof, a lemma is required to extend the latter isomorphism to one between the algebraic closures $\overline{\mathcal{F}(S_1)}$ and $\overline{\mathcal{F}(S_2)}$. By Exercise 7.41, since, for $i \in \{1, 2\}$, $\mathcal{K}_i$ is an algebraically closed, algebraic extension of $\mathcal{F}(S_i)$, $\mathcal{K}_i = \overline{\mathcal{F}(S_i)}$.

We now turn to details and begin with:

Lemma 7.59 *Let $\mathcal{K}_1, \mathcal{K}_2$ be isomorphic fields and $\mathcal{C}_1, \mathcal{C}_2$ their respective algebraic closures. If $f : \mathcal{K}_1 \longrightarrow \mathcal{K}_2$ is a field isomorphism, then f can be extended to a field isomorphism $\overline{f} : \mathcal{C}_1 \longrightarrow \mathcal{C}_2$.*

Proof Since $\mathcal{K}_i$ is included in $\mathcal{C}_i$, for $i = 1, 2$, we refer to the relevant inclusion mapping as ι_i. The function $\iota_2 \circ f$ isomorphically embeds $\mathcal{K}_1$ into $\mathcal{C}_2$ and $\mathcal{C}_1$ is an algebraic extension of $\mathcal{K}_1$. Theorem 6.32 now applies to guarantee the existence of a field homomorphism $\overline{f} : \mathcal{C}_1 \longrightarrow \mathcal{C}_2$ that agrees with $\iota_2 \circ f$ over $\mathcal{K}_1$. In particular $\overline{f}[\mathcal{K}_1] = \mathcal{K}_2$. Since all field homomorphisms are injective, we only have to verify that $\overline{f}$ is surjective. To this end, it suffices to note that $K_2 \subseteq \overline{f}[\mathcal{C}_1]$ and that $\overline{f}[\mathcal{C}_1] \subseteq \mathcal{C}_2$ is included in an algebraic extension of $\mathcal{K}_2$. Because $\overline{f}$ is an embedding, $\overline{f}[\mathcal{C}_1]$ is the domain of an algebraically closed field extension of $\mathcal{K}_2$, but this extension is also algebraic, so it is an algebraic closure of $\mathcal{K}_2$. Because all algebraic closures are isomorphic, $\overline{f}[\mathcal{C}_1] \subseteq \mathcal{C}_2$ implies equality and $\overline{f}$ is the required field isomorphism. □

Theorem 7.60 *If $\mathcal{K}_1, \mathcal{K}_2$ are models of ACF, then they are isomorphic iff they have the same characteristic and transcendence degree.*

Proof If $\mathcal{K}_1, \mathcal{K}_2$ are isomorphic models of ACF, we leave it as an exercise to verify that they must have the same characteristic and transcendence degree. If, on the other hand $\mathcal{K}_1, \mathcal{K}_2$ have the same characteristic and transcendence degree, let F be their common, prime subfield (in general, their prime subfields will be distinct and isomorphic: there is no harm here in identifying them) and let S_1, S_2 be the respective transcendence bases. Since $|S_1| = |S_2|$, there is a bijection $f : S_1 \longrightarrow S_2$. Thus, $S_2 = f[S_1]$. The bijection f can be extended to a ring isomorphism between $F[S_1]$ and $F[S_2]$. The elements of $F[S_1]$ are polynomials with coefficients in S, whose variables $x_1, \ldots, x_n$ are evaluated over $s_1, \ldots, s_n \in S_1$. The elements of $F[S_2]$ have the obvious structure. To obtain the desired isomorphic extension f_1, it suffices to send any evaluation of a polynomial in $x_1, \ldots, x_n$ over $s_1, \ldots, s_n$ into an evaluation of the same polynomial over $f(s_1), \ldots, f(s_n)$. Polynomial addition and multiplication are preserved by construction.

7.7 Completeness of ACF$_0$ and ACF$_p$

Because $F[S_1]$ and $F[S_2]$ are both integral domains, they can be embedded into the corresponding fields of fractions $F(S_1)$ and $F(S_2)$. The isomorphism f_1 is easily extended to an isomorphism f_2 between $\mathcal{F}(S_1)$ and $\mathcal{F}(S_2)$. Lemma 7.59 finally applies, allowing us to extend f_2 to an isomorphism f_3 from $\overline{\mathcal{F}(S_1)}$ to $\overline{\mathcal{F}(S_2)}$, i.e. between $\mathcal{K}_1$ and $\mathcal{K}_2$. □

Exercise 7.61 Describe the action of f_1, f_2 from the proof of Theorem 7.60.

Lemma 7.62 *Let $\mathcal{K}$ be an uncountably large, algebraically closed field and S a transcendence base for $\mathcal{K}$ over its prime subfield $\mathcal{F}$. Then $|\mathcal{K}| = |S|$.*

Proof We suppose that $|\mathcal{K}| = \mu$ and want to show that $|S| = \mu$. First we note that, because $\mathcal{K}$ is an algebraic extension of $\mathcal{F}(S)$, $|F(S)| = |\mathcal{K}|$. To see this, we overestimate the size of an algebraic extension of $\mathcal{F}(S)$. To this end, it suffices to consider the set of all polynomials with coefficients in $\mathcal{F}(S)$, denoted by $F(S)[x]$, and assign to each $f \in F(S)[x]$ a countably infinite set (certainly larger then the set of f's roots in $\mathcal{K}$). We leave the argument's details as an exercise.

The next size estimate we need concerns the field $\mathcal{F}(S)$. Using the fact that F is countable and the hypothesis that $|\mathcal{K}| = \mu$ is uncountable, we are going to show that $|S| = \kappa < \mu$ implies $|F(S)| < \mu$, contradicting $|F(S)| = |\mathcal{K}|$. Because obviously $|F(S)| \leq \mu$, it will follow that $|F(S)| = \mu = |\mathcal{K}|$.

To evaluate $|F(S)|$ when $|S| = \kappa$, we restrict attention to the set $F[S]$. An element of $F[S]$ is a finite sum of terms, each of which is a finite product of an element of F with finitely many powers of elements of S, also known as a *monomial*. We therefore have to count the possible monomials arising from elements of S. It is convenient to do so by attending to the set of monomials whose length is some fixed, positive integer m. A monomial of length m is uniquely determined by a sequence of m choices from S, followed by a sequence of m choices from $\mathbb{N}$ (the exponents of the monomial's factors).

There are κ sequences of m terms without repetitions[9] and $\aleph_0$ distinct choices of exponents for each of them. Overall, there are $\aleph_0 \cdot \kappa = \kappa$ possibilities. This is true for each positive integer m: we conclude that there must be $\aleph_0 \cdot \kappa$ monomials based on elements of S. This value does not change if we consider monomials preceded by an element of F.

An element of $F[S]$ may now be identified with a finite sequence of distinct monomials. For any $n \geq 1$ is an arbitrary integer, there are $\underbrace{\kappa + \ldots \kappa}_{n\ times} = \kappa$ elements of $F[S]$ of length n. The size of $F[S]$ is therefore $\aleph_0 \kappa = \kappa$. Because $F(S)$ is the field of fractions of $F[S]$ its size is the same as the size of the set $F[S] \times F[S]$, i.e. $\kappa \times \kappa = \kappa$ (as noted in Sect. 6.5, if $\mathcal{R}$ is an integral domain, the set $R \times R$ can be used as the domain of its field of fractions).

We arrive at the chain of relations $|\mathcal{K}| = |F(S)| = \kappa < \mu = |\mathcal{K}|$, which cannot hold. It follows that $|S| = \mu = |\mathcal{K}|$, as desired. □

---

[9] We have $\kappa(\kappa - 1) \cdots (\kappa - m + 1)$ choices but each factor equals κ because κ is infinite so cardinal arithmetic yields κ as the value of the last product.

Note the similarity between Lemma 7.62 and the result that an uncountably large vector space $\mathcal{V}$ over a countable field K has the same size as any of its bases. We are now only one step away from proving completeness for ACF_0 and ACF_p, with p a prime.

Theorem 7.63 *Let $\kappa > \aleph_0$ be an uncountable cardinal and $\mathcal{K}_1, \mathcal{K}_2$ be algebraically closed fields with the same characteristic and the same size κ. Then $\mathcal{K}_1, \mathcal{K}_2$ are isomorphic.*

Proof By Theorem 7.60 $\mathcal{K}_1, \mathcal{K}_2$ are isomorphic iff they have the same characteristic and transcendence degree. By hypothesis, they have the same characteristic and their transcendence degree is κ in virtue of the hypothesis and Lemma 7.62. Thus, $\mathcal{K}_1$ and $\mathcal{K}_2$ are isomorphic. □

By Vaught's test, we obtain:

Theorem 7.64 *The theory ACF_0 is complete. Moreover, for any prime p, the theory ACF_p is complete.*

7.8 Model-Theoretic Algebraic Closure

In Sect. 3.2.2 and Exercise 7.58 we have adopted the notation $acl_\mathcal{M}(A)$ to indicate, respectively, a subspace of a vector space and an algebraically closed subfield of a model of ACF.

These kinds of substructures, i.e. linear closures and algebraic closures, specify in the respective contexts an abstract model-theoretic notion of algebraic closure. When this notion satisfies a key condition, known as the *exchange property*, it gives rise to a well-defined notion of dimension, which specialises to linear dimension in vector spaces and transcendence degree in algebraically closed fields.

A full account of abstract dimension theory will be offered at end of Chap. 8, where it can be more explicitly connected with the model-theoretic properties of VS_K^∞ and ACF. Here we focus on defining the key concepts of closure, basis, independence and dimension.

Definition 7.65 Let $\mathcal{M}$ be a $\mathcal{L}$-structure and $A \subseteq M$. The **algebraic closure** of A in $\mathcal{M}$, in symbols $acl_\mathcal{M}(A)$, is the union of the finite subsets of M defined by $\mathcal{L}$-formulae with parameters in A.

By the last definition, $acl_\mathcal{M}(A)$ is obtained by adjoining to A all the elements of M that satisfy a $\mathcal{L}$-formula $\varphi(x, a_1, \ldots, a_n)$ such that $a_1, \ldots, a_n \in A$ and $\mathcal{M} \models \exists^{=k} x \varphi(x, a_1, \ldots, a_n)$, where the last expression abbreviates a longer formula saying that there are *exactly k* elements of M satisfying it. We shall resort to abbreviations of this kind systematically.

It follows from the definition of algebraic closure that $acl_\mathcal{M}(A)$ is the domain of a substructure of $\mathcal{M}$. By an abuse of notation, we shall refer to this substructure and its domain indifferently as $acl_\mathcal{M}(A)$.

7.8 Model-Theoretic Algebraic Closure

Because we have defined the algebraic closure of a set with reference to formulae that are satisfied by finitely many elements of a fixed domain, it is helpful have a distinctive name for them.

Definition 7.66 Let $\mathcal{M}$ be a $\mathcal{L}$-structure, $A \subseteq M$ and $\varphi(x, a_1, \ldots, a_n)$ a $\mathcal{L}$-formula with parameters $a_1, \ldots, a_n \in A$. We say that $\varphi(x, a_1, \ldots, a_n)$ is an **algebraic formula over A in $\mathcal{M}$** iff for some $k \in \mathbb{N}$, $\mathcal{M} \models \exists^{=k} x \varphi(x)$ (we allow the possibility that $A = \emptyset$).

Remark 7.67 From now on, whenever convenient and unambiguously practicable, we shall refer to a finite n-tuple of parameters like $(a_0, \ldots, a_{n-1})$ and $(f(a_0), \ldots, f(a_{n-1}))$ by $\bar{a}$, $f(\bar{a})$ respectively, and to a finite sequence of variables like $(x_0, \ldots, x_{n-1})$ by $\bar{x}$.

Exercise 7.68

(i) Show that, if $\mathcal{M} \preceq \mathcal{N}$, then $\varphi(x)$ is algebraic over A in $\mathcal{M}$ iff $\varphi(x)$ is algebraic over A in $\mathcal{N}$ as well;
(ii) Deduce that, if $\mathcal{M} \preceq \mathcal{N}$, then $acl_\mathcal{M}(A) = acl_\mathcal{N}(A)$;
(iii) Show that, if $A \subseteq M$, then $A \subseteq acl_\mathcal{M}(A)$;
(iv) Show that, if $A, B \subseteq M$, then $A \subseteq B$ implies $acl_\mathcal{M}(A) \subseteq acl_\mathcal{M}(B)$.

Exercise 7.69 Let $\mathcal{M} \models \mathsf{ACF}_0$. Any atomic $\mathcal{L}_r$-formula $\varphi(x)$ without parameters is, up to equivalence, of the form $a_n x^n + a_{n-1} x^{n-1} + \ldots + a_1 x + a_0 = 0$, where $a_n, \ldots, a_0$ are closed $\mathcal{L}_r$-terms that denote elements of $\mathbb{Z}$. Because any polynomial in $\mathbb{Z}[x]$ has finitely many roots in $\mathcal{M}$, $\varphi(x)$ is certainly algebraic in $\mathcal{L}$.

(i) Show that the set of roots of any polynomial in $\mathbb{Q}[x]$ can be defined by means of a $\mathcal{L}_r$-formula without parameters.
(ii) Deduce that $acl_\mathcal{M}(\emptyset)$ contains every element of M that is algebraic over the prime subfield $\mathbb{Q}$. In other words $\overline{\mathbb{Q}} \subseteq acl_\mathcal{M}(\emptyset)$.
(iii) Suppose that every $\mathcal{L}_r$-formula is equivalent to a quantifier-free $\mathcal{L}_r$-formula modulo ACF_0. This is to say that, if $\psi(x, \bar{y})$ is a $\mathcal{L}_r$-formula, then there is a quantifier-free $\mathcal{L}_r$-formula $\varphi(x, \bar{y})$ such that:

$$\mathsf{ACF}_0 \models \forall \bar{y} (\psi(x, \bar{y}) \leftrightarrow \varphi(x, \bar{y})).$$

Show that every definable subset of M is either finite or has a finite complement (*Hint*: by Exercise 2.47 in chapter two, $\varphi(x, \bar{a})$ may be written as a disjunctive normal form).
(iv) For any $A \subseteq M$, let $Q(A)$ be the fraction field of $\mathcal{M}$, the substructure of $\mathcal{M}$ generated by A. Show that $acl_\mathcal{M}(A) = \overline{Q(A)}$.
(v) Show that $\overline{\mathbb{Q}} = acl_\mathcal{M}(\emptyset)$.

Model-theoretic algebraic closure satisfies some important properties, foreshadowed by Exercise 7.58.

Lemma 7.70 Let $\mathcal{M}$ be a $\mathcal{L}$-structure and $A \subseteq M$. Then:

(1) $A \subseteq acl_{\mathcal{M}}(A)$;
(2) $A, B \subseteq M$, then $A \subseteq B$ implies $acl_{\mathcal{M}}(A) \subseteq acl_{\mathcal{M}}(B)$;
(3) $acl_{\mathcal{M}}(acl_{\mathcal{M}}(A)) = acl_{\mathcal{M}}(A)$;
(4) if $A \subseteq acl_{\mathcal{M}}(B)$ and $B \subseteq acl_{\mathcal{M}}(C)$, then $A \subseteq acl_{\mathcal{M}}(C)$;
(5) $acl_{\mathcal{M}}(A)$ is the union of all sets of the form $acl_{\mathcal{M}}(A_0)$, where A_0 is a finite subset of A.

Proof

(1) see Exercise 7.68-(iii);
(2) see Exercise 7.68-(iv);
(3) by (i) and (ii), $acl_M(A) \subseteq acl_M(acl_M(A))$. We verify the converse inclusion. Suppose $b \in acl_M(acl_M(A))$. Then there are parameters $b_1, \ldots, b_m \in acl_M(A)$ and an algebraic formula $\varphi(x, b_0, \ldots, b_{m-1})$ such that $\mathcal{M} \models \varphi(b, b_0, \ldots, b_{m-1})$. Suppose that $\varphi(x, b_0, \ldots, b_{m-1})$ defines a subset of M of size l.

Since $b_i \in acl_M(A)$ $i = 0, \ldots, m-1$, for each i there is a formula $\psi_i(y_i, \overline{a}_i)$ over A that is satisfied by b_i and, in addition, can only be satisfied by finitely many, say l_i, other elements of M. As a result, the conjunction:

$$\bigwedge_{i=0}^{m-1} \psi_i(y_i, \overline{a}_i))$$

defines a finite set of m-tuples containing $n = l_0 \cdots l_{m-1}$ elements. Let us abbreviate the last conjunction by $\eta(\overline{y}, \overline{a})$. We next consider the formula:

$$\exists \overline{y}(\varphi(x, \overline{y}) \wedge \eta(\overline{y}, \overline{a}) \wedge \exists^{=l} z \varphi(z, \overline{y})),$$

which we shall refer to as $\Phi(x)$. Our goal is to show that $\Phi(x)$ is algebraic and $\mathcal{M} \models \Phi(b)$.

When $\overline{y}$ is evaluated on a tuple of M^m other than the n choices satisfying $\eta(\overline{y}, \overline{a})$ in $\mathcal{M}$, $\Phi(x)$ cannot be satisfied by any assignment to x because one of its conjuncts is not satisfied. We can therefore restrict attention to n possible selections from M^m for $\overline{y}$. If the selection $\overline{c}$ determines a conjunct $\varphi(x, \overline{c})$ that defines an infinite subset of M, then the parametric formula $\exists^{=l} z \varphi(z, \overline{c})$ cannot be satisfied in $\mathcal{M}$.

It now follows that $\Phi(x)$ defines a set of size at most ln, since $\varphi(x, \overline{c})$ may, in principle, be satisfied by l distinct elements of M for each of n choices of $\overline{c}$. In any case, $\Phi(x)$ is algebraic over A and it is satisfied by b. We conclude that $b \in acl_M(A)$.

(4) This statement follows easily from (2) and (3).
(5) If $b \in acl_M(A)$, then b belongs to the finite set defined by a formula $\varphi(x, \overline{a})$, with $\overline{a} \in A$. Thus, by definition of algebraic closure, $b \in acl_M(\overline{a})$. It follows that $acl_M(A)$ is included in the union of algebraic closures of finite subsets of A. The converse inclusion follows from (2).

7.8 Model-Theoretic Algebraic Closure

□

With the notion of algebraic closure at our disposal, we have a natural way of defining independence and the corresponding notion of a basis.

Definition 7.71 Let $\mathcal{M}$ be a $\mathcal{L}$-structure with $A, B \subseteq M$. We say that B is **independent over** A iff for any $b \in B$, $b \notin acl_\mathcal{M}((A \cup B) - \{b\})$. We say that B is **independent** *simpliciter* iff B is independent over $\emptyset$.

Exercise 7.72 Show that independence reduces to linear independence when specialised to vector spaces over a fixed field $\mathcal{K}$.

We next define bases in terms of independence.

Definition 7.73 Let $\mathcal{M}$ be a $\mathcal{L}$-structure and $A \subseteq M$. A **basis** for A is an independent set $B \subseteq A$ such that $acl_\mathcal{M}(B) = acl_\mathcal{M}(A)$. If $C \subseteq M$, a basis for A over C is a set $B \subseteq A$ that is independent over C and satisfies the equality $acl_\mathcal{M}(A \cup C) = acl_\mathcal{M}(B \cup C)$.

Exercise 7.74 Under the assumptions of Exercise 7.69, show that, if $\mathcal{M} \models \mathsf{ACF}_0$, a model-theoretic basis for $A \subseteq M$ is a transcendence base for $\overline{\mathbb{Q}(A)}$ over $\mathbb{Q}$.

With independence and bases at our disposal, we are one step away from a theory of dimension. What keeps us from it is the fact that we could define *the* dimension of A over C to be the cardinality of a basis for A over C only if this cardinality was not affected by a choice of basis, i.e. only if every basis had the same cardinality.

We cannot ensure this for a notion of closure that only satisfies the conditions listed in Lemma 7.70. We do know, however, that if one further condition, known as the exchange property, holds, then dimension, defined in terms of bases, is uniquely determined.

We have verified the field-theoretic version of the exchange property in Exercise 7.51. In purely model-theoretic terms, the exchange property amounts to the fact that, if $\mathcal{M}$ is a $\mathcal{L}$-structure and $A \subseteq M$, then $c \in acl_\mathcal{M}(A \cup \{b\})$ and $c \notin acl_\mathcal{M}(A)$ jointly imply $b \in acl_\mathcal{M}(A \cup \{c\})$.

Even though the exchange property does not follow from the definition of $acl_\mathcal{M}$, it is satisfied by the models of **strongly minimal theories**, to be discussed in Sect. 8.8. Here we suppose that it is satisfied and exhibit its connection with the uniqueness of dimension. The following theorem prepares us to do so with respect to finite bases.

Theorem 7.75 (Finite Exchange) *Suppose that $\mathcal{M}$ is a $\mathcal{L}$-structure with the exchange property. Let $E, F \subseteq M$ be finite independent sets such that $|E| = |F|$ and $E \subset acl_\mathcal{M}(F)$. Then $F \subset acl_\mathcal{M}(E)$.*

Proof In the following argument we freely employ the notation $acl_\mathcal{M}(e_2, \ldots, e_{m+1})$ instead of the formally correct $acl_\mathcal{M}(\{e_2, \ldots, e_{m+1}\})$.

We proceed by induction on the common size n of E, F. If $n = 1$, set $E = \{e_1\}, F = \{f_1\}$. By hypothesis $e_1 \in acl_\mathcal{M}(\emptyset \cup \{f_1\})$. By independence

$e_1 \notin acl_{\mathcal{M}}(E - \{e_1\}) = acl_M(\emptyset)$. The Exchange Property applies and we obtain $f_1 \in acl_M(\emptyset \cup \{e_1\}) = acl_{\mathcal{M}}(\{e_1\})$.

Next, suppose that the lemma holds for any m independent elements and let $n = m+1$. Then:

$$E = \{e_1, \ldots, e_{m+1}\}, F = \{f_1, \ldots, f_{m+1}\}.$$

We shall now show that $acl_{\mathcal{M}}(\{f_2, \ldots, f_{m+1}\})$ cannot include the whole of E. If it did, we should have $E \subset acl_{\mathcal{M}}(f_2, \ldots, f_{m+1})$ and, *a fortiori*, $\{e_2, \ldots, e_{m+1}\} \subset acl_{\mathcal{M}}(f_2, \ldots, f_{m+1})$. By the inductive hypothesis, Finite Exchange implies:

$$\{f_2, \ldots, f_{m+1}\} \subset acl_{\mathcal{M}}(e_2, \ldots, e_{m+1}).$$

Thus:

$$acl_{\mathcal{M}}(f_2, \ldots, f_{m+1}) \subset acl_{\mathcal{M}}(e_2, \ldots, e_{m+1}).$$

If, therefore $e_1 \in E - \{e_2, \ldots, e_{m+1}\}$, the hypothesis $e_1 \in acl_{\mathcal{M}}(f_2, \ldots, f_{m+1})$ implies $e_1 \in acl_{\mathcal{M}}(e_2, \ldots, e_{m+1})$, which contradicts the independence of E. We conclude that $e_1 \notin acl_{\mathcal{M}}(f_2, \ldots, f_{m+1})$.

Having obtained the conditions:

$$e_1 \notin acl_{\mathcal{M}}(f_2, \ldots, f_{m+1}) \text{ and } e_1 \in acl_{\mathcal{M}}(\{f_2, \ldots, f_{m+1}\} \cup \{f_1\}).$$

we can apply the exchange property to deduce:

$$f_1 \in acl_{\mathcal{M}}(\{f_2, \ldots, f_{m+1}\} \cup \{e_1\}).$$

We now turn to the set $\{e_1, f_2, \ldots, f_{m+1}\} = \{e_1, f_3, \ldots, f_{m+1}\} \cup \{f_2\}$. Because the set $\{e_1, f_3, \ldots, f_{m+1}\}$ has m elements, its algebraic closure cannot contain the whole of E. If it did, we would reach the contradiction $e_1 \in acl_{\mathcal{M}}(e_2, \ldots, e_{m+1})$ again (since $\{e_2, \ldots, e_{m+1}\} \subseteq acl_{\mathcal{M}}(e_1, f_3, \ldots, f_{m+1})$ and Finite Exchange holds by the inductive hypothesis). We can therefore select $e_2 \in E$ such that $e_2 \notin acl_M(\{e_1, f_3, \ldots, f_{m+1}\})$.

We next claim that $e_2 \in acl_{\mathcal{M}}(\{e_1, f_3, \ldots, f_{m+1}\} \cup \{f_2\})$. Because $e_2 \in acl_{\mathcal{M}}(F)$, there is an algebraic formula $\varphi(x, f_1, \ldots, f_{m+1})$ satisfied in M by e_2 (the parameters occurring in φ belong to F but do not have to exhaust F). Let l_0 be the finite size of the set defined by $\varphi(x, f_1, \ldots, f_{m+1})$ in M.

Since $f_1 \in acl_{\mathcal{M}}(\{f_2, \ldots, f_{m+1}\} \cup \{e_1\})$, there is a formula $\psi(y, e_1, f_2, \ldots, f_{m+1})$ satisfied in M by f_1 and defining a finite subset of M. If the latter set has l_1 elements, the formula:

$$\exists^{=l_0} z \exists^{=l_1} y (x = z \wedge \varphi(x, y, f_2, \ldots, f_{m+1}) \wedge \psi(y, e_1, f_2, \ldots, f_{m+1}))$$

7.8 Model-Theoretic Algebraic Closure

is algebraic with parameters in $\{e_1, f_3, \ldots, f_{m+1}\} \cup \{f_2\}$ and it is satisfied by $e_2 \in M$.

We have now established:

$$e_2 \notin acl_{\mathcal{M}}(\{e_1, f_3, \ldots, f_{m+1}\}) \text{ and } e_2 \in acl_{\mathcal{M}}(\{e_1, f_3, \ldots, f_{m+1}\} \cup \{f_2\})$$

and we can use the exchange property again to deduce:

$$f_2 \in acl_{\mathcal{M}}(\{e_1, e_2, f_3, \ldots, f_{m+1}\}).$$

Imitating the argument just described, we arrive at:

$$f_1 \in acl_{\mathcal{M}}(\{e_1, e_2, f_3, \ldots, f_{m+1}\}).$$

Thus, $\{f_1, f_2\} \subset acl_{\mathcal{M}}(e_1, e_2, f_3, \ldots, f_{m+1})$. A finite iteration of the procedure described so far yields the inclusion $F \subset acl_{\mathcal{M}}(E)$. □

Lemma 7.76 *Let $\mathcal{M}$ be a $\mathcal{L}$-structure in which the exchange property holds for $acl_{\mathcal{M}}$ and $A, C \subseteq M$. If A has a finite basis over C, then any two bases over C have the same size.*

Proof Let B_1, B_2 be two bases of A over C. Without loss of generality, we may suppose $|B_1| \leq |B_2|$. The latter hypothesis naturally induces us to consider $B_3 \subseteq B_2$ such that $|B_1| = |B_3|$. We shall show that B_3 is in fact the set B_2. By the definitions of B_1 and B_3, we have:

$$B_3 \subseteq acl_{\mathcal{M}}(A \cup C) = acl_{\mathcal{M}}(B_1).$$

By Finite Exchange, $B_1 \subseteq acl_{\mathcal{M}}(B_3)$. By Lemma 7.70 we further deduce $acl_{\mathcal{M}}(B_1) \subseteq acl_{\mathcal{M}}(B_3)$. Since B_1 is a basis of A over C, we conclude $acl_{\mathcal{M}}(A \cup C) \subseteq acl_{\mathcal{M}}(B_3)$. Applying $acl_{\mathcal{M}}$ to $B_3 \subseteq B_2$, we obtain the converse inclusion. Thus, $acl_{\mathcal{M}}(B_2) = acl_{\mathcal{M}}(B_3)$. If $b \in B_2 - B_3$, then $b \notin acl_{\mathcal{M}}(B_3)$ by the independence of B_2. Since, however, $acl_{\mathcal{M}}(B_3) = acl_{\mathcal{M}}(B_2)$, we arrive at the contradiction $b \notin acl_{\mathcal{M}}(B_2)$. It follows that $B_3 = B_2$. As a consequence $|B_1| = |B_2|$. □

We have now shown that algebraic closure determines a well-defined notion of finite dimension. As noted in Sect. 7.6, the proof of Theorem 7.46 may in essence be replicated to generalise the last theorem to arbitrary dimension.

Theorem 7.77 *Let $\mathcal{M}$ be a $\mathcal{L}$-structure in which $acl_{\mathcal{M}}$ has the exchange property and $A, C, \subseteq M$. If A has a basis B_1 over C, then any two bases over C have the same size.*

Proof If A has a finite basis, Lemma 7.76 yields the result. We may thus suppose that B_1, B_2 are infinite bases of A over C. It is convenient to expand the language by adding constant symbols that name the elements of C, in order to regard B_1, B_2 as bases of A over $\emptyset$ and forget the additional parameters.

By Lemma 7.76, B_2 cannot be finite. By the Axiom of Choice, $|B_1| \leq |B_2|$ or $|B_2| \leq |B_1|$. Suppose the former inequality holds. We conclude the proof by constructing an injection of B_2 into B_1 and applying the Cantor-Bernstein theorem.

Since B_2 is a basis for A (in the expanded language), each $b \in B_1 \subseteq A$ is dependent on B_2 and, in particular, on a finite set $S_b \subseteq B_2$ by part (v) of Lemma 7.70. By the Axiom of Choice again B_1 may be well-ordered as $\{b_\delta : \delta \in \alpha\}$ for some ordinal α.

We induce a well-ordering on the finite sets associated with each element of B_1 in the way described in Sect. 7.6. Let these sets be enumerated as $\{S_{b_\delta} : \delta \in \alpha\}$. The given enumeration is again reducible to an enumeration of disjoint, nonempty sets by the exact same approach adopted in the proof of Theorem 7.46. It suffices to set $S'_{b_\delta} = S'_{b_\delta} - \bigcup_{\gamma \in \delta} S_{b_\gamma}$.

As in the proof involving transcendence bases, the union of the S'_{b_δ}, which are pairwise disjoint, coincides with B_2. Since each S'_{b_δ} is finite, we deduce $|B_2| \leq \aleph_0 |B_1| = |B_1|$. The equality $|B_2| = |B_1|$ follows. □

We shall return to the dimension theory associated with algebraic closure at the end of Chap. 8.

Chapter 8
Model-Completeness and Quantifier Elimination

Abstract A $\mathcal{L}$-theory T is model-complete if $T \cup \mathcal{D}(\mathcal{M})$ is complete, where $\mathcal{D}(\mathcal{M})$ is the diagram of $\mathcal{M} \models T$. If $\mathcal{M} \models T$ and $\mathcal{A} \sqsubseteq \mathcal{M}$ imply that $T \cup \mathcal{D}(\mathcal{A})$ is a complete theory, then T is substructure-complete.

When the extensions of T we have described are complete, important results follow concerning the complexity of the first-order information codified by a model of T. If, for instance, T is model-complete, then every formula is T-equivalent to an existential $\mathcal{L}$-formula and, if T is substructure-complete, every formula is T-equivalent to a quantifier-free $\mathcal{L}$-formula.

In this chapter we pursue a comprehensive study of model-completeness and of substructure-completeness, starting with the former, weaker property. In Sect. 8.1, model-completeness is defined and simple theories of linear orders are tested for it. In Sect. 8.2 we study the preservation of formulae under substructures and extensions in order to obtain important characterisations of model-completeness. Section 8.3 covers essential material on unions of chains needed to establish Lindström's test for model-completeness, which immediately yields this property for the algebraic theories discussed in the preceding chapters.

From Sect. 8.4 we turn to quantifier elimination and its consequences. After showing how quantifier elimination may be enforced by suitable expansions of a first-order language, we prove a few different quantifier elimination tests. We focus on finite relational languages in Sect. 8.5 and on the connection between quantifier elimination and extensions of embeddings in Sects. 8.6 and 8.7.

The final sections of this chapter discuss the impact of quantifier elimination on definability. In Sect. 8.8 we focus on strongly minimal theories, whose models have the least possible amount of parametrically definable sets, and show that, in particular, the theories of torsion-free, divisible abelian groups, of infinite vector spaces and of algebraically closed fields are strongly minimal. The models of strongly minimal theories support the model-theoretic dimension theory developed in Chap. 7 because they have the exchange property, a fact we prove in Sect. 8.8.

In Sect. 8.9 we offer a brief discussion of o-minimal theories, linear orders in which the only parametrically definable subsets are the finite unions of points and intervals.

8.1 Model-Completeness

Part (ii) of Exercise 4.36, directly implies that IS is model-complete. This is clear in view of the following definition of model-completeness.

Definition 8.1 Let T be a $\mathcal{L}$-theory. We say that T is **model-complete** iff for any $\mathcal{M}, \mathcal{N} \models T$, $\mathcal{M} \sqsubseteq \mathcal{N}$ implies $\mathcal{M} \preceq \mathcal{N}$.

In other words, T is model-complete iff the submodels of its models are not only substructures but also elementary substructures. The next lemma shows that Definition 8.1 is equivalent to the definition mentioned in the opening of the present chapter.

Lemma 8.2 Let T be a $\mathcal{L}$-theory. Then T is model-complete (in the sense of Definition 8.1) iff for any $\mathcal{M} \models T$, $T \cup \mathcal{D}(\mathcal{M})$ is a complete $\mathcal{L}(C)$-theory, where $\mathcal{L}(C)$ is the language obtained from $\mathcal{L}$ by adding a set C of new constant symbols to name the elements of M not already named by the constant symbols in $\mathcal{L}$.

Proof Suppose that T is model-complete. To show that $T \cup \mathcal{D}(\mathcal{M})$ is complete, we prove that any two $\mathcal{L}(C)$-structures $\mathcal{N}_1^+, \mathcal{N}_2^+ \models T \cup \mathcal{D}(\mathcal{M})$ are elementarily equivalent.

Because, in particular, $\mathcal{N}_1^+, \mathcal{N}_2^+ \models \mathcal{D}(\mathcal{M})$, the structure $\mathcal{M}$, and thus its expansion $\mathcal{M}^+$ to $\mathcal{L}(C)$, can be embedded into each of $\mathcal{N}_1^+, \mathcal{N}_2^+$. In view of Theorem 4.46, we can identify $\mathcal{N}_1^+, \mathcal{N}_2^+$ with their isomorphic copies extending $\mathcal{M}^+$. Thus, we regard $\mathcal{M}^+$ as a common substructure of $\mathcal{N}_1^+, \mathcal{N}_2^+$.

We now consider an arbitrary $\mathcal{L}(C)$-sentence $\varphi(c_1, \ldots, c_k)$, in which only the constant symbols $c_1, \ldots, c_k \in \mathcal{L}(C)$ occur. Since these symbols have interpretations in M, we may assume that they name $a_1, \ldots, a_k \in M$ respectively. Thus $\varphi(a_1, \ldots, a_n)$ is a $\mathcal{L}$-formula with parameters in M corresponding to the sentence $\varphi(c_1, \ldots, c_k)$.

Let $\mathcal{N}_1, \mathcal{N}_2$ the $\mathcal{L}$-reducts of $\mathcal{N}_1^+, \mathcal{N}_2^+$ respectively. The model-completeness of T yields $\mathcal{M} \preceq \mathcal{N}_1, \mathcal{N}_2$, from which we are able to deduce the following chain of equivalences:

$$\mathcal{N}_1^+ \models \varphi(c_1, \ldots, c_k) \iff \mathcal{N}_1 \models \varphi(a_1, \ldots, a_k)$$
$$\iff \mathcal{M} \models \varphi(a_1, \ldots, a_k)$$
$$\iff \mathcal{N}_2 \models \varphi(a_1, \ldots, a_k)$$
$$\iff \mathcal{N}_2^+ \models \varphi(c_1, \ldots, c_k).$$

We have proved $\mathcal{N}_1^+ \equiv \mathcal{N}_2^+$. Consequently, $T \cup \mathcal{D}(\mathcal{M})$ is complete.

We next suppose that, for any $\mathcal{M} \models T$, $T \cup \mathcal{D}(\mathcal{M})$ is complete. Consider $\mathcal{M}, \mathcal{N} \models T$ such that $\mathcal{M} \sqsubseteq \mathcal{N}$. The last condition implies that an expansion of $\mathcal{N}$ is a model of $T \cup \mathcal{D}(\mathcal{M})$. Completeness implies that, relative to the language

8.1 Model-Completeness

$\mathcal{L}(C)$, the expanded structures $\mathcal{N}^+, \mathcal{M}^+$ are elementarily equivalent. It is easy to see that the last condition implies $\mathcal{M} \preceq \mathcal{N}$. □

Definition 8.1 tells us that, when $\mathcal{M}, \mathcal{N} \models T$ and $\mathcal{M}$ is a submodel of $\mathcal{N}$, the first-order information we can obtain from $\mathcal{N}$ over M is already available in $\mathcal{M}$. In particular, if $\mathcal{N} \models \exists x_1 \ldots \exists x_n \varphi(x_1, \ldots, x_n, \overline{a})$, where $\overline{a} \in M$ and φ is a quantifier-free formula, then $\mathcal{M} \models \exists x_1 \ldots \exists x_n \varphi(x_1, \ldots, x_n, \overline{a})$.

In other words, existential sentences modelled by $\mathcal{N}$ already have witnesses in $\mathcal{M}$. The existential information carried by $\mathcal{N}$ over M is thus reflected by $\mathcal{M}$. This reflection property is equivalent to model-completeness, as we shall now prove.

Recall that the formula $\varphi(\overline{x})$, whose free variables occur among those listed in the tuple $\overline{x}$, is T-equivalent to the formula $\psi(\overline{x})$ iff:

$$T \models \forall \overline{x}(\varphi(\overline{x}) \leftrightarrow \psi(\overline{x})).$$

If ψ is existential (respectively, universal), then we say that φ is T-equivalent to an existential (respectively, universal) formula.

Exercise 8.3 Show that, for each integer $n \geq 1$, T-equivalence is an equivalence relation on the set of formulae whose free variables belong to the set $\{x_1, \ldots, x_n\}$.

Theorem 8.4 *Let T be a $\mathcal{L}$-theory. The following statements are equivalent:*

(1) T is model-complete.
(2) If $\mathcal{M} \models T$ and $\exists \overline{y} \psi(\overline{x} \overline{y})$ is an existential $\mathcal{L}$-formula with $\overline{x}$ a n-tuple of variables, then, for any supermodel $\mathcal{N} \supseteq \mathcal{M}$ and any n-tuple $\overline{a} \in M$, $\mathcal{N} \models \exists \overline{y} \psi(\overline{a}, \overline{y})$ implies $\mathcal{M} \models \exists \overline{y} \psi(\overline{a}, \overline{y})$.
(3) Every $\mathcal{L}$-formula it T-equivalent to an existential formula.
(4) Every $\mathcal{L}$-formula is T-equivalent to a universal formula.

Proof By Definition 8.1, (1) implies (2).

We next prove that (2) implies (3) by an induction on the complexity of formulae. Atomic formulae are existential and the disjunction of two existential formulae is equivalent (thus, T-equivalent) to an existential formula (verify). Moreover, if $\varphi(y, \overline{x})$ is T-equivalent to an existential formula, certainly $\exists y \varphi(y, \overline{x})$ is, too. Thus, the only nontrivial step in the induction involves negation. The inductive hypothesis is that $\varphi(\overline{x})$ is T-equivalent to an existential formula $\psi(\overline{x})$.

In this case $\neg \varphi(\overline{x})$ is T-equivalent to a universal formula $\psi'(\overline{x})$. By (2), for any models $\mathcal{N} \supseteq \mathcal{M}$ of T and any n-tuple $\overline{a} \in M^n$:

$$\mathcal{N} \models \psi(\overline{a}) \text{ implies } \mathcal{M} \models \psi(\overline{a}).$$

It follows that:

$$\mathcal{M} \models \psi'(\overline{a}) \text{ implies } \mathcal{N} \models \psi'(\overline{a}).$$

We know from Chap. 4 that every existential formula is preserved by supermodels. Suppose we can prove the converse of this statement, i.e. that every formula preserved by supermodels is T-equivalent to an existential formula. This preservation result together with the last implication shows that $\psi'(\overline{x})$ must be T-equivalent to an existential formula. Thus (2) implies (3), given a preservation theorem to be proved in Sect. 8.2.

By Exercise 2.62, (3) implies (4).

If (4) holds and $\mathcal{M}, \mathcal{N}$ are models of T such that $\mathcal{M} \sqsubseteq \mathcal{N}$, the Tarski-Vaught test yields (1) (details left to the reader). It follows that $\mathcal{M} \preceq \mathcal{N}$ and T is model-complete. □

Exercise 8.5 Verify that a disjunction of existential formulae is equivalent to an existential formula.

In view of the last theorem, we introduce the important:

Definition 8.6 Let T be a $\mathcal{L}$-theory and $\mathcal{M}$ a model of T. Then $\mathcal{M}$ is said to be **existentially closed** in the class of structures $Mod T$ iff condition (2) from the statement of Theorem 8.4 holds relative to any extension of $\mathcal{M}$ in $Mod T$. More generally, if **K** is a class of $\mathcal{L}$-structures, then $\mathcal{M}$ is said to be existentially closed in **K** if, for any extension $\mathcal{N}$ of $\mathcal{M}$ in the class **K**, existential $\mathcal{L}$-formulae with parameters in M and modelled by $\mathcal{N}$ are already modelled by $\mathcal{M}$.

It follows from the last definition that T is model-complete iff all of its models are existentially closed in the class $Mod T$. A slightly simpler equivalent of existential closure can be obtained by restricting attention to existential formulae whose matrices are conjunctions of literals.

Definition 8.7 Let ψ be a $\mathcal{L}$-formula. If ψ is existential and its matrix consists of a conjunction of literals, then ψ is said to be a **primitive** $\mathcal{L}$-formula.

Exercise 8.8

(i) Show that condition (2) from the statement of Theorem 8.4 is equivalent to condition (2) restricted to primitive $\mathcal{L}$-formulae. This latter condition is known as **Robinson's test** for model-completeness.
(ii) Use Robinson's test to show that EQ^∞ from Exercise 7.17 is model-complete.

Exercise 8.9 Let T be the theory of fields. Show that, if $\mathcal{M}$ is existentially closed in $Mod T$, then $\mathcal{M}$ is an algebraically closed field.

Remark 8.10 Suppose $\mathcal{K} \models \mathsf{ACF}$ and $\mathcal{M}$ is a field that extends $\mathcal{K}$. Let $\psi := \exists x_1 \ldots \exists x_n \varphi(x_1, \ldots, x_n)$ be a primitive $\mathcal{L}_r$-formula with parameters in K. Then ψ is a conjunction of polynomial equations of the form $p(x_1, \ldots, x_n) = 0$, with $p(x_1, \ldots, x_n) \in K[x_1, \ldots, x_n]$, and of negated polynomial equations.

Using substructure completeness, we shall see in Sect. 8.7 that, if $M \models \psi$, $\mathcal{K} \models \psi$. It follows that an algebraically closed field is existentially closed in the class of fields. By Exercise 8.9, the field-theoretic notion 'algebraically closed' is a specialisation of the model-theoretic notion 'existentially closed'.

8.1 Model-Completeness

We have not yet given examples of model-complete theories: we remedy this omission in the next two subsections, which focus on dense and discrete linear orders respectively.

8.1.1 Applications to Dense Linear Orders

Robinson's test provides a useful method to check that model-completeness holds. A very simple application of the test shows that the theory of dense linear orders without endpoints is model-complete. Our proof relies on the convenient fact that a quantifier-free $\mathcal{L}_<$-formula is $\mathsf{DLO}_{-,-}$-equivalent to a quantifier-free $\mathcal{L}_<$-formula in which negations does not occur. This is because trichotomy enables us uniformly to replace every literal of the form $\neg(x < y)$ with the $\mathcal{L}_<$-formula $x = y \vee y < x$ and every literal of the form $\neg(x = y)$ with the $\mathcal{L}_<$-formula $x < y \vee y < x$.

Robinson's test may thus be applied to primitive formulae in whose matrices only positive literals figure.

Lemma 8.11 $\mathsf{DLO}_{-,-}$ is model-complete.

Proof Suppose $\mathcal{M}, \mathcal{N} \models \mathsf{DLO}_{-,-}$ and $\mathcal{M} \sqsubseteq \mathcal{N}$. If ψ is a primitive formula whose matrix consists of a conjunction of positive literals, then ψ specifies the relative position of $x_1, \ldots, x_n$ relative to one another and to a finite set of parameters $a_1, \ldots, a_m \in M$. The resulting system of inequalities (equalities may be eliminated by substitution) is consistent because it has a solution in $\mathcal{N}$. Because $\mathcal{M} \models \mathsf{DLO}_{-,-}$, the same system of inequalities can be satisfied in M. For instance, we may select $b_1 \in M$ to satisfy the inequalities involving x_1 and a parameter in the matrix of ψ. Next, we select $b_2 \in M$ to satisfy the inequalities involving $b_1, a_1, \ldots, a_m$. Proceeding in this manner, we arrive at a simultaneous solution of the original system of inequalities in n unknowns. As a result, $\mathcal{M} \models \psi$. □

By Exercise 7.65, the $\mathcal{L}_<$-theory $\mathsf{DLO}_{+,-}$ of dense linear orders with a least element but no greatest element, is not model-complete. If $\mathcal{N} \models \mathsf{DLO}_{+,-}$ and $a \in M$ is not the least element of N, the primitive formula $\exists y (x < a)$ holds in $\mathcal{N}$ but not in its submodel $\mathcal{M}$ with domain parametrically defined in N by the literal $a \leq x$.

We can nonetheless reinstate model-completeness by enriching our language.[1] Since we lose it because we cannot single out the least element of a model, we introduce a unary relation symbol P and extend $\mathsf{DLO}_{+,-}$ to a $\mathcal{L}_< \cup \{P\}$-theory by the addition of the single sentence:

$$\forall x (P(x) \leftrightarrow \forall y (\neg(y < x))).$$

We set $\mathcal{L}_< \cup \{P\} = \mathcal{L}^*$ and call the $\mathcal{L}^*$-extension of $\mathsf{DLO}_{+,-}$ just described $\mathsf{DLO}^*_{+,-}$.

---

[1] Here and below we follow the discussion of Robinson and Zakon from [63], pp.225–231.

Exercise 8.12 Show that any model of $\mathsf{DLO}_{+,-}$ can be expanded to a model of $\mathsf{DLO}^*_{+,-}$ and that $\mathsf{DLO}_{+,-}$ is complete if $\mathsf{DLO}^*_{+,-}$ is. Finally, show that, if $\mathcal{N} \models \mathsf{DLO}^*_{+,-}$, the least element of N is a_0 and $\mathcal{M}$ is a submodel of $\mathcal{N}$, then $a_0 \in M$.

An application of Theorem 8.4 and Definition 8.7 leads us to the following result.

Lemma 8.13 *The $\mathcal{L}^*$-theory $\mathsf{DLO}^*_{+,-}$ is model-complete.*

Proof Suppose $\mathcal{N}, \mathcal{M} \models \mathsf{DLO}^*_{+,-}$ with $\mathcal{M} \sqsubseteq \mathcal{N}$. Let a_0 be the least element of N. We only have to show that, if $\exists \overline{y} \phi(\overline{yx})$ is a primitive $\mathcal{L}^*$-formula, $\overline{a} \in M^n$ and $\mathcal{N} \models \exists \overline{y} \phi(\overline{y}, \overline{a})$, then $\mathcal{M} \models \phi(\overline{y}, \overline{a})$.

We now examine the possible literals occurring in $\phi(\overline{y}, \overline{a})$, using y_h, y_k as generic elements of the tuple $\overline{y}$ and a_i, a_j as generic elements of $\overline{a}$.

Because we can eliminate negations from equalities and inequalities, and then eliminate equalities by substitution, $\phi(\overline{b}, \overline{a})$ can be assumed to contain only inequalities of the form:

$$a_i < a_j, y_k < a_j, a_i < y_k, y_k < y_h.$$

Together with these inequalities, $\phi(\overline{y}, \overline{a})$ may also contain atomic formulae of the form:

$$P(a_0), \neg P(y_k), \neg P(a_i).$$

Literals without parameters, in which the unary relation symbol P occurs are easily satisfied in M. Negative literals can always be satisfied because only finitely many of them occur in $\phi(\overline{b}, \overline{a})$ and $\neg P(x)$ defines an infinite subset of M, namely $M - \{a_0\}$. The positive literal $P(x)$ is only satisfied by $a_0 \in M$ (see Exercise 8.12).

It is, furthermore, clear that $\mathcal{M}$ satisfies the inequalities in $\phi(\overline{y}, \overline{a})$. Any finite set of inequalities determines a finite linear ordering that can be realised in M using its density or the absence of a greatest element. Note that, if y_i is involved in any inequalities and $\neg P(y_i)$ is a literal in ϕ, there is no problem choosing a value for y_i in M. If instead the positive literal $P(y_i)$ occurs, we already know that it can only be satisfied by a_0 in M. We conclude that $\mathcal{M} \models \exists \overline{y} \varphi(\overline{y}, \overline{a})$, i.e. $\mathsf{DLO}^*_{+,-}$ is model-complete. □

We refer to $\mathsf{DLO}^*_{+,-}$ as a *model-completion* of $\mathsf{DLO}_{+,-}$

Exercise 8.14 Show that $\mathsf{DLO}^*_{+,-}$ has a model that can be embedded into every model of the same theory. We call this model an **algebraically prime** model of $\mathsf{DLO}^*_{+,-}$. Use the existence of an algebraically prime model to deduce that $\mathsf{DLO}^*_{+,-}$ is a complete theory. Infer that $\mathsf{DLO}_{+,-}$ is also complete.

Exercise 8.15 Find a model-completion of $\mathsf{DLO}_{+,+}$, show that it has an algebraically prime model and deduce that $\mathsf{DLO}_{+,+}$ is complete.

8.1 Model-Completeness

Exercise 8.16 The following exercise is based on [64], pp.148–149. Let $\mathcal{L}_R$ be a first-order language whose only extra-logical symbol is the binary relation symbol R.

(i) Call T the $\mathcal{L}_R$-theory that is finitely axiomatised by the following sentences:

(a) $\forall x \forall y (R(x, y) \to R(y, x))$;
(b) $\exists^{=1} x \forall y \neg R(x, y)$;
(c) $\exists^{=3} x \neg R(x, x)$;
(d) $\forall x \forall y (R(x, x) \to (R(x, y) \to x = y))$;
(e) $\exists x \exists y (\neg(x = y) \land R(x, y))$.

State in words what each axiom asserts.

(ii) Use the Tarski-Vaught criterion to prove that T is model-complete.

(iii) If $\mathcal{M} \models T$, let $a, b, c \in M$ be the three objects that satisfy axiom (c) and a the unique object satisfying axiom (b). Verify that $\{a\}, \{b\}$ determine isomorphic substructures of $\mathcal{M}$. Call $\mathcal{D}(\mathcal{A})$ the common diagram of these substructures and show that $(T \cup \mathcal{D}(\mathcal{A}))$ is not a complete theory. Deduce that model-completeness does not imply substructure-completeness.

By definition, substructure-completeness implies model-completeness. By the last exercise, the converse implication fails.

Exercise 8.17 In Sect. 8.7 we shall prove that a $\mathcal{L}$-theory T is substructure complete iff every $\mathcal{L}$-formula is T-equivalent to a quantifier-free formula. Use Exercise 8.16 to show that not every formula that is simultaneously preserved by submodels and supermodels of T is T-equivalent to a quantifier-free formula.

8.1.2 Applications to Discrete Linear Orders

Our approach to the study of model-completeness for dense linear orders extends to discrete linear orders. The distinctive feature of a discrete linear order is that each one of its elements has an immediate predecessor and an immediate successor, unless it is an endpoint.

In order to express the notions of predecessor and successor we add one binary predicate S to $\mathcal{L}_<$ and set $\mathcal{L}_< \cup \{S\} = \mathcal{L}_d$. The theory of discrete linear orders is obtained from the theory of linear orders by adding to it the following $\mathcal{L}_d$-sentences:

(a) $\forall x \forall y (S(x, y) \leftrightarrow \forall z ((z < x \lor z = x) \lor (y < z \lor z = y)))$;
(b) $\forall x (\exists y (x < y) \to \exists z (x < z \land S(x, z)))$;
(c) $\forall x (\exists y (y < x) \to \exists z (z < x \land S(z, x)))$.

Sentence (a) says that, if y immediately succeeds x, then there is a gap between x and y, i.e. no other elements lie between them. Sentence (b) guarantees the existence of the immediate successor for any element with a strict upper bound. Sentence (c) guarantees the existence of the immediate predecessor for any element with a strict lower bound.

Let dLO be the theory of discrete linear orders. We adopt the notation $\mathsf{dLO}_{-,-}$ to designate the theory of discrete linear orders without endpoints. Then $\mathsf{dLO}_{+,-}$ and $\mathsf{dLO}_{-,+}$ are, respectively, the theory of discrete linear orders with a least element and the theory of discrete linear orders with a greatest element. Each of the last three theories has only infinite models.

In the next exercises we prove that these theories are model-complete.

Exercise 8.18 Suppose $\mathcal{M}, \mathcal{N} \models \mathsf{dLO}_{-,-}$. Let $\exists \bar{y} \varphi(\bar{y}, \bar{a})$ be a primitive $\mathcal{L}_d$-formula with parameters $\bar{a}$ in M and such that $\mathcal{N} \models \exists \bar{y} \varphi(\bar{y}, \bar{a})$.

(i) Describe the kinds of literals that may occur in $\varphi(\bar{b}, \bar{a})$ and show that they can be $\mathsf{dLO}_{-,-}$-reduced to quantifier-free formulae in which only inequalities and atomic formulae of the form $S(b, a)$ occur (to this end, show in particular that $\neg S(b, a)$ can be reduced to an equality or to inequalities).

(ii) Consider the linear ordering of $\bar{b} = (b_1, \ldots, b_k), \bar{a} = (a_1, \ldots, a_n)$ in N and suppose that $a_1 < a_2 < \ldots < a_n$ (this can always be arranged by relabelling, if necessary). If every element of $\bar{b}$ precedes every element of $\bar{a}$ or every element of $\bar{b}$ follows every element of $\bar{a}$, then $\mathcal{M} \models \exists \bar{y}(\bar{y}, \bar{a})$. The nontrivial cases arise when some of the elements of $\bar{b}$ lie between a_i, a_{i+1}, for some $1 \le i \le n-1$. Show that, if only finitely elements of N lie between a_i and a_{i+1}, then all of these elements already lie in M. Deduce that $\mathcal{M} \models \exists \bar{y}(\bar{y}, \bar{a})$. Show that the same conclusion follows when infinitely many elements of N lie between a_i and a_{i+1}.

(iii) Deduce that $\mathsf{dLO}_{-,-}$ is model-complete. Find an algebraically prime model for this theory and prove that it is complete.

Exercise 8.19 How might the strategy described in the previous exercise be adapted to prove the completeness of $\mathsf{dLO}_{+,-}$?

Exercise 8.20 Prove that $\mathsf{dLO}_{+,+}$ is incomplete but that the theory of infinite, discrete linear orders with endpoints is complete. Describe a model of the latter theory. Finally, classify the infinite, discrete linear orders up to elementary equivalence (i.e. count the equivalence classes into which the class of infinite models of dLO is partitioned relative to the relation of elementary equivalence).

8.1.3 Model-Completeness and Completeness

Although we can sometimes use model-completeness to prove that certain theories are complete, the properties of completeness and model-completeness are independent of each other. We have already encountered complete theories that are not model-complete, e.g. $\mathsf{DLO}_{+,-}$. There also are incomplete theories that are model-complete. A simple, if artificial, example is provided by a variation of Exercise 8.16. Consider the sentences:

(a) $\forall x \forall y (R(x,y) \to R(y,x))$;
(b) $\exists! x \exists! y ((\neg(x=y) \wedge Rxy \wedge \neg Rxx \wedge \neg Ryy) \vee (\neg(x=y) \wedge Rxy \wedge Rxx \wedge Ryy))$;
(c) $\exists x (Rxx \wedge \forall y (Rxy \to x=y))$;
(d) $\forall x \forall y (R(x,x) \to (R(x,y) \to x=y))$;

Let Rel^∞ be the $\mathcal{L}$-theory obtained by adding the sentences just listed to IS. The next exercise shows that Rel^∞ is an incomplete but model-complete $\mathcal{L}_R$-theory.

Exercise 8.21

(i) Show that Rel^∞ cannot be complete.
(ii) Show that Rel^∞ is model-complete.

Exercise 8.22 Prove that if T is complete and has a finite model, then it must also be model-complete.

8.2 Preservation and Axiomatisability

The proof of Theorem 8.4 was made to rest on a preservation result to the effect that, for any $\mathcal{L}$-theory T, a $\mathcal{L}$-formula is T-equivalent to an existential $\mathcal{L}$-formula iff it is preserved by supermodels, i.e. extensions of models of T that are themselves models of T. We derive the dual result, which will immediately imply the one we have stated.

Instead of directly focussing on existential sentences, we work with universal sentences. The following definitions single out two notions that will play a central role in the proof of the preservation theorems we are after. First, we recall an idea already used in Chap. 4.

Definition 8.23 Let T be a $\mathcal{L}$-theory. The $\mathcal{L}$-formula $\varphi(x_1, \ldots, x_n)$ is T-**hereditary** iff $\mathcal{M}, \mathcal{N} \models T$, $a_1, \ldots, a_n \in M$, $\mathcal{M} \sqsubseteq \mathcal{N}$ and $\mathcal{N} \models \varphi(a_1, \ldots, a_n)$ jointly imply $\mathcal{M} \models \varphi(a_1, \ldots, a_n)$.

Since we focus on universal sentences, we are especially interested in the 'universal part' of a given theory.

Definition 8.24 Let T be a $\mathcal{L}$-theory. The **universal part** of T, denoted by $T_\forall$, is the set of consequences of T that are also universal $\mathcal{L}$-sentences.

A useful characterisation of $T_\forall$ follows directly from the Łos-Tarski theorem, which says that a theory is universal, i.e. axiomatised by a set of universal sentences, exactly when the substructures of its models are submodels.

Theorem 8.25 (Łos-Tarski) *Let T be a $\mathcal{L}$-theory. Then T has a universal axiomatisation iff $\mathcal{M} \models T$ and $\mathcal{A} \sqsubseteq \mathcal{M}$ jointly imply $\mathcal{A} \models T$.*

Proof Suppose that T has a universal axiomatisation S. Obviously, $T \models S$ and, since every sentence in S is universal, $S \subseteq T_\forall$. If $\mathcal{M} \models T$ and $\mathcal{A} \subset \mathcal{M}$, then

$\mathcal{A} \models T_\forall$ because universal sentences are preserved by substructures (Lemma 4.31): in particular $\mathcal{A} \models S$. Because every sentence of T follows from S, $\mathcal{A} \models T$.

To establish the converse implication, we consider $T_\forall$. It suffices to show that we can use this set of universal sentences as an axiomatisation of T. In other words, if $\mathcal{M} \models T_\forall$, we have to prove that $\mathcal{M} \models T$. In order to make use of the hypothesis, we need an extension $\mathcal{N} \sqsupseteq \mathcal{M}$ that models T.

If we had $\mathcal{N} \models T$ and $\mathcal{M} \sqsubseteq \mathcal{N}$, we could deduce $\mathcal{M} \models T$ from our hypothesis. Towards this goal, we show that $T \cup \mathcal{D}(\mathcal{M})$ has a model. We assume that the new constant symbols occurring in the sentences contained in $\mathcal{D}(\mathcal{M})$ do not figure in the signature of $\mathcal{L}$.

If the set $\Sigma = T \cup \mathcal{D}(\mathcal{M})$ has no model, then some finite subset $\Delta \subseteq \Sigma$ has no model by compactness. We derive a contradiction from the antecedent of this conditional, taken as a hypothesis.

Note that T has a model because it is a theory and $\mathcal{M} \models \mathcal{D}(\mathcal{M})$. The set Δ must therefore contain finitely many sentences $\tau_1, \ldots, \tau_n$ from T and finitely many sentences $\delta_1, \ldots, \delta_m$ from $\mathcal{D}(\mathcal{M})$, or it would have a model.

Let $\bar{c}$ be a list of the new constant symbols occurring in the conjunction $\delta_1 \wedge \ldots \wedge \delta_m$. We refer to this conjunction as $\delta(\bar{c})$ and to $\tau_1 \wedge \ldots \wedge \tau_n$ simply as τ. Then:

$$\tau \models \neg \delta(\bar{c})$$

Since none of the constant symbols in the tuple $\bar{c}$ occurs in $\mathcal{L}$ and T is a $\mathcal{L}$-theory, we can deduce (exercise):

$$\tau \models \forall \bar{x} \neg \delta(\bar{x}).$$

The sentence $\forall \bar{x} \neg \delta(\bar{x})$ is universal because $\neg \delta(\bar{x})$ is quantifier-free. Since $T \models \tau$, $T \models \forall \bar{x} \neg \delta(\bar{x})$. Using the fact that $\mathcal{M}$ models T as well as its own diagram, we arrive at the contradiction:

$$\mathcal{M} \models \exists \bar{x} \delta(\bar{x}) \wedge \forall \bar{x} \neg \delta(\bar{x}).$$

It follows that Σ has a model $\mathcal{N}$. The proof is complete. □

Exercise 8.26

(i) Show that, if $T \models \phi(c)$, where c is a constant symbol not occurring in T, then $T \models \forall x \phi(x)$.
(ii) Show that a conjunction of universal $\mathcal{L}$-formulae is equivalent to a single, universal $\mathcal{L}$-formula.

The strategy we have adopted to prove the Łos-Tarski theorem could have been used to prove:

Lemma 8.27 *Let T be a $\mathcal{L}$-theory. Then $\mathcal{M} \models T_\forall$ iff $\mathcal{M}$ is a substructure of a model $\mathcal{N}$ of T.*

8.2 Preservation and Axiomatisability

Proof Exercise. □

Exercise 8.28 Prove Lemma 8.27.

We are now in a position to prove the preservation theorem announced at the beginning of this section.

Theorem 8.29 *Let T be a $\mathcal{L}$-theory and $\varphi(\overline{x})$ be a $\mathcal{L}$-formula. Then $\varphi(\overline{x})$ is T-hereditary iff it is T-equivalent to a universal $\mathcal{L}$-formula $\psi(\overline{x})$.*

Proof If $\varphi(\overline{x})$ is T-equivalent to a universal $\mathcal{L}$-formula $\psi(\overline{x})$, the fact that the latter is preserved by submodels (recall Lemma 4.31, easily relativised to models of T) implies that $\varphi(\overline{x})$ is T-hereditary.

Conversely, suppose that $\varphi(\overline{x})$ is T-hereditary. We apply compactness to complete the proof. To this end, we expand $\mathcal{L}$ by adding to it a finite tuple of new constant symbols $\overline{c}$. Let $\mathcal{L}_c$ be the expanded language. We consider the $\mathcal{L}_c$-theories $T' = T \cup \{\varphi(\overline{c})\}$ and $T'' = T \cup \{\neg\varphi(\overline{c})\}$. Clearly, T', T'' cannot both fail to have models, or T would not be a theory.

It follows that the set $T'_\forall \cup T''$ has no models. If $\mathcal{M} \models T'_\forall \cup T''$ then, in particular, $\mathcal{M} \models T'_\forall$ is a substructure of some $\mathcal{N} \models T'$, by the Tarski-Łoś' theorem. Since $\mathcal{M} \models T''$, $\mathcal{M} \models T$, the T-hereditariness of $\varphi(x)$ implies $\mathcal{M} \models T'$. Thus, $\mathcal{M}$ is a submodel of a model $\mathcal{N}$ of T'. Since $\mathcal{M} \models T''$, we reach the contradiction $\mathcal{M} \models \varphi(\overline{c}) \wedge \neg\varphi(\overline{c})$.

Compactness implies that, if $T'_\forall \cup T''$ has no models, a finite subset Δ of $T'_\forall \cup T''$ has no models. In general, Δ will contain finitely many universal sentences from $T'_\forall$, whose conjunction we call $\psi'(\overline{c})$. The last sentence is equivalent to a universal $\mathcal{L}_c$-sentence $\psi(\overline{c})$, by Exercise 8.26-(ii).

Because a finite part of T'' implies $\neg\psi(\overline{c})$, we deduce $T \cup \{\neg\varphi(\overline{c})\} \models \neg\psi(\overline{c})$. On the other hand, $T'_\forall \models \psi(\overline{c})$ and $T' \models T'_\forall$ imply $T \cup \{\varphi(\overline{c})\} \models \psi(\overline{c})$. We conclude

$$T \models \varphi(\overline{c}) \leftrightarrow \psi(\overline{c}).$$

But T is a $\mathcal{L}$-theory in which the constant symbols in the tuple $\overline{c}$ do not occur. Exercise 8.26-(i) leads to the consequence relation:

$$T \models \forall x (\varphi(\overline{x}) \leftrightarrow \psi(\overline{x})),$$

in which $\psi(\overline{x})$ is a universal $\mathcal{L}$-formula. □

In the special case $T = \emptyset$, Theorem 8.29 reduces to the statement that every formula preserved by substructures is equivalent to a universal formula. An immediate consequence of what we have proved is the following:

Theorem 8.30 *Let T be a $\mathcal{L}$-theory. The $\mathcal{L}$-formula $\varphi(x_1, \ldots, x_n)$ is preserved by supermodels of T iff it is T-equivalent to an existential $\mathcal{L}$-formula $\psi(x_1, \ldots, x_n)$.*

Proof Since existential formulae are preserved by supermodels (recall Lemma 4.31), only one direction of the biconditional needs to be proved. If $\varphi(x_1, \ldots, x_n)$ is preserved by supermodels of T and $\mathcal{M} \subseteq \mathcal{N}$ are models of T, then, for any $a_1, \ldots, a_n \in M$:

$$\mathcal{M} \models \varphi(a_1, \ldots, a_n) \implies \mathcal{N} \models \varphi(a_1, \ldots, a_n),$$

which is equivalent to the condition:

$$\mathcal{N} \models \neg\varphi(a_1, \ldots, a_n) \implies \mathcal{M} \models \neg\varphi(a_1, \ldots, a_n).$$

We deduce that $\neg\varphi(\bar{x})$ is T-hereditary and, by Theorem 8.29, T-equivalent to a universal formula $\psi(\bar{x})$. Then $\neg\psi(\bar{x})$ is T-equivalent to an existential formula and to $\varphi(\bar{x})$. □

Exercise 8.31 In this exercise, we shall prove the dual of the Łos-Tarski theorem, which characterises existential axiomatisability. For the purpose of stating this result we define the **existential part** of a theory T, denoted by $T_\exists$, as the set of existential consequences of T. We aim to prove:

Theorem 8.32 *Let T be a $\mathcal{L}$-theory. Then T has an existential axiomatisation iff for any $\mathcal{M} \models T$ and any embedding $f : M \longrightarrow N$, with N the domain of a $\mathcal{L}$-structure $\mathcal{N}$, we also have $\mathcal{N} \models T$.*

(i) Verify that, if T has an existential axiomatisation, then the condition given in the statement of the above theorem follows.
(ii) Now assume that, if $\mathcal{M} \models T$ is embeddable in $\mathcal{N}$, then $\mathcal{N} \models T$. We are going to show that, under this hypothesis, $T_\exists$ provides an existential axiomatisation of T. Explain why we obtain this result if, assuming $\mathcal{N} \models T_\exists$, we can embed a model $\mathcal{M}$ of T into an elementary extension of $\mathcal{N}$.
(iii) Consider the theory $T \cup T_\neg$, where $T_\neg$ is the set of $\mathcal{L}$-sentences of the form $\neg\varphi$ where φ is existential and $\mathcal{N} \not\models \varphi$. Use compactness to show that $T \cup T_\neg$ has a model (*Hint*: set up a proof by contradiction and show that, given your hypotheses, a finite part of T implies $\neg\delta_1 \vee \ldots \vee \neg\delta_n$, where $\delta_1, \ldots, \delta_n \in T_\neg$. Use the fact that $\neg\delta_1 \vee \ldots \vee \neg\delta_n$ is equivalent to a formula in $T_\exists$ and the hypothesis $\mathcal{N} \models T_\exists$ to obtain a contradiction).
(iv) Let $\mathcal{M}$ be a model of $T \cup T_\neg$. Use compactness to show that $\mathcal{ED}(\mathcal{N}) \cup \mathcal{D}(\mathcal{M})$ has a model (*Hint*: if not, note that $\mathcal{ED}(\mathcal{N})$ implies the negation of an existential sentence that must also hold in $\mathcal{M}$ because $\mathcal{M} \models T_\neg$. It then follows that $\mathcal{M}$ models at once an existential sentence and its negation.).
(v) Conclude the proof of Theorem 8.32.

8.3 Chains and Lindström's Test

One way to determine whether a theory is model-complete is, as we have seen, to apply Robinson's test. This test works on a case-by-case basis, because it looks at the form of literals determined by the language of a theory T.

An alternative way of testing for model-completeness consists in providing sufficient conditions that imply it. These conditions may apply uniformly to linguistically distinct scenarios, thus circumventing the need for local analyses. In this section we identify sufficient conditions for model-completeness and use them to prove at once that DAG^+, VS_K^∞, ACF_0 and ACF_p, for each prime p, are model-complete theories.

We take advantage of the fact that these theories satisfy the hypotheses of Vaught's test. All of them have only infinite models and they are κ-categorical, with κ larger than the cardinality of their language. If, for any of these theories, we can find an existentially closed model that has size κ, then every model of the same theory and of the same size κ is existentially closed (verify). As the next lemma shows, this is enough to ensure model-completeness.

Lemma 8.33 *Let T be a $\mathcal{L}$-theory and $\kappa \geq |\mathcal{L}|$ a cardinal. If T has only infinite models, then T is model-complete iff for any models $\mathcal{M} \sqsubseteq \mathcal{N}$ of size κ, and any existential formula $\exists \overline{x} \varphi(\overline{x}, \overline{a})$ with parameters in M, we have:*

$$\mathcal{N} \models \exists \overline{x} \varphi(\overline{x}, \overline{a}) \implies \mathcal{M} \models \exists \overline{x} \varphi(\overline{x}, \overline{a}).$$

Proof If T is model-complete then $\mathcal{M} \preceq \mathcal{N}$ and existential formulae with parameters in M transfer from $\mathcal{N}$ to $\mathcal{M}$.

Conversely, suppose that, for any $\mathcal{M} \sqsubseteq \mathcal{N}$ with $\mathcal{M}, \mathcal{N} \models T$ and $|M| = \kappa = |N|$, existential formulae with parameters in M transfer from $\mathcal{N}$ to $\mathcal{M}$. We are going to show that every existential $\mathcal{L}$-formula is T-equivalent to a universal one: model-completeness follows by Theorem 8.4.

We consider an arbitrary existential $\mathcal{L}$-formula $\varphi(x_1, \ldots, x_n)$ and expand the language to $\mathcal{L}_c$ by adding a n-tuple of new constant symbols $c_1, \ldots, c_n$ or, in short, $\overline{c}$. We let Γ be the set of all universal $\mathcal{L}$-sentences that are T-consequences of $\varphi(\overline{c})$. In other words:

$$\gamma(\overline{c}) \in \Gamma \text{ iff } T \models \varphi(\overline{c}) \to \gamma(\overline{c}).$$

Note that, not all (or, indeed, any) of the symbols in $\overline{c}$ have to occur in $\gamma(\overline{c})$. It is easy to see that $\Gamma \neq \emptyset$. If $\varphi(\overline{x})$ cannot be satisfied by any assignment in any model of T, then it is T-equivalent to the universal $\mathcal{L}$-formula $\neg(x_1 = x_1)$. Thus, we may assume that there is a $\mathcal{L}_c$-structure $\mathcal{M}_0$ that models $T \cup \{\varphi(\overline{c})\}$. The Löwenheim-Skolem theorems allow us to choose an elementary extension or an elementary substructure $\mathcal{M}$ of $\mathcal{M}_0$, as the case may be, with $|M| = \kappa$.

By definition of Γ, $\mathcal{M} \models T \cup \Gamma$. In order to use our hypothesis, we need a model of T that is also an extension of $\mathcal{M}$. We look for models of $T \cup \{\varphi(\overline{c})\} \cup$

$\mathcal{D}(\mathcal{M})$. Since $T \cup \{\bar{c}\}$ is consistent and $\mathcal{D}(\mathcal{M})$ is closed under conjunction, the form of compactness given in Corollary 5.63 implies that, if $T \cup \{\varphi(\bar{c})\} \cup \mathcal{D}(\mathcal{M})$ is inconsistent:

$$T \cup \{\varphi(\bar{c})\} \models \neg \theta(\bar{d}, \bar{c}),$$

where θ is a conjunction of literals from the diagram of $\mathcal{M}$ and the tuple $\bar{d}$ consists of constant symbols not occurring in $\mathcal{L}_c$. Since these symbols do not occur in T or φ, we deduce:

$$T \cup \{\varphi(\bar{c})\} \models \forall \bar{x} \neg \theta(\bar{x}, \bar{c}).$$

Then $\forall \bar{x} \neg \theta(\bar{x}, \bar{c}) \in \Gamma$ and $\mathcal{M}$ is a model of $\forall \bar{x} \neg \theta(\bar{x}, \bar{c})$ as well as $\exists \bar{x} \theta(\bar{x}, \bar{c})$. This contradiction yields a $\mathcal{L}_c$-extension $\mathcal{N} \sqsupseteq \mathcal{M}$ such that $\mathcal{N} \models T$. Again, by the Löwenheim-Skolem theorems, we may assume $|N| = \kappa$.

Note that $\mathcal{M}, \mathcal{N}$ assign identically interpret each symbol in the tuple $\bar{c}$. Let the common interpretation determine the tuple $\bar{a}$. We refer to $\mathcal{M}^-, \mathcal{N}^-$ as the $\mathcal{L}$-reducts of $\mathcal{M}, \mathcal{N}$ respectively. Since these reducts are models of T, we can use our assumption concerning $\mathcal{M}_0$ to deduce $\mathcal{M} \models \varphi(\bar{c})$.

Irrespective of whether $\mathcal{M} \preceq \mathcal{M}_0$ or $\mathcal{M} \preceq \mathcal{M}_0$, we have $\mathcal{M}_0, \mathcal{M} \models \varphi(\bar{c})$). Since $\mathcal{M}_0$ is an arbitrary model of $T \cup \Gamma$, we conclude $T \cup \Gamma \models \varphi(\bar{c})$. By compactness, a finite conjunction of sentences of Γ implies $\varphi(\bar{c})$ *modulo* T (this is to say that the implication is a consequence of T). Such a conjunction of universal $\mathcal{L}_c$-sentences is itself equivalent to a universal $\mathcal{L}_c$-sentence $\gamma(\bar{c})$. We thus obtain:

$$T \models \gamma(\bar{c}) \to \varphi(\bar{c}).$$

It is now easy to complete the proof. □

Exercise 8.34 Show that we may allow $\mathcal{M}, \mathcal{N}$ in the statement of Lemma 8.33 to be either finite and of the same size or infinite and of the same, sufficiently large, size.

The problem we now need to solve in order to arrive at sufficient conditions for model-completeness is that of building existentially closed models of a theory T. To this end, we introduce a model-building technique that progresses through successive stages. At each stage we obtain an extension of the preceding stage, which is also a substructure of the successive stage: the resulting chain of structures determines a single structure, obtained by taking the chain's union, in a sense we proceed to clarify.

Definition 8.35 Let α be an ordinal. A **chain** of $\mathcal{L}$-structures of length α is a set $\{\mathcal{M}_\beta : \beta \in \alpha\}$ of $\mathcal{L}$-structures such that $\gamma \in \beta \in \alpha$ implies $\mathcal{M}_\gamma \sqsubseteq \mathcal{M}_\beta$. The **union of the chain** $\{\mathcal{M}_\beta : \beta \in \alpha\}$ is the $\mathcal{L}$-structure $\mathcal{M}$ defined by the following three conditions:

8.3 Chains and Lindström's Test

(a) The domain of $\mathcal{M}$ is the set $\bigcup_{\beta \in \alpha} M_\beta$;
(b) for any constant symbol c in $\mathcal{L}$, $c^\mathcal{M} = c^{\mathcal{M}_0}$;
(c) for any m-ary relation symbol R in $\mathcal{L}$, $R^\mathcal{M} = \bigcup_{\beta \in \alpha} R^{\mathcal{M}_\beta}$;
(d) for any k-ary function symbol f in $\mathcal{L}$ and k-tuple $\bar{a} \in M^k$, $f^\mathcal{M}(\bar{a}) = f^{\mathcal{M}_\beta}(\bar{a})$ where $\bar{a} \in M_\beta^k$.

If, in the definition of a chain of structures, we require that $\gamma \in \beta \in \alpha$ imply $\mathcal{M}_\gamma \preceq \mathcal{M}_\beta$, we obtain an **elementary chain**. The union of an elementary chain is defined exactly as the union of a chain.

Exercise 8.36 Show that condition (d) from Definition 8.35 determines well-defined functions (i.e. their values are unique). Prove that the union of a chain is an extension of each structure in the chain.

Exercise 8.37 Let $\{\mathcal{M}_\beta : \beta \in \alpha\}$ an elementary chain of length α and $\mathcal{M}$ its union. Prove by induction on the complexity of formulae that, for any $\beta \in \alpha$, $\mathcal{M}_\beta \preceq \mathcal{M}$.

We now characterise the set of first-order sentences that are modelled by the union of a chain if they are modelled by every structure in the chain.

Definition 8.38 A $\mathcal{L}$-sentence φ is said to be **preserved by unions of chains** iff, for any chain of $\mathcal{L}$-structures $\{\mathcal{M}_\beta : \beta \in \alpha\}$, $\mathcal{M}_\beta \models \varphi$ for each $\beta \in \alpha$ implies $\mathcal{M} \models \varphi$.

Existential sentences are certainly preserved by unions of chains because they are preserved by extensions.

Exercise 8.39
  (i) Show that universal sentences are preserved by unions of chains.
  (ii) Show that, if $\varphi(x, y)$ is a quantifier-free $\mathcal{L}$-formula, then the $\mathcal{L}$-sentence $\forall x \exists y \varphi(x, y)$ is preserved by unions of chains.

Definition 8.40 Let $\mathcal{L}$ be a first-order language. A **universal-existential** $\mathcal{L}$-sentence is a sentence of the form $\forall x_1 \ldots \forall x_m \exists y_1 \ldots \exists y_n \varphi(x_1, \ldots, x_m, y_1, \ldots, y_n)$, where $\varphi(x_1, \ldots, x_m, y_1, \ldots, y_n)$ is quantifier-free. Note that any existential $\mathcal{L}$-sentence is universal-existential, since one may add universal quantifiers vacuously. The set of universal-existential consequences of a theory T is denoted by $T_{\forall\exists}$.

Inspection shows that the axioms of DAG^+, VS_K^∞, ACF_0, ACF_p are universal or universal-existential sentences in a countable language. A trivial generalisation of Exercise 8.37 (obtained by replacing x, y with $\bar{x}$, $\bar{y}$ throughout) shows that all of these sentences are preserved by chains.

This condition suffices to guarantee that we can extend sufficiently large models to existentially closed models of the same size.

Theorem 8.41 *Let T be a $\mathcal{L}$-theory whose sentences are preserved by unions of chains. For any cardinal $\kappa \geq |\mathcal{L}|$, every model of T of size κ can be extended to an existentially closed model of the same size.*

Proof Let $\mathcal{M}$ be a model of T of size κ. We add to $\mathcal{L}$ enough constant symbols to name every element of M and call $\mathcal{L}_M$ the language so expanded. Since there are κ

existential $\mathcal{L}_M$-sentences (corresponding to existential $\mathcal{L}$-sentences with parameters over M), we can index them over κ and enumerate them as the set $\{\varphi_\alpha : \alpha \in \kappa\}$.

We use transfinite recursion and the enumeration just described to build a chain of length κ (transfinite recursion works on the whole class **On**, so we can truncate it at any ordinal stage). The relevant recursion clauses are:

(a) $\mathcal{M}_0 = \mathcal{M}$;
(b) if $\beta = \alpha + 1$ is the successor of α, we consider φ_α. If any extensions of $\mathcal{M}_\alpha$ model $T \cup \{\varphi_\alpha\}$, we choose one of them to be $\mathcal{M}_\beta$. Otherwise $\mathcal{M}_\beta = \mathcal{M}_\alpha$;
(c) if λ is a limit ordinal, we let $\mathcal{M}_\lambda$ be the union of the chain $\{\mathcal{M}_\delta : \delta \in \lambda\}$.

Note that we obtain a model of T at limit steps in the above recursion because T is preserved by unions of chains. Let $\mathcal{N}_0$ be the union of the chain $\{\mathcal{M}_\alpha : \alpha \in \kappa\}$. Then $\mathcal{N}_0 \models T$. We can make such arrangements as guarantee $|\mathcal{N}_0| = \kappa$ (exercise).

Moreover, if $\exists \overline{x} \varphi(\overline{x}, \overline{a})$ is φ_δ in our enumeration of existential $\mathcal{L}_M$-sentences and, and for some $\mathcal{N} \models T, \mathcal{N} \sqsupseteq \mathcal{N}_0$ and $\mathcal{N} \models \exists \overline{x} \varphi(\overline{x}, \overline{a})$, then $\mathcal{N}_0 \models \exists \overline{x} \varphi(\overline{x}, \overline{a})$ by construction (the details are left to the reader).

Nonetheless, $\mathcal{N}_0$ is not in general existentially closed because it may have extensions modelling $\mathcal{L}_M$-formulae with parameters not in M. We therefore enrich the language by adding no more than κ new constant symbols to make sure that every element of $\mathcal{N}_0$ is named, enumerate the existential sentences in the expanded language and proceed as above. We obtain $\mathcal{N}_1 \sqsupseteq \mathcal{N}_0$ such that $\mathcal{N}_1 \models T$ and $|\mathcal{N}_1| = \kappa$.

A countably infinite iteration of the procedure we have described produces the countable chain:

$$\mathcal{N}_0 \sqsubseteq \mathcal{N}_1 \sqsubseteq \mathcal{N}_2 \sqsubseteq \ldots \sqsubseteq \mathcal{N}_m \sqsubseteq \mathcal{N}_{m+1} \sqsubseteq \ldots$$

Let $\mathcal{N}$ be the union of the last chain. Then $\mathcal{N} \models T$ and $|\mathcal{N}| = \kappa$. It is readily verified that $\mathcal{N}$ is also existentially closed in $Mod\, T$ (see the next exercise). □

Exercise 8.42

(i) Show that it is possible to force the equality $|\mathcal{N}_i| = \kappa$ for each $i \in \mathbb{N}$ and deduce that $|\mathcal{N}| = \kappa$.
(ii) Show that $\mathcal{N}_0$ models every existential $\mathcal{L}_M$-sentence modelled by any one of its extensions.
(iii) Show that $\mathcal{N}$ is existentially closed.

We can now prove:

Theorem 8.43 (Lindström's Test) *Let T be a $\mathcal{L}$-theory preserved by unions of chains. If T has only infinite models and it is κ-categorical for $\kappa \geq |\mathcal{L}|$, then T is model-complete.*

Proof Let $\mathcal{M} \models T$. Since $\mathcal{M}$ is infinite, it has, depending on its size, an elementary substructure or an elementary extension $\mathcal{M}_1$ of size κ (by the Löwenheim-Skolem

8.3 Chains and Lindström's Test

theorems). By Theorem 8.41, $\mathcal{M}_1$ can be extended to an existentially closed model of T, say $\mathcal{N}_1$, itself of size κ. Every $\mathcal{N} \models T$ of size κ is isomorphic to $\mathcal{N}_1$ and thus existentially closed. Lemma 8.27 applies and T is model-complete. □

Corollary 8.44 *The theories* DAG^+, VS_K^∞, ACF_0 *and* ACF_p, *for each prime* p, *are model-complete.*

Exercise 8.45 Let T be a $\mathcal{L}$-theory with $\mathcal{L}$ a countable language. Suppose that T is model-complete and that any two countable models of T can be embedded into a joint extension that is itself a model of T. Show that T is complete.

Although we could not prove that ACF is model-complete by Lindström's test, because this theory is not categorical in any infinite cardinality, we can obtain its model-completeness easily, as the next exercise shows.

Exercise 8.46 Show that the completions of ACF are ACF_0 and, for each prime p, ACF_p. Use this fact to show that ACF is model-complete.

8.3.1 Universal-Existential Axiomatisability

Using a suitable chain construction we can obtain a theorem about universal-existential axiomatisability that corresponds to the Łos-Tarski theorem for universal axiomatisability. Roughly speaking, we can show that universal-existential axiomatisability is equivalent to the fact that unions of chains are models.

To prove the last equivalence, we require a preliminary characterisation of existential closure *in a structure*. If $\mathcal{M}, \mathcal{N}$ are $\mathcal{L}$-structures and $\mathcal{M} \sqsubseteq \mathcal{N}$, we say that $\mathcal{M}$ is **existentially closed in** $\mathcal{N}$ iff every existential sentence with parameters in $\mathcal{M}$ and modelled by $\mathcal{N}$ is already modelled by $\mathcal{M}$.

Lemma 8.47 *Let* $\mathcal{M}, \mathcal{N}$ *be* $\mathcal{L}$-*structures. If* $\mathcal{M} \sqsubseteq \mathcal{N}$, *then* $\mathcal{M}$ *is existentially closed in* $\mathcal{N}$ *iff there are a* $\mathcal{L}$-*structure* $\mathcal{B}$, *an elementary embedding* $g : \mathcal{M} \longrightarrow \mathcal{B}$ *and an embedding* $f : \mathcal{N} \longrightarrow \mathcal{B}$ *such thtat, for any* $a \in \mathcal{M}$, $f(a) = g(a)$.

Proof We assume that the expanded languages of $\mathcal{ED}(\mathcal{M}), \mathcal{D}(\mathcal{N})$ respectively share the constant symbols naming the elements of $\mathcal{M}$. To prove one half of the lemma it suffices to show that, if $\mathcal{M}$ is existentially closed in $\mathcal{N}$, the set $\mathcal{ED}(\mathcal{M}) \cup \mathcal{D}(\mathcal{N})$ is consistent.

Toward a contradiction, we may assume that $\mathcal{ED}(\mathcal{M}) \cup \mathcal{D}(\mathcal{N})$ has no models. Since $\mathcal{ED}(\mathcal{M})$ is consistent and $\mathcal{D}(\mathcal{N})$ is closed under conjunction, we can apply compactness in the form of Corollary 5.63 to reach the desired contradiction. We leave it for the reader to complete the argument.

If, on the other hand, the embeddings mentioned in the statement of the theorem exist, it is clear that $\mathcal{M}$ is existentially closed in $\mathcal{N}$. □

Exercise 8.48 Fill the gaps in the proof of Lemma 8.47.

Theorem 8.49 *Let T be a $\mathcal{L}$-theory. Then T has a universal-existential axiomatisation iff T is preserved by unions of chains.*

Proof If T has a universal-existential axiomatisation, Exercise 8.37-(ii) guarantees its preservation by unions of chains.

To prove the converse, it suffices to show that, if $\mathcal{M} \models T_{\forall\exists}$, then $\mathcal{M} \models T$. The proof strategy consists in the construction of an elementary chain starting from $\mathcal{M} = \mathcal{M}_0$, whose union we can ensure to be a model of T. More precisely, we elementarily embed $\mathcal{M}_0$ into some $\mathcal{M}_1$, but do so via Lemma 8.47, so that a model o T, say $\mathcal{N}_0$, also embeds into $\mathcal{M}_1$.

By means of this trick, we produce a chain that alternates between $\mathcal{M}_i$ and $\mathcal{N}_i$, for each $i \in \mathbb{N}$. When we regard the union of this chain as the union of the $\mathcal{N}_i$, we obtain a model of T. When we regard it as the union of the $\mathcal{M}_i$, we obtain an elementary extension of $\mathcal{M}_0$, so $\mathcal{M}_0$ models T.

The details of our chain construction are as follows. Since $\mathcal{M}_0 \models T_{\forall\exists}$, we expand the language to $\mathcal{L}_{M_0}$, so that we can name each element of M_0, and consider $T^{\forall}$, the set of universal sentences that hold in $\mathcal{M}_0$. If $\mathcal{N} \models T \cup T^{\forall}$, then $\mathcal{N} \models T$ and $\mathcal{M}_0$ can be embedded into $\mathcal{N}_0$ because $\mathcal{D}(\mathcal{M}_0) \subseteq T^{\forall}$. It is easy to see that $\mathcal{M}_0$ is existentially closed in $\mathcal{N}$ (argue by contradiction).

Because $T^{\forall}$ is closed under conjunction and T has a model, we can apply compactness in the form of Corollary 5.63 to obtain a contradiction from the inconsistency of $T \cup T^{\forall}$ (the contradiction in question requires the assumption $\mathcal{M}_0 \models T_{\forall\exists}$: see Exercise 8.50). We conclude that there is a $\mathcal{L}_M$-structure $\mathcal{N} \models T \cup T'$.

Let $\mathcal{N}_0$ be the $\mathcal{L}$-reduct of the last structure: it continues to be existentially closed in $\mathcal{M}_0$ and a model of T. Lemma 8.47 applies and it guarantees the existence of an extension $\mathcal{M}_1$ of $\mathcal{N}_0$ that also elementarily extends $\mathcal{M}_0$.

Clearly, $\mathcal{M}_1 \models T_{\forall\exists}$: we can therefore repeat the construction just described, starting from $\mathcal{M}_1$. By a transfinite iteration, we arrive at the chain:

$$\mathcal{M}_0 \sqsubseteq \mathcal{N}_0 \sqsubseteq \mathcal{M}_1 \sqsubseteq \mathcal{N}_1 \sqsubseteq \ldots \sqsubseteq \mathcal{M}_k \sqsubseteq \mathcal{N}_k \sqsubseteq \ldots$$

Since $\bigcup_{i \in \mathbb{N}} M_i = \bigcup_{i \in \mathbb{N}} N_i$ and these two sets are domains of the same structure, the proof is complete. □

Exercise 8.50 Use compactness to show that, if $\mathcal{M} \models T_{\forall\exists}$, then $T \cup T^{\forall}$, as defined in the proof of Theorem 8.49, has a model.

Exercise 8.51 Let T be a $\mathcal{L}$-theory. The chain construction used in the proof of Theorem 8.49 can be adapted to deliver a preservation theorem. We shall prove that $\mathcal{L}$-formula is T-equivalent to a universal-existential formula iff it is preserved by unions of chains whose links are models of T. One direction of the equivalence is Exercise 8.37-(ii). The other direction can be established as follows:

(i) Let $\varphi(\overline{x})$ be a $\mathcal{L}$-formula that is preserved by unions of chains whose links are models of T. We add a tuple $\overline{c}$ of new constant symbols to $\mathcal{L}$ and focus on the sentence $\varphi(\overline{c})$. Consider the set of sentences $T_{\varphi,\forall\exists}$, defined by the condition:

$\psi(\bar{c}) \in T_{\varphi,\forall\exists}$ iff $\psi(\bar{x})$ is universal-existential and $T \models \varphi(\bar{c}) \to \psi(\bar{c})$. Show that if $T \cup T_{\varphi,\forall\exists}$ happens to be inconsistent, the proof becomes trivial. We may thus assume that $T \cup T_{\varphi,\forall\exists}$ has a model.

(ii) Show that the conjunction of two formulae in $T_{\varphi,\forall\exists}$ is equivalent to a formula in $T_{\varphi,\forall\exists}$.

(iii) Derive a contradiction from the hypothesis that $T \cup T_{\varphi,\forall\exists} \cup \neg\varphi(\bar{c})$ is consistent. To do so, imitate the chain construction from the proof of Theorem 8.49, starting from a model $\mathcal{M}_0$ of $T \cup \{\neg\varphi(\bar{c})\} \cup T_{\varphi,\forall\exists}$. Use Compactness to show that $T \cup T^\forall \cup \{\varphi(\bar{c})\}$ (where $T^\forall$ is defined as in the proof of Theorem 8.49) has a model $\mathcal{N}_0$. Iterate the construction until you get a chain whose union models both $\varphi(\bar{c})$ and $\neg\varphi(\bar{c})$.

(iv) Deduce from (ii) that $\varphi(c)$ is T-equivalent to a universal-existential sentence and complete the proof of the theorem.

8.4 Quantifier Elimination

When a $\mathcal{L}$-theory is model-complete, it reduces $\mathcal{L}$-formulae to T-equivalents that are at most prefixed by a string of existential quantifiers. In other words, model-completeness implies that we can eliminate from a $\mathcal{L}$-formula all but at most a string of existential quantifiers. In this section we consider a stronger property of theories, which implies the elimination of all quantifiers.

Definition 8.52 Let T be a $\mathcal{L}$-theory. We say that T has **quantifier elimination** iff for any $\mathcal{L}$-formula $\varphi(\bar{x})$ there is a quantifier-free $\mathcal{L}$-formula $\theta(\bar{x})$ such that:

$$T \models \forall \bar{x}(\varphi(\bar{x}) \leftrightarrow \theta(\bar{x})).$$

Remark 8.53 In Definition 8.52, the free variables occurring in $\theta(\bar{x})$ are among the free variables occurring in $\varphi(\bar{x})$. Thus, if a $\mathcal{L}$-theory T has quantifier elimination, we expect each $\mathcal{L}$-sentence to be equivalent to a *quantifier-free* sentence. If a language has no constant symbols, the last condition cannot be fulfilled *a priori*, since there are no quantifier-free sentences. In this case we resort to a small trick and identify a sentence φ with the equivalent formula $\varphi \wedge (x = x)$.

If a theory T is complete, however, we may simply decide to supplement the language with a pair constant symbols $\top, \bot$ (top and bottom or *verum* and *falsum*), which are to be viewed as 0-ary propositional connectives.

By definition, $\top$ is true in any structure and $\bot$ is false in any structure. In any complete theory T, $T \models \varphi$ implies $T \models \varphi \leftrightarrow \top$. If $T \not\models \varphi$, completeness implies $T \models \neg\varphi$: consequently $T \models \varphi \leftrightarrow \bot$.

If the language of T contains constant symbols, we may obviously dispense with $\top$ and $\bot$. For instance, any consequence of ACF_0 is ACF_0-equivalent to the sentence $0 = 0$.

Remark 8.54 We shall see in Sect. 8.7 that quantifier elimination is equivalent to substructure-completeness (see Exercise 8.79).

Since quantifier-free formulae are existential, our definition implies that a theory with quantifier elimination is model-complete. We have already seen that the opposite implication does not hold. Relative to the language $\mathcal{L}_< \cup \{P\}$, discussed in Sect. 8.1.1, the theory of dense linear orders with a least element is model-complete but it cannot have quantifier elimination because the formula $\exists y(y < x)$ does not have a quantifier-free equivalent.

We can isolate a sufficient condition for quantifier elimination if we attempt to prove it by an induction on the complexity of formulae and observe how far we can go without relying on any special assumptions. This line of argument leads to the statement and proof of the next theorem.

Theorem 8.55 *Let T be $\mathcal{L}$-theory. If, for any primitive $\mathcal{L}$-formula $\exists y \eta(y, \overline{x})$ with $\overline{x}$ of length ≥ 1, $T \models \forall \overline{x}(\exists y \varphi(y, \overline{x}) \leftrightarrow \theta(\overline{x}))$, with $\theta(\overline{x}))$ quantifier-free, then T has quantifier elimination.*

Proof We require $\overline{x}$ of length ≥ 1 because zero length yields sentences, for which we may have no quantifier-free equivalents if e.g. we have no constant. We proceed by induction on the complexity of formulae. Since every atomic formula is quantifier-free, the inductive base is easily established.

The inductive hypothesis for $\neg, \vee$ is that the formulae $\varphi_1(\overline{x}), \varphi_2(\overline{x})$ are T-equivalent to quantifier-free $\mathcal{L}$-formulae $\theta_1(\overline{x}), \theta_2(\overline{x})$. The reader may easily complete the inductive step for $\neg, \vee$.

The inductive hypothesis for $\exists$ is that, given a formula of the form $\exists y \varphi(y, \overline{x})$, the formula $\varphi(y, \overline{x})$ is T-equivalent to a quantifier-free formula $\theta(y, \overline{x})$. Thus, $\exists y \varphi(y, \overline{x})$ is T-equivalent to $\exists y \theta(y, \overline{x})$, where the latter formula is existential and its matrix may be assumed in disjunctive normal form.

We may rewrite the last formula as:

$$\exists y(\theta_1(y, \overline{x}) \vee \ldots \vee \theta_k(y, \overline{x}))$$

where each $\theta_i(\overline{x})$ is a conjunction of literals. The last formula is equivalent to a disjunction of primitive formulae, namely:

$$\exists y \theta_1(y, \overline{x}) \vee \ldots \vee \exists y \theta_k(y, \overline{x}).$$

By hypothesis, each disjunct is T-equivalent to a quantifier-free formula. Consequently, the whole disjunction is. The induction is complete. □

We have seen that model-completeness is not a syntactically robust property of theories in the sense that it may be affected by small changes in the signature of a language, e.g. the deletion or addition of one symbol. The same is true of quantifier elimination. Nonetheless, both properties can always be forced to hold by enriching a language's signature. The next subsections illustrate two distinct ways in which this enrichment may be effected.

8.4 Quantifier Elimination

8.4.1 Morleysation

Let T be a $\mathcal{L}$-theory. If T does not have quantifier elimination, we can always find a $\mathcal{L}^m$-theory T^m, called the **Morleysation** of T, that extends T and has it. The language $\mathcal{L}^m$ is obtained from $\mathcal{L}$ by adding one n-ary relation symbol R_φ for each $\mathcal{L}$-formula. The theory T^m is obtained from T by adding to it one sentence of the form:

$$\forall \overline{x}(R_\varphi \overline{x} \leftrightarrow \varphi(\overline{x}))$$

for each new relation symbol R_φ. By construction, every $\mathcal{L}$-formula is T^m-equivalent to an atomic $\mathcal{L}^m$-formula. Quantifier elimination holds for T^m by the next exercise.

Exercise 8.56

(i) Prove that every $\mathcal{L}^m$-formula is T^m-equivalent to a $\mathcal{L}$-formula. Deduce that T^m has quantifier elimination.
(ii) Let $\mathcal{M}, \mathcal{N} \models T^m$. Show that any homomorphism $f : M \longrightarrow N$ is an elementary embedding.

Exercise 8.57

(i) Let T be a $\mathcal{L}$-theory. A set Σ of $\mathcal{L}$-formulae is an **elimination set** for T iff every $\mathcal{L}$-formula $\phi(\overline{x})$ is T-equivalent to a formula in Σ. Let Θ be a set of $\mathcal{L}$-formulae with the following properties:

(a) Θ contains every atomic $\mathcal{L}$-formula.
(b) If $\varphi(\overline{x}), \psi(\overline{x}) \in \Theta$, then $\neg \varphi(\overline{x}), \varphi(\overline{x}) \vee \psi(\overline{x})$ are in Θ.
(c) If $\varphi(\overline{x}, y)$ is in Θ, then $\exists y \varphi(\overline{x}, y)$ is T-equivalent to a formula in Θ.

Show that Θ is an elimination set for T.
(ii) Let T be a $\mathcal{L}$-theory and Σ be an elimination set for T. Show that a map between two models of T is elementary iff it strongly preserves the formulae in Σ.

8.4.2 Skolemisation

By Theorem 8.55, quantifier elimination is achieved if only single existential quantifiers can be eliminated. If T implies, for each formula of the form $\exists y \varphi(\overline{x}, y)$, its T-equivalence to a formula of the form $\psi(\overline{x}, t(\overline{x}))$, with $t(\overline{x})$ a term, we can immediately deduce that T has quantifier elimination.

If this is not the case, we can nonetheless add new sentences to T in order to achieve the same result relative to a richer theory T^s. To illustrate this approach in greater detail, we require an important new idea.

Definition 8.58 Let T be a $\mathcal{L}$-theory. We say that T **has in-built Skolem functions** iff for every $\mathcal{L}$-formula of the form $\exists y \varphi(\overline{x}, y)$, there is a $\mathcal{L}$-term $t(\overline{x})$ such that $T \models \forall \overline{x} (\exists y \varphi(\overline{x}, y) \to \varphi(\overline{x}, t(\overline{x})))$.

The reference to functions in the last definition follows from the fact that, whenever $\mathcal{M} \models T$, the term $t(\overline{x})$ names a unique object $b \in M$ for each assignment of $\overline{a}$ to the tuple of variables $\overline{x}$. In other words, the term t can be regarded as the name of a n-ary function $t^{\mathcal{M}} : M^n \longrightarrow M$.

Exercise 8.59

(i) Use the Tarski-Vaught test to show that, if T has Skolem functions, then it is model-complete.
(ii) Show that, if T has Skolem functions, then it has quantifier elimination.

If T lacks Skolem functions, we can always expand its language $\mathcal{L}$ to $\mathcal{L}^s$ and add suitable sentences to T until we obtain an extension $T^s \supseteq T$ with Skolem functions.

Theorem 8.60 Let T be a $\mathcal{L}$-theory. There are an expansion $\mathcal{L}^s$ of $\mathcal{L}$ and an extension T^s of T such that:

(1) if $\mathcal{M} \models T$, then $\mathcal{M}$ can be expanded to a model $\mathcal{M}^s$ of T^s;
(2) the $\mathcal{L}^s$-theory T^s has Skolem functions.

Proof Consider the set of all $\mathcal{L}$-formulae of the form $\exists y \varphi(\overline{x}, y)$, with $\overline{x}$ a n-tuple of variables. For each of these formulae we add a corresponding n-ary function symbol $F_{\varphi(\overline{x}, y)}$ to $\mathcal{L}$.

We set $\mathcal{L} = \mathcal{L}_0$ and let $\mathcal{L}_1 \supseteq \mathcal{L}_0$ be the expanded language obtained from $\mathcal{L}_0$ by adding the function symbols just described. We also set $T = T_0$ and let $T_1 \supseteq T_0$ be the $\mathcal{L}_1$-theory obtained from T_0 by adding to it one sentence of the form:

$$\forall \overline{x} (\exists y \varphi(\overline{x}, y) \to \varphi(\overline{x}, F_{\varphi(\overline{x}, y)}(\overline{x}))), \tag{8.1}$$

for each function symbol $F_{\varphi(\overline{x}, y)}$. We iterate these additions until we obtain respectively a chain of languages $\{\mathcal{L}_n\}_{n \in \mathbb{N}}$ and a chain of theories $\{T_n\}_{n \in \mathbb{N}}$ such that T_n is a $\mathcal{L}_n$-theory. We finally set:

$$\mathcal{L}^s = \bigcup_{n \in \mathbb{N}} \mathcal{L}_n \text{ and } T^s = \bigcup_{n \in \mathbb{N}} T_n.$$

If $\mathcal{M} \models T$, the $\mathcal{L}$-structure $\mathcal{M}$ can be expanded to a $\mathcal{L}^s$-model of T^s. It suffices to find an interpretation $F^s_{\varphi(\overline{x}, y)}$ for each function symbol $F_{\varphi(\overline{x}, y)}$ that satisfies condition (8.1). To this end, we observe that, for each $\overline{a} \in M^n$, $\mathcal{M} \models \exists y \varphi(\overline{a}, y)$ or $\mathcal{M} \not\models \exists y \varphi(\overline{a}, y)$.

Let us suppose that there are tuples $\overline{a}$ for which $\mathcal{M} \models \exists y \varphi(\overline{a}, y)$. Each of these tuples determines a nonempty subset of M defined by the formula $\varphi(\overline{a}, y)$. We therefore have at our disposal a nonempty family of nonempty sets. By a version of the Axiom of Choice (i.e. (AC1) from Sect. A.6) there is a function F^s that selects one element from each set in the family.

The function F^s is not defined on all of M^n: there may be a tuple $\bar{b}$ such that $\mathcal{M} \not\models \exists y \varphi(\bar{b}, y)$. In this case we fix $c \in M$ and extend F^s to $F^s_{\varphi(\bar{x},y)}$ by setting:

$$F^s_{\varphi(\bar{x},y)}(\bar{a}) = \begin{cases} F^s(\bar{a}) & \text{if } \mathcal{M} \models \exists y \varphi(\bar{a}, y) \\ c & \text{otherwise} \end{cases}$$

We leave it as an exercise for the reader to verify that conditions (1) and (2) are satisfied by the interpretations we have constructed. □

Exercise 8.61 Show that the $\mathcal{L}^s$-theory T^s satisfies conditions (1) and (2) from the statement of Theorem 8.60.

Exercise 8.62 Let T be a $\mathcal{L}$-theory with infinite models. Show that both T^m and T^s have models of size $|\mathcal{L}|$.

Although it is possible artificially to expand the signature of a language to force model-completeness and quantifier elimination, these properties are especially significant when they hold relative to the natural signatures of significant mathematical theories. For this reason it is helpful to develop general tests for quantifier elimination that can be used on given theories. A few basic ones constitute the main subject of the next three sections.

8.5 Finite Relational Languages

Since every theory can be completed, it is reasonable to look for a quantifier elimination test that applies to complete theories. In this section we obtain one that works for complete $\mathcal{L}$-theories under the restriction that, in the signature of $\mathcal{L}$, $\mathbf{F} \cup \mathbf{C} = \emptyset$ and $\mathbf{R}$ be finite.

When the last two conditions are fulfilled we say that T has a *finite relational language*. The test we will prove relies on the following characterisation of T-equivalence to a quantifier-free formula.

Theorem 8.63 *Let T be a complete $\mathcal{L}$-theory and $\bar{x}$ a n-tuple of free variables with $n \geq 1$. The following conditions are equivalent:*

(1) *the $\mathcal{L}$-formula $\varphi(\bar{x})$ is T-equivalent to a quantifier-free $\mathcal{L}$-formula $\theta(\bar{x})$;*
(2) *if $\mathcal{M} \models T$, if $\bar{a}, \bar{b} \in M^n$ satisfy exactly the same atomic $\mathcal{L}$-formulae, then $\mathcal{M} \models \varphi(\bar{a}) \iff \mathcal{M} \models \varphi(\bar{b})$.*

Proof The implication from condition (1) to (2) is left as an exercise. We assume that (2) holds and expand the language $\mathcal{L}$ to $\mathcal{L}_{\bar{c}}$ by adding a n-tuple of new constant symbols. Let Q be the set of quantifier-free $\mathcal{L}_c$-sentences implied by $T \cup \varphi(\bar{c})$. Thus, for any quantifier-free sentence $\theta(\overline{c})$, we have $\theta(\overline{c}) \in Q$ iff $T \models \varphi(\overline{c}) \to \theta(\overline{c})$. Our goal is to prove $T \cup Q \models \varphi(\overline{c})$, which, together with the definition of $\theta(\overline{c})$, implies (1).

Toward a contradiction, we suppose $\mathcal{M}_{\overline{c}} \models T \cup Q \cup \{\neg \varphi(\overline{c})\}$. Since we are given the $\mathcal{L}_{\overline{c}}$-structure $\mathcal{M}_{\overline{c}}$, we let P be the set of quantifier-free $\mathcal{L}_{\overline{c}}$-sentences modelled by $\mathcal{M}_{\overline{c}}$. By definition, $Q \subseteq P$.

It now follows that the set $T \cup P \cup \{\varphi(\overline{c})\}$ has a model $\mathcal{N}_{\overline{c}}$. If not, $T \cup P \models \neg \varphi(\overline{c})$ by Corollary 5.63 and, by compactness again, there is a conjunction $\theta(\overline{c})$ of elements of P, itself in P and such that $T \cup \theta(\overline{c}) \models \neg \varphi(\overline{c})$. Since $T \models \varphi(\overline{c}) \to \neg \theta(\overline{c})$, we deduce $\neg \theta(\overline{c}) \in Q \subseteq P$. It follows that P has no models, which is impossible because $\mathcal{M}_{\overline{c}} \models P$.

We now work with the $\mathcal{L}$-reducts of $\mathcal{M}_{\overline{c}}, \mathcal{N}_{\overline{c}}$, which we call $\mathcal{M}, \mathcal{N}$ respectively. There are $\overline{a} \in M^n$ and $\overline{b} \in N^n$ such that $\mathcal{M} \models \neg \varphi(\overline{a})$ and $\mathcal{N} \models \varphi(\overline{b})$.

Because T is complete and $\mathcal{M}, \mathcal{N}$ are models of T, they can be elementarily embedded into a common elementary extension $\mathcal{M}^*$ (recall Theorem 7.15). We identify them with two elementary substructures of this common extension by Theorem 4.31. Thus $\mathcal{M}^* \models T$ and $\overline{a}, \overline{b} \in M^*$.

Because $\mathcal{D}(\mathcal{M})$ and $\mathcal{D}(\mathcal{N})$ may be identified with subsets of P, $\overline{a}, \overline{b}$ satisfy the same atomic formulae on account of the fact that their respective expansions $\mathcal{M}_{\overline{c}}, \mathcal{N}_{\overline{c}}$ model P. We can now use condition (2) to deduce:

$$\mathcal{M}^* \models \varphi(\overline{a}) \iff \mathcal{M}^* \models \varphi(\overline{b}),$$

which is impossible by the way $\mathcal{M}^*$ was defined. The contradiction we have arrived at implies $T \cup Q \models \varphi(\overline{c})$, as desired. □

Exercise 8.64 Prove the implication from (1) to (2) in the statement of Theorem 8.63.

The restriction to finite relational vocabularies did not come into play in the proof of the preceding theorem but is needed in the following quantifier elimination test:

Theorem 8.65 *Let T be a complete $\mathcal{L}$-theory in a finite relational vocabulary. The following conditions are equivalent:*

(1) T has quantifier elimination;
(2) let $At(\overline{x})$ be the set of atomic $\mathcal{L}$-formulae whose free variables occur among $\overline{x}$. Suppose $\mathcal{M} \models T$, $\overline{a}, \overline{b} \in M^n$ and, for any $\eta(\overline{x}) \in At(\overline{x})$, $\mathcal{M} \models \eta(\overline{a}) \leftrightarrow \eta(\overline{b})$. Then, for any $(n+1)$-tuple $\overline{a}a_{n+1} = (a_1, \ldots, a_n, a_{n+1})$ there is a $(n+1)$-tuple $\overline{b}b_{n+1}$ such that $\mathcal{M} \models \theta(\overline{a}a_{n+1}) \leftrightarrow \theta(\overline{b}b_{n+1})$ for each $\theta \in At(\overline{x}, x_{n+1})$.

Proof To prove that (1) implies (2), we note that, by our restriction to finite relational languages, $At(\overline{x})$ is finite. We may therefore set $At(\overline{x}) = \{\eta_1(\overline{x}), \ldots, \eta_m(\overline{x})\}$ for some $m \in \mathbb{N}$. When we assign $\overline{a} \in M^n$ to $\overline{x}$ we either satisfy $\eta_i(\overline{x})$ in $\mathcal{M}$ or we satisfy $\neg \eta_i(\overline{x})$, for each $i \in \{1, \ldots, m\}$.

We can thus determine a finite sequence of literals, whose conjunction we call $\psi(\overline{x})$. By hypothesis $\mathcal{M} \models \psi(\overline{a}), \psi(\overline{b})$. For any $a_{n+1} \in M$, since $At(\overline{x}x_{n+1})$

is finite, we may include all the literals satisfied by $\bar{a}a_{n+1}$ in $\mathcal{M}$ into a single conjunction $\theta(\bar{x}, x_{n+1})$. It follows that $\mathcal{M} \models \exists x_{n+1}\theta(\bar{a}, x_{n+1})$ and, by (1), $T \models \forall \bar{x}(\exists x_{n+1}\theta(\bar{a}, x_{n+1}) \leftrightarrow \xi(\bar{x}))$, where $\xi(\bar{x})$ is quantifier-free. We conclude $\mathcal{M} \models \xi(\bar{a})$.

Because $\bar{a}, \bar{b}$ satisfy exactly the same atomic formulae, they satisfy exactly the same quantifier-free formulae by an easy induction. Consequently, $\mathcal{M} \models \xi(\bar{b})$, from which we immediately obtain $\mathcal{M} \models \exists x_{n+1}\theta(\bar{b}, x_{n+1})$. It is therefore possible to pick b_{n+1} in such a way that the $(n+1)$-tuples $\bar{a}a_{n+1}$ and $\bar{b}b_{n+1}$ satisfy exactly the same atomic formulae in $At(\bar{x}, x_{n+1})$.

To prove that (2) implies (1), we rely on Theorem 8.55 and consider a primitive formula $\exists y \varphi(\bar{x}, y)$, where $\varphi(\bar{x}, y)$ is a conjunction of $\mathcal{L}$-literals. Fix $\mathcal{M} \models T$. If $\bar{a}, \bar{b}$ satisfy the same atomic formulae and $\mathcal{M} \models \exists y \varphi(\bar{x}, y)$, we deduce $\mathcal{M} \models \varphi(\bar{a}, a_{n+1})$ for some $a_{n+1} \in M$. We can therefore find $b_{n+1} \in M$ such that $\mathcal{M} \models \varphi(\bar{b}, b_{n+1})$ and, conclude $\mathcal{M} \models \exists y \varphi(\bar{b}, y)$. The argument is reversible and it shows that:

$$\mathcal{M} \models \exists y \varphi(\bar{a}, y) \leftrightarrow \exists y \varphi(\bar{b}, y).$$

Because $\mathcal{M}$ is an arbitrary model of T, Theorem 8.63 applies and we conclude that $\exists y \varphi(\bar{x}, y)$ is T-equivalent to a quantifier-free formula $\zeta(\bar{x})$. Since we can reach this conclusion for every primitive formula, T has quantifier elimination. □

Exercise 8.66

(i) Show that IS has quantifier elimination.
(ii) Show that DLO$_{-,-}$ has quantifier elimination. Use this result to deduce $(\mathbb{Q}, <^\mathbb{Q}) \preceq (\mathbb{R}, <^\mathbb{Q})$. Finally, if $\mathcal{M} \models$ DLO$_{-,-}$, describe the subsets of M definable by a $\mathcal{L}$-formula with parameters over M.
(iii) Show that the theory of equivalence classes axiomatised in Exercise 7.17 has quantifier elimination.
(iv) Show that dLO$_{+,-}$, regarded as a $\mathcal{L}_d$-theory (see Sect. 8.1.2) does not have quantifier elimination. Does anything change if we add to $\mathcal{L}_d$ a unary predicate P to dLO$_{+,-}$ the sentence $\forall x(Px \leftrightarrow \forall y(\neg(y < x)))$?

8.6 T-Closures

Although model-completeness does not imply quantifier elimination, many of the theories we have studied so far satisfy a condition under which the two properties are equivalent.

Definition 8.67 Let T be a $\mathcal{L}$-theory, $\mathcal{M}$ a model of T and $\mathcal{A} \sqsubseteq \mathcal{M}$. We say that $\mathcal{M}$ is a T-**closure** of $\mathcal{A}$ iff for any embedding f of $\mathcal{A}$ into a model $\mathcal{N}$ of T, there is an embedding $g : \mathcal{M} \longrightarrow \mathcal{N}$ such that $f = g \circ \iota$, where ι is the inclusion of $\mathcal{A}$ into $\mathcal{M}$.

Note that, by Lemma 8.27, $\mathcal{A} \models T_\forall$. An example of T-closure we have repeatedly discussed is offered by ACF, regarded as a $\mathcal{L}_r$-theory. Any substructure of a model of ACF is an integral domain. We have seen in Sect. 6.5 that $\mathcal{A}$ can be embedded in its field of fractions $Q(A)$ and, ultimately, in the algebraic closure $\overline{Q(A)}$ of this field, which is a model of ACF.

Exercise 8.68 Verify that $\overline{Q(\mathcal{A})}$ is an ACF-closure of the integral domain $\mathcal{A}$.

Exercise 8.69

(i) Let A be a finite set. Describe a IS-closure of A and show that any two IS-closures of finite sets are isomorphic.
(ii) Set $T = \mathsf{VS}_K^\infty$ and, for any $\mathcal{L}_v$-substructure $\mathcal{A}$ of an infinite vector space over K, determine a T-closure of $\mathcal{A}$ (*Hint*: since K may be finite or infinite, it is convenient to consider these cases separately).

If T-closures exist, compactness yields a useful characterisation of T-equivalence with a quantifier-free formula.

Lemma 8.70 *Let T be a $\mathcal{L}$-theory and $\overline{x}$ be a n-tuple of variables with $n > 0$. If every model of $T_\forall$ has a T-closure, the following conditions are equivalent:*

(1) *the formula $\varphi(x_1, \ldots, x_n)$ is T-equivalent to a quantifier-free $\mathcal{L}$-formula $\eta(\overline{x})$;*
(2) *for any $\mathcal{M}, \mathcal{N} \models T$ and any tuple $\overline{a} \in M^n$, if $f : \mathcal{M} \longrightarrow \mathcal{N}$ is an embedding, then:*

$$\mathcal{M} \models \varphi(\overline{a}) \iff \mathcal{N} \models \varphi(f(\overline{a})).$$

Proof Since embeddings strongly preserve quantifier-free formulae, it is clear that (1) implies (2). To prove the converse, we essentially rely on compactness. Let Γ be the set of quantifier-free $\mathcal{L}$-formulae $\psi(\overline{x})$ such that:

$$T \models \forall \overline{x}(\varphi(\overline{x}) \to \psi(\overline{x})).$$

Let us expand $\mathcal{L}$ to $\mathcal{L}_{\overline{c}}$, where $\overline{c}$ is a tuple of new constant symbols whose length is the length of $\overline{x}$. It suffices to prove $T \cup \Gamma_{\overline{c}} \models \varphi(\overline{c})$, where $\Gamma_{\overline{c}}$ is obtained from Γ by substituting the new constant symbols for the respective variables in $\overline{x}$.

Toward a contradiction, we suppose the existence of a model $\mathcal{M}' \models T \cup \Gamma_{\overline{c}} \cup \{\neg \varphi(\overline{c})\}$. Let $\overline{a}$ be the interpretation of $\overline{c}$ in $\mathcal{M}'$ and $\mathcal{M}$ be the $\mathcal{L}$-reduct of $\mathcal{M}'$. Moreover, let A be the set containing $t^{\mathcal{M}_1}(\overline{a})$ for each $\mathcal{L}$-term $t(\overline{x})$. Then A is the domain of a structure $\mathcal{A} \sqsubseteq \mathcal{M}_1$. More precisely, $\mathcal{A}$ is the substructure generated by $\overline{a}$.

By hypothesis, $\mathcal{A}$ has a T-closure $\mathcal{B}$, which can be embedded in $\mathcal{M}$ and identified with an extension of $\mathcal{A}$ that is also a submodel of $\mathcal{M}$. We now consider the set Δ of quantifier-free $\mathcal{L}$-formulae of the form $\delta(\overline{x})$ such that $\mathcal{M} \models \delta(\overline{a})$.

Let Δ_c be obtained from Δ by substituting the new constant symbols for the respective variables in $\overline{x}$. Toward a contradiction, let $\mathcal{N}'$ be a model of $T \cup \Delta_{\overline{c}} \cup$

8.6 T-Closures

$\{\varphi(\bar{c})\}$ and $\bar{b}$ the interpretation of $\bar{c}$ in N^n. Because $\mathcal{N}$, the $\mathcal{L}$-reduct of $\mathcal{N}'$, models Δ, any atomic formula satisfied by $\bar{a}$ in $\mathcal{M}$ is satisfied by $\bar{b}$ in $\mathcal{N}$.

In other words, the function f^- that sends $\bar{a}$ into $\bar{b}$ can be extended to an embedding f from $\mathcal{A}$ into $\mathcal{N}$. Because $\mathcal{B}$ is a T-closure of $\mathcal{A}$, the embedding f can in turn be extended to an embedding f^+ of $\mathcal{B}$ into $\mathcal{N}$.

From $f^+(\bar{a}) = \bar{b}$, the hypothesis (2) and the fact that $\mathcal{B} \models \neg\varphi(\bar{a})$, we deduce $\mathcal{N} \models \neg\varphi(\bar{b})$. A contradiction follows because, by construction, $\mathcal{N} \models \varphi(\bar{b})$. This contradiction implies that $T \cup \Delta_{\bar{c}} \cup \{\varphi(\bar{c})\}$ has no model. Since T is consistent and $\Delta_{\bar{c}}$ is closed under conjunction, Corollary 5.63 applies and there is a sentence $\delta(\bar{c}) \in \Delta_{\bar{c}}$ such that, using contraposition once:

$$T \models \varphi(\bar{c}) \to \neg\delta(\bar{c}).$$

Because the constant symbols $\bar{c}$ do not occur in any sentence of T, the last condition implies $\neg\delta(\bar{x}) \in \Gamma$. Thus $\mathcal{M} \models \neg\delta(\bar{a})$. A contradiction follows again because, by the definition of Δ, $\mathcal{M} \models \delta(\bar{a})$. We conclude:

$$T \cup \Gamma_c \models \varphi(\bar{c}).$$

By compactness, a finite part of Γ_c entails $\varphi(\bar{c})$ in any model of T. Because Γ_c is closed under conjunction, a quantifier-free sentence entails $\varphi(\bar{c})$ in any model of T. Condition (1) follows. □

We can now easily prove a theorem that yields at once a criterion for quantifier elimination and a sufficient condition for the equivalence between model-completeness and quantifier elimination. The straightforward proof requires a preliminary definition.

Definition 8.71 Let $\mathcal{M}, \mathcal{N}$ be $\mathcal{L}$-structures and $f : M \longrightarrow N$ an embedding. We say that f is **simple** iff, for any primitive formula $\exists y \varphi(\bar{x}, y)$ and any matching tuple $\bar{a}$ of elements of M, we have:

$$\mathcal{N} \models \exists y \varphi(f(\bar{a}), y) \implies \mathcal{M} \models \exists y \varphi(\bar{a}, y).$$

Theorem 8.72 (Shoenfield-Robinson) *Let T be a $\mathcal{L}$-theory. If any model of $T_\forall$ has a T-closure, then the following conditions are equivalent:*

(1) T has quantifier elimination;
(2) T is model-complete;
(3) if $\mathcal{A}, \mathcal{B} \models T$ and $f : A \longrightarrow B$ is an embedding, then f is simple.

Proof See Exercise 8.73. □

Exercise 8.73 Give a round robin proof of Theorem 8.76. Use induction on the complexity of formulae and rely on Lemma 8.70 to obtain the implication from (3) to (1).

We immediately obtain:

Corollary 8.74 *The theories* ACF, ACF_0, ACF_p *for any prime p and* VS_K^∞ *have quantifier elimination.*

Proof Let T be any one of the theories listed in the statement of the corollary. If $\mathcal{A} \models T_\forall$, then $\mathcal{A}$ has a T-closure by Exercise 8.61. Moreover, T is model-complete by Corollary 8.44. Theorem 8.72 applies. □

Because ACF is an incomplete theory (e.g. the sentence $1+1 = 0$ is true in some of its models and false in others), the last corollary shows that quantifier elimination does not imply completeness.

Exercise 8.75 Consider the language $\mathcal{L}_{0,s} = \mathcal{L}_= \cup \{0, s\}$, where 0 is a constant symbol and s a unary function symbol. We abbreviate the n-th iteration of the function s on the variable x (the constant symbol 0, respectively) by $s^n(x)$ ($s^n(0)$, respectively). Let $T_{0,s}$ be the $\mathcal{L}_{0,s}$-theory axiomatised by the sentences:

(1) $\forall x(\neg(s(x) = 0))$;
(2) $\forall x(\neg(x = 0) \to \exists y(s(y) = x))$;
(3) $\forall x \forall y((s(x) = s(y) \to x = y))$;
(4) for each integer $n \geq 1$, $\forall x \neg(s^n(x) = x)$.

Note that, the $\mathcal{L}_{0,s}$-structure $\mathcal{N} = (\mathbb{N}, s^\mathbb{N}, 0^\mathbb{N})$, where $s^\mathbb{N}$ is the successor function sending n into $n + 1$, and $0^\mathbb{N}$ the number zero, is a model of $T_{0,s}$.

(i) Show that $\mathcal{N}$ can be embedded in any model of $T_{0,s}$.
(ii) Show that any model of T_s contains an isomorphic copy of the natural numbers as well as a set of disjoint isomorphic copies of $\mathbb{Z}$. Use this information to prove that $T_{0,s}$ is complete.
(iii) Deduce that $T_{0,s} = Th(\mathcal{N})$ and conclude that the theory of the natural numbers with zero and the successor function is axiomatisable.
(iv) Prove that no finite subset of $T_{0,s}$ axiomatises $Th(\mathcal{N})$. Deduce that the latter theory is not finitely axiomatisable.
(v) Show that $T_{0,s}$ has quantifier elimination (*Hint*: if $\mathcal{A} \sqsubseteq \mathcal{M}$ and $\mathcal{M} \models T_{0,s}$, then $\mathcal{A}$ has a $T_{0,s}$-closure).
(vi) Let $\mathcal{N}_s = (\mathbb{N}, s)$ be the reduct of $\mathcal{N}$ to $\mathcal{L}_s = \mathcal{L}_= \cup \{s\}$. Show that $Th(\mathcal{N}_s)$ does not have quantifier elimination.

8.7 Algebraically Prime Models

We opened this chapter by stating that quantifier elimination is equivalent to substructure-completeness. The equivalence easily follows from the next theorem, which provides one further test for quantifier elimination.

Theorem 8.76 *Let T be a $\mathcal{L}$-theory and $\varphi(\overline{x})$ be a $\mathcal{L}$-formula where $\overline{x}$ is a n-tuple of variables with $n > 0$. The following conditions are equivalent:*

8.7 Algebraically Prime Models

(1) $\varphi(\bar{x})$ *is T-equivalent to a quantifier-free formula;*
(2) if $\mathcal{M}, \mathcal{N} \models T$, $\mathcal{A}$ *is a common substructure of* $\mathcal{M}, \mathcal{N}$ *and* $\bar{a} \in A^n$, *then:*

$$\mathcal{M} \models \varphi(\bar{a}) \iff \mathcal{N} \models \varphi(\bar{a});$$

(3) the restriction of (2) to finitely generated, common substructures.

Proof It is easy to see that (1) implies (2) and (3) is just a special case of (2). To prove that (3) implies (1), we adapt the proof of Lemma 8.70. Working with an expansion of the language that includes the new constant symbols $\bar{c}$, the goal is again to show that $T \cup \Gamma_c \models \varphi(\bar{c})$, where $\Gamma_{\bar{c}}$ is defined as in Lemma 8.70.

The proof proceeds by contradiction. Once a $\mathcal{L}$-structure $\mathcal{M} \models T \cup \Gamma_c \cup \{\neg \varphi(\bar{c})\}$ is postulated, the interpretations $\bar{a} \in M^n$ are used to identify a finitely generated substructure of $\mathcal{M}$, i.e. the smallest substructure $\mathcal{A} \sqsubseteq \mathcal{M}$ containing $\bar{a}$. The existence of $\mathcal{M}$ entails the existence of a model $\mathcal{N} \models T \cup \mathcal{D}(\mathcal{A}) \cup \{\varphi(\bar{c})\}$.

It is thus possible to regard $\mathcal{A}$ as a common substructure of $\mathcal{M}, \mathcal{N}$ and to deduce $\mathcal{M} \models \neg\varphi(\bar{a})$, $\mathcal{N} \models \varphi(\bar{a})$, against (3). Full details are left to the reader. □

Exercise 8.77 Fill the gaps in the proof of Theorem 8.76.

Exercise 8.78 Show that, in the statement of Theorem 8.76, it is enough to require that conditions (2) and (3) hold for primitive formulae.

Exercise 8.79 Show that T has quantifier elimination iff for any $\mathcal{A} \models T_\forall$, $T \cup \mathcal{D}(\mathcal{A})$ is a complete theory. Deduce that **substructure-completeness** in the class $Mod\,T$ is equivalent to quantifier elimination.

Exercise 8.80 We shall prove that algebraically closed fields are existentially closed in the class of fields, as noted in Remark 8.10. Let $\mathcal{K}$ be a $\mathcal{L}_r$-structure that models the field axioms and $\sigma(x_1, \ldots, x_n)$ an existential formula with parameters in K. Show that, if $\sigma(x_1, \ldots, x_n)$ is modelled by a field extension of $\mathcal{K}$, then it is modelled by any algebraically closed field extension (*Hint*: $ACF \cup \mathcal{D}(\mathcal{K}) \cup \{\sigma(x_1, \ldots, x_n)\}$ has a model).

We next turn to a quantifier elimination test closely related to the notion of T-closure.

Definition 8.81 A $\mathcal{L}$-theory T is said to have **algebraically prime models** iff any model of $T_\forall$ has a T-closure. Whenever $\mathcal{A} \models T_\forall$, we call the T-closure of $\mathcal{A}$ an algebraically prime model over $\mathcal{A}$.

Theorem 8.82 *Let T be a $\mathcal{L}$-theory with algebraically prime models. If, for any models $\mathcal{M} \sqsubseteq \mathcal{N}$ of T, any n-tuple $\bar{a} \in M^n$, and any primitive formula $\exists y \varphi(\bar{x}, y)$:*

$$\mathcal{N} \models \exists y \varphi(\bar{a}, y) \implies \mathcal{M} \models \exists y \varphi(\bar{a}, y),$$

then T has quantifier elimination.

Proof Fix a primitive formula $\exists y \varphi(\overline{x}, y)$. It suffices to show that condition (2) of Theorem 8.76 holds relative to this fixed formula. Thus, we suppose that $\mathcal{M}, \mathcal{N} \models T$ have a common substructure $\mathcal{A}$. Because T has algebraically prime models, we can extend $\mathcal{A}$ to $\mathcal{A}_1 \sqsubseteq \mathcal{M}$ and to $\mathcal{A}_2 \sqsubseteq \mathcal{M}$, where $\mathcal{A}_1, \mathcal{A}_2 \models T$ are algebraically prime over $\mathcal{A}$.

Suppose $\mathcal{N} \models \exists y \varphi(\overline{a}, y)$. By the theorem's hypothesis and the fact that $\mathcal{A}_2$ is a submodel of $\mathcal{M}$, $\mathcal{A}_2 \models \exists y \varphi(\overline{a}, y)$. Note that, because $\mathcal{A} \sqsubseteq \mathcal{A}_1$, there is an embedding of $\mathcal{A}_1$ into $\mathcal{A}_2$ that keeps $\mathcal{A}$ fixed (by definition of prime model over $\mathcal{A}$).

Because embeddings preserve existential formulae, we deduce $\mathcal{A}_1 \models \exists y \varphi(\overline{a}, y)$. Because extensions preserve existential formulae, we conclude $\mathcal{M} \models \exists y \varphi(\overline{a}, y)$. The argument is reversible and leads to the equivalence:

$$\mathcal{M} \models \exists y \varphi(\overline{a}, y) \iff \mathcal{N} \models \exists y \varphi(\overline{a}, y).$$

By condition (2) from Theorem 8.76, $\exists y \varphi(\overline{a}, y)$ is T-equivalent to a quantifier-free formula. Exercise 8.75 implies that T has quantifier elimination. □

We turn to some basic applications of Theorem 8.82.

Example 8.83 Consider the theory DAG^+ of non-degenerate, divisible abelian groups. If $\mathcal{G} \models \mathsf{DAG}^+$, then a substructure $\mathcal{A} \sqsubseteq \mathcal{G}$ is a torsion-free, but not necessarily divisible, group. It can even be the degenerate group whose domain is $\{0^G\}$. If $\mathcal{A}$ is non-degenerate, its divisible hull is an algebraically prime model over $\mathcal{A}$. If $\mathcal{A}$ is degenerate, we consider $a \neq 0^G$ in G: an algebraically prime model over $\mathcal{A}$ is in this case provided by the divisible hull of the subgroup of $\mathcal{G}$ generated by $\{0, a\}$.

Thus, DAG^+ has algebraically prime models. In order to apply Theorem 8.82 to DAG^+, we consider the matrix of a generic, primitive formula $\exists y \varphi(x_1, \ldots, x_n, y)$. Its conjuncts are literals of the form $t_1(x_1, \ldots, x_n, y) = t_2(x_1, \ldots, x_n, y)$.

It is clear that, if $a_1, \ldots, a_n, b \in G$ and $\mathcal{G} \models t_1(a_1, \ldots, a_n, b) = t_2(a_1, \ldots, a_n, b)$, the last equality holds iff, after suitable algebraic manipulations:

$$\mathcal{G} \models m_1 a_1 + \ldots + m_n a_n = mb, \quad \text{where } m_1, m_n, m \text{ are integer coefficients.}$$

Similarly, a negated literal holds in $\mathcal{G}$ under the same assignment iff:

$$\mathcal{G} \models \neg(m_1 a_1 + \ldots + m_n a_n = mb).$$

We can now easily verify that DAG^+ satisfies the hypotheses of Theorem 8.82. If $\mathcal{G}, \mathcal{H} \models \mathsf{DAG}^+, \mathcal{H} \sqsubseteq \mathcal{G}$ and $a_1, \ldots, a_n \in H$, suppose:

$$\mathcal{G} \models \exists y \varphi(a_1, \ldots, a_n, y)$$

where $\varphi(a_1, \ldots, a_n, y)$ is a conjunction of literals. Either one of the conjuncts is a positive literal (i.e. not negated), or none is.

8.7 Algebraically Prime Models

In the former case, let $b \in G$ solve an equality of the form $m_1 a_1 + \ldots + m_n a_n = mb$. Since $m_1 a_1 + \ldots + m_n a_n$ designates an element of H, and $\mathcal{H}$ is divisible, there is $a' \in H$ such that $ma' = m_1 a_1 + \ldots + m_n a_n = mb$. It follows that $a' = b \in H$. Thus $\mathcal{G} \models m_1 a_1 + \ldots + m_n a_n = ma'$. The fact that equalities are preserved by substructures implies $\mathcal{H} \models m_1 a_1 + \ldots + m_n a_n = ma'$. We conclude:

$$\mathcal{H} \models \exists y \varphi(a_1, \ldots, a_n, y).$$

The negative literals in φ require that y differ from finitely many elements of H, which is automatically the case, if the formula is at all satisfiable, since H is infinite and a' is uniquely determined.

If, on the other hand, no positive literal occurs in $\varphi(a_1, \ldots, a_n, y)$, then the negative literals in which y occurs constrain a solution to differ from finitely many elements of the domain, as noted. Because H is infinite, the requirement can certainly be met in H. We have again:

$$\mathcal{G} \models \exists y \varphi(a_1, \ldots, a_n, y) \implies \mathcal{H} \models \exists y \varphi(a_1, \ldots, a_n, y).$$

It follows that DAG^+ has quantifier elimination.

Exercise 8.84 Show that the sentence $\exists x(\neg(x = 0))$ is not DAG-equivalent to a quantifier-free $\mathcal{L}_g$-sentence. Conclude that DAG does not have quantifier elimination.

Exercise 8.85 Show that the theory of the structure $(\mathbb{Q}, +^{\mathbb{Q}}, 0^{\mathbb{Q}})$ has quantifier elimination.

Example 8.86 We can use the key ideas from Example 8.83 to show that the theory VS_K^∞ of infinite vector spaces over a fixed, infinite field K has quantifier elimination. If $\mathcal{A}$ is an arbitrary substructure of a model of VS_K^∞, then $\mathcal{A}$ is certainly a vector space over K and, if it is not 0-dimensional, it is also a model of VS_K^∞.

This is to say that, unless $A = \{\mathbf{0}\}$, where $\mathbf{0}$ is the null vector, $\mathcal{A}$ is an algebraically prime model over A. If A only contains the null vector, we consider any other vector v, which must exist in the model including $\mathcal{A}$, and take the linear closure of $\{\mathbf{0}, v\}$ as a prime model over A. Thus, in any case, algebraically prime models exist.

The reader may check that the other hypothesis of Theorem 8.82 is also satisfied. It suffices, as in the case of DAG^+, to examine the literals that may occur in the matrix of a primitive $\mathcal{L}_v$-formula $\exists y \varphi(a_1, \ldots, a_n, y)$ which holds in $\mathcal{V}$ with parameters $a_1, \ldots, a_n$ from a subspace $\mathcal{W} \sqsubseteq \mathcal{V}$. The positive literals say that y is a linear combination of the parameters and the negative literals say that y must differ from some (at most finitely many) linear combinations of the parameters.

This information suffices to replicate the argument given in the previous example.

Exercise 8.87 Suppose that $\mathcal{V}_1, \mathcal{V}_2$ model VS_K^∞ and that they are both proper extensions of a common subspace $\mathcal{V}$. Show that, if $\bar{a}$ is a tuple of parameters from

V and if $b \in V_1 - V_2$ satisfies the conjunction of literals $\varphi(\bar{a}, y)$ in V_1, then there is $c \in V_2 - V$ such that $V_2 \models \varphi(\bar{a}, c)$.

Exercise 8.88 Use the previous exercise and Theorem 8.76 to show that VS_K^∞ has quantifier elimination.

Exercise 8.89 Let $\mathsf{DAG}_<^+$ be the theory of non-degenerate, ordered, divisible abelian groups. This theory is obtained from $\mathsf{DAG}_<$, described in Sect. 3.2, by the addition of the sentence $\exists x(\neg(x = 0))$. Using the fact that, if $\mathcal{G} \models \mathsf{DAG}_<^+$, the $\mathcal{L}_<$-reduct of $\mathcal{G}$ is a dense, linear order without endpoints, prove that $\mathsf{DAG}_<^+$ has quantifier elimination.

Quantifier elimination entails that all embeddings between two models are elementary. For instance, all algebraic closed field embeddings are elementary. In particular, $\overline{\mathbb{Q}}$ can be elementarily embedded into any algebraically closed field of 0 characteristic and $\overline{\mathbb{Z}_p}$ can be elementarily embedded into any model of ACF_p. It is also easy to see that $(\mathbb{Q}, <^\mathbb{Q})$ has elementary embeddings into any dense linear orders.

We call models that can be elementarily embedded into any model of a theory T **prime models** of T. In the next chapter we shall be able to say more about the existence and fine structure of prime models.

Exercise 8.90 Let T be a $\mathcal{L}$-theory. Suppose that there is a $\mathcal{L}$-structure $\mathcal{A}$ that can be embedded into any model of T. Show that, if T has quantifier elimination, then T is complete. Provide an illustration of this result in which $\mathcal{A}$ is not a model of T.

Besides forcing all embeddings to be elementary, quantifier elimination has a dramatic impact on definability. If T is a $\mathcal{L}$-theory with quantifier elimination and $\mathcal{M} \models T$, a careful look at the $\mathcal{L}$-literals is sufficient to characterise the parametrically definable subsets of M. This possibility has momentous consequences, which we begin to explore in the next section.

8.8 Strongly Minimal Theories

The definable subsets of a $\mathcal{L}$-structure $\mathcal{M}$ are in general determined by formulae in prenex forms that can have arbitrarily many alternations of quantifiers. Such complex hierarchy (which arises, for instance, in logical number theory) is wiped out by quantifier elimination.

The simplest illustration of its effect is afforded by IS. This theory has quantifier elimination. It follows that, if M is an infinite set, any parametrically $\mathcal{L}_=$-definable subset of M is defined by a disjunctive normal form $\varphi(x)$ whose disjuncts are conjunctions of $\mathcal{L}_=$-literals. If $\varphi(x)$ contains the parameters $a_1, \ldots, a_n$, any one of its disjuncts says that x equals some parameters and that x is distinct from other parameters. Because the disjuncts of $\varphi(x)$ in which positive literals occur define individual elements of M, $\varphi(x)$ defines a finite subset of M if its disjuncts contain

8.8 Strongly Minimal Theories

any positive literals. If only negative literals occur in any one disjunct, it defines the complement of a finite subset of M, also known as a **cofinite** set, and so does $\varphi(x)$.

In general, $\phi(x)$ defines a union of one-element sets and cofinite sets. This is to say that it defines either a finite or a cofinite subset of M. Most of the theories we have studied so far possess the same special property on account of quantifier elimination. Its importance call for a special name.

Definition 8.91 Let $\varphi(x)$ be a $\mathcal{L}$-formula and $\mathcal{M}$ be a $\mathcal{L}$-structure. We say that $\varphi(x)$ is **minimal** in $\mathcal{M}$ iff $\varphi(x)$ has parameters in M, $\varphi(M)$ is an infinite subset of M and, for any set $\psi(M) \subseteq$ defined by a formula $\psi(x)$ with parameters in M, either $\varphi(M) \cap \psi(M)$ is finite or $\varphi(M) \cap \neg\psi(M)$ is finite.

We say that $\varphi(x)$ is **strongly minimal** in $\mathcal{M}$ if it is minimal in $\mathcal{M}$ and in every elementary extension of $\mathcal{N}$.

A $\mathcal{L}$-structure $\mathcal{M}$ is minimal if the $\mathcal{L}$-formula $x = x$ is minimal in $\mathcal{M}$. If $x = x$ is minimal in every elementary extension of $\mathcal{M}$, then $\mathcal{M}$ is said to be a strongly minimal structure.

Finally, we say that a $\mathcal{L}$-theory T is strongly minimal if every model of T is a minimal structure. It follows that, in this case, every model of T is in fact a strongly minimal structure.

As observed, interesting examples of strongly minimal theories are not hard to find.

Example 8.92 Let $\mathcal{V}$ be a model of VS_K^∞. By quantifier elimination, every definable subset of V is in particular defined by a quantifier-free formula $\eta(x)$ with parameters from V. We may assume $\eta(x)$ to be in disjunctive normal form. We examine the kinds of literals that may occur in the disjuncts of $\eta(x)$.

A positive literal can only be an equality between two terms and, given the axioms of vector spaces, it is reducible to the form $x = \lambda_1 v_1 + \ldots + \lambda_m v_m$, where $v_1, \ldots, v_m \in V$ are parameters. The last literal defines a single vector in the linear closure of the parameters $v_1, \ldots, v_m$.

A negative literal can only be a negated inequality and it is reducible to the formula $\neg(x = \lambda_1 v_1 + \ldots + \lambda_m v_m)$, which defines the infinite set of vectors that differ from the linear combination of $\lambda_1 v_1 + \ldots + \lambda_m v_m$.

We conclude that positive literals define finite sets (in fact, singletons) whereas negative literals define cofinite sets. Thus, if each disjunct of $\eta(x)$ contains a positive literal, then $\eta(x)$ defines a finite set of vectors. By contrast, if at least one disjunct of $\eta(x)$ contains only negative literals, then $\eta(x)$ defines a cofinite set.

It follows that VS_K^∞ is a strongly minimal theory.

Exercise 8.93 Show that DAG^+ is a strongly minimal theory.

Exercise 8.94 Let $\mathcal{V}$ be a model of VS_K^∞, where K is an infinite field.

(i) Show that the only definable *subspaces* of $\mathcal{V}$ are $\mathcal{V}$ itself and the subspace consisting of the null vector alone.
(ii) If $\mathcal{V}$ is at least two-dimensional, show that there is no $\mathcal{L}_v$-formula $\varphi(x_1, x_2)$ that holds iff x_1, x_2 are linearly independent.

Exercise 8.95 Let T be a $\mathcal{L}$-theory with quantifier elimination. The following statements are equivalent:

(1) T is strongly minimal;
(2) for any $\mathcal{M} \models T$, any atomic formula $\varphi(x, y_1, \ldots, y_n)$ and any $a_1, \ldots, a_n \in M$, the parametric formula $\varphi(x, a_1, \ldots, a_n)$ defines a finite or cofinite subset of M.

Exercise 8.96 Use the previous exercise to show that ACF is a strongly minimal theory. It follows that ACF_0 and, for each prime p, ACF_p are strongly minimal theories.

8.8.1 Minimality and Dimension

If $\mathcal{M}$ is a $\mathcal{L}$-structure and $\varphi(x)$ a $\mathcal{L}$-formula with parameters in M, we refer to the set defined by $\varphi(x)$ in M as $\varphi(M) = \{a \in M : \mathcal{M} \models \varphi(a)\}$ and call it a **minimal set**. If the parameters of $\varphi(x)$ all come from a subset $B \subseteq M$, then we say that the set $\varphi(M)$ is **definable over** B. Minimal sets have the exchange property and thus support the dimension theory developed at the end of Chap. 7.

Theorem 8.97 (Exchange Lemma) *Let $\mathcal{M}$ be a $\mathcal{L}$-structure and $C \subseteq M$ a minimal set definable over $B \subseteq M$. If $a, b \in C$, then:*

$$a \in acl_{\mathcal{M}}(B \cup \{b\}) \text{ and } a \notin acl_{\mathcal{M}}(B) \text{ jointly imply } b \in acl_{\mathcal{M}}(B \cup \{a\}).$$

Proof We expand the language to include constant symbols naming every element of B. In this case our hypothesis reduces to:

$$a \in acl_{\mathcal{M}}(b) \text{ and } a \notin acl_{\mathcal{M}}(\emptyset).$$

Instead of showing $b \in acl_{\mathcal{M}}(a)$ directly, we derive a contradiction from the additional assumption $b \notin acl_{\mathcal{M}}(a)$.

Because $a \in acl_{\mathcal{M}}(b)$, there is a formula $\varphi(x, b)$ that is algebraic over $\{b\}$. We may suppose that $\varphi(x, b)$ is satisfied by n distinct elements of M. Thus:

$$\mathcal{M} \models \varphi(a, b) \wedge \exists^{=n} x \varphi(x, b)$$

We next call $\psi(a, y)$ the formula:

$$\varphi(a, y) \wedge \exists^{=n} x \varphi(x, y)$$

in which a occurs as a parameter. Because $b \notin acl_{\mathcal{M}}(a)$ and $\mathcal{M} \models \psi(a, b)$, the formula $\psi(a, y)$ cannot be algebraic over $\{a\}$. Now suppose C is defined by the formula $\theta(x)$. Because C is minimal, either $\theta(M) \cap \psi(a, M)$ is finite or $\theta(M) \cap \neg\psi(a, M)$ is. Since, however $\mathcal{M} \models \theta(b) \wedge \psi(a, b)$, the formula $\theta(y) \wedge \psi(a, y)$

8.8 Strongly Minimal Theories

cannot be algebraic. Consequently $\theta(M) \cap \psi(a, M)$ is cofinite. This is to say that, for all but finitely many $c \in C$:

$$M \models \varphi(a, c) \wedge \exists^{=n} x \varphi(x, c).$$

The assumption $a \notin acl_M(\emptyset)$ implies that the formula:

$$\forall y((\theta(y) \wedge \exists u \psi(u, y)) \to \psi(x, y))$$

cannot be algebraic, because it holds in M when a is assigned to x (check the last statement). It follows there is an infinite sequence $\{a_i\}_{i \in \mathbb{N}} \subseteq M$ such that, for any $i \in \mathbb{N}$ and any $c \in C \cap \psi(a, M)$, $\psi(a_i, c)$ holds. In particular, $M \models \psi(a_0, c) \wedge \ldots \wedge \psi(a_n, c)$, which implies:

$$M \models \varphi(a_0, c) \wedge \ldots \wedge \varphi(a_n, c) \wedge \exists^{=n} x \varphi(x, c),$$

i.e. that at least $n + 1$ elements of M satisfy a condition that holds for exactly n elements of M. This contradiction proves the exchange property. □

When M models a $\mathcal{L}$-theory T that is strongly minimal, acl_M has the properties we have deduced in Chap. 7 as well as the exchange property. Thus, models of strongly minimal theory have a well-defined, model-theoretic dimension theory, which generalises the vectorial and field-theoretic notions of dimension.

An interesting consequence of this outcome is that we can obtain certain algebraic theorems as specialisations of purely model-theoretic results, thus uncovering their logical structure.

To illustrate this possibility, we reprove Steinitz's theorem, which we had arrived at by algebraic means in Chap. 7. The proof relies on two lemmas of independent interest.

Lemma 8.98 *Let M, N be $\mathcal{L}$-structures. If $X \subseteq M, Y \subseteq N$, then any elementary bijection $f : X \longrightarrow Y$ can be extended to an elementary embedding of $acl_M(X)$ into N whose image is $acl_N(Y)$.*

Proof Let $\{a_\delta : \delta \in \alpha\}$ be an enumeration of $acl_M(X) - X$ over the ordinal α. We use the latter enumeration to extend f to $acl_M(X)$ by transfinite recursion. Clearly, $f_0 = f$ and, if λ is a limit ordinal $f_\lambda = \bigcup_{\delta \in \lambda} f_\delta$. We only have to select an extending operation for successor ordinals.

Let us therefore assume that f_γ has been defined and that $\beta = \gamma + 1$ is the successor of α. Because $X \subseteq X \cup \{a_\delta : \delta \in \gamma\}$, $acl_M(X) \subseteq acl_M(X \cup \{a_\delta : \delta \in \gamma\})$ and there is an algebraic formula with parameters in $X \cup \{a_\delta : \delta \in \gamma\}$ that is satisfied by finitely many elements of M, including a_γ.

We choose an algebraic formula $\varphi(x, a_1, \ldots, a_m)(a_i \in X \cup \{a_\delta : \delta \in \gamma\})$ that is satisfied by the *least* finite number of elements of M, including a_γ. Let n be this least number. Since f_γ is elementary by hypothesis:

$$\mathcal{N} \models \exists^{=n} x \varphi(x, f_\gamma(a_1), \ldots, f_\gamma(a_n)).$$

Among the n elements that satisfy $\varphi(x, f_\gamma(a_1), \ldots, f_\gamma(a_n))$ in $\mathcal{N}$, we select any $b \in N$ that has not been used at an earlier stage of our construction and define $f_\beta = f_\gamma \cup \{(a_\gamma, b)\}$. It remains to show that f_β is elementary.

To this end, let $\psi(x, \overline{y})$ any $\mathcal{L}$-formula and $\overline{c}$ any matching tuple of parameters from M. Suppose $\mathcal{M} \models \psi(a, \overline{c})$: we want to show that $\mathcal{N} \models \psi(b, f_\gamma(\overline{c}))$.

Note that:

$$\mathcal{M} \models \forall x (\varphi(x, \overline{a}) \to \psi(x, \overline{c})),$$

where $\overline{a}$ is the tuple $a_1, \ldots, a_n$.

If not, there is $a' \in M$ such that $\mathcal{M} \models \varphi(a', \overline{a}) \to \neg \psi(a', \overline{c})$, so the parametric formula $\varphi(x, \overline{a}) \wedge \psi(x, \overline{c})$ is satisfied by $n-1$ elements of M including a_γ, against the choice of $\varphi(x, \overline{a})$. This contradiction implies:

$$\mathcal{M} \models \forall x (\varphi(x, \overline{a}) \to \psi(x, \overline{c})) \text{ and } \mathcal{N} \models \forall x (\varphi(x, f_\gamma(\overline{a})) \to \psi(x, f_\gamma(\overline{c}))).$$

Since $\mathcal{N} \models \varphi(b, f_\gamma(\overline{a}))$ by our choice of b, we conclude $\mathcal{N} \models \psi(b, f_\gamma(\overline{c}))$. The argument is reversible and shows that f_β is elementary.

We have obtained an elementary function at each stage of our transfinite recursion. The union of all stages:

$$f^* = \bigcup_{\beta \in \alpha} f_\beta$$

is therefore elementary and thus injective. Furthermore, its domain is $acl_\mathcal{M}(X)$. As for its image, it certainly is included in $acl_\mathcal{N}(Y)$. It remains to prove that f^* is surjective. If $b \in acl_\mathcal{N}(Y)$, there is a formula $\theta(x, f^*(\overline{a}))$, with $\overline{a}$ a tuple of parameters in X, that is satisfied by b and has m solutions. Consequently, $\mathcal{M} \models \exists^{=m} x \theta(x, \overline{a})$.

Because there are m elements of $acl_\mathcal{M}(X)$ that take f^*-values in $acl_\mathcal{N}(Y)$ and these elements are the only ones to satisfy $\theta(x, \overline{a})$, one of them must take the value b under f^*. Thus f^* is surjective. □

The last lemma almost enables us to prove that, if two models of a strongly minimal theory T have the same dimension, then they are isomorphic. The proof strategy that suggests itself is to use Lemma 8.98 to extend a bijection between two bases to an elementary map between their algebraic closures. For this strategy to work, we require an elementary bijection between two bases. The next lemma, which rests on the notion of an independent sequence, guarantees its existence.

Definition 8.99 Let $\mathcal{M}$ be a $\mathcal{L}$-structure, A be a minimal subset of M and $X \subseteq M$. An **independent sequence** of A over X is an ordinal sequence $\{a_\beta : \beta \in \alpha\} \subseteq A$ such that, for each $\gamma \in \alpha$, $a_\gamma \notin acl_\mathcal{M}(X \cup \{a_\delta : \delta \in \gamma\})$. When $X = \emptyset$, we speak of an independent sequence *simpliciter*.

8.8 Strongly Minimal Theories

Lemma 8.100 *Let $\mathcal{M} \equiv \mathcal{N}$ be $\mathcal{L}$-structures with A minimal in $\mathcal{M}$ and B minimal in $\mathcal{N}$. Suppose, moreover, that $\{a_\delta : \delta \in \beta\} \subseteq A$ is independent over $X \subseteq A$ and $\{b_\delta : \delta \in \beta\}$ is independent over $Y \subseteq B$.*

Any elementary bijection f from X to Y can be extended to an elementary bijection from $X \cup \{a_\delta : \delta \in \beta\}$ to $Y \cup \{b_\delta : \delta \in \beta\}$.

Proof Note that, if $X = \emptyset$, then $Y = \emptyset$ and the hypothesis $\mathcal{M} \equiv \mathcal{N}$ implies that the empty map $\emptyset$ is elementary (it preserves all formulae in which no free variables occur). Thus, the lemma works even if X, Y are empty. We set:

$$g(a_\delta) = b_\delta \text{ for each } \delta \in \beta,$$

and use the notation g_γ to designate the restriction $g \restriction \gamma$, with $\gamma \in \beta$. The proof will be completed when we have shown by transfinite induction up to β that each g_γ is elementary. This is clear for $g_0 = f$ and for g_λ with λ a limit ordinal, because a union of elementary maps is elementary.

The only nontrivial case concerns successor ordinals. We thus suppose g_γ elementary and consider $g_{\gamma+1} = g_\gamma \cup \{(a_\gamma, b_\gamma)\}$. Given an arbitrary formula $\psi(x, \bar{y}, \bar{z})$, our goal is to show that, for any tuple $\bar{c}$ from X and tuple $\bar{d}$ from $\{a_\delta : \delta \in \gamma\}$:

$$\mathcal{M} \models \psi(a_\gamma, \bar{c}, \bar{d}) \iff \mathcal{N} \models \psi(b_\gamma, g_\gamma(\bar{c}), g_\gamma(\bar{d})). \tag{8.2}$$

Suppose $\mathcal{M} \models \psi(a_\gamma, \bar{c}, \bar{d})$ and let $\theta(x)$ define A (we omit parameters for notational convenience). Then $\theta(x) \wedge \psi(x, \bar{c}, \bar{d})$ can only define a finite or a cofinite subset of A. But $a_\gamma \in A$ and a_γ is independent of $X \cup \{a_\delta : \delta \in \gamma\}$: since a_γ cannot lie in the algebraic closure of the latter set, it follows that $\theta(x) \wedge \psi(x, \bar{c}, \bar{d})$ is cofinite.

Let $A^* \subseteq A$ be the set defined by $\theta(x) \wedge \psi(x, \bar{c}, \bar{d})$. Set $g_\gamma[A^*] = B^* \subseteq B$. Since g_γ is elementary and thus injective, B^* is an infinite subset of B.

If $b \in B^*$, then b satisfies the formula $\eta(x) \wedge \psi(x, g_\gamma(\bar{c}), g_\gamma(\bar{d}))$, where $\eta(x)$ defines B. It follows that $\eta(x) \wedge \psi(x, g_\gamma(\bar{c}), g_\gamma(\bar{d}))$ defines a cofinite subset of B. Consequently, $\eta(x) \wedge \neg\psi(x, g_\gamma(\bar{c}), g_\gamma(\bar{d}))$ defines a finite subset of B, which cannot contain b_γ, or b_γ would belong to the algebraic closure of $Y \cup \{b_\delta : \delta \in \gamma\}$ in $\mathcal{N}$.

Thus:

$$\mathcal{N} \models \psi(b_\gamma, g_\gamma(\bar{c}), g_\gamma(\bar{d})).$$

We can run exactly the same argument from the initial assumption $\mathcal{M} \models \neg\psi(a_\gamma, \bar{c}, \bar{d})$ to obtain the right-to-left implication in (8.2) via contraposition. The proof is complete. □

We can now abstractly prove what we have severally verified for vector spaces and algebraically closed fields, namely that bijections between bases and transcendence bases respectively extend to isomorphisms.

Theorem 8.101 *Let $\mathcal{M}, \mathcal{N}$ be models of a complete, strongly minimal theory T. $\mathcal{M}, \mathcal{N}$ are isomorphic iff $dim(\mathcal{M}) = dim(\mathcal{N})$.*

Proof Because T is complete, $\mathcal{M} \equiv \mathcal{N}$. Thus, $\emptyset$ is an elementary bijection from M to N. Let B_M be a basis for $\mathcal{M}$ and B_N a basis for $\mathcal{N}$. Once we fix an enumeration for each basis, we obtain two independent sequences of the same length. Lemma 8.100 with $X = Y = \emptyset$ now applies to yield an elementary bijection f between B_M and B_N.

Since $acl_\mathcal{M}(B_M)$ is $\mathcal{M}$ and $acl_\mathcal{M}(B_M)$ is $\mathcal{N}$, Lemma 8.98 ensures that f can be extended to of an elementary bijection between $\mathcal{M}, \mathcal{N}$. It follows that these structures are isomorphic.

If, on the other hand, $\mathcal{M}, \mathcal{N}$ are isomorphic, it is easy to see that they must have the same dimension: a proof is left to the reader. □

Exercise 8.102 Complete the proof of Theorem 8.101

When specialised to vector spaces, Theorem 8.101 shows that two vector spaces of the same dimension are isomorphic. Steinitz's theorem, to the effect that algebraically closed fields with the same characteristic and transcendence bases of the same size are isomorphic, is also a consequence of Theorem 8.101, via the next result.

Theorem 8.103 *Let T be a countable, complete and strongly minimal $\mathcal{L}$-theory. For any $\lambda > \aleph_0$, T is λ-categorical.*

Proof Let $\mathcal{M}, \mathcal{N}$ be models of T such that $|M| = \lambda = |N|$. If S is a basis for $\mathcal{M}$, then $acl_\mathcal{M}(S) = M$. Because the language of T is countable, if $|S| = \kappa$ there are at most κ $\mathcal{L}$-formulae with parameters from S. Thus, in particular, there are at most κ algebraic $\mathcal{L}$-formulae, each of which defines a finite subset of M. It follows that $acl_\mathcal{M}(S)$ is a set of size κ and, since it coincides with M, we must have $|S| = \kappa = \lambda$. The same argument shows that the dimension of $\mathcal{N}$ must be λ as well. By Theorem 8.101, $\mathcal{M} \simeq \mathcal{N}$. □

Corollary 8.104 *The theories ACF_0 and ACF_p, for each prime p, are uncountably categorical.*

In Chaps. 10 and 11 we shall encounter further examples of model-theoretic results that have mathematically significant algebraic corollaries or specialisations.

Exercise 8.105 Let $\mathcal{L}$ be the language of signature $\{E\}$, where E is a binary relation symbol, and $\mathcal{M}$ a $\mathcal{L}$-structure in which E^M is an equivalence relation. Suppose that each equivalence class determined by E^M is finite and that there is exactly one equivalence class of size n for each $n \geq 1$. Show that $\mathcal{M}$ is minimal but not strongly minimal.

8.9 o-Minimal Theories

If the $\mathcal{L}_<$-structure $\mathcal{M}$ is a linear order without endpoints, its atomic formulae must define more than the finite and cofinite subsets of M. For instance, the formula $a < x$, with $a \in M$, defines a subset of M that is infinite but not cofinite.

Thus, we cannot in general expect strongly minimal theories in presence of quantifier elimination for a theory of linearly ordered structures. The definable subsets admit nonetheless a simple characterisation.

Definition 8.106 Let $\mathcal{M}$ be a linear order. A subset of M defined by a parametric equality of the form $x = a$, with $a \in M$, is a **point** of M. A subset of M is an **interval** iff it is defined by one of the following parametric formulae: $x < a, a < x$, $a < x \wedge x < b$, for any $a, b \in M$, or, alternatively, their counterparts in which not all inequalities are strict.[2] A $\mathcal{L}$-theory is said to be **o-minimal** iff, for any $\mathcal{M} \models T$ the parametrically definable subsets of $\mathcal{M}$ are finite unions of points and intervals.

Example 8.107 Let $\mathcal{M}$ be a model of $\mathsf{DLO}_{-,-}$. If $\varphi(x)$ is a $\mathcal{L}_<$-formula with parameters over M, we know that it is $\mathsf{DLO}_{-,-}$-equivalent to a quantifier-free $\mathcal{L}_<$-formula $\eta(x)$ with the same parameters. The formula $\eta(x)$ may be assumed in disjunctive normal form: let its disjuncts be $\eta_1(x), \ldots, \eta_k(x)$. Then $\eta(x)$ defines the union of the subsets of M defined by each $\eta_j(x)$, with $j = 1, \ldots, k$.

Each disjunct $\eta_j(x)$ defines, in turn, the intersection of the subsets of M defined by its conjuncts. By trichotomy, we may suppose that only positive literals occur in $\eta_j(x)$. If $a \in M$ is a parameter, a positive literal is of the form $x = a$, in which case it defines $\{a\}$, or of the form $a < x$ or $x < a$, in which case it defines an interval.

The intersection of sets defined by positive literals is itself a point or an interval (with or without endpoints). It follows that the definable subsets of M are exactly the finite unions of points or intervals. This is to say that $\mathsf{DLO}_{-,-}$ is o-minimal.

The theory $\mathsf{DAG}^+_<$ is also o-minimal by quantifier elimination, as the reader may verify. In fact, o-minimality characterises the ordered, divisible abelian groups. This result depends on the penury of the definable subgroups included in any o-minimal, ordered group.

Theorem 8.108 Let $\mathcal{G}$ be an o-minimal ordered group and $\mathcal{H} \sqsubseteq \mathcal{G}$ a definable subgroup. Then $H = \{0\}$ or $\mathcal{G} = \mathcal{H}$.

Proof In the following proof, we drop superscripts and write $<, -, +$ instead of $<^G, -^G, +^G$. We also use $x > y$ as a variant of $y < x$. Suppose $H \neq \{0\}$. It follows that H is infinite. By o-minimality, it is possible to represent H as a finite union of points and intervals. Since unions of overlapping intervals are intervals, we

---

[2] Any one strict inequality may thus be replaced by a formula of the form $x = r \vee x < r$ or of the form $x = r \vee r < x$, where $r \in \{a, b\}$.

may represent H as a minimal union of pairwise disjoint intervals. For convenience, we take points to be degenerate intervals. In symbols, H has a representation of the following form:

$$H = H_1 \cup \ldots \cup H_n, \text{ where } H_1, \ldots, H_n \text{ are intervals and n is least.}$$

Not every one of the H_i can be a point, or H would be finite. In other words, at least one of the intervals partitioning H is non-degenerate. We next observe that, since $0 \in H$ and H is the domain of a subgroup of $\mathcal{G}$, H must be symmetrically distributed relative to 0, since $0 < a \in H$ implies $0 > -a \in H$ and vice versa.

Toward a contradiction, we suppose $H \neq G$.

If H_j is a non-degenerate interval containing 0, then it is of the form $[-a, a]$ or $(-a, a)$. In principle, we might have no interval of this form but e.g. a pair of non-degenerate intervals H_j, H_k of the form $[-d, -c]$, $[c, d]$ respectively, with $-c < 0 < c$, or a pair of intervals of a similar form but lacking one or both endpoints, e.g. $(-d, -c)$, (c, d). By symmetry, it suffices to focus only on the right interval of each such pair.

We focus on the case $H_k = (c, d]$. Because $(c, d]$ is non-degenerate, there is $a \in (c, d)$. We consider any $b \in (d, d + (d - a)]$ and show that $b \in H$. Clearly, $b - (d - a) < (d + (d - a)) - (d - a) = d$ and $b - (d - a) > d - (d - a) = a > c$. In other words $b - (d - a) \in H_k \subseteq H$. By the closure properties of H, $b - (d - a) + (d - a) = b \in H$. It now follows that at least the whole segment $(c, d + (d - a)]$ lies in H. Because this segment strictly includes H_k, it must overlap some other segment in the given partition of H. It follows that the given partition of H can be reduced to fewer than n components, against the choice of n as least.

If $H_k = (c, d)$, the fact that (c, d) is non-degenerate implies the existence of distinct $a, e \in (c, d)$. We can reduce ourselves to the previous case by showing that $d \in H$. We may assume $a < e$, in which case we observe that $(d - e) + a < (d - a) + a = d$ and that, since $c < a$ and $0 < d - e, c < (d - e) + a$. We conclude again that $(d - e) + a \in H_j \subseteq H$. The fact that H is the domain of a subgroup containing a, e implies $d \in H$.

A similar approach handles bounded intervals containing 0. If H_j is unbounded of the form $(-\infty, d)$ with $d < 0$ (otherwise $H = G$ trivially), we can deduce $d \in H$ as we did before. If H_j is unbounded of the form $(d, +\infty)$ with $0 < d$, we may consider $(-\infty, -d) \subseteq H$, prove $-d \in H$, which implies $d \in H$. In each possible case we arrive at a contradiction by showing that the partition chosen to describe H is not minimal. It follows that $H = G$. □

Theorem 8.109 *Let $\mathcal{G}$ be a non-degenerate, ordered group: $\mathcal{G}$ is o-minimal iff $\mathcal{G} \models$ DAG$_<^+$.*

Proof It suffices to show that $\mathcal{G}$ is abelian and divisible. To prove that $\mathcal{G}$ is abelian, we consider, for each $a \in G$, the formula $y + a = a + y$, which defines the set of group elements commuting with a. It is easy to see that the latter set is a subgroup $\mathcal{H}_a$ of $\mathcal{G}$. When $a \neq 0$, Theorem 8.108 applies and $\mathcal{H}_a = \mathcal{G}$ for each $a \in G$. It follows that any two $a, b \in G$ commute (because e.g. $b \in H_a$), i.e. $\mathcal{G}$ is abelian. We

8.9 o-Minimal Theories

now show divisibility. If $n \geq 1$, the formula $\exists y(ny = x)$ defines the subgroup $\mathcal{H}_n$ of 'n-th parts'. Because $\mathcal{G}$ is non-degenerate, there is $a \neq 0$ in G. Thus $0 \neq na \in \mathcal{H}_n$ for each $n \geq 1$ and Theorem 8.108 implies $\mathcal{G} = \mathcal{H}_n$ for each $n \geq 1$. In other words, $\mathcal{G}$ is divisible. □

Our study of o-minimal structures will continue in Chap. 11, where we discuss the theory of ordered, real closed fields.

Chapter 9
Types

Abstract If $\mathcal{M}$ be a $\mathcal{L}$-structure, $Th\mathcal{M}$ is a complete $\mathcal{L}$-theory that carries global information about $\mathcal{M}$. A finer analysis of $\mathcal{M}$ may be effected by including the first-order information carried by the elements of M or M^n. For instance, if Φ is the set of all $\mathcal{L}$-formulae, by fixing $a \in M$ we determine the set $p(x) = \{\varphi(x) \in \Phi : \mathcal{M} \models \varphi(a)\}$.

Note that $p(x) \supseteq Th\mathcal{M}$ and, for any $\mathcal{L}$-formula $\psi(x)$ in which at most x occurs free, either $\psi(x) \in p(x)$ or $\neg\psi(x) \in p(x)$. Because of this, we may regard $p(x)$ as a generalisation of $Th\mathcal{M}$ that records first-order information at the level of a single element of $\mathcal{M}$.

We call this generalisation a (complete) *type*. In this chapter we use types to refine our study of $\mathcal{L}$-structures. Sections 9.1 and 9.2 introduce key definitions and single out types with special properties. In Sect. 9.3 we briefly look at an application of types outside model theory, which is closely connected with the final sections of Chap. 12.

Section 9.4 is devoted to the connection between types and elementary maps, which prepares the ground for the study of homogeneity in Sect. 9.5. Section 9.6 focusses on atomic structures: it includes the proofs of two fundamental results, i.e. the Omitting Types theorem and the Ryll-Nardzewski theorem. Section 9.7 discusses saturated structures and closes with a discussion of large model-theoretic frameworks known as monster models.

9.1 Key Definitions

We generalise the notion of a $\mathcal{L}$-theory by considering, instead of a satisfiable, or, equivalently, finitely satisfiable set of sentences, a finitely satisfiable set of formulae in a fixed tuple $\overline{x}$ of distinct variables.

Definition 9.1 Let $\overline{x}$ be a tuple of distinct variables and $p(\overline{x})$ be a set of $\mathcal{L}$-formulae whose free variables figure in the list $\overline{x}$. We say that $p(\overline{x})$ is a **type** iff it is finitely satisfiable, i.e., for any finite subset $S \subseteq p(\overline{x})$, there are a $\mathcal{L}$-structure $\mathcal{M}$ and $\overline{a} \in M$ such that $\mathcal{M} \models \psi(\overline{a})$, for each $\psi(\overline{x}) \in S$.

If $\bar{x}$ is a n-tuple, we refer to $\bar{x}$ as a n-**type**. If, moreover, for every $\mathcal{L}$-formula $\varphi(\bar{x})$ whose free variables occur in the list $\bar{x}$, $\varphi(\bar{x}) \in p(\bar{x})$ or $\neg \varphi(\bar{x}) \in p(\bar{x})$, then we say that $p(\bar{x})$ is a **complete** n-**type**.

Exercise 9.2 Let $\mathcal{M}$ be a $\mathcal{L}$-structure and $a \in M$. Show that the set $\{\varphi(x) : \mathcal{M} \models \varphi(a)\}$, where $\varphi(x)$ is a $\mathcal{L}$-formula in which at most x occurs free, is a complete 1-type.

Exercise 9.3 Show that, for any $n \geq 1$, every n-type can be extended to a complete n-type.

When we determine a complete n-type relative to a particular structure and an element from the domain of that structure, we replace the notation $p(\bar{x})$ with the more specific alternative $tp_\mathcal{M}(a)$. If we had been selecting a n-tuple $\bar{a}$ from M, we would have used the obvious variant $tp_\mathcal{M}(\bar{a})$ for the corresponding, complete n-type.

Compactness guarantees that any n-type $p(\bar{x})$ is satisfied in some $\mathcal{L}$-structure. When working with the types associated with a given $\mathcal{L}$-structure, we do not restrict attention to $\mathcal{L}$-formulae alone but allow parameters, too.

Definition 9.4 Let $\mathcal{M}$ be a $\mathcal{L}$-structure, $A \subseteq M$ and $\bar{x}$ a n-tuple of distinct variables. A n-**type** $p(\bar{x})$ **of** $\mathcal{M}$ **over** A is a set of $\mathcal{L}$-formulae with parameters in A and free variables among $\bar{x}$ such that any finite subset of $p(\bar{x})$ is satisfied by $\mathcal{M}$.

It follows from Definition 9.4 that, if $p(\bar{x})$ is a n-type $p(\bar{x})$ of $\mathcal{M}$ over A and $\psi_1(\bar{x}), \ldots, \psi_k(\bar{x}) \in p(\bar{x})$, then $\mathcal{M} \models \exists \bar{x}(\psi_1(\bar{x}) \wedge \ldots \wedge \psi_k(\bar{x}))$. Conversely, if the last condition holds for every finite subset of $p(\bar{x})$, the latter is a n-type of $\mathcal{M}$ over A.

It follows that $p(\bar{x})$ is a type of $\mathcal{M}$ over A iff $p(\bar{x})$ is finitely consistent with $\mathcal{ED}(\mathcal{M})$. Definition 9.4 thus invites us to consider the elementary extensions of a fixed structure $\mathcal{M}$. We also allow types to depend on a theory, rather than a structure.

Definition 9.5 Let T be a $\mathcal{L}$-theory, $\bar{x}$ a n-tuple of variables. A n-type $p(\bar{x})$ of T is any set of $\mathcal{L}$-formulae finitely consistent with T (i.e. the union of T and any finite subset of $p(\bar{x})$ has a model).

Since a type $p(\bar{x})$ may or may not be satisfied by a structure $\mathcal{M}$, we introduce a special terminology to highlight these distinct possibilities.

Definition 9.6 Let $p(\bar{x})$ be a n-type and $\mathcal{M}$ be a structure for the language of $p(\bar{x})$. If, there is a n-tuple $\bar{a} \in M$ such that, for every $\psi(\bar{x})$, $\mathcal{M} \models \psi(\bar{a})$, we say that $\mathcal{M}$ **realises** $p(\bar{x})$. Otherwise, we say that $\mathcal{M}$ **omits** $p(\bar{x})$.

Exercise 9.7 Exhibit a $\mathcal{L}$-structure $\mathcal{M}$ and a 1-type that $\mathcal{M}$ omits.

Before turning to a general discussion of types, we provide an example that will play a significant role in Sect. 9.3.

9.1 Key Definitions

Example 9.8 Consider the $\mathcal{L}_<$-structure $(\mathbb{Q}, <^\mathbb{Q})$. If we allow parameters over $\mathbb{Q}$, we can easily describe 1-types that are not realised in $\mathbb{Q}$. Consider for instance the set $p(x)$ of inequalities of the form $a < x, x < b$, where $a, b \in \mathbb{Q}$ are such that $a^2 <^\mathbb{Q} 2$ and $2 <^\mathbb{Q} b^2$. Density implies that any finite subset of $p(x)$ is realised in $\mathbb{Q}$, i.e. that $p(x)$ is a type. A realisation c of $p(x)$ should satisfy the equality $c^2 = 2$, so it cannot be a rational number. It is easy to see that $p(x)$ is realised in the elementary extension $(\mathbb{R}, <^\mathbb{R})$ of $(\mathbb{Q}, <^\mathbb{Q})$. A similar argument works for any set of inequalities that describes upper and lower rational approximations of a fixed irrational number.

Exercise 9.9 Find infinitely many 1-types in $(\mathbb{Q}, <^\mathbb{Q})$ over $\mathbb{Q}$ that are not realised in the elementary extension $(\mathbb{R}, <^\mathbb{Q})$.

Exercise 9.10 Suppose that $\mathcal{M} \models \mathsf{DLO}_{-,-}$ and that $p(x)$ is a complete 1-type in the language $\mathcal{L}_<$. Show that, if $\mathcal{M}$ realises $p(x)$ and $q(x)$ is a 1-type containing the same inequalities as $p(x)$, then $\mathcal{M}$ realises $q(x)$.

Exercise 9.11 Given a $\mathcal{L}$-structure $\mathcal{M}$ and $A \subseteq M$, use compactness and a suitable expansion of $\mathcal{L}_A$ to prove the following statement: there is an elementary $\mathcal{L}_A$-extension $\mathcal{N}$ of $\mathcal{M}^A$, the $\mathcal{L}_A$-expansion of $\mathcal{M}$, in which every complete n-type of $\mathcal{M}$ over A is realised.

We use the notation $S_n(\mathcal{M}/A)$ to denote the set of all complete n-types over A in $\mathcal{M}$: when $A = \emptyset$, we adopt the notation $S_n(\mathcal{M})$.

Exercise 9.12 Let $[\varphi(\overline{x})] \subseteq S_n(\mathcal{M}/A)$ be the set of all complete n-types of $\mathcal{M}$ over A that contain the formula $\varphi(\overline{x})$. We shall refer to subsets of $S_n(\mathcal{M}/A)$ of the form $[\varphi(\overline{x})] \subseteq S_n(\mathcal{M}/A)$ as **basic open sets**.

(i) Show that $\emptyset$ and $S_n(\mathcal{M}/A)$ are basic open sets.
(ii) Show that the intersection of two basic open sets is a basic open set and that the union of two basic open sets is a basic open set.
(iii) Define an **open set** of $S_n(\mathcal{M}/A)$ to be a union of basic open sets. Show that the intersection of two open sets is an open set and that, if $\{U_i\}_{i \in I}$ is a collection of open sets, then their union is an open set.
(iv) The complement of an open set is called a **closed set**. Show that any basic open set is both open and closed, i.e. it is a **clopen** set.
(v) Let $\mathsf{U} = \{U_i\}_{i \in I}$ be a family of open subsets of $S_n(\mathcal{M}/A)$ whose union equals $S_n(\mathcal{M}/A)$. Use compactness to show that $S_n(\mathcal{M}/A)$ is already the union of finitely many open sets from U (*Hint*: argue by contradiction, using the fact that each U_i is a union of basic open sets).

Given a set S, a **topological space** over S is a pair of sets (S, τ), with $\tau \subseteq \mathcal{P}(S)$, such that $\emptyset, S \in \tau$ and τ contains the finite intersections as well as the arbitrary unions of its elements. The elements of τ are called **open sets**.

By the previous exercise, $(S_n(\mathcal{M}/A), \tau)$ is a topological space. Part (v) of that exercise establishes a fundamental property of this particular topological space, namely the fact that it has the finite subcover property. A **cover** of $S_n(\mathcal{M}/A)$ is a family $\mathsf{U} = \{U_i\}_{i \in I}$ of open sets such that $\bigcup_{i \in I} U_i = S_n(M/A)$. A **finite subcover** of U is a finite subset of U whose union is still $S_n(M/A)$. We have shown that any cover U of $S_n(M/A)$ contains a finite subcover. This topological property is also known as compactness (and, if adopted as a hypothesis, could be used to prove the compactness theorem).

Moreover, for any 'point' $p(\overline{x})$ of $S_n(M/A)$, it is possible to find a partition of the space into disjoint open sets U, V such that $x \in U$ and $x \notin V$ (if $p(\overline{x}) \in [\varphi(\overline{x})]$, consider $[\varphi(\overline{x})]$ and $[\neg\varphi(\overline{x})]$): we say that $S_n(M/A)$ is **totally disconnected**. Compact, totally disconnected topological spaces are known as **Stone spaces**. Spaces of complete types are thus concrete examples of Stone spaces.

9.1.1 The Space $S_n(T)$

Let T be a first-order theory. The Stone space $S_n(T)$ may be regarded as a topological space exactly as $S_n(M/A)$, since we can work with the same notion of basic open set (we leave the details to the reader). When T is a complete theory and $\mathcal{M} \models T$, the spaces $S_n(T)$ and $S_n(M)$ are closely related. If $p(\overline{x}) \in S_n(M)$, then $p(\overline{x})$ is finitely consistent with $Th\mathcal{M} = T$, so $p(\overline{x}) \in S_n(T)$. Conversely, if $p(\overline{x}) \in S_n(T)$, then there is $\mathcal{N} \models T$ that realises $p(\overline{x})$. By completeness, $\mathcal{M}$ and $\mathcal{N}$ can be embedded into a joint elementary extension (recall Theorem 7.15), which realises $p(\overline{x})$. It follows that $p(\overline{x})$ is realised in an elementary extension of $\mathcal{M}$, i.e. $p(\overline{x}) \in S_n(M)$. We conclude that $S_n(T) = S_n(M)$.

We next turn to types that can be uniquely determined by a formula, which will play an important role in the sections to follow.

9.2 Isolated and Algebraic Types

Suppose that $\mathcal{F} \models \mathsf{ACF}$ and that $\mathcal{K}$ is a subfield of $\mathcal{F}$. Thus, in particular, $\mathcal{K}, \mathcal{F}$ are $\mathcal{L}_r$-structures. Let $a \in F$ be algebraic over $\mathcal{K}$. We use the fact that ACF has quantifier elimination to describe the complete 1-type $tp_\mathcal{F}(a/K)$.

First, we note that every $\mathcal{L}_r$-formula $\varphi(x)$ with parameters over K is equivalent to a quantifier-free formula in $\mathcal{F}$ with the same parameters over K. Moreover, if $\varphi(x)$ is an atomic $\mathcal{L}_r$-formula other than a sentence, then $\varphi(x)$ is a polynomial equation of the form $0 = a_0 + a_1 x + \ldots + a_n x^n$.

The complete 1-type $tp_\mathcal{F}(a/K)$ is uniquely determined by its polynomial equations, which fix its quantifier-free elements and thus, up to equivalence, the entire type. We can thus reduce the problem of determining which formulae occur in $tp_\mathcal{F}(a/K)$ to the problem of determining which polynomial equations occur in this type. The problem admits a further reduction, as the next exercise shows.

9.2 Isolated and Algebraic Types

Exercise 9.13 Let $\mathcal{F} \models \mathsf{ACF}$, $\mathcal{K} \sqsubseteq \mathcal{F}$, $a \in F$ algebraic over K with minmum polynomial $m(x)$.

  (i) Show that, if $b \neq a$ is a root of $m(x)$ in F, then $m(x)$ is the minimum polynomial of b over K.
  (ii) Show that, if $f(x) = 0$ is a polynomial equation in $tp_\mathcal{F}(a/K)$, then $\mathcal{F} \models \forall x(m(x) = 0 \to f(x) = 0)$.
  (iii) Show that, if $f(x) = 0$ is a polynomial equation not in $tp_\mathcal{F}(a/K)$, then $\mathcal{F} \models \forall x(m(x) = 0 \to \neg(f(x) = 0))$.
  (iv) Deduce from part (iii) that, for any quantifier-free formula $\psi(x) \in tp_\mathcal{F}(a/K)$, $\mathcal{F} \models \forall x(m(x) = 0 \to \psi(x))$.
  (v) Conclude from (iv) that $tp_\mathcal{F}(a/K)$ contains an algebraic formula that implies every other formula in the same 1-type.

In view of the last Exercise, we can say two significant things about $tp_\mathcal{F}(a/K)$: (i) $tp_\mathcal{F}(a/K)$ contains a formula $\varphi(x)$ (namely $m(x) = 0$) such that, for every $\psi(x) \in tp_\mathcal{F}(a/K)$, $\mathcal{F} \models \forall x(\varphi(x) \to \psi(x))$; (ii) the formula $\varphi(x)$ is algebraic. Conditions (i) and (ii) motivate the next general definitions.

Definition 9.14 Let $\mathcal{M}$ be a $\mathcal{L}$-structure, $A \subseteq M$ and $p(\overline{x}) \in S_n(M/A)$. We say that $p(\overline{x})$ is an **isolated type** of $S_n(M/A)$ iff, for any $\psi(\overline{x}) \in p(\overline{x})$, $\mathcal{M} \models \forall \overline{x}(\varphi(\overline{x}) \to \psi(\overline{x}))$.

Definition 9.15 Let $\mathcal{M}$ be a $\mathcal{L}$-structure, $A \subseteq M$ and $p(\overline{x}) \in S_n(M/A)$. We say that $p(\overline{x})$ is an **algebraic type** of $S_n(M/A)$ iff $p(\overline{x})$ contains an algebraic formula.

Algebraic types are isolated but, as we shall shortly see, isolated types may not be algebraic.

Exercise 9.16 Show that any algebraic type $p(\overline{x}) \in S_n(M/A)$ is isolated (*Hint*: consider the algebraic formula in $p(\overline{x})$ with the smallest number of solutions in M).

Exercise 9.17 Let $p(\overline{x})$ be an isolated n-type in $S_n(M/A)$ and $\varphi(\overline{x}) \in p(\overline{x})$ a formula such that, for each $\psi(\overline{x}) \in p(\overline{x})$, $\mathcal{M} \models \forall \overline{x}(\varphi(\overline{x}) \to \psi(\overline{x}))$. Show that $[\varphi(\overline{x})] = \{p(\overline{x})\}$. The last equality says that $p(\overline{x})$ is the only element of the basic open set $[\varphi(\overline{x})]$. We refer to $p(\overline{x})$ as an **isolated point** of the Stone space $S_n(M/A)$. Show that, if $p(\overline{x})$ is an isolated point of $S_n(M/A)$, then it is an isolated type.

Example 9.18 We describe an isolated type that is not algebraic. To this end, we consider the language $\mathcal{L}$ with signature $\{R\}$, where R is a binary relation symbol. Let T^∞ be the theory obtained by adding the following $\mathcal{L}$-sentences to IS:

(1) $\forall x\, Rxx$;
(2) $\forall x \forall y (\neg(x = y) \to (\neg Rxy \wedge \neg Ryx))$.

Every model of T^∞ consists of an infinite set of disjoint loops. It is easy to see that T^∞ is complete because any two models of the same size must be isomorphic. Since T^∞ has a finite relational vocabulary, it easily follows that it has quantifier elimination (the verification is left to the reader).

Fix $\mathcal{M} \models T^\infty$. By quantifier elimination, a complete 1-type $p(x) \in S_1(M)$ is uniquely determined by the atomic formulae it contains. The only atomic formulae are Rxx and $x = x$. If one of them were not to occur in $p(x)$, then $p(x)$ would not be finitely satisfiable in $\mathcal{M}$, against the fact that it is a 1-type in $\mathcal{M}$. It follows that $S_1 = \{p(x)\}$, where $p(x)$ contains both $x = x$ and $R(x,x)$.

It is clear that $p(x)$ is not algebraic because every element of M realises $p(x)$ and $|M| \geq \aleph_0$. By contrast, $p(x)$ is isolated in light of Exercise 9.17.

Exercise 9.19 Let $p(\bar{x}) \in S_n(M/A)$. Show that the following statements are equivalent:

(1) $\{p(\bar{x})\}$ is the basic open set $[\varphi(\bar{x})]$ (i,e. $p(\bar{x})$ is isolated);
(2) $\psi(\bar{x}) \in p(\bar{x}) \implies \mathcal{M} \models \forall \bar{x}(\varphi(\bar{x}) \to \psi(\bar{x}))$;
(3) $\psi(\bar{x}) \in p(\bar{x}) \iff \mathcal{M} \models \forall \bar{x}(\varphi(\bar{x}) \to \psi(\bar{x}))$.

Since, in view of the last exercise, a n-type is isolated exactly when it is the only n-type to contain a formula $\varphi(\bar{x})$, we say that $\varphi(\bar{x})$ **isolates** $p(\bar{x})$. Note that, if $\varphi(\bar{x}), \phi(\bar{x})$ isolate $p(\bar{x})$, then $\mathcal{M} \models \forall \bar{x}(\varphi(\bar{x}) \leftrightarrow \phi(\bar{x}))$.

Exercise 9.20 Let $\mathcal{M}$ be a $\mathcal{L}$-structure and $A \subseteq M$. Show that $\mathcal{M}$ realises each isolated type in $S_n(M/A)$. Deduce that $\mathcal{M}$ realises its algebraic types.

We conclude this section with a few simple results about isolated types that will be used later in the chapter. We adopt the notation $\bar{a}, \bar{b}$ to designate a n-tuple followed by a m-tuple, for some $n, m \in \mathbb{N}$.

Lemma 9.21 *Let $\mathcal{M}$ be a $\mathcal{L}$-structure and $A \subseteq M$. Then $tp_\mathcal{M}(\bar{a}, \bar{b}/A)$ is isolated iff $tp_\mathcal{M}(\bar{a}/A \cup \{\bar{b}\})$ and $tp_\mathcal{M}(\bar{b}/A)$ are isolated.*

Proof Suppose $tp_\mathcal{M}(\bar{a}/A \cup \{\bar{b}\})$ and $tp_\mathcal{M}(\bar{b}/A)$ are isolated by $\psi(\bar{x}, \bar{b}, \bar{c})$, with $\bar{c} \in A$, and by $\theta(\bar{y}, \bar{d})$, with $\bar{d} \in A$, respectively.

Then the formula $\psi(\bar{x}, \bar{y}, \bar{c}) \wedge \theta(\bar{y}, \bar{d})$ is in $tp_\mathcal{M}(\bar{a}, \bar{b}/A)$. It suffices to show that this formula isolates $tp_\mathcal{M}(\bar{a}, \bar{b}/A)$. To this end, we consider a formula $\zeta(\bar{x}, \bar{y}, \bar{e}) \in tp_\mathcal{M}(\bar{a}, \bar{b}/A)$, with $\bar{e} \in A$. We have the following chain of equivalences, which follow from Exercise 9.17 or the substitution theorem:

$$\zeta(\bar{x}, \bar{y}, \bar{e}) \in tp_\mathcal{M}(\bar{a}, \bar{b}/A) \iff \mathcal{M} \models \zeta(\bar{a}, \bar{b}, \bar{e})$$
$$\iff \mathcal{M} \models \forall \bar{x}(\psi(\bar{x}, \bar{b}, \bar{e}) \to \zeta(\bar{x}, \bar{b}, \bar{e}))$$
$$\iff \mathcal{M} \models \forall \bar{y}(\theta(\bar{y}, \bar{d}) \to \forall \bar{x}(\psi(\bar{x}, \bar{b}, \bar{e}) \to \zeta(\bar{x}, \bar{b}, \bar{e})))$$
$$\iff \mathcal{M} \models \forall \bar{y} \forall \bar{x}(\theta(\bar{y}, \bar{d}) \to \forall \bar{x}(\psi(\bar{x}, \bar{b}, \bar{e}) \to \zeta(\bar{x}, \bar{b}, \bar{e})))$$
$$\iff \mathcal{M} \models \forall \bar{x} \forall \bar{y}(\theta(\bar{y}, \bar{d}) \to (\psi(\bar{x}, \bar{b}, \bar{e}) \to \zeta(\bar{x}, \bar{b}, \bar{e})))$$
$$\iff \mathcal{M} \models \forall \bar{x} \forall \bar{y}((\theta(\bar{y}, \bar{d}) \wedge \psi(\bar{x}, \bar{b}, \bar{e})) \to \zeta(\bar{x}, \bar{b}, \bar{e}))$$
$$\iff \mathcal{M} \models \forall \bar{x} \forall \bar{y}((\psi(\bar{x}, \bar{b}, \bar{e}) \wedge \theta(\bar{y}, \bar{d})) \to \zeta(\bar{x}, \bar{b}, \bar{e}))$$

The converse implication is established by a similar approach (see Exercise 9.22-(iii)). □

Exercise 9.22

(i) Let $\mathcal{M}$ be a $\mathcal{L}$-structure and $A \subseteq B \subseteq M$. Show that, if $\varphi(\overline{x})$ isolates $tp_{\mathcal{M}}(\overline{a}/B)$, then $\varphi(\overline{x})$ isolates $tp_{\mathcal{M}}(\overline{a}/A)$.
(ii) Show that if $tp_{\mathcal{M}}(\overline{a}, \overline{b}/A)$ is isolated, then $tp_{\mathcal{M}}(\overline{a}/A)$ is.
(iii) Complete the proof of Lemma 9.21.
(iv) Show that, if $a \in acl_{\mathcal{M}}(A)$, then $tp_{\mathcal{M}}(a/A)$ is isolated.

We define the isolated types of a theory T as the isolated points of $S_n(T)$. This definition yields a characterisation that produces analogues of the results obtained in Exercise 9.17.

Lemma 9.23 *Let T be a $\mathcal{L}$-theory. The following statements are equivalent:*

(1) $p(\overline{x})$ is an isolated point of $S_n(T)$.
(2) There is $\varphi(\overline{x})$ such that $\psi(\overline{x}) \in p(\overline{x}) \implies T \models \forall \overline{x}(\varphi(\overline{x}) \rightarrow \psi(\overline{x}))$.
(3) There is $\varphi(\overline{x})$ such that $\psi(\overline{x}) \in p(\overline{x}) \iff T \models \forall \overline{x}(\varphi(\overline{x}) \rightarrow \psi(\overline{x}))$.

Exercise 9.24 Prove Lemma 9.23.

Suppose that $\varphi(\overline{x})$ isolates the n-type $p(\overline{x}) \in S_n(T)$. If $T \not\models \exists \overline{x} \varphi(\overline{x})$, $p(\overline{x})$ may not be realised in some model of T. Thus, isolated types do not necessarily have realisations in every model of an incomplete theory. If, however, T is complete, $\varphi(\overline{x})$ is satisfiable in some model of T and, consequently $T \models \exists \overline{x} \varphi(\overline{x})$, which in turn implies that $p(\overline{x})$ is realised in any model of T.

Exercise 9.25 Use the fact that $S_n(M/A)$ is a compact topological space to show that $|S_n(M/A)| < \aleph_0$ iff every point of $S_n(M/A)$ is isolated.

Exercise 9.26 Let $\mathcal{M}$ be a $\mathcal{L}$-structure with $A, B \subseteq M$.

(i) Show that, if $A \subseteq B$, then every non-algebraic n-type in $S_n(M/A)$ can be extended to a non-algebraic n-type in $S_n(M/B)$.
(ii) Show that $p(\overline{x}) \in S_n(M/A)$ is algebraic iff all of its realisations lie in $\mathcal{M}$ (*Hint*: algebraic formulae are preserved by elementary extensions).

Exercise 9.27 Suppose $\mathcal{M} \models \mathsf{IS}$ and $A \subseteq M$. How many non-algebraic types are there in $S_1(M/A)$? Answer the same question for $\mathcal{M} \models \mathsf{VS}_K^\infty$.

9.3 An Application to Measurement Theory

In this section we briefly pause to consider a little explored, yet very interesting, application of types to measurement theory. Our discussion is entirely based on [53] and [54]. Measurement theory is an abstract study of scaling processes,[1] which are typically, though not always, portrayed as embeddings of $\mathcal{L}$-structures codifying an empirical variable into a $\mathcal{L}$-structure with domain $\mathbb{R}$.

Example 9.28 Traditionally, pharmacists used two types of equal-arm balance, one of which weighed up to 2 ounces or approximately 56 grams, while the other weighed up to 2 pounds or approximately 907 grams. In order to weigh a prescription, a pharmacist would counterbalance it using a set of standard weights.

It is not difficult to associate the pharmacist's operations with a first-order language $\mathcal{L}_w$ of signature $\{<, \circ\}$, where $<$ is a binary relation symbol and $\circ$ a binary function symbol. Basic comparison of substances and standard weights may be naturally described by atomic $\mathcal{L}_w$-formulae. For instance the formula $x \circ x < y \circ y \circ y$ might codify the observation that twice the weight of x does not evenly balance three times the weight of y.

Since a classic contribution of Otto Hölder's (see [28]), it is known that each $\mathcal{L}_w$-structure from a class **M** we shall now describe has embeddings into the $\mathcal{L}_w$-structure $(\mathbb{R}, <^{\mathbb{R}}, +^{\mathbb{R}})$.

If $\mathcal{M} \in \mathbf{M}$, the $\mathcal{L}_<$-reduct of $\mathcal{M}$ is a linear order. Moreover, $\mathcal{M} \models \forall x \forall y \forall z (x < y \to (x \circ z < y \circ z))$ and the operation $\circ^M$ is associative. Thus $\mathcal{M}$ is an ordered semigroup. In addition, $\mathcal{M}$ models two first-order sentences known to measurement theorists as *positivity* and *right solvability*. The former sentence is $\forall x \forall y (x < x \circ y \land y < x \circ y)$ and the latter is $\forall x \forall y (x < y \to \exists z (x \circ z = y))$.

It follows that $\mathcal{M}$ is not a monoid, or positivity would fail, and therefore not a group. It might, however, be embedded in an ordered group of which it would constitute the positive part.

In addition to the first-order sentences mentioned so far, $\mathcal{M}$ satisfies a non-elementary condition that is necessary for embeddability into an Archimedean structure like $(\mathbb{R}, <^{\mathbb{R}}, +^{\mathbb{R}})$: this condition is obviously a version of the Archimedean property discussed in Chap. 5. It states that $x < y$ implies the existence of a positive integer n such that n successive concatenations of x with itself exceed y, i.e. $y < nx$.

Hölder essentially proved that any Archimedean ordered, positive semigroup can be embedded into $(\mathbb{R}, <^{\mathbb{R}}, +^{\mathbb{R}})$. This result may be viewed as an abstract account of scalability for empirical variables that behave like the pharmacist's standard weights. It is worth noting that Archimedean, ordered semigroups have a commutative operation so there is no need of including the commutativity of $\circ$ among the conditions that single out **M**.

In view of Example 9.30, we think of an empirical variable simply as a $\mathcal{L}$-structure $\mathcal{M}$ with $\mathcal{L} \supseteq \mathcal{L}_<$ and such that the $\mathcal{L}_<$-reduct of $\mathcal{M}$ is a linear order.[2] An empirical variable is **real scalable** iff there is an embedding of $\mathcal{M}$ into a $\mathcal{L}$-structure whose $\mathcal{L}_<$-reduct is $(\mathbb{R}, <^{\mathbb{R}})$.

---

[1] See [36] for survey.

[2] This requirement can be slightly weakened since a complete preorder suffices, but the weakening is not of special interest to our discussion and adds a small, unnecessary complication.

9.3 An Application to Measurement Theory

Reinhard Niederée has observed that this notion of scalability is to a certain extent unhelpful because it is equivalent to the single size requirement $|M| \leq 2^{\aleph_0} = |\mathbb{R}|$, a constraint that exhibits no natural connection with measurement. In [53], Niederée proposed to look at measurement values in an 'intrinsic' way, i.e. not as the numerical values of an embedding but as types determined by the $\mathcal{L}$-structure to be scaled.

If measurement values are symbolic codes of information arising from experimental procedures, then types provide a promising, abstract notion of measurement value because they contain formulae that can be interpreted as linguistic expressions of experimental information. A further advantage of using types as measurement values is that they can be made to depend on a finite number of parameters, interpretable as the reference points of a scale. For instance, types over a singleton may codify scale values relative to a unit of measure; types over a pair, scale values relative to a choice of origin and unit (as in the case of the Fahrenheit scale), and so on.

Because, in general, complete types may contain information that is too complex to relate in any easy way to experimental interaction, a focus on types that are not complete, but can nonetheless suffice to discriminate the 'levels' of an empirical variable is reasonable. The next definition will clarify the last remarks.

Definition 9.29 Let $\mathcal{L}$ be a first-order language and $\mathfrak{E}$ the set of existential $\mathcal{L}$-formulae in the free variables $x, y_1, \ldots, y_n$. If $\mathcal{M}$ is a $\mathcal{L}$-structure, $A = \{a_1, \ldots, a_n\}$ a subset of M, we let $\mathcal{E}_A$ be the set obtained from $\mathcal{E}$ by assigning the parameter a_i to the variable y_i, for each $i \in \{1, \ldots, n\}$.

For any $\mathcal{E} \subseteq \mathfrak{E}$, a $\mathcal{E}$-**type** $s(x)$ in $\mathcal{M}$ over A is the intersection $\mathcal{E}_A \cap p(x)$ for some $p(x) \in S_1(M/A)$.

We say that $\mathcal{M}$ is $\mathcal{E}$-**separable** over A iff the function $f : M \longrightarrow \mathcal{P}(\mathcal{E}_A)$ determined by the condition $a \mapsto \mathcal{E}_A \cap tp_{\mathcal{M}}(a/A) = s_a(x)$ is an injection.

When $\mathcal{M}$ is $\mathcal{E}$-separable over A, it induces a measuring $\mathcal{L}$-structure $\mathcal{M}_S$. The domain of $\mathcal{M}_S$ is $f[M]$. The relations, functions and constants of $f[M]$ are defined in the trivial way. For instance, if the m-ary relation symbol R figures in the signature of $\mathcal{L}$, we define:

$$R^{\mathcal{M}_S}(s_{a_1}(x), \ldots, s_{a_m}(x)) \iff R^{\mathcal{M}}(a_1, \ldots, a_m).$$

We refer to $\mathcal{M}_S$ as a **scaling structure for** $\mathcal{M}$ **over** A. The fact that we induce the structure of $\mathcal{M}$ on its scaling structure over A automatically ensures that f is a strong homomorphism. $\mathcal{E}$-separability implies that f is also an embedding. This embedding is also surjective, i.e. it is an isomorphism.

There is a simple way of characterising the $\mathcal{L}$-structures with a domain of $\mathcal{E}$-types that are scaling structures for some $\mathcal{L}$-structure $\mathcal{M}$.

Exercise 9.30 Let $\mathcal{E}$ be a set of existential $\mathcal{L}$-formulae in the free variables $x, y_1, \ldots, y_n$, $\{a_1, \ldots, a_n\} = A$ a finite set of parameters and $\mathcal{E}_A$ the set of existential $\mathcal{L}$-formulae in the free variable x from Definition 9.29. If $\mathcal{N}$ is a structure whose elements are sets of $\mathcal{L}$-formulae from $\mathcal{E}_A$, show that the following statements are equivalent:

(1) there is a $\mathcal{L}$-structure $\mathcal{M}$ such that $A \subseteq M$, $\mathcal{M}$ is $\mathcal{E}$-separable over A and $\mathcal{N}$ is a scaling structure for $\mathcal{M}$ over A;
(2) there are is a subset $C = \{c_1, \ldots, c_n\}$ of N such that, for each $s(x) \in N$, $tp_{\mathcal{N}}(s(x)/C) = s(x)$.

The type-theoretic approach to measurement does not only provide a natural way of seeing scales as codes of experimental information, but also yields an alternative outlook on key aspects of measurement. As an illustration, we characterise the Archimedean property in terms of separability.

Lemma 9.31 *Let $\mathcal{L}$ be the language discussed in Example 9.28, $\mathcal{E}_0$ the set of atomic $\mathcal{L}$-formulae in x, y. Let $\mathcal{M}$ be a positive, right-solvable ordered semigroup. The following statements are equivalent:*

(1) $\mathcal{M}$ is an Archimedean ordered semigroup;
(2) for any $a \in M$, $\mathcal{M}$ is $\mathcal{E}_0$-separable over $\{a\}$.

Proof Suppose that $\mathcal{M}$ is not Archimedean ordered. It follows that there are $a, b \in M$ such that $a <^M b$ and, for any $n \geq 1$ $na <^M b$. We deduce that $\mathcal{M}$ is not $\mathcal{E}_0$-separable over $\{a\}$ because the $\mathcal{E}_0$-types of b, $b \circ^M b$ over $\{a\}$ are identical. Let us call these types s_b and s_{2b} respectively.

An atomic formula over $\{a\}$ is of the form $ma < nx$, $nx < ma$ or $ma = nx$. Neither of s_b, s_{2b} contains equalities of the given form. As for the inequalities, none of the form $nx < ma$ can belong to s_b or s_{2b} because any multiple of a lies below b and, thus below nb and $n(b \circ^M b)$. Finally, since $\mathcal{M} \models ma < nb$ and $\mathcal{M} \models b < bob$ by positivity, it follows that $ma < nx \in s_b$ and $ma < nx \in s_{2b}$.

We conclude that $s_b = s_{2b}$ while $b \neq 2b$. This conclusion shows that $\mathcal{E}_0$-separability implies the Archimedean property.

We obtain the converse inclusion by contraposition. Suppose that $\mathcal{M}$ is not $\mathcal{E}_0$-separable but that it satisfies the Archimedean property. We shall derive a contradiction from these hypotheses.

Because separability fails, there must be distinct $b, c \in M$ such that $s_b = s_c$, where s_b, s_c are the $\mathcal{E}_0$-types of b, c over $\{a\}$. We use the Archimedean property to show that $s_b \neq s_c$. Because a tedious survey of cases is required, we only discuss two, leaving the remaining ones to the reader as an exercise.

Since $b \neq c$, trichotomy implies $b <^M c$ or $c <^M b$. Whichever holds, if a belongs to the interval with endpoints b, c, then it is clear that each of the formulae $x <^M a$ and $a <^M x$ belongs to exactly one of s_b, s_c, against the assumption that they are equal.

We may thus assume that $a <^M b <^M c$. The Archimedean property implies the existence of a largest positive integer n such that $nb \leq^M c <^M (n+1)b$ (where $x \leq^M y$ is an abbreviation of $x <^M y \vee x = y$). There also exists a largest positive integer m such that $ma \leq^M nb <^M (m+1)a$.

We cannot have $(m+1)a = c$ because, in this case, $(m+1)a = x \in s_c$ and $(m+1)a = x \notin s_b$. We cannot have $(m+1)a <^M c$ either because, in this latter case, $(m+1)a$ lies between b and c and their $\mathcal{E}_0$-types must consequently differ. Finally, if $c <^M (m+1)a$ $a <^M b$ implies $(m+1)a <^M nb \circ^M b = (n+1)b$ by monotonicity. A further application of monotonicity yields $nb <^M c$. The inequality $(n+1)b <^M 2c$, obtained again by monotonicity, implies that $nx <^M (m+1)a$ is in s_b but not in s_c.

Because we reach the same contradiction under the assumption $a <^M c <^M b$, simply by interchanging the roles of b, c in the previous argument, we must suppose that a is greater than b, c. This assumption leads to contradiction by arguments similar to those we have used and left to the reader.

It follows that the Archimedean property must fail if separability fails. □

We conclude this section with a couple of remarks concerning countable languages, whose size is appropriate for measurement settings.

Lemma 9.32 *Let $\mathcal{L}$ be a countable language.*

(1) If the $\mathcal{L}$-structure $\mathcal{M}$ is $\mathcal{E}$-separable over a finite subset A of M, then $|M| \leq 2^{\aleph_0}$.
(2) Let T be a $\mathcal{L}$-theory with infinite models. Then T has a countably infinite model $\mathcal{M}^*$ that is not $\mathcal{E}$-separable over A, for any subset $\mathcal{E}$ of $\mathfrak{E}$ and any finite subset A of M^*.

Proof Part (1) is straightforward and left to the reader. It provides an independent motivation for the cardinality restriction imposed by measurement theorists when they require scaling structures to have $\mathbb{R}$ as their domain. We only sketch the proof of part (2). Since T has an infinite model, by the Downward Löwenheim-Skolem theorem we may fix a countably infinite model $\mathcal{M}$ of T. We say that $\{c_i\}_{i \in \omega}$ is a sequence of indiscernibles over M iff, for any $\mathcal{L}$-formula $\varphi(x)$ with parameters in M, we have $\mathcal{M} \models \varphi(c_i) \leftrightarrow \varphi(c_j)$. The existence of a countable model $\mathcal{M}$ of T with an infinite sequence of indiscernibles over M is a consequence of Theorem 5.1.3 from [70], p.154. It is clear that no type, and in particular no $\mathcal{E}$-type over a finite subset of N will distinguish two indiscernibles. Separability thus fails, even for a larger set of formulae than $\mathcal{E}$. □

The type-theoretic approach to measurement we have outlined is studied in greater detail in Niederée's thesis [55]. We shall return to the interaction between model theory and measurement theory in Chap. 12, where we look at a model-theoretic approach to a fundamental question concerning the classification of measurement scales.

9.4 Types and Elementary Maps

We are already familiar with a few mathematically interesting instances of complete, countable theories. We now begin a systematic study of $S_n(T)$ for a countable, complete theory, which will provide further insight into the examples we know.

Our first objective is to relate types and elementary maps. A connection between them is not hard to see once we observe that, if $\mathcal{M}$ realises a complete n-type $p(\overline{x})$ over $A \subseteq M$, this type has to be $tp_{\mathcal{M}}(\overline{a})$ for some n-tuple $\overline{a}$ and it has to be realised in any elementary extension $\mathcal{N}$ of $\mathcal{M}$. In this case, $\mathcal{M}$ and $\mathcal{N}$ are trivially related by an elementary map, namely inclusion. What we have observed is however a general fact. If $f : \mathcal{M} \longrightarrow \mathcal{N}$ is elementary, any type over A that is realised by $\overline{a}$ in $\mathcal{M}$ is realised by $f(\overline{a})$ in $\mathcal{N}$ over the parameter set $f[A]$.

Conversely, if $\overline{a}$ realises the same complete n-type in $\mathcal{M}$ as $\overline{b}$ does in $\mathcal{N}$, i.e. if $tp_{\mathcal{M}}(\overline{a}) = tp_{\mathcal{N}}(\overline{b})$, then (the verification is left to the reader), for any $\mathcal{L}$-formula $\varphi(\overline{x})$, whose free variables occur among $x_1, \ldots, x_n$:

$$\mathcal{M} \models \varphi(\overline{a}) \iff \mathcal{N} \models \varphi(\overline{b}).$$

The last condition says that the function $f : \overline{a} \longrightarrow \mathcal{N}$ is a **partial elementary map**, where 'partial' alludes to the fact that its domain is not the whole of M (it does not have to be the domain of a substructure of $\mathcal{M}$ either). Note that, when $\overline{a}, \overline{b}$ realise the same type in $\mathcal{M}$, we have a partial elementary self-map of $\mathcal{M}$.

An interesting question is whether this map can be extended, ideally to an elementary bijection, i.e. an automorphism of $\mathcal{M}$. If we move to a suitable elementary extension $\mathcal{N}$ of $\mathcal{M}$, then this is indeed the case: any partial elementary map of the kind described can be extended to an automorphism of $\mathcal{N}$.

In order to prove this result, we require two lemmas on extensions of elementary maps.

Lemma 9.33 *Let $\mathcal{M}, \mathcal{N}$ be $\mathcal{L}$-structures and $A \subseteq M$. If $f : A \longrightarrow \mathcal{N}$ is a partial elementary map and $a \in M - A$, then there is an elementary extension $\mathcal{N}_1$ of $\mathcal{N}$ such that $f' : A \cup \{a\} \longrightarrow \mathcal{N}_1$ is an elementary map extending f.*

Proof We consider $tp_{\mathcal{M}}(a/A)$ and define $p(x)$ to be the set of formulae obtained by replacing each $\varphi(x, \overline{c}) \in tp_{\mathcal{M}}(a/A)$ with $\varphi(x, f(\overline{c}))$. We leave it to the reader to show that $\mathcal{ED}(\mathcal{N}) \cup p(x)$ is finitely consistent. It follows that $p(x) \in S_n(N/f[A])$, i.e. that $p(x)$ is realised in some $\mathcal{N}_1 \succeq \mathcal{N}$, say by $b \in N_1$.

Consider an arbitrary $\mathcal{L}$-formula $\psi(x, \overline{c})$ with parameters $\overline{c}$ over A. Then $\mathcal{M} \models \psi(a, \overline{c})$ iff $\psi(x, \overline{c}) \in tp_{\mathcal{M}}(a/A)$ iff $\psi(x, f(\overline{c})) \in tp_{\mathcal{N}_1}(b/f[A])$ iff $\mathcal{N}_1 \models \psi(b, f(\overline{c}))$.

It follows that the function $f' = f \cup \{(a, b)\}$ is a partial elementary map extending f. □

Exercise 9.34 Show that $\mathcal{ED}(\mathcal{N}) \cup p(x)$ is finitely consistent.

Lemma 9.33 allows us to deal with one-point extensions of elementary maps. It is plausible to think that a transfinite iteration of the lemma should allow us to extend partial elementary maps to total elementary maps. The next lemma confirms the conjecture.

9.4 Types and Elementary Maps

Lemma 9.35 *Let $\mathcal{M}, \mathcal{N}_0$ be $\mathcal{L}$-structures and $A \subseteq M$. If $f_0 : A \longrightarrow N_0$ is an elementary map, then there is an elementary extension $\mathcal{N}$ of $\mathcal{N}_0$ such that $f : M \longrightarrow N$ is an elementary map extending f.*

Proof Let $\{a_\beta : \beta \in \alpha\}$ be an enumeration of $M - A$. We use transfinite recursion to obtain a sequence of pairs $(\mathcal{N}_\beta, f_\beta)$ such that: (i) $\mathcal{N}_\beta \succeq \mathcal{N}_\gamma$ for each ordinal $\gamma \in \beta$; (ii) f_β is an elementary map with domain $A \cup \{a_\gamma : \gamma \in \beta\}$ and codomain N_β and $f_\beta \supseteq f_\gamma$ for each ordinal $\gamma \in \beta$.

The lemma's hypothesis gives us $(\mathcal{N}_0, f_0)$ and, at an ordinal stage λ, we simply take $\mathcal{N}_\lambda$ to be the union of the elementary chain $\{\mathcal{N}_\delta : \delta \in \lambda\}$ and f_λ to be the union $\bigcup_{\delta \in \lambda} f_\delta$.

As usual, the nontrivial extending operation we have to define involves successor ordinals. Thus, suppose the pair $(\mathcal{N}_\gamma, f_\gamma)$ has been defined. We consider a_γ in the enumeration of $M - A$. If a_γ is in the domain of f_γ, then we set $f_\gamma = f_{\gamma+1}$ and $\mathcal{N}_\gamma = \mathcal{N}_{\gamma+1}$. Otherwise we apply Lemma 9.33 to obtain the required one-point extension of f_γ and a corresponding elementary extension $\mathcal{N}_{\gamma+1}$ of $\mathcal{N}_\gamma$.

Let $\mathcal{N}$ be the union of the elementary chain $\{\mathcal{N}_\beta : \beta \in \alpha\}$. Then the union $\bigcup_{\beta \in \alpha} f_\beta$ is an elementary map with domain M and codomain N. □

Exercise 9.36 Verify the last assertion of Lemma 9.35.

We can finally use Lemma 9.35 to set up a chain construction that allows us to prove:

Theorem 9.37 *Let $\mathcal{M}$ be a $\mathcal{L}$-structure, $\overline{a}, \overline{b} \in M$. The equality $tp_\mathcal{M}(\overline{a}) = tp_\mathcal{M}(\overline{b})$ holds iff there are $\mathcal{N} \succeq \mathcal{M}$ and an automorphism $f : N \longrightarrow N$ such that $f(\overline{a}) = \overline{b}$.*

Proof Set $\mathcal{M} = \mathcal{M}_0$. Since the function sending $\overline{a}$ into $\overline{b}$ is a partial elementary map from $\overline{a}$ to $\mathcal{M}_0$, we use Lemma 9.35 to extend this map to an elementary map f_0 from $\mathcal{M}_0$ to a suitable elementary extension $\mathcal{N}_0 \succeq \mathcal{M}_0$.

Because f_0 is elementary, it is injective. Thus, f_0^{-1} is a function and, in particular, a partial elementary map from some $B \subseteq N_0$ to $\mathcal{M}_0$. Using Lemma 9.35 again, we can extend f_0^{-1} to an elementary map g_0 from $\mathcal{N}_0$ to $\mathcal{M}_1 \succeq \mathcal{M}_0$.

We can now consider the inverse of g_0, which extends f_0, and apply Lemma 9.35 once more, to obtain an elementary map $f_1 \supseteq f_0$ from $\mathcal{M}_1$ to $\mathcal{N}_1 \succeq \mathcal{N}_0$. Proceeding in this manner (i.e. by a recursion truncated at ω), we obtain the elementary chain:

$$\mathcal{M}_0 \preceq \mathcal{N}_0 \preceq \mathcal{M}_1 \preceq \mathcal{N}_1 \preceq \ldots \mathcal{M}_k \preceq \mathcal{M}_k \preceq \ldots$$

For each $k \in \mathbb{N}$, $f_k : \mathcal{M}_k \longrightarrow \mathcal{N}_k$ is an elementary map. Let $\mathcal{N}$ be the union of the above chain. Then $\mathcal{N}$ is both the union of $\{\mathcal{N}_k\}_{k \in \mathbb{N}}$ and of $\{\mathcal{M}_k\}_{k \in \mathbb{N}}$. Furthermore, $f = \bigcup \{f_k\}_{k \in \mathbb{N}}$ is an elementary map from N to N, as the reader is asked to verify.

It only remains to show that $f : N \longrightarrow N$ is surjective. Consider $b \in N$: there is some integer $i \geq 0$ such that $b \in N_i$. For convenience, let us suppose $i = 0$: the argument can then be generalised to an arbitrary index by the uniformity of our construction. If $i = 0$, we observe that, in the course of our chain construction, we have considered an extension g_0 of f_0^{-1}, which maps $\mathcal{N}_0$ into $\mathcal{M}_1$. Because this extension is invertible, we may consider the map g_0^{-1} from $A_1 \subseteq M_1$ onto N_0. Because g_0^{-1} is surjective, when we extend this map to obtain the injection f_1 from the whole of M_1 into N_1, we guarantee that $f_1(M_1) \supseteq N_0$. This is because f_1 behaves exactly like g_0^{-1} over A_1 and thus includes N_0 in its image.

The inclusion $f_1(M_1) \supseteq N_0$ allows us to conclude that any $b \in N_0$ is a value of f_1 and thus of f. If we had had $b \in N_i$, we could have run exactly the same argument on $f_{i+1}, M_{i+1}, N_{i+1}$, concluding that $f_{i+1}(M_{i+1}) \supseteq N_i$.

We have shown that f is surjective: it follows that it is an automorphism of $\mathcal{N}$ sending $\bar{a}$ into $\bar{b}$. The proof is complete. □

Exercise 9.38 Show that, in the proof of Theorem 9.37, $f = \bigcup \{f_k\}_{k \in \mathbb{N}}$ is an elementary map from N to N.

Exercise 9.39 Let $\mathcal{M}, \mathcal{N}$ be $\mathcal{L}$-structures and $f : M \longrightarrow N$ be an elementary map. Given $A \subseteq M$ and $a \in M$, define $q(x)$ to be the set of formulae over $f[A]$ obtained from the 1-type $tp_{\mathcal{M}}(a/A)$ by replacing each formula $\varphi(x, \bar{b}) \in tp_{\mathcal{M}}(a/A)$ with $\varphi(x, f(\bar{b}))$. Show that $q(x) \in S_1(N/f[A])$ and that $q(x)$ is isolated iff $tp_{\mathcal{M}}(a/A)$ is isolated.

Theorem 9.37 allows us to deduce the existence of an automorphism in an elementary extension of $\mathcal{M}$ from an equality between types. Under certain conditions, which are satisfied by the main algebraic examples we have considered, we may extend partial elementary self-maps of $\mathcal{M}$ to automorphisms of $\mathcal{M}$.

In order to carry out the analysis needed to establish the last claim, we turn to a central notion concerning one-point extensions of elementary maps, namely homogeneity.

9.5 Homogeneity

Lemma 9.33 guaranteed the existence of one-point extensions of partial elementary maps from a structure $\mathcal{M}$ into a structure $\mathcal{N}$. We focus attention on the structures that have the same property relative to partial elementary self-maps. To this end, we introduce:

Definition 9.40 Let κ be an infinite cardinal. A $\mathcal{L}$-structure $\mathcal{M}$ is said to be κ-**homogeneous** iff for any $A \subseteq M$ such that $|A| < \kappa$ and any $a \in M$, every partial self-map $f : A \longrightarrow M$ has a one-point extension $\overline{f} : A \cup \{a\} \longrightarrow M$.

A $\mathcal{L}$-structure $\mathcal{M}$ is said to be **homogeneous** iff it is $|M|$-homogeneous.

A connection with the theme of Theorem 9.37 is provided by a stronger version of homogeneity:

Definition 9.41 Let κ be an infinite cardinal. A $\mathcal{L}$-structure $\mathcal{M}$ is said to be **strongly κ-homogeneous** iff for any $A \subseteq M$ such that $|A| < \kappa$, every partial self-map $f : A \longrightarrow M$ can be extended to an automorphism of $\mathcal{M}$.

9.5 Homogeneity

An infinite $\mathcal{L}$-structure $\mathcal{M}$ is said to be **strongly homogeneous** iff it is strongly $|M|$-homogeneous.

We can now prove:

Theorem 9.42 *An infinite $\mathcal{L}$-structure $\mathcal{M}$ is homogeneous iff it is strongly homogeneous.*

Proof We only have to show that homogeneity implies strong homogeneity. To this end, set $|M| = \kappa$. We consider $A \subseteq M$ such that $|A| < \kappa$ and an elementary map $f : A \longrightarrow M$. Our goal is to extend f to a surjective, elementary map from $\overline{f} : M \longrightarrow M$, i.e. an automorphism of $\mathcal{M}$.

The proof proceeds, as customary, by transfinite recursion. We consider an enumeration $\{a_\gamma : \gamma \in \kappa\}$ of $M - A$ and obtain a chain of elementary extensions $\{f_\gamma\}_{\gamma \in \kappa}$ of f such that a_γ is both an argument and a value of $f_{\gamma+1}$. This will ensure that $\overline{f} = \bigcup_{\gamma \in \kappa} f_\gamma$ is an elementary function with the whole of M as domain and the whole of M as image, i.e. that $\overline{f}$ is an automorphism of $\mathcal{M}$.

The recursion is easy at the base step and at limit steps. We set $f = f_0$ and, if λ is a limit ordinal, $f_\lambda = \bigcup_{\delta \in \lambda} f_\delta$. The successor step requires two appeals to homogeneity. Suppose f_γ has been defined. We consider a_γ. If a_γ is already in the domain of f_γ, we set $g = f_\gamma$. Otherwise, we let g be a one-point extension of f_γ whose domain contains a_γ.

Once g has been defined, we consider the elementary map g^{-1} and repeat the above construction to guarantee that g^{-1} or a one-point extension of g^{-1} has a_γ in its domain. We let $f_{\gamma+1}$ be the inverse of the resulting elementary map. By construction, a_γ belongs to the domain of $f_{\gamma+1}$ and, since a_γ also belongs to the domain of its inverse, it is a value of $f_{\gamma+1}$.

It is now easy to see that, by construction, each element of M is in both the domain and image of $\overline{f} = \bigcup_{\gamma \in \kappa} f_\gamma$. □

Example 9.43 Consider an algebraically closed field $\mathcal{F}$ and $A \subseteq F$ such that $|A| < |F|$. Let $f_0 : A \longrightarrow F$ be an elementary self-map of F. We are going to show that f_0 has elementary one-point extensions, thus proving the homogeneity of $\mathcal{F}$.

Recall that f extends to an embedding f_1 of the substructure $\mathcal{B}$ generated by A within $\mathcal{F}$. Using the theorems on field embeddings from Chap. 4, we can also extend the embedding f_1 to an embedding of the fraction field $\mathcal{Q}(\mathcal{B})$ of $\mathcal{B}$ into $\mathcal{F}$. A further extension $f_2 = f$ embeds the algebraic closure of $\mathcal{Q}(\mathcal{B})$ into $\mathcal{F}$. By quantifier elimination for **ACF**, all embeddings are elementary.

This familiar chain of extensions enables us to restrict attention to embeddings with domain an algebraically closed field $\mathcal{K} \sqsubseteq \mathcal{F}$ such that $|K| < |F|$.

We now consider any $a \in F - K$. Then a is transcendental over $\mathcal{K}$ (verify) and $\mathcal{K}$ does not contain a transcendence base for $\mathcal{F}$. Because $f : K \longrightarrow f[K]$ is an isomorphism of algebraically closed fields, if $f[K]$ contained a transcendence base for $\mathcal{F}$, the image of $\mathcal{K}$ under f should coincide with $\mathcal{F}$, against the hypothesis $|K| < |F|$. It follows that there is $b \in F$ transcendental over $f(K)$.

Because the complete type of b contains the quantifier-free formula $\neg(q(x) = 0)$ for each polynomial $q(x)$ with coefficients in $f[K]$, the function $f \cup \{(a,b)\}$ is an elementary one-point extension of f (verify).

We conclude that any algebraically closed field is homogeneous and thus, by Theorem 9.42, strongly homogeneous.

Exercise 9.44 Complete the argument from Example 9.43 by carrying out the required verifications.

Exercise 9.45 Show that infinite vector spaces are strongly homogeneous.

Homogeneity does not only provide information about the automorphisms of a single structure. It also allows us to characterise isomorphisms between structures, under suitable hypotheses.

Theorem 9.46 *Let $\mathcal{M}, \mathcal{N}$ be countable, homogeneous structures. $\mathcal{M}, \mathcal{N}$ are isomorphic if, and only if, for each $n \geq 1$, $\{tp_{\mathcal{M}}(a_1, \ldots, a_n) : (a_1, \ldots, a_n) \in M^n\} = \{tp_{\mathcal{N}}(b_1, \ldots, b_n) : (b_1, \ldots, b_n) \in N^n\}$, i.e. $\mathcal{M}, \mathcal{N}$ realise the same complete n-types over $\emptyset$.*

Proof If $\mathcal{M}, \mathcal{N}$ are isomorphic, the associated Stone spaces of n-types over the empty set are easily seen to be identical. If, on the other hand, $\mathcal{M}, \mathcal{N}$ have identical Stone spaces, then $\mathcal{M} \equiv \mathcal{N}$. Thus, if one of them is finite, they are isomorphic. We may therefore assume that $\mathcal{M}, \mathcal{N}$ are both infinite.

Our goal is to construct an isomorphism in stages, using simultaneously two fixed enumerations $M = \{a_1, a_2, \ldots\}$ and $N = \{b_1, b_2, \ldots\}$ (we shall essentially repeat the argument given in the proof of Theorem 7.30). At stage 0, we set $f_0 = \emptyset$, which is partial elementary because $\mathcal{M} \equiv \mathcal{N}$.

If $0 < n = 2k - 1$ and the partial elementary map f_{2k-2} has been defined. Let the $(2k-2)$-tuple $\bar{a}$ be the domain of f_{2k-2} and let the $(2k-2)$-tuple $\bar{b}$ be the image of f_{2k-2}. Because f_{2k-2} is elementary, we have $tp_{\mathcal{M}}(\bar{a}) = tp_{\mathcal{M}}(\bar{b})$.

We consider the least $i \in \mathbb{N}$ such that, in the given enumeration of M, a_i differs from any object in $\bar{a}$. By hypothesis, $tp_{\mathcal{M}}(a_i, \bar{a}) \in S_{2k-2}(M) = S_{2k-2}(N)$: thus, for a suitable choice of $c, \bar{c} \in N$, $tp_{\mathcal{M}}(a_i, \bar{a}) = tp_{\mathcal{N}}(c, \bar{c})$.

By looking at the restrictions of the last types to the terminal $(2k-2)$-tuples, we conclude that $tp_{\mathcal{N}}(\bar{b}) = tp_{\mathcal{M}}(\bar{a}) = tp_{\mathcal{N}}(\bar{c})$. It now follows that there is a partial elementary map $\bar{c} \longrightarrow \bar{b}$.

Because $\mathcal{N}$ is homogeneous, we can extend this map to $c, \bar{c}$ and thus find $b_j \in N$ such that $tp_{\mathcal{M}}(a_i, \bar{a}) = tp_{\mathcal{N}}(c, \bar{c}) = tp_{\mathcal{N}}(b_j, \bar{b})$. Note that, since $c, \bar{c}$ are all distinct, so are $b_j, \bar{b}$. The selection of b_j is uniquely determined if we look for the least index j such that, in the given enumeration of N, b_j satisfies the last chain of equalities. We finally set $f_{2k-1} = f_{2k-2} \cup \{(a_i, b_j)\}$.

If $0 < n = 2k$, we suppose that f_{2k-1}, sending the $2k$-tuple $\bar{a}$ into $\bar{b}$, has been defined. We extend it to f_{2k} essentially as we have done earlier, starting from the least j such that, in the given enumeration of N, b_j is not in the image of f_{2k-1}.

Since b_j is not in the domain of f_{2k}^{-1}, we consider $tp_{\mathcal{N}}(b_j/\bar{b})$ and note again that, by hypothesis, the last type is the same as $tp_{\mathcal{M}}(d, \bar{d})$, with $d, \bar{d} \in M$. The

9.5 Homogeneity

construction then proceeds almost exactly as in the odd-numbered steps (bearing in mind that what is extended is in the first instance f_{2k}^{-1}), this time relying on the homogeneity of $\mathcal{M}$. We invite the reader to supply the missing details.

By means of the back-and-forth construction just described, we obtain a countable chain of partial elementary maps $\{f_i : i \in \mathbb{N}\}$. Their union f is an elementary map with domain $\mathcal{M}$ (since a_i is added at the latest to the domain of f_{2i-1}) and image $\mathcal{N}$ (since b_j is added at the latest to the image of f_{2j}), i.e. an isomorphism between $\mathcal{M}$ and $\mathcal{N}$. □

If we add $\mathcal{M} \equiv \mathcal{N}$ to our hypotheses, the conclusion of the last theorem does not require any restriction on the size of $\mathcal{M}, \mathcal{N}$.

Theorem 9.47 *Let $\mathcal{M}, \mathcal{N}$ be homogeneous $\mathcal{L}$-structures. If $\mathcal{M} \equiv \mathcal{N}$ and $|\mathcal{M}| = \kappa = |\mathcal{N}|$, the following conditions are equivalent:*

(1) for any $n \geq 1$, $\{tp_{\mathcal{M}}(a_1, \ldots, a_n) : a_1, \ldots, a_n \in \mathcal{M}\} = \{tp_{\mathcal{N}}(b_1, \ldots, b_n) : b_1, \ldots, b_n \in \mathcal{N}\}$.
(2) $\mathcal{M}, \mathcal{N}$ are isomorphic.

Proof The implication from (2) to (1) is unproblematic. In order to show that (1) implies (2), we make use of Lemma 9.48, proved below. This lemma guarantees that, for any cardinal κ, if $A \subseteq \mathcal{M}$ is of size $\leq \kappa$, there is a partial elementary map $f : A \longrightarrow \mathcal{N}$ and that, if $B \subseteq \mathcal{N}$ is of size $\leq \kappa$, there is a partial elementary map $g : B \longrightarrow \mathcal{M}$.

With this lemma in place we run a transfinite version of the back-and-forth argument used in the proof of Theorem 9.46. Instead, however, of carrying out a single operation at each stage, we carry out two.[3]

We work with two fixed enumerations of M, N, say $\{a_\alpha : \alpha \in \kappa\} = M$ and $\{b_\alpha : \alpha \in \kappa\} = N$. Note that, since κ is a cardinal, it is an initial ordinal, i.e., for any $\alpha \in \kappa$, $|\alpha| < \kappa$. Our goal, as in the previous theorem, is to build a chain of elementary functions $\{f_\alpha : \alpha \in \kappa\}$, such that the domain A_α of f_α has size $|\alpha|$.

Our starting point is $\emptyset = f_0$. At limit ordinals, we only have to take the union of the preceding stages. The nontrivial part of the back-and-forth argument concerns successor ordinals. Thus, suppose that f_α has been constructed and consider the least γ such that a_γ is not in the domain A_α of f_α. Since the size of A_α is the same as the size of $A_\alpha \cup \{a_\gamma\}$ and it is smaller than κ, there is an elementary map f from $A_\alpha \cup \{a_\gamma\}$ to $\mathcal{N}$.

Because we eventually want to extend f_α but we do not know whether f does extend it, our immediate objective is to overcome this uncertainty. We consider the restriction of f to A_α and note that the function $g = f_\alpha \circ f^{-1}$ sends $f[A_\alpha]$ into $f_\alpha[A_\alpha]$, i.e. it is an elementary, partial self-map of $\mathcal{N}$. We can therefore use homogeneity to extend g to the domain $f[A_\alpha] \cup \{f(a_\gamma)\}$. Calling the extension h and setting $g(f(a_\alpha)) = b$, we see that, for any $a \in A_\alpha$, $g(f(a)) = f_\alpha(a)$. Thus $h \circ f = f_\alpha^*$ is an elementary one-point extension of f_α.

---

[3] This saves us a digression into the distinction between even and odd ordinals. The distinction can in fact be drawn: if an ordinal α is a sequence of β pairs, then α is even, otherwise it is odd.

We finally consider the least δ such that, in the given enumeration of N, b_δ does not belong to the image of f_α^*. Reasoning exactly as above and making use of the homogeneity of $\mathcal{M}$, we arrive at a one-point extension of f_α^* that contains b_δ in its image. We call this one-point extension $f_{\alpha+1}$.

It is clear that, by stage $\alpha + 1$ in our construction, we arrive at a partial elementary map that has a_α in its domain and b_α in its image. Thus, setting $\overline{f} = \bigcup_{\alpha \in \kappa} f_\alpha$, we see that $\overline{f}$ is an isomorphism between $\mathcal{M}$ anbd $\mathcal{N}$. □

Lemma 9.48 *Let $\mathcal{N}$ be a homogeneous $\mathcal{L}$-structure. Suppose $\mathcal{M} \equiv \mathcal{N}$, $|N| = \kappa$ and, for any $n \geq 1$, $\{tp_\mathcal{M}(a_1, \ldots, a_n) : a_1, \ldots, a_n \in M\} = \{tp_\mathcal{N}(b_1, \ldots, b_n) : b_1, \ldots, b_n \in N\}$. Then, if $A \subseteq M$ is of size $\leq \kappa$ there is a partial elementary map $f : A \longrightarrow N$.*

Proof We proceed as in the proof of Theorem 9.47. If A is finite, the result follows immediately from elementary equivalence on the hypothesis that $\mathcal{M}, \mathcal{N}$ realise the same complete n-types, for any $n \geq 1$. The rest of the proof proceeds as a transfinite induction on the infinite size of A.

Given a cardinal $\lambda < \kappa$, the inductive hypothesis is that, if $C \subseteq M$ and $|C| < \lambda$, there is an elementary map from A to N. Our goal is to obtain the desired map as the union of a chain whose existence is in turn guaranteed by an application of transfinite induction. Each element of the chain will be a function f_α, for some ordinal $\alpha \in \lambda$, with domain A_α of size $|\alpha|$. Recall that, since λ is an initial ordinal, for each $\alpha \in \lambda$ we have $|\alpha| < \lambda$.

We fix an enumeration $\{a_\alpha : \alpha \in \lambda\}$ of A. The inductive base is given by $f_0 = \emptyset$ and, at limit points, we take unions of the preceding stages, which are elementary maps. Let us therefore suppose that f_α has been constructed. Since $|\alpha| = |\alpha \cup \{\alpha\}| < |\lambda|$, the transfinite inductive hypothesis implies the existence of an elementary map $f : A_\alpha \cup \{a_\alpha\} \longrightarrow N$.

We consider the restriction of f to A_α and note that the function $g = f_\alpha \circ f^{-1}$ sends $f[A_\alpha]$ into $f_\alpha[A_\alpha]$, i.e. it is an elementary, partial self-map of $\mathcal{N}$. We obtain a one-point extension of f_α exactly as in the proof of Theorem 9.47, using the homogeneity of $\mathcal{N}$. The union $\bigcup_{\alpha \in \lambda} f_\alpha$ is an elementary map from A into N, as desired. □

The proof just given implies that, if $\mathcal{M}$ is homogeneous, there are elementary maps from subsets of N of size $\leq |M|$ into M.

What we have learnt about homogeneity will help us with the study of structures that realise only a few types and structures that realise many types. We shall focus on the former in the next section and on the latter in Sect. 9.7.

9.6 Atomic Structures

We have already noted that any $\mathcal{L}$-structure $\mathcal{M}$ realises all of the isolated types in $S_n(M)$. We are now interested in structures that realise as few types as possible over $\emptyset$, i.e. none other than the isolated types, which cannot be omitted.

9.6 Atomic Structures

Definition 9.49 A $\mathcal{L}$-structure $\mathcal{M}$ is said to be **atomic** iff for any tuple $\bar{a} \in \mathcal{M}$, $tp_{\mathcal{M}}(\bar{a})$ is isolated.

If $\bar{a}$ is the list $a_1, \ldots, a_n$ and $tp_{\mathcal{M}}(a_n, a_{n-1}, \ldots, a_1)$ is isolated, Lemma 9.21 implies that $tp_{\mathcal{M}}(a_n/\{a_{n-1}, \ldots, a_1\})$ is isolated. Using this observation, it is easy to see that $\mathcal{M}$ is atomic iff any 1-type over a finite subset of M is isolated.

Example 9.50 Consider the complete, countable theory $\mathsf{DLO}_{-,-}$ of dense linear orders without endpoints. Since $\mathsf{Q} = (\mathbb{Q}, <) \models \mathsf{DLO}_{-,-}$, $S_n(T) = S_n(\mathsf{Q})$ for each $n \geq 1$. It follows that Q realises all of the isolated types of $S_n(T)$. We shall now prove that Q realises no other types, i.e. that it is an atomic $\mathcal{L}_<$-structure.

To this end, consider an arbitrary complete n-type in of Q, say $tp_{\mathsf{Q}}(a_1, \ldots, a_n)$. We may assume that $a_1, \ldots, a_n$ are listed in order of magnitude (otherwise, it suffices to reassign numerical indices). We shall verify that the formula:

$$x_1 < x_2 \wedge \ldots \wedge x_{n-1} < x_n,$$

which we are going to call $\varphi(x_1, \ldots, x_{n-1})$, isolates $tp_{\mathsf{Q}}(a_1, \ldots, a_n)$. To this end, we consider a $\mathcal{L}_<$-formula $\psi(x_1, \ldots, x_n) \in tp_{\mathsf{Q}}(a_1, \ldots, a_n)$ and suppose $\mathsf{Q} \models \varphi(b_1, \ldots, b_n)$.

By the definition of φ, we have $b_1 < b_2 \wedge \ldots \wedge b_{n-1} < b_n$ i.e. the n-tuples $a_1, \ldots, a_n$ and $b_1, \ldots, b_n$ are ordered in the same way. Thus, the function sending a_i into b_i is a partial, order-preserving injection. Let $A = \mathbb{Q} - \{a_1, \ldots, a_n\}$ and $B = \mathbb{Q} - \{b_1, \ldots, b_n\}$. Clearly, we can enumerate A and B, which are countably infinite, dense linear orders without endpoints.

By the $\aleph_0$-categoricity of $\mathsf{DLO}_{-,-}$, there is an order-preserving isomorphism f from A to B. It follows that $f \cup \{(a_1, b_1), \ldots, (a_n, b_n)\}$ is an automorphism of Q. Because automorphisms preserve first-order information, we conclude that:

$$\mathsf{Q} \models \varphi(b_1, \ldots, b_n) \implies \mathsf{Q} \models \psi(b_1, \ldots, b_n)$$

for any $\psi(x_1, \ldots, x_n) \in tp_{\mathsf{Q}}(a_1, \ldots, a_n)$. Since $b_0, \ldots, b_n$ are arbitrary rational numbers satisfying φ, we conclude that:

$$\mathsf{Q} \models \forall x_1 \ldots \forall x_n (\varphi(x_1, \ldots, x_n) \to \psi(x_1, \ldots, x_n)).$$

In other words $tp_{\mathsf{Q}}(a_1, \ldots, a_n)$ is an isolated type. Thus, Q is atomic.

We know from the previous chapter that Q is a prime model of $\mathsf{DLO}_{-,-}$. We can reach the same conclusion using the last example and the following, general theorem concerning countable, atomic models of countable, complete theories.

Theorem 9.51 *Let T be a countable, complete theory without finite models. Every countable, atomic model of T is a prime model of T.*

Proof Let $\mathcal{M}, \mathcal{N} \models T$ with $\mathcal{M}$ a countable, atomic model of T. We build an elementary embedding of $\mathcal{M}$ into $\mathcal{N}$ in stages, using an enumeration $\{a_i\}_{i \in \mathbb{N}}$ of $\mathcal{M}$.

Given a_0, we note that $tp_\mathcal{M}(a_0)$ is an isolated 1-type of T. Because any model of T realises the isolated types of T, there is $b_0 \in N$ such that $tp_\mathcal{M}(a_0) = tp_\mathcal{N}(b_0)$. As a result, the map sending a_0 into b_0 is a partial elementary map f_0. Note that, although we already have a theorem that allows us to extend partial elementary maps to elementary maps, that theorem will not in general keep $\mathcal{N}$ fixed but might require an elementary extension of $\mathcal{N}$.

It is to overcome this issue that we make essential use of the atomicity of $\mathcal{M}$. We therefore consider a_1 and the type $tp_\mathcal{M}(a_1/a_0)$. Because the types $tp_\mathcal{M}(a_1, a_0)$ and $tp_\mathcal{M}(a_0)$ are both isolated by atomicity, Lemma 9.21 implies that $tp_\mathcal{M}(a_1/a_0)$ is isolated.

Let $\varphi(x, a_0)$ an isolating formula for $tp_\mathcal{M}(a_1/a_0)$. For any $\psi(x, a_0) \in tp_\mathcal{M}(a_1/a_0)$, isolation implies $\mathcal{M} \models \forall x(\varphi(x, a_0) \rightarrow \psi(x, a_0))$. Since f_0 is partial elementary, we deduce:

$$\mathcal{N} \models \forall x(\varphi(x, b_0) \rightarrow \psi(x, b_0)).$$

The last condition says that the type obtained from $tp_\mathcal{M}(a_1/a_0)$ by replacing each occurrence of the parameter a_0 with an occurrence of b_0, is an isolated type of $\mathcal{N}$ over b_0. As such, it is realised by some $b_1 \in N$.

We now have a partial elementary map f_1 that sends a_0, a_1 into b_0, b_1 respectively, as the reader should verify. This construction can obviously be iterated: when we arrive at elementary map that sends $a_0, \ldots, a_{n-1}$ into $b_0, \ldots, b_{n-1}$ respectively, we consider the next element of M in the enumeration, namely a_n, and note that the type obtained from $tp_\mathcal{M}(a_n/a_{n-1}, \ldots, a_0)$ by replacing its parameters with $b_{n-1}, \ldots, b_0$ respectively is isolated.

It then suffices to select a realisation $b_n \in N$ to obtain an extended, partial elementary map f_n sending $a_0, \ldots, a_n$ into $b_0, \ldots, b_n$ respectively. The map $f = \bigcup_{i \in \mathbb{N}} f_i$ is an elementary embedding of $\mathcal{M}$ into $\mathcal{N}$. □

Exercise 9.52 Show that a countable atomic structure is homogeneous.

Exercise 9.53 Show that any two countable, atomic structures are isomorphic (*Hint*: use the previous exercise and Theorem 9.47).

Exercise 9.54 Show that any finite structure is atomic.

What we know so far about atomic structures only enables us to prove that, if T is a countable, complete theory, any countable atomic model of T is a prime model. The reason why the converse statement might fail is that, *a priori*, we do not yet know whether there are any countable and complete theories whose models always realise a non-isolated type. In this case, a prime model could not be atomic. This possibility is ruled out by an important theorem, known as the Omitting Types Theorem, which we are going to prove in a separate subsection.

9.6.1 The Omitting Types Theorem

Theorem 9.55 (Omitting Types) *Let T be a countable $\mathcal{L}$-theory and $p \in S_n(T)$ a non-isolated n-type, with $n \geq 1$. There is a countable model of T that omits p.*

Proof We first expand $\mathcal{L}$ to $\mathcal{L}(C)$ by adding to the former language an infinite sequence of new constant symbols. Our goal is to obtain a complete $\mathcal{L}(C)$-extension of T in two stages. First we build a consistent extension $T^* \supseteq T$ such that, for any n-tuple $\bar{c}$ of new constant symbols, T^* contains a sentence of the form $\neg \psi(\bar{c})$, for some $\psi(\bar{x}) \in p$.

In the second stage, we use Zorn's lemma to obtain a completion of T^*, which we shall call T^+. There is a canonical way of building a countable $\mathcal{L}(C)$-structure $\mathcal{M}^+$ that T^+ and whose domain consists entirely of constants interpreting the new symbols in $\mathcal{L}(C)$. By the way we have defined T^+, no n-tuple $\bar{c}^{\mathcal{M}^+}$ of elements of $\mathcal{M}^+$ can realise p.

In order to extend T to T^*, we set up a recursion up to ω. Let $\{\varphi_i\}_{i \in \mathbb{N}}$ be a fixed enumeration of the $\mathcal{L}(C)$-sentences and $\{\bar{c}_i\}_{i \in \mathbb{N}}$ a fixed enumeration of the n-tuples of new constant symbols in $\mathcal{L}(C)$.

We start from $T_0 = T$ and use the first enumeration to build the odd-numbered stages of our recursion, while the second enumeration is used to build the even-numbered stages other than 0. Each stage T_n is obtained by adding only finitely many sentences to T_0.

Thus, suppose the even stage T_{2k} of our construction, with $k \geq 0$, has been defined in such a way that it contains only finitely many sentences in addition to those in T_0. We consider the sentence φ_k from the given enumeration and proceed as follows:

(1) if φ_k is not of the form $\exists x \varphi(x)$ and it does not follow from T_{2k}, then we set $T_{2k+1} = T_{2k}$;
(2) if φ_k is of the form $\exists x \varphi(x)$ and it does not follow from T_{2k}, then we select the first constant symbol c_j from the given enumeration that does not occur in any sentence of T_{2k} and set $T_{2k+1} = T_{2k} \cup \{\exists x \varphi(x) \to \varphi(c_j)\}$;
(3) if $\varphi(k)$ is of the form $\exists x \varphi(x)$ and $T_{2k} \models \exists x \varphi(x)$, then we select the first constant symbol c_j from the given enumeration that does not occur in any sentence of T_{2k} and set $T_{2k+1} = T_{2k} \cup \{\exists x \varphi(x), \varphi(c_j)\}$.

We leave it for the reader to check that, if T_{2k} is consistent, each of (1) to (3) produces a consistent extension of T_{2k}.

We next turn to the even-numbered stages of our recursion: thus, let us suppose that T_{2k+1} has been defined and that it contains only finitely many sentences in addition to those in T_0. We consider $\bar{c}_k$, the k-th n-tuple in our fixed enumeration. The conjunction $\theta(\bar{c}, \bar{c}_k)$ of the sentences in $T_{2k+1} - T_0$ is a $\mathcal{L}(C)$-sentence because the last set is finite. In general, $\theta(\bar{c}, \bar{c}_k)$ may contain new constant symbols $\bar{c}$ not in tuple $\bar{c}_k$ as well as constant symbols in that n-tuple.

Consider the $\mathcal{L}$-formula $\theta(\bar{y}, \bar{x})$ and let $\varphi(\bar{x})$ be $\exists \bar{y} \theta(\bar{y}, \bar{x})$. Because p is non-isolated, the last $\mathcal{L}$-formula does not isolate it. This is to say that, for some $\psi(\bar{x}) \in p$, we have:

$$T_0 \not\models \forall \bar{x}(\varphi(\bar{x}) \to \psi(\bar{x})).$$

In other words, there is a $\mathcal{L}$-structure $\mathcal{M}$ such that $\mathcal{M} \models \varphi(\bar{x}) \wedge \neg\psi(\bar{x})$. It is easy to see that a suitable expansion of $\mathcal{M}$ models $\theta(\bar{c}, \bar{c}_k) \wedge \neg\psi(\bar{c}_k)$. By the way θ has been defined, this is to say that $T_{2k+1} \cup \{\neg\psi(\bar{c}_k)\}$ is consistent. In view of this last fact, we set $T_{2k+2} = T_{2k+1} \cup \{\neg\psi(\bar{c}_k)\}$.

It is now easy to prove by induction that $T^* = \bigcup_{n \in \mathbb{N}} T_n$ is a consistent extension of T_0. By Zorn's lemma, T^* has a maximal consistent extension T^+, i.e. a completion. By the way the odd-numbered steps in the construction of T^* have been defined, if a sentence of the form $\exists x \varphi(x)$ is in T^+, then $\varphi(c_j) \in T^+$, for some constant symbol c_j from the infinite sequence added to $\mathcal{L}$.

We can easily construct a model $\mathcal{M}^+$ of T^+ whose domain M^+ consists of constant symbols from our expanded language and is therefore countable (see Exercise 9.56). By construction, no n-tuple of elements of M^+ realises p. □

Exercise 9.56 We shall construct a model $\mathcal{M}^+$ of T^+ using the fact that T^+ is a complete theory. Let M be the set of closed terms in $\mathcal{L}_C$. We define M^+ to be a quotient of M, obtained by identifying two closed terms t_1, t_2 in M iff the sentence $t_1 = t_2$ is in T^+ (by completeness, this sentence or its negation must lie in T^+).

(i) show that each equivalence class of M contains a constant symbol;
(ii) if R is a k-ary relation symbol in $\mathcal{L}_C$, we define the k-ary relation $R^{\mathcal{M}^+}$ on M^+ by the condition:

$$R^{\mathcal{M}^+} c_1 \cdots c_k \text{ iff } R c_1 \cdots c_k \in T^+,$$

where we take advantage of part (i) to identify the elements of M^+ with fixed constant symbols from the corresponding equivalence classes. Explain how functions and constants are to be defined on M^+.
(iii) Prove by induction on the complexity of formulae that $\mathcal{M}^+ \models T^+$.

Exercise 9.57 How can the proof of the Omitting Types theorem be modified to ensure the existence of a countable model that omits two non-isolated types p_1, p_2 simultaneously?

The attentive reader will have noticed that the last exercise lends itself to an easy generalisation. By differentiating the operations to be carried out at different stages, the proof of Theorem 9.55 can be generalised to yield countable models of T that simultaneously omit any given finite number of non-isolated types. It is in fact possible to omit a countably infinite set $\{p_i\}_{i \in \mathbb{N}}$ of non-isolated n-types. The transitions from even-numbered steps to odd-numbered ones used in the proof of the

9.6 Atomic Structures

Omitting Types theorem remains the same but the transitions from odd-numbered to even-numbered steps have to be modified.

Since every odd number is of the form $2^q(2r+1) - 1$ (by prime factor decomposition, any even number is the product of a power of two times an odd number), at the odd stage $2^q(2r+1) - 1$ the r-th n-tuple $\bar{c}_r$ of new constant symbols in a fixed enumeration is considered and, using a familiar argument, a formula from the q-th type is found, whose negation is consistent with $T_{2^q(2r+1)-1}$. Since this is done for each positive power of 2, the interpretation of $\bar{c}_r$ does not realise any type from the list $\{p_i\}_{i \in \mathbb{N}}$ in the extension T^+ of T described earlier. Because the last assertion holds for each $r \in \mathbb{N}$, infinitely many non-isolated types are omitted.

Example 9.58 The restriction to theories in a countable language is necessary to the proof of the Omitting Types Theorem. A standard way of showing it makes use of a theory in the uncountable language $\mathcal{L}^*$, obtained from $\mathcal{L}_=$ by adding a countably infinite sets C of new constant symbols and a distinct, uncountable set D of new constant symbols.

Suppose $|D| = \lambda > \aleph_0$ and set $T = \{\neg(d = d') : d, d' \in D\}$. We consider the 1-type $p(x) = \{\neg(x = c) : c \in C\}$. Since any model $\mathcal{M}$ of T is uncountable, $\mathcal{M}$ realises $p(x)$. Nevertheless, $p(x)$ is not an isolated type. To see this, we note that T has quantifier elimination (see Exercise 9.59). Given a $\mathcal{L}^*$-formula $\varphi(x)$ such that $T \cup \varphi(x)$ is consistent, $\varphi(x)$ is T-equivalent to a quantifier-free, disjunctive normal form. We may thus identify $\varphi(x)$ with such a formula.

Consider $\mathcal{M} \models T$ such that $\mathcal{M}$ realises $\varphi(x)$. Since $\mathcal{M}$ must satisfy at least one disjunct of $\varphi(x)$, we may take $\varphi(x)$ to be a conjunction of literals. Either all of its literals are negative or at least one is positive. In the former case, $\varphi(x)$ is of the form $\neg(x = c_1) \wedge \ldots \wedge \neg(x = c_k) \wedge \neg(x = d_1) \wedge \ldots \wedge \neg(x = d_m)$, with $c_1, \ldots, c_k \in C$, $d_1, \ldots, d_m \in D$.

Since C is countably infinite, it is clear that, for any $c \in C - \{c_1, \ldots, c_k\}$, we have $\mathcal{M} \models \varphi(c)$, i.e. $\mathcal{M} \models \varphi(c) \wedge (x = c)$ or $\mathcal{M} \models \neg(\varphi(c) \to \neg(x = c))$. It follows that $T \not\models \forall x(\varphi(x) \to \neg(x = c))$. In other words, $\varphi(x)$ does not isolate $p(x)$.

If, on the other hand, a positive literal occurs in $\varphi(x)$, it is either of the form $x = c$ for some $c \in C$ or of the form $x = d$ for some $d \in D$. In the former case it is obvious that $\varphi(x)$ does not isolate $p(x)$. In the latter case, we consider a model of T, which may or may not be $\mathcal{M}$, that assigns the same interpretation to the constant symbol d and to a suitably chosen symbol $c' \in C$ (we only have to make sure that c' differs from any c_i such that $\neg(x = c_i)$ is a literal in $\varphi(x)$).

In any case we can conclude $T \not\models \forall x(\varphi(x) \to \neg(x = c))$, i.e. $\varphi(x)$ does not isolate $p(x)$. Because $\varphi(x)$ was chosen arbitrarily, it follows that $p(x)$ is a non-isolated 1-type realised in every model of T.

Exercise 9.59 Show that the theory T from Example 9.51 is not λ-categorical and that it has quantifier elimination. Is T a complete theory?

An immediate consequence of the Omitting Types theorem, which we had already adumbrated, is:

Lemma 9.60 *Let T be a countable, complete theory: $\mathcal{M}$ is a prime model of T iff it is countable and atomic.*

Exercise 9.61 Prove the last lemma and provide at least two algebraic examples of countable, atomic structures.

Exercise 9.62 Is the linear order $(\mathbb{R}, <)$ atomic? What about the ordered, divisible abelian group $(\mathbb{Q}, <, +, -, 0)$?

A remarkable consequence of the Omitting Types theorem is the following type-theoretic characterisation of countable and complete $\aleph_0$-categorical theories.

9.6.2 The Ryll-Nardzewski Theorem

Theorem 9.63 (Ryll-Nardzewski) *Let T be a countable and complete theory. The following conditions are equivalent:*

(1) T is $\aleph_0$-categorical;
(2) for each $n \geq 1$, and every $p \in S_n(T)$, p is isolated;
(3) for each $n \geq 1$, $S_n(T)$ is finite;
(4) for each $n \geq 1$, every $\mathcal{L}$-formula is T-equivalent to one among finitely many $\mathcal{L}$-formulae $\varphi_1(\overline{x}), \ldots, \varphi_k(\overline{x})$;
(5) if $\mathcal{M} \models T$, then $\mathcal{M}$ is atomic;
(6) if $\mathcal{M}$ is a countable model of T, then $\mathcal{M}$ is atomic.

Proof If (1) holds, T has only infinite models. Suppose $q \in S_n(T)$ is a non-isolated type. By the Omitting Types theorem, q is omitted by a countable model of T, say $\mathcal{M}$. On the other hand, There is a model of T that realises q. By the Downward-Löwenheim Skolem, the latter model has a countable, elementary submodel $\mathcal{N}$ that realises q (exercise). We may suppose that $\overline{b}$ realises q in N, i.e. $\mathcal{N} \models q(\overline{b})$. Since $\mathcal{M}, \mathcal{N}$ are countably infinite, they must be isomorphic. But if $g : N \longrightarrow M$ is an isomorphism, then $f(\overline{b})$ must realise q in $\mathcal{M}$, a contradiction. We conclude that (2) holds.

By Exercise 9.25, (2) is equivalent to (3).

Let us next assume (3) and suppose that there are finitely many n-types of T, say $p_1, \ldots, p_m$, all of them isolated. For any $i = 1, \ldots, m$, we call $\theta_i(\overline{x})$ the formula isolating p_i. We consider a formula $\zeta(\overline{x})$ in the language of T that belongs to the types $p_{i_1}, \ldots, p_{i_k} \in \{p_1, \ldots, p_m\}$.

Let $\varphi(\overline{x})$ be $\theta_{i_1}(\overline{x}) \vee \ldots \vee \theta_{i_k}(\overline{x})$ and consider an arbitrary model $\mathcal{M}$ of T. If $\mathcal{M} \models \varphi(\overline{b})$, there is $j \in \{1, \ldots, k\}$ such that $\mathcal{M} \models \theta_{i_j}(\overline{b})$, which implies $\mathcal{M} \models \zeta(\overline{b})$. Conversely, if $\mathcal{M} \models \zeta(\overline{b})$, the n-tuple $\overline{b}$ realises a type in $S_n(M) = S_n(T)$. This type clearly cannot contain $\neg \zeta(\overline{b})$: as a result, it must be one of the types isolated by $\theta_{i_1}(\overline{x}), \ldots, \theta_{i_k}(\overline{x})$ respectively. Thus, $\mathcal{M} \models \varphi(\overline{b})$ and, by the arbitrary choice of $\mathcal{M}, \overline{b}$, we have $T \models \forall \overline{x}(\varphi(\overline{x}) \leftrightarrow \zeta(\overline{x}))$. Since there are only 2^m subsets

9.6 Atomic Structures

of $\{p_1, \ldots, p_m\}$, the argument just given shows that any formula in the language of T is T-equivalent to one of 2^m disjunctions.

Let us next assume (4) and consider an arbitrary model $\mathcal{M}$ of T and $\overline{a} \in M^n$, with $n \geq 1$. Our goal is to show that $tp_{\mathcal{M}}(\overline{a})$ is isolated. The set $S = \{\varphi_1(\overline{x}), \ldots, \varphi_{2^m}(\overline{x})\}$ of formulae whose existence is guaranteed by (4) is partitioned by $\overline{a}$ into two subsets, namely:

$$S_1 = \{\varphi_i(\overline{x}) \in S : \mathcal{M} \models \varphi_i(\overline{a})\} \text{ and } S_2 = \{\varphi_i(\overline{x}) \in S : \mathcal{M} \not\models \varphi_i(\overline{a})\}.$$

Let $\eta(\overline{x})$ be the conjunction of the elements of S_1 and $\theta(\overline{x})$ be the conjunction of the elements of S_2. We claim that $\eta(\overline{x}) \wedge \theta(\overline{x})$ isolates $tp_{\mathcal{M}}(\overline{a})$. To verify this, suppose $\mathcal{M} \models \eta(\overline{b}) \wedge \theta(\overline{b})$ for some $\overline{b} \in M^n$ and consider any formula $\zeta(\overline{x}) \in tp_{\mathcal{M}}(\overline{a})$. By (4), $\zeta(\overline{x})$ is T-equivalent to a formula from the set S. Because $\mathcal{M} \models \zeta(\overline{a})$, the formula $\zeta(\overline{x})$ cannot be T-equivalent to any formula in S_2 and must therefore be T-equivalent to a formula in S_1. As such it is T-implied by $\eta(\overline{x})$, which contains among its conjuncts a T-equivalent of $\zeta(\overline{x})$. It follows in particular that $\mathcal{M} \models \zeta(\overline{b})$. If $\mathcal{M} \not\models \zeta(\overline{a})$, the argument just given continues to work with reference to $\eta(\overline{x})$.

By the arbitrary choice of $\overline{b}$ in M^n, we finally deduce $\mathcal{M} \models \forall \overline{x}((\eta(\overline{x}) \wedge \theta(\overline{x})) \to \zeta(\overline{x}))$. Since $\zeta(\overline{x})$ is an arbitrary formula in $tp_{\mathcal{M}}(\overline{a})$, we have shown that the latter type is isolated. The argument applies to any type in $S_n(M)$, with $n \geq 1$. It follows that $\mathcal{M}$ is atomic.

If (5) holds, every model of T is atomic. This is certainly true of any countable model, i.e. (6) holds.

If (6) holds, (1) follows by Exercise 9.53. □

Exercise 9.64 What happens in the proof that (3) implies (4) if every type in $\{p_1, \ldots, p_m\}$ contains $\neg \zeta(\overline{x})$?

Exercise 9.65 Let $\mathcal{M}$ be a $\mathcal{L}$-structure and $\mathcal{A}$ the group of automorphisms of $\mathcal{M}$.

(1) if $\overline{a} \in M^n$, the **orbit** of $\overline{a}$ relative to $\mathcal{A}$ is the set of n-tuples $\{\overline{b} \in M^n : \overline{b} = \alpha(\overline{a}),$ for some automorphsm α of $\mathcal{A}\}$. Show that the relation $\overline{a} \sim \overline{b}$, defined to hold iff $\overline{a}, \overline{b}$ belong to the same orbit relative to $\mathcal{A}$, is an equivalence relation.
(2) By part (i), M is partitioned by its automorphism group into orbits. Suppose that $\mathcal{M}$ is a countable model of a countable, complete and $\aleph_0$-categorical theory T. Show that, if $\mathcal{M} \models T$, then M^n is partitioned by $\mathcal{A}$ into finitely many orbits. We say that $\mathcal{A}$ is an **oligomorphic** automorphism group.
(3) Give two examples of structures with an oligomorphic automorphism group.

Although we have introduced atomic structures by looking for the smallest amount of types that models of complete theories can realise, the Ryll-Nardzewski theorem tells us that some atomic structures realise all that there is to realise. Because $S_n(T)$ is finite when T is countable, complete and $\aleph_0$-categorical, any model of T realises *every* n-type over the empty set. It is thus natural to ask whether we can depart from the special case of $\aleph_0$-categorical theories and find models that realise many types even when the space of types is not small, i.e. countable. We turn to this question in the next section.

9.7 Saturated Structures

The central idea we work with is that of saturation, defined below.

Definition 9.66 Let $\mathcal{M}$ be a $\mathcal{L}$-structure and κ an infinite cardinal. We say that $\mathcal{M}$ is κ-**saturated** iff, for any $A \subseteq M$ of size $< \kappa$, $\mathcal{M}$ realises every type in $S_1(\mathcal{M}/A)$. We say that $\mathcal{M}$ is **saturated** iff it is $|M|$-saturated.

Exercise 9.67

(i) Show that, if $\mathcal{M}$ is κ-saturated, $A \subseteq M$ and $|A| < \kappa$, then $\mathcal{M}$ realises every type in $S_n(\mathcal{M}/A)$.
(ii) Show that, if $\mathcal{M}$ is finite, then it is κ-saturated for every infinite cardinal κ.

Saturated structures have three distinctive properties, which we are going to establish in the next three lemmas.

Lemma 9.68 Let $\mathcal{M}$ be a κ-saturated $\mathcal{L}$-structure. Then $\mathcal{M}$ is κ-homogeneous.

Proof Let $A \subseteq M$ be of size $< \kappa$ and $f : A \longrightarrow M$ be a partial elementary function. We fix $a \in M - A$ and consider $tp_{\mathcal{M}}(a/A)$. Set:

$$p(x) = \{\varphi(x, f(\overline{a})) : \varphi(x, \overline{a}) \in tp_{\mathcal{M}}(a/A)\}.$$

We leave it as an exercise to show that $p(x)$ is a complete 1-type over $f[A]$. Because f is injective, $|f[A]| = |A| < \kappa$ and, by κ-saturation, some $b \in M$ realises $p(x)$. We set $g = f \cup \{(a, b)\}$. It follows that g is an elementary one-point extension of f. □

Exercise 9.69 Supply the missing details in the proof of the last lemma.

In order to prove the next lemma, we need the definition of a new model-theoretic property.

Definition 9.70 Let $\mathcal{M}$ be a model of a $\mathcal{L}$-theory T and κ an infinite cardinal. We say that $\mathcal{M}$ is κ-**universal** iff $\mathcal{N} \models T$ and $|N| < \kappa$ imply that $\mathcal{N}$ can be elementarily embedded into $\mathcal{M}$. When elementary embeddability extends to models of size smaller than or equal to κ, we say that $\mathcal{M}$ is κ^+-universal.

Lemma 9.71 Let $\mathcal{M}$ be a κ-saturated model of a complete theory T. Then $\mathcal{M}$ is κ^+-universal.

Proof Suppose $\mathcal{N} \models T$ and $|N| = \kappa$ (if $|N| < \kappa$, essentially the same proof can be used). We determine a chain of embeddings by transfinite recursion, given a fixed

9.7 Saturated Structures

enumeration of N, say $\{b_\gamma : \gamma \in \kappa\}$. Because T is complete, $\mathcal{M} \equiv \mathcal{N}$ and $f_0 = \emptyset$ is elementary. At limit ordinals, we only have to take unions of the previously defined embeddings.

Finally, if f_γ has been defined as an elementary map from $\{b_\delta : \delta \in \gamma\}$ into $\mathcal{M}$, clearly the domain of f_γ has size smaller than κ. We can therefore apply Lemma 9.68 to obtain a one-point extension $f_{\gamma+1}$ of γ that includes b_γ into its domain.

The union $\bigcup_{\alpha \in \kappa} f_\alpha$ embeds $\mathcal{N}$ into $\mathcal{M}$. □

The last two lemmas lead to a single converse, which requires a constraint on the cardinality of the language:

Lemma 9.72 *Let T be a $\mathcal{L}$-theory with $|\mathcal{L}| \leq \kappa$. If the $\mathcal{L}$-structure $\mathcal{M} \models T$ is is κ-homogeneous and κ^+-universal, then it is κ-saturated.*

Proof Fix $A \subseteq M$ such that $|A| < \kappa$ and choose an arbitrary 1-type $p \in S_1(M/A)$. There is $\mathcal{N} \succeq \mathcal{M}$ such that $\mathcal{N}$ realises p. Fix a realisation $a \in N$. By the Downward Löwenheim-Skolem theorem, there is $\mathcal{N}' \preceq \mathcal{N}$ with $A \cup \{a\} \subseteq N'$ and $|N'| \leq \kappa$, given the hypothesis on $|\mathcal{L}|$.

Since $\mathcal{M}$ is κ^+-universal, there is an elementary embedding $f : \mathcal{N}' \longrightarrow \mathcal{M}$. The inverse f^{-1} is also elementary and its restriction g to $f[A]$ sends $f[A] \subseteq M$ into $A \subseteq M$. We can therefore exploit the κ-homogeneity of $\mathcal{M}$ and extend g to the domain $f[A] \cup \{f(a)\}$ (note that $f(a) \in M$). Because g is elementary, for any formula $\varphi(x/A) \in p$, we have:

$$\mathcal{N}' \models \varphi(a/A) \text{ iff } \mathcal{M} \models \varphi(f(a)/f[A]) \text{ iff } \mathcal{M} \models \varphi(g(a)/A).$$

We conclude that $\mathcal{M}$ realises p. Because p was chosen arbitrarily, $\mathcal{M}$ is κ-saturated. □

Exercise 9.73 Let $\mathcal{M}, \mathcal{N}$ be countably infinite, elementarily equivalent and saturated structures. Show that they are isomorphic.

So far, we have only proved that the κ-saturated models of complete theories are exactly the κ-homogeneous and κ^+-universal models but we have not proved that any such models exist. Our current aim is therefore to show the existence of structures that have certain saturation properties. We begin with:

Lemma 9.74 *Let $\mathcal{M}$ be a $\mathcal{L}$-structure, $|\mathcal{L}| = \aleph_0$ and κ be an infinite cardinal. Furthermore, let S be the set of 1-types over subsets of M whose cardinality is $\leq \kappa$. Then $|S| \leq |M|^\kappa$ and there is $\mathcal{N} \succeq \mathcal{M}$ such that $\mathcal{N}$ realises every type in S and $|N| \leq |M|^\kappa$.*

Proof First, note that $|M|^\kappa$ determines the size of the set of functions from κ to M. Among these functions, the injections single out the subsets of M of size exactly κ. The remaining functions include those mapping κ into a subset of M of cardinality strictly smaller than κ. Thus, $|M|^\kappa$ provides an overestimate of, and certainly an upper bound for, the set of subsets of M that have size at most κ.

Next, fix $A \subseteq M$ such that $|A| = \kappa$. Since the language is countable, its expansion to $\mathcal{L}_A$, which contains one new constant symbol for each parameter in A, is of size κ, i.e. there are κ distinct $\mathcal{L}_A$-formulae. Each 1-type over A is a subset of a set of κ formulae, so there are at most 2^κ types. The product of the upper bound for the number of complete 1-types and the upper bound for the number of parameter sets yields an upper bound for the number of 1-types over any subset of size $< \kappa$, namely: $2^\kappa \cdot |M|^\kappa = |M|^\kappa$.

Fix an enumeration of the 1-types over the subsets of M of size $\leq \kappa$, say $\{p_\beta : \beta \in |M|^\kappa\}$ (if we run out of types at a cardinality strictly lower than our bound, we may e.g. repeat p_0 until we reach the bound). We set $\mathcal{M} = \mathcal{N}_0$ and run a transfinite recursion of length $|M|^\kappa$. At limit ordinals, we only have to take unions. If γ is a successor ordinal and $\mathcal{N}_\gamma$ such that $|\mathcal{N}_\gamma| \leq |M|^\kappa$ has been defined, we look for an elementary extension $\mathcal{N}_\gamma^* \succeq \mathcal{N}_\gamma$ that realises p_γ.

By the Downward Löwenheim-Skolem theorem we can, if necessary, find an elementary substructure of $\mathcal{N}_\gamma^*$ that includes $N_\gamma \cup \{a_\gamma\}$, where a_γ realises p_γ. Let $\mathcal{N}_{\gamma+1}$ be such an elementary substructure. It is clear that $|N_{\gamma+1}| \leq |M|^\kappa$.

The union $\mathcal{N}$ of the elementary chain $\{\mathcal{N}_\beta : \beta \in |M|^\kappa\}$ is an elementary extension of $\mathcal{M}$ that realises all 1-types over subsets of M whose size is at most κ. Moreover $|N| \leq |M|^\kappa$. □

An iteration of Lemma 9.74 yields the existence of κ^+-saturated, elementary extensions.

Theorem 9.75 *Let $\mathcal{M}$ be a $\mathcal{L}$-structure with $|\mathcal{L}| = \aleph_0$ and κ an infinite cardinal. There is a κ^+-saturated $\mathcal{N} \succeq \mathcal{M}$ such that $|N| \leq |M|^\kappa$.*

Proof Set $\mathcal{M} = \mathcal{N}_0$ and call a set 'small' if its size is smaller than or equal to κ. We use Lemma 9.74 to obtain, by transfinite recursion, an elementary chain of length $|M|^\kappa$. The base structure is $\mathcal{N}_0$ and at limit ordinals we take unions of elementary chains.

Now suppose $\mathcal{N}_\beta$ has been defined, with $|N_\beta| \leq |M|^\kappa$. Let S_β be the set of 1-types over a small subset of N_0. By Lemma 9.74, there is $\mathcal{N}_{\beta+1} \succeq \mathcal{N}_\beta$ that realises every type in S_β and is of size at most $|N_\beta|^\kappa \leq (|M|^\kappa)^\kappa = |M|^{\kappa \cdot \kappa} = |M|^\kappa$.

We let $\mathcal{N}$ be the union of the chain $\{\mathcal{N}_\alpha : \alpha \in |M|^\kappa\}$. Let A be a small subset of N: because $|A| < \kappa$ and the successor k^+ is a regular cardinal,[4] there cannot be a function from $|A|$ into k^+ whose image is unbounded (see Lemma A.79 from the Appendix). In particular, if A is small, the function assigning to each $a \in A$ the least γ such that $a \in N_\gamma$ has its full image included in some N_α, with $\alpha \in k^+$. Because the entirety of A lies in N_α and $\mathcal{N}_{\alpha+1}$ realises every 1-type over A, all such types are also realised in the elementary extension $\mathcal{N}$ of $\mathcal{N}_{\alpha+1}$.

We conclude that $\mathcal{N}$ is κ^+-saturated. Since N is the union of $|M|^\kappa$ sets each of cardinality $\leq |M|^\kappa$, $|N| \leq |M|^\kappa \cdot |M|^\kappa = |M|^\kappa$. □

---

[4] See Appendix, Sect. A.5.2, for a definition.

9.7 Saturated Structures

Exercise 9.76 Let T be a $\mathcal{L}$-theory. Suppose that, for any $\mathcal{M} \models T$, for each finite $A \subseteq M$, $|S_1(M/A)| \leq \kappa$, where κ is an infinite cardinal. Fix $\mathcal{M} \models T$ of size $|\mathcal{L}|$.

(i) Show that there is $\mathcal{N}_0 \succeq \mathcal{M}$ such that $\mathcal{N}_0$ realises every complete 1-type over a finite subset of $\mathcal{N}_0$ and $|N| \leq |\mathcal{L}| + \kappa$.
(ii) Use part (i) to show that there is a $\aleph_0$-saturated, elementary extension $\mathcal{N}$ of $\mathcal{M}$ that has size $\leq |\mathcal{L}| + \kappa$.

9.7.1 Small Theories

A theory T is said to be **small** when it realises only countably many types, more precisely when $|S_n(T)| \leq \aleph_0$ for each $n \geq 1$. Countable, complete theories are small iff they have a $\aleph_0$-saturated model. In order to prove this result, we need the following characterisation of $\aleph_0$-saturated models of complete theories.

Lemma 9.77 Let T be a complete $\mathcal{L}$-theory and $\mathcal{M} \models T$: $\mathcal{M}$ is $\aleph_0$-saturated iff it is $\aleph_0$-homogeneous and it realises every type in $S_n(T)$, for all $n \geq 1$.

Proof If $\mathcal{M}$ is $\aleph_0$-saturated, then it is $\aleph_0$-homogeneous by Lemma 9.68. We only have to show that every $p \in S_n(T)$ is realised in $\mathcal{M}$. Since $S_n(T) = S_n(M)$, p is a complete n-type of $\mathcal{M}$ over $\emptyset$. By hypothesis, $\mathcal{M}$ realises all complete n-types over a parameter set A with $|A| < \aleph_0$. The last condition obviously holds if $A = \emptyset$, so p is realised by $\mathcal{M}$.

To complete the proof, we suppose that $\mathcal{M}$ is $\aleph_0$-homogeneous and realises every type in $S_n(T)$, with $n \geq 0$. It suffices to show that $\mathcal{M}$ realises every 1-type in $S_1(M/\bar{a})$, with $\bar{a}$ a finite n-tuple of parameters. To this end, we associate $p \in S_1(M/\bar{a})$ with the the $(n+1)$-type $q \in S_{n+1}(T)$, defined as follows:

$$q = \{\psi(x, \bar{y}) : \psi(x, \bar{a}) \in p\}.$$

By hypothesis, $\mathcal{M}$ realises q. We fix a realisation $b\bar{c}$ and note that $tp_\mathcal{M}(\bar{c}) = tp_\mathcal{M}(\bar{a})$ by the definition of q. It follows that there is a partial elementary map f sending $\bar{c}$ into $\bar{a}$. By $\aleph_0$-homogeneity, we can extend f to g with domain $b\bar{c}$. Then $g(b)$ realises p and $\mathcal{M}$ is $\aleph_0$-saturated. □

We now characterise small theories. In the sequel, a countable theory is just a theory in a countable language.

Theorem 9.78 Let T be a countable, complete $\mathcal{L}$-theory. T is small iff it has a countable, $\aleph_0$-saturated model.

Proof If $\mathcal{M} \models T$ is $\aleph_0$-saturated, then it realises every type in $S_n(M/A)$ where A is finite. We may in particular choose $A = \emptyset$. It follows that every n-type in $S_n(M)$ is realised by $\mathcal{M}$. Since $S_n(T) = S_n(M)$ and M is countable, $|S_n(T)| \leq \aleph_0$, i.e. T is small.

Conversely, suppose T is small and let $\mathcal{M}$ be a countable model of T (which exists by the Downward Löwenheim-Skolem theorem). Given an enumeration of $\bigcup_{n\in\mathbb{N}} S_n(T)$, we can build a countably infinite, elementary chain $\mathcal{M} = \mathcal{M}_0 \preceq \mathcal{M}_1 \preceq \ldots \preceq \mathcal{M}_k \preceq \ldots$ of countable models ot T, each realising one further type along the enumeration. The union $\mathcal{M}^\omega$ of the chain just described is countable and it realises every type in $S_n(T)$, for each $n \geq 1$.

As shown in the next lemma, we can find a countable, elementary extension $\mathcal{N} \succeq \mathcal{M}^\omega$ that is also $\aleph_0$-homogeneous. Lemma 9.77 implies that $\mathcal{N}$ is $\aleph_0$-saturated. □

Lemma 9.79 *Let T be a countable, complete theory with infinite models and $\mathcal{M} \models T$. There is $\mathcal{N} \succeq \mathcal{M}$ such that $|M| = |N|$ and $\mathcal{N}$ is $\aleph_0$-homogeneous.*

Proof We may assume $\mathcal{M}$ infinite since finite models are κ-saturated for any infinite cardinal κ (Exercise 9.67-(ii)) and thus also κ-homogeneous.

The proof proceeds by means of a double chain construction. We first use transfinite recursion to build an elementary chain starting from $\mathcal{M} = \mathcal{M}_0$. In particular, we consider an enumeration A of all triples of the form $(\bar{a}, \bar{b}, c)$ such that $\bar{a}, \bar{b} \in M^n$ for some $n \geq 1$ and $tp_\mathcal{M}(\bar{a}) = tp_\mathcal{M}(\bar{b})$. Since $|A| < |M|$, we may set $A = \{(\bar{a}_\alpha, \bar{b}_\alpha, c) : \alpha \in |M|\}$.

In order to define the recursion, we only need to specify the extending operation required at successor ordinals. If $\mathcal{M}_\alpha$ has been constructed, we consider the triple $(\bar{a}_\alpha, \bar{b}_\alpha, c)$. Since $tp_\mathcal{M}(\bar{a}_\alpha) = tp_\mathcal{M}(\bar{b}_\alpha)$, there is a partial elementary map sending $\bar{a}_\alpha$ into $\bar{b}_\alpha$.

Our goal is to add to $\mathcal{M}_\alpha$ an object d such that $tp_{\mathcal{M}_{\alpha+1}}(c_\alpha \bar{a}_\alpha) = tp_{\mathcal{M}_{\alpha+1}}(d\bar{b}_\alpha)$ and $|M_{\alpha+1}| = |M_\alpha|$. By doing so, we obtain an approximation of $\aleph_0$-homogeneity relative to the triple considered.

Note that, since $\mathcal{M} \preceq \mathcal{M}_\alpha$, we have $tp_{\mathcal{M}_\alpha}(\bar{a}_\alpha) = tp_{\mathcal{M}_\alpha}(\bar{b}_\alpha)$. Set $p(y, \bar{x}) = tp_{\mathcal{M}_\alpha}(c_\alpha \bar{a}_\alpha) = tp_\mathcal{M}(c_\alpha \bar{a}_\alpha)$. It is easy to check that $p(y/\bar{b})$ is a 1-type over $\bar{b}$, which is therefore realised by an object d in an elementary extension $\mathcal{M}^*$ of $\mathcal{M}_\alpha$.

By the Downward Löwenheim-Skolem theorem, we may restrict attention to the elementary substructure of $\mathcal{M}^*$ generated by $M_\alpha \cup \{d\}$ and call it $\mathcal{M}_{\alpha+1}$. Clearly, $|M_{\alpha+1}| = |M_\alpha|$.

We now let $\mathcal{N}_0$ be the union of the elementary chain $\{\mathcal{M}_\alpha : \alpha \in |M|\}$. It can be shown by transfinite induction that, for each $\alpha \in |M|$, $|M_\alpha| = |M|$. Consequently, $|N_0| = |M|$, since N_0 is the union of $|M|$ sets of size $|M|$. We iterate the chain construction just described ω times.

The first iteration leads from $\mathcal{N}_0$ to $\mathcal{N}_1$, where $\mathcal{N}_1$ satisfies the condition that, for each triple $(\bar{a}, \bar{b}, c)$, with $\bar{a}, \bar{b} \in N_0^n$ and $tp_{\mathcal{N}_0}(\bar{a}) = tp_{\mathcal{N}_0}(\bar{b})$, any partial elementary map from $\bar{a}$ to $\bar{b}$ can be extended to a partial elementary map with domain $\bar{a}c$.

For any $k \geq 1$, we apply the same construction to obtain the desired elementary extension $\mathcal{N}_{k+1}$ of $\mathcal{N}_k$. We obtain the elementary chain $\{\mathcal{N}_k : k \in \mathbb{N}\}$, whose union we call $\mathcal{N}$. A simple inductive proof shows that $|N_0| = |N_k|$ for each $k \in \mathbb{N}$. Thus $|N| = |N_0| = |M|$.

Since $\mathcal{N} \succeq \mathcal{N}_0 \succeq \mathcal{M}$, $\mathcal{N} \succeq \mathcal{M}$ and $|N| = |M|$. Moreover $\mathcal{N}$ is $\aleph_0$-homogeneous. To see why, consider an elementary self-map of N sending $\bar{a}$ into $\bar{b}$ and pick $c \in N$.

Clearly, we have $tp_{\mathcal{N}}(\bar{a}) = tp_{\mathcal{N}}(\bar{b})$. Moreover, since $\bar{a}, \bar{b}$ and c are finite lists of elements, each occurring in some N_i, there is a least $k \in \mathbb{N}$ such that N_k contains all of the elements listed. It follows that there is $d \in N_{k+1}$ such that $tp_{\mathcal{N}_{k+1}}(c\bar{a}) = tp_{\mathcal{N}_{k+1}}(d\bar{b})$. Thus $tp_{\mathcal{N}}(c\bar{a}) = tp_{\mathcal{N}}(d\bar{b})$ and the given partial elementary map can be extended to c. □

Example 9.80 By the Ryll-Nardzewski's theorem, $\mathsf{DLO}_{-,-}$ is a small theory. Since it is $\aleph_0$-categorical, it has only one countable, $\aleph_0$-saturated model up to isomorphism. Thus, in particular, $(\mathbb{Q}, <)$ is both atomic and $\aleph_0$-saturated.

Exercise 9.81 Let T be a complete theory and $\mathcal{M} \models T$ be an atomic model.

(i) Show that, if $A = \{a_1 \ldots, a_n\} \subseteq M^n$, the expansion $\mathcal{M}_A$ of $\mathcal{M}$ to a language containing constant symbols for each of $a_1, \ldots, a_n$ is still atomic.
(ii) Use part (i) to show that an atomic structure is $\aleph_0$-saturated.
(iii) Use part (ii) to show that ACF_0 and, for any prime p, ACF_p, are small theories.

9.7.2 Monster Models

In certain model-theoretic investigations we may not wish to consider structures of arbitrarily large size. For instance, when studying the extensive structures discussed in Sect. 9.3, we are in effect only interested in models of size $\leq 2^{\aleph_0}$.

Even independently of special restrictions, we may operate at a sufficient level of generality by looking at the models of a theory T whose size lies below some 'large' infinite cardinal κ. In such cases it may be convenient to work with a single ambient structure inside which all models of interest live. We can do this if T has a model $\mathbb{M}$ that is κ-saturated and κ^+-strongly homogeneous. By Lemma 9.71, $\mathbb{M}$ is κ^+-universal, so any model of T of size $\leq \kappa$ is elementarily embeddable in $\mathbb{M}$. By κ^+-strong homogeneity, the elementary maps between models of T may also be viewed as restrictions of automorphisms of $\mathbb{M}$. Large models like $\mathbb{M}$ are sometimes called **monster models**. We close this chapter by proving their existence.

Theorem 9.82 *Let T be a $\mathcal{L}$-theory and κ be an infinite cardinal. Then T has a κ-saturated and κ^+-strongly homogeneous model $\mathbb{M}$.*

Proof Fix $\mathcal{M} \models T$. We may assume M infinite, since finite structures are κ-saturated for any infinite cardinal κ. We use transfinite recursion of length κ^+ to determine an elementary chain $\{\mathcal{M}_\alpha : \alpha \in \kappa^+\}$ of models of T such that $\mathcal{M}_0 = \mathcal{M}$ and, for any $\alpha \in \kappa^+$, $\mathcal{M}_{\alpha+1} = \mathcal{M}$ is $|M_\alpha|^+$-saturated.

While we take unions of elementary chains at limit steps, the successor step follows the construction discussed in Theorem 9.75. Since we have now dropped the hypothesis that $\mathcal{L}$ is countable, we can no longer provide bounds for model size similar to those established in that theorem. Everything else works.

Let $\mathbb{M}$ be the union of the elementary chain $\{\mathcal{M}_\alpha : \alpha \in \kappa^+\}$, with domain $D_\mathbb{M}$. We first check that $\mathbb{M}$ is κ^+-saturated. For any $A \subseteq D_\mathbb{M}$ such that $|A| \leq \kappa$, the regularity of κ^+ implies that A is already included in some $\mathcal{M}_\alpha$, with α an ordinal of cardinality strictly smaller than κ^+.

Since $\mathcal{M}_{\alpha+1}$ is $|M_\alpha|^+$-saturated, any 1-type over A in $S_1(D_\mathbb{M}/A)$ is already realised in $\mathcal{M}_{\alpha+1}$ and thus in $\mathbb{M} \succeq \mathcal{M}_{\alpha+1}$. This establishes the κ^+-saturation of $\mathbb{M}$.

We use transfinite recursion again to prove that $\mathbb{M}$ is κ^+-strongly homogeneous, relying on the fact that $\mathbb{M}$ has been constructed as the union of a particular elementary chain. Suppose that $f : A \longrightarrow D_\mathbb{M}$ is a partial elementary self-map of $\mathbb{M}$. Since $D_\mathbb{M} = \bigcup_{\alpha \in \kappa^+} M_\alpha$, an argument already given and resting on the regularity of κ^+ guarantees that both A and $f[A]$ are included in some M_α. In particular, we may regard f as a partial elementary function with domain $A \subseteq M_\alpha$ and codomain $M_{\alpha+1}$.

Because $A \subseteq M_\alpha \subseteq M_{\alpha+1}$, we may also regard f as a partial elementary self-map of $\mathcal{M}_{\alpha+1}$. The the last structure is $|M_\alpha|^+$-saturated so it is also $|M_\alpha|^+$-homogeneous: we can thus iterate one-point extensions to extend the domain of f to the entirety of M_α. We let $f_0 : M_\alpha \longrightarrow M_{\alpha+1}$ be the resulting elementary map.

The remainder of the proof requires two applications of homogeneity in the fashion just described to obtain extensions of elementary maps at successor steps in a transfinite recursion of length κ^+. The recursion will yield a chain of elementary functions of the form $f_\beta : M_{\alpha+\beta} \longrightarrow M_{\alpha+\beta+1}$, with $\beta \in \kappa^+$. Since we already have f_0 and we only take unions at limit ordinals, the nontrivial part of the recursion occurs at successor steps. We wish to define an extending operation that will eventually support a proof by transfinite induction of the condition:

$$M_{\alpha+\beta} \subseteq f_{\beta+1}[M_{\alpha+\beta+1}], \text{ with } \beta \in \kappa^+.$$

Because the union of a chain of elementary functions is an elementary function, the last condition guarantees its surjectivity, thus showing that we can indeed extend f_0 to an automorphism of $\mathbb{M}$, i.e. that κ^+-strong homogeneity holds.

Let us suppose $f_\gamma : M_{\alpha+\gamma} \longrightarrow M_{\alpha+\gamma+1}$ has been defined. We now explain how f_γ is to be extended to $f_{\gamma+1}$. Because f_γ is elementary and has image $f_\gamma[M_{\alpha+\gamma}]$, f_γ^{-1} may be regarded as a partial elementary function with a domain of size $|M_{\alpha+\gamma}| = \mu$ included in $M_{\alpha+\gamma+1}$ and with codomain $M_{\alpha+\gamma+1} \supseteq M_{\alpha+\gamma}$. In short, f_γ^{-1} may be regarded as a partial elementary self-map of $M_{\alpha+\gamma+1}$ with domain of size $\mu < |M_{\alpha+\gamma}|^+$.

Note $\mu = |M_{\alpha+\gamma}| \leq |M_{\alpha+\gamma} \cup f_\gamma[M_{\alpha+\gamma}]| \leq \mu + \mu = \mu < \kappa^+$. This means that, because $\mathcal{M}_{\alpha+\gamma+1}$ is κ^+-saturated, and thus κ^+-homogeneous, we can extend f_γ^{-1} to:

$$g : f[M_{\alpha+\gamma}] \cup M_{\alpha+\gamma} \longrightarrow M_{\alpha+\gamma} \subseteq M_{\alpha+\gamma+1}.$$

9.7 Saturated Structures

We now extend g^{-1}, proceeding essentially as we have done to obtain g. The image of g^{-1} is $M_{\alpha+\gamma} \cup g[M_{\alpha+\gamma}]$ and g^{-1} itself may be viewed as a partial, elementary self-map of $\mathcal{M}_{\alpha+\gamma+2}$ with a domain included in $\mathcal{M}_{\alpha+\gamma+1}$. Using the $|M_{\alpha+\gamma+1}|^+$-saturation of $\mathcal{M}_{\alpha+\gamma+2}$, we extend g^{-1} to:

$$f_{\gamma+1} : \mathcal{M}_{\alpha+\gamma+1} \longrightarrow \mathcal{M}_{\alpha+\gamma+2}.$$

It is not too difficult to see that $M_{\alpha+\beta} \subseteq f_{\beta+1}[M_{\alpha+\beta+1}]$ for any $\beta \in \kappa^+$. The function $f^* = \bigcup_{\alpha \in \kappa^+} f_\alpha$ is an automorphism of $\mathbb{M}$. As a consequence $\mathcal{M}$ is a κ^+-strongly homogeneous model of T. □

Chapter 10
Algebraically Closed Fields and Algebraic Geometry

Abstract In this chapter we pursue the interplay between field-theoretic and model-theoretic concepts already explored in Chaps. 7 and 8 and expand it to algebraic geometry. Our main goal is to offer a concrete illustration of the fact that, in one respect, model theory provides a 'logical analysis' of certain algebraic results, framing in an abstract form the conditions or properties that, within specific algebraic settings, coincide with independently studied mathematical content.

We certainly do not claim that model theory is to be viewed mainly or primarily as an instrument of logical analysis, but we find it important to emphasise this role, which is both mathematically and philosophically interesting. It is mathematically interesting because it shows that model-theoretic ideas and techniques provide a new outlook on familiar objects and results. It is philosophically interesting because it shows that model-theoretic results govern a reconstruction of mathematical content along novel, distinctively logical, lines.

Section 10.1 gives two model-theoretic characterisation of field-theoretic notions. The first is a simple observation to the effect that 'algebraic element' in the model-theoretic sense specialises to 'algebraic element' in the field-theoretic sense. The second and central result of this section is a beautiful theorem proved by McKenna, McIntyre and Van den Dries, to the effect that the infinite fields with quantifier elimination are exactly the models of ACF. This theorem provides a purely model-theoretic characterisation of 'algebraically closed'.

In Sect. 10.2 we introduce the rudiments of algebraic geometry. Our central goal here is to obtain a proof of Hilbert's *Nullstellensatz* based on the model-completeness of ACF. In the process we illustrate some natural model-theoretic translations of geometric notions like that of a variety.

Section 10.3 concludes our brief exploration of the interplay between model theory and algebraic geometry. This section contains a first introduction to Morley rank and a proof that this notion coincides with that of dimension of a variety when set in the context of algebraic geometry.

10.1 Specialisations

If $\mathcal{K} \sqsubseteq \mathcal{M}$ are fields, we know that, whenever $a \in K$ is algebraic over K, a satisfies an algebraic $\mathcal{L}_r$-formula of the form $m(x) = 0$, where $m(x)$ is the minimal polynomial of a in $K[x]$. Thus, $a \in acl_{\mathcal{M}}(K)$. The converse is true, i.e. algebraic elements in the model-theoretic sense are algebraic in the field-theoretic sense.

Although this result implicitly follows from remarks made in Chap. 7, we can offer a type-theoretic argument for it, which depends on the following characterisation of 1-types in models of strongly minimal theories.

Lemma 10.1 *Let T be a complete theory with an infinite model. The following conditions are equivalent:*

(1) T is strongly minimal;
(2) For any $\mathcal{M} \models T$ and any $A \subseteq M$, exactly one type in $S_1(A)$ is not algebraic.

Proof We first show that (1) implies (2) by contraposition, i.e. we suppose $\mathcal{M} \models T$ but $S_1(A)$ contains two distinct types, p, q, none of which is algebraic. There is, in this case, a formula $\varphi(x)$ such that $\varphi(x) \in p$ and $\neg\varphi(x) \in q$. The formula $\varphi(x)$ cannot be algebraic, i.e. it does not define a finite set. Because this is also true of $\neg\varphi(x)$, $\varphi(x)$ does not define a cofinite set either. We conclude that $\mathcal{M}$ is not a minimal structure and T is not a strongly minimal theory because not all of its models are minimal.

If, on the other hand T is not strongly minimal, then there are $\mathcal{M} \models T$, a parameter set A (possibly empty) and a formula $\varphi(x)$ with parameters in A such that $\varphi(M)$ is neither finite nor cofinite. It follows that $\varphi(x)$, $\neg\varphi(x)$ are not algebraic. Each of these formulae can be extended to a non-algebraic type over A, so we obtain a counterexample to (2). To see why e.g. $\varphi(x)$ can be extended to a non-algebraic type, consider the set $p = \{\neg\psi(x) : \psi(x) \text{ is algebraic over } A\}$, where $\psi(x)$ is a formula in the language of $\varphi(x)$. A finite conjunction of elements of S is equivalent to $\neg \bigvee_{i=1}^{m} \psi_i(x)$, with $\psi_i(x)$ algebraic for each $i \in \{1, \ldots, m\}$. Because $\bigvee_{i=1}^{m} \psi_i(x)$ defines a finite subset of M and M is infinite, its negation must have a solution in M. In short, p is a type. Note, furthermore, that, for each $\mathcal{L}$-formula $\psi(x)$, either $\psi(x)$ is algebraic (in which case $\neg\psi(x) \in p$) or it is not (in which case $\neg\neg\psi(x) \in p$), i.e. p is a complete type. $\square$

The last result entails:

Lemma 10.2 *For any $\mathcal{M} \models \mathsf{ACF}$ and any $A \subseteq M$, the following conditions are equivalent.*

(1) $a \in acl_{\mathcal{M}}(A)$;
(2) a is algebraic over $\mathcal{A}$, the subfield of M generated by A.

Proof The lemma is clear in light of our earlier verification that model-theoretic algebraic closure specialises to the field-theoretic notion in models of ACF. We can, however, provide an alternative argument that relies on strong minimality. Since it is clear that (2) implies (1), we only need to show that the converse implication holds. We do so by contraposition. Suppose a is not algebraic over $\mathcal{A}$. Then it is transcendental over $\mathcal{A}$. For each equality of the form $f(x) = 0$, where $f(x)$ is a polynomial with coefficients in the domain of $\mathcal{A}$, $tp(a/A)$ contains the atomic formula $\neg(f(x) = 0)$. Quantifier elimination ensures that $tp(a/A)$ is uniquely determined by its literals. Because every other type is algebraic, $tp(a/A)$ must be the unique non-algebraic type in $S_1(A)$. Since $a \in acl_\mathcal{M}(A)$ implies that $tp(a/A)$ is algebraic, we conclude $a \notin acl_\mathcal{M}(A)$. □

Exercise 10.3 In this exercise we describe the complex field C in two different ways. In Chap. 11 we shall be able to prove that this field is a model of ACF_0.

(i) Let $\mathbb{R}^2$ be the set of ordered pairs of reals. Define two binary operations $\oplus, \odot$ on $\mathbb{R}^2$ by means of the following stipulations: $(a,b) \oplus (c,d) = (a+c, b+d)$ and $(a,b) \odot (c,d) = (ac-bd, ad+bc)$. Show that the $\mathcal{L}_r$-structure $(\mathbb{R}^2, \oplus, \odot)$ is a field.
(ii) Consider the polynomial ring $\mathbb{R}[x]$. The quotient ring $\mathbb{C} = \mathbb{R}[x]/(x^2 + 1)$ is a field. Show that the $\mathcal{L}_r$-structure $(\mathbb{C}, +^\mathbb{C}, \cdot^\mathbb{C}, 0^\mathbb{C}, 1^\mathbb{C})$ is isomorphic to the field constructed in part (i).

10.1.1 Infinite Fields with Quantifier Elimination

In a classic study [48], Angus McIntyre, Kenneth Gerard McKenna and Lou van den Dries systematically explored the algebraic content of quantifier elimination relative to several classes of fields. The first theorem in their article shows that quantifier elimination characterises algebraically closed fields.

More precisely, they show that an infinite field with quantifier elimination is a model of ACF. This is to say that, over the class of infinite fields, the model-theoretic property of having quantifier elimination specialises to the algebraic property of being algebraically closed (relative to the language of rings).

In order to establish this result, we make use of a few ideas and results from field theory not previously covered. For present purposes, it is sufficient to state key definitions and results. The reader interested in more detail or a slower-paced and more informative discussion may consult e.g. [30], pp.79–84 and pp.110–115.

Recall that, if $\mathcal{K}$ is a field and $f(x) \in K[x]$ does not have some of its roots in K, we can always build a field extension $\mathcal{M}$ of $\mathcal{K}$ in which $f(x)$ can be expressed as a product of linear factors (whose exponents could in general be greater than 1). In this case we say that $f(x)$ **splits completely** over $\mathcal{M}$. The smallest field extension of $\mathcal{K}$ (relative to inclusion) over which $f(x)$ splits completely is a **splitting field** for $f(x)$. It can be shown that, up to isomorphism, the splitting field is unique.

Definition 10.4 Let $\mathcal{K}$ be a field. An irreducible polynomial $f(x) \in K[x]$ is said to be **separable over** K iff its roots in a splitting field for $\mathcal{K}$ are all distinct. A polynomial $f(x) \in K[x]$ is said to be **separable over** K iff its irreducible factors are. If $\mathcal{M} \supseteq \mathcal{K}$ and $\alpha \in M$ is algebraic over K, we say that α is **separable over** K iff the minimal polynomial of α is separable over K.

We require two more field-theoretic definitions.

Definition 10.5 Let $\mathcal{K}$ be a field, $\overline{\mathcal{K}}$ its algebraic closure. We say that $\mathcal{K}$ is **separably closed** iff every separable element of $\overline{K}$ is already an element of K.

Definition 10.6 Let $\mathcal{K}$ be a field. We say that $\mathcal{K}$ is **perfect** iff every irreducible polynomial in $K[x]$ is separable.

Every field of characteristic 0 is perfect. We shall shortly see next exercise that any separably closed field of characteristic 0 is a model of ACF_0. If, on the other hand, we have a separably closed field $\mathcal{K}$ of prime characteristic $p > 0$, we cannot immediately deduce from this fact that $\mathcal{K} \models \mathsf{ACF}_p$. To this end, we have to verify the equality $K = K^p = \{a^p : a \in K\}$ (note that one inclusion is trivial). We record the last remark as a lemma:

Lemma 10.7 *Let $\mathcal{K}$ be a field of positive characteristic p. Then $\mathcal{K}$ is perfect iff $K^p = K$.*

Exercise 10.8 Show that a separably closed, perfect field is algebraically closed.

The final notion we need before we can embark on a proof of our main theorem is that of a **symmetric polynomial** in n variables $s_1, \ldots, s_n$ over a fixed field $\mathcal{K}$. A polynomial $f(\overline{s}) \in K[\overline{s}]$ is symmetric iff it is invariant under every permutation of the set $\{1, \ldots, n\}$. It is easy to see that the polynomials $s_1 + s_2 + s_3$ and $s_1 s_2 s_3$, regarded as elements of $\mathbb{Q}[x_1, x_2, x_3]$, are symmetric in the sense just described.

We record one easy lemma:

Lemma 10.9 *Let $\mathcal{K}$ be a field. The coefficients of the polynomial:*

$$\Pi_{i=1}^n (x - s_i) = (x - s_1)(x - s_2) \cdots (x - s_n) \in K[s_1, \ldots, s_n]$$

are symmetric polynomials in $s_1, \ldots, s_n$.

Proof By induction on the number of variables. □

The coefficients of $\Pi_{i=1}^n (x - s_i)$ distinct from 1 are polynomial functions of $s_1, \ldots, s_n$: let $\sigma_k(\overline{s})$ be the coefficient of x^{k-1} ($n \leq k \leq 1$). The polynomials $\sigma_i(\overline{s})$ ($i = 1, \ldots, n$) are called the **elementary symmetric polynomials** in $s_1, \ldots, s_n$. We shall appeal to two fundamental properties of elementary symmetric polynomials. The first is declared in the statement of:

Theorem 10.10 *Let $\mathcal{K}$ be a field not containing the variables $s_1, \ldots, s_n$. The elementary symmetric polynomials are algebraically independent over K.*

A proof may be found in [30], p.180. The next, fundamental property of elementary symmetric polynomials, which will also be invoked once in Chap. 11, to prove the Artin-Schreier theorem, is stated below:

10.1 Specialisations

Theorem 10.11 *Every symmetric polynomial in the variables $s_1, \ldots, s_n$ with coefficients in K can be expressed as a polynomial function of $\sigma_1(\bar{s}), \ldots, \sigma_n(\bar{s})$ with coefficients in K (i.e. as an element of the ring $K[\sigma_1(\bar{s}), \ldots, \sigma_n(\bar{s})]$.*

A proof of Theorem 10.11 may be found in [12], pp.347-348.

We have now covered the background needed for the long proof of our main result, namely:

Theorem 10.12 (McIntyre, McKenna, van den Dries) *Let $\mathcal{K}$ be an infinite field with quantifier elimination. Then $\mathcal{K} \models$ ACF.*

Proof The proof's strategy is to show that $\mathcal{K}$ is separably closed. In 0 characteristic, there is nothing left to prove. In positive characteristic, we have to check that $K^p = K$ holds.

Let us assume that $\mathcal{K}(\alpha)$, is a nontrivial algebraic extension of $\mathcal{K}$ (i.e. $[K(\alpha) : K] > 1$) and that α is separable over K. It follows that $\alpha \notin K$, i.e. $\mathcal{K}$ is not separably closed. Our goal is to derive a contradiction from this assumption.

We consider the atomic formula $\theta(y, \bar{x})$:

$$y^n + x_{n-1}y^{n-1} + \ldots + x_1 y + x_0 = 0$$

and note that, by quantifier elimination, there is a quantifier-free formula $\varphi(\bar{x})$ in the language of rings such that:

$$\mathcal{K} \models \forall \bar{x}[\forall y(\neg \theta(y, \bar{x})) \leftrightarrow \varphi(\bar{x}))].$$

We may assume that $\varphi(\bar{x})$ is in disjunctive normal form. Our immediate aim is to prove that $\varphi(\bar{x})$ has a disjunct without positive literals, which can be written as $\neg(q(\bar{x}) = 0)$ (if more than one negated polynomial equality occurs, we take the product of the polynomials involved and reduce ourselves to one negated equality).

To this end we introduce an auxiliary polynomial $F(y, \bar{z})$, presently to be described. Because α is separable, its minimal polynomial, of degree n, has n distinct roots $\alpha = \alpha_1, \ldots, \alpha_n$ in a splitting field. We associate α_i with the term $z_1 + z_2 \alpha_i + \ldots + z_n \alpha_i^{n-1}$, where $z_1, \ldots, z_n$ are new variables, different from $\bar{x}$ and y.

The auxiliary object we focus on is the polynomial:

$$F(y, \bar{z}) = \prod_{i=1}^{n}(y - (z_1 + z_2 \alpha_i + \ldots z_n \alpha_i^{n-1})),$$

which, once we carry out the long multiplication, may be rewritten as:

$$F(y, \bar{z}) = F_0(\bar{z}) + F_1(\bar{z})y + \ldots + F_{n-1}(\bar{z})y^{n-1} + y^n.$$

It is crucial to observe that $F(y, \bar{z})$ is, by construction, a symmetric function of $\alpha_1, \ldots, \alpha_n$. In a suitable field extension of $\mathcal{K}$, the minimal polynomial of α can be written as the product $(y - \alpha_1) \cdots (y - \alpha_n)$, whose coefficients are the elementary symmetric polynomials in $\alpha_1, \ldots, \alpha_n$. Because these coefficients are in K, any symmetric polynomial in $\alpha_1, \ldots, \alpha_n$ is an element of K by Theorem 10.11.

It follows that $F(y, \bar{z})$, a symmetric function of $\alpha_1, \ldots, \alpha_n$, has coefficients in K, i.e. is a polynomial in the ring $K[y, \bar{z}]$.

We now show that the set $\{F_0(\bar{z}), \ldots, F_{n-1}(\bar{z})\}$ is algebraically independent over K. We first observe that the set $\{z_1, \ldots, z_n\}$ is algebraically independent over K and the transcendence degree of $\mathcal{K}(z_1, \ldots, z_n)$ over K is n. Moreover, if each z_i ($i = 1, \ldots, n$) is algebraic over $\mathcal{K}(F_0(\bar{z}), \ldots, F_{n-1}(\bar{z}))$, the field $\mathcal{K}(z_1, \ldots, z_n)$ is an algebraic extension of $\mathcal{K}(F_0(\bar{z}), \ldots, F_{n-1}(\bar{z}))$. Consequently, the transcendence degree of $K(F_0(\bar{z}), \ldots, F_{n-1}(\bar{z}))$, which must be $\leq n$, cannot be strictly smaller than n, which is what we eventually wish to conclude. In order to reach this conclusion, it suffices to verify the hypothesis on which it rests, i.e. that each z_i ($i = 1, \ldots, n$) is algebraic over $\mathcal{K}(F_0(\bar{z}), \ldots, F_{n-1}(\bar{z}))$.

We now regard $F(y, \bar{z})$ as a polynomial with coefficients in $K(F_0(\bar{z}), \ldots, F_{n-1}(\bar{z}))$, whose roots in a splitting field are algebraic over the field $K(F_0(\bar{z}), \ldots, F_{n-1}(\bar{z}))$. By the definition of $F(y, \bar{z})$ as a product of factors linear in y, its roots $r_1, \ldots, r_n$ are:

$$r_1 = z_1 + z_2\alpha_1 + \ldots + z_n\alpha_1^{n-1}; \ldots; r_n = z_1 + z_2\alpha_n + \ldots + z_n\alpha_n^{n-1}.$$

Using matrix notation, we can write the list of roots $r_1, \ldots, r_n$ as the matrix product:

$$\begin{pmatrix} z_1 & z_2 & \cdots & z_n \end{pmatrix} \cdot \begin{pmatrix} 1 & 1 & \cdots & 1 \\ \alpha_1 & \alpha_2 & \cdots & \alpha_n \\ \vdots & \vdots & \ddots & \vdots \\ \alpha_1^{n-1} & \alpha_2^{n-1} & \cdots & \alpha_n^{n-1} \end{pmatrix} = \begin{pmatrix} r_1 & r_2 & \cdots & r_n \end{pmatrix}$$

in which r_i is obtained by component-wise multiplication of the first factor times the i-th column of the second factor. The $n \times n$ matrix above is known as a Vandermonde matrix, whose determinant is nonzero because $\alpha_1, \ldots, \alpha_n$ are all distinct. We omit a definition of determinant (for which the reader is directed to e.g. chapter 8 of [7]) and a proof of the last assertion (which may be found e.g. in [72], pp.276–277). What matters to our present purposes is simply that the $n \times n$ matrix M is invertible.

This means that, if rewrite the above product as $\bar{z} \cdot M = \bar{r}$, there is a $n \times n$ matrix M^{-1} such that $\bar{z} = \bar{r}M^{-1}$. Because the entries of M^{-1} only depend on $\alpha_1, \ldots, \alpha_n$ and these field elements are algebraic over K, while each r_i is, as we have already noted, algebraic over $K(F_0(\bar{z}), \ldots, F_{n-1}(\bar{z}))$, we deduce that $z_1, \ldots, z_n$ are algebraic over $K(F_0(\bar{z}), \ldots, F_{n-1}(\bar{z}))$, as desired.

We now claim that, for any n-tuple $\bar{k} \in K^n$ different from the constant n-tuple $\bar{0}$:

$$\mathcal{K} \models \varphi(F_0(\bar{k}), \ldots, F_{n-1}(\bar{k})), \tag{10.1}$$

10.1 Specialisations

where $\varphi(x_1, \ldots, x_n)$ was defined at the beginning of our proof. To establish the claim, we observe that, for any fixed $\bar{k}$ other than $\bar{0}$, $k_1 + k_2\alpha_j + \ldots + k_n\alpha_j^{n-1} \notin K$. If not, the polynomial:

$$k_n x^{n-1} + \ldots + k_2 x + (k_1 - (k_1 + k_2\alpha_j + \ldots + k_n\alpha_j^{n-1})),$$

of degree $n-1$, would be in $K[x]$ and have α_j as a root, against the hypothesis that the minimal polynomial of α (and thus of each α_j) is of degree n. Now given the form of r_j, the j-th root of $F(y, \bar{z})$, it follows from what we have just observed that $F(y, \bar{k})$ cannot have any root in K. In other words:

$$\mathcal{K} \models \forall y (\neg(F_0(\bar{k}) + F_1(\bar{k})y + \ldots + F_{n-1}(\bar{k})y^{n-1} + y^n = 0)).$$

Because we have chosen $\varphi(y, \bar{x})$ to be a quantifier-free equivalent of the last formula, we have established (10.1) for an infinitely large set of n-tuples. We now consider the set:

$$T = \{(F_0(\bar{k}), \ldots, F_{n-1}(\bar{k})) : \bar{k} \in K - \{\bar{0}\}\}.$$

The set T cannot be the set of roots of a polynomial $f(\bar{x}) \in K[\bar{x}]$ or it should be possible to describe a nontrivial polynomial function of $F_0(\bar{z}), \ldots, F_{n-1}(\bar{z})$ that takes the constant value 0 over K^n (i.e. $f(F_0(\bar{x}), \ldots, F_{n-1}(\bar{x})) \cdot (F_0(\bar{x}) + \ldots + F_{n-1}(\bar{x}) - (F_0(\bar{0}) + \ldots + F_{n-1}(\bar{0}))))$, against the fact that $F_0(\bar{k}), \ldots, F_{n-1}(\bar{k})$ are algebraically independent over K.

In fact, the argument just given shows that T cannot even be included in the set of roots of one or finitely many polynomials in $K[\bar{x}]$ (the reduction from finitely many polynomials to one is obtained by summing the given polynomials).

As a consequence, $\varphi(y, \bar{x})$ must contain a disjunct in which positive literals do not occur. If not, the formula $\varphi(F_0(\bar{x}) + \ldots + F_{n-1}(\bar{x}))$ defines a subset of polynomial roots that includes T. We have shown that $\varphi(y, \bar{x})$ must contain a disjunct of the form $\neg(q(\bar{x}) = 0)$ (if we had more than one negative literal, we could reduce ourselves to a single negative literal by taking the product of the polynomials occurring in each literal and setting it to be different from zero). In other words, the formula $\neg(q(\bar{x}) = 0)$ defines a nonempty set.

Pick $\bar{k}$ such that $\mathcal{K} \models \neg(q(\bar{k}) = 0)$. If we satisfy one disjunct of $\varphi(\bar{x})$, we obviously satisfy the formula as a whole, so $\mathcal{K} \models \varphi(\bar{k})$. The last condition is equivalent to the fact that the polynomial $f(y, \bar{k})$ has no roots in K. We are now in a position to derive the contradiction we are after. To this end, we introduce the polynomial:

$$G(y, \bar{z}) = \prod_{i=1}^{n}(y - z_i) = \sigma_1(\bar{z}) + \sigma_2(\bar{z})y + \ldots + \sigma_n(\bar{z})y^{n-1} + y^n,$$

where $\sigma_i(\bar{z})$ is the i-th elementary, symmetric polynomial in $\bar{z}$. Clearly, for any $\bar{k} \in K^n$, the solutions of $G(y, \bar{k})$ all lie in K. By Theorem 10.10, $\{\sigma_i(\bar{z})\}_{1 \leq i \leq n}$ is algebraically independent over K. It follows that $q(\sigma_1(\bar{z}), \ldots, \sigma_n(\bar{z}))$ cannot be constant and equal to zero. In other words, there is a n-tuple $\bar{k} \in K^n$ such that $q(\sigma_1(\bar{k}), \ldots, \sigma_n(\bar{k})) \neq 0$.

We deduce:

$$\mathcal{K} \models \varphi(\sigma_1(\bar{k}), \ldots, \sigma_n(\bar{k})),$$

which implies:

$$\mathcal{K} \models \neg(f(y, \sigma_1(\bar{k}), \ldots, \sigma_n(\bar{k})) = 0).$$

The last condition contradicts the fact that $f(y, \sigma_1(\bar{k}), \ldots, \sigma_n(\bar{k}))$ is $G(y, \bar{k})$, whose roots all lie in K. The contradiction we have reached was based on the assumption that the field element α, separable over K, was not in K. We conclude that $\mathcal{K}$ is separably closed. If its characteristic is zero, then it is also a perfect field and, by Exercise 10.8, $\mathcal{K} \models \mathsf{ACF}_0$. If its characteristic is $p > 0$, it suffices to show $K^p = K$ by Lemma 10.7. We do so by proving that $K - K^p$ must be empty.

Since $K - K^p$ is defined by the formula $\neg \exists x(x^p - y = 0)$ and quantifier elimination holds, $\mathcal{K}$ is minimal. In particular, $K - K^p$ must be either finite or cofinite. If the last set were nonempty, it could be neither. To see why, view $\mathcal{K}$ as a vector space over the subfield with domain K^p. Then K^p is infinite because K is infinite by hypothesis, the function sending x into x^p is a field homomorphism[1] and field homomorphisms are injective.

If $a \in K - K^p$, then, for any two distinct, nonzero scalars $\lambda, \mu \in K^p$, $\lambda a \neq \mu a$, so $K - K^p$ is infinite and has an infinite complement in K, against the minimality of $\mathcal{K}$. We conclude that $K - K^p = \emptyset$, i.e. $K = K^p$. Once again, $\mathcal{K}$ is separably closed and perfect: consequently, $\mathcal{K} \models \mathsf{ACF}_p$. □

Remark 10.13 The last theorem implies that the $\mathcal{L}_r$-structure $(\mathbb{Q}, +, -, \cdot, 0, 1)$ does not admit quantifier elimination (i.e. its $\mathcal{L}_r$-theory does not have quantifier elimination). We shall prove in the next chapter that the relative algebraic closure of $\mathbb{Q}$ in $\mathbb{R}$, also known as the set of real algebraic numbers $\mathbb{R}_{alg}$, is not algebraically closed. It follows that the $\mathcal{L}_r$-structure $(\mathbb{R}_{alg}, +, -, \cdot, 0, 1)$ does not have quantifier elimination.

---

[1] It is clear that this function preserves field multiplication. Moreover, since, in characteristic $p \neq 0$, $(x + y)^p = x^p + x^p$ (as may be shown by noting that, once we expand the p-th power of the given binomial, the only nonzero terms are those with coefficients that are not multiples of p), field addition is preserved as well.

10.2 Elements of Algebraic Geometry

In this section we consider a fixed model $\mathcal{K}$ of **ACF** and associate with it the **affine n-space** $\mathbb{A}^n$, which is simply the Cartesian product K^n. We call the parametrically definable subsets of K^n (in the language $\mathcal{L}_r$) **affine algebraic** or simply **algebraic** sets (the adjective **constructible** is also used with the same meaning). Because of quantifier elimination, an algebraic set is a finite union of sets defined by conjunctions of literals. When we only have one disjunct and only positive literals in it, we reduce ourselves to the important notion of an algebraic variety.

Definition 10.14 Let $V \subseteq \mathbb{A}^n$. We say that V is an **algebraic variety** iff there is a conjunction of positive $\mathcal{L}_r$-literals $\psi(x_1, \ldots, x_n)$ with parameters in K such that $\psi(\mathbb{A}^n) = V$.

The notion of algebraic variety that we have introduced in model-theoretic terms has an equivalent algebraic formulation in terms of ideals. A conjunction of positive $\mathcal{L}_r$-literals in $x_1, \ldots, x_n$ can always be reduced to the parametric form:

$$p_1(x_1, \ldots, x_n) = 0 \wedge \ldots \wedge p_k(x_1, \ldots, x_n) = 0.$$

where $p_j(x_1, \ldots, x_n) \in K[x_1, \ldots, x_n]$, $j = 1, \ldots, k$. The set V defined by the last quantifier-free formula is just the set of points in affine n-space on which the polynomials $p_1, \ldots, p_k$ simultaneously vanish.

Exercise 10.15 Let S be a nonempty set of polynomials in $K[x_1, \ldots, x_n]$.

(a) Let J be the smallest ideal in $K[x_1, \ldots, x_n]$ that contains S. Show that, if V is the set of points on which every polynomial in S vanishes, then V is also the set of points on which every polynomial in J vanishes.
(b) Let V be as in part (1). Use Hilbert's basis theorem to explain why V is a variety in the model-theoretic sense.

We can indifferently make use of the model-theoretic definition of a variety in terms of definability and of the ring-theoretic one, on which a variety is the set of points on which the polynomials that belong to an ideal of $K[x_1, \ldots, x_n]$ vanish. At a very basic level, algebraic geometry studies the interplay between the algebraic properties of the ideals in $K[x_1, \ldots, x_n]$ and the geometrical properties of the sets of 'zeros' of polynomials associated with these ideals (which may be e.g. lines, planes, conics, or more complicated surfaces).

The model-theoretic properties of algebraically closed fields allow a distinctively model-theoretic study of this interplay. In the remainder of this section we shall show, in particular, that a fundamental theorem of algebraic geometry known as Hilbert's *Nullstellensatz* can be elegantly obtained by model-theoretic means. It can be proved by algebraic means that an infinite field $\mathcal{K}$ is algebraically closed iff it satisfies the *Nullstellensatz*. Thus quantifier elimination for **ACF**, or indeed model completeness, is in fact equivalent to Hilbert's fundamental result.

In order to state the *Nullstellensatz* and to understand what it actually says, we need to acquire a little familiarity with a few basic facts concerning the connection between ideals of polynomials and sets of zeros associated with them. To survey such facts is the main goal of the next subsection. In closing this one, we note that quantifier elimination for ACF is equivalent to a result known as Chevalley's theorem.

Theorem 10.16 *The projection of an algebraic set is algebraic.*

Proof Let $U \subseteq \mathbb{A}^n$ be an algebraic set in n-affine space. Then U is defined by a quantifier-free formula $\varphi(x_1, \ldots, x_n)$. A projection U_1 of U, say on the axes $x_2, \ldots, x_n$, is defined by the formula $\exists x_1 \varphi(x_1, x_2, \ldots, x_n)$. Because U_1 is definable over K, it is defined by a quantifier-free formula, i.e. it is an algebraic set. □

10.2.1 The Fundamental Correspondence

In Sect. 3.1 we considered two operations s, t that linked sets of sentences and classes of structures, thus bridging first-order syntax and semantics. In the context now at hand we encounter a similar correspondence, which bridges algebra and geometry. The operations we now have to do with assign ideals to varieties and varieties to ideals respectively. We first describe an operation $\mathcal{Z}$, which associates an ideal of polynomials with the variety on which its elements vanish. In symbols, if J is an ideal of $K[x_1, \ldots, x_n]$:

$$\mathcal{Z}(J) = \{(a_1, \ldots, a_n) \in K^n : \text{for each } f \in J, f(a_1, \ldots, a_n) = 0\}$$

Conversely, when $V \subseteq \mathbb{A}^n$ is variety, we define, we define a second operation $\mathcal{I}$, defined by the equality:

$$\mathcal{I}(V) = \{f \in K[x_1, \ldots, x_n] : f(a_1, \ldots, a_n) = 0 \text{ for each } (a_1, \ldots, a_n) \in V\}$$

It is instructive to check some easy properties of the operations of $\mathcal{Z}, \mathcal{I}$, given in the next exercise:

Exercise 10.17 Let $\mathcal{Z}, \mathcal{I}$ be as above.

(i) If $I \subseteq J$ are ideals of $K[x_1, \ldots, x_n]$, then $\mathcal{Z}(J) \subseteq \mathcal{Z}(I)$.
(ii) If $U \subseteq V$ are varieties of $\mathbb{A}^n$, then $\mathcal{I}(V) \subseteq \mathcal{I}(U)$.
(iii) $\mathcal{Z}(0) = \mathbb{A}^n$ (here 0 stands for the principal ideal (0) generated by 0).
(iv) $\mathcal{Z}(1) = \emptyset$ (here 1 stands for the principal ideal $(1) = K[x_1, \ldots, x_n]$).
(v) If J is an ideal of $K[x_1, \ldots, x_n]$, then $J \subseteq \mathcal{I}(\mathcal{Z}(J))$.
(vi) If V is a variety of $\mathbb{A}^n$, then $V \subseteq \mathcal{Z}(\mathcal{I}(V))$.
(vii) For any field $\mathcal{K}$, $\mathcal{I}$ is injective and $\mathcal{Z}$ is surjective (as we shall see, when $\mathcal{K}$ is algebraically closed, we obtain two one-to-one correspondences).

10.2 Elements of Algebraic Geometry

If (v), (vi) were equalities, we could view the operations $\mathcal{Z}, \mathcal{I}$ as inverses of each other. At present we are only able to prove the following:

Lemma 10.18 *If V is a variety, $V = \mathcal{Z}(\mathcal{I}(V))$. Moreover, if J is an ideal of the form $\mathcal{I}(V)$, for some variety V, then $J = \mathcal{I}(\mathcal{Z}(J))$.*

Proof For each part of the lemma we only have to establish one inclusion. The first inclusion is $\mathcal{Z}(\mathcal{I}(V)) \subseteq V$. We appeal to our definition of a variety to note that $V = \mathcal{Z}(J)$ for some ideal J of $K[x_1, \ldots, x_n]$. By Exercise 10.17-(v), $J \subseteq \mathcal{I}(\mathcal{Z}(J)) = \mathcal{I}(V)$, i.e. $J \subseteq \mathcal{I}(V)$. Every polynomial in $\mathcal{I}(V)$ vanishes at each $(a_1, \ldots, a_n) \in \mathcal{Z}(\mathcal{I}(V))$: *a fortiori*, this must be the case for every polynomial in J. It follows that $a \in \mathcal{Z}(J) = V$.

The second inclusion we wish to establish is $\mathcal{I}(\mathcal{Z}(J)) \subseteq J$, under the hypothesis $J = \mathcal{I}(V)$. By the previous part of the lemma we can deduce $\mathcal{Z}(J) = \mathcal{Z}(\mathcal{I}(V)) = V$. The conclusion immediately follows. □

Note that the equality $V = \mathcal{Z}(\mathcal{I}(V))$ could be proved by a simple appeal to the definition of variety whereas the equality $J = \mathcal{I}(\mathcal{Z}(J))$ required a special hypothesis on the ideal J, namely that it was of the form $\mathcal{I}(V)$. Relative to an algebraically closed field, Hilbert's *Nullstellensatz* isolates a key algebraic property of ideals that corresponds to the geometric property of arising as the 'zero' set of a variety in affine n-space. A separate discussion of the relevant property, given in the next subsection, is a necessary preliminary to our model-theoretic proof of the *Nullstellensatz*.

10.2.2 Radical and Primary Ideals

Let V be a variety in $\mathbb{A}^n$. The ideal $\mathcal{I}(V)$ has one distinctive feature, i.e. it contains the polynomial f iff it contains any power f^m of f. If $f^m(a_1, \ldots, a_n) = 0$, the fact that $\mathcal{K}$ is, in particular, an integral domain, ensures that $f(a_1, \ldots, a_n) = 0$. The converse implication is clear. As we shall see, this condition on powers of polynomials characterises the ideals of the form $\mathcal{I}(V)$. We isolate it by means of:

Definition 10.19 Let $\mathcal{R}$ be a commutative ring with unity and $I \subseteq R$ be an ideal. The **radical** of I is the set $\sqrt{I} = \{a \in R : a^m \in I \text{ for some } m \geq 1\}$. An ideal I is said to be a **radical ideal** iff it coincides with its radical.

Because a general commutative ring with unity, which is not necessarily an integral domain, may well contain non-zero elements with an integer power equal to zero, it is important to single them out by an appropriate definition.

Definition 10.20 Let $\mathcal{R}$ be a commutative ring with unity and $I \subseteq R$ be an ideal. An element $a \in R$ is said to be **nilpotent** iff $a^m = 0$ for some integer $m \geq 1$. The set of nilpotent elements of $\mathcal{R}$ is called the **nilradical** of $\mathcal{R}$.

We shall now verify that, if I is an ideal in a commutative ring with unity, $\sqrt{I} \supseteq I$ is in fact an ideal (that it extends I follows from the definition of radical).

Lemma 10.21 *Let $\mathcal{R}$ be a commutative ring with unity, $I \subseteq R$ an ideal. Then $\sqrt{I}$ is an ideal.*

Proof It suffices to prove that the nilradical of $\mathcal{R}$ is an ideal. To see why, note that that, once the quotient ring R/I is taken, the nilradical of this quotient ring contains the cosets of the form $a + I$ and such that $a^m + I = I$ iff $a^m \in I$. In other words, $a \in rad\,I$ iff a is in the nilradical of R/I, where I plays the role of 0 in the additive group of $\mathcal{R}$. If we could prove that the nilradical of the ring R/I was an ideal extending the ring's zero element, we would have proved that $\sqrt{I}$ was a radical extending I, on account of the correspondence between ideals in R/I and ideals extending $I \subseteq R$ (see Remark 3.41).

In view of the above remarks, we show that, given an arbitrary ring $\mathcal{R}$, its nilradical N is an ideal. We leave it as an exercise to verify that, if $a \in N$ and $r \in R$, then $ra \in N$. If $a, b \in N$, it is clear that $-b = (-1)b \in N$ so it remains to show that $a + b \in N$. Suppose $a^n = 0 = b^m$. Because the Binomial Theorem holds for commutative rings, we can write:

$$(a+b)^{m+n} = \sum_{i=1}^{n+m} r_i a^i b^{n+m-1}.$$

Our goal is to show that the last expression equals zero, which implies $a+b \in N$. The product $a^i b^{n+m-i}$ is such that $i \geq n$ or $i < n$. In the former case, $a^i = 0$ and the product equals zero. In the latter case, $n + m - i = (n-i) + m \geq m$ and $b^{n+m-i} = 0$. The whole sum must therefore equal zero. □

Hilbert's *Nullstellensatz* amounts to the fact that, for a radical ideal $J \in K[x_1, \ldots, x_n]$, with $\mathcal{K}$ algebraically closed, the equality $\mathcal{I}(\mathcal{Z})(J) = J$ holds. Its model-theoretic proof rests on an important decomposition theorem for ideals. In order to clarify what a decomposition is in this context, we introduce:

Definition 10.22 Let $\mathcal{R}$ be a commutative ring with unity and $Q \subseteq R$ be an ideal. We say that Q is **primary** iff $\sqrt{Q}$ is a prime ideal.[2]

Exercise 10.23 Let $\mathcal{R}$ be a commutative ring with unity and $Q \subseteq R$ be an ideal.

(i) Show that Q is primary iff $ab \in Q$ implies $a \in Q$ or $b^m \in Q$ for some integer $m \geq 1$.
(ii) Show that if Q is prime, then Q is primary.
(iii) Show that Q is primary iff every zero divisor in the quotient R/Q is nilpotent.
(iv) Verify that, in the $\mathcal{L}_r$-structure $(\mathbb{Z}, +, -, \cdot, 0, 1)$, the ideal (p^m), with p a prime number, is primary, and that $\sqrt{(p^m)} = p$.

---

[2] See Definition 3.34 from Sect. 3.4.

10.2 Elements of Algebraic Geometry

Example 10.24 We know that, in the $\mathcal{L}_r$-structure $(\mathbb{Z}, +, -, \cdot, 0, 1)$, every integer n has a prime factor decomposition. In other words $n = p_1^{k_1} \cdots p_m^{k_m}$, where $p_1, \ldots, p_m$ are distinct primes. In terms of ideals, this is to say that $n \in (p^{k_1}) \cap \ldots \cap (p^{k_m})$ and, since any integer multiple of n is an integer multiple of its decomposition as a product of prime factors, $(n) = (p^{k_1}) \cap \ldots \cap (p^{k_m})$. By exercise 10.23-(iv), every ideal that appears in the last intersection is primary. Thus, in the Noetherian ring $(\mathbb{Z}, +, -, \cdot, 0, 1)$, prime factor decomposition may be viewed as a special case of the decomposition of an ideal into primary ideals.

In view of the last example, the following definition should not be surprising:

Definition 10.25 Let $\mathcal{R}$ be a commutative ring with unity and $I \subseteq R$ an ideal. We say that I has a **primary decomposition** if there are primary ideals $Q_1, \ldots, Q_m$ such that:

$$I = Q_1 \cap \ldots \cap Q_m.g$$

We say, moreover, that the given primary decomposition is **minimal** iff:

- no primary ideal Q_j ($1 \leq j \leq m$) includes the intersection of the remaining ideals;
- for any $i \neq j \in \{1, \ldots, m\}$ $\sqrt{Q_i} \neq \sqrt{Q_j}$.

The ideals $Q_1, \ldots, Q_m$ determining a minimal primary decomposition of I are called the **primary components** of I.

Any ideal in a Noetherian ring has a minimal primary decomposition (see e.g. [14], pp.681–685). Since $K[x_1, \ldots, x_n]$ is Noetherian by Hilbert's basis theorem, ideals of polynomials have minimal primary decompositions. We rely on this fact to obtain a characterisation of radical ideals in terms of their primary components.

Theorem 10.26 *Let $\mathcal{R}$ be a Noetherian ring. An ideal I is radical iff the primary components of a minimal primary decomposition of I are all prime ideals.*

Proof Suppose $I = \bigcap_{i=1}^m Q_i$, where $Q_1, \ldots, Q_m$ are prime ideals determining a minimal primary decomposition of I. Using the fact that prime ideals are radical (see the parenthetical remark in Definition 10.19), we obtain the following chain of equalities:

$$\begin{aligned}\sqrt{I} &= \sqrt{Q_1 \cap \ldots \cap Q_m} \\ &= \sqrt{Q_1} \cap \ldots \cap \sqrt{Q_m} \\ &= Q_1 \cap \ldots \cap Q_m \\ &= I\end{aligned}$$

which shows I to be a radical ideal. We leave it as an exercise to check that the radical of an intersection of ideals is the intersection of the corresponding radicals.

To prove the converse, we suppose I to be a radical ideal and consider a minimal primary decomposition $I = \bigcap_{i=1}^{m} Q_i$ with primary components $Q_1, \ldots, Q_m$. By the definition of primary ideal, radicals of primary ideals are prime: we thus focus on the prime ideals $\sqrt{Q_i} = P_i$, $i = 1, \ldots, m$. Because radicals extend their corresponding ideals, we have the inclusion:

$$I = \bigcap_{i=1}^{m} Q_i \subseteq \bigcap_{i=1}^{m} P_i.$$

We only need to check that the converse inclusion holds. To this end, we suppose $a \in \bigcap_{i=1}^{m} P_i$, from which it follows that $a^{k_i} \in Q_i$ with $i = 1, \ldots, m$. We next set $k = max\{k_1, \ldots, k_m\}$. It is easy to see that $a^k \in Q_i$ with $i = 1, \ldots, m$, which in turn implies $a^k \in I$ and thus $a \in \sqrt{I} = I$. Using the original decomposition of I, we conclude $a \in Q_i$ for each $i \in \{1, \ldots, m\}$. We have proved:

$$I = \bigcap_{i=1}^{m} Q_i = \bigcap_{i=1}^{m} P_i.$$

The set $\{P_1, \ldots, P_m\}$ determines a minimal primary decomposition of I (see Exercise 10.27). □

It can be proved that the decomposition of I into prime ideals is unique (it can more precisely be shown that $P_i = Q_i$ for each $i \in \{1, \ldots, m\}$ and a general uniqueness result for minimal primary decompositions can be invoked to complete the argument), but we do not need this additional piece of information for the *Nullstellensatz*, which is our central goal.

Exercise 10.27 Complete the proof of Theorem 10.26. In particular:

(i) show that, if Q_1, Q_2 are ideals in a commutative ring, then $\sqrt{Q_1 \cap Q_2} = \sqrt{Q_1} \cap \sqrt{Q_2}$;
(ii) show that $P_i \neq P_j$ for any $i \neq j \in \{1, \ldots, m\}$ and note that each P_i is a radical ideal;
(iii) suppose that, for some $i \in \{1, \ldots, m\}$, $P_i \supseteq \bigcap_{j \neq i} P_j$ and deduce $Q_i \supseteq \bigcap_{j \neq i} Q_j$;
(iv) conclude from parts (ii) and (iii) that the set $\{P_1, \ldots, P_m\}$ determines a minimal primary decomposition of I.

10.2.3 Hilbert's Nullstellensatz

We can now state and prove Hilbert's *Nullstellensatz*.

Theorem 10.28 (Hilbert's Nullstellensatz) *Let $\mathcal{K}$ be an algebraically closed field and J a radical ideal in $\mathcal{K}[x_1, \ldots, x_n]$. Then:*

$$\mathcal{I}(\mathcal{Z}(J)) = J.$$

10.2 Elements of Algebraic Geometry

Proof As noted earlier, we only have to establish the inclusion $\mathcal{I}(\mathcal{Z}(J)) \subseteq J$. Toward a contradiction, we suppose $f \in \mathcal{I}(\mathcal{Z}(J)) - J$. By Theorem 10.26, J has a minimal primary decomposition determined by a set of prime ideals $\{P_1, \ldots, P_m\}$. Because $f \notin J$, f must fail to belong to at least one P_i from the given decomposition. We set $P_i = P$. Because P is a prime ideal, the quotient: $K[x_1, \ldots, x_n]/P$ is an integral domain (recall Sect. 3.4). We construct, as we have many times done, the field of fractions $\mathcal{F}$ of $K[x_1, \ldots, x_n]/P$ and then take its algebraic closure $\overline{\mathcal{F}}$.

Because $K[x_1, \ldots, x_n]$ is Noetherian, J is finitely generated: let us fix a set of generators $\{g_1, \ldots, g_r\}$. The generators determine corresponding polynomial functions of the cosets $x_i + P \in K[x_1, \ldots, x_n]/P$ $(i = 1, \ldots, r)$. Because $J \subseteq P$, for each $i \in \{1, \ldots, r\}$ we have $g_i \in P$, which, translated into cosets, becomes the equality $g(x_1, \ldots, x_n) + P = P$. In an analogous manner we obtain $f(x_1, \ldots, x_n) + P \neq P$.

The fact that we can embed $K[x_1, \ldots, x_n]/P$ into $\overline{\mathcal{F}}$ guarantees that the equalities and the negated equality just considered continue to hold in $\overline{\mathcal{F}}$. Bearing in mind that P plays the role of 0 in the quotient ring $K[x_1, \ldots, x_n]/P$ and the fields that extend it, the equalities we have obtained correspond to the condition:

$$\overline{\mathcal{F}} \models \bigwedge_{i=1}^{r} g_i(x_1 + P, \ldots, x_n + P) = 0 \wedge f(x_1 + P, \ldots, x_n + P) = 0)$$

in which the parameters $x_i + P$ $(i = 1, \ldots, n)$ occur. We eliminate these parameters by repeated existential quantifications and obtain:

$$\overline{\mathcal{F}} \models \exists x_i \ldots \exists x_n \left(\bigwedge_{i=1}^{r} g_i(x_1, \ldots, x_n) = 0 \wedge f(x_1, \ldots, x_n) = 0) \right).$$

Since $\mathcal{K}, \overline{\mathcal{F}} \models \mathbf{ACF}$ and $\mathbf{ACF}$ is model complete, we deduce:

$$\mathcal{K} \models \exists x_i \ldots \exists x_n \left(\bigwedge_{i=1}^{r} g_i(x_1, \ldots, x_n) = 0 \wedge f(x_1, \ldots, x_n) = 0) \right).$$

By the semantics of the existential quantifier, K^n contains at least one n-tuple $(a_1, \ldots, a_n)$ at which each g_i $(i = 1, \ldots, r)$ vanishes, while f does not. The hypothesis $f \in \mathcal{I}(\mathcal{Z}(J))$ however implies that f vanishes on the zero set determined by the ideal J, which contains $(a_1, \ldots, a_n)$ because a set of generators of J vanishes at that point. This contradiction concludes the proof. □

Exercise 10.29 Let $\mathcal{K}$ be an algebraically closed field. Use Hilbert's *Nullstellensatz* to prove the following statements:

(i) For any ideal J of the polynomial ring $K[x_1, \ldots, x_n]$, $\mathcal{I}(\mathcal{Z}(J)) = \sqrt{J}$;
(ii) $\mathcal{Z}$, viewed as an operation that sends radical ideals into varieties, is injective;
(iii) $\mathcal{I}$, viewed as an operation that sends varieties into radical ideals, is surjective;
(iv) Let $\mathcal{M}$ be an algebraically closed extension of $\mathcal{K}$, $g, f_1, \ldots, f_m \in K[x_1, \ldots, x_n]$. If the joint roots of the f_j $(j = 1, \ldots, m)$ are also roots

of g in $\mathcal{M}$, then there is an integer $k > 0$ such that g^k belongs to the ideal of $K[x_1, \ldots, x_n]$ generated by $f_1, \ldots, f_m$.

We conclude this section with a proof of the *Nullstellensatz* in an alternative form, which relies on quantifier elimination for ACF, as opposed to the weaker (albeit, in the present context, equivalent) property of model completeness used in the proof of Theorem 10.28.

Theorem 10.30 *Let $\mathcal{K}$ be a field, $f_1, \ldots, f_m \in K[x_1, \ldots, x_n]$. If the ideal $(f_1, \ldots, f_m)$, generated by the polynomials $f_1, \ldots, f_m$, is not the ring $K[x_1, \ldots, x_n]$, then $f_1, \ldots, f_m$ have a joint root in every algebraically closed extension of $\mathcal{K}$.*

Proof Suppose the field extension $\mathcal{K}^*$ of $\mathcal{K}$ contains a joint root of $f_1, \ldots, f_m$. Thus:

$$\mathcal{K}^* \models \exists \overline{x}(f_1(\overline{x}) = 0 \wedge \ldots \wedge f_m(\overline{x}) = 0).$$

Because $\mathcal{K}$ is included in $\mathcal{K}^*$, a suitable expansion of the latter structure models the diagram of $\mathcal{K}$ as well as the last formula, now viewed as a sentence (since there are constant symbols to name the parameters from K). Let $\mathcal{M}$ be the algebraic closure of this structure. Because existential sentences are preserved by extensions:

$$\mathcal{M} \models \mathsf{ACF} \cup \mathcal{D}(\mathcal{K}) \cup \{\exists \overline{x}(f_1(\overline{x}) = 0 \wedge \ldots \wedge f_m(\overline{x}) = 0)\}.$$

By quantifier elimination, the theory $\mathsf{ACF} \cup \mathcal{D}(\mathcal{K})$ is complete. Because $\exists \overline{x}(f_1(\overline{x}) = 0 \wedge \ldots \wedge f_m(\overline{x}) = 0)$ is a sentence in the language of $\mathsf{ACF} \cup \mathcal{D}(\mathcal{K})$ true in a model $\mathcal{M}$ the latter theory, it follows that each one of its models contains a joint root of $f_1, \ldots, f_m$. Furthermore, each such model can be identified with an algebraically closed extension of $\mathcal{K}$.

What we have proved so far is that, if one algebraically closed field extension $\mathcal{K}^*$ of $\mathcal{K}$ contains a joint root of the given polynomials, then all algebraically closed field extensions do. We must now show that one field extension with the properties of $\mathcal{K}^*$ actually exists. To this end, we note that the ideal $(f_1, \ldots, f_m)$ can be extended to a maximal ideal M in $K[x_1, \ldots, x_n]$ (without any appeal to Zorn's lemma, since $K[x_1, \ldots, x_n]$ is Noetherian). The quotient $K[x_1, \ldots, x_n]/M = \mathcal{K}^*$ is a field and, as the reader should check, using an argument in the proof of Theorem 10.28, $\mathcal{K}^*$ does contain a joint root of the polynomials $f_1, \ldots, f_m$. □

Exercise 10.31 We assume a few basic results concerning finite fields. First, letting $\mathbb{F}_p$ be the finite field $\mathbb{Z}/p\mathbb{Z}$, we note that every finite field of characteristic p has p^n elements, for some $n \geq 1$. The field with p^n elements, denoted by $\mathbb{F}_{p^n}$, is unique up to isomorphism and $\bigcup_{n \geq 1} \mathbb{F}_{p^n}$ is the algebraic closure $\overline{\mathbb{F}_p}$ of $\mathbb{F}_p$. It is clear that any finitely generated subfield of $\overline{\mathbb{F}_p}$ is a finite field, since it must be included in $\mathbb{F}_{p^n}$ for sufficiently large n.

(i) Suppose $f_1, \ldots, f_n \in \overline{\mathbb{F}_p}[x_1, \ldots, x_n]$ and consider the polynomial map $G = (f_1, \ldots, f_n)$ from $\overline{\mathbb{F}_p}^n$ to $\overline{\mathbb{F}_p}^n$. Show that, if G is injective, then it must be surjective (*Hint*: argue by contradiction. Consider $\bar{a}$ not in the image of G and work with a suitable, finitely generated subfield of $\overline{\mathbb{F}_p}^n$ containing $\bar{a}$).
(ii) Consider a polynomial map $G = (f_1, \ldots, f_n)$ where $f_1, \ldots, f_n \in K[x_1, \ldots, x_n]$ and $\mathcal{K}$ an algebraically closed field. Write a $\mathcal{L}_r$-sentence φ saying that, if G is injective, then G is surjective.
(iii) If $\mathcal{K}$ is an algebraically closed field of positive characteristic p, show that $\mathcal{K} \models \varphi$. Deduce $\mathsf{ACF}_p \models \varphi$.
(iv) Deduce $\mathsf{ACF}_0 \models \varphi$ (*Hint*: suppose $\mathsf{ACF}_0 \not\models \varphi$ and apply compactness to show that the latter condition implies $\mathsf{ACF}_p \models \neg\varphi$ for infinitely many primes).

10.3 Morley Rank

In the previous chapters we have developed an abstract notion of dimension that applies to the models of strongly minimal theories. In this section we introduce the fundamental notion of Morley rank, which, in the restricted context of strongly minimal theories, may be regarded as a generalisation of this notion. Our limited task in the remainder of this chapter is precisely to prove the last statement. We first define the Morley rank of a formula and a of a type, as well as the allied notion of Morley degree. Once in possessions of a few basic consequences of these definitions, we shall prove that, in models of a strongly minimal theory, Morley rank coincides with the model-theoretic notion of dimension we already know.

This result will enable us to show, in the next and final section, that, in the context of algebraic geometry, Morley rank coincides, in a sense to be clarified, with the dimension of a variety.

Given a $\mathcal{L}$-structure $\mathcal{M}$ and the expanded language $\mathcal{L}_M$ of $\mathcal{D}(\mathcal{M})$ (in which every element of M is named by a constant symbol), we now define the Morley rank of a $\mathcal{L}_M$-formula $\varphi(\bar{x})$ or, equivalently, the Morley rank of the set it defines (which will be a subset of M^n, with $n \geq 1$). The notation $MR^M(\varphi(\bar{x}))$ refers to the Morley rank of $\varphi(\bar{x})$ in $\mathcal{M}$. We shall systematically use the abbreviation $MR^M(\varphi)$ for $MR^M(\varphi(\bar{x}))$. We do not define Morley rank directly: instead, we define the meaning of the inequality $MR^M(\varphi) \geq \alpha$ for each ordinal α.

Definition 10.32 Let $\mathcal{M}$ be a $\mathcal{L}$-structure and φ be a $\mathcal{L}_M$-formula that defines a subset of M^n. Then:

(i) $MR^M(\varphi) \geq 0$ iff φ defines a nonempty subset of M^n;
(ii) $MR^M(\varphi) \geq \alpha + 1$ iff there is a sequence $\{\psi_i\}_{i \in \mathbb{N}}$ such that: (i) $MR^M(\psi_i) \geq \alpha$ and (ii) $\mathcal{M} \models \forall \bar{x}(\psi(\bar{x}) \to \varphi(\bar{x}))$; (iii) for any $i \neq j \in \mathbb{N}$, ψ_i, ψ_j define mutually disjoint subsets of M^n;
(iii) whenever λ is a limit ordinal $MR^M(\varphi) \geq \lambda$ iff, for any $\alpha \in \lambda$, $MR^M(\varphi) \geq \alpha$.

Let X be the subset of M^n defined by φ. By Definition 10.32, the Morley rank of φ in M grows to the successor of an ordinal α if it is possible to find an infinite sequence sequence $\{Y_1, Y_2, \ldots\}$ of disjoint, definable subsets of X, each of Morley rank at least α. In view of the last definition we can make sense of equalities in which the Morley rank of a formula takes a specific ordinal value.

Definition 10.33 Let $\mathcal{M}$ be a $\mathcal{L}$-structure and φ be a $\mathcal{L}_M$-formula that defines a subset of M^n. Then:

(i) $MR^M(\varphi) = \alpha$ iff $MR^M(\varphi) \geq \alpha$ and $MR^M(\varphi) \not\geq \alpha + 1$;
(ii) $MR^M(\varphi) = -\infty$ iff φ defines the empty set;
(iii) $MR^M(\varphi) = \infty$ iff, for any ordinal α, $MR^M(\varphi) \geq \alpha$.

Exercise 10.34 Show that φ defines an infinite subset of M^n iff $MR^M(\varphi) \geq 1$. Deduce that $MR^M(\psi) = 0$ iff $\psi(M)$ is a finite subset of M. Moreover, show that, if ψ defines a minimal subset of M, then $MR^M(\psi) = 1$. Deduce that the Morley rank of an algebraically closed field is 1.

10.3.1 Basic Properties of Morley Rank

In this section we establish some simple but important properties of Morley rank. Proofs are inevitably by transfinite induction. Here is a simple illustration:

Lemma 10.35 Let $\mathcal{M}$ be a $\mathcal{L}$-structure, φ, ψ $\mathcal{L}_M$-formulae such that a suitable expansion of $\mathcal{M}$ models $\forall \overline{x}(\varphi(\overline{x}) \to \psi(\overline{x}))$. Then $MR^M(\varphi) \leq MR^M(\psi)$.

Proof We prove by transfinite induction that the Morley rank of ψ is never strictly bounded by the Morley rank of φ. In other words, we show that, for any ordinal α, $MR^M(\varphi(\overline{x})) \geq \alpha$ implies $MR^M(\psi(\overline{x})) \geq \alpha$. If $\alpha = 0$, then φ defines a finite set. Consequently, the set defined by ψ is nonempty and $MR^M(\psi) \geq 0$. If α is the successor ordinal $\beta + 1$ and $MR^M(\varphi) \geq \beta + 1$, call X, Y the sets defined by $\varphi(\overline{x}), \psi(\overline{x})$ respectively in the relevant Cartesian power of M. Then $X \subseteq Y$ and X includes an infinite sequence of pairwise disjoint, definable subsets of Morley rank β. These are also definable subsets of Y so $MR^M(\psi) \geq \beta + 1 = \alpha$. If α is a limit ordinal, the inductive hypothesis immediately yields the sought result. □

An immediate consequence of the last lemma is that two formulae defining the same set have the same Morley rank.

Exercise 10.36 Let $\mathcal{M}$ be a $\mathcal{L}$-structure, $\varphi(x_1, \ldots, x_n)$ a $\mathcal{L}_M$-formula and X the set of M^n defined by φ. Show that, if $MR^M(\varphi) \geq \alpha$ and $\beta < \alpha$, there is a $\mathcal{L}_M$-formula ψ such that $MR^M(\psi) = \beta$ and ψ defines a subset of X.

Given two parametrically definable subsets of a $\mathcal{L}$-structure M, the next lemma allows us to compute the Morley rank of their union in terms of their Morley ranks:

10.3 Morley Rank

Lemma 10.37 *Let $\mathcal{M}$ be a $\mathcal{L}$-structure, φ, ψ $\mathcal{L}_M$-formulae. Then: $MR^M(\varphi \vee \psi) = max\{MR^M(\varphi), MR^M(\psi)\}$.*

Proof Since each of φ, ψ implies the disjunction $\varphi \vee \psi$, we immediately have $MR^M(\varphi \vee \psi) \geq max\{MR^M(\varphi), MR^M(\psi)\}$. To establish the converse inequality, we use transfinite induction as in Lemma 10.35. In other words, if α is an ordinal, we want to show:

$$MR^M(\varphi \vee \psi) \geq \alpha \implies max\{MR^M(\varphi), MR^M(\psi)\} \geq \alpha.$$

The last condition, together with the inequality we have just established, implies the equality we want. If $\alpha = 0$ or α is a limit ordinal, the inductive proof does not pose any special problems. We turn to a successor ordinal $\alpha = \beta + 1$. Let X, Y be the sets defined by φ, ψ respectively. If $MR^M(\varphi \vee \psi) \geq \alpha$, then there is a sequence of pairwise disjoint, definable subsets of $X \cup Y$, call it $\{Z_i\}_{i \geq 1}$, such that $MR^M(Z_i) \geq \beta$. We consider the sequences $\{X \cap Z_i\}_{i \geq 1}$ and $\{Y \cap Z_i\}_{i \geq 1}$. Both are sequences of pairwise disjoint sets. Since $Z_i \subseteq X \cup Y$, at least one of these two sequences contains a subsequence of nonempty sets. Suppose it is $\{Y \cap Z_i\}_{i \geq 1}$. We may in fact assume that the whole sequence consists of nonempty sets (e.g. by deleting the possible empty elements and relabelling the remaining ones). If θ_i defines Z_i, then $Y \cap Z_i$ is defined by $\psi \wedge \theta_i$. We can conclude $MR^M(Y) \geq \alpha$ if we are able to show that $MR^M(\psi \wedge \theta_i) \geq \beta$. The last inequality follows from an application of Lemma 10.35, which yields:

$$\beta \leq MR^M(\theta_i) = MR^M((\varphi \wedge \theta_i) \vee (\psi \wedge \theta_i)).$$

Because β is smaller than $\beta + 1$, the inductive hypothesis applies and we obtain:

$$\beta \leq max\{MR^M(\varphi \wedge \theta_i), MR^M(\psi \wedge \theta_i)\}.$$

If $\beta \leq MR^M(\varphi \wedge \theta_i)$ for infinitely many values of i, then $MR^M(\varphi) \geq \beta + 1$. Otherwise $\beta \leq MR^M(\psi \wedge \theta_i)$ for infinitely many values of i and $MR^M(\psi) \geq \beta+1$. The induction is complete. □

Exercise 10.38 Let $\varphi(x_1, \ldots, x_n)$ be a formula whose free variables occur among $x_1, \ldots, x_n$. A fortiori, we may view φ as a formula $\psi(x_1, \ldots, x_n, y_1, \ldots, y_m)$ whose free variables occur among $x_1, \ldots, x_n, y_1, \ldots, y_n$. Show that $MR^M(\psi) \geq MR^M(\varphi)$.

We rely on the last exercise to obtain a result about the Morley rank of projections.

Lemma 10.39 *Let $\mathcal{M}$ be a $\mathcal{L}$-structure, $\psi(\overline{x}, \overline{y})$ a $\mathcal{L}_c$-formula. Set $\varphi(\overline{x}) := \exists \overline{y} \psi(\overline{x}, \overline{y})$. Then: $MR^M(\psi) \geq MR^M(\varphi)$.*

Proof We use transfinite induction again to prove that $MR^M(\varphi) \geq \alpha$ implies $MR^M(\psi) \geq \alpha$. If $\alpha = 0$ or α is a limit ordinal, the proof does not pose any problems and is left to the reader. Suppose $\alpha = \beta + 1$ is a successor ordinal and $MR^M(\varphi) \geq \alpha, \beta$. By the inductive hypothesis, $MR^M(\psi) \geq \beta$ holds and there is an infinite sequence of formulae $\{\varphi_i(\overline{x})\}_{i \geq 1}$, defining pairwise disjoint subsets of the set defined by $\varphi(\overline{x})$, each of Morley rank at least β.

Since $\mathcal{M} \models \forall \overline{x}(\varphi_i(\overline{x}) \to \varphi(\overline{x}))$ for each $i \geq 1$, $MR^M(\varphi \wedge \varphi_i) \geq MR^M(\varphi_i) \geq \beta$ or, equivalently $MR^M(\exists \overline{y}(\psi \wedge \psi_i)) \geq \beta$.

The inductive hypothesis holds at β for any two formulae of the relevant form. In particular: $MR^M(\psi \wedge \varphi_i) \geq \beta$. By Exercise 10.38, we may regard φ_i as a formula ψ_i in the variables $\overline{x}, \overline{y}$, thus obtaining $MR^M(\psi \wedge \psi_i) \geq \beta$. Each conjunction $\psi \wedge \psi_i$ defines a subset of the set defined by ψ. By the properties of ψ_i, any two such sets are disjoint. Consequently $MR^M(\psi) \geq \beta + 1 = \alpha$. The induction is complete. $\square$

10.3.2 Morley Degree

Let $\mathcal{M}$ be a $\mathcal{L}$-structure and $A \subseteq M$ a parameter set such that $|A| < |M|$. If the $\mathcal{L}_A$-formula φ has Morley rank $MR^M(\varphi) = \alpha < \infty$, the set X defined by φ admits a decomposition into a uniquely determined, finite number of disjoint parts. This number is called the **Morley degree** of φ, or, equivalently, of X. The decomposition we are interested in consists of definable sets with a distinctive property, defined below.

Definition 10.40 Let $\mathcal{M}$ be a $\mathcal{L}$-structure, $A \subseteq M$ and $\varphi(\overline{x})$ a $\mathcal{L}_A$-formula. We say that $\varphi(\overline{x})$ is α-**strongly minimal over** A iff $MR^M(\varphi) = \alpha$ and, for any $\mathcal{L}_A$-formula $\psi(\overline{x})$, either $MR^M(\varphi \wedge \psi) < \alpha$ or $MR^M(\varphi \wedge \neg \psi) < \alpha$. A definable subset of M^n, with $n \geq 1$, is α-strongly minimal over A iff its defining formula is.

When $\varphi(\overline{x})$ defines $X \subseteq M^n$, we shall from now on take the notations $MR^M(\varphi)$ and $MR^M(X)$ to have the same meaning and freely switch between them.

Exercise 10.41 Let $\mathcal{M}$ be a $\mathcal{L}$-structure, $A \subseteq M$.

(i) Show that X is a 1-strongly minimal subset of M iff X is minimal.
(ii) Let X, Y be sets defined by $\mathcal{L}_M$-formulae. We introduce a binary relation $X \sim_\alpha Y$, defined by the condition $MR^M((X - Y) \cup (Y - X)) < \alpha$: show that $\sim_\alpha$ is an equivalence relation. We shall call two definable sets α-**equivalent** when they satisfy the equivalence relation we have just defined.
(iii) Let $\varphi(x_1, \ldots, x_n)$ be a $\mathcal{L}_A$-formula of Morley rank α in $\mathcal{M}$. Show that the set of all formulae ψ such that $MR^M(\varphi \wedge \neg \psi) < \alpha$ is a n-type.
(iv) If $\varphi(x_1, \ldots, x_n)$ is α-strongly minimal in $\mathcal{M}$, show that the set of all formulae ψ such that $MR^M(\varphi \wedge \neg \psi) < \alpha$ is a complete n-type.

Theorem 10.42 *Let $\mathcal{M}$ be a $\mathcal{L}$-structure, $A \subseteq M$, $\varphi(\overline{x})$ a $\mathcal{L}_A$-formula and X the set defined by φ. If $MR^M(X) = \alpha < \infty$ then X can be represented as a disjoint union of a uniquely determined, finite family of pairwise disjoint, α-strongly minimal sets.*

10.3 Morley Rank

Proof By hypothesis $MR^M(X) = \alpha$. We proceed by contradiction and show that, if the theorem fails, then $MR^M(X) \geq \alpha + 1$. To this end, we need to find an infinite sequence of definable, pairwise disjoints subset of X. If X does not admit the representation described in the theorem's statement, then it cannot itself be α-strongly minimal, or it would determine a trivial partition. This is to say that X can be partitioned into definable subsets X_1, Y_1 (using a suitable formula ψ and the conjunctions $\varphi \wedge \psi, \varphi \wedge \neg \psi$ respectively), none of which is of Morley rank smaller than α. Because these sets cannot be of Morley rank strictly greater than α, they are both of Morley rank exactly α. At least one of them, say X_1, must fail to be α-strongly minimal (if both were, the representation of X we have ruled out would exist). We may therefore iterate the argument just given and find two definable subsets of Y_1, say X_2, Y_2, one of which at least is not α-strongly minimal. If X_2 is the set in question, then we have obtained two definable subsets of X, namely X_1, X_2, with $X_1 \cap X_2 = \emptyset$ and $MR^M(X_i) \geq \alpha$ for $i = 1, 2$. Focussing on X_2, we may again iterate our construction. It is clear that, since we do not have to stop after finitely many steps, we generate a sequence of pairwise disjoint, definable subsets of X, each of Morley rank at least α. We arrive at the conclusion $MR^M(X) \geq \alpha + 1$, which produces a contradiction.

What the contradiction we have reached shows is that, for some positive $k \in \mathbb{N}$, there are α-strongly minimal subsets $X_1, \ldots, X_k$ of X such that:

For any $1 \leq i \neq j \leq k$, $X_i \cap X_j = \emptyset$ and $X = X_1 \cup \ldots \cup X_k$.

It only remains to show that k is uniquely determined, i.e. that, if $X = Y_1 \cup \ldots \cup Y_s$, where $Y_1, \ldots, Y_s$ are definable, pairwise disjoint and α-strongly minimal, then $k = s$. It may happen that the X_i and Y_j are not the same sets but we shall show that they are pairwise α-equivalent.

Fix Y_{r_1}, Y_{r_2} with $1 \leq r_1 \neq r_2 \leq s$. Each of these sets admits a definable partition into the cells $Y_{r_k} \cap X_1, \ldots, Y_{r_k} \cap X_k$, with $k \in \{1, 2\}$. There are X_i, X_j such that $MR^M(Y_{r_1} \cap X_i) = \alpha = MR^M(Y_{r_2} \cap X_j)$ (otherwise the Morley rank of Y_{r_k} would be $< \alpha$). We must have $i \neq j$, or we could partition the same set X_i into two definable, pairwise disjoint subsets each of Morley rank α, against the α-strong minimality of X_i). This shows that there is an injection from $\{Y_1, \ldots, Y_s\}$ into $\{X_1, \ldots, X_k\}$, i.e. that $k \leq s$. The roles of the last two sets in the argument just given can be reversed. We conclude that s equals k: in other words, there is a bijection f from $\{Y_1, \ldots, Y_s\}$ onto $\{X_1, \ldots, X_k\}$. We leave it as an exercise for the reader to show that $f(Y_i) \sim_\alpha Y_i$, for each $i \in \{1, \ldots, k\}$. □

Exercise 10.43 With reference to the proof of Theorem 10.48, show that $f(Y_i) \sim_\alpha Y_i$, for each $i \in \{1, \ldots, k\}$.

If X is defined by $\varphi(\overline{x})$ and $MR^M(X) = \alpha < \infty$, Theorem 10.48 guarantees the existence of formulae $\varphi_1(\overline{x}), \ldots, \varphi_k(\overline{x})$ defining the α-strongly minimal components of a partition of X. These formulae are called the *components* of φ and their number, which is uniquely determined, is called the **Morley degree** of φ, in symbols $MD^M(\varphi)$.

In the next subsection we shall make use of Morley degree and of one further notion, the Morley rank of a type. Let $\mathcal{M}$ be a $\mathcal{L}$-structure, $A \subseteq M$ and p a type over A. We define the **Morley rank of the type** p, in symbols $MR^M(p)$, as $min\{MR^M(\varphi) : \varphi \in p\}$. Because any nonempty collection of ordinals has a least element, if $MR^M(p) = \alpha$, there is a formula $\varphi \in p$ such that $MR^M(\varphi) = \alpha$. We have the following:

Lemma 10.44 *Let $\mathcal{M}$ be a $\mathcal{L}$-structure, $A \subseteq M$ and φ a $\mathcal{L}_A$-formula of Morley rank ≥ 0. Then:*

$$MR^M(\varphi) = max\{MR^M(p) : \varphi \in p \text{ and } p \in S_n(A)\}.$$

Exercise 10.45 Prove Lemma 10.44 as follows:

(i) suppose $MR^M(\varphi) = \infty$ and consider the set containing each $\mathcal{L}_A$-formula ψ such that $MR^M(\varphi \wedge \neg\psi < \infty)$. By Exercise 10.41-(iii), the latter set is a type q. Let p be a complete type extending q: show that $\varphi \in p$ and $MR^M(p) = \infty$.
(ii) Suppose $MR^M(\varphi) = \alpha < \infty$ and adapt the argument from part (i).

10.3.3 Morley Rank in Strongly Minimal Structures

In the next section we shall be interested in the rank of types for the sake of characterising the dimension of a variety in model-theoretic terms. Here we establish one important result, which is of independent interest and shows Morley rank to generalise the notion of dimension that we have introduced for strongly minimal theories. In order to reach this theorem, we require a technical lemma to the effect that Morley rank is invariant under the adjunction of an algebraic element.

Lemma 10.46 *Let $\mathcal{M}$ be a $\mathcal{L}$-structure, $A \subseteq M$ with $|A| < |M|$, b algebraic over A. Then $MR^M(tp^M(\bar{a}, b/A)) = MR^M(tp^M(\bar{a}/A))$.*

Proof Consider any formula $\psi(\bar{x}, \bar{y}) \in tp^M(\bar{a}, b/A))$. Since $\exists\bar{y}\psi(\bar{x}, \bar{y}) \in tp^M(\bar{a}/A)$, by Lemma 10.39, $MR^M(\psi) \geq MR^M(\exists\bar{y}\psi)$, which shows the Morley rank of $tp^M(\bar{a}, b/A)$ to be bounded below by the Morley rank of $tp^M(\bar{a}/A)$. In symbols:

$$MR(tp^M(\bar{a}, b/A)) \geq MR(tp^M(\bar{a}/A)).$$

We only need to prove by transfinite induction that:

$$MR(tp^M(\bar{a}, b/A) \geq \alpha \implies MR(tp^M(\bar{a}/A) \geq \alpha.$$

10.3 Morley Rank

The only nontrivial step is the successor step. Our hypotheses are that b is algebraic over $A \cup \{\bar{a}\}$ and that $MR^{\mathcal{M}}(tp^{\mathcal{M}}(\bar{a}, b/A)) \geq \beta + 1$. Since, in this case, $MR^{\mathcal{M}}(tp^{\mathcal{M}}(\bar{a}, b/A)) \geq \beta$, the inductive hypothesis yields $MR^{\mathcal{M}}(tp^{\mathcal{M}}(\bar{a}/A)) \geq \beta$.

Toward a contradiction, we suppose $MR(tp^{\mathcal{M}}(\bar{a}/A)) = \beta$. The existence of a formula $\varphi \in tp^{\mathcal{M}}(\bar{a}/A)$ such that $MR^{\mathcal{M}}(\varphi) = \beta$ follows. Theorem 10.48 moreover guarantees the existence of certain β-strongly minimal formulae $\varphi_1, \ldots, \varphi_d$ that partition the set defined by φ.

Because b is algebraic over $A \cup \{\bar{a}\}$, there is a formula $\psi(\bar{x}, \bar{y})$ such that:

$$\mathcal{M} \models \psi(\bar{a}, b) \wedge \exists^{=m} y \psi(\bar{a}, y).$$

Set $\theta(\bar{x}, y) = \psi(\bar{x}) \wedge \psi(\bar{x}, y) \wedge \exists^{=m} y \psi(\bar{x}, y)$. We have $\theta(\bar{x}, y) \in tp^{\mathcal{M}}(\bar{a}, b/A)$, i.e. $MR(\theta) \geq \beta + 1$. By definition of Morley rank, there is a sequence of formulae $\theta_i(\bar{x}, y)$, $i \in \mathbb{N}$ such that they define pairwise disjoint subsets of the set defined by θ and each is of Morley rank β. We let:

$$\xi_i(\bar{x}) = \exists y \theta_i(\bar{x}, y), \text{ for each } i \in \mathbb{N}.$$

Suppose we had proved the following two claims:

(1) $MR(\xi_k) \geq \beta$ for $k \geq 1$;
(2) $MR(\varphi_1 \wedge \bigwedge_{i=1}^{k} \xi_i) \geq \beta$, where φ_1 is one of the β-strongly minimal components of φ.

Using these claims we could quickly reach the contradiction needed to complete our proof by transfinite induction. To see how, note that the claims imply the inequalities:

$$\beta \leq MR^{\mathcal{M}}(\varphi_1 \wedge \bigwedge_{i=1}^{k} \xi_i) \leq MR(\bigwedge_{i=1}^{k} \xi_i)$$

Since $\beta \geq 0$, it follows that the sequence $\{\xi_k\}_{k \in \mathbb{N}}$ is a n-type. As such, it is included in a complete n-type, which must be realised by some elementary extension $\mathcal{N}$ of $\mathcal{M}$. Let $\bar{c}$ be a realisation in $\mathcal{N}$. By definition of ξ_k, we can determine, for each $k \geq 1$, an element d_k such that $\mathcal{N} \models \theta_k(\bar{c}, d_k)$. Because we know that the sets defined by the θ_i are all disjoint in $\mathcal{M}$, and continue to be so in $\mathcal{N}$, it follows that $\{d_k\}_{k \in \mathbb{N}}$ is an infinite sequence of distinct elements. But θ_k implies θ, which is a conjunction including the formula $\psi(\bar{x}, y)$, so $\mathcal{M} \models \psi(\bar{c}, d_k) \wedge \exists_{=m} y \psi(\bar{c}, y)$, which is the contradiction we wanted.

It now remains to prove that (1) and (2) actually hold. As for (1), we first note that $MR(\theta_k(\bar{x}, y)) \geq \beta \geq 0$, which implies the existence of $\bar{c}, d$ such that $\mathcal{M} \models \theta_k(\bar{c}, d)$. By Lemma 10.44, the Morley rank of $\theta_k(\bar{x}, y)$ is the maximum among the Morley ranks of types containing θ_k. We may thus choose $\bar{c}, d$ in such a way that the Morley rank of $tp^{\mathcal{M}}(\bar{c}, d/A \cup \bar{b}_k)$, where $\bar{b}_k$ are possible parameters occurring in θ_k, is at least β. Because $\mathcal{M} \models \forall \bar{x} \forall y (\theta_k(\bar{x}, y) \to \theta(\bar{x}, y))$, we see that d is algebraic over $A\cup\{\bar{c},\bar{b}_k\}$. The inductive hypothesis applies and we deduce $MR(\bar{c}/A\cup\{\bar{b}_k\}) \geq \beta$. Since $\mathcal{M} \models \xi_k(\bar{c})$, we conclude $MR^M(\xi_k) \geq \beta$.

In order to establish (2), we observe that:

$$\mathcal{M} \models \xi_k(\bar{x}) := \exists y \theta_k(\bar{x}, y) \implies \mathcal{M} \models \exists y \theta(\bar{x}, y)$$
$$\implies \mathcal{M} \models \varphi(\bar{x})$$
$$\implies \mathcal{M} \models \varphi_1(\bar{x}) \vee \ldots \vee \varphi_d(\bar{x}).$$

It follows that $\beta \leq MR^M(\xi_k) \leq MR^M(\xi_k \wedge \varphi)$. Furthermore, we have:

$$\beta \leq MR^M(\xi_k \wedge \varphi) = max\{MR^M(\xi_k \wedge \varphi_1), \ldots, MR(\xi_k \wedge \varphi_d)\}.$$

It follows that there is at least one $i \in \{1, \ldots, d\}$ such that $MR^M(\xi_k \wedge \varphi_i) \geq \alpha$. Because this is true for every $k \geq 1$ and there are only finitely many choices of i, there is j such that $MR^M(\xi_k \wedge \varphi_j) \geq \beta$ for fixed j and infinitely many values of k. We restrict attention to the infinite subsequence determined by these values and, by an abuse of notation, refer to it as $\{\xi_k\}_{k\in\mathbb{N}}$. The corresponding sequence of formulae satisfies (2). If not, there should be a least k such that, setting $j = 1$ (which can be done, possibly by relabelling):

$$MR^M(\varphi_1 \wedge \bigwedge_{i=1}^{k-1} \xi_i) \geq \beta \text{ and } MR^M(\varphi_1 \wedge \bigwedge_{i=1}^{k} \xi_i) \not\geq \beta.$$

In this case, since:

$$\beta \leq MR^M(\varphi_1 \wedge \xi_k) = max\{MR^M(\varphi_1 \wedge \bigwedge_{i=1}^{k} \xi_i), MR^M(\varphi_1 \wedge \xi_k \wedge \neg \bigwedge_{i=1}^{k-1} \xi_k)\}$$

and the Morley rank of $\varphi_1 \wedge \bigwedge_{i=1}^{k} \xi_i$ lies below β, we must have the inequality $MR^M(\varphi_1 \wedge \xi_k \wedge \neg \bigwedge_{i=1}^{k-1} \xi_k) \geq \beta$. The sets defined by the formulae $\varphi_1 \wedge \bigwedge_{i=1}^{k-1} \xi_i$ and $(\varphi_1 \wedge \xi_k \wedge \neg \bigwedge_{i=1}^{k-1} \xi_i) \vee (\varphi_1 \wedge \neg \xi_k \wedge \neg \bigwedge_{i=1}^{k-1} \xi_i)$ determine a partition of $\varphi_1(\mathbb{M})$ into two subsets, each of Morley rank at least β, contradicting the β-strong minimality of φ_1. This contradiction establishes the second assertion.

The proof is now complete. □

We can now explicitly relate Morley rank to the notion of dimension supported by strongly minimal theories. In the sequel we focus on sufficiently large models of a countable, strongly minimal theory T, i.e. on uncountable ones, because their saturation can be taken advantage of in the inductive argument we are about to propose. By Lemma 10.1, if $\mathcal{M}$ is a strongly minimal $\mathcal{L}$-structure and $A \subseteq M$ with $|A| < |M|$, there is only one non-algebraic 1-type over A. Since the algebraic 1-types over A are all realised by the elements of $acl_\mathcal{M}(A)$, if this algebraic closure

10.3 Morley Rank

does not fill M, its complement must contain a realisation of the non-algebraic 1-type. It follows that $\mathcal{M}$ is saturated. These remarks lead to:

Lemma 10.47 *Let T be a countable, strongly minimal theory, $\mathcal{M} \models T$ a model of size $|M| > \aleph_0$. Then $\mathcal{M}$ is saturated.*

Proof If $|A| < |M|$, the size of $acl_\mathcal{M}(A)$ is the larger of $|A|$ and $\aleph_0$. It follows that the complement of $acl_\mathcal{M}(A)$ in M is nonempty, i.e. that every 1-type over A is realised in $\mathcal{M}$. Since A is an arbitrary subset of M of cardinality smaller than $|M|$, it follows that $\mathcal{M}$ is saturated. □

Theorem 10.48 *Let $\mathcal{M}$ be model of a countable, strongly minimal theory T such that $|M| > |T|$. If $A \subseteq M$ and $|A| < |M|$, then:*

$$MR^M(tp^M(a_1, \ldots, a_n/A)) = dim(acl_\mathcal{M}(\{a_1, \ldots a_n\}/A)).$$

Proof The right-hand side of the last equality denotes a number $k \leq n$ such that the (model-theoretic) algebraic closure in $\mathcal{M}$ over A of a k-element subset of $\{a_1, \ldots, a_n\}$ is identical with the algebraic closure of $\{a_1, \ldots, a_n\}$ over A. In view of Lemma 10.46, we may assume that $\{a_1, \ldots, a_n\}$ is an independent set over A, in which case $dim(acl_\mathcal{M}(\{a_1, \ldots a_n\}/A)) = n$. With this simplifying assumption in place, we proceed by induction on the dimension n.

If $n = 1$, $MR^M(tp^M(a_1, \ldots, a_n/A)) = MR^M(tp^M(a_1/A))$. Because a_1 is not algebraic over A, $\varphi \in tp^M(a_1/A) \geq 1$. Let ψ be a formula whose Morley rank is the Morley rank of the type $tp^M(a_1/A)$. Then $MR^M(\psi) \geq 1$ and, because T is strongly minimal, there is no definable sequence of pairwise disjoint, infinite subsets of $\{c \in M : \mathcal{M} \models \psi(c)\}$. It follows that $MR^M(\psi) \not\geq 2$, i.e. $MR^M(\psi) = 1$.

Suppose the theorem holds for $n = m - 1$ and consider the independent set $\{a_1, \ldots, a_m\}$. We wish to evaluate $MR^M(tp^M(a_1, \ldots, a_m/A))$. To this end, we focus on $\mathcal{N} = acl_\mathcal{M}(A \cup \{a_1, \ldots, a_m\})$. By construction, each $b \in M - N$ is independent over A. We next take $\varphi(x_1, \ldots, x_m) \in tp^M(a_1, \ldots, a_m/A)$ with Morley rank equal to the Morley rank of $tp^M(a_1, \ldots, a_m/A)$.

Consider the parametrically definable set:

$$B = \{b \in M : \mathcal{M} \models \varphi(a_1, \ldots, a_{m-1}, b_i)\}.$$

By the strong minimality of T, B is either finite or cofinite. If $\varphi(a_1, \ldots, a_{m-1}, x_m)$ had only finitely many solutions in M, since a_m has to be one of them, the independence of $\{a_1, \ldots, a_m\}$ over A would be contradicted. It follows that B is cofinite. In particular, there is a sequence $\{b_i\}_{i \geq 1}$ (of elements of $M - N$) such that, for each positive i, $a_1, \ldots, a_{m-1}, b_i$ is independent over A and $\mathcal{M} \models \varphi(a_1, \ldots, a_{m-1}, b_i)$.

For each $i \geq 1$, we next consider the formula $\varphi_i(x_1, \ldots, x_m) := \varphi(x1, \ldots, x_m) \wedge x_m = b_i$. We shall now show that $MR^M(\varphi_i) \geq m - 1$, which will imply, by the definition of Morley rank, $MR^M(\varphi) \geq m$.

By Lemma 10.39, $MR^M(\varphi_i) \geq MR^M(\exists x_m \varphi_i)$. The last formula belongs to $tp^{\mathcal{M}}(a_1, \ldots, a_{m-1}/A \cup \{b_i\})$ so its Morley rank is $\geq MR^M(a_1, \ldots, a_{m-1}/A \cup \{b_i\})$. We can use the inductive hypothesis to determine the last Morley rank if we are able to show that $a_1, \ldots, a_{m-1}$ continue to be independent over $A \cup b_i$. Because b_i is independent of the $a_1, \ldots, a_{m-1}$ over A, we can use the contrapositive of the exchange lemma and the fact that $b_i \notin acl_{\mathcal{M}}(\{a_1\} \cup (A \cup \{a_2, \ldots, a_{m-1}\})$ to deduce $a_1 \notin acl_{\mathcal{M}}(\{b_i\} \cup acl_{\mathcal{M}}(A \cup \{a_1, \ldots, a_{m-1}\}) - acl_{\mathcal{M}}(A \cup \{a_2, \ldots, a_{m-1}\})$. Because the $a_1, \ldots, a_{m-1}$ are independent over A, we have $a_1 \notin acl_{\mathcal{M}}(A \cup \{a_2, \ldots, a_{m-1}\})$. We conclude:

$$a_1 \notin acl_{\mathcal{M}}(A \cup \{a_2, \ldots, a_{m-1}\} \cup \{b_i\}).$$

Thus a_1 is independent of $a_2, \ldots, a_{m-1}$ over $A \cup \{b_i\}$. The same argument holds for $i = 2, \ldots, m - 1$. If $|A| < |M|$, then $|A \cup \{b_i\}| < |M|$. The inductive hypothesis applies and we conclude:

$$MR^M(\varphi_i) \geq MR^M(tp^{\mathcal{M}}(a_1, \ldots, a_{m-1}/A \cup \{b_i\})) = m - 1.$$

Call X_i the subset of M defined by φ_i. The sequence $\{X_i\}_{i \geq 1}$ consists of pairwise disjoint, definable subsets of the set defined by φ, each of Morley rank $\geq m - 1$. As a consequence, $MR^M(tp^{\mathcal{M}}(a_1, \ldots, a_m/A)) \geq m$. It only remains to show that the Morley rank of $tp^{\mathcal{M}}(a_1, \ldots, a_m/A)$ is not greater than m. Let X be the set defined by φ. If $MR^M(\varphi) \geq m + 1$, we must be able to determine an infinite sequence of pairwise disjoint, definable subsets of X, each defined by a formula ψ of Morley rank $\geq m$.

Because these formulae define pairwise disjoint sets, at most one of them will be satisfied by the m-tuple $a_1, \ldots, a_m$. We are going to show that, for any formula ψ such that $\mathcal{M} \models \neg \psi(a_1, \ldots, a_m)$, we have $MR^M(\psi) < m$. The last inequality suffices to rule out the possibility that the Morley rank of φ, and thus of $tp^{\mathcal{M}}(a_1, \ldots, a_m/A)$, should grow above m.

Suppose that $\mathcal{M} \models \psi(b_1, \ldots, b_m)$. If $b_1, \ldots, b_m$ were algebraically independent over A, the properties of independence in strongly minimal theories (by an easy adaptation of Lemma 8.100) would guarantee the existence of a partial elementary map sending a_i into b_i, $i = 1, \ldots, m$. It would thus follow that $tp^{\mathcal{M}}(a_1, \ldots, a_m/A) = tp^{\mathcal{M}}(b_1, \ldots, b_m/A)$ but this is impossible because ψ is satisfied in $\mathcal{M}$ by the tuple $b_1, \ldots, b_m$ but not by the tuple $a_1, \ldots, a_m$. We conclude that $b_1, \ldots, b_m$ are not algebraically independent over A. We can therefore reduce them to a strictly smaller, independent set with the same Morley rank. Our inductive hypothesis now implies:

$$MR^M(tp^{\mathcal{M}}(b_1, \ldots, b_m/A) < m.$$

By Lemma 10.44, $MR^M(\psi) = max\{MR^M(p) : \psi \in p \text{ and } p \in S_m(A)\}$. Because, however, M is saturated, each $p \in S_m(A)$ is realised by a m-tuple $b_1, \ldots, b_m$ in M^m. The last condition can therefore be rewritten as:

$$MR^{\mathcal{M}}(\psi) = max\{MR^{\mathcal{M}}(tp^{\mathcal{M}}(b_1,\ldots,b_m/A) : \mathcal{M} \models \psi(b_1,\ldots,b_m)\}.$$

By what we have shown, the maximum mentioned in the last equality must lie below m. The induction is now complete. □

We know from Exercise 7.39 that, given a field $\mathcal{K}$ and an extension $\mathcal{M}$, a set $\{a_1,\ldots,a_m\} \subseteq \mathcal{M}$ is algebraically independent over $\mathcal{K}$ iff the transcendence degree of $\mathcal{K}(a_1,\ldots,a_m)$ over the field $\mathcal{K}$ equals m. Because, when T is the theory of algebraically closed fields in a fixed characteristic, dimension specialises to transcendence degree, Theorem 10.48 shows[3] that, when $\mathcal{K} \models T$, the Morley rank of $tp^{\mathcal{M}}(a_1,\ldots,a_n/\mathcal{K})$ coincides with the transcendence degree of $\mathcal{K}(a_1,\ldots,a_m)$ over the field $\mathcal{K}$. We shall rely directly on this fact to establish a connection between Morley rank and the dimension of an algebraic variety.

Corollary 10.49 *Let $\mathcal{M} \models T$ be as in the statement of Theorem 10.48 and let φ be the formula $\bigwedge_{i=1}^n x_i = x_i$, which defines M^n. Then $MR^{\mathcal{M}}(\varphi) = n$.*

Proof By Lemma 10.44, $MR^{\mathcal{M}}(\psi) = max\{MR^{\mathcal{M}}(p) : \psi \in p \text{ and } p \in S_n(\emptyset)\}$. The saturation of $\mathcal{M}$ allows us to reduce the last condition to:

$$MR^{\mathcal{M}}(\varphi) = max\{MR^{\mathcal{M}}(tp^{\mathcal{M}}(a_1,\ldots,a_n) : \mathcal{M} \models \varphi(a_1,\ldots,a_m)\}.$$

Because Morley rank and dimension coincide for a strongly minimal theory T, the maximum value we can attain is achieved when $\{a_1,\ldots,a_n\}$ is an independent set, in which case the Morley rank of the corresponding type is n. □

10.4 Dimension of an Algebraic Variety

An algebraic variety V is said to be **irreducible** if it cannot be expressed as a union of varieties $V_1 \cup V_2$ in any non-trivial way. In other words, if V is irreducible and $V = V_1 \cup V_2$, we must have $V_1 = V$ or $V_2 = V$.

Before introducing any model-theoretic considerations, we prove a fundamental result about varieties and introduce one among many equivalent ways of defining the dimension of a variety. Our first result concerns the connection between the irreducibility of V and the algebraic properties of the ideal $\mathcal{I}(V)$.

Theorem 10.50 *The variety V is irreducible iff $\mathcal{I}(V)$ is a prime ideal.*

Proof We prove this theorem by two applications of contraposition. We first suppose V to be reducible. It follows that $V = V_1 \cup V_2$, where $V_1, V_2 \neq V$.

---

[3] When $\mathcal{K}$ is the parameter set, the theorem as we have stated applies only if we embed $\mathcal{K}$ into a sufficiently large model of T. Since both models are saturated and uncountable, they are both $\aleph_0$-homogeneous and this ensures (see [70], p.136 for a proof) Morley rank to be invariant as we move between $\mathcal{K}$ and its elementary extension.

Because $V_i \subset V$ (inclusion is strict because the decomposition of V is nontrivial) for $i = 1, 2$, we can deduce $\mathcal{I}(V_i) \supset \mathcal{I}(V)$ (with strict inclusion because $\mathcal{I}$ is a bijection). As a consequence, there is $f_i \in \mathcal{I}(V_i) - \mathcal{I}(V)$, for $i = 1, 2$. It now follows that the polynomial $f_1 f_2$ is in $\mathcal{I}(V)$ while neither of f_1, f_2 is. In other words, $\mathcal{I}(V)$ is not a prime ideal.

We next show that, if $\mathcal{I}(V)$ is not a prime ideal, then we can decompose V into two nontrivial varieties V_1, V_2. We pick f_1, f_2 such that $f_1 f_2 \in \mathcal{I}(V)$ and $f_1, f_2 \notin \mathcal{I}(V)$. Consider $V_i = \mathcal{Z}(f_i) \cap V$, with $i = 1, 2$. Because $V = \mathcal{Z}(J)$ for some ideal, and J is finitely generated by, say, $g_1, \ldots, g_m$, we can represent V as follows:

$$V = \mathcal{Z}(g_1) \cap \ldots \cap \mathcal{Z}(g_m) = \mathcal{Z}(J)$$

Consequently, we can represent V_i as a value of $\mathcal{Z}$, namely $\mathcal{Z}(f_i) \cap \mathcal{Z}(g_1) \cap \ldots \cap \mathcal{Z}(g_m) = \mathcal{Z}((f_i, g_1, \ldots, g_m))$, where $(f_i, g_1, \ldots, g_m)$ is the ideal generated by the set $\{f_1, g_1, \ldots, g_m\}$. This is to say that both V_1, V_2 are values of $\mathcal{Z}$, i.e. they are both varieties. Note that f_i does not vanish on V so the set of points at which $f_i, g_1, \ldots, g_m$ vanish simultaneously cannot be V itself. In other words, V_i is a proper subset of V. Because $f_1 f_2 \in \mathcal{I}(V)$, it is easy to deduce that $V = V_1 \cup V_2$. In other words, V is reducible. □

Suppose that V is an irreducible variety in affine n-space $\mathbb{A}$ over the algebraically closed field $\mathcal{K}$. Let $K[\mathbb{A}^n]$ be the ring of polynomials in $K[x_1, \ldots, x_n]$, viewed as functions from $\mathbb{A}^n$ to $\mathbb{A}$. By the last theorem $\mathcal{I}(V)$ is prime so $K[\mathbb{A}^n]/\mathcal{I}(V)$ is an integral domain. Let us call $\mathcal{K}(V)$ the associated field of fractions.

Definition 10.51 Let V be an irreducible variety. The **dimension** of V is the transcendence degree of $\mathcal{K}(V)$ over the field $\mathcal{K}$.

Example 10.52 Let the algebraically closed field of complex numbers be the affine 1-space $\mathbb{A}^1$. In this case, because polynomials in $\mathbb{C}[x]$ have only finitely many roots, the varieties are the finite subsets of $\mathbb{C}$ or the whole of $\mathbb{C} = \mathbb{A}^1$ (when we consider the ideal generated by the constant polynomial 0). It is easy to see that the whole of $\mathbb{A}^1$ and its points are the irreducible varieties. For any fixed $a \in \mathbb{C}$, the ideal generated by $x - a$ is a maximal ideal of $\mathbb{C}[x]$ and thus a prime ideal. The quotient $\mathbb{C}[x]/(x-a)$ is clearly isomorphic to $\mathbb{C}$. Since the degree of transcendence of $\mathbb{C}$ over itself is zero, the dimension of a point of $\mathbb{A}^1$ is 0. If we turn to the whole space $\mathbb{A}^1$, we obtain the quotient $\mathbb{C}[\mathbb{A}^1]/(0)$, which is isomorphic to the integral domain $\mathbb{C}[x]$. The associated field of fractions is $\mathbb{C}(x)$, which has transcendence degree 1 over $\mathbb{C}$. As expected, the dimension of $\mathbb{A}^1$ is 1. We define the dimension of a reducible variety simply as the largest dimension of an irreducible component. In the present example, we only encounter the trivial situation in which the dimension of a finite set of points (a union of 0-dimensional, irreducible varieties) is zero.

Our next result (which is mirrored by the decomposition of radical ideals into prime components) concerns the decomposability of any variety into a uniquely determined number of irreducible components.

10.4 Dimension of an Algebraic Variety

Theorem 10.53 *Let V be a nonempty variety. Then V has a unique representation of the form:*

$$V = V_1 \cup \ldots \cup V_m$$

where $m \in \mathbb{N}$, $V_1, \ldots, V_m$ are irreducible varieties.

Proof Toward a contradiction, suppose that the family $\mathcal{R}$ of nonempty varieties that cannot be written as finite unions of irreducible varieties is nonempty. Applying $\mathcal{I}$ to the elements of $\mathcal{R}$ determines a family $\mathcal{S}$ of ideals (of the ring $K[x_1, \ldots, x_n]$, with $n \geq 1$ and K algebraically closed), which must contain a maximal element because the underlying ring is Noetherian.

When we apply $\mathcal{Z}$ to the elements of $\mathcal{S}$, we obtain the elements of $\mathcal{R}$ back. Because, however, inclusions are reversed, the maximal element of $\mathcal{S}$ now corresponds to a minimal element V^* of $\mathcal{R}$. Because V^* cannot be expressed as a union of irreducible varieties, it is in particular not an irreducible variety itself. Consequently, there are varieties $V_1 \cup V_2$, such that $V^* = V_1 \cup V_2$ and, by the minimality of V^*, $V_1, V_2 \notin \mathcal{R}$. Each of V_1, V_2 admits a decomposition into finitely many irreducible varieties. The union of these decompositions is a decomposition of V^* into irreducible varieties: this result contradicts the condition $V^* \in \mathcal{R}$. We conclude that $\mathcal{R} = \emptyset$, i.e. that every nonempty variety has a representation as a union of irreducible varieties.

To prove uniqueness, suppose V can be represented both as the union $U_1 \cup \ldots \cup U_q$ and as the union $W_1 \cup \ldots \cup W_s$, where U_i, W_j ($1 \leq i \leq q$, $1 \leq j \leq s$) are irreducible varieties. We take these representations to be irredundant: it is never the case that $U_i \subseteq U_j$ when $i \neq j$, and similarly for the $W_1, \ldots, W_s$ (we simply expunge redundant subsets, if any occur).

Because $U_1 \cup \ldots \cup U_q$ covers V, it will cover W_1, which may therefore be expressed as:

$$W_1 = (W_1 \cap U_1) \cup \ldots \cup (W_1 \cap U_q).$$

Because W_1 is irreducible, no nontrivial union of varieties can represent it, i.e. $W_1 = W_1 \cap U_j$ for some $j \in \{1, \ldots, q\}$. We deduce $W_1 \subseteq U_j$. We can repeat the argument just given for U_j and conclude that $U_j \subseteq W_k$ for some $k \in \{1, \ldots, s\}$. It now follows that $W_1 \subseteq W_k$: because we have expunged redundant subsets, $W_1 = W_k$ and, consequently, $W_1 = U_j$. Continuing in this manner, we identify each W_i with some U_j: because the U_j determine an irredundant decomposition of V, we must have $q = s$. □

In order to establish a model-theoretic connection between Morley rank and the dimension of a variety, we make use of the correspondence between irreducible varieties and prime ideals established by Theorem 10.50 to relate certain complete n-types to irreducible varieties. We do so by associating prime ideals of polynomials in n variables uniquely with such n-types. What we know about the Morley rank of

n-types in the context of a strongly minimal theory will then allow us to draw a connection between the Morley rank of n-types and the geometric dimension of varieties.

We begin by introducing two operations that link prime ideals and n-types. Let $\mathcal{K}$ be an algebraically closed field and $p \in S_n(K)$. The operation $\mathfrak{P}$ sends p into a prime ideal $J = \mathfrak{P}(p)$ of $K[x_1, \ldots, x_n]$, according to the rule:

$$f(x_1, \ldots, x_n) \in \mathfrak{P}(p) \text{ iff the formula } f(x_1, \ldots, x_n) = 0 \text{ is in } p.$$

The operation $\mathfrak{T}$ sends a prime ideal J of $K[x_1, \ldots, x_n]$ into the unique complete n-type $\mathfrak{T}(J)$ extending the n-type given by the conditions:

(1) $f(x_1, \ldots, x_n) = 0$ is in $\mathfrak{T}(J)$ if $f(x_1, \ldots, x_n) \in J$;
(2) $\neg(f(x_1, \ldots, x_n) = 0)$ is in $\mathfrak{T}(J)$ if $f(x_1, \ldots, x_n) \notin J$.

Exercise 10.54

(i) Show that $\mathfrak{P}(p)$ is a prime ideal of $K[x_1, \ldots, x_n]$.
(ii) Show that conditions (1), (2) determine a n-type by verifying that they can be realised in an algebraically closed field extending $K[x_1, \ldots, x_n]/J$.
(iii) Explain why the n-type discussed in part (ii) has a unique complete extension.
(iv) Show that $\mathfrak{P}(\mathfrak{T})(J) = J$ and that $\mathfrak{T}(\mathfrak{P})(p) = p$.

Definition 10.55 Let V be an irreducible variety in affine n-space $\mathbb{A}^n$. Then $\mathcal{I}(V)$ is a prime ideal and $p_V = \mathfrak{T}(\mathcal{I}(V))$, a complete n-type over K, is called the **generic point** of V. If V is not irreducible, its generic points are the generic points of its irreducible components.

If p_V is the generic point of $V \subseteq \mathbb{A}^n$, then:

$$f(\overline{a}) = 0 \text{ for every } \overline{a} \in V \text{ iff } f(\overline{x}) \in \mathcal{I}(V) \text{ iff } f(\overline{x}) = 0 \in p_V.$$

Given an irreducible variety V, the field of fractions $K(V)$ of $K[\overline{x}]/\mathcal{I}(V)$ contains a realisation of p_V given by the n-tuple of cosets $a_i = x_i + \mathcal{I}(V)$, $i = 1, \ldots, n$. The field of fractions associated with the integral domain $K[\overline{x}]/\mathcal{I}(V)$ is, by definition, $K(V)$. By construction, it is the field $\mathcal{K}(a_1, \ldots, a_n)$. Within a suitable algebraically closed field, the Morley rank of p_V coincides, by Theorem 10.48, with the dimension of $\{a_1, \ldots, a_n\}$ over K, i.e. with the transcendence degree of $\mathcal{K}(V)$ over the field $\mathcal{K}$ or, equivalently, with the dimension of the irreducible variety V, which we shall denote by $dim(V)$. We record the last observation as:

Lemma 10.56 *Let $\mathcal{K}$ be an uncountable, algebraically closed field of fixed characteristic, $V \subseteq K^n$ be an irreducible variety and p_V its generic point. Then $dim(V) = MR^K(p_V)$.*

10.4 Dimension of an Algebraic Variety

Our goal is now within reach since we only need the equality $MR^K(p_V) = MR^K(V)$ to show that the geometric dimension $dim(V)$ is a specialisation of Morley rank. We shall see a proof presently, which relies on an evaluation of Morley rank for types to be established in the next lemma.

Theorem 10.57 *Let K be an uncountable, algebraically closed field of fixed characteristic, $V \subseteq K^n$ be an irreducible variety. Then $dim(V) = MR^K(V)$.*

Proof The generic point p_V contains a formula that defines V and satisfies the equality $\mathcal{Z}(\mathfrak{P}(p_V)) = V$. To see why the first claim holds, we note that $\mathcal{I}(V)$ is finitely generated by polynomials $g_1, \ldots, g_m \in K[x_1, \ldots, x_n]$, each of which vanishes over V. By the definition of p_V, the formula $g_i(\overline{x}) = 0$ is in p_V for each $i = 1, \ldots, m$. Consequently, the conjunction φ of these formulae also belongs to p_V. Clearly, φ defines V.

As for the second claim, it suffices to note that $\mathfrak{P}(p_V) = \mathcal{I}(V)$ and to recall that $\mathcal{Z}, \mathcal{I}$ are inverses by the *Nullstellensatz*. Consider an arbitrary type $q \in S_n(K)$ such that $\varphi \in q$ and $\mathcal{Z}(\mathfrak{P}(q)) = V$. Because $\mathcal{Z}$ is a bijection we obtain $\mathfrak{P}(q) = \mathfrak{P}(p_V)$ and the fact that $\mathfrak{P}$ is also a bijection yields $q = p_V$. In other words, p_V is the only type with the properties declared in the statement of the theorem.

It follows that, if $\varphi \in q \in S_n(K)$ and $q \neq p_V$, then $\mathfrak{P}(q) \supset \mathcal{I}(V) = \mathfrak{P}(p_v)$. Because $\mathcal{Z}$ reverses inclusion, we deduce:

$$V_0 = \mathcal{Z}(\mathfrak{P}(q)) \subset \mathcal{Z}(\mathfrak{P}(p_V)) = V.$$

It will be proved in the next lemma that, given the last inclusion, $MR^K(q) < MR^K(p_V)$. By Lemma 10.44, $MR^K(V) = MR^K(\varphi)$ is the maximum among the Morley ranks of n-types in $S_n(K)$ containing φ. Since p_V is one of these types and its Morley rank is greater than that of any other such type, we can invoke Lemma 10.58 to conclude:

$$MR^K(V) = MR^K(p_V) = dim(V).$$

□

The proof of Theorem 10.57 relies on the inequality $MR^K(q) < MR^K(p_V)$, which follows from the lemma below.

Lemma 10.58 *Let $V \subset W$ be irreducible varieties of K^n, with K algebraically closed. Then $MR^K(p_V) < MR^K(p_W)$.*

Proof The hypothesis and the fact that $\mathcal{I}$ reverses inclusions jointly imply $\mathcal{I}(W) \subseteq \mathcal{I}(V)$. Consider an arbitrary polynomial $f(\overline{x}) \in K[\overline{x}]$ and suppose that $f(\overline{a}_W) = 0$, where $\overline{a}_W$ realises p_W. This is to say that $f \in \mathcal{I}(W)$ (by the definition of generic point) and the inclusion we have already established yields $f \in \mathcal{I}(V)$, which is equivalent to $f(\overline{x}) \in p_V$. In this case, if $\overline{a}_V$ realises p_V, we have $f(\overline{a}_V) = 0$. In sum, for any polynomial $f(\overline{x}) \in K[\overline{x}]$:

$$f(\bar{a}_W) = 0 \implies f(\bar{a}_V) = 0.$$

It follows that, if the n-tuple of field elements $\bar{a}_V$ is algebraically independent over K, so must the n-tuple $\bar{a}_W$ be (otherwise the latter n-tuple makes a non-zero polynomial with coefficients in K vanish, and so does the former n-tuple, by what we have proved). In other words, the transcendence degree of $\mathcal{K}(W)$ over the field $\mathcal{K}$ is not smaller than the transcendence degree of $\mathcal{K}(V)$ over the field $\mathcal{K}$. Because, by Theorem 10.48 (supposing we work with sufficiently large algebraically closed fields), this transcendence degree is the Morley rank of the respective generic point, we obtain the inequality:

$$MR^K(p_V) \leq MR^K(p_W).$$

In order to show that, when V is a strict subset of W, strict inequality holds, it will suffice to assume $MR^K(p_V) = MR^K(p_W)$ and deduce that, in this case, $V = W$. If we have equality, it is clear that, given a transcendence basis for $\mathcal{K}(W)$ over $\mathcal{K}$, which is a sublist of the n-tuple $\bar{a}_W$, the corresponding sublist of $\bar{a}_V$ determines a transcendence basis for $\mathcal{K}(V)$ over $\mathcal{K}$.

Let the respective transcendence bases be $a_{1W}, \ldots, a_{kW}, a_{1V}, \ldots, a_{kV}$, with $k \leq n$. Since the fields $\mathcal{K}(V) = \mathcal{K}(a_{1V}, \ldots, a_{kV})$ and $\mathcal{K}(W) = \mathcal{K}(a_{1W}, \ldots, a_{kW})$ are generated from $\mathcal{K}$ by finite transcendence bases, we can describe their elements as ratios of polynomials in $K[x_1, \ldots, x_k]$ evaluated on $a_{1W}, \ldots, a_{kW}$ and $a_{1V}, \ldots, a_{kV}$ respectively.

It is therefore easy to see that there is an isomorphism ρ from $\mathcal{K}(W)$ onto $\mathcal{K}(V)$ that fixes K and sends a_{iW} into a_{iV}. If k, the common transcendence degree of $\mathcal{K}(W)$ and $\mathcal{K}(V)$ over $\mathcal{K}$, equals n, this isomorphism sends $\bar{a}_W$ into $\bar{a}_V$. It follows that $p_V = p_W$, since they must contain the same formulae over K. We noted in the proof of Theorem 10.57 that generic points are uniquely determined by the corresponding varieties. We may therefore conclude $V = W$.

A problem arises when the transcendence degree k is strictly smaller than n. In this case $n - k$ of the a_{iW} are algebraically dependent over the remaining ones and the same is true of the a_{iV}. If a_{iW} is algebraically dependent on $a_{1W}, \ldots, a_{kW}$, there is a nonzero polynomial $h(x)$ in $\mathcal{K}(W)[x]$ such that $h(a_{iW}) = 0$. Note that $h(a_{iW})$ may also be regarded as a polynomial $h'(x_1, \ldots, x_n)$ evaluated on $a_{1W}, \ldots, a_{nW}$ and equal to 0 at such arguments. It follows that the same polynomial vanishes at $a_{1V}, \ldots, a_{nV}$.

Now we note that the isomorphism ρ can be canonically extended to a ring isomorphism between $\mathcal{K}(W)[x]$ and $\mathcal{K}(V)[x]$. If $h(x) \in \mathcal{K}(W)[x]$ is the minimal polynomial of a_{iW}, it follows that its ρ-image is also monic, irreducible and, by what we have noted, vanishing at a_{iV}. We deduce that the ρ-image of $h(x)$ must be the minimal polynomial of a_{iV} in $\mathcal{K}(V)[x]$. The isomorphism ρ sends a_{iW} into a root of the minimal polynomial of a_{iV}. Even though this root is not a_{iV} exactly, the corresponding algebraic extension fields are isomorphic.

10.4 Dimension of an Algebraic Variety

In view of this fact, we know that, even when $k < n$, we can find an isomorphism ρ' that fixes K and sends $\bar{a}_W$ into $\bar{a}_V$. It again follows that $p_V = p_W$ and $V = W$. □

Exercise 10.59 We have defined the dimension of a variety as the largest dimension attained by its irreducible components. Show that, if $W \subseteq K^n$ is a variety (not necessarily irreducible), $MR^K(W) = dim(V)$.

10.4.1 A Note on ω-Stable Theories

Our brief discussion of Morley rank has focussed on showing how an abstract model-theoretic notion can be specialised to more concrete algebraic and geometric notions, which in turn acquire a novel, logical character. Although this result is of independent significance, in that it provides an abstract 'logical analysis' of the notion of dimension, the scope of Morley rank extends well beyond the mathematical contexts we have restricted attention to.

Although we have no space to pursue this theme here, it is at least worth noting that Morley rank determines an abstract independence relation for a class of theories that exceeds the strongly minimal ones we have been focussing on in this chapter. The theories in question are singled out by a key combinatorial property concerning their types.

Definition 10.60 Let T be a theory, κ an infinite cardinal. We say that T is κ-**stable** iff, for any $\mathcal{M} \models T$ and any $A \subseteq M$ with $|A| \leq \kappa$, $|S_1^{\mathcal{M}}(A)| \leq \kappa$.

When $\kappa = \aleph_0$ we say that the theory T is ω-**stable**.

Exercise 10.61 Show that the theory ACF is ω-stable.

It can be proved (see [70], p.139–140) that, if T is countable and complete, then T is ω-stable iff every formula (possibly with parameters over a model of T) has an ordinal Morley rank (i.e. $< \infty$). The connection between Morley rank and abstract independence in ω-stable theories would lead us to the beginnings of stability theory. The reader interested in pursuing this topic will find some pertinent references among the suggestions for further reading at the end of the book.

Chapter 11
Real Closed Fields

Abstract Relative to the language of rings $\mathcal{L}_r$, the algebraically closed fields are exactly the existentially closed fields. The existentially closed *ordered* fields in the language $\mathcal{L}_{r<}$ are algebraically known as real closed ordered fields, while their $\mathcal{L}_r$-reducts are the real closed fields.

In this chapter we provide first-order axiomatisations of the above classes of fields, study their key model-theoretic properties and discuss some of their basic applications to real algebraic geometry.

Section 11.1 covers essential background on ordered fields. The algebraic characterisation of real closed fields introduced in Sect. 10.2 leads to a formulation of the first-order $\mathcal{L}_r$-theory RCF, which axiomatises the class of real closed fields, and of the $\mathcal{L}_{r<}$-theory RCF$_<$, which extends RCF and axiomatises the class of real closed ordered fields.

In Sect. 11.3 we prove quantifier elimination for RCF$_<$ and deduce from it the model-completeness of RCF. Although the latter theory does not have quantifier elimination, it is, like its extension, complete and decidable. We also obtain an analogue of the main theorem proved in Sect. 10.1, to the effect that an ordered field has quantifier elimination iff it is a model of RCF$_<$.

Section 11.4 contains some basic applications of the model-theoretic properties of RCF$_<$, in particular quantifier elimination and o-minimality, to real algebraic geometry.

11.1 Ordered Fields

An ordered field is a $\mathcal{L}_{r<}$-structure that models the sentences axiomatising the class of fields, the sentences axiomatising the class of linear orders, and the following $\mathcal{L}_{r<}$-sentences that govern the interaction between order and the field operations:

(a) $\forall x \forall y \forall z (x < y \rightarrow x + z < y + z)$;
(b) $\forall x \forall y \forall z ((0 < z \land x < y) \rightarrow xz < yz)$.

Any ordered field $\mathcal{F}$ has characteristic 0 and is infinite. If we expand the language by adding a new unary predicate symbol P and introduce the new axiom

$\forall x (P(x) \leftrightarrow 0 < x)$, $\mathcal{F}$ expands to an ordered field $\mathcal{F}^+$ with a distinguished subset of positive elements, which in turn models the following three conditions in the expanded language:

(c) $\neg P(0)$;
(d) $\forall x \forall y ((P(x) \wedge P(y)) \rightarrow (P(x+y) \wedge P(xy)))$;
(e) $\forall x (x = 0 \vee P(x) \vee \neg P(x))$.

If we add conditions (c) to (e) to the axioms of a field, a model of the resulting set of sentences is a $\mathcal{L}_r \cup P$-structure $\mathcal{F}^-$, which carries a linear order defined by the nested atomic formula $P(x - y)$. In this case, if we add to our language the binary relation symbol $<$, defined as follows:

$$\forall x \forall y (y < x \leftrightarrow P(x - y))$$

and expand $\mathcal{F}^-$ to a $(\mathcal{L}_{r<} \cup P)$-structure $\mathcal{F}^+$, the $\mathcal{L}_{r<}$-reduct of $\mathcal{F}^+$ is an ordered field $\mathcal{F}$.

In some contexts it is convenient to think about ordered fields as fields with a distinguished set of positive elements, while in other cases it is convenient to think about them simply as $\mathcal{L}_{r<}$-structures. The two approaches are equivalent by the above observations. We mainly adopt the former in the remainder of this section and in the next.

The reader may think of an ordered field $\mathcal{F} = (F, P^F, +^F, -^F, \cdot^F, 0^F, 1^F)$ as the $(\mathcal{L}_r \cup P)$-reduct of a $(\mathcal{L}_{r<} \cup \{P\})$-structure, obtained in turn by expanding an ordered field as we have shown above.

Exercise 11.1 Verify that, if $\mathcal{F}$ is an ordered field with P^F the associated set of positive elements, the expansion of $\mathcal{F}$ to $(\mathcal{L}_{r<} \cup \{P\})$, obtained by defining the relation symbol $<$ as shown above, models the following sentences:

(i) $\neg P(-1)$;
(ii) $\forall x ((\neg (x = 0) \wedge \exists y (x = y \cdot y)) \rightarrow 0 < x)$;
(iii) $\forall x \forall y \forall u \forall z ((\neg (x = 0) \wedge \neg (y = 0) \wedge x < y \wedge xu = 1 \wedge yz = 1) \rightarrow z < y)$.

Sentence (ii) implies that nonzero squares are positive and (iii) says that, if $a, b \in K$, then $a < b$ implies $1/a < 1/b$, where $1/a, 1/b$ are the multiplicative inverses of a, b respectively.

We are interested in isolating conditions under which a field can be ordered, i.e. expanded to a $(\mathcal{L}_r \cup \{P\})$-structure that models (c) to (e). The last exercise provides some useful clues. For instance it tells us that, in an ordered field, every square is positive. Moreover, condition (d) implies that a sum of two squares is positive. It follows that, in an ordered field, -1 cannot be expressed as a sum of squares.

In view of the last remark, we introduce the following:

Definition 11.2 Let $\mathcal{F}$ be a field, conceived as a $\mathcal{L}_r$-structure. We say that $\mathcal{F}$ is **formally real** iff $-1 \in F$ cannot be expressed as a sum of squares of elements of F.

11.1 Ordered Fields

Exercise 11.3 Prove that a field $\mathcal{F}$ is formally real iff the equality $0 = a_1^2 + \ldots + a_k^2$, where $a_i \in F$, $(i = 1, \ldots, k)$ holds exactly when $a_1 = \ldots = a_k = 0$.

Formally real fields can be ordered. To see this, we note that, if a field $\mathcal{F}$ can be ordered, the subset of F that interprets P, i.e. the set of positive elements of F, must contain the (finite) sums of squares of nonzero elements of F.

Exercise 11.4 Let $\mathcal{F}$ be a formally real field and call $\Sigma(F^2)$ the set of sums of squares of nonzero elements of F. This is to say that $a \in \Sigma(F^2)$ implies the existence of a positive $n \in \mathbb{N}$ and of $a_1, \ldots, a_n \in F-\{0\}$ such that $a = a_1^2 + \ldots + a_n^2$. Show that:

(i) $0 \notin \Sigma(F^2)$;
(ii) $a, b \in \Sigma(F^2)$ implies $a + b, ab \in \Sigma(F^2)$;
(iii) $a \neq 0$ and $a \in \Sigma(F^2)$ implies $1/a \in \Sigma(F^2)$.

We refer to $\Sigma(F^2)$ as a **division cone** over F and to $\Sigma(F^2) \cup \{0\}$ as a **nonnegative division cone** over F. By the last exercise, if $\mathcal{F}$ is formally real, then the division cone $\Sigma(F^2)$ satisfies two of the three conditions that define the positive elements of a field. It is therefore quite natural to try and extend the division cone $\Sigma(F^2)$ to a set of positive elements for F. We can do it by means of Zorn's lemma if we work with the nonnegative division cone $\Sigma(F^2) \cup \{0\}$.

Theorem 11.5 *Let $\mathcal{F}$ be a formally real field, $\Sigma(F^2)$ its associated division cone. There is an extension P^F of $\Sigma(F^2)$ such that the expansion of $\mathcal{F}$ to $\mathcal{L}_r \cup \{P\}$ interpreting P on P^F is an ordered field.*

Proof We consider the nonnegative division cone $\Sigma(F^2) \cup \{0\}$ and a special collection $\mathcal{S}$ of subsets of F. The set $X \subseteq F$ is in $\mathcal{S}$ iff:

(1) $-1 \notin X$;
(2) $\Sigma(F^2) \cup \{0\} \subseteq X$;
(3) $a, b \in X$ implies $a + b, ab \in X$.

Note that, by (2) and (3), $a \in X$ and $a \neq 0$ jointly imply $1/a \in X$. The set $\mathcal{S}$ is nonempty and partially ordered by inclusion. Furthermore, $\subseteq$-chains in $\mathcal{S}$ have upper bounds. Zorn's Lemma applies and yields the existence of a maximal element P_0^F of $\mathcal{S}$. We show that, for any $a \neq 0$ in F, $a \in P_0^F$ or $-a \in P_0^F$ (we cannot have both $a, -a \in P_0^F$ or $1/a \in P_0^F$ and, by (3), $-1 \in P_0^F$).

Toward a contradiction, we suppose that, for some $b \in F - \{0\}$, $b, -b \notin P_0^F$. In this case it suffices to verify that the set $T = \{c + bd : c, d \in P_0^F\}$ is a proper extension of P_0^F containing b, against the maximality of P_0^F. We leave the verification to the reader. The contradiction arrived at shows that $P^F = P_0^F - \{0\}$ expands $\mathcal{F}$ to an ordered field. □

Exercise 11.6 Let $\mathcal{S}, T$ be as in the previous theorem. Show that the following conditions hold:

(i) $b \in T$;
(ii) $u, v \in T$ implies $u + v, uv \in T$;
(iii) $\Sigma(F^2) \cup \{0\} \subseteq T$;
(iv) $-1 \notin T$.

Exercise 11.7

(i) Let $\mathcal{F}$ be an ordered field and $\mathcal{K}$ be an ordered extension of $\mathcal{K}$, i.e. $P^F \subseteq P^K$. If $a_1, \ldots, a_m$ are any nonzero elements of K and $b_1, \ldots, b_m \in P^F$, show that the field element:

$$b_1 a_1^2 + \ldots + b_m a_m^2$$

is in P^K. Moreover, show that -1 cannot be expressed as a field element of the form $b_1 a_1^2 + \ldots + b_m a_m^2$.

(ii) Let $\mathcal{F}$ be an ordered field and $\mathcal{K}$ a field extension of $\mathcal{F}$. Set:

$$\Sigma_p = \{b_1 a_1^2 + \ldots + b_m a_m^2 : b_1, \ldots, b_m \in P^F \text{ and } a_1, \ldots, a_m \in K\}$$

The proof of Theorem 11.5 implies that a field has an ordering iff it is formally real. Adapt this proof to show that $\mathcal{K}$ is an ordered field extension of $\mathcal{F}$ iff $-1 \notin \Sigma_p$.

As we shall see, real closed ordered fields do not admit of proper algebraic extensions that are also ordered extensions. For this reason, before we can introduce them, as well as real closed fields, we need some information on the orderability of algebraic field extensions. The next lemma provides sufficient conditions for a simple algebraic extension to be an ordered field extension.

Lemma 11.8 *Let $\mathcal{F}$ be an ordered field, viewed as a $(\mathcal{L}_r \cup \{P\})$-structure. If the polynomial $p(x) \in F[x]$ is irreducible over F and, moreover, for some $a, b \in F$, $b - a \in P^F$ and $\mathcal{F} \models \neg P(p(a)p(b)) \land \neg (p(a)p(b) = 0)$, then the quotient $F[x]/(p(x))$, regarded as a field-extension $\mathcal{K}$ of $\mathcal{F}$, can be ordered in such a way that $P^K \supseteq P^F$.*

To make sense of the proof to follow, recall that $b - a \in P^F$ is the same condition as $a <^F b$ and note that $\neg P(p(a)p(b)) \land \neg (p(a)p(b) = 0)$ is simply a circuitous way of writing $p(a)p(b) < 0$.

From now on we drop the superscript on the relation $<^F$ and, with an abuse of notation, we let $<$ designate both a relation symbol and the relation it denotes. Context will make apparent which one is meant.

Proof Let $p(x)$ be a counterexample to the lemma that is also of least degree. We are going to construct a counterexample of lower degree. The ensuing contradiction will prove the lemma.

We begin with the quotient $F[x]/(p(x))$, viewed as a field, and suppose that, even though $p(a)p(b) < 0$, we cannot extend the ordering of F to $F[x]/(p(x))$. Exercise 11.7 implies that $F[x]/(p(x))$ is an ordered field extension of $\mathcal{F}$ iff $-1 \notin$

11.1 Ordered Fields

Σ_p, where Σ_p is defined as in that exercise. The last condition implies that 0 cannot be expressed as a field element of the form:

$$b_1 a_1^2 + \ldots + b_m a_m^2$$

where $b_1, \ldots, b_m \in P^F$ and $a_1, \ldots, a_m$ are equivalence classes of $F[x]/(p(x))$ other than 0, i.e. the principal ideal $(p(x))$. Because the equivalence classes in question may be identified with the residues of division by $p(x)$, we may set $a_i = r_i(x)$, where r_i is a polynomial of degree strictly smaller than $\partial p(x)$, for each $i \in \{1, \ldots, m\}$.

Because, by hypothesis, $F[x]/(p(x))$ cannot be expanded to an ordered extension of $\mathcal{F}$, the condition just described holds, i.e. there are m residues $r_i(x)$ and m positive elements of F, say b_i ($i = 1, \ldots, m$), such that:

$$b_1 r_1^2(x) + \ldots + b_m r_m^2(x) = h(x) p(x). \tag{11.1}$$

In other words, the polynomial on the left-hand side of the last equality is equivalent to the zero of $F[x]/(p(x))$. Note that, because the degree of each $r_i(x)$ is smaller than the degree of $p(x)$, this is true of $h(x)$ as well.

We may suppose that, for any $c \in F$, $b_1 r_1^2(c) + \ldots + b_m r_m^2(c) \in P^F$. If not, the fact that $\mathcal{F}$ is formally real implies $b_i r_i^2(c) = 0$ for each $i \in \{1, \ldots, m\}$ and, since $b_i \neq 0$, we must have $r_i(c) = 0$. In other words, we can cancel the factor $(x - c)^2$ on either side of equation (11.1). We may thus assume that all possible cancellations have been made and that, for any $c \in F$, $b_1 r_1^2(c) + \ldots + b_m r_m^2(c)$ is positive in F.

Let us rewrite the right-hand side of (11.1) as:

$$h_1(x) \cdots h_s(x) p(x), \tag{11.2}$$

where $h_1(x)$, $i = 1, \ldots, s$ is irreducible over F. If $h_j(x)$ divided each $r_i(x)$, we could cancel the factor $h_j^2(x)$ out in (11.1). We may thus assume that, for each $j \in \{1, \ldots, s\}$, there is $k \in \{1, \ldots, m\}$ such that $h_j(x)$ does not divide $r_k(x)$.

We now have sufficient information to conclude the proof. Because the left-hand side of (11.1) is positive when evaluated in F, we deduce $h(a) p(a) > 0$, $h(b) p(b) > 0$. The hypothesis $p(a) p(b) < 0$ further implies $h(a) h(b) < 0$. By (11.2), the last inequality may be rewritten as:

$$[h_1(a) h_1(b)][h_2(a) h_2(b)] \cdots [h_s(a) h_s(b)] < 0. \tag{11.3}$$

Because not every product within square brackets in (11.3) can be positive, we may assume $h_1(a) h_1(b) < 0$. The degree of $h_1(x)$ is certainly smaller than the degree of $p(x)$, which is the least violating our lemma. As a consequence, the lemma must hold for $F[x]/h_1(x)$ and this field can be expanded to an ordered extension of $\mathcal{F}$. Note, however, that, by (11.2):

$$b_1 r_1^2(x) + \ldots + b_m r_m^2(x) = h_1(x) [h_2(x) \cdots p(x)], \tag{11.4}$$

where the right-hand side is the zero of $F[x]/h_1(x)$. Part (ii) of exercise 11.7 implies that each term on the left-hand side of the last equality should be zero. This cannot happen because, as we have noted earlier, one of the m polynomials $r_i(x)$ is not divided by $h_1(x)$. We have reached a contradiction, which concludes our proof. □

The last lemma entails that, if a polynomial $p(x) \in F[x]$ has opposite signs at two points a, b, then it must have a root in some ordered extension of $\mathcal{F}$. This is the content of the next theorem, which will play an essential role in proving quantifier elimination for $\mathsf{RCF}_<$.

Theorem 11.9 *Let $\mathcal{F}$ be an ordered field, viewed as a $(\mathcal{L}_r \cup \{P\})$-structure. If the polynomial $p(x) \in F[x]$ is irreducible over F and, moreover, for some $a, b \in F$, $b - a \in P^F$ and $\mathcal{F} \models \neg P(p(a)p(b)) \wedge \neg(p(a)p(b) = 0)$. There exists an ordered extension $\mathcal{M} \sqsupseteq \mathcal{F}$ such that, for some $c \in M$, $\mathcal{M} \models P(c-a) \wedge P(b-c) \wedge p(c) = 0$.*

Proof We set up a proof by induction on the degree of $p(x)$. If $\partial p(x) = 1$, then $p(x) = cx + d$, with $c, d \in F$ and $p(a), p(b)$ have opposite signs. If $p(a)$ is negative and $p(b)$ positive, then $ca + d < 0$ and $cb + d > 0$ imply $a < -d/c < b$ and we are done.

We next suppose that the theorem holds up to degree n and we suppose $p(x)$ to be of degree $n + 1$.

Because we can construct successive field extensions of $\mathcal{F}$ that contain roots of $p(x)$, we continue doing so until we reach an extension of the form $F(\alpha_1, \ldots, \alpha_m)$, in which $p(x)$ splits into a product of linear factors $(x - \alpha_1) \cdots (x - \alpha_m)$.

We next consider the set $\mathcal{S}$ of field extensions of $\mathcal{F}$ included in $F(\alpha_1, \ldots, \alpha_m)$ and carrying sets of positive elements that include P^F. Since $\mathcal{F}$ responds to this description, $\mathcal{S}$ is nonempty. We temporarily adopt the notation (K, P^K) to designate a generic element of $\mathcal{S}$. Let the binary relation $(K, P^K) \leq (N, P^N)$ hold iff $K \subseteq N$ and $P^K \subseteq P^N$. It is easy to see that $\leq$ is a partial order and that $\mathcal{S}$ is closed under unions of $\leq$-chains. Zorn's lemma applies and yields a $\leq$-maximal element $\mathcal{M}$ of $\mathcal{S}$.

Note that $p(x)$ cannot be irreducible over $\mathcal{M}$. If it were, then $\mathcal{M}$ could be extended by the adjunction of any α_i and the resulting field, which has an ordering by Lemma 11.8, would provide a proper ordered extension of $\mathcal{M}$ in $\mathcal{S}$, against the maximality of $\mathcal{M}$.

It follows that, in $M[x]$, we can represent $p(x)$ as a product of the following form:

$$p(x) = (x - \alpha_{i_1}) \cdots (x - \alpha_{i_k}) p_1(x) \cdots p_j(x),$$

where there is at least one linear factor, and each p_i ($i = 1, \ldots, j$) is irreducible. It follows that $p_i(x)$ has degree at least 2 and at most n. By the inductive hypothesis, the theorem holds for each $p_i(x)$. Thus $p_i(a)p_i(b) > 0$, or we could find a root β of $p_i(x)$ between $a, b \in M$ and deduce $\beta \in M$, using once more the maximality of $\mathcal{M}$ and Lemma 11.8.

11.2 Real Closed Fields

It remains to evaluate a few inequalities. First, relative to the ordering of M, which extends that of F, $p(a)p(b) < 0$. Second, by the argument just given, $p_1(a) \cdots p_j(a) > 0$ and $p_1(b) \cdots p_j(b) > 0$. It now follows that the products $(a - \alpha_{i_1}) \cdots (a - \alpha_{i_k})$ and $(b - \alpha_{i_1}) \cdots (b - \alpha_{i_k})$ must have opposite signs, i.e.:

$$(a - \alpha_{i_1}) \cdots (b - \alpha_{i_1})(a - \alpha_{i_k}) \cdots (b - \alpha_{i_k}) < 0.$$

If, for each $j = 1, \ldots k$, $(a - \alpha_{i_j}) \cdots (b - \alpha_{i_j}) > 0$ the previous inequality would fail. We conclude that, for some $r \in \{1, \ldots, k\}$, $(a - \alpha_{i_r}) \cdots (b - \alpha_{i_r}) < 0$.

Because $a < b$, we must have $a - \alpha_{i_r} < 0$ and $0 < b - \alpha_{i_r}$. Setting $c = \alpha_{i_r}$, we see that the conditions given in the statement of the theorem are satisfied. □

11.2 Real Closed Fields

Algebraically closed fields do not admit proper algebraic extensions and, in this respect, they are maximal relative to algebraic extensions. We now adapt this idea to ordered fields by means of the following two definitions.

Definition 11.10 An ordered field $\mathcal{F}$, viewed as a $(\mathcal{L}_r \cup \{P\})$-structure, is **order closed** iff no proper, finite extension of the $\mathcal{L}_r$-reduct of $\mathcal{F}$ can be expanded to an ordered extension of $\mathcal{F}$.

Definition 11.11 A field $\mathcal{F}$, viewed as a $\mathcal{L}_r$-structure, is **real closed** iff it is formally real and no proper, algebraic extension of $\mathcal{F}$ can be ordered.

The notions of an order closed and a real closed field look rather different since one refers to finite extensions only, whereas the other refers to arbitrary algebraic extensions. In fact, the only significant difference between these two notions concerns the signatures occurring in the respective definitions.

Lemma 11.12 *Let $\mathcal{F}$ be a field, viewed as a $\mathcal{L}_r$-structure and $\mathcal{F}_P$ be an ordered expansion of $\mathcal{F}$ to the language $\mathcal{L}_r \cup \{P\}$. Then $\mathcal{F}$ is real closed iff $\mathcal{F}_P$ is order closed. Moreover, the expansion of $\mathcal{F}$ to $\mathcal{F}_P$ is unique.*

Proof Suppose that $\mathcal{F}$ is real closed. Because $\mathcal{F}$ can be ordered, it has characteristic 0. It follows that $\mathcal{F}$ includes an isomorphic copy of $\mathbb{Q}$. Because 1 is positive under any ordering of F, so is $\mathbb{Q}^+$, the positive part of $\mathbb{Q}$, under the usual ordering of the rational numbers (which is also their only ordering, as the reader may check).

We next consider the set:

$$\Sigma'_{\mathbb{Q}} = \{q_1 a_1^2 + \ldots + q_m a_m^2 : q_1, \ldots, q_m \in \mathbb{Q}^+ \cup \{0\}, a_1, \ldots, a_m \in F - \{0\}\}$$

The set $\Sigma_{\mathbb{Q}} = \Sigma'_{\mathbb{Q}} - \{0\}$ must be included in the set of positive elements of F for any choice of ordered expansion. Let us fix one, say $\mathcal{F}_P$, which must exist because $\mathcal{F}$ is formally real. Clearly $\Sigma_{\mathbb{Q}} \subseteq P^F$. We show that the opposite inclusion holds.

If not, there is $a \in P^F$ such that $a \notin \Sigma_{\mathbb{Q}}$. It this event, a is not a square in F and, consequently, that the polynomial $x^2 - a$ is irreducible over F. By Lemma 11.8,[1] can be expanded to $F[x]/(x^2 - a)$ an ordered extension of $\mathcal{F}_P$. In other words, $F[x]/(x^2 - a)$ is an orderable, algebraic extension of the real closed field $\mathcal{F}$, a contradiction. We conclude $\Sigma_{\mathbb{Q}} = P^F$.

Thus $\mathcal{F}$ has a unique order, determined by $\Sigma_{\mathbb{Q}}$. Endowed with this unique order, $\mathcal{F}$ can only be expanded to $\mathcal{F}_{\Sigma_{\mathbb{Q}}} = \mathcal{F}_P$. Its unique expansion must be order closed or $\mathcal{F}$ would not be real closed.

If $\mathcal{F}_P$ is order closed, we can essentially repeat the previous argument to show that $P^F = \Sigma_{\mathbb{Q}}$. If $\mathcal{F}$ were not real closed, it would nonetheless have the unique ordering $\Sigma_{\mathbb{Q}}$. Toward a contradiction, we suppose that some proper algebraic extension $\mathcal{M}$ of $\mathcal{F}$ can be ordered. The restriction of the ordering of M to F is $\Sigma_{\mathbb{Q}}$. Because, moreover, $\mathcal{M}$ is proper, $M - F$ is nonempty. For any $b \in M - F$, the field $\mathcal{F}(b)$, ordered by a restriction of the ordering of M, is a finite, ordered extension of $\mathcal{F}_P$, against the fact that the latter field is order closed. This contradiction shows that $\mathcal{F}$ is real closed. □

Exercise 11.13 We shall show that, if $\mathcal{F}$ is real closed, then no polynomial of odd degree over $\mathcal{F}$ is irreducible. Let $p(x) = a_0 + a_1 x + \ldots + a_{n-1} x^{n-1} + x^n$ be a polynomial with coefficients in F and with n odd. Expand $\mathcal{F}$ to the uniquely determined ordered field $\mathcal{F}_P$.

(i) Set $b = max\{|a_0|, \ldots, |a_{n-1}|\}$ and $c = max\{1, nb\}$ (here $|a_i|$ is the absolute value of a_i, defined as in Chap. 5. The definition makes sense in every ordered field). Show that, if $d \in F$ and $d > c$, then $p(d) > 0$.
(ii) Show that, for the same choice of $d \in F$, $p(-d) < 0$.
(iii) Apply Lemma 11.8 to deduce that $p(x)$ has a root in $\mathcal{F}$. Deduce that any polynomial of odd degree in $F[x]$ has a root in F.

Exercise 11.14 Let $\mathcal{F}$ be a real closed field. Prove that, for any $a \in F$, exactly one of a, $-a$ is a square, i.e. there is $b \in F$ such that either $a = b^2$ or a is not a square and $-a = b^2$.

In two classic articles (see [3, 4]), Emil Artin and Otto Schreier showed that the properties of real closed fields established in the last two exercises suffice to characterise them. Because these properties can be expressed in the language of rings, a first-order axiomatisation RCF of the class of real closed fields is thereby shown to exist.

Before stating the axioms of RCF, we prove the Artin-Schreier characterisation:[2]

---

[1] The hypotheses of the lemma are satisfied because, setting $x^2 - a = p(x)$, we have $p(0)p(1 + a) < 0$.

[2] Because many proofs of this result make use of Galois theory, we find it helpful to include the elementary proof from [8] in full, adding a little detail to the terser original. This also enables us very easily to see that the complex field C models ACF_0.

11.2 Real Closed Fields

Theorem 11.15 (Artin-Schreier) *Let $\mathcal{F}$ be a field. The following statements are equivalent:*

(1) $\mathcal{F}$ is real closed;
(2) $\mathcal{F}$ is formally real, every polynomial of odd degree in $F[x]$ has a root in F and, for any $a \in F$, either a or $-a$ is a square;
(3) $\mathcal{F}$ is not algebraically closed but the field extension $F[x]/(x^2+1) = F(i)$ is algebraically closed.

Proof Exercises 11.13 11.14 show that (1) implies (2). To prove that (2) implies (3), note that -1 cannot be a square because $\mathcal{F}$ is formally real. Consequently the polynomial $x^2 + 1 \in F[x]$ has no roots in F and $\mathcal{F}$ is not algebraically closed. It follows that $x^2 + 1$ is irreducible and $F(i)$ is a field including F. The nontrivial part of the proof consists in showing that $F(i)$ is in fact algebraically closed. We arrive at this result by first showing that each polynomial $p(x) \in F[x]$ has a root in $F(i)$. Because the degree of $p(x)$ can be decomposed into prime factors, it can always be represented as a product of the form $2^m k$, with k an odd number.

We proceed by induction on m. If $m = 1$, the degree of $p(x)$ is odd and condition (2) implies the existence of a root of $p(x)$ in $F \subseteq F(i)$. The inductive hypothesis is that any polynomial in $F[x]$ of degree $2^{m-1}k$, with k odd, has a root in $F(i)$. We suppose the degree of $p(x)$ to be $n = 2^m k$ and consider the roots $b_1, \ldots, b_n$ of $p(x)$ in an algebraic closure $\overline{F}$ of F, which we may assume to include $F(i)$ (we know from Chap. 4 that algebraic extensions of fields can be embedded into their algebraic closures).

The product $(x - b_1) \cdots (x - b_n)$ thus equals $p(x)$. Viewed as a product of linear factors, $p(x)$ is obtained by carrying out polynomial multiplication repeatedly in such a way that the resulting coefficients are (up to sign) the elementary symmetric polynomials in n variables, evaluated on $b_1, \ldots, b_n$ respectively.

Because $(x - b_1) \cdots (x - b_n) = p(x) \in F[x]$, it follows that the elementary symmetric polynomials $s_1, \ldots, s_n$ in $b_1, \ldots, b_n$ are elements of the field F. Consequently, any symmetric polynomial in $b_1, \ldots, b_n$ is an element of F.

We now consider an infinite family of auxiliary polynomials $\{g_h\}_{h \in \mathbb{Z}}$. For each integer $h \in \mathbb{Z}$, we define:

$$g_h = \Pi_{1 \le r < s \le n}(x - b_r - b_s - h b_r b_s)$$

The coefficients of g_h are symmetric polynomials in $b_1, \ldots, b_n$ (checking this for $n = 3$ may provide some intuition). By Theorem 10.11, every coefficient of g_h can be expressed as a polynomial in $F[s_1, \ldots, s_n]$. Because the s_j are elements of F, so are the coefficients of g_h, i.e. $g_h \in F[x]$.

We next observe that the degree of g_h is $n(n-1)/2$, the number of possible ways of selecting two distinct elements from the set $\{b_1, \ldots, b_n\}$, i.e. the set of roots of $p(x)$ in $\overline{F}$. Because $n = 2^m k$ is even, $n - 1$ is odd and the degree of g_h is $2^{m-1}k(n-1)$, with $k(n-1)$ odd. By the inductive hypothesis, g_h has a root $b_r + b_s + h b_r b_s \in F(i)$. Because there are only finitely many possible choices of

b_r, b_s for each $h \in \mathbb{Z}$, at least one fixed pair b_r, b_s must figure in a root of g_h for infinitely many values of $h \in \mathbb{Z}$.

Suppose that, among such values, we find two distinct positive values h_1, h_2. Then $b_r + b_s + h_1 b_r b_s \in F(i)$, $b_r + b_s + h_2 b_r b_s \in F(i)$ imply $(h_2 - h_1)(b_r + b_s) \in F(i)$ and, since $(h_2 - h_1) \neq 0$ is an invertible element of F, $(b_r + b_s) \in F(i)$. The same result can be arrived at if h_1, h_2 are both negative, have opposite sign, or if one of them equals zero. We conclude that $a = b_r + b_s$ and $c = b_r b_s$ are elements of $F(i)$.

The quadratic equation $x^2 - ay + c = 0$ has the two roots b_r, b_s, which must also lie in $F(i)$ by an application of the quadratic formula from secondary school. Because these field elements are also roots of $p(x)$, we finally deduce that $p(x)$ has roots in $F(i)$, thus completing the induction.

To show that $F(i)$ is algebraically closed, it only suffices to remind ourselves that any element of $F(i)$ can be written in the form $a + ic$, with $a, c \in F$. The function α sending each $a + ic$ into its conjugate $a - ic$ is a ring homomorphism of $F(i)$ onto itself that naturally extends to the ring of polynomials $F(i)[x]$ (it suffices to switch the coefficients of any given polynomial to the respective conjugates). Note that $a + ic$ is its own conjugate exactly if $c = 0$ iff $a + ic \in F$. In a similar vein, $q(x) \in F(i)[x]$ is its own conjugate exactly if $q(x) \in F[x]$.

Calling α^* the extension of α to $F(i)[x]$ and noting that, for any $p(x) \in F(i)[x]$, $p(x)\alpha^*(p(x))$ is always its own conjugate, we can easily deduce that $p(x)\alpha^*(p(x)) \in F[x]$ always has a root $a + ic$ in $F(i)$. If $a + ic$ is also a root of $p(x)$, we are done. If not, then $a + ic$ must be a root of $\alpha^*(p(x))$ and this implies that $a - ic$ is a root of $p(x)$. In any case, a polynomial with coefficients in $F(i)$ has a root in $F(i)$. In other words, $F(i)$ is algebraically closed.

It only remains to show that (3) implies (1). We have to show that $\mathcal{F}$ is formally real and that it has no nontrivial algebraic extensions that can be ordered. Suppose $\mathcal{F}$ is formally real. The only nontrival, algebraic extension of $\mathcal{F}$ is $\mathcal{F}(i)$, because $[F(i) : F] = 2$ and any algebraic extension of $\mathcal{F}$ can be embedded into the algebraically closed extension $\mathcal{F}(i)$. Certainly $\mathcal{F}(i)$ is not formally real, so $\mathcal{F}$ is real closed.

It remains to show that $\mathcal{F}$ is formally real. Clearly -1 is not a square in F. It only remains to show that any sum of squares of F is a square of F. Consider $a, b \in F$. The field $F(i)$ contains a square root $c + id$ of $a + ib$, i.e. a root of $x^2 - (a + ib)$. The conjugate of this root is a root of $x^2 - (a - ib)$ (this property of conjugates was used in the proof of (3) from (2)). It follows that:

$$a^2 + b^2 = (a + ib)(a - ib) = (c + id)^2 (c - id)^2 \text{ or } a^2 + b^2$$
$$= (c^2 + d^2)^2 \in F.$$

□

Using (2) from the last theorem, we axiomatise the class of real closed fields. The **theory of real closed fields** or RCF is a $\mathcal{L}_r$-theory consisting of the axioms that isolate the class of fields together with the infinite set of sentences consisting of:

11.2 Real Closed Fields

(a) the sentence $\forall x_1 \ldots \forall x_n \neg(-1 = x_1^2 + \ldots + x_n^2)$ for each $n \geq 1$;
(b) the sentence $\forall x_0 \ldots \forall x_n \exists y (x_0 + x_1 y + \ldots + x_n y^n = 0)$, for each $n = 2k + 1$ ($k \geq 0$);
(c) the sentence $\forall x \exists y (x = y^2 \vee x + y^2 = 0)$.

If we expand the language to $\mathcal{L}_{r<}$ and add to RCF only the axioms isolating the class of ordered fields in which the symbol $<$ occurs, we obtain the **theory of real closed ordered fields** or RCF$_<$.

Example 11.16 Using only basic analysis, in particular the Intermediate Value Theorem for polynomials (see e.g. [1] pp.120-121), it is easy to verify that the real field $(\mathbb{R}, +, -, \cdot, 0, 1)$ is a model of RCF and its ordered expansion a model of RCF$_<$. Theorem 11.15 then implies that the complex field $\mathbf{C}$ is algebraically closed. In particular, $\mathbf{C} \models \mathsf{ACF}_0$.

The argument from Example 11.16 could be used to show that the set of real algebraic numbers, i.e. the relative algebraic closure of the rational numbers in the reals, is real closed. We arrive at the same conclusion by more abstract means.

Lemma 11.17 *Let $\mathcal{F}_P$ be an ordered field and $\mathcal{R}$ a real closed extension of the $\mathcal{L}_r$-reduct $\mathcal{F}$ of $\mathcal{F}_P$ whose unique order includes F^P. The relative algebraic closure of $\mathcal{F}$ in $\mathcal{R}$ is real closed.*

Proof Let $\mathcal{T}$ be the relative algebraic closure of $\mathcal{F}$ in $\mathcal{R}$. By Theorem 11.15, $\mathcal{R}$ has an algebraically closed extension $\mathcal{R}_i$ with domain $R(i)$. Let $\overline{\mathcal{F}}$ be the algebraic closure of $\mathcal{F}$ in $\mathcal{R}_i$.

It suffices to show that, for any irreducible polynomial $p(x) \in T[x]$, $T[x]/p(x)$ contains i. It will follow that no algebraic extension of $\mathcal{T}$ can be formally real, i.e. can be expanded to an ordered extension of $\mathcal{T}$.

We therefore suppose $p(x)$ irreducible over T and note that $p(x)$ has a root of the form $\alpha = a + bi$ ($a, b \in R, b \neq 0$) in $\overline{F}$. Set $\alpha' = a - bi$. Then α, α' are two roots of $p(x)$, both lying in $\overline{F}$: this is because the function sending $a + bi$ into $a - bi$ is an automorphism of $\overline{F}$ that fixes T and sends roots of polynomials over T into roots of the same polynomials over T.

The extension of $\mathcal{T}$ determined by the adjunction of α is, up to isomorphism, the field with domain $T[\alpha]$. We shall show that $i \in T[\alpha]$ using the closure properties of fields. First we can deduce that $i \in \overline{F}$ because $\alpha^2 - \alpha'^2 = 4ib \in \overline{F}$, which implies $ib \in \overline{F}$ and $-b^2 = ib^2 \in \overline{F}$. It then follows that $-ib/b^2 = -i/b \in \overline{F}$, from which we conclude that $1/b \in \overline{F}$. Finally $i = ib \cdot 1/b \in \overline{F}$.

Next we note that $2a = \alpha + \alpha' \in \overline{F}$, so a is algebraic over F. Since $2a \in R$ and T includes the elements of R that are algebraic over F, $a \in T$. The same argument relative to $2b = i(\alpha' - \alpha)$ implies $b \in T$.

We conclude:

$$i = \frac{\alpha - a}{b} \in T[\alpha].$$

□

Example 11.18 Consider the ordered field of rational numbers $\mathsf{Q}_<$. Let Q_{alg} its relative algebraic closure in the real closed field R, which must extend the unique ordering of the rationals. By the previous lemma Q_{alg} is a model of RCF and its expansion to an ordered field is a model of $\mathsf{RCF}_<$.

The last example shows that RCF, $\mathsf{RCF}_<$ have countably infinite models. In the next section we shall prove that the ordered field $\mathsf{Q}_{alg,<}$ is in fact embeddable into any model of $\mathsf{RCF}_<$. More generally, we shall prove that $\mathsf{RCF}_<$ has algebraically prime models. This fact, together with Theorem 11.5, yields a proof of quantifier elimination for $\mathsf{RCF}_<$, discussed in the next section.

11.3 Quantifier Elimination of $\mathsf{RCF}_<$ and Its First Consequences

We prove quantifier elimination for $\mathsf{RCF}_<$ using the test established in Theorem 8.82. For the test to apply, two conditions must be satisfied. First, $\mathsf{RCF}_<$ must have algebraically prime models. Second, certain primitive formulae must transfer to submodels. We now prove that both conditions hold.

Let us consider the existence of algebraically prime models first. We want to show that, if $\mathcal{A}$ models the universal part of $\mathsf{RCF}_<$, i.e. if it is a substructure of a model of $\mathsf{RCF}_<$, then any embedding of $\mathcal{A}$ into a model $\mathcal{N} \models \mathsf{RCF}_<$ extends to an embedding of some $\mathcal{L}_{<r}$-structure $\mathcal{M} \sqsupseteq \mathcal{A}$ such that $\mathcal{M} \models \mathsf{RCF}_<$.

Exercise 11.19 Show that if $\mathcal{A} \sqsubseteq \mathcal{M} \models \mathsf{RCF}_<$, then $\mathcal{A}$ includes a copy of the ordered ring of integers.

The key notion we shall be using to prove that embeddings of substructures can be extended to models of $\mathsf{RCF}_<$ is that of a real closure.

Definition 11.20 Let $\mathcal{F}_P$ be an ordered field. Then a field $\mathcal{E} \supseteq \mathcal{F}$ is said to be a **real closure** of $\mathcal{F}$ iff:

(1) $\mathcal{E}$ is an algebraic extension of $\mathcal{F}$;
(2) $\mathcal{E} \models \mathsf{RCF}$;
(3) the unique ordering of $\mathcal{E}_P$ extends the ordering of $\mathcal{F}$.

Note that, in order to guarantee the existence of a real closure of $\mathcal{F}$, it suffices to find an ordered closure of $\mathcal{F}_P$, i.e. a model $\mathcal{M}$ of $\mathsf{RCF}_<$ whose $\mathcal{L}_r$-reduct is an algebraic extension of $\mathcal{F}$ and that, expanded to $\mathcal{L}_r \cup \{P\}$, is an ordered extension of $\mathcal{F}_P$. The $\mathcal{L}_r$-reduct of $\mathcal{M}$ is a real closure of $\mathcal{F}$.

An application of Zorn's lemma yields:

Lemma 11.21 *Every ordered field $\mathcal{F}_P$ has an ordered closure.*

Proof Consider the $\mathcal{L}_r$-reduct $\mathcal{F}$ of the given ordered field and its algebraic closure $\overline{\mathcal{F}}$. We focus on the class $\mathcal{S}$ whose elements are ordered extensions of $\mathcal{F}$ with

11.3 Quantifier Elimination of RCF$_<$ and Its First Consequences

domains included in $\overline{F}$. A generic element of S may be described by means of the pair (E, P^E), where $F \subseteq E \subseteq \overline{F}$ and $P^F \subseteq P^E$. It is clear that the elements of S are partially ordered by a simultaneous double inclusion involving fields and their associated orderings. Chains of elements of S have upper bounds (it suffices to take unions) so Zorn's lemma applies.

Let $\mathcal{M}_P$ be a maximal element and suppose that there is an algebraic, ordered extension $\mathcal{N}_P$ of $\mathcal{M}_P$. The $\mathcal{L}_r$-reduct $\mathcal{N}$ of $\mathcal{N}_P$ is an algebraic extension of $\mathcal{F}$, which we may identify wih a subfield of $\overline{\mathcal{F}}$. In this case the expansion $\mathcal{N}_P$ is an ordered extension of $\mathcal{F}_P$ with domain included in $\overline{F}$ and strictly including M, against the maximality of $\mathcal{M}$. It follows that $\mathcal{M}_P$ is an ordered closure of $\mathcal{F}_P$ and $\mathcal{M}$ a real closure of $\mathcal{F}$. □

The last lemma, together with the remarks preceding it, ensures the existence of a real closure for any formally real field. Our next goal is to prove that, exactly like algebraic closures, real closures are unique up to isomorphism. We can then easily show that RCF$_<$ has algebraically prime models. To see why, observe that $\mathcal{A} = (\mathbb{Z}, <^{\mathbb{Z}}, +^{\mathbb{Z}}, -^{\mathbb{Z}}, .^{\mathbb{Z}}, 0^{\mathbb{Z}}, 1^{\mathbb{Z}})$ is embeddable in every real closed, ordered field $\mathcal{N}$. This is true of $\mathcal{A}$'s ordered field of fractions $\mathbb{Q}_<$.

We may therefore view the field $\mathbb{Q}_<$ as a common substructure of the ordered, real field $\mathbb{R}_<$ and of $\mathcal{N}$. By the last lemma and Exercise 11.22, $\mathcal{Q}$ has distinct, isomorphic real closures $\mathcal{R}_1, \mathcal{R}_2$ in the real field and in (a reduct of) $\mathcal{N}$ respectively.

Since $\mathcal{R}_1, \mathcal{R}_2$ are isomorphic, the expansion of $\mathcal{R}_1$ to $\mathcal{L}_{r<}$ is a model of RCF$_<$ that can be embedded into any model of RCF$_<$.

Because the set of real algebraic numbers $\mathbb{Q}_{alg}$ is a real closure of $\mathbb{Q}$ in the real field, it follows that $\mathbb{Q}_{alg}$ is embeddable in any model of RCF and that its $\mathcal{L}_{r<}$-expansion induced by the ordering of the reals is an algebraically prime model of RCF$_<$. It will follow from quantifier elimination that the latter structure is in fact a prime model of RCF$_<$.

Exercise 11.22 Show that any substructure $\mathcal{A}$ of a model $\mathcal{M}$ of RCF$_<$ can be extended to a model of RCF$_<$ whose $\mathcal{L}_r$-reduct is a real closure of the fraction field determined by $\mathcal{A}$.

It remains to prove that real closures are unique. The following proposition is a stepping stone to the theorem we are after.

Lemma 11.23 *Let $\mathcal{F}_1, \mathcal{F}_2$ be ordered fields (in the signature $\mathcal{L}_{r<}$) with real closures $\mathcal{R}_1, \mathcal{R}_2$ (in the signature $\mathcal{L}_r$) respectively. Let $\sigma : F_1 \longrightarrow F_2$ be an order isomorphism. Then σ naturally extends to a ring isomorphism $\overline{\sigma} : F_1[x] \longrightarrow F_2[x]$. For any $p(x) \in F_1[x]$, the number of roots of $p(x)$ in R_1 is the same as the number of roots of $\overline{\sigma}(p(x))$ in R_2.*

The 'natural' extension of σ alluded to sends the polynomial $a_0 + a_1 x + \ldots + a_n x^n \in F_1[x]$ into the polynomial $\sigma(a_0) + \sigma(a_1)x + \ldots + \sigma(a_n)x^n \in F_2[x]$. The proof of Lemma 11.23 requires a generalisation to real closed fields of a theorem originally due to Sturm, which we omit. A proof may be found in [8], pp.12–13.

The lemma we have stated allows us to obtain a 'finite approximation' of the result we eventually want to prove.

Lemma 11.24 *Let $\mathcal{F}_1', \mathcal{F}_2'$ be ordered fields (in the signature $\mathcal{L}_{r<}$) with the respective real closures $\mathcal{R}_1, \mathcal{R}_2$ (in the signature $\mathcal{L}_r$). Let $\mathcal{F}_1, \mathcal{F}_2$ be the $\mathcal{L}_r$-reducts of $\mathcal{F}_1', \mathcal{F}_2'$. If $\sigma : F_1' \longrightarrow F_2'$ is an order isomorphism and $A = \{a_1, \ldots, a_n\} \subseteq R_1$, there are a subfield $\mathcal{E} \sqsubseteq \mathcal{R}_1$ with domain $E \supseteq F_1 \cup A$ and an extension σ^* of σ to E such that, relative to the unique orderings of $R_1, R_2, a_i <^{R_1} a_j$ iff $\sigma(a_i) <^{R_2} \sigma(a_j)$.*

Proof We first identify the domain of $\mathcal{E}$. To this end, we suppose that the elements of A are listed in increasing order[3] as $a_1 <^{R_1} \ldots <^{R_1} a_m$. It follows that, for any $i \in \{1, \ldots, m-1\}$, $a_{j+1} - a_j$ is positive and thus a square. It follows that the positive square root of $a_{j+1} - a_j$, call it b_j, is an element of R_1.

We extend the field $\mathcal{F}_1$ by successively adjoining to it $a_1, \ldots, a_m, b_1, \ldots, b_{m-1}$. By Theorem 6.35, the field $\mathcal{F}_1(a_1, \ldots, a_m, b_1, \ldots, b_{m-1})$ is a simple algebraic extension $\mathcal{F}_1(\alpha)$: consequently, $\alpha \in R_1$.

If $p(x) \in F_1[x]$ is the minimum polynomial of α, the polynomial $\overline{\sigma}(p(x)) \in F_2[x]$, determined by the extension of σ we have already described, has a root $\beta \in R_2$ by Lemma 11.23.

Note that $p(x)$ and $\overline{\sigma}(p(x))$ have the same degree. Thus, the fields determined by $F_1(\alpha)$ and $F_2(\beta)$ are extensions of the same degree, say k, over F_1, F_2 respectively. As a result, each element of $F_1(\alpha)$ can be expressed as a linear combination of $1, \alpha, \ldots, \alpha^{k-1}$ and that each element of $F_1(\beta)$ can be expressed as a linear combination of $1, \beta, \ldots, \beta^{k-1}$.

It is now easy to extend σ to a field isomorphism $\sigma^* : F_1(\alpha) \longrightarrow F_2(\beta)$ that satisfies the equality $\sigma^*(\alpha) = \beta$. Because σ^* is a field isomorphism, for each $j \in \{1, \ldots, m-1\}$ we have:

$$\sigma^*(a_{j+1}) - \sigma^*(a_j) = \sigma^*(a_{j+1} - a_j) = \sigma^*(b_j^2) = (\sigma^*(b_j))^2.$$

It follows that, relative to the unique ordering of $\mathcal{R}_2$, b_j is positive, i.e. $\sigma^*(a_j) <^{R_2} \sigma^*(a_{j+1})$. The proof is complete. □

Roughly speaking, Lemma 11.24 allows us to extend certain isomorphisms between ordered fields to finite parts of their real closures. We want extensions to the entirety of their real closures. Because finite parts are already taken care of, it is plausible to conjecture that the compactness theorem suffices to complete the argument we want. This conjecture is confirmed by the next theorem.

Theorem 11.25 *Let $\mathcal{F}_1', \mathcal{F}_2'$ be ordered fields (in the signature $\mathcal{L}_{r<}$) with the respective real closures $\mathcal{R}_1, \mathcal{R}_2$ (in the signature $\mathcal{L}_r$). If σ is an order isomorphism between $\mathcal{F}_1'$ and $\mathcal{F}_2'$, then there is an extension of σ, restricted to $\mathcal{L}_r$-reducts, to an isomorphism between $\mathcal{R}_1, \mathcal{R}_2$, which is already an order isomorphism between the $\mathcal{L}_{r<}$-expansions of $\mathcal{R}_1, \mathcal{R}_2$.*

---

[3] Relative to the unique ordering of R_1. In the remainder of the proof, we freely help ourselves to expansions or reducts of real closed fields as needed.

11.3 Quantifier Elimination of RCF$_<$ and Its First Consequences

Proof Without loss of generality, we may assume that R_1, R_2 are disjoint (if not, we can produce disjoint, isomorphic copies of the real closures $\mathcal{R}_1$, $\mathcal{R}_2$).

We introduce a sufficiently rich first-order language $\mathcal{L}^*$ and a set of $\mathcal{L}^*$-sentences Σ that in essence formalise the statement of the theorem. Lemma 11.24 will play a key role in proving the finite satisfiability of Σ.

The vocabulary of $\mathcal{L}^*$ consists of:

(1) the unary relation symbols A_1, A_2, B_1, B_2;
(2) the binary relation symbols $Inv_1, Inv_2, <_1, <_2, S$;
(3) the ternary relation symbols $Add_1, Mult_1, Add_2, Mult_2$;
(4) the constant symbols $0_1, 1_1, 0_2, 1_2$;
(5) a set $C = \{c_i\}_{i \in I}$ of constant symbols such that $|C| = |R_1|$;
(6) a set $D = \{d_j\}_{j \in J}$ of constant symbols such that $|D| = |R_2|$.

Note that $\mathcal{L}^*$ is a relational vocabulary. Intuitively, we use A_i to describe the field F_i and B_i to describe its real closure R_i ($i = 1, 2$). The relation symbols $Inv_i, Add_i, Mult_i$ ($i = 1, 2$) are intended to describe the field operations in each domain F_i separately.

The reason to turn to a relational vocabulary is that it allows us to regard $\mathcal{R}_1, \mathcal{R}_2$ and their respective subfields as parts of the same structure $\mathcal{M}^*$, without having to define functions that can take arguments from both fields simultaneously.

The binary relation symbol S is intended to provide a description of the sought isomorphism between $\mathcal{R}_1, \mathcal{R}_2$. Our next step is to list a set Σ of $\mathcal{L}^*$-sentences that codifies the statement we want to prove. The set Σ consists of:

(a) the $\mathcal{L}^*$-diagrams $\mathcal{D}(\mathcal{R}_i)$, $i = 1, 2$.
(b) the sentences $\forall x(A_i(x) \to B_i(x))$, $i = 1, 2$;
(c) sentences of the form $S(c_i, d_j)$ with c_i, d_j naming $a_i \in F_1, b_j \in F_2$ respectively and $\sigma(a_i) = b_j$;
(d) sentences of the form $S(c_i, d_j) \land S(c_i, d_k) \to d_j = d_k$;
(e) for each $c_i \in C$, the sentence $\exists x(B_2(x) \land S(c_i, x))$;
(f) sentences of the form $(c_i <_1 c_k \land S(c_i, d_j) \land S(c_k, d_h)) \to d_j <_2 d_h$;
(g) sentences of the form $(Add_1(c_i, c_j, c_k) \land S(c_i, d_r) \land S(c_j, d_s) \land S(c_k, d_t)) \to Add_2(d_r, d_s, d_t)$;
(h) sentences of the form $(Mult_1(c_i, c_j, c_k) \land S(c_i, d_r) \land S(c_j, d_s) \land S(c_k, d_t)) \to Mult_2(d_r, d_s, d_t)$;
(i) sentences of the form $(Inv_1(c_i, c_j) \land S(c_i, d_r) \land S(c_j, d_s)) \to Inv_2(d_r, d_s)$;
(j) for each $d_k \in D$, the sentence $\exists x(B_1(x) \land S(x, d_k))$.

The sentences in (c) describe the action of σ in terms of S. The sentences in (d) and (e) say that S is the graph of a single-valued relation that is total on the domain that B_1 names. The sentences in (f) say that S is order-preserving (and thus injective), while the sentences in (g)-(i) say that S is a field-embedding. Finally, the sentences in (j) say that S maps B_1 onto B_2. We next show that Σ is finitely satisfiable using $\mathcal{M}^*$ as the model of reference for its finite parts.

It will follow that Σ has a model $\mathcal{N}^*$, in which we can embed $\mathcal{R}_1, \mathcal{R}_2$ by (a). Within $\mathcal{N}^*$, the embedded copies of $\mathcal{R}_1, \mathcal{R}_2$ are isomorphic, so we can deduce that $\mathcal{R}_1, \mathcal{R}_2$ must also be isomorphic.

It only remains to consider a finite part Φ of Σ and verify that $\mathcal{M}^* \models \Phi$. Because $\mathcal{M}^*$ models (a) and (b) anyway, we may assume that Φ contains a finite part of (c) to (j). Suppose, in particular, that Φ contains the finite list:

$$\exists x_1 (B(x_1) \wedge S(x_1, d_1)), \ldots, \exists x_1 (B_1(x_n) \wedge S(x_n, d_n)). \tag{11.5}$$

Let $\overline{q}_i(x) \in F_2[x]$ the minimum polynomial of d_i. Using the isomorphism between $F_1[x]$ and $F_2[x]$ induced by σ, the isomorphism between F_1, F_2, we can associate with each $\overline{q}_i(x) \in F_2[x]$ the unique polynomial $q_i(x) \in F_1[x]$.

By Lemma 11.23, $\overline{q}_i(x)$ and $q_i(x)$ have the same number of roots in the respective real closed fields. Let the roots in R_1 of the n polynomials $q_1(x), \ldots, qn(x)$ be $a_1, \ldots, a_m$, named by the constant symbols $c_1, \ldots, c_m$ in the signature of $\mathcal{L}^*$. We may assume $F_1 \cap \{a_1, \ldots, a_m\} = \emptyset$.

We next consider the ordered list $a_1 <^{R_1} a_2 <^{R_1} \ldots <^{R_1} a_m$. An application of Lemma 11.24 ensures the existence of a field-isomorphism σ^* from a subfield of $\mathcal{R}_1$ to a subfield of $\mathcal{R}_2$. In $\mathcal{M}^*$, we interpret the symbol S as the graph of σ^*. The finite part of (c), (d), (e) is automatically satisfied under this interpretation. The constant symbols $d_1, \ldots, d_n$ from (11.5) are interpreted on suitable σ^*-images of the elements of R_1 named by $c_1, \ldots, c_m$. As for the remaining constant symbols d_j from (f), (g), (h), we can straightforwardly interpret them on the σ^*-images of the interpretations of the c_i occurring in the same sentences.

It follows that Σ has a model and, as suggested earlier, we can reduce this result, via diagrams, to the existence of an isomorphism between $\mathcal{R}_1$ and $\mathcal{R}_2$. In other words, real closures are unique up to isomorphism. □

Exercise 11.26 Prove that the extension of σ referred to in the statement of Theorem 11.25 is unique.

Theorem 11.27 *The theory* $\mathsf{RCF}_<$ *has quantifier elimination. Consequently, it is model-complete.*

Proof We recall the statement of Theorem 8.82. According to that result, a theory T has quantifier elimination if it satisfies the following conditions:

(1) T has algebraically prime models.
(2) For any models $\mathcal{M} \sqsubseteq \mathcal{N}$ of T, any n-tuple $\overline{a} \in M^n$, and any primitive formula $\exists y \varphi(\overline{x}, y)$:

$$\mathcal{N} \models \exists y \varphi(\overline{a}, y) \implies \mathcal{M} \models \exists y \varphi(\overline{a}, y),$$

Setting $T = \mathsf{RCF}_<$, we note that (1) holds by earlier remarks. It remains to prove that (2) holds. We consider $\mathcal{M}, \mathcal{N} \models \mathsf{RCF}_<$ with $\mathcal{M} \sqsubseteq \mathcal{N}$ and a positive, primitive

11.3 Quantifier Elimination of RCF$_<$ and Its First Consequences

formula $\exists y \varphi(\bar{a}, y)$ with parameters $\bar{a}$ in M. By 'positive' we mean that $\varphi(\bar{x}, y)$ is a conjunction of positive literals, in which the negation symbol does not occur.

We wish to show that $\mathcal{N} \models \exists y \varphi(\bar{a}, y)$ implies $\mathcal{M} \models \exists y \varphi(\bar{a}, y)$. To this end we note that there is $b \in N$ such that $\mathcal{N} \models \varphi(\bar{a}, b)$, where $\varphi(\bar{a}, x)$ is a conjunction of positive literals.

Such literals are polynomial equalities or inequalities of the form $p(\bar{a}, y) = 0$ or $p(\bar{a}, y) > 0$ respectively (any inequality of the form $p(\bar{a}, y) < 0$ may be replaced by the equivalent $- p(\bar{a}, y) > 0$).

If any equalities occur, then $\mathcal{N} \models p(\bar{a}, c) = 0$ for some polynomial with coefficients in M. This is because the parameters of $\varphi(\bar{a}, y)$ are from M and the possible terms occurring in it denote elements of M. Since $\mathcal{M}$ is real closed, $c \in M$, otherwise the field extension obtained by adding c to M would be proper, algebraic and ordered by a suitable restriction of the ordering of N.

If, on the other hand, $\varphi(\bar{a}, y)$ is a conjunction of polynomial inequalities, we have to study the satisfiability of a formula of the form $p_1(\bar{a}, y) > 0 \wedge \ldots \wedge p_m(\bar{a}, y) > 0$ in $\mathcal{M}$. If none of the polynomials involved has any root in N, it follows from Theorem 11.15[4] that their signs never vary over N and, thus, over M. In this case, any $b \in M$ can satisfy $\varphi(\bar{a}, y)$.

The only possibility left to examine is the scenario in which $p_1(\bar{a}, y), \ldots p_m(\bar{a}, y)$ have roots in N. Because there are only finitely many of them and they can be ordered, we may represent them by means of the ordered list:

$$b_1 < b_2 < \ldots < b_k,$$

where $b_i \in M$ ($i = 1, \ldots, k$), since $\mathcal{M}$ is real closed. By trichotomy, c lies in an open interval bounded by two of the b_i, precedes b_1 or follows b_k. By Theorem 11.15, the sign of each $p_j(\bar{a}, y)$ within the interval containing c must be positive. Because this interval also contains elements of M, any $b \in M$ that belongs to it satisfies the conjunction $p_1(\bar{a}, y) > 0 \wedge \ldots \wedge p_m(\bar{a}, y) > 0$.

We conclude that, under all possible circumstances, $\mathcal{M} \models \exists y \varphi(\bar{a}, y)$. By the next exercise, what we have proved for conjunctions of positive literals carries over to conjunctions of literals. Theorem 8.82 now applies and the proof is complete. □

Exercise 11.28 Let $\varphi(\bar{x}, y)$ be a conjunction of literals in the language of RCF$_<$.

(i) Show that, *modulo* RCF$_<$, any quantifier-free formula $\varphi(\bar{x}, y)$ is equivalent to a quantifier-free formula in which the negation symbol does not occur;
(ii) deduce that, if $\mathcal{N} \models \exists y \varphi(\bar{a}, y)$ implies $\mathcal{M} \models \exists y \varphi(\bar{a}, y)$ for any choice of parameters $\bar{a}$ from M and any conjunction of positive literals $\varphi(\bar{a}, y)$, then the same implication must hold for any conjunction of literals.

---

[4] This theorem guarantees the existence of roots in certain ordered algebraic extensions of an ordered field. Because we are working with real closed fields, the only possible extensions in the present context are the fields themselves.

Exercise 11.29 Show that if RCF had quantifier elimination, then it would be a strongly minimal theory. Find $\mathcal{M} \models$ RCF and a formula $\varphi(x)$ such that $\varphi(M)$ is neither finite nor cofinite. Deduce that RCF does not have quantifier elimination.

Quantifier elimination has two immediate consequences, stated in the next theorems.

Theorem 11.30 *The theory of ordered, real closed fields* $\mathsf{RCF}_<$ *is complete. Furthermore, the theory of real closed fields* RCF *is complete.*

Exercise 11.31 Prove the last theorem.

Exercise 11.32 Prove that a field is real closed iff it is elementarily equivalent to the real field R.

Theorem 11.33 *The theory* $\mathsf{RCF}_<$ *is o-minimal.*

Proof If $\mathcal{M} \models \mathsf{RCF}_<$, let $X \subseteq M$ be defined by the formula $\varphi(x, \overline{a})$ with parameters. By quantifier elimination, $\varphi(x, \overline{a})$ is equivalent to a disjunctive normal form $\psi(x, \overline{a})$ modulo $\mathsf{RCF}_<$. Because negations can be eliminated, we may assume that the disjuncts of $\psi(x, \overline{a})$ are conjunctions of positive literals, i.e. formulae of the form:

$$p_1(x, \overline{a}) = 0 \wedge \ldots \wedge p_m(x, \overline{a} = 0) \wedge q_1(x, \overline{a}) > 0 \wedge \ldots \wedge q_n(x, \overline{a}) > 0,$$

where it is allowed that $m = 0$ or $n = 0$. If $m \neq 0$, the whole conjunction defines a point. If $m = 0$, then $n \neq 0$. The set $p(x, \overline{a}) > 0$ defines a union of intervals in M. This is because $p(x, \overline{a})$ has finitely many solutions in M, say $a_1 < \ldots < a_k$ and, by Theorem 11.53, its sign is constant before a_1, after a_k and over each open interval (a_i, a_{i+1}) with $i = 1, \ldots, k-1$.

The formula $p(x, \overline{a}) > 0$ singles out those intervals over which $p(x, \overline{a})$ is positive, if any.

In sum, a conjunction of literals without equalities defines an intersection of unions of open intervals, which must itself be a union of open intervals.

It now follows that $\psi(x, \overline{a})$, and thus its equivalent $\varphi(x, \overline{a})$, defines a finite union of points and open intervals in M. □

Theorem 11.34 *The theory of real closed fields* RCF *is model-complete.*

Proof Suppose $\mathcal{M}, \mathcal{N} \models \mathsf{RCF}$ with $\mathcal{M} \sqsubseteq \mathcal{N}$. It suffices to show that the inclusion map can be expanded to an order-preserving inclusion. To this end, we expand $\mathcal{M}, \mathcal{N}$ to models $\mathcal{M}_<, \mathcal{N}_<$ of $\mathsf{RCF}_<$ and note $\mathcal{M}_< \sqsubseteq \mathcal{N}_<$ (we leave the verification to the reader). It follows from quantifier elimination for $\mathsf{RCF}_<$ that $\mathcal{M}_< \prec \mathcal{N}_<$. As a consequence, all $\mathcal{L}_r$-formulae with parameters in M transfer from $\mathcal{N}_<$ to $\mathcal{M}_<$. Because these formulae do not depend for their satisfaction on how the relation symbol $<$ is interpreted, the inclusion of $\mathcal{M}$ into $\mathcal{N}$ is elementary. Consequently, RCF is a model-complete theory. □

Exercise 11.35 Complete the proof of the preceding theorem.

11.3.1 o-Minimal Fields

By Theorem 11.33, ordered real closed fields are o-minimal. It makes sense to ask which other ordered fields, if any, are o-minimal. In this subsection we shall prove that, among ordered fields, only the models of RCF$_<$ are o-minimal. While offering an interesting model-theoretic characterisation of real closed fields in terms of definability, this result shows that the abstract notion of o-minimality specialises to real closedness in the context of ordered fields.

Our goal is to show that, if $\mathcal{M}$ is a $\mathcal{L}_{r<}$-structure that models the axioms of ordered fields and is, in addition, o-minimal, then $\mathcal{M} \models$ RCF$_<$. To arrive at this result we look at $\mathcal{M}$ as a topological space, a notion we have already briefly discussed in Exercise 9.12. In the present context the topology of $\mathcal{M}$ is determined by its order: we take intervals of the form (a, b), defined by $a < x \wedge x < b$, to be the basic open sets (this includes the empty interval defined by $a < x \wedge x < a$). Every open set is a union of basic open sets. The family of open sets determine a topology over M. The complement of an open set in M is called a closed set.

Exercise 11.36 Let $\mathcal{M}$ be an o-minimal field. Show that, for any $a \in M$, $\{a\}$ is closed. Show that a finite union of closed set is closed and deduce that any finite subset of M is closed. Thus, the finite subsets of M are definable, closed subsets.

Exercise 11.37 Let $\mathcal{M}$ be an o-minimal field with $a, b, c \in M$ and $a < b, b < c$. Is $\{a\} \cup (b, c)$ open? What about $\{b\} \cup (b, c)$? Is any of these sets closed?

By 'definable' we henceforth mean $\mathcal{L}_{r<}$-definable with parameters from a fixed o-minimal field $\mathcal{M}$.

Exercise 11.38 Let $\mathcal{M}$ be an o-minimal field. Show that any *definable*, open subset of M can be represented as a finite union of disjoint intervals.

Definable intervals are indecomposable, in a sense presently to be specified. We say that a subset of $V \subseteq M$ is **definably connected** if V cannot be represented as $U_1 \cup U_2$, where U_1, U_2 are definable, nonempty and open subsets of M without elements in common (i.e. $U_1 \cap U_2 = \emptyset$).

Lemma 11.39 *Let $\mathcal{M}$ be an o-minimal field. For any $a, b \in M$ such that $a < b$, the intervals (a, b) and $[a, b]$ are definably connected.*

Proof We proceed by contradiction and suppose that the interval V (which may indifferently be (a, b) or $[a, b]$) can in fact be represented as the disjoint union of two nonempty, definable open sets U_1, U_2. We must therefore have $c \in U_1, d \in U_2$, with $c, d \in V$ and we may assume $c < d$ without loss of generality.

By Exercise 12.10, each of U_1, U_2 can be represented as a finite union of disjoint intervals. Focussing on one such representation of U_1, we can find an interval V_1 containing c, whose right endpoint is some $s \in M$. Then $c < s$ and we cannot have $d < s$ because $c < d$ implies $d \in V_1$ (by definition, an interval contains any point lying between its endpoints), against the fact that U_1, U_2 are disjoint. Thus $s \leq d$. Because V is an interval, it follows that $s \in V$. If $s \in d$, then $s \in U_2$. If

$s < d$, the fact that s is an endpoint of a partition of U_1 into finitely many disjoint, open intervals implies $s \notin U_1$. Because, by hypothesis, $V = U_1 \cup U_2$, we conclude $s \in U_2$.

Thus, in any case, there is $V_2 \subseteq U_2$, such that V_2 is an interval of the form (r_1, r_2) and $s \in (r_1, r_2)$. Because $\mathcal{M}$ is dense, we can find a point $r \in V_1$ such that $r_1 < r$ and $r < s$ (note that we can do this because o-minimal fields are dense: given $u < z \in M$, we can always pick the midpoint $u + (z - u)/2 \in (u, z)$, since $u + (z - u)/2$ is a field element). It follows that $r \in (c, s)$, i.e. $r \in U_1$, against the fact that $r \in V_2 \subseteq U_2$. This contradiction concludes the proof. □

In order to use the lemma just obtained we need to know how definable functions interact with open sets. We say that a n-ary function $f(x_1, \ldots, x_n)$ is definable in M if there is a formula (possibly with parameters) $\psi(x_1, \ldots, x_n, y)$ such that, for each n-tuple $a_1, \ldots, a_n \in M^n$, $\mathcal{M} \models \psi(a_1, \ldots, a_n, b)$ exactly when $f(a_1, \ldots, a_n) = b$.

Definable functions are remarkably well-behaved on definably connected sets when they are continuous. We define f to be a **continuous function** from M to M iff for any basic open set $(a, b) \subseteq M$, $f^{-1}[(a, b)] = \{c \in M : f(c) \in (a, b)\}$ is an open subset of M. Note that, if f is a definable function, then $f^{-1}[(a, b)]$ is also definable.

Exercise 11.40 Let $\mathcal{M}$ be an o-minimal field. Show that a definable function $f : M \longrightarrow M$ is continuous iff, for any open interval $(a, b) \subseteq M$, $f^{-1}[(a, b)]$ is a finite union of open intervals.

Exercise 11.41 Let $\mathcal{M}$ be an o-minimal field. Show that, if $f : M \longrightarrow M$ is a definable function and $U \subseteq M$ a definably connected subset of M, then $f[U]$ is definably connected (*Hint*: reason by contraposition).

We can now establish a theorem that will immediately lead to the model-theoretic characterisation of o-minimal, ordered fields announced above.

Theorem 11.42 *Let $\mathcal{M}$ be an o-minimal field, $f : [a, b] \longrightarrow M$ a definable, continuous function with $f(a) < f(b)$. If $d \in (f(a), f(b))$, there is $c \in (a, b)$ such that $f(c) = d$. Equivalently, $f[[a, b]] = [f(a), f(b)]$.*

Proof By Lemma 11.39 and Exercise 11.41, both $[a, b]$ and $f([a, b])$ are definably connected. Toward a contradiction, we suppose that $d \in (f(a), f(b))$ but $d \notin f[[a, b]]$. Consider the definable, open sets $(-\infty, d)$ and $(d, +\infty)$. Because $f([a, b])$ is definable, so are the sets $U_1 = (-\infty, d) \cap f([a, b])$ and $U_2 = (d, +\infty) \cap f([a, b])$. By trichotomy, $U_1 \cup U_2 = f([a, b])$. Moreover, $f(a) \in U_1$, $f(b) \in U_2$. Since U_1, U_2 are also disjoint by construction, we have shown that $f([a, b])$ is not definably connected, contradicting Exercise 11.41. It follows that $d = f(c)$ for some $c \in (a, b)$. □

Using the Artin-Schreier Theorem, we can finally prove:

Theorem 11.43 *Let $\mathcal{M}$ be an ordered field. Then $\mathcal{M}$ is o-minimal iff it is real closed.*

11.3 Quantifier Elimination of RCF$_<$ and Its First Consequences

Proof A model RCF$_<$ is o-minimal by Theorem 11.33. If, conversely, $\mathcal{M}$ is o-minimal, then it satisfies Theorem 11.42. We use the latter theorem to show that, for every $a \in M$, either a or $-a$ is a square. We may assume $0 < a$. The polynomial $p(x) = x^2 - a$ is, like every other polynomial in $M[x]$, a definable, continuous function. We have the chain of inequalities $p(0) = -a < 0 < 1+a^2+a = p(1+a)$. Because $0 \in p[[0, 1+a]]$, it follows that $p(x)$ has a root in M, i.e. that a is a square. By Exercise 11.13, the same argument shows that every polynomial of odd degree with coefficients in M must have a root in M. The Artin-Schreier theorem now implies that $\mathcal{M} \models$ RCF$_<$. □

We have taken for granted that polynomials are continuous functions. This fact follows directly from the continuity of addition and multiplication in M. Because the composition of two continuous functions is continuous, iterated multiplication and addition continuously generate all polynomials.

When verifying their continuity, we have to bear in mind that they are *binary* functions from $M \times M$ to M. In order to prove their continuity, we therefore have to view $M \times M$ as a topological space: the basic open sets are just Cartesian products of basic open subsets of M, of the form $(a, b) \times (c, d)$.

Exercise 11.44 Let $\mathcal{M}$ be an ordered field, $f : M \times M \longrightarrow M$ a binary function and $U \subseteq M$ an open set. Suppose that, for any $a \in U$ such that $a = f(u, v)$, there are intervals $I = (u_1, u_2)$, $J = (v_1, v_2)$ such that $u \in I$, $v \in J$ and $f[I \times J] \subseteq U$. Show that f is continuous.

Exercise 11.45 Let $\mathcal{M}$ be an ordered field, $f, g : M \times M \longrightarrow M$ continuous functions. Show that the function $h : M \times M \times M \longrightarrow M$, determined by the condition $h(x, y, z) = g(f(x, y), z)$, is continuous.

Lemma 11.46 *Let $\mathcal{M}$ be an o-minimal field. Any polynomial function $p(x) : M \longrightarrow M$ is continuous.*

Proof We first note that the projection function $\iota : M \times M \longrightarrow M$, defined by the condition $\iota(x, y) = x$, is certainly continuous. Let $+, \cdot$ denote addition and multiplication in M and suppose these binary functions to be continuous. Exercise 11.45 implies that $\cdot(\iota(x, y), x) = x^2$. By iteration, the higher powers of x are continuous, too. Because, moreover, constant functions, sending each element of M into a fixed value $a \in M$, are continuous, ax^n is continuous, for any $a \in M$. The continuity of addition and Exercise 11.45 now imply that every polynomial function is continuous.

It remains to show that addition and multiplication are indeed continuous. We can restrict attention to basic open subsets of M of the form $(a - \epsilon, a + \epsilon)$, whose elements x satisfy the inequality $|a - x| < \epsilon$. Recall that open sets are unions of basic open sets, which, in this case, are unions of open intervals.

We focus on addition first and consider an open set $U \subseteq M$ with $a = c+d \in M$. For some $\epsilon > 0$, we must have $(a - \epsilon, a + \epsilon) \subseteq U$ because U is a union of basic open sets. By Exercise 11.44, it is enough to show that there are basic open intervals

$I, J \subseteq M$ with $c \in I, d \in J$ and $\cdot[I \times J] \subseteq (a - \epsilon, a + \epsilon)$. The last conditions are satisfied by setting $I = (c_1 - \epsilon/2, c_1 + \epsilon/2)$ and $J = (d_1 - \epsilon/2, d_1 + \epsilon/2)$.

As for multiplication, we similarly consider an open set $U \subseteq M$ with $a = cd \in M$. For some $\epsilon > 0$, we must have $(a - \epsilon, a + \epsilon) \subseteq U$ because U is a union of basic open sets. We can in particular choose ϵ to be smaller than 1. We now set:

$$\delta = \frac{\epsilon}{2}\left(\frac{1}{1 + |c| + |d|}\right) < 1.$$

Let I be the interval $(c - \delta, c + \delta)$ and J be the interval $(d - \delta, d + \delta)$. If $x \in I, y \in J$ we obtain:

$$\begin{aligned}
|\cdot(x, y) - \cdot(c, d)| &= |xy - cd| \\
&= |xy - xd + xd - cd| \\
&= |x(y - d) + d(x - c)| \\
&\leq |x||y - d| + |d||x - c| \\
&\leq |x|\delta + |d|\delta
\end{aligned}$$

where the last inequality follows from the definition of I, J and the fact that these intervals have 'radius' δ. Because $\delta < 1$ and $|x - c| < \delta$, it follows that $|x| - |c| < |x - c| < 1$, i.e. $|x| < 1 + |c| < 1 + |c| + |d|$. As a result:

$$|x|\delta < (1 + |c| + |d|)\delta < \epsilon/2.$$

We conclude $|x|\delta + |d|\delta < \epsilon$, which implies $\cdot[I \times J] \subseteq U$. Exercise 11.44 again yields the continuity of multiplication. □

Exercise 11.47 Let M be an ordered field. We say that M has quantifier elimination iff for every $\mathcal{L}_{r<}$-formula $\phi(\overline{x})$ there is a quantifier-free $\mathcal{L}_{r<}$-formula $\psi(\overline{x})$ such that $M \models \forall \overline{x}(\phi(\overline{x}) \leftrightarrow \psi(\overline{x}))$. Show that M has quantifier elimination iff $M \models \mathsf{RCF}_<$.

11.3.2 Decidability

A $\mathcal{L}$-theory T is said to be **decidable** iff there is a uniform, mechanical procedure to determine, for any $\mathcal{L}$-sentence φ, whether or not $T \models \varphi$. Our definition is rather loose, because the notion of a mechanical procedure cannot do much mathematical work unless sharpened.

Although there are standard ways of spelling it out, all studied in the context of computability theory (see e.g. chapter 1 of [18] for a quick survey of key ideas), our loose notion suffices for the sake of the present digression. In fact, we can convince

11.3 Quantifier Elimination of RCF$_<$ and Its First Consequences

ourselves that it should be possible mechanically to check whether a $\mathcal{L}_r$-formula or a $\mathcal{L}_{r<}$-formula is an axiom of RCF or of RCF$_<$ respectively.

For this reason, we say that RCF and RCF$_<$ have decidable axiomatisations. By the completeness theorem of first-order logic, when a theory has a decidable axiomatisation it is possible to generate an enumeration of its theorems, possibly with repetitions.[5] In our case, if φ is a theorem of RCF, then we can generate a list of $\mathcal{L}_r$-sentences that will produce φ after finitely many steps. The same is true of RCF$_<$ with respect to $\mathcal{L}_{r<}$-sentences.

It now follows from the completeness of RCF and RCF$_<$ that these theories are decidable. To see this, e.g. for RCF$_<$, note that, if φ follows from RCF$_<$, then it is going to occur, after finitely many other sentences, in a fixed enumeration of the theorems of RCF$_<$. If φ does not follow from RCF$_<$, then RCF$_< \models \neg\varphi$ and $\neg\varphi$ will occur in the same enumeration. In short, given the question 'is φ a theorem of RCF$_<$?', the occurrence of φ in a fixed enumeration of the theorems of RCF$_<$ is equivalent to the answer 'yes', while the occurrence of $\neg\varphi$ in the same enumeration is equivalent to the answer 'no'.

Decidability is not an obvious property of first-order theories. By a fundamental theorem of logic due to K. Göedel, the complete theory $Th(\mathbb{N})$ of the structure $(\mathbb{N}, +, \cdot, 0, 1)$ (and, *a fortiori*, of any expansion of this structure) is undecidable. It follows that any decidable set of axioms for the arithmetic of the natural numbers fails to imply the complete theory of $(\mathbb{N}, +, \cdot, 0, 1)$.

This result is sensitive to the choice of a first-order language. If, instead of adopting the notion of formula defined in Chap. 2, we work with a more restrictive notion, we can recover decidability and a form of quantifier elimination for arithmetic. This surprising result has been obtained by Marker and Slaman in [46]. We introduce it here because it essentially rests on the model-theoretic properties of RCF$_<$. Its proof depends on the following exercise:

Exercise 11.48 Let $\mathcal{L}_{r<}^-$ be the language with vocabulary $\{<, +, \cdot, 0, 1\}$ and $\mathcal{R}^-$ be the $\mathcal{L}_{r<}^-$-structure $(\mathbb{R}, <, +, \cdot, 0, 1)$. Show that, for every $\mathcal{L}_{r<}^-$-formula $\varphi(\overline{x})$ there is a quantifier-free $\mathcal{L}_{r<}^-$-formula $\psi(\overline{x})$ such that:

$$\mathcal{R}^- \models \forall \overline{x}(\varphi(\overline{x}) \leftrightarrow \psi(\overline{x})).$$

In addition, prove that $\mathcal{R}^-$ is o-minimal.

We introduce a new quantifier $\mathfrak{Q}$, which will be called 'for all but finitely many' when referred to $\mathbb{N}$. More concretely, we take the string $\mathfrak{Q}x$ to abbreviate the string $\exists z \forall x(x > z \to \ldots)$. In other words, we introduce a quantifier saying that a certain condition holds from some point on. Because of this, we also abandon the standard notion of $\mathcal{L}_{r<}^-$-formula in favour of a variant based on our new quantifier. The set of formulae we now consider is the set $\mathfrak{F}$ of strings inductively generated by the

---
[5] For both this result and the completeness theorem, see [17], pp.135–143.

atomic $\mathcal{L}_{r<}^-$-formulae under the application of the propositional connectives and the quantifier $\mathfrak{Q}$.

Thus, for instance, if $\varphi(x, y)$ is the atomic formula $x < y$, then $\mathfrak{Q}x\varphi(x, y)$ is the $\mathcal{L}_{r<}^-$-formula $\exists z \forall x (x > z \rightarrow x < y)$, which belongs to $\mathfrak{F}$. By contrast, $\forall x \exists y (x < y)$ is not a formula in the set $\mathfrak{F}$. We say that two $\mathcal{L}_{r<}^-$-structures are elementarily $\mathfrak{F}$-equivalent iff they model the same formulae in $\mathfrak{F}$. We also say that, in this case, two structures have the same $\mathfrak{F}$-theory. The notion of a $\mathfrak{F}$-elementary substructure is defined in the obvious way.

It is clear that $\mathcal{N} = (\mathbb{N}, <, +, \cdot, 0, 1)$ is a substructure of $\mathcal{R}^-$. The o-minimality of $\mathcal{R}^-$ implies:

Lemma 11.49 $\mathcal{N}$ *is a $\mathfrak{F}$-elementary substructure of $\mathcal{R}^-$. In particular $\mathcal{R}^-$ and $\mathcal{N}$ have the same $\mathfrak{F}$-theory.*

Proof Our goal is to show that, for any formula $\varphi(x_1, \ldots, x_k) \in \mathfrak{F}$ and any parameters $n_1, \ldots, n_k \in \mathbb{N}$:

$$\mathcal{N} \models \varphi(n_1, \ldots, n_k) \iff \mathcal{R}^- \models \varphi(n_1, \ldots, n_k).$$

We proceed by induction on the complexity of formulae relative to the standard connectives and the quantifier $\mathfrak{Q}$. The inductive base and the inductive step for connectives are left to the reader. The only part of the inductive step left involves $\mathfrak{Q}$. When $\varphi(x_1, \ldots, x_k)$ is of the form $\mathfrak{Q}x\psi(x, x_1, \ldots, x_k)$, the inductive hypothesis is that the lemma holds for $\psi(x, x_1, \ldots, x_k)$.

If $\mathcal{R}^- \models \mathfrak{Q}x\psi(x, n_1, \ldots, n_k)$, it is clear that $\psi(x, n_1, \ldots, n_k)$ holds for every real number greater than some r. Because the reals are an Archimedean field, there is a least $n \in \mathbb{N}$ such that $r < n$. Then $\mathcal{R}^- \models \psi(m, n_1, \ldots, n_k)$ for each $m > n$. The inductive hypothesis implies the same result for $\mathcal{N}$, i.e. $\mathcal{N} \models \mathfrak{Q}x\psi(x, n_1, \ldots, n_k)$.

If $\mathcal{N} \models \mathfrak{Q}x\psi(x, n_1, \ldots, n_k)$, then $\mathcal{N} \models \psi(m, n_1, \ldots, n_k)$ for every m greater than some fixed $n \in \mathbb{N}$. By the inductive hypothesis, this is true of $\mathcal{R}^-$. Because the set X defined by $\psi(x, n_1, \ldots, n_k)$ is unbounded above and $\mathcal{R}^-$ is o-minimal, X must include an unbounded interval of the form $(r, +\infty)$. It follows that $\mathcal{R}^- \models \psi(s, n_1, \ldots, n_k)$ for every real $s > r$. We conclude that $\mathcal{R}^- \models \mathfrak{Q}x\psi(x, n_1, \ldots, n_k)$. The induction is complete. □

We can finally prove:

Theorem 11.50 *The $\mathfrak{F}$-theory of $\mathcal{N}$ is decidable.*

Proof Because $\mathsf{RCF}_<$ is decidable, there is in particular a mechanical procedure to determine whether or not a $\mathcal{L}_{r<}^-$-sentence $\varphi \in \mathfrak{F}$ follows from $\mathsf{RCF}_<$. Because $\mathsf{RCF}_<$ is complete, $\mathsf{RCF}_< \models \varphi$ iff $\mathcal{R}^- \models \varphi$. In short, we can decide the $\mathfrak{F}$-theory of $\mathcal{R}^-$. By the previous lemma, this is the same as the $\mathfrak{F}$-theory of $\mathcal{N}$. □

Exercise 11.51 Show that, for every $\mathcal{L}_{r<}^-$-formula $\varphi(\overline{x})$ in $\mathfrak{F}$, there is a quantifier-free $\mathcal{L}_{r<}^-$-formula $\psi(\overline{x})$ such that:

$$\mathcal{N} \models \forall \overline{x}(\varphi(\overline{x}) \leftrightarrow \psi(\overline{x})).$$

Exercise 11.52 Is is true that a theory with quantifier elimination must be decidable?

11.4 Applications to Real Geometry and Algebra

The notion of an affine algebraic set from algebraic geometry corresponds to the logical notion of a definable subset of affine space K^n, where $\mathcal{K}$ is an algebraically closed field.

By quantifier elimination, definable subsets are isolated by a formula obtained from polynomial equalities of the form $p(\overline{x}) = 0$ through repeated application of the propositional connectives. Such a formula, which is none other than a quantifier-free $\mathcal{L}_r$-formula, is sometimes referred to as a **Boolean combination** of atomic $\mathcal{L}_r$-formulae.

Although classical algebraic geometry is based on an underlying algebraically closed field, it is possible to consider a real closed field as the environment of reference and develop a rather different study of *real* algebraic geometry. In this context, the definable sets are called **semi-algebraic** and they are Boolean combinations of polynomial equalities and inequalities.

The model-theoretic properties of $\mathsf{RCF}_<$ afford useful proof techniques in the context of real algebraic geometry. For instance, model-completeness licenses the transfer of first-order information from a real closed field to a real closed subfield. Furthermore, completeness provides a useful transfer principle: anything that can be proved about the ordered reals and formulated as a $\mathcal{L}_{r<}$-sentence generalises to arbitrary ordered real closed fields.

In the next three subsections we shall look at some basic uses of model-theoretic properties as instruments of proof in real algebraic geometry. The first property we consider is quantifier elimination for $\mathsf{RCF}_<$ often referred to by geometers as the Tarski-Seidenberg theorem. Let $\mathcal{R}$ be a model of $\mathsf{RCF}_<$ with $A \subseteq R^{n+1}$ a definable set. If the A is defined by $\phi(x_1, \ldots, x_n, x_{n+1})$, the formula $\exists x_{n+1} \phi(x_1, \ldots, x_n, x_{n+1})$ is said to define a projection of A on the first n coordinates.

Theorem 11.53 (Tarski-Seidenberg) *Let $\mathcal{R}$ be a real closed field and $A \subseteq R^{n+1}$ a semi-algebraic set. The projection of a A on the first n coordinates is semi-algebraic.*

Proof Let $\mathcal{R}$ be an ordered, real closed field and $A \subseteq R^{n+1}$ a semi-algebraic set. Let the $\mathcal{L}_{r<}$-formula $\phi(x_1, \ldots, x_n, x_{n+1})$ define A. By quantifier elimination, the formula $\exists x_{n+1} \phi(x_1, \ldots, x_i, \ldots, x_n, x_{n+1})$ is equivalent to a quantifier-free formula $\psi(x_1, \ldots, x_n)$ in every model of $\mathsf{RCF}_<$. Clearly, $\psi(x_1, \ldots, x_n)$ defines a semi-algebraic set. □

Exercise 11.54 Explain why quantifier elimination for RCF$_<$ yields a strengthening of the Tarski-Seidenberg theorem.

Exercise 11.55 Let $\mathcal{R}$ be an ordered, real closed field, $a \in R$. Define:

$$sign(a) = \begin{cases} 0 & \text{if } a = 0, \\ 1 & \text{if } a > 0, \\ -1 & \text{if } a < 0. \end{cases}$$

Prove the following equivalent of quantifier elimination for RCF$_<$:

Let $\bar{y}$ be a n-tuple of variables and $p_i(x, \bar{y}) = h_{i,m_i}(\bar{y})x^{m_i} + \ldots + h_{i,0}(\bar{y})$ be a sequence of s polynomials (i.e. $i = 1, \ldots, s$) in the variables $\bar{y}, x$ with coefficients in $\mathbb{Z}$. Furthermore, let f be a function from $\{1, \ldots, s\}$ to $\{-1, 0, 1\}$. There is a Boolean combination $\mathsf{B}(\bar{y})$ of polynomial equations and inequalities in the variables $\bar{y}$ with coefficients in $\mathbb{Z}$ such that for every real closed field $\mathcal{R}$ and every $\bar{b} \in R^n$, the system:

$$\begin{cases} sign(p_1(x, \bar{b})) = f(1) \\ \vdots \\ sign(p_s(x, \bar{b})) = f(s). \end{cases}$$

has a solution $a \in R$ iff $\mathcal{R}$ satisfies $\mathsf{B}(\bar{b})$.

11.4.1 Extensions

Real algebraic geometers are often interested in studying extensions of semi-algebraic sets and functions between real closed fields. Extensions naturally arise by fixing a $\mathcal{L}_{r<}$-formula and considering which sets of tuples it defines in a real closed field and in its extensions respectively. More concretely, we consider $\mathcal{R}, \mathcal{R}_1 \models \mathsf{RCF}_<$ with $\mathcal{R} \preceq \mathcal{R}_1$. If $S \subseteq R^m$ is semi-algebraic, there is a $\mathcal{L}_{r<}$-formula $\varphi(\bar{x})$, possibly with parameters in R, such that $S = \{\bar{a} \in R^m : \mathcal{R} \models \varphi(\bar{a})\}$. We abbreviate the last equality by the notation $S = \varphi(R^m)$.

The extension of S to $\mathcal{R}_1$ is the set $S_1 = \varphi(R_1^m)$. The next exercise describes some basic properties of extensions, which already hold for real closed fields, without taking order into account.

Exercise 11.56 Let $\mathcal{R}, \mathcal{R}_1 \models \mathsf{RCF}$, $\varphi(\bar{x})$ a $\mathcal{L}_r$-formula such that $\varphi(R^m) = S$ and $\varphi(R_1^m) = S_1$.

(i) Show that, if $\psi(\bar{x})$ is equivalent to $\varphi(\bar{x})$ in $\mathcal{R}$, then $S_1 = \psi(R_1^m)$.
(ii) Show that, if θ defines $T \supseteq S$, then $S_1 \subseteq T_1 = \theta(R_1^m)$.
(iii) Show that $S \subseteq S_1$ and that, if S is finite, equality holds.

11.4 Applications to Real Geometry and Algebra

(iv) Show that $(R^m - S)_1 = R^m_1 - S_1$ and that, if $\theta(R^m) = T$, $(S \cup T)_1 = S_1 \cup T_1$.

Exercise 11.57 Let S, $\mathcal{R}$ be as in the previous exercise. Prove that, if S is infinite, $|S| = \lambda$, and $\kappa > max\{|R_1|, \lambda\}$, there is an elementary extension $\mathcal{R}_\kappa$ of $\mathcal{R}$ such that $|S_1| = \kappa$.

If $S \subseteq R^{n+1}$ is semi-algebraic, then the function Π that projects $(a_1, \ldots, a_{n+1}) \in S$ onto $(a_1, \ldots, a_n)$ is semi-algebraic. To see this, it suffices to consider a defining formula for S, say $\eta(x_1, \ldots, x_{n+1})$ and to note that the formula:

$$\zeta(x_1, \ldots, x_{n+1}, y_1, \ldots, y_b) = \eta(x_1, \ldots, x_{n+1}) \wedge \bigwedge_{i=1}^{n} x_i = y_i$$

defines the graph of Π. We can use the last formula to extend Π to Π_1, the projection with domain S_1.

Exercise 11.58 Show that $\Pi_1[S_1] = (\Pi[S])_1$.

As the discussion of projections indicated, we say that a function is semi-algebraic when its graph can be defined by a $\mathcal{L}_{r<}$-formula. It follows that, if f is semi-algebraic, its domain and image are. If, moreover, $f \subseteq R^{m+1}$ is a m-ary function with domain a semi-algebraic set $S \subseteq R^m$, then $f_1 \subseteq R_1^{m+1}$ is a m-ary function with domain the semi-algebraic set S_1. As we extend f to f_1, the first-order properties of (the graph of) f are preserved.

Exercise 11.59 Show that f is injective iff f_1 is and that f is surjective iff f_1 is.

Exercise 11.60 If $\mathcal{R} \models \mathsf{RCF}_<$, show that the absolute value function $|\cdot| : R \longrightarrow R$, given by the conditions:

$$|x| = \begin{cases} x & \text{if } x \geq 0, \\ -x & \text{if } x < 0. \end{cases}$$

is semi-algebraic.

Because the absolute value is semi-algebraic, we can express the continuity of a semi-algebraic function $f : A \longrightarrow R$ with a semi-algebraic domain $A \subseteq R$ at a point of A by a $\mathcal{L}_{r<}$-formula. Let $\theta(x, y)$ be a defining formula for f and $\eta(x)$ a defining formula for A. Abbreviating strings like $\forall x(x > 0 \to \ldots)$ and $\exists x(x > 0 \wedge \ldots)$ by $\forall x > 0$ and $\exists x > 0$ respectively, we obtain the formula:

$$\forall x(\eta(x) \to \forall \epsilon > 0 \exists \delta > 0 \forall u(\eta(u) \to (|x - u| < \delta \to$$
$$(\forall y \forall z((\theta(x, y) \wedge \theta(u, z)) \to |y - z| < \epsilon)))),$$

which says that, for any x in the domain of f, f is continuous at x, i.e. it is possible to chose points so close to x that there f-values differ from $y = f(x)$ by less than any prescribed, positive quantity ϵ. A function is continuous on a set S iff it is continuous at every point of S.

We have shown:

Lemma 11.61 *Let $\mathcal{R}$ be an ordered real closed field, $S \subseteq R$ a semi-algebraic set and $f : A \longrightarrow R$ a semi-algebraic function. If $\mathcal{R}_1 \succeq \mathcal{R}$, then f_1 is continuous on S_1 iff f is continuous on S.*

The last result continues to hold if we consider semi-algebraic m-ary functions with $m > 1$, provided we use a suitable notion of 'distance' between two elements of R^m in place of the absolute value function.

We close this section with a lemma that will play an important role in the proof of Milnor's Curve Selection theorem at the end of this chapter. We obtain the lemma for semi-algebraic functions on the ordered real field and then transfer it to arbitrary, ordered real closed fields by the completeness of RCF$_<$.

Lemma 11.62 *Let $\mathcal{R}$ be an ordered, real closed field, $f : R \longrightarrow R$ semi-algebraic. For any open interval $(a, b) \subseteq R$ there is $x \in (a, b)$ such that f is continuous at x.*

Proof It is not too difficult (albeit tedious) to verify that the Lemma's statement is a $\mathcal{L}_{r<}$-sentence. As we have observed, it suffices to prove the lemma for the ordered field R of real numbers. Because every other model of RCF$_<$ is elementarily equivalent to it, the lemma generalises.

In view of what we have observed, we proceed to consider a semi-algebraic function $f : \mathbb{R} \longrightarrow \mathbb{R}$ and an open real interval $(a, b) \subseteq \mathbb{R}$. Following the proof strategy in [45], p.100, we apply the law of excluded middle and consider two cases.

In case 1, there is an open subset V of (a, b) such that $f[V]$ is a finite set. In case 2, there is no such open subset of (a, b). Case 1 is easily dealt with: it suffices to fix $b \in f[V]$, which must be the f-value of infinitely many elements of (a, b). In other words, the parametrically definable set $f^{-1}[b]$ is infinite: by o-minimality, it contains an interval, obviously included in (a, b). Because f is constant on the latter interval, it is also continuous on it.

In case 2, for any open set $V \subseteq (a, b)$, $f[V]$ is not finite. Moreover, no open set $W \subseteq V$ can have a finite f-image because $W \subseteq (a, b)$. We use the last facts to build an infinite descending chain:

$$(a, b) = V_0 \supseteq V_1 \supseteq V_2 \supseteq \ldots \supseteq V_n \supseteq V_{n+1} \supseteq \ldots$$

such that, for $i \geq 1$, V_i is an open interval of length at most $1/n$ and, in addition, its topological closure,[6] denoted by $\overline{V_i}$, is also included in V_{i-1}. Our construction essentially relies on the o-minimality of the structure R. Suppose we have already defined V_n. Because we are in case 2, the definable set $f[V_n]$ is infinite and, as a consequence, it must include an interval (c, d). Note that we can choose this interval in such a way that $d - c < 1/n$ (if the endpoints are too far away from each other, we can e.g. bisect the given interval sufficiently many times).

---

[6] The closure of a set $U \subseteq \mathbb{R}$ is the $\subseteq$-least closed subset of $\mathbb{R}$ to include U. We refer here to the topology determined by the ordering of $\mathbb{R}$, whose basic open sets are open intervals.

The definable set $f^{-1}[(c,d)]$ is certainly infinite so, by o-minimality again, it contains an interval $(a'_n, b'_n) \subseteq V_n$. It now suffices to fix two distinct points a_n, b_n inside the interval (a'_n, b'_n) in order to guarantee that $[a_n, b_n]$ (i.e. the interval (a_n, b_n) together with its endpoints) should be included in (a'_n, b'_n) and, thus, in V_n. We set $(a_n, b_n) = V_{n+1}$. The closure of V_{n+1} is $[a_n, b_n] \subseteq V_n$, as desired.

Next we consider the intersection $\bigcap_{i \geq 1} V_i$. Because we are working with a descending chain of sets, this intersection must be included in $V_2, V_3, \ldots$. Note that, by the properties of closure, $V_n \subseteq \overline{V_n}$ and, by our construction, $\overline{V_{n+1}} \subseteq V_n$. In other words, for each $n \geq 1$, each V_n is squeezed between two closed sets. It follows that:

$$\bigcap_{i \geq 1} V_i = \bigcap_{i \geq 1} \overline{V_i}.$$

In the real field, any intersection of nonempty, nested intervals is nonempty (see e.g. [1], pp.18–19). We conclude that there is a real number r in $\bigcap_{i \geq 1} V_i$. We next observe that, for any n, the set V_{n+1} is such that $f[V_{n+1}]$ is included in an interval of length at most $1/n$. In other words, for any real number sufficiently close to r (i.e. in the interval $V_{n+1} = (a_{n+1}, b_{n+1})$), we can make sure that its f-value differs from $f(r)$ by less than $1/n$. Because we can choose n as large as we please, f is continuous at r. □

11.4.2 The Artin-Lang Theorem and Hilbert's 17th Problem

A $\mathcal{L}$-theory T with quantifier elimination possesses a useful transfer principle. Because every $\mathcal{L}$-formula φ is T-equivalent to a quantifier-free formula, it is possible to preserve φ, if satisfied by a model of T, in any substructure or extension that models T. Model-completeness affords a similar transfer principle: since every $\mathcal{L}$-formula φ is T-equivalent to an existential formula, extensions that model T preserve φ. Downward preservation holds in virtue of the fact that, if T is model-complete and $\mathcal{M} \models T$, any submodel of T is an elementary substructure, i.e. it inherits any φ that holds in $\mathcal{M}$ with parameters in the submodel's domain.

Because both $\mathsf{RCF}_<$ and RCF are model-complete, they support the transfer principle just described. A central consequence of this principle is that, if a certain set of polynomial equalities or inequalities has a solution in particular real closed field, then a solution must also exist in its real closed subfields including the same parameters.

We are going to make use of this fact to obtain an important theorem of real algebraic geometry known as the Artin-Lang homomorphism theorem. The form in which we state this theorem and the proof we provide are due to Abraham Robinson.

For any real closed field $\mathcal{R}$, Robinson's version of the Artin-Lang theorem concerns a R-algebra of finite type, which is, up to isomorphism, a quotient ring $R[x_1, \ldots, x_n]/I$, where I is an ideal in $R[x_1, \ldots, x_n]$.

Theorem 11.63 (Artin-Lang) *Let $\mathcal{R}$ be a real closed field, A a R-algebra of finite type. If $\mathcal{R}_1$ is a real closed field extending $\mathcal{R}$ and there is a R-algebra homomorphism $f : A \longrightarrow R_1$, then there is a R-algebra homomorphism $h : A \longrightarrow R$.*

Proof Because R is a field, it is Noetherian (it only has two ideals). By Hilbert's Basis theorem, $R[x_1, \ldots, x_n]$ is also Noetherian and, consequently, the ideal I is finitely generated, say by the polynomials $p_i(x_1, \ldots, x_n), i = 1, \ldots, k$.

The homomorphism f sends the equivalence class of x_j into some $r_j \in R_1$, with $j = 1, \ldots, n$. Because f preserves the ring operations, the coefficients in R, and sends the elements of I into the field element 0, we obtain:

$$0 = f(p_i(x_1, \ldots, x_n)) = p_i(r_1, \ldots, r_n) \text{ with } i = 1, \ldots, k.$$

It follows that:

$$\mathcal{R}_1 \models \exists y_1 \ldots \exists y_n \left(\bigwedge_{i=1}^{k} p_i(y_1, \ldots, y_n) = 0 \right).$$

By the model-completeness of RCF, $\mathcal{R}_1 \succeq \mathcal{R}$ and the last sentence transfers to $\mathcal{R}$. Let $(a_1, \ldots, a_n) \in R^n$ be a fixed n-tuple that satisfies the matrix of the last formula.

We introduce the partial map defined by:

$$h(x_i) = a_i$$

and extend h to A by setting, for any $q(x_1, \ldots, x_n)$ in an equivalence class of A, $\varphi(q(x_1, \ldots, x_n)) = q(a_1, \ldots, a_n) \in R$. By construction, $h : A \longrightarrow R$ is a R-algebra homomorphism. □

Beyond a proof of the Artin-Lang theorem, Robinson's classic contributions to real algebra famously include a remarkably concise model-theoretic solution to Hilbert's 17th problem. Its number locates it in the list of twenty-three problems published by Hilbert on the proceedings of the 1900 Congress of Mathematicians held in Paris. The problem asks to determine whether a rational function $f(x_1, \ldots, x_n) \in \mathbb{R}(x_1, \ldots, x_n)$ such that $f(a_1, \ldots, a_n) \geq 0$ for each $(a_1, \ldots, a_n) \in \mathbb{R}^n$ can always be expressed as a sum of squares of finitely many rational functions.

Hilbert's 17th problem had provided motivation for Artin's construction of the theory of real closed fields and his eventual positive solution. Robinson's solution for an arbitrary, real closed field, follows.

Theorem 11.64 (Hilbert's 17th Problem) *Let $\mathcal{R}$ be a real closed field, $f \in R(x_1, \ldots, x_n)$ be such that $f(a_1, \ldots, a_n) \geq 0$ for every n-tuple $(a_1, \ldots, a_n) \in R^n$. Then there are $g_1, \ldots, g_k \in R(x_1, \ldots, x_n)$ such that $f = g_1^2 + \ldots + g_k^2$.*

Proof We proceed by contradiction and suppose that f cannot be represented as a sum of squares of rational functions in $R(x_1, \ldots, x_n)$. Because $\mathcal{R}$ is formally real, so is $R(x_1, \ldots, x_n)$ (see Exercise 11.65). Because f is not in the non-negative division

11.4 Applications to Real Geometry and Algebra 317

cone of $R(x_1, \ldots, x_n)$, we can adapt an argument from Theorem 11.5 to show that $R(x_1, \ldots, x_n)$ can be ordered in such a way that $f < 0$.

The proof is completed by taking the real closure $\mathcal{R}_n$ of the ordered field $R(x_1, \ldots, x_n)$. Because $\mathcal{R}_n$ is an ordered extension of $R(x_1, \ldots, x_n)$, f is a negative element of R_n and it is trivially true that:

$$\mathcal{R}_n \models \exists x_1 \ldots \exists x_n (f(x_1, \ldots, x_n) < 0),$$

since $x_1, \ldots, x_n$ are themselves solutions. By model-completeness, $\mathcal{R} \preceq \mathcal{R}_n$. We deduce:

$$\mathcal{R} \models \exists x_1 \ldots \exists x_n (f(x_1, \ldots, x_n) < 0)$$

and reach a contradiction, which concludes the proof. □

Exercise 11.65 Let $\mathcal{R}$ be a formally real field, $R[x]$ the associated polynomial ring. Show that the field of fractions $R(x_1)$ is formally real. Deduce from this result that $R(x_1, \ldots, x_n)$ is formally real.

11.4.3 Milnor's Curve Selection

In Chap. 8, Definition 8.58, we introduced the notion of in-built Skolem functions. A $\mathcal{L}$-theory T has in-built Skolem functions if, for any $\mathcal{L}$-formula $\varphi(y_1, \ldots, y_n, y)$, there are a formula $\psi(y_1, \ldots, y_n, y)$ and a term $t(y_1, \ldots, y_n)$ such that, if $\mathcal{M} \models T$ and $\mathcal{M} \models \exists y \varphi(a_1, \ldots, a_n, y)$, then $\mathcal{M} \models \varphi(a_1, \ldots, a_n, t(a_1, \ldots, a_n))$.

If instead of a $\mathcal{L}$-term $t(y_1, \ldots, y_n)$ denoting a n-ary function we had a formula $\psi(y_1, \ldots, y_n, y)$ defining the graph of a n-ary function $f(y_1, \ldots, y_n)$, we would say that T has **definable Skolem functions**. More precisely, abbreviating $(y_1, \ldots, y_n)$ by $\overline{y}$, we say that T has definable Skolem functions iff for any $\mathcal{L}$-formula $\varphi(\overline{y}, x)$ there is $\psi(\overline{y}, x)$ such that:

$$T \models \forall \overline{y}(\exists_{=1} x \psi(\overline{y}, x) \wedge (\exists x \varphi(\overline{y}, x) \rightarrow \forall x (\psi(\overline{y}, x) \rightarrow \varphi(\overline{y}, x)))).$$

What the last formula says is that ψ 'selects' a function out of the correspondence defined by φ. Note that nothing stops us from considering vector-valued functions. Given a formula $\varphi(\overline{y}, \overline{x})$, with $\overline{x}$ a m-tuple of variables, the existence of definable Skolem functions implies the existence of $\psi(\overline{y}, \overline{x})$ satisfying the above condition with the added constraint that a choice of arguments $\overline{y}$ fixes a unique m-tuple $\overline{x}$ of values for ψ.

Lou van den Dries has proved that $\mathsf{RCF}_<$ has definable Skolem functions, an important property leading to a very simple model-theoretic proof of Milnor's Curve Selection theorem, a result originally obtained by geometric means.

Theorem 11.66 (van den Dries) $\mathsf{RCF}_<$ *has definable Skolem functions.*

Proof In order to devise a proof strategy we first need to unpack the theorem's statement. To this end, we fix a model $\mathcal{R} \models \mathsf{RCF}_<$. One way to do so is in terms of semi-algebraic sets and functions, because these objects are definable. Thus, instead of a $\mathcal{L}_{r<}$-formula $\varphi(\bar{y}, \bar{x})$, we refer to the corresponding semi-algebraic set $S = \varphi(R^{n+m})$. Instead of $\psi(\bar{y}, \bar{x})$ we consider a semi-algebraic function $f : R^n \longrightarrow R^m$, whose graph is defined by ψ. If $\mathsf{RCF}_<$ has definable Skolem functions, then for any semi-algebraic set S there is a semi-algebraic function f such that, if, given the n-tuple $\bar{a}$, there is a m-tuple $\bar{b}$ such that $(\bar{a}, \bar{b}) \in S$, then $(\bar{a}, f(\bar{a})) \in S$.

We leave it as an exercise to show that the condition we have just stated is equivalent to the formal condition preceding this proof. The reason for introducing a restatement of the result we want is that we can now obtain it by an induction on m.

If $m = 1$, S is a set of $(n + 1)$-tuples and, when we fix $\bar{a}$, $\varphi(\bar{a}, x)$ defines a subset of $\mathcal{R}$. By o-minimality, such a subset is a finite union of points and intervals. Whatever set $\varphi(\bar{a}, R)$ we determine, we can single out a unique $b \in R$ such that $\mathcal{R} \models \varphi(\bar{a}, b)$, as follows:

(1) if $\varphi(\bar{a}, R) = \emptyset$, we set $f(\bar{a}) = 0$;
(2) if $\varphi(\bar{a}, R)$ has a least element b, we set $f(\bar{a}) = 0$;
(3) if the leftmost interval of $\varphi(\bar{a}, R)$ is (c, d), we set $f(\bar{a}) = (d - c)/2$;
(4) if the leftmost interval of $\varphi(\bar{a}, R)$ is $(c, +\infty)$, we set $f(\bar{a}) = c + 1$;
(5) if the leftmost interval of $\varphi(\bar{a}, R)$ is $(-\infty, c)$, we set $f(\bar{a}) = c - 1$.

Note that case (1) arises in particular when no $(n+m)$-tuple of S begins with the n-tuple $\bar{a}$. We include case (1) to ensure that f has domain R^n.

The function f we have just described is definable. To see it, observe that e.g. condition (1) can be expressed by the formula $\neg \exists u \varphi(\bar{y}, u) \land x = 0$ and condition (2) is expressed by the formula $\exists u (\varphi(\bar{y}, u) \land \forall z (\varphi(\bar{y}, z) \to u \le z) \land u = x)$. We can express the remaining conditions in a similar manner. In particular, each of them is expressed by a formula $\varphi_i(\bar{y}, x)$, with $i = 1, \ldots, 5$. The disjunction of the φ_i defines the desired semi-algebraic function. This concludes the inductive base.

We now turn to the inductive step and suppose that if $m \ge 1$ and $T \subseteq R^{n+m}$, we can definably select a semi-algebraic function from T. We now consider $S \subseteq R^{n+m+1}$ and note that, by the inductive hypothesis, there is a semi-algebraic function $g_1 : R^{n+1} \longrightarrow R^m$ satisfying the following condition: if, for any $(\bar{a}, b) \in S$, there is $\bar{c} \in R^m$ such that $\mathcal{R} \models \varphi(\bar{a}, b, \bar{c}) \in R^{n+m+1}$, then $\mathcal{R} \models \varphi(\bar{a}, b, g_1(\bar{a}, b))$.

A further application of the inductive hypothesis guarantees the existence of $g_2 : R^n \longrightarrow R$ satisfying the following condition: if, for any $(\bar{a}) \in S$ and some fixed $\bar{c} \in R^m$ there is $d \in R$ such that $\mathcal{R} \models \varphi(\bar{a}, d, \bar{c}) \in R^{n+m+1}$, then $\mathcal{R} \models \varphi(\bar{a}, g_2(\bar{a}), \bar{c}))$.

Now a semi-algebraic function $h : R^n \longrightarrow R^{m+1}$ that does what we want can be defined by the condition:

$$h(a_1, \ldots, a_n) = (g_1(g_2(a_1, \ldots, a_n), a_1, \ldots, a_n), g_2(a_1, \ldots, a_n)).$$

$\square$

11.4 Applications to Real Geometry and Algebra 319

In order to prove Milnor's theorem, we require the following lemma:

Lemma 11.67 *Let $\mathcal{R}$ be an ordered, real closed field, $f : R \longrightarrow R$ a semi-algebraic function. Then R can be expressed as a partition of the form $I_1 \cup \ldots \cup I_k \cup E$, where $I_1, \ldots, I_k$ are disjoint open intervals on which f is continuous and E is a finite set of points.*

Proof The remarks immediately before Lemma 11.61 imply (by the suppression of a single universal quantifier) that the set D of points in R at which f is discontinuous is semi-algebraic. If D were infinite, it would contain an open interval I. Since Lemma 11.62 implies that f should be continuous at some point $x \in I$, D must be finite. Thus $R - D$ is infinite and f is continuous at each point in $R - D$. Because $R - D$ is, by o-minimality, a finite union of points and intervals, we can take E to contain D and the points in the union that determines $R - D$. The remaining intervals are disjoint or may be represented as disjoint (taking overlaps as distinct intervals). The desired partition of R results. □

Exercise 11.68 Generalise the last Lemma to (vector-valued) semi-algebraic functions from R to R^n (*Hint*: projections are semi-algebraic).

Theorem 11.69 (Milnor's Curve Selection) *Let $\mathcal{R}$ be a real closed ordered field, $S \subseteq R^n$ be a semi-algebraic set and $\overline{a} \in R^n$ be a tuple in the closure of S. For some $r \in R$ there is a continuous, semi-algebraic function $f : (0, r) \longrightarrow R^n$ such that, for any $s \in (0, r)$, $f(s) \in S$ and, as s approaches 0, $f(s)$ approaches $\overline{a}$.*

Proof Let $\varphi(\overline{x})$ be a defining formula for S. We consider the definable set:

$$Y = \{(s, \overline{y}) : 0 < s \wedge \varphi(\overline{y}) \wedge (y_1 - a_1)^2 + \ldots + (y_n - a_n)^2 < s\}.$$

Because $\mathsf{RCF}_<$ has definable Skolem functions, there is a semi-algebraic function $f : R \longrightarrow R^n$ such that $(s, f(s)) \in Y$ for each positive s, by choosing y_i sufficiently close to a_i. By Exercise 11.68, we can partition R into disjoint open intervals on which f is continuous and a finite set of points. Whether or not the latter set contains 0, there certainly is an open interval of the form $(0, r)$ on which f is continuous. By the definition of Y, for any $s \in (0, r)$ we have $f(s) \in S$. The same definition implies that, if we pick s sufficiently close to 0, $f(s)$ must be as close to $\overline{a}$. □

Chapter 12
Fraïssé Limits and Measurement Scales

Abstract In this chapter we present a fundamental model-theoretic construction due to Roland Fraïssé, which produces countable $\mathcal{L}$-structures by amalgamating finite $\mathcal{L}$-structures. The resulting amalgams are unique up to isomorphism and possess a rich automorphism group: they are known as Fraïssé' limits.

The first four sections of this chapter provide an introduction to Fraïssé' limits. After some preliminaries on embeddings, covered in Sect. 12.1, Fraïssé' limits are defined, proved to exist and to be unique in Sect. 12.2.

Section 12.3 describes two very simple examples of Fraïssé' limit. Section 12.4 proves quantifier elimination for the theory of a Fraïssé' limit in a finite relational language, from which the limit's saturation easily follows.

The second half of this chapter discusses a fascinating application of Fraïssé limits to measurement theory due to Peter Jephson Cameron. Section 12.5 provides the necessary measurement-theoretic background.

Sections 12.6 and 12.7 discuss, respectively, the model-theoretic and group-theoretic results needed to prove the main theorem from Sect. 12.8. This theorem provides information about the possible, abstract scale types supported by the ordered rational numbers: it turns out that there are infinitely many more than can be supported by the ordered reals!

Throughout this chapter, $\mathcal{L}$ is a countable language.

12.1 Classes of Finitely Generated Substructures

Recall that, given a first-order language $\mathcal{L}$ and a $\mathcal{L}$-structure $\mathcal{M}$, the subset $\{a_1, \ldots, a_m\}$ determines a substructure $\mathcal{A} \sqsubseteq \mathcal{M}$ whose elements are simply the interpretations of fhe $\mathcal{L}$-terms $t(x_1, \ldots, x_n)$ (it follows from our discussion of terms in Chap. 2 that it is enough to restrict attention to unnested ones). We call $\mathcal{A}$ the substructure of $\mathcal{M}$ finitely generated by $\{a_1, \ldots, a_m\}$, which we also denote by $\langle a_1, \ldots, a_m \rangle$ or, when the length of the ordered list is immaterial, by $\langle \overline{a} \rangle$.

Exercise 12.1 Let $\mathcal{M}$ be a $\mathcal{L}$-structure, $\mathcal{A}, \mathcal{B} \sqsubseteq \mathcal{M}$ finitely generated substructures, with generator sets $\{a_0, \ldots, a_{m-1}\}, \{b_0, \ldots, b_{n-1}\}$ respectively. Describe a finitely generated substructure of $\mathcal{M}$ into which $\mathcal{A}, \mathcal{B}$ can both be embedded.

Once we have a $\mathcal{L}$-structure $\mathcal{M}$, the class of its finitely generated substructures is determined. We shall be interested in the opposite question, i.e. in finding conditions under which a class K of finitely generated structures may be viewed as the collection of the finitely generated substructures of a uniquely determined, countable structure $\mathcal{M}$. Equivalently, we look for conditions on K under which its elements are all embeddable into a single countable structure $\mathcal{M}$. The following terminology is due to Roland Fraïssé:

Definition 12.2 Let $\mathcal{M}$ be a $\mathcal{L}$-structure. The **age** of $\mathcal{M}$, denoted $Age(\mathcal{M})$ is the class of finitely generated structures embeddable into $\mathcal{M}$.

It is clear that, if $\mathcal{A} \in Age(\mathcal{M})$ and $\mathcal{A}'$ is isomorphic to $\mathcal{A}$, then $\mathcal{A}' \in Age(\mathcal{M})$. Thus $Age(\mathcal{M})$ is closed under isomorphism. Note that, if $\mathcal{M}$ is a $\mathcal{L}$-structure and $\mathcal{L}$ only contains finitely many relation symbols, then $Age(\mathcal{M})$ will only have countably many isomorphism types (i.e. there are only countably many ways in which two elements of $Age(\mathcal{M})$ can fail to be isomorphic).

Exercise 12.3 Prove the last statement.

In what follows, *we assume that, whenever we consider a class $\mathcal{K}$ of finitely generated $\mathcal{L}$-structures, the class exhibits only countably many isomorphism types*. So far, we have not yet isolated conditions that guarantee a class $\mathcal{K}$ to be the age of some structure $\mathcal{M}$. Since we are looking for necessary conditions, they must already be satisfied by any class of the form $Age(\mathcal{M})$.

Suppose $\mathcal{A} \in Age(\mathcal{B})$ and $\mathcal{B} \in Age(\mathcal{M})$. Then clearly $\mathcal{A} \in Age(\mathcal{M})$. In other words, if $\mathcal{A}$ is embeddable in $\mathcal{B}$ and $\mathcal{B}$ is embeddable in $\mathcal{M}$, then $\mathcal{A}$ is so embeddable, by a composition of the given embeddings. Moreover, in view of Exercise 12.1, $\mathcal{A}, \mathcal{B} \in Age(\mathcal{M})$ implies the existence of a finitely generated substructure in $Age(\mathcal{M})$ such that $\mathcal{A}, \mathcal{B}$ can be simultaneously embedded in it. We now restate the properties we have identified relative to a class K of finitely generated $\mathcal{L}$-structures.

Definition 12.4 Let K be a class of finitely generated $\mathcal{L}$-structures. We say that:

(1) K is **closed under isomorphisms** iff, when $\mathcal{A}$ is in K and $\mathcal{A}'$ is isomorphic to $\mathcal{A}$, the structure $\mathcal{A}'$ is in K;
(2) K satisfies the **Hereditary Property** (**HP**) iff, for any $\mathcal{B}$ in K, if $\mathcal{A}$ is a finitely generated substructure of $\mathcal{B}$, then $\mathcal{A}$ is in K;
(3) K satisfies the **Joint Embedding Property** (**JEP**) iff, for any $\mathcal{A}, \mathcal{B}$ in K, there are a structure $\mathcal{D}$ in K and embeddings f, g such that $f : A \to D, g : B \to D$. The Joint Embedding Property is illustrated by the diagram below:

12.1 Classes of Finitely Generated Substructures

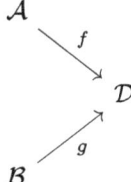

We have enough information to prove the following:

Theorem 12.5 *Let $\mathcal{L}$ be a countable language and K be a class of finitely generated $\mathcal{L}$-structures that is closed under isomorphism. The following statements are equivalent:*

(1) $\mathsf{K} = Age(\mathcal{M})$ for some countable structure $\mathcal{M}$;
*(2) K satisfies **HP** and **JEP**.*

Proof If (1) holds, then (2) does by remarks already made. The nontrivial part of the proof consists in building a suitable, countable $\mathcal{M}$ given (2). We do so by means of a chain construction that essentially relies on our assumption to the effect that K has only countably many isomorphism types. As a consequence, we can enumerate chosen representatives of each isomorphism type as a sequence $\{\mathcal{A}_i\}_{i\in\mathbb{N}}$, with $\mathcal{A}_i$ a finitely generated structure in K for each $i \geq 0$.

Our chain construction proceeds recursively, with $\mathcal{A}_0 = \mathcal{B}_0$ as its starting point. Our goal is to define a finitely generated structure $\mathcal{B}_{n+1}$ given $\mathcal{B}_n$. To this end, we simply consider $\mathcal{A}_n$ and apply **JEP** to find, in K, a finitely generated structure $\mathcal{B}_{n+1}$ into which both $\mathcal{A}_n$ and $\mathcal{B}_n$ can be embedded. In particular $\mathcal{B}_n \sqsubseteq \mathcal{B}_{n+1}$, since we can identify $\mathcal{B}_n$ with its isomorphic copy inside $\mathcal{B}_{n+1}$.

Note that, because $\mathcal{L}$ is countable, any finitely generated $\mathcal{L}$-structure is countable, although not necessarily finite. It follows that $\mathcal{B} = \bigcup_{i\in\mathbb{N}} \mathcal{B}_i$ is a countable extension of each $\mathcal{B}_i$. It remains to verify that K is $Age(\mathcal{B})$. If $\mathcal{C}$ is in K, then $\mathcal{C}$ may be identified, up to isomorphism, with $\mathcal{A}_i$, for some $i \geq 0$. Since $\mathcal{A}_i$ is embedded into $\mathcal{B}$, $\mathcal{C}$ is also embeddable into $\mathcal{B}$.

If, conversely, $\mathcal{C}$ is a finitely generated structure in $Age(\mathcal{B})$, then it may be identified with a substructure $\mathcal{A}$ of $\mathcal{B}$, in which case its finitely many generators are included in $\mathcal{B}_k$, for some sufficiently large k. By **JEP**, $\mathcal{B}_k$ is in K. By **HP**, $\mathcal{A}$ is also in K. Because $\mathcal{C}, \mathcal{A}$ are isomorphic and K is closed under isomorphisms, we conclude that $\mathcal{C}$ is in K. We conclude that K is the age of the countable $\mathcal{L}$-structure $\mathcal{B}$.
□

Consider the finite relational language $\mathcal{L}_<$. The class K of finitely generated $\mathcal{L}_<$-structures that are models of the theory of linear orders consists of the finite linear orders (there is only one of size n for each $n \in \mathbb{N}$, up to isomorphism). It is clear that this class is both $Age(\mathbb{N})$ and $Age(\mathbb{Z})$, but $\mathbb{N}, \mathbb{Z}$ are not isomorphic $\mathcal{L}_<$-structures. We require further conditions on classes of finitely generated structures to view them as the ages of uniquely determined structures. We discuss uniqueness conditions in the next section.

12.2 Fraïssé Limits

Let $\mathcal{M}, \mathcal{N}$ be $\mathcal{L}$-structures. If $Age(\mathcal{M}) \subseteq Age(\mathcal{N})$, we say that $\mathcal{M}$ is **younger** than $\mathcal{N}$. When we restrict attention to countable $\mathcal{L}$-structures, we can specify a condition under which one particular $\mathcal{L}$-structure $\mathcal{N}$ may be viewed as an extension of any younger structure. With this result in hand, we can easily prove that two countable structures of the same age are isomorphic.

The key property we rely on is given in the next definition:

Definition 12.6 Let $\mathcal{M}$ be a $\mathcal{L}$-structure. We say that $\mathcal{M}$ is **algebraically ω-homogeneous** iff, for any finitely generated $\langle \overline{a} \rangle, \langle \overline{b} \rangle \sqsubseteq \mathcal{M}$, if g is an isomorphism from $\langle \overline{a} \rangle$ to $\langle \overline{b} \rangle$ and $c \in M$, then there are $d \in M$ and an isomorphism h from $\langle \overline{a}c \rangle$ to $\langle \overline{b}d \rangle$ such that h extends g.

Algebraic ω-homogeneity can be stated in various equivalent forms, two of which are given in the next exercise.

Exercise 12.7 Let $\mathcal{M}$ be a $\mathcal{L}$-structure and K be $Age(\mathcal{M})$. Show that the following statements are equivalent:

(1) $\mathcal{M}$ is algebraically ω-homogeneous;
(2) for any $\mathcal{A}, \mathcal{B}$ in K and embeddings $f_0 : A \longrightarrow B$, $f_1 : A \longrightarrow M$, there is an embedding $g : B \longrightarrow M$ such that $g \circ f_0 = f_1$.
(3) for any finitely generated $\mathcal{A} \sqsubseteq \mathcal{M}$, any $\mathcal{B}$ in K and any embedding $f_1 : A \longrightarrow B$, there is $g : B \longrightarrow M$ such that $g \circ f_1 = \iota_A$, where ι_A is the inclusion of A into M.

Algebraic ω-homogeneity plays a key role in the following:

Lemma 12.8 *Let $\mathcal{M}, \mathcal{N}$ be countable $\mathcal{L}$-structures with $\mathcal{M}$ younger than $\mathcal{N}$. If $\mathcal{N}$ is algebraically ω-homogeneous, there is an embedding $f : M \longrightarrow N$.*

Proof Fix a finitely generated $\mathcal{L}$-structure $\mathcal{A}_0$ in the age of $\mathcal{M}$. Since $\mathcal{N}$ is older, there is an embedding $f_0 : A_0 \longrightarrow N$. All we have to do is extend f_0 to the desired embedding f.

We view $\mathcal{A}_0$ as a finitely generated substructure of $\mathcal{M}$. Because $\mathcal{M}$ is countable, $M - A_0 = \{b_i\}_{i \in \mathbb{N}}$. Starting with $\mathcal{A}_0 = \langle \overline{a} \rangle$, we consider the sequence of finitely generated structures $\langle \overline{a}b_0 \rangle, \langle \overline{a}b_0 b_1 \rangle, \ldots$, whose union is exactly $\mathcal{M}$. We shall refer to this sequence as $\{\mathcal{A}_i\}_{i \in \mathbb{N}}$. Our next step in the proof is to construct a chain of embeddings $\{f_i\}_{i \in \mathbb{N}}$ such that $f_i : A_i \longrightarrow N$. The union of this chain will be the embedding we want.

Because we already have f_0, we suppose $f_n : A_n \longrightarrow N$ given (with $n \geq 0$) and show how f_{n+1} is constructed. Because $\mathcal{A}_{n+1}$ is a finitely generated substructure of $\mathcal{M}$ and $\mathcal{M}$ is younger than $\mathcal{N}$, there is an embedding of $\mathcal{A}_{n+1}$ into $\mathcal{N}$. The restriction of this embedding to its image is an isomorphism g between $\mathcal{A}_{n+1}$ and a substructure $\mathcal{B}$ of $\mathcal{N}$. Because $\mathcal{A}_n \sqsubseteq \mathcal{A}_{n+1}$, the isomorphism g further restricts to an embedding $g_0 : A_n \longrightarrow B$. As the configuration below shows, we can use the algebraic ω-homogeneity of $\mathcal{N}$ (version (2) from Exercise 12.7) to obtain $h : B \longrightarrow N$.

12.2 Fraïssé Limits

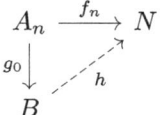

Note that, by construction, $h \circ g_0 = f_n$. We set $f_{n+1} = h \circ g$. When g is restricted to A_n, it equals g_0, so it is clear that $f_{n+1} \supseteq f_n$. The recursive construction just described produces a chain of embeddings $\{f_i\}_{i \in \mathbb{N}}$. Its union $f : \mathcal{M} \longrightarrow \mathcal{N}$ is itself an embedding. □

When $\mathcal{M}, \mathcal{N}$ have the same age and they are both algebraically ω-homogeneous, we obtain:

Theorem 12.9 *Let $\mathcal{M}, \mathcal{N}$ be countable $\mathcal{L}$-structures with the same age, both algebraically ω-homogeneous. Then $\mathcal{M}, \mathcal{N}$ are isomorphic.*

Exercise 12.10 Prove Theorem 12.9 by adapting the proof of Lemma 12.8. In particular, construct a chain of finite isomorphisms of the form $f_i : \mathcal{M}_i \longrightarrow \mathcal{N}_i$ such that, if $j < i$, $M_i \supseteq \{m_0, \ldots, m_j\}$ and $N_i = \{n_0, \ldots, n_j\}$.

Suppose that $\mathcal{M}$ is a countable, algebraically ω-homogeneous structure and that there is an isomorphism h between two of its finitely generated substructures $\mathcal{A}_0, \mathcal{B}_0$. Then the proof of Theorem 12.9, with $\mathcal{M} = \mathcal{N}$, allows us to extend h to an automorphism of $\mathcal{M}$.

Definition 12.11 Let $\mathcal{M}$ be a $\mathcal{L}$-structure. We say that $\mathcal{M}$ is **ultrahomogeneous** when any isomorphism between finitely generated substructures of $\mathcal{M}$ can be extended to an automorphism of $\mathcal{M}$.

Exercise 12.12 We have shown that a countable, algebraically ω-homogeneous $\mathcal{L}$-structure is ultrahomogeneous. Show that ultrahomogeneity implies algebraic ω-homogeneity.

In order to solve the problem that motivated this section, we still have to isolate a property on a class **K** of finitely generated structures that, together with **HP** and **JEP**, characterises **K** not simply as the age of a countable structure $\mathcal{M}$, but in fact as the age of a countable, ultrahomogeneous structure $\mathcal{M}$, uniquely determined up to isomorphism. We shall refer to such a structure, whose existence we are about to prove, as the **Fraïssé limit** of **K**. By Theorem 12.9 we know that, if a Fraïssé limit exists, it must be unique up to isomorphism, as well as ultrahomogeneous. The condition on **K** that guarantees its existence is an amalgamation property of **K**.

Definition 12.13 Let **K** be a class of finitely generated $\mathcal{L}$-structures. We say that **K** has the **Amalgamation Property** (**AP**) iff for any $\mathcal{A}, \mathcal{B}, \mathcal{C}$ in **K** such that $f_1 : \mathcal{A} \longrightarrow \mathcal{B}$, $g_1 : \mathcal{A} \longrightarrow \mathcal{C}$, there are a finitely generated structure $\mathcal{D}$ in **K** and embeddings $f_2 : \mathcal{B} \longrightarrow \mathcal{D}$, $g_2 : \mathcal{C} \longrightarrow \mathcal{D}$ such that $f_2 \circ f_1 = g_2 \circ g_1$. The Amalgamation Property is illustrated by the diagram below:

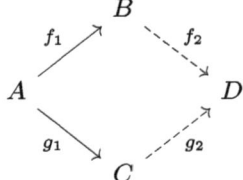

We are now ready to prove the existence of Fraïssé limits.

Theorem 12.14 *Let* K *be a class of finitely generated* L-*structures, closed under isomorphism and with countably many isomorphism types. The following statements are equivalent:*

(1) there is a Fraïssé limit M *of* K*;*
(2) K *satisfies* **EP**, **JEP** *and* **AP**.

Proof Showing that (1) implies (2) is relatively easy. Because K is the age of M, we already know that it satisfies **EP** and **JEP**. To show that **AP** holds, suppose we are given A, B, C and the embeddings f_1, g_1 from Definition 12.23.

Because A is in the age of M, there is an embedding $h : A \longrightarrow M$. If we restrict f_1 to its image, it is easy to see that $h \circ f_1^{-1}$ is an embedding of $f_1[A]$ into M. Using the inclusion $f_1[A] \subseteq B$ and appealing to algebraic ω-homogeneity, we obtain f_2 in the manner illustrated by the diagram below.

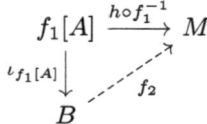

In order to construct g_2, we invoke algebraic ω-homogeneity again, this time relying on $g_1 \circ f_1^{-1}$ to embed $f_1[A]$ into C and on the restriction of f_2 to $f_1[A]$ as the required embedding of $f_1[A]$ into M. The diagram below shows how we obtain g_2.

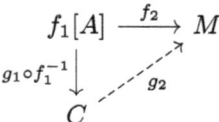

Since B, C are finitely generated, they are $\langle \overline{b} \rangle$, $\langle \overline{c} \rangle$ respectively, for suitable, finite tuples of generators. We set $\mathcal{D} = \langle f_2(\overline{b}), g_2(\overline{c}) \rangle$. It is clear from the last diagram that $g_2 \circ (g_1 \circ f_1^{-1}) = f_2$. Since embeddings are invertible, 'multiplying on the right' by f_1 we obtain $g_2 \circ g_1 = f_2 \circ f_1$.

We now turn to proving that (2) implies (1). Our starting point is a sequence $\{A_i\}_{i \in \mathbb{N}}$ of finitely generated structures, each representing one isomorphism type

12.2 Fraïssé Limits

of **K**. Our goal is recursively to define a chain $\{\mathcal{M}_i\}_{i \in \mathbb{N}}$ of structures from **K** whose union $\mathcal{M}$ is going to be the Fraïssé limit of **K**. Although the stages of our construction are indexed by $\mathbb{N}$, it is useful to describe them by ordered pairs of natural numbers and then use a suitable bijection $\tau : \mathbb{N} \longrightarrow \mathbb{N}^2$ to decode each stage as a specific task.

Following this approach, we first describe the tasks and the way they are labelled by pairs of natural numbers. The central idea is that a task is either a structure from the sequence $\{\mathcal{A}_i\}_{i \in \mathbb{N}}$ or an embedding from a substructure of $\mathcal{M}_n$ into $\mathcal{A}_i$, for certain specified values of n, i. If the task is a structure, then we carry it out by an application of **JEP**. If it is an embedding, we carry it out by an application of **AP**.

We take $\mathcal{A}_i$ to be the task labelled by $(i, 0)$. It follows that any task (i, j) with $j \neq 0$ is an embedding task. Suppose we are given the structure $\mathcal{M}_i$. The set of embeddings from a finitely generated substructure $\mathcal{C}$ of $\mathcal{M}_i$ into some $\mathcal{A}_k$ can be enumerated. To see this, note that, since $\mathcal{M}_i$ is at most countable, it has only countably many finite subsets, i.e. possible sets of generators of a finite substructure. The set of all embeddings from the domain of a finitely generated substructure of $\mathcal{M}_i$ into a fixed $\mathcal{A}_k$ is countable because any such embedding is fixed by its action on a finite set of generators. Because there are countably many structures of the form $\mathcal{A}_k$, the set of embeddings from a finitely generated substructure of $\mathcal{M}_i$ into a structure from the sequence $\{\mathcal{A}_i\}_{i \in \mathbb{N}}$ is itself countable. We may therefore fix an enumeration of this last set. Task (i, j) is the j-th embedding in the fixed enumeration.

Because we want the tasks we have described to be carried out sequentially, we rely on a 'decoding function' τ that, at each stage n, 'unpacks' it into a pair (i, j). Not every function will do: in particular we want to ensure $\tau(0) = (0, 0)$ and $\tau(n) = (i, j)$ with $i < n$ so that, by the n-th stage, we can take as constructed enough structures of the form $\mathcal{M}_i$. A bijection τ that meets the *desiderata* we have specified may be readily found: we can for instance list the pairs of natural numbers according to their weight (where the weight of (i, j) is $i + j$ and we fix a lexicographic ordering for pairs of equal weight, listing those with a smaller first component first).

We now explicitly describe how each task is carried out. The 0-th stage is $(0, 0)$: at this stage we fix $\mathcal{A}_0$ and set $\mathcal{A}_0 = \mathcal{M}_0$.

At the n-th stage we suppose $\mathcal{M}_{n-1}$ given. If $\tau(n) = (i, 0)$, we use **JEP** to embed both $\mathcal{M}_{n-1}$ and $\mathcal{A}_i$ into a finitely generated structure $\mathcal{M}_n$ in **K**. By repeatedly carrying out tasks of type $(i, 0)$, we ensure that $\mathcal{M}$ has **K** as its age, exactly as we did in the proof of Theorem 12.5.

If $\tau(n) = (i, j)$, then $\mathcal{M}_i$ has been constructed. In fact, we have at our disposal a specific chain of finitely generated substructures from **K**, namely $\mathcal{M}_i \sqsubseteq \ldots \sqsubseteq \mathcal{M}_{n-1}$. The index j singles out an embedding from a finitely generated substructure $\mathcal{C} \sqsubseteq \mathcal{M}_i$ into some $\mathcal{A}_k$. We view $\mathcal{C}$ as a finitely generated substructure of $\mathcal{M}_{n-1}$ and apply **AP** in the form depicted below.

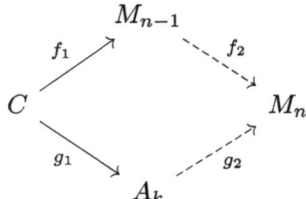

The function g_1 is determined by j in the pair (i, j), while f_1 is an inclusion embedding. As we go through each stage along the sequence $\tau(n)$ we carry out each of the $|\mathbb{N} \times \mathbb{N}|$ tasks we set ourselves. At the end of this process we obtain a countable structure $\mathcal{M} = \bigcup_{n \in \mathbb{N}} \mathcal{M}_n$, which has age K and, as we shall now prove, is also algebraically ω-homogeneous.

Suppose $\mathcal{C}, \mathcal{B}$ are in K and $f_0 : C \longrightarrow B$, $f_1 : C \longrightarrow M$ are embeddings. Because $\mathcal{C}$ is finitely generated, the construction of $\mathcal{M}$ implies that there is an embedding $h : C \longrightarrow M_i$, for sufficiently large i. Because $\mathcal{B}$ is in K, it is isomorphic to $\mathcal{A}_k$, for some $k \in \mathbb{N}$. We view $\mathcal{C}$ as a finitely generated substructure of $\mathcal{M}_i$ and compose f_0 with a suitable isomorphism to obtain $h : C \longrightarrow A_k$. By the way we labelled tasks, the embedding h is (i, j) for some $j > 0$. Since (i, j) is carried out at some stage $\tau(n)$, the application of **AP** made at this stage ensures the existence of an embedding $g : A_k \longrightarrow M_n$ such that, when we regard h as an embedding from C into $M_n \supseteq M_i$ (since $i < n$ by the choice of τ), $f_1 = g \cdot h$. Composing with the relevant isomorphisms shows that algebraic ω-homogeneity holds. □

12.3 Two Concrete Limits

In this section we describe two concrete classes of structures and their respective Fraïssé limits. Our first example is the class L of finite linear orders and our second example is the class V of finite vector spaces over a fixed finite field $\mathbb{F}_q$ with q elements.

12.3.1 Finite Linear Orders

It is easy to see that the class L of finite linear orders has countably many isomorphism types and that it also has **HP** and **JEP**.

Exercise 12.15 Show that L has **HP** and **JEP**.

In order to guarantee the existence of a Fraïssé limit for L, we only need to verify the Amalgamation Property.

Lemma 12.16 *The class* L *has* **AP**.

12.3 Two Concrete Limits

Proof Suppose $\mathcal{A}$ is a finite linear order embedded into the distinct, finite linear orders $\mathcal{B}, \mathcal{C}$ via f_1, g_1 respectively. Taking isomorphic copies, we can arrange A to be a common subset of B, C. We next consider the set $B \cup C$. The sets B, C are each linearly ordered and they share a common linear suborder on A. Suppose the ordering of A is given by:

$$a_0 < a_1 < \ldots < a_{m-2} < a_{m-1}.$$

We extend this linear suborder to an ordering of $B \cup C$. If $(b, c) \in B \times C$ and $b, c \notin A$, the relative position of b, c with respect to the linear order of A may vary.[1] If $b, c < a_0$ or $a_{m-1} < b, c$, we set $b < c$. Otherwise, at least one of b, c lies between two consecutive elements of A, say a_i, a_{i+1}. If both lie between a_i, a_{i+1}, we again set $b < c$. Otherwise, one of them lies below a_i or above a_{i+1}, in which case we stipulate that it is respectively smaller or greater than the element of $B \cup C$ between a_i, a_{i+1}. It is clear that, with the stipulations we have described, any two elements of $B \cup C$ are comparable: nothing changes when we compare two elements of B or two elements of C and we know how to compare any element of B with any element of C. It is clear that the comparability relation we have defined is actually irreflexive and transitive. We have thus obtained a finite linear order in which $\mathcal{B}, \mathcal{C}$ can be embedded and that, in addition, embeds $\mathcal{A}$ identically via $\mathcal{B}$ or, alternatively, $\mathcal{C}$. □

In view of Theorem 12.14, L has a Fraïssé limit. If we can exhibit a linear order that is algebraically ω-homogeneous and has age L, this is, up to isomorphism, the limit in question.

Theorem 12.17 *The dense linear order without endpoints* $\mathcal{Q} = (\mathbb{Q}, <)$ *is the Fraïssé limit of* L.

Proof It is clear that $Age(\mathcal{Q})$ is L. We only have to check that $\mathcal{Q}$ is algebraically ω-homogeneous. The embeddings we are given by hypothesis determine the configuration:

$$\begin{array}{ccc} L_0 & \xrightarrow{f_0} & \mathbb{Q} \\ {\scriptstyle f_1}\downarrow & & \\ L_1 & & \end{array}$$

where L_0, L_1 are the domains of two finite linear orderings and f_0, f_1 are embeddings. We need to show that an embedding $g : L_1 \longrightarrow \mathbb{Q}$ exists such that $g \circ f_1 = f_0$. We apply **AP** to obtain the configuration:

---

[1] If $b, c \in A$, there is nothing to do. If only one of b, c is in A, we can easily adapt the argument for $b, c \notin A$.

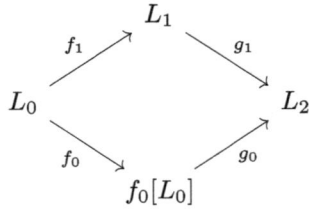

where $g_0 \circ f_0 = g_1 \circ f_1$. Because embeddings are invertible, it follows that $g_0^{-1} \circ g_1 \circ f_1 = f_0$. We thus set $g = g_0^{-1} \circ g_1$. □

12.3.2 Finite Vector Spaces over $\mathbb{F}_q$

We use the notation $\mathbb{F}_q$ to indicate the (unique up to isomorphism) finite field with q elements, where q is a prime power. Finitely generated vector spaces over $\mathbb{F}_q$ are finite and viceversa. Thus, we may take the class of structures we are interested in to be the class of finite vector spaces over $\mathbb{F}_q$.

Because a vector space is determined up to isomorphism by a basis, it is clear that the isomorphism types in V are represented by the finite vector spaces $\mathbb{F}_q, \mathbb{F}_q^2, \ldots, \mathbb{F}_q^n, \ldots$. In other words, we have countably many isomorphism types.

It is clear that **HP** holds. We shall now verify that **JEP** and **AP** also holds.

Lemma 12.18 *The class* V *satisfies* **JEP**.

Proof We may work with representatives of isomorphism types alone. In this case, suppose $\mathbb{F}_q^n$ and $\mathbb{F}_q^k$ are given. These vector spaces can be embedded into $\mathbb{F}_q^{n+k}$. Consider the linear map f_i, with $i \in \{n, k\}$, which sends any vector v of $\mathbb{F}_q^n$ to the vector $f_i(v) \in \mathbb{F}_q^{n+k}$, whose first i components agree with those of v and whose remaining components equal 0. □

Lemma 12.19 *The class* V *satisfies* **AP**.

Proof We are given three vector spaces $\mathbb{F}_q^j, \mathbb{F}_q^k, \mathbb{F}_q^n$, with $\mathbb{F}_q^j$ embeddable into $\mathbb{F}_q^k$ via f_0 and into $\mathbb{F}_q^n$ via f_1. The image of $\mathbb{F}_q^j$ under these embeddings is a j-th dimensional subspace of F_q^k or F_q^n. Using a suitable automorphism g_i' of $\mathbb{F}_q^i$, with $i \in \{n, k\}$, we can always embed $\mathbb{F}_q^j$ into F_q^i as the set of i-tuples whose components of index greater than j equal 0 (this automorphism extends an isomorphism between two j-th dimensional subspaces of the same space).

Proceeding as in the proof of Lemma 12.18, we embed $\mathbb{F}_q^k$ and $\mathbb{F}_q^n$ into $\mathbb{F}_q^{n+k}$ via linear maps g_0', g_1' respectively. Suppose $a \in F_q^j$: then $g_k' \circ f_0(a)$, once composed with g', is a vector of length k whose components of index greater then j equal 0. Furthermore $g_0' \circ g_k' \circ f_0(a)$ is the vector of $\mathbb{F}_q^{n+k}$ that coincides with a in the first j components and equals 0 in the remaining $(n+k) - j$ components.

12.4 Saturation and Quantifier Elimination

The argument we have just given can be repeated with minor modifications to show that $g'_1 \circ g'_n \circ f_1(a) = g'_0 \circ g'_k \circ f_0(a)$. We may thus simply set $g_0 = g'_0 \circ g'_k$ and $g_1 = g'_1 \circ g'_n$.

$\square$

The last two lemmas imply that $\mathbf{V}$ has a Fraïssé limit. Consider the set M of all infinite sequences of elements of $\mathbb{F}_q$. Relative to componentwise addition and scalar multiplication, M determines a vector space M of dimension $\aleph_0$. It is clear that the age of M is $\mathbf{V}$.

The properties of vector spaces make it easy to show that $\mathcal{M}$ is ultrahomogeneous. We conclude that $\mathcal{M}$ is the Fraïssé limit of $\mathbf{V}$.

Exercise 12.20 Show that $\mathcal{M}$ is ultrahomogeneous.

12.4 Saturation and Quantifier Elimination

In this section we introduce some additional results on Fraïssé limits, restricting attention to finite relational languages. These results will play a crucial role in the discussion of measurement scales that concludes the chapter.

If $\mathcal{L}$ is a finite relational language, any $\mathcal{L}$-structure $\mathcal{M}$ is **locally finite**, i.e. any finitely generated substructure is finite. In particular, if $A \subseteq M$ is a set of n generators, the substructure of $\mathcal{M}$ generated by A has n elements, since it arises from restricting the relations on $\mathcal{M}$ to the relevant Cartesian products of A.

Suppose $A = \{a_0, \ldots, a_{n-1}\}$. The conjunction of all literals in the variables $x_0, \ldots, x_n$ that are satisfied by $\bar{a}$ is a $\mathcal{L}$-formula $\varphi_{\bar{a}}(\bar{x})$ that describes $\langle \bar{a} \rangle$ up to isomorphism. Moreover, given a fixed number of generators, a finitely generated, and thus finite, $\mathcal{L}$-structure realises only one of finitely many isomorphism types.

In order to shorten the statement of our next result, we refer to a class of finite structures that has **HP**, **JEP** and **AP** as a **Fraïssé class**.

Theorem 12.21 *Let $\mathcal{L}$ be a finite, relational language and K be a Fraïssé class of $\mathcal{L}$-structures. If $\mathcal{M}$ is the Fraïssé limit of K, then $Th(\mathcal{M})$ is $\aleph_0$-categorical and it has quantifier elimination.*

Proof Following [25], p.349, we first provide an explicit axiomatisation T_M of $Th(\mathcal{M})$. Let $\varphi_{\bar{a}}(\bar{x})$ be the (quantifier-free) description of the structure in K generated by $\bar{a}$. The set T_M is the union of two sets of sentences. The first set, T_M^1, consists of infinitely many $\mathcal{L}$-sentences of the form:

$$\forall \bar{x}(\varphi_{\bar{a}}(\bar{x}) \to \exists y \varphi_{\bar{a},b}(\bar{x}, y)).$$

The second set, T_M^2, consists of infinitely many $\mathcal{L}$-sentences of the form:

$$\forall \overline{x} \bigvee_{\overline{a}} (\varphi_{\overline{a}}(\overline{x})).$$

Suppose $\mathcal{N} \models T_M$ and $\mathcal{N}$ is countable. Then, in particular, $\mathcal{N} \models T_M^1$. When we consider the empty set of generators, the corresponding sentence of T_M^1 reduces to $\exists y \varphi_b(y)$, which says that any structure in K generated by one element (e.g. b) is isomorphic to a substructure of $\mathcal{N}$ or, equivalently, can be embedded in $\mathcal{N}$.

Consider a fixed structure $\mathcal{A} = \langle \overline{a} \rangle$ in K, generated by n elements, with $n \geq 1$: the corresponding formula in T_M^1 is a statement of algebraic ω-homogeneity restricted to 'one-generator extensions' of finitely generated structures. To see this, note that the formulae $\varphi_{\overline{a}}(\overline{x})$ and $\varphi_{\overline{a},b}(\overline{x},y))$ hold in $\mathcal{N}$ if there are substructures $\mathcal{N}_{\overline{a}}, \mathcal{N}_{\overline{a},b} \sqsubseteq \mathcal{N}$ isomorphic to $\langle \overline{a} \rangle$ and $\langle \overline{a}, b \rangle$ respectively. Equivalently, $\langle \overline{a} \rangle$ and its extension $\langle \overline{a}, b \rangle$ are embeddable in $\mathcal{N}$.

For any n-tuple $\overline{c}$ of elements of N we have:

$$\mathcal{N} \models \varphi_{\overline{a}}(\overline{c}) \to \exists y \varphi_{\overline{a},b}(\overline{c}, y)).$$

In other words, given $\langle \overline{a} \rangle$, and an embedding f of this substructure into $\mathcal{N}$, the structure $\langle \overline{a}, b \rangle \sqsupseteq \langle \overline{a} \rangle$ can also be embedded in $\mathcal{N}$ by a map g that extends f (since it agrees with f on the generators of the embedded structure $\langle \overline{c} \rangle$).

What we have observed so far amounts to two statements: (i) the structures in K with exactly one generator can be embedded in $\mathcal{N}$; (ii) if a r structure from K with n generators is embeddable via the map f into $\mathcal{N}$, so is any one of its extensions in K with $n+1$ generators, via an embedding g extending f. Statements (i) and (ii) clearly support an inductive argument to the effect that $K \subseteq Age(N)$.

The sentences in T_M^2 enable us to obtain the converse inclusion. If $\mathcal{C} \in Age(N)$ and $\overline{c}$ is a n-tuple generating $\mathcal{C}$, then this n-tuple must satisfy a sentence in T_M^2 that forces $\mathcal{C}$ to realise the isomorphism type of a structure in K. Because K is closed under isomorphism, it follows that $\mathcal{C}$ is in K. In other words, $Age(N)$ is K.

This alone allows us to conclude that $\mathcal{N}$ can be embedded in the Fraïssé limit $\mathcal{M}$ of K but we can guarantee that $\mathcal{M}, \mathcal{N}$ are isomorphic because we can repeatedly apply algebraic ω-homogeneity for one-generator extensions to obtain the same property for k-generator extensions, with k an arbitrary positive integer. In other words, algebraic ω-homogeneity holds in $\mathcal{N}$, which is, consequently, a Fraïssé limit for K. Since Fraïssé limits are unique up to isomorphism, $\mathcal{N}$ is isomorphic to $\mathcal{M}$.

Because $\mathcal{N}$ is an arbitrary model of T_M, this theory is $\aleph_0$-categorical. Since T_M has only infinite models (because they embed finite structures with an arbitrary number of generators), it is complete. Clearly, $T_M \subseteq Th(M)$ and the former theory's completeness implies equality. We have shown that $Th(M)$ is $\aleph_0$-categorical.

To obtain quantifier elimination we consider a $\mathcal{L}$-formula $\phi(\overline{x})$, with $\overline{x}$ a n-tuple of variables, and the set $X \subseteq M^n$ defined by $\phi(\overline{x})$. If $\overline{a} \in X$ and, for any $\overline{b}$, there is an isomorphism from $\langle \overline{a} \rangle \longrightarrow \langle \overline{b} \rangle$, ultrahomogeneity implies that his isomorphism extends to an automorphism of $\mathcal{M}$. Because automorphisms preserve all first-order formulae, $\overline{b} \in X$.

12.5 Measurement Scales over $\mathbb{Q}$ 333

It follows that $\bar{b}$ satisfies $\phi(\bar{x})$ in $\mathcal{M}$ iff $\langle \bar{b} \rangle$ determines an isomorphism type realised by some n-tuple $\bar{c}$ in X. A single n-tuple can only realise one among finitely many isomorphism types. Moreover, each isomorphism type is in turn realised by a n-tuple $\bar{c}$ iff $\bar{c}$ satisfies a suitable, quantifier-free formula $\varphi_{\bar{a}}(\bar{x})$ in $\mathcal{M}$. A finite disjunction of quantifier-free formulae $\bigvee_{\bar{a}} \varphi_{\bar{a}}(\bar{x})$ is thus equivalent to $\phi(\bar{x})$ in $\mathcal{M}$.

In other words:

$$\mathcal{M} \models \forall \bar{x} \left(\bigvee_{\bar{a}} \varphi_{\bar{a}}(\bar{x}) \leftrightarrow \phi(\bar{x}) \right).$$

Because $Th(\mathcal{M})$ is complete, the last sentence transfers to every model of $Th(\mathcal{M}) = T_\mathcal{M}$ and $T_\mathcal{M}$ has quantifier elimination (sentences can be treated as formulae by adding to them the conjunct $x = x$, as noted in Chap. 8). □

The last theorem makes it very easy to show that Fraïssé limits of classes of relational structures (in a finite language) are saturated.

Theorem 12.22 *Let $\mathcal{L}$ be a finite, relational language, $\mathbf{K}$ a Fraïssé class of $\mathcal{L}$-structures and $\mathcal{M}$ the Fraïssé limit of $\mathbf{K}$. Then $\mathcal{M}$ is saturated.*

Proof Because $\mathcal{M}$ is countable, we only have to show that, if A is a finite subset of M, every type in $S_1(M/A)$ is realised in $\mathcal{M}$. Suppose $p \in S_1(M/A)$. Because $Th(\mathcal{M})$ has quantifier elimination, p is uniquely determined by its positive literals. Since the language contains only finitely many relation symbols, p is uniquely determined by finitely many positive literals or, equivalently, by their conjunction. Because types in $S_1(M/A)$ are finitely satisfiable, the latter conjunction is satisfied by some $c \in M$. It follows that c realises p in $\mathcal{M}$. □

12.5 Measurement Scales over $\mathbb{Q}$

We conclude this chapter with an application of Fraïssé limits to measurement theory. As explained in Sect. 9.3, a measurement scale is in this context a function that encodes a structure $\mathcal{M}$ into a suitable expansion of $(\mathbb{R}, <)$. In the simplest cases the encoding is carried out by a strong homomorphism. Any $a \in M$ may be regarded as an object expressing a 'level' of an empirical variable like mass, length, subjective loudness etc.

Objects expressing the same level of a fixed empirical variable are identified and, relative to this equivalence, a strong homomorphism $f : M \longrightarrow \mathbb{R}$ induces an embedding whose domain is a suitable quotient of $\mathcal{M}$.

In applied contexts like modelling with differential equations, the empirical variables involved are typically viewed as continua. Measurement theorists portray this situation by formally describing an empirical variable as a particular $\mathcal{L}$-structure $\mathcal{M}$, which is a linear order isomorphic to the ordered reals relative to the reduct $\mathcal{L}_<$

of $\mathcal{L}$. The automorphisms of $\mathcal{M}$, even though they are not automorphisms of the ordered reals, are obviously associated with a subgroup of $Aut(\mathbb{R})$, the group of automorphisms of $(\mathbb{R}, <)$.

The study of certain subgroups of $Aut(\mathbb{R})$ has led to deep measurement-theoretic results. To see how, let us consider a $\mathcal{L}$-structure that formally describes the empirical variable 'mass' and that is isomorphic to the positive semigroup of the additive reals R. A measurement scale is, in this special context, an isomorphism h from $\mathcal{M}$ onto R. Suppose $h(x) = 1$, so that x is a unit of measure.

The fact that any level of 'mass' may be adopted as a unit of measure corresponds to the fact that, for any $x, y \in M$, there is an automorphism $\alpha : M \longrightarrow M$ that satisfies the equality $\alpha(x) = y$. In this case $h \cdot \alpha^{-1}(y) = 1$. Scale changes track property of $Aut(M)$, i.e. the existence of an automorphism relating any two distinct elements of M. Furthermore, the fact that, once a unit of measure is fixed, a unique measurement scale is fixed, corresponds to the fact the only automorphism $\alpha \in Aut(M)$ with one fixed point, i.e. $x \in M$ such that $\alpha(x) = x$, is the identity.

In the very special case we have just described, fundamental features of a measurement scale reduce to properties of a particular automorphism group, which may be identified with a subgroup of the group of automorphisms of the ordered reals (since $\mathcal{M}$, R are isomorphic, as noted). These properties concern the maximum length of tuples joined by an automorphism and the minimum length of tuples of fixed points that force an automorphism to coincide with the identity. In the case we have illustrated these lengths take the same value, namely 1. We say that $\mathcal{M}$ is of scale type $(1, 1)$.

With these ideas as a starting point, we can envisage a classification of scale types over a continuum based on an enumeration of the possible pairs of positive integer values associated with groups of order-preserving automorphisms of $(\mathbb{R}, <)$ or, alternatively, an abstract continuum. To make the suggestion work, we introduce the following:

Definition 12.23 Let $\mathcal{L}$ be a relational vocabulary including $<$ and $\mathcal{M}$ a $\mathcal{L}$-structure linearly ordered by $<$. The group $Aut(M)$ is m-**point homogeneous** iff, for any two m-tuples $\langle a_0, \ldots, a_{m-1} \rangle$, $\langle b_0, \ldots, b_{m-1} \rangle$ in M^m such that $a_0 <^M \ldots <^M a_{m-1}$ and $b_0 <^M \ldots <^M b_{m-1}$, there is $\alpha \in Aut(M)$ such that, for each $i \in \{0, \ldots, m-1\}$, $\alpha(a_i) = b_i$.

The group $Aut(M)$ is n-**point unique** iff for any $\alpha, \beta \in Aut(M)$ and any $a_0, \ldots, a_{n-1}$ such that $a_0 <^M \ldots <^M a_{n-1}$, the equalities $\alpha(a_i) = \beta(a_i)$ with $i = 0, \ldots, n-1$ imply $\alpha = \beta$. Equivalently, any $\alpha \in Aut(M)$ with n fixed points is the identity.

Exercise 12.24 Show that, if $Aut(M)$ is m-point homogeneous and $k < m$, then $Aut(M)$ is k-point homogeneous. Dually, if $Aut(M)$ is n-point unique and $n < k$, then it is k-point unique. Moreover, show that, if $Aut(M)$ is m-point homogeneous and n-point unique, then $m \leq n$.

In view of the last exercise, we define the **finite homogeneity degree** of $Aut(M)$ as the largest m for which $Aut(M)$ is m-point homogeneous and the **finite**

uniqueness degree of $Aut(M)$ as the least n for which $Aut(M)$ is n-point unique. We are now in a position to define 'finite scale type'.

Definition 12.25 The **finite scale type** of a linearly ordered $\mathcal{L}$-structure $\mathcal{M}$ is the pair $(m, n) \in \mathbb{N}^2$, where m is the finite homogeneity degree of $Aut(M)$ and n is the finite uniqueness degree of $Aut(M)$.

A rather striking theorem of measurement theory shows that, if $\mathcal{M}$ is a continuum of finite scale type and homogeneity degree at least 1 then $n \leq 2$ (see [40], pp.120–122 for a more detailed statement of this result, which includes an explicit description of the automorphism groups that can occur, and for a proof). This result restricts the possible homogeneous, finite scale types to $(1, 1)$, $(1, 2)$ and $(2, 2)$.

If we turn from scale types over a continuum to scale types over the ordered rationals, i.e. if we turn to subgroups of $Aut(\mathbb{Q})$, the options suddenly escalate. Any pair $(m, n) \in \mathbb{N}^2$ determines a possible finite scale type over the ordered rational. When $m \neq n$, this result is due to Peter Jephson Cameron, who proved in [11] with the help of Fraïssé limits.[2]

In what follows we shall not prove Cameron's results in full generality but will go far enough to show that there are infinitely many distinct, 1-point homogeneous scale types over $(\mathbb{Q}, <)$.

12.6 Model-Theoretic Preliminaries

Our eventual aim is to associate n-point unique subgroups of $Aut(\mathbb{Q})$ with relational expansions of $(\mathbb{Q}, <)$. There is a general way of reconstructing a permutation group as the automorphism group of a suitable relational structure. The next lemma, which rests on the group-theoretic notion of orbit defined below, describes it.

Definition 12.26 Let X be an infinite set, $\mathcal{G}$ a group of permutations of X (i.e. bijections from X onto X). For any $\bar{a} \in X^n$, the n-orbit of $\bar{a}$ is $Y = \{\bar{b} \in X^n : \text{for some } g \in G, g(\bar{a}) = \bar{b}\}$.

Lemma 12.27 Let X be an infinite set, $\mathcal{G}$ a group of permutations of X. If G is n-point unique, there is a relational structure $\mathcal{M} = \langle X, R^{\mathcal{M}} \rangle$ such that $G = Aut(M)$.

Proof Let X^{n+1} be partitioned into the sequence of orbits $\{O_i\}_{i \in \mathbb{N}}$, given by a fixed enumeration. We introduce a first-order language $\mathcal{L}$ whose signature only contains one $2(n + 1)$-ary relation symbol R. We focus on a specific $\mathcal{L}$-structure $\mathcal{M} = \langle X, R^{\mathcal{M}} \rangle$, determined by the following condition:

for any $\bar{a}, \bar{b} \in X^{n+1} \mathcal{M} \models R(\bar{a}, \bar{b})$ iff there are two $(n + 1)$-orbits O_i, O_{i+1}

---

[2] Dugald McPherson proved in [49] that the ordered rationals also support every scale type of the form (m, m), with m a positive integer.

in the given enumeration such that $\bar{a} \in O_i$ and $\bar{b} \in O_{i+1}$.

If $g \in G$, then g sends the elements of a $(n+1)$-orbit into elements of the same orbit. In model-theoretic terms:

$$\mathcal{M} \models R\bar{a}, \bar{b} \iff \mathcal{M} \models Rg(\bar{a}), g(\bar{b}).$$

It follows that $g \in Aut(M)$. More precisely, $\mathcal{G}$ is a subgroup of $Aut(M)$. We only have to show that any automorphism of $\mathcal{M}$ is already an element of $\mathcal{G}$.

To this end, we note that the formula $\varphi_1(\bar{x}) := \forall y \neg (\bar{y}, \bar{x})$ defines O_1. It is then easy to see that, for each $i \geq 1$, there is a formula $\varphi_i(\bar{x})$ defining O_i. Because every $h \in Aut(M)$ preserves the definable subsets of M^{n+1}, it follows that $Aut(M)$ determines the same $(n+1)$-orbits on X^{n+1} as does $\mathcal{G}$.

Suppose $h \in Aut(M)$. We select any $\bar{a} \in O_i$ and observe that $h(\bar{a}) \in O_i$ because h preserves the definable subsets of M^{n+1}. Because $Aut(M)$ and $\mathcal{G}$ determine the same $(n+1)$-orbits, there must be $g \in G$ such that $h(\bar{a}) = g(\bar{a})$. Since g is invertible, we deduce $g^{-1}h(\bar{a}) = \bar{a}$. Because $\mathcal{G} \sqsubseteq Aut(M)$, $g^{-1}h$ is an automorphism of $\mathcal{M}$.

We claim that $g^{-1}h$ is the identity on M. This is because $\mathcal{G}$ is n-point unique: as a consequence, if $b \neq c$, the n-tuples $(a_0, \ldots, a_{n-1}, b)$ and $(a_0, \ldots, a_{n-1}, c)$ must belong to different $(n+1)$-orbits. Because $g^{-1}h$ preserves orbits and it fixes $a_0, \ldots, a_{n-1}$, it must fix every other point of X. It follows that $g^{-1}h$ is the identity, i.e. that h is the inverse of g^{-1} and thus an element of G. It now follows that $\mathcal{G}$ is the group $Aut(M)$. $\square$

In order to study the scale types over the ordered rationals, we shall make use of certain Fraïssé limits of relational structures, which we discuss in this section. These limits satisfy a strong version of **AP**, the **strong amalgamation property** (**SAP**). We say that a class **C** of finite relational structures has **SAP** if, for any $\mathcal{A}, \mathcal{B}, \mathcal{C} \in \mathbf{C}$, whenever $\mathcal{A}$ is a common substructure of $\mathcal{B}, \mathcal{C}$, there are embeddings g_0, g_1 from $\mathcal{B}, \mathcal{C}$ respectively into the domain D of a structure $\mathcal{D}$ in **C** such that $g_0(d) = g_1(d)$ implies $d \in A$. In other words, the images of $\mathcal{B}, \mathcal{C}$ in $\mathcal{D}$ intersect exactly on A.

We conclude this section with a verification that L, the class of finite linear orders, has **SAP**, as does the class obtained from L by expanding its elements to structures that interpret a $2n$-ary relation symbol on an equivalence relation between n-element sets (as distinct from n-tuples). What we prove here will play an important role in the concluding section of the present chapter.

To verify that strong amalgamation holds in L, we may, without loss of generality, restrict attention to finite linear orders of rational numbers. We thus focus on the class $\mathbf{Q}_n$ of finite $\mathcal{L}$-structures, where $\mathcal{L}$ consists only of two relation symbols $<, E$, of arities 2 and $2n$ respectively: the symbol $<$ designates a linear order and the symbol E designates an equivalence relation on n-element sets. Since the equivalence relation may be carried by a structure independently of a linear ordering, the $\mathcal{L}_<$-reducts of the elements of $\mathbf{Q}_n$ determine, up to isomorphism, L.

We leave it as an exercise for the reader to check that Q_n has **HP** and **JEP** (the latter property can be verified by adapting the argument from the next lemma). By Exercise 12.3, Q_n has only countably many isomorphism types. It also has **SAP**.

Lemma 12.28 *The class Q_n has the strong amalgamation property.*

Proof Let $\mathcal{A}, \mathcal{B}, \mathcal{C}$ be elements of Q_n and $f_0 : A \longrightarrow B$, $f_1 : A \longrightarrow C$ be embeddings. We consider the $\mathcal{L}_<$-reducts $\mathcal{A}_<, \mathcal{B}_<, \mathcal{C}_<$ of $\mathcal{A}, \mathcal{B}, \mathcal{C}$ respectively. Let $\mathcal{D}_<$ be a strong amalgam of $\mathcal{B}_<, \mathcal{C}_<$ and $g_0 : B \longrightarrow D$, $g_1 : C \longrightarrow D$ the relevant embeddings.

We first extend the given embeddings to the interpretations of E^B, E^C: we simply stipulate that two n-element subsets of B be related by E^B iff their g_0-images are related by E^D, and introduce a similar stipulation relative to E^C and g_1. We then endow $\mathcal{D}_<$ with an equivalence relation by partitioning D into equivalence classes.

It remains to expand $\mathcal{D}_<$ to a full $\mathcal{L}$-structure. The n-element subsets of D that only contain images of elements of B or only contain images of elements of C are already partitioned into equivalence classes. We stipulate that the remaining n-element subsets of D belong to the same equivalence class. □

We record an immediate consequence of Theorem 12.14 as a further lemma.

Lemma 12.29 *The class Q_n has a Fraïssé limit $\mathcal{M}_n$ whose $\mathcal{L}_<$-reduct is (up to isomorphism) the set of ordered rational numbers $\mathcal{Q}$.*

Proof The structure $\mathcal{M}_n$ is saturated by Theorem 12.35. By the next exercise, its reducts are also saturated. In particular, its $\mathcal{L}_<$-reduct is a countable saturated structure of age $\mathbf{Q}$. Since $\mathcal{Q}$ is a Fraïssé limit, it is itself a countable saturated structure. By Lemma 9.71, $\mathcal{M}_n$ can be elementarily embedded into $\mathcal{Q}$, so these two structures are elementary equivalent and saturated. It follows from Exercise 9.73 that they must be isomorphic. □

Exercise 12.30 Prove that, if a $\mathcal{L}$-structure $\mathcal{M}$ is saturated and the language $\mathcal{L}_-$ is a subset of $\mathcal{L}$, then the $\mathcal{L}_-$-reduct of $\mathcal{M}$ is also saturated.

Exercise 12.31 Show that $Aut(\mathcal{M}_n)$ is a subgroup of $Aut(\mathbb{Q})$.

12.7 Group-Theoretic Preliminaries

In order to isolate groups that have suitable finite degrees of homogeneity and uniqueness, we focus on groups generated by an infinite sequence of symbols.

Let $\{f_i\}_{i \in \mathbb{N}}$ be such a sequence. Let $\mathcal{L} = \{\cdot, ^{-1}, 1\}$ be a multiplicative variant of $\mathcal{L}_g$: we can view the $\mathcal{L}$-terms (in any number of variables) as strings of 'integer powers' of the symbols $\{f_i\}_{i \in \mathbb{N}}$, among which we include the empty string e. We turn the $\mathcal{L}$-terms, which we shall refer to as *words*, into a group by taking the group operation $\cdot$ to be the concatenation of strings, by treating f^{-1} as the inverse of f^1

and e as the group's neutral element. We take words to be given in their reduced form (e.g. we identify the string $f_1 f_2 f_2^{-1}$ with the word f_1).

We say that a countably infinite group $\mathcal{H}$ formally generated by the sequence $\{f_i\}_{i \in \mathbb{N}}$ is a **free group of countable rank**. The preliminary lemma to follow ensures the existence of certain free groups of countable rank as subgroups of permutation groups.

Lemma 12.32 *Let $\mathcal{G}$ be a permutation group on a countably infinite set, $\mathcal{N}$ a normal subgroup of $\mathcal{G}$ with the same n-orbits on X as $\mathcal{G}$, for each $n \geq 1$. If $\mathcal{N} \sqsubseteq \mathcal{K} \sqsubseteq \mathcal{G}$ and the quotient K/N is a free group of countable rank, then there is $\mathcal{H} \sqsubseteq \mathcal{G}$ such that:*

(1) $HN = H$ and $H \cap N = \{e\}$;
(2) $\mathcal{H}$ is a free group of countable rank;
(3) $\mathcal{H}$ has the same 1-orbits on X as does $\mathcal{G}$.

Proof Let $\{Nf_i\}_{i \in \mathbb{N}}$ be a sequence of generating cosets for the quotient K/N. Moreover, let $\{(x_i, y_i)\}_{i \in \mathbb{N}}$ be a fixed enumeration of the pairs in X^2 whose components lie in the same orbit of $\mathcal{G}$. Note that $f_i(x_i)$ lies in the same orbit as x_i. Because $\mathcal{N}$ has the same orbits as $\mathcal{G}$, there is $h_i \in N$ such that $h_i(f_i(x_i)) = y_i$.

We let $\mathcal{H}$ be the group generated by the sequence $\{h_i f_i\}_{i \geq 1} \subseteq G$. By construction if x, y lie within the same orbit of $\mathcal{G}$, they appear as the pair (x_m, y_m) in our fixed enumeration. It follows that $h_m f_m(x_m) = y_m$, i.e. that x, y lie within the same orbit of $\mathcal{H}$. The converse is certainly true so (iii) holds.

We next turn to (i). Because $\mathcal{N}$ is a normal[3] subgroup of $\mathcal{G}$, $HN = NH$ is the domain of a subgroup of $\mathcal{G}$. It is easy to see that $HN \subseteq K$. On the other hand, if $k \in K$, Nk is a right coset[4] of K/N and, as such, of the form:

$$Nf_{i_1}^{t_1} \cdots Nf_{i_n}^{t_m},$$

since the cosets $\{Nf_i\}_{i \in \mathbb{N}}$ generate K/N. Note that, for any $a \in N$, $Na = N$. Using the latter fact we obtain, for each $j \in \{1, \ldots, m\}$ the following chain of equalities:

$$Nf_{i_1}^{t_1} = Nf_{i_1} \cdots Nf_{i_1} = Nh_{i_j} f_{i_j} \cdots Nh_{i_j} f_{i_j} = Nh_{i_j} f_{i_j} \cdots h_{i_j} f_{i_j},$$

where the term multiplying N on the right is in H. Using the last equalities repeatedly, we find that $Nk = Nh$, with $h \in H$. We deduce that $k \in Nh$, i.e. there is $n \in N$ such that $k = nh$. Thus $K \subseteq NH = HN$ and, in view of the previously established inclusion, $K = HN$.

In order to complete the proof of (i), we have to show $H \cap N = \{e\}$. To this end we consider an arbitrary word $w \in H \cap N$. The word w is a product of integer

---

[3] This means that, for any $f \in G$, the set $fNf^{-1} = \{fnf^{-1} : n \in N\}$ is N. It is not hard to verify that, in this case, $HN = NH$ and that either term of the last equality is a group.

[4] Because we are not working with abelian groups, we draw a distinction between left and right cosets, depending on which side of the subgroup N multiplication takes place. Nevertheless, every right coset of a normal subgroup is a left coset, and vice versa.

12.7 Group-Theoretic Preliminaries

powers of generators of $\mathcal{H}$ (with the relevant reductions carried out). Suppose w is a product of the group elements in $\{h_i f_i\}_{1 \leq i \leq s}$.

If we substitute each of the generators occurring in w with the corresponding right cosets $N f_i$, $1 \leq i \leq s$, the resulting word w designates the coset N from the quotient K/N. To see why, consider a simple example in which w is $h_1 f_1 h_2 f_2$. Bearing in mind that, for any h_i, $N h_i = N$ and that $w \in N$, we obtain:

$$N f_1 N f_2 = N f_1 f_2 = N h_1 f_1 N h_2 f_2 = N h_1 f_1 h_2 f_2 = N.$$

The same outcome occurs with more complicated words. The only way for a product of powers of the generators of K/N to determine the group identity N is for w to be the trivial word, corresponding to the term 1 in the language $\mathcal{L}$ described before the proof of this lemma. It follows that w is the neutral element of $\mathcal{H}$.

We finally apply the second isomorphism theorem for groups[5] to conclude that the quotient $H/(H \cap N)$ is isomorphic to the quotient K/N. Because H, N only share their common neutral element, the first quotient is $\mathcal{H}$. The latter group is therefore isomorphic to K/N, a free group of countable rank by hypothesis, and thus must itself be a free group of countable rank. Condition (ii) is now established and the proof concluded. □

Remark 12.33 For every $n \geq 1$, it is possible to enumerate the pair of tuples in the same n-orbit of $\mathcal{G}$. A modification of Lemma 12.32 thus ensures that $\mathcal{H}$ has the same n-orbits as $\mathcal{G}$ (this is because $\mathcal{G}$ is rich enough to connect any two n-tuples of distinct elements).

The existence of a free group of countable rank is closely connected with the problem of classifying scale types, in particular with finding n-point homogeneous subgroups of an automorphism group. The next theorem, which requires a preliminary definition, spells out the relevant connection.

Definition 12.34 Let X be an infinite set and $\mathcal{G}$ be a group of permutations of X. We say that $\mathcal{G}$ is **highly transitive** on X iff for any $n \geq 1$, any two n-tuples of distinct points lie in the same n-orbit of $\mathcal{G}$.

Theorem 12.35 *Let X be an infinite set. The free group of countable rank $\mathcal{H}$, viewed as a group of permutations of X, is highly transitive on X. Moreover, any element of H other than e fixes finitely many points of X but the number of fixed points is unbounded.*

Proof We may refer to 'the' free group of countable rank because it is unique up to isomorphism. We shall locate an isomorphic copy of this group inside $\mathcal{G}$, the group of permutations from X onto X. To this end, we observe that, Since both X and the domain M of the free group of countable rank $\mathcal{M}$ are countable, we are allowed to identify them by means of a fixed bijection. With this identification in place, there

---

[5] The second isomorphism theorem is exactly the statement that concludes the sentence. A proof of this result may be found in [33], p.72.

is a canonical way of embedding $\mathcal{M}$ into $\mathcal{G}$. We associate any $g \in M$ with the permutation f_g of M defined by the condition: $f_g(x) = xg$ (note multiplication on the right), for every $x \in M$. We refer to these as the *right regular representation* of M.

An argument that we omit, but is nothing other than the proof of Cayley's theorem (see e.g. [23], pp.71–73) ensures that $\mathcal{M}$ is isomorphic to a subgroup of $\mathcal{G}$, which we shall keep calling $\mathcal{M}$, by an abuse of notation. We observe that, by the way we have represented $\mathcal{M}$ as a group of permutations, $f_g(x) = x$ implies $x = xg$, which, multiplying on the left by x^{-1}, yields $g = e$. It follows that $f_g = f_e$, i.e. that the only element of M with fixed points at all is the identity.

We next single out $\mathcal{N}$, the normal subgroup of $\mathcal{G}$ consisting of all permutations leaving all but finitely many points fixed (the reader may check that $\mathcal{N}$ is normal). It is clear that each $h \in N$ has countably many fixed points. By the previous observation we made concerning $\mathcal{M}$, $M \cap N = \{e\}$. We now set $K = MN$: the second isomorphism theorem implies that $\mathcal{M}$ is isomorphic to the quotient K/N. In other words K/N is a free group of countable rank.

It is easy to see that $\mathcal{N}$ has the same n-orbits as $\mathcal{G}$, for each $n \geq 1$. If e.g. there is $g \in G$ such that $g(x_i) = y_i$ for $i = 1, \ldots, n$, there certainly is a permutation in N that fixes every other element of X and sends x_i into y_i. The hypotheses of Lemma 12.32 are satisfied and, by Remark 12.33 we can deduce that there is a free group of countable rank $\mathcal{H} \sqsubseteq \mathcal{G}$ with the same n-orbits as $\mathcal{G}$. Because of this, for any two n-tuples of distinct points $\overline{x}, \overline{y} \in X^n$, there is $h \in H$ such that $h(\overline{x}) = \overline{y}$, i.e. $\mathcal{H}$ is highly transitive.

Moreover, because $H \subseteq HN = K = MN$, every $h \in H$ is the product of some $m \in M$, without fixed points, and some $n \in N$, which fixes all but finitely many points. Thus any $h \in H$ fixes only finitely many points of X. Suppose h fixes, in particular $x_1, \ldots, x_m \in X$ and select $x' \notin \{x_1, \ldots, x_m\}$. Because $\mathcal{H}$ has the same n-orbits as $\mathcal{G}$, there is $k \in H$ such that $k(x_i) = x_i$ and $k(x') = h(x')$. It is easy to check that $h^{-1} \circ k \circ h$ fixes at least $m+1$ points, namely $x_1, \ldots, x_m, x'$. Because we can repeat, if necessary, this construction, the number of fixed points of non-identity elements of H is unbounded. $\square$

With Theorem 12.35 we have covered the algebraic background needed to reach one of Cameron's theorems on scale types over the ordered rationals. In the next section we shall break its proof into three preliminary lemmas and one main theorem.

12.8 A Theorem on Rational Scale Types

We begin with the announced preliminary lemmas.

Lemma 12.36 *$Aut(M_n)$ has the same orbits on M^{n-1} as $Aut(\mathbb{Q})$.*

12.8 A Theorem on Rational Scale Types

Proof Consider two $(n-1)$-element sets. We may view them as linear orders but also, vacuously, as finite substructures of $\mathcal{M}_n$, since they determine no n-element set (such sets must contain n distinct elements). Thus, two isomorphic $n-1$-element linear orders determine two isomorphic finite substructures of $\mathcal{M}_n$. By ultrahomogeneity, any given finite isomorphism extends to an automorphism of $\mathcal{M}_n$. □

The following two lemmas lemma concern a structure $\mathcal{M}^*$ obtained from $\mathcal{M}_n$ by focussing on the set of equivalence classes determined by $E^{\mathcal{M}_n}$. Let $A = \{A_i\}_{i \in \mathbb{N}}$ be this set, given under a fixed enumeration. We first expand $\mathcal{M}_n$ to a countable language with an infinite sequence of n-ary relation symbols $\{R_i\}_{i \in \mathbb{N}}$ in such a way that R_i names the equivalence class A_i. The structure $\mathcal{M}^*$ is the reduct of $\mathcal{M}_n$'s expansion obtained by dropping the relation symbol E.

We establish two properties of $\mathcal{M}^*$, which will enable us to use Theorem 12.35 in the proof of our main result.

Lemma 12.37 *$\mathcal{M}^*$ is an ultrahomogeneous structure.*

Proof The expansion of $\mathcal{M}_n$ to a finite relational language containing the symbols $\{R_i\}_{i<k}$ is a Fraïssé limit of the class of finite structures obtained by suitably expanding the structures in $Age(\mathcal{M}_n)$. In particular, it is algebraically ω-homogeneous. We exploit this fact to extend finite isomorphisms between substructures of $\mathcal{M}^*$. Suppose $\langle \overline{a} \rangle, \langle \overline{b} \rangle \sqsubseteq \mathcal{M}^*$ are isomorphic and pick $c \in M$. If $R_1, \ldots, R_m$ name the equivalence classes partitioning the n-sets in $\langle \overline{a}, c \rangle$, using the reduct of $\mathcal{M}^*$ to the language $<, \{R_i\}_{i<m+1}$, we can find $d \in M$ such that $\langle \overline{a}, c \rangle$ and $\langle \overline{b}, d \rangle$ are isomorphic. The latter isomorphism is not affected by the way we interpret R_i for $i \geq m+1$. Thus, it continues to be a finite isomorphism in $\mathcal{M}^*$. Since (we omit a verification) the age of $\mathcal{M}^*$ can be obtained from the age of $\mathcal{M}_n$ by expanding its elements to the language with a name for each equivalence relation and then omitting E, it is easy to see that $Age(\mathcal{M}^*)$ is a Fraïssé class. We conclude that $\mathcal{M}^*$, being algebraically ω-homogeneous, must also be ultrahomogeneous. □

Lemma 12.38 *Let $A = \{A_i\}_{i \in \mathbb{N}}$ be the set of equivalence classes determined by $E^{\mathcal{M}_n}$ in $\mathcal{M}_n$. Then $Aut(\mathcal{M}_n)$ induces the full symmetric group on A.*

Proof Consider a permutation σ of $\mathbb{N}$ and the resulting enumeration $\{A_{\sigma(i)}\}$ of A. We can associate with the latter enumeration a structure $\mathcal{M}^{**}$ constructed exactly as $\mathcal{M}^*$, with the sole exception that the relation symbol R_i names $A_{\sigma(i)}$. By Lemma 12.37, $\mathcal{M}^*$ and $\mathcal{M}^{**}$ are Fraïssé limits of the same Fraïssé class, obtained from the structures in $\mathbf{Q}_n$ by dropping their equivalence relation and naming the corresponding equivalence classes by means of the n-ary relation symbols in $\{R_i\}_{i \in \mathbb{N}}$.

It follows that $\mathcal{M}^*$ and $\mathcal{M}^{**}$ are isomorphic. Let h be an isomorphism between them: then $h[A_i] = h[A_{\sigma(i)}]$. Moreover, h is order-preserving and also preserves E, because we have $E^{\mathcal{M}_n}(\overline{a}, \overline{b})$ iff $\overline{a}, \overline{b} \in R_i^{\mathcal{M}^*}$ for some $i \in \mathbb{N}$ iff $h(\overline{a}), h(\overline{b}) \in R_i^{\mathcal{M}^{**}}$ for some $i \in \mathbb{N}$ iff $E^{\mathcal{M}_n}(h(\overline{a}), h(\overline{b}))$.

If we expand both $\mathcal{M}^*$ and $\mathcal{M}^{**}$ by interpreting E in the same way on each, we see that h is an order-preserving permutation of $\mathbb{Q}$ that also preserves E. In other words $h \in Aut(M_n)$. We can determine one such automorphism for each permutation of A. As a result, $Aut(M_n)$ induces the full symmetric group over A.
□

The above preparatory lemmas directly lead to our main result.

Theorem 12.39 *Let $\mathcal{Q}$ be the $\mathcal{L}_<$-structure $(\mathbb{Q}, <)$, $Aut(\mathbb{Q})$ the group of automorphisms of $\mathcal{Q}$. For any integer $n \geq 1$ there is a subgroup $\mathcal{H} \sqsubseteq Aut(\mathbb{Q})$ such that:*

(1) the degree of homogeneity of $\mathcal{H}$ is $n - 1$;
(2) the degree of uniqueness of $\mathcal{H}$ is n.

Proof We let $\mathcal{M}_n$ be the Fraïssé limit previously described and A the set of E^{M_n}-equivalence classes. The automorphisms in $Aut(M_n)$ may or may not permute distinct equivalence classes determined by E^{M_n}. When they do not permute them, they at most permute n-tuples within the same equivalence class. Automorphisms acting in the manner described determine a subgroup $\mathcal{N}$ of $Aut(M_n)$, which is also normal in $Aut(M_n)$ (if $h \in N$ and $k \in Aut(M_n)$, consider any n-tuple $\bar{a}$ in the equivalence class A_i. If k sends $\bar{a}$ into A_j, then it is clear that, since h leaves it in A_j, the automorphism $k^{-1}hk(\bar{a})$ must leave $\bar{a}$ in A_i, i.e. $k^{-1}hk \in N$).

A crucial feature of $\mathcal{N}$ is that it has the same orbits on M^{n-1} as $Aut(M_n)$. To see this, note that we may identify N with $Aut(\mathcal{M}^*)$ ($\mathcal{M}^*$ was defined immediately before Lemma 12.38). If $g \in Aut(M_n)$ relates the $(n-1)$-tuples $\bar{a}, \bar{b}$, there is a finite isomorphism between the finite structures generated by the latter tuples. This isomorphism leaves the equivalence classes of n-sets unaffected because a tuple consisting of $n-1$ elements can have no subsets of n distinct elements. It follows that the given finite isomorphism is also a finite isomorphism in $\mathcal{M}^*$. By Lemma 12.37, this isomorphism extends to an automorphism of $\mathcal{M}^*$, i.e. to some $h \in N$ (which evidently need not be g).

We now proceed as in the proof of Theorem 12.35 and take $\mathcal{H}'$ the subgroup of $Aut(M_n)$ that induces the regular representation of the free group of countable rank on the set A (this representation was described in the proof of Theorem 12.35: such a subgroup exists because $Aut(M_n)$ induces the full symmetric group on A, as shown in Lemma 12.38).

We can now apply Theorem 12.21 to N and $K = H'N$ and obtain the existence of a subgroup $\mathcal{H}$ of $Aut(M_n)$ with the same orbits as the latter group of automorphisms on M^{n-1}. By Lemma 12.36 these orbits coincide with those of $Aut(\mathbb{Q})$, which sends any ordered list $q_0 < \ldots < q_{n-1}$ of rationals into any other ordered list $s_0 < \ldots < s_{n-1}$. It now follows that $\mathcal{H}$ is $(n-1)$-point homogeneous.

To see that $\mathcal{H}$ is n-point unique, suppose $h(\bar{a}) = \bar{a}$ for some n-tuple $\bar{a} \in \mathbb{Q}^n$. Then h preserves one equivalence class and, because it acts regularly on A, it preserves all of them (by an argument given in the proof of Theorem 12.35). Since $\mathcal{H}$ is isomorphic to K/N, it identifies the elements of N (which preserve every

12.8 A Theorem on Rational Scale Types

equivalence class but may permute n-tuples inside the same equivalence class) with its neutral element. It follows that h must be the identity of $\mathcal{H}$.

The proof is not yet complete because we have not ruled out the possibility that $\mathcal{H}$ is in fact $(n-1)$-point unique (because of $(n-1)$-point homogeneity, m-point uniqueness cannot hold for any $m < n-1$ but it might hold for $m = n-1$). In order to do so we must show that there are $a_1, \ldots, a_{n-1} \in \mathbb{Q}$ and $h \in \mathcal{H}$ such that $h(a_i) = a_i$ while h is not the neutral element of $\mathcal{H}$.

To this end we observe that K/N has an infinite sequence of generators $\{Nf_i\}_{i \in \mathbb{N}}$. As in the proof of Lemma 12.32, we focus on the infinite sequence $\{f_i\}_{i \in \mathbb{N}}$. The lemma's proof shows that we can use the countable sequence $\{f_i\}_{i>0}$ to ensure that $(n-1)$-point homogeneity holds.

At the same time, we can use f_0 to ensure that n-point uniqueness holds, as follows. Since f_0 is not in N, it does not leave all equivalence classes fixed. Suppose f_0 sends the equivalence class A_1 into the equivalence class A_2.

We now use Lemma 12.37 and consider the Fraïssé limit $\mathcal{M}^*$, whose age is obtained from the age of $\mathcal{M}_n$ by expanding the relevant, finite structures to a language containing $\{R_i\}_{i \in \mathbb{N}}$ (as discussed above) and then taking reducts that omit the interpretations of E.

Let $\{a_1, \ldots, a_{n-1}\}$ be a $(n-1)$-element set. Consider the n-element sets $\{a_1, \ldots, a_{n-1}, b\}$ and $\{a_1, \ldots, a_{n-1}, c\}$, chosen to be isomorphic as linear orders. We can have two finite n-element structures with the latter sets as domains and modelling the formulae $\neg(y = z)$, $R_1(x_1, \ldots, x_{n-1}, y)$ and $R_2(x_1, \ldots, x_{n-1}, z)$ under the obvious assignments. A strong amalgam of these structures over $\{a_1, \ldots, a_{n-1}\}$ in $Age(\mathcal{M}^*)$ exists. Because this amalgam can be identified with a substructure of $\mathcal{M}^*$, the following conditions are satisfied in $\mathcal{M}^*$:

(1) $\{a_1, \ldots, a_{n-1}, b\} \in A_1$ and $\{a_1, \ldots, a_{n-1}, c\} \in A_2$;
(2) there is a $\mathcal{L}_<$-isomorphism between the finite structures just described.

Note that the finite structure generated by $\{a_1, \ldots, a_{n-1}, y\}$ is isomorphic to the finite structure generated by $\{f_0(a_1), \ldots, f_0(a_{n-1}), f_0(b)\}$ because f_0 is an automorphism. Because, however f_0 sends A_1 into A_2, it follows that the finite structures generated by $\{f_0(a_1), \ldots, f_0(a_{n-1}), f_0(b)\}$ and $\{a_1, \ldots, a_{n-1}, c\}$ respectively are isomorphic with respect to the expansion of $\mathcal{L}_<$ to $\{R_i\}_{i \in \mathbb{N}}$.

Lemma 12.37 now implies that this isomorphism extends to an automorphism of $\mathcal{M}^*$. Note that the automorphisms of $\mathcal{M}^*$ are elements of N. In other words, there is $h_0 \in N$ such that:

$$h_0 f_0(a_i) = a_i \, (i = 1, \ldots, n-1) \text{ and } h_0 f_0(b) = c.$$

The function $h_0 f_0$ fixes $n-1$ points but differs from the identity. Thus n-point uniqueness holds. □

As an immediate consequence of the preceding theorem, we obtain:

Theorem 12.40 *Let $\mathcal{Q}$ be as in the statement of the previous theorem. For each positive $n \in \mathbb{N}$, $\mathcal{Q}$ can be expanded to a relational structure of scale type $(n-1, n)$.*

Proof By Theorem 12.39, there is a subgroup $\mathcal{H} \sqsubseteq Aut(\mathbb{Q})$ that is $(n-1)$-point homogeneous and n-point unique. Lemma 12.27 ensures that, after adding one $2(n+1)$-ary relation symbol to the language, we can expand $\mathcal{Q}$ to a relational structure $\mathcal{Q}^+$ whose automorphism group is $\mathcal{H}$. □

Further Reading

The material covered in this book offers a basic introduction to model theory. Readers who wish to extend their study beyond the basics may find the following, concise suggestions of help.

Intermediate Model Theory

For the sake of consolidation, it may be useful to read, in part or wholly, other introductory texts. We especially recommend [35, 58] (at least its first three chapters and the two opening sections of the fourth) and [64]. It should also be noted that a much deeper and more extensive treatment of all the topics discussed here can be found in Hodges' classic text [25], a significantly more concise, but still profitable, version of which is [26].

The main directions of study that go immediately beyond the scope of this book concern further applications of model theory to algebra and analysis, as well as the proof of Morley's theorem and the more abstract developments of stability theory. A book that does both well and concisely is [70]. Most of the material covered at a rather fast pace in the first four chapters of the latter book has been covered more leisurely in our first 11 chapters. It is thus possible to tackle the early chapters with relative ease and move quickly to the discussion of indiscernibles and Vaughtian pairs that leads to a proof of Morley's theorem in the fifth chapter. The last two sections of Chap. 4 (pp.128–132) provide essential information on ω-stability, a topic that may be more extensively pursued in the sixth chapter of [44]. A more detailed introduction to stability theory, which is readable even directly after this book, is [9].

The reader with a specific interest in applications of model theory could again turn to [70], whose seventh chapter, together with Appendix C, contains a self-contained, if demanding, introduction to the model theory of valued fields, including a full proof of the Ax-Kochen-Ershov theorem. It is advisable to gain some familiarity with valued fields in the final chapter of [58] before turning to [70]. Alternatively, since the algebraic background covered in Appendix C is presented in a rather condensed form, it may be profitable to study the first four chapters of [20] by way of preparation.

Another important area of algebra in which model theory has proved fruitful is the theory of modules (roughly speaking, vector spaces in which 'scalars' come from a ring that may not be a field). An excellent and standard reference for the model theory of modules is [57].

Finally, o-minimality has led to important applications of model theory, rather more closely related to topology and analysis than algebra. A standard reference on such applications is [76], which should be readily accessible after studying this book, even though it requires some knowledge of linear algebra and topology, which could be picked up as needed.

Historical and Philosophical Themes

A study of philosophical themes specifically connected with model theory cannot prescind from [10] and [6]. The first book tends to focus on problems closer to mainstream philosophy of mathematics, whereas the latter focusses on the articulation of an overall view of mathematical practice from the standpoint of model theory. A significant feature of [10] is its final chapter on the history of model theory, contributed by Wilfrid Hodges. The reader with a decidedly historical interest in model theory will find a large number of articles, monographs, and collections of classic papers, from which we select a small fraction, which may constitute a reasonable starting point.

The selection of Tarski's papers in [75] and the comprehensive collection of Mal'tsev's papers in [43] are historically fundamental. So is Robinson's monograph [59]. Among research monographs, we mention the extremely thorough study of the model-theoretic treatment of real algebra and real closed fields in [71]. Among articles, we single out the illuminating discussion of model-completeness in [47] and the insightful and entertaining [27].

Appendix A
Set-Theoretic Background

This Appendix provides, in a largely self-contained manner, the set-theoretic background needed for Chaps. 7 to 11. The reader unfamiliar with basic set theory should study the Appendix before turning to Chap. 7.

A.1 Well-Ordered Sets

The central set-theoretic notion we will study is that of an ordinal. We arrive at it through a specialisation of a more general order-theoretic concept, that of a well-ordered set.

Definition A.1 The structure $(M, \leq)$, where $\leq$ is a binary relation, is a partially ordered set or **poset** iff $\leq$ is reflexive, transitive and antisymmetric.[1] We say that the poset $(M, \leq)$ is a **well-ordered set** iff, for every nonempty $S \subseteq M$, S has a least element relative to $\leq$. In symbols, there is $s \in S$ such that, for each $a \in S$, $s \leq a$.

It follows from the last definition that a well-ordered set is linearly ordered: to see why, note that, for any $a, b \in M$, $\{a, b\}$ is a nonempty subset of M with a least element. Thus, either $a \leq b$ (if a is least) or $b \leq a$ (if b is least). If both inequalities hold, then $a = b$. Thus, if $a \neq b$, we must have $a < b$ or $b < a$ (by definition $x < y$ holds iff $x \leq y$ and $x \neq y$). We may therefore regard the given well-ordered set also as the linear order $(M, <)$.

Exercise A.2 Let (M, R) be a nonempty set endowed with an irreflexive, binary relation R. Suppose that, if $\emptyset \neq S \subseteq M$, there is $s \in S$ such that, for every $a \in S - \{s\}$, sRa. Show that (M, R) is a linear order.

---
[1] This is to say that, for any $a, b \in M$, $a \leq b$ an $b \leq a$ jointly imply $a = b$.

Exercise A.3 Let $\mathcal{M} = (M, <)$ be an infinite, linearly ordered set. Show that $\mathcal{M}$ is a well-ordered set iff it does not include an infinite sequence $\{a_i\}_{i \in \mathbb{N}}$ such that, for each $i \in \mathbb{N}$, $a_i > a_{i+1}$.

In view of the previous exercise, we have an easy way of proving that the class of well-ordered sets, regarded as $\mathcal{L}_<$-structures, is not elementary.

Exercise A.4 Prove by contradiction that the class **W** of well-ordered sets is not an elementary class of $\mathcal{L}_<$-structures. (*Hint*: expand the language and apply Compactness).

Given a well-ordered set $(M, <)$, its elements fall under exactly one of three categories. Clearly, M has one least element relative to $<$, which we may call c_0. If $x \in M$ is not c_0, then we consider $L_x = \{a \in M : a < x\}$, the **initial segment** of M determined by x. The initial segment L_x may or may not have a maximum. If it does, let the maximum be b: then $b < x$ and, if $b < z$, it follows that $z \notin L_x$, i.e. $x \leq z$, by trichotomy. In other words, x is the least element greater than b. We call x the *successor* of b or a *successor point* of M. If L_x has no maximum, $M - L_x$ has a least element, namely x ($x \leq b$ for each $b \in M - L_x$). Since x is not the successor of any element of M, we call it a *limit point* of M.

In sum, when $(M, <)$ is a well-ordered set, any $a \in M$ can only be: (i) the least element of M; (ii) a successor point of M; (iii) a limit point of M. Well-ordered sets may not contain representatives of each category, as the following example shows.

Example A.5 When conceived as a $\mathcal{L}_<$-structure, the set of natural numbers $\mathbb{N}$ is well-ordered by $<^{\mathbb{N}}$. Its least element is 0 and every other element of $\mathbb{N}$ is a successor point. There are no limit points. The $\mathcal{L}_<$-structure $(\mathbb{N} \cup \{-1\}, <^{\mathbb{N}}_{-1})$ is a well-ordered set when the relation $<^{\mathbb{N}}_{-1}$ is defined by the following condition: $x <^{\mathbb{N}}_{-1} y$ iff $x, y \in \mathbb{N}$ and $x < y$ or $x \in \mathbb{N}$ and $y = -1$. The well-ordered set $(\mathbb{N} \cup \{-1\}, <^{\mathbb{N}}_{-1})$ has exactly one limit point.

By the previous example, a well-ordered set may be infinite without limit points. If, however, a well-ordered set $(M, <)$ has at least one limit point, then M must be infinite.

Exercise A.6 Prove the last statement.

Well-ordered sets support a proof technique known as transfinite induction. This form of induction depends on the next result:

Theorem A.7 Let $(M, <)$ be a well-ordered set and $S \subseteq M$. Suppose that $L_x \subseteq S$ implies $x \in S$. Then $S = M$.

Proof Toward a contradiction, suppose $S \neq M$, which implies $M - S \neq \emptyset$. Because M is well-ordered, let a be least in $M - S$. In this case, $L_a \subseteq S$ and, by hypothesis, $a \in S$. Thus $a \in S \cap (M - S)$, which is impossible. □

Exercise A.8 Let $(M, <)$ be a well-ordered set with least element c. Suppose that $S \subseteq M$ satisfies the following three conditions:

(1) $c \in S$;
(2) if $x \in S$, then the successor of x is in S;
(3) if x is a limit point of M and $y < x$ implies $y \in S$, then $x \in S$.

Show that $S = M$.

A well-ordered set may be thought of as a sequence that begins with an initial stage and goes through successor and limit stages, achieving a certain 'length'. Thus, it is natural to suppose that, if two well-ordered sets are distinct, they essentially differ only by their length. In other words, one of them should be isomorphic to an initial segment of the other. This will be proved in the next section. We first need a way of comparing lengths.

Definition A.9 Let $(M, <^M)$, $(N, <^N)$ be well-ordered sets. A **similarity** between $(M, <^N)$, $(N, <^N)$ is a bijection $f : M \longrightarrow N$ such that, for any $a, b \in M$, $a <^M b$ iff $f(a) <^N f(b)$.

Lemma A.10 *Let $(M, <)$ be a well-ordered set. The only similarity of $(M, <)$ onto itself is the identity.*

Proof First, we note that, if f is a similarity of $(M, <)$ onto itself, then $a \leq f(a)$ for any $a \in M$. If this were not the case, the set $A = \{c \in M : f(c) < c\}$ would be nonempty and have a least element a. Then $f(a) < a$ and, since f preserves $<$, we could deduce $ff(a) < f(a)$, against the choice of a as least in A. Next, we note that, if f is a similarity of $(M, <)$ onto itself, so is f^{-1}. Consequently, for any $a \in M$ we have $a \leq f(a)$ and $a \leq f^{-1}(a)$. Since f is order-preserving, the last inequality implies $f(a) \leq a$. Now antisymmetry yields $f(a) = a$. □

Exercise A.11 Suppose that $(M, <^M)$, $(N, <^N)$ are well-ordered sets and that f is a similarity between $(M, <^M)$ and initial segment of $(N, <^N)$. Show that f is unique (*Hint*: suppose there is a distinct similarity g and use Lemma A.10).

Exercise A.12 Suppose that $(M, <^M)$, $(N, <^N)$ are well-ordered sets and that f is a similarity between them. Show that f has the following properties (relative to the respective orderings):

(i) If c is the least element of M, then $f(c)$ is the least element of N.
(ii) If b is the successor of a in M, then $f(b)$ is the successor of $f(a)$ in N.
(iii) If a is a limit point of M, then $f(a)$ is a limit point of N.

Lemma A.13 *Suppose that $(M, <^M)$, $(N, <^N)$ are well-ordered sets and let f, g be similarities between a proper initial segment of $(M, <^M)$ and a proper initial segment of $(N, <^N)$. Either f extends g or g extends f, i.e. $f \subseteq g$ or $g \subseteq f$.*

Proof Let L_a, L_b be the domains of f, g respectively. We may assume $a \neq b$. By trichotomy, $a < b$ or $b < a$. In the former case $L_a \subset L_b$ while, in the latter case, $L_b \subset L_a$. Suppose the former. In view of Exercise A.11, there is a unique similarity from L_b to $g[L_b]$. It is clear that the restriction of this similarity to L_a is a similarity. But f is, again by Exercise A.11, the unique similarity between L_a and $f[L_a]$. It must therefore coincide with the restriction of g to L_a. In other words, $f \subset g$. If we had assumed $L_b \subset L_a$, we would have obtained $g \subset f$. □

We are primarily interested in a special kind of well-ordered set, endowed with the property stated in:

Definition A.14 A set X is **transitive** iff $y \in X$ implies $y \subseteq X$.

Exercise A.15

(i) Show that a set X is transitive iff $Y \subseteq X$ implies $\bigcup Y \subseteq \bigcup X$.
(ii) Show that the following sets are transitive:

$$\emptyset, \{\emptyset\}, \{\emptyset, \{\emptyset\}\}, \{\emptyset, \{\emptyset\}, \{\emptyset, \{\emptyset\}\}\}.$$

We can now define ordinal numbers or ordinals.

Definition A.16 An **ordinal** is a transitive set that is well-ordered by $\in$.

Exercise A.17

(i) Show that the sets from Exercise A.15-(ii) are ordinals.
(ii) Show that any initial segment of an ordinal is an ordinal.
(iii) Show that any element of an ordinal α is an initial segment of α.
(iv) Show that an ordinal is well-ordered by inclusion.

Definition A.16 has one important consequence, recorded in:

Lemma A.18 *If α, β are distinct ordinals, then $\alpha \in \beta$ iff $\alpha \subset \beta$.*

Proof If $\alpha \in \beta$, the transitivity of β implies $\alpha \subseteq \beta$ and the hypothesis $\alpha \neq \beta$ implies $\alpha \subset \beta$. To prove the converse implication, suppose $\alpha \subset \beta$. Since $\beta - \alpha \neq \emptyset$, there is a $\in$-least $\gamma \in \beta$ such that $\gamma \notin \alpha$. We show that $\gamma = \alpha$. On the one hand, if $\delta \in \gamma$, $\delta \in \alpha$ because γ is the least element of β not in α. We deduce $\gamma \subseteq \alpha$. If $\delta \in \alpha$, trichotomy yields three possibilities: $\delta = \gamma$, $\delta \in \gamma$ or, finally, $\gamma \in \delta$. We rule out the first, since it implies $\gamma \in \alpha$, against the choice of γ. If $\gamma \in \delta$, then $\delta \in \alpha$ and the transitivity of $\in$ in β imply $\gamma \in \alpha$, which cannot hold. We conclude that $\delta \in \gamma$, i.e. $\alpha \subseteq \gamma$. It now follows that $\alpha = \gamma$, as desired. □

Exercise A.19

(i) Show that, if α is an ordinal, then $\alpha \notin \alpha$.
(ii) Show that, if α is an ordinal other than $\emptyset$, then $\emptyset \in \alpha$. Deduce that $\emptyset$ is the least element of every nonempty ordinal.

By Exercise A.19, the least element of every ordinal other than $\emptyset$ is $\emptyset$. It is also easy to see that, if α is an ordinal, then $\alpha \cup \{\alpha\}$ is, too.

Exercise A.20 Establish the last claim.

We next characterise the successor and limit points of an ordinal α.

Lemma A.21 *Let α be an ordinal, $\gamma \in \alpha$. If L_γ has a maximum δ, then $\gamma = \delta \cup \{\delta\}$. Otherwise $\gamma = \bigcup_{\beta \in \gamma} \beta$.*

Proof Suppose that L_γ has a maximum δ and consider $A = \{\beta \in \alpha : \delta \in \beta\}$. The set A is nonempty because it contains γ. Thus $\delta \in \gamma$ and, since γ itself is an ordinal, its transitivity yields $\delta \subseteq \gamma$. Thus $\delta \cup \{\delta\} \subseteq \gamma$. By Exercise A.20, $\delta \cup \{\delta\}$ is an ordinal. By Lemma A.13, $\delta \cup \{\delta\} \subset \gamma$ is equivalent to $\delta \cup \{\delta\} \in \gamma$, against the fact that δ is the maximum of γ. Thus $\gamma = \delta \cup \{\delta\}$.

If L_γ has no maximum then, for any $\beta \in \gamma$, the transitivity of γ implies $\beta \subseteq \gamma$. Consequently:

$$\bigcup \{\beta \in \alpha : \beta \in \gamma\} \subseteq \gamma.$$

The union $\bigcup \{\beta \in \alpha : \beta \in \gamma\}$ is an ordinal (by Exercise A.22 below) and it cannot be strictly included in γ or it would be an element of L_γ and the latter set would have a maximum. Thus:

$$\gamma = \bigcup \{\beta \in \alpha : \beta \in \gamma\}.$$

By definition, the last union is $\bigcup_{\beta \in \gamma} \beta$. □

Exercise A.22 Show that, if α is an ordinal and $S \subseteq \alpha$, then $\bigcup S$ is an ordinal.

Similarities do not relate isomorphic ordinals: they specify individual ordinals.

Theorem A.23 *Suppose that α, β are ordinals and that $f : \alpha \longrightarrow \beta$ is a similarity. Then $\alpha = \beta$.*

Proof We proceed by transfinite induction and consider the set $S = \{\delta \in \alpha : f(\delta) = \delta\} \subseteq \alpha$. Our goal is to show that $S = \alpha$. If this is the case, then f is the identity over α and its range is β, i.e. $\alpha = \beta$. If α, β are both empty, there is nothing to prove. Thus, we may assume that they are both nonempty. Since they have the same least element $\emptyset$ and $f(\emptyset) = \emptyset$, we deduce $\emptyset \in S$.

Next, we show that, if $\delta \in S$, then $\delta \cup \{\delta\} \in S$. We know from Exercise A.12-(ii) that $f(\delta \cup \{\delta\}) = f(\delta) \cup \{f(\delta)\}$. By the inductive hypothesis, $f(\delta) = \delta$ so $f(\delta \cup \{\delta\}) = \delta \cup \{\delta\}$, as desired.

Finally, suppose that $\lambda \in S$ is a limit point of α. The inductive hypothesis is that, for any $\beta \in \lambda$, $\beta \in S$, i.e. $f(\beta) = \beta$. Since f is a similarity, $\beta \in \lambda$ implies $f(\beta) \in f(\lambda)$ (ordinals are well-ordered by $\in$), so $\beta \in f(\lambda)$, which entails $\lambda \subseteq f(\lambda)$. The reader may establish the converse inclusion. It follows that $\lambda = f(\lambda)$.

We can now conclude that $S = \alpha$. □

Corollary A.24 *If α, β are ordinals and α is similar to an initial segment of β, then $\alpha \in \beta$.*

Proof By Theorem A.23, the given similarity strictly includes α in β. Thus $\alpha \subset \beta$ and, in view of Lemma A.13, $\alpha \in \beta$. □

By the last corollary and Lemma A.13, if $\alpha \neq \beta$, the alternative $\alpha \in \beta$ or $\beta \in \alpha$ holds if there is a similarity to compare α, β. In Sect. A.3 we shall see that there are enough similarities to guarantee the comparability of any two ordinals. Exercise A.19-(i) implies that set-theoretic membership is irreflexive on ordinals. Membership is also transitive since $\alpha \in \beta$ and $\beta \in \gamma$ imply $\beta \subseteq \gamma$ (by the transitivity of γ) and, consequently, $\alpha \in \gamma$. Certainly, if we could single out a nonempty subcollection of the collection of ordinals, it should contain a least element: since the given subcollection must contain some ordinal α, its least element lies in $L_{\alpha \cup \{\alpha\}}$.

These remarks open the prospect of treating the collection of ordinals as a well-ordered aggregate. We may not, however, treat it as a set, for reasons to be clarified in the next section.

A.2 Classes and Sets

Structures, as we have defined them, are set-theoretic objects. We have drawn upon set-theoretic resources to supply the semantic content of model theory. In this book, these resources presuppose the axioms of Zermelo-Fraenkel set theory together with the Axiom of Choice, in short **ZFC** (for a full list of the axioms, see e.g. [31], pp.267–268).

Because, however, the axioms of **ZFC** can be stated as first-order sentences in the language $\mathcal{L}_\in = \mathcal{L}_= \cup \{\in\}$, where '$\in$' is a binary relation symbol, we may regard **ZFC** as a first-order theory among many and study it by model-theoretic means. If we were to do it in our present context, we should presuppose the same theory at a 'higher level', as providing the semantic metatheory for **ZFC**.

Here we do not engage in such a task but only note that, once we take a structure $\mathcal{M}$ to be a model of **ZFC**, we refer to the elements of its domain as sets, much as we refer to the elements of a vector space as vectors. Exactly as we would not ask whether the collection of all points in a plane is a point, we do not *a priori* accept it as a meaningful question to ask whether the collection of all sets in $\mathcal{M}$ is a set. It turns out, however, that certain parametrically $\mathcal{L}_=$-definable subcollections of M may be viewed as sets, i.e. elements of M, while others do not. It is therefore important to draw a distinction between the definable collections that 'are' sets and those that are not: adhering to standard terminology, we refer to the latter as **proper classes** or classes that are not sets.[2]

The collection **On** of ordinals provides an example of a class. To check that this collection is $\mathcal{L}_\in$-definable, note that the formula $\forall s (s \in x \rightarrow s \subseteq x)$ (where

---

[2] It may be helpful to bear in mind the following, intuitive way of distinguishing between sets and classes. Let the relation $a \in b$ be depicted by an arrow pointing from a to b. A collection of sets K could be identified with a set $c \in M$ if there were enough arrows pointing from the elements of K to c. When arrows are insufficient K cannot be identified with a point of M.

$s \subseteq x$ abbreviates $\forall z(z \in s \rightarrow z \in x))$, says that x is a transitive set. Call this formula ψ_1. There are formulae ψ_2, ψ_3 saying respectively that x is linearly ordered by $\in$ and that any nonempty subset of x has a least element. Thus, the formula $\psi_1(x) \wedge \psi_2(x) \wedge \psi_3(x)$ defines the collection **On**.

Exercise A.25 Write ψ_2 and ψ_3 explicitly.

Even though the collection of ordinals is definable, it is not a set.

Theorem A.26 *The collection **On** is not a set.*

Proof Toward a contradiction, suppose **On** is the set β. By the remark closing Sect. A.1, β is well-ordered by $\in$ and transitive. It follows that β is an ordinal, which implies $\beta \in \beta$, against Exercise A.19-(i). □

The distinction between class and set will play a role in some later proofs (see e.g. Theorems A.28 and A.58): we sometimes have to rely on the fact that certain classes are also sets or, by contrast, on the fact that they cannot be sets. Furthermore, some of the proofs we are about to encounter explicitly rely on one axiom schema (i.e. an infinite family of axioms of the same form) known as the Replacement axiom. Such axiom refers to a *functional correspondence* F, i.e. single-valued collections of ordered pairs parametrically defined by a $\mathcal{L}_\in$-formula over M. Any correspondence F is a definable collection of sets, whose domain may be restricted to a set. Replacement says that

if F is a unary functional correspondence and X a set, then F[X] is a set.

where we abuse the notation F[X], since F is not in general a function. More formally, the Replacement axiom schema is:

$$(\forall x \forall y \forall z \phi(x,y) \wedge \phi(x,z) \rightarrow y = z) \rightarrow \forall t \exists u (\forall v (v \in u \leftrightarrow \exists w (w \in t \wedge \phi(w,v)))),$$

where ϕ is any formula that defines a functional correspondence. Replacement implies that definable subcollections of sets are sets, a result on which we are also going to rely.[3]

Exercise A.27 Show that, if the formula $x = x$ defined a set, then the formula $\neg(x \in x)$ would define a set. Derive a contradiction from this conclusion and deduce that $x = x$ defines a class that is not a set.

---

[3] To see why this result holds, consider a $\mathcal{L}_\in$-formula $\varphi(x)$ and suppose that $X, \varphi(X)$ are nonempty. Fix $c \in \varphi(X)$. The formula $(\varphi(x) \wedge x = y) \vee (\neg \varphi(x) \wedge y = c)$, which is of the form $\phi(x, y)$, defines a functional correspondence F that, when restricted to the set X, precisely determines the image $\{x \in X : \varphi(x)\}$. The Replacement axiom for $\phi(x, y)$ ensures that the last subcollection of X is actually a set.

A.3 Comparability and Towers

We now prove that any two well-ordered sets are comparable, i.e. that one of them is similar to an initial segment of the other. It follows that both membership and inclusion linearly order the class **On**.

In what follows we shall refer to a well-ordered collection whose initial segments are sets as a *proper, well-ordered class*.

Theorem A.28 *Let* C, D *be proper, well-ordered classes. The following three possibilities are mutually exclusive and exhaustive: (i) there is a class similarity*[4] *between the whole of* C *and an initial segment of* D *(since* D *is a proper class, its initial segments are sets); (ii) there is a class similarity between the whole of* D *and an initial segment of* C*; (iii) there is a class similarity between the whole of* C *and the whole of* D.

Proof Let L_a, L_b be initial segments of C, D respectively. Because they are both sets, they are included in longer initial segments $L_{a'}, L_{b'}$ respectively. Thus, whenever we consider a similarity between two initial segments of C, D, we may think of it as a similarity between proper initial segments of two well-ordered sets, i.e. as a function f. Any two such similarities are comparable by inclusion in view of Lemma A.13. The collection S of similarities between proper initial segments of C, D is definable by a formula $\psi(x)$ in the language of ZFC and thus it is a class.

The class defined by $\exists y \psi(y) \wedge x \in y$ is, intuitively, the 'union' of S, which we denote by $\bigcup S$ for the sake of notational convenience.[5] Because, for any two functions f, g from the class $\bigcup S$, one extends the other, it is clear that the class of ordered pairs $\bigcup S$ is a functional correspondence (a given argument in C determines a unique value under a similarity f and this value stays the same under the similarities extending f).

It now suffices to show that either the first components of the ordered pairs in $\bigcup S$ exhaust C or the second components of these ordered pairs exhaust D (or both). If neither happened, there would be least elements c, d in C, D respectively such that the ordered pairs in $\bigcup S$ were all contained in $L_c \times L_d$. In this case $\bigcup S$ would be a function f and, more precisely, a similarity between an initial segment of C and an initial segment of D that does not contain the ordered pair (c, d). Because, however, the function $f \cup \{(c, d)\}$ is a similarity, it must be included in $\bigcup S$, i.e. (c, d) at once is and is not in $\bigcup S$. This contradiction concludes the proof. □

When C, D are sets, we obtain:

Corollary A.29 *Let* $(M, <^M), (N, <^N)$ *be well-ordered sets. Then either they are similar to each other or one of them is similar to an initial segment of the other.*

---

[4] This is simply a collection of ordered pairs that provides the class-equivalent of a similarity between well-ordered sets.

[5] Insofar as 'union' is a set-theoretic operation, it may not apply to classes that are not sets.

A Set theory

Exercise A.30 Show that no well-ordered set can be similar to any of its proper initial segments.

If we specialise to ordinals, Corollary A.24 yields:

Corollary A.31 *Let α, β be distinct ordinals. Either $\alpha \in \beta$ or $\beta \in \alpha$.*

In view of the last Corollary, we conclude that **On** is a proper, well-ordered class. Its centrality among proper, well-ordered classes depends on the following:

Theorem A.32 *Let C be a proper, well-ordered class. If C is a set, then it is similar to an ordinal. If it is not a set, then there is a class similarity between C and **On**.*

Proof Suppose C was a set. Then **On** could not be similar to an initial segment of C or, by Replacement (relative to the inverse of the given similarity), **On** would be itself a set, against Theorem A.26. It follows that C must be similar to an initial segment of **On**, say L_α. Since L_α is the set of ordinals smaller than α (relative to $\in$), it is α: thus, C is similar to the ordinal α. If C was not a set, it could not be similar to an initial segment of **On**, i.e. an ordinal, by Replacement again. Thus C must be similar to the class of ordinals by Theorem A.28. □

In view of the last theorem, ordinals provide similarity standards for any well-ordered set and the class of ordinals itself provides a single standard for proper, well-ordered classes. We may describe such classes in abstract terms. To do so, we first need to single out certain features of the class **On**, which we shall later generalise.

Our starting point is the functional correspondence σ that assigns to each ordinal its successor. This correspondence σ is *strictly increasing* in the sense that, if α is an ordinal, then $\alpha \subset \sigma(\alpha)$. We may therefore describe **On** as a class containing the empty set, closed under σ (i.e. if α is in **On**, so is $\sigma(\alpha)$) and such that every element other than $\emptyset$ is either of the form $\sigma(\alpha)$, for some ordinal α, or it is the union of the chain of preceding ordinals.

In virtue of these features **On** supports proof by transfinite induction. In order to clarify what this means, we fix a class P with defining $\mathcal{L}_\in$-formula $\phi(x)$.

Theorem A.33 *Suppose that the collection P satisfies the following conditions, where α, β are ordinals:*

(1) *$\emptyset$ is in P;*
(2) *If α is in P, then $\sigma(\alpha)$ is in P;*
(3) *If λ is a limit ordinal and, for every $\delta \in \lambda$, δ is in P, then $\lambda = \bigcup_{\delta \in \lambda} \delta$ is in P.*

Then every ordinal is in P.

Exercise A.34 Prove Theorem A.33.

Exercise A.35

(i) By an axiom of ZFC called the axiom of infinity, there is a set N such that:

   (a) $\emptyset \in N$;
   (b) $x \in N$ implies $x \cup \{x\} \in N$.

Let S be the family of subsets of N that satisfy conditions (a) and (b). Call ω the intersection of the elements of S. Show that ω is the least set (relative to inclusion) to satisfy conditions (a), (b).
(ii) Show that, if a subset of ω satisfies (a) and (b), then it coincides with ω. Use this result to show that ω is a set of ordinals. List the first four elements of ω.
(iii) Show that no element of ω is a limit ordinal and deduce that ω is the least limit ordinal.

The elements of ω are the **finite ordinals** and ω is standardly used as the set-theoretic representation of $\mathbb{N}$. For this reason, its least element is identified with 0, the successor of 0 is identified with 1, the successor of 1 is identified with 2, and so on. Note that, under these conventions, $0 = \emptyset$, $1 = \{0\}$, $2 = \{0, 1\}$ and so on.

We may regard **On** as a hierarchy starting at 0 and increasing through applications of successor and unions (at limit points). By the way it grows, this structure admits transfinite induction. The abstract type of hierarchy exemplified by **On** can now be explicitly defined.

Definition A.36 Let T be a class and F a unary functional correspondence such that, for any x in T, $x \subset F(x)$. Then T is said to be a F-tower iff it satisfies the following conditions:

(1) 0 is in T;
(2) if x is in T, then $F(x)$ is in T;
(3) if C is a chain whose elements are in T, then $\bigcup C$ is in T;
(4) T admits transfinite induction.

Exercise A.37 State a transfinite induction theorem for T and show that, if S and T are F-towers, then they coincide.

Definition A.36 is sufficient to retrieve order-theoretic properties that we have already established for **On**. In particular, we can show that a F-tower T is well-ordered by inclusion and that it has no maximum. The point of obtaining these results is that they lead to a transparent proof of the transfinite recursion theorem, a few equivalent forms of which we shall establish in the next section. This theorem plays a major role in Chaps. 7 to 9. In the following subsection, we establish a key feature of towers (our discussion closely follows [73], pp.52–57).

A.3.1 Order Properties of Towers

We wish to show that, if T is a F-tower, then any two sets in this class are comparable relative to inclusion. If the comparability could be reduced to a unary property, we could prove it by transfinite induction. Since, however, comparability depends on the binary property defined by the formula $x \subseteq y \vee y \subseteq x$, we must adopt a variant of transfinite induction that works for classes of ordered pairs, as opposed to classes *simpliciter*. Let $\mathsf{R}(x, y)$ be a class of ordered pairs. We can prove that, for any $x, y \in \mathsf{T}$, $\mathsf{R}(x, y)$ holds if T satisfies the following three conditions:

(1) For any x in T, $\mathsf{R}(x, 0)$;
(2) $\mathsf{R}(x, y)$ and $\mathsf{R}(y, x)$ jointly imply $\mathsf{R}(x, \mathsf{F}(y))$;
(3) if C is a chain such that, for any $y \in C$, y is in T and $\mathsf{R}(x, y)$, then $\mathsf{R}(x, \bigcup C)$.

Exercise A.38 Follow the steps below to prove that, if T satisfies conditions (1) to (3) above, then $\mathsf{R}(x, y)$ holds for any x, y in T.

(i) First, call x in T right-normal iff, for any y in T, $\mathsf{R}(y, x)$. Define left-normal analogously.
(ii) Show by transfinite induction for T that, if x is right-normal, then it must also be left-normal (to this end, use conditions (1) to (3)).
(iii) Show by transfinite induction on T that, for every x in T, x is right normal (note that what (1) says is that 0 is right-normal).
(iv) Conclude that, for any x, y in T, $\mathsf{R}(x, y)$ must hold.

Now let R be the class defined by the formula $\mathsf{F}(x) \subseteq y \vee y \subseteq x$. If a F-tower satisfies (1) to (3) relative to $\mathsf{F}(x)$, then it is linearly ordered by inclusion. This is because the formula implies $x \subseteq y \vee y \subseteq x$ (due to the fact that F is strictly increasing).

Lemma A.39 *Let* T *be a F-tower and* R *the class defined by* $\mathsf{F}(x) \subseteq y \vee y \subseteq x$. *Then* T *satisfies conditions (1) to (3).*

Proof Condition (1) is left to the reader. To verify condition (2), it suffices to note that its hypothesis amounts to the fact that $\mathsf{F}(x) \subseteq y \vee y \subseteq x$ and $\mathsf{F}(y) \subseteq x \vee x \subseteq y$ hold. The last two disjunctions determine four combinations of conditions. One of them, namely $\mathsf{F}(x) \subseteq y$ and $\mathsf{F}(y) \subseteq x$, can be ruled out because $x \subset \mathsf{F}(x)$ and $y \subset \mathsf{F}(y)$. If $x \subseteq y$ and $y \subseteq x$ hold simultaneously, then $x = y$, which implies $\mathsf{F}(x) = \mathsf{F}(y)$ and thus $\mathsf{F}(x) \subseteq \mathsf{F}(y) \vee \mathsf{F}(y) \subseteq x$, i.e. $\mathsf{R}(x, y)$. The remaining cases are left to the reader.

As for condition (3), suppose that $\mathsf{R}(x, y)$ holds for each $c \in C$, where C is a chain. If $y \subseteq x$ for each $y \in C$, it follows that $\bigcup C \subseteq x$, which is what we want. Otherwise there is z such that $\mathsf{F}(x) \subseteq z$. Because $z \subseteq \bigcup C$, we obtain $\mathsf{F}(x) \subseteq \bigcup C$ and condition (3) still holds. □

Exercise A.40 Complete the proof of Lemma A.39.

Exercise A.41 Show that no F-tower T can be a set (*Hint*: if it were, then it would be a chain by Lemma A.39).

We now know that any F-tower T is linearly ordered by inclusion. We want to show that T is in fact well-ordered by inclusion. To this end, we consider a subclass U of T and look at the sets in T that are strictly included in every set from U. As we shall see, by doing so we can identify the least element of U in a way that brings to light a close, though perhaps unsurprising, parallel between F-towers and **On**.

Theorem A.42 *Let* T *be a* F*-tower and* U *a subclass of* T *that is not the empty set. Then* U *has a least element relative to inclusion. More precisely, if L is the collection of sets from* T *that are strictly included in every set from* U*, we have the following, exhaustive list of mutually exclusive cases:*

(a) $L = \emptyset$ *and* $\emptyset$ *is the least set in* U*;*
(b) *L has a maximum x and* F(x) *is the least set in* U*;*
(c) *L has no maximum and* $\bigcup L$ *is the least set in* U*.*

Proof In case (a), $\emptyset$ is not strictly included in every set from U. Because $\emptyset$ is strictly included in every nonempty set, we must have $\emptyset$ in U.

In case (b), since $x \subset F(x)$, it follows that $F(x)$ cannot be a strict subset of every set in U, i.e. it must itself be a set in U because U is linearly ordered by inclusion. Moreover, we cannot have any set y in U such that $x \subset y \subset F(x)$. This is because, by Lemma A.39, $R(x, y)$ must hold and, if $y \subseteq x$ does not hold (since, by hypothesis, $x \subset y$), then $F(x) \subseteq y$. In other words, $F(x)$ is least in U.

In case (c), we observe that $\bigcup L$ is a set. This is because a set in L is included in all, and thus in any set x from U. It follows that the elements of L are all subsets of x, i.e. elements of $\mathcal{P}(x)$. Their collection is therefore an element of the set $\mathcal{PP}(x)$ and thus a set. If $\bigcup L \in L$, then $\bigcup L$ would be the maximum of L relative to set-theoretic inclusion. It follows that $\bigcup L$ is in U. If y was in U and $y \subset \bigcup L$, then we could easily deduce $x \subseteq y$ for each $x \in L$. Consequently $\bigcup L \subseteq y \subset \bigcup L$, which is impossible. We conclude that $\bigcup L$ is least in U.

□

Although we have proved a few results about F-towers, we have not proved that any exist, unless F is σ. That F-towers exist for any strictly increasing functional correspondence F follows from a result known as Cowen's theorem. We omit a proof of this theorem, a detailed discussion of which appears in [73], pp.66–69. It is however important to remark that Cowen's theorem shows that, given F, there is a first-order formula in the language of ZFC that defines a F-tower T (see in particular the remark after Definition 7.3 on p.66 of [73]). By Exercise A.37, T is unique. Thus, we simply refer to *the* F-tower T. Its existence is appealed to in the proof of Theorem A.46 below.

A.4 Transfinite Recursion

By the last subsection, a F-tower is a class well-ordered by inclusion, which we may regard as a counterpart to **On** in which F plays the role of σ. It is thus plausible to suppose that there should be a systematic matching between the sets in a F-tower and the sets in the class **On**. The existence of a matching is the content of the transfinite recursion theorem.

Theorem A.43 (Transfinite Recursion) *Let* T *be a* F*-tower. Then* **On** *is similar to* T *and the similarity can be used to index the sets in* T *as follows:*

A Set theory

(1) $T_0 = 0$;
(2) $T_{\sigma(\alpha)} = \mathsf{F}(T_\alpha)$;
(3) $T_\lambda = \bigcup_{\delta \in \lambda} T_\delta$, with λ a limit ordinal.

Proof Note that T is a proper, well-ordered class. By Theorem A.32 there is a class similarity On between **On** and T. We describe the effect of On by setting:

$$On(\alpha) = T_\alpha, \text{ for each } \alpha \text{ in } \mathbf{On}.$$

Since On must associate the least set in **On** with the least set in T, we obtain $T_0 = 0$. Because On preserves successors (by an easy variant of Exercise A.12 for classes), it associates $T_{\sigma(\alpha)}$ with the successor of T_α. By Theorem A.42, this successor is $\mathsf{F}(T_\alpha)$. Finally, because On preserves limits (by Exercise A.12 again), it must associate T_λ with the limit of the set L_{T_λ}. By Theorem A.42, this limit is the union of the T_δ with $T_\delta \subseteq T_\lambda$. Because On is a similarity, the T_δ in question are exactly the On-values of each $\delta \in \lambda$. Consequently, $T_\lambda = \bigcup_{\delta \in \lambda} T_\delta$. □

Theorem A.43 does not state transfinite recursion in the form most suitable for model-theoretic applications. To reach the desired form of transfinite induction, we rely on two important definitions.

Definition A.44 Let α be an ordinal. A function f with domain α is said to be an **ordinal sequence of length** α.

Definition A.45 Let E be a functional correspondence defined on ordinal sequences. Then E is said to be an **extending operation** iff, for any ordinal sequence f of length α, there is a set x such that $\mathsf{E}(f) = f \cup \{(\alpha, x)\}$.

An extending operation simply takes an ordinal sequence and adds one further step to it. With the last two definitions in place, we reach the following variant of transfinite recursion:

Theorem A.46 *If* E *is an extending operation, there is a functional correspondence* F *such that, for any ordinal α, the following equality holds:*

$$\mathsf{F} \upharpoonright \sigma(\alpha) = \mathsf{E}(\mathsf{F} \upharpoonright \alpha).$$

Proof In view of Cowen's theorem, we consider the E-tower T. By transfinite recursion (Theorem A.43), the sets in T determine a linear order relative to inclusion, whose stages are kept track of by the ordinals. We use transfinite induction on the class **On** to show that each set T_α is an ordinal sequence of length α.

It is clear that $T_0 = \emptyset$ is the empty function, i.e. an ordinal sequence of length 0. Next, consider $\sigma(\alpha)$ and suppose that T_α is an ordinal sequence of length α. By Theorem A.43, $T_{\sigma(\alpha)} = \mathsf{E}(T_\alpha)$ and, since E is an extending operation applied to an ordinal sequence of length α, it follows that $T_{\sigma(\alpha)}$ is an ordinal sequence of length $\sigma(\alpha)$.

Finally, consider T_λ with λ a limit ordinal and suppose that, for each $\delta \in \lambda$, T_δ is an ordinal sequence of length δ. By Theorem A.43, $T_\lambda = \bigcup_{\delta \in \lambda} T_\delta$. The last union is a function and, in particular, an ordinal sequence, since its domain is a union of ordinals and thus an ordinal. More precisely, T_λ is an ordinal sequence of length λ.

Since we are looking for a functional correspondence F defined on every ordinal, this correspondence must be 'longer' than any ordinal sequence. Because T is definable by a first-order formula $\phi(x)$, we define F by the formula $\exists y (\phi(y) \wedge x \in y)$. With an abuse of notation, we set $F = \bigcup T$, simply to indicate that F is obtained by joining together longer and longer ordinal sequences. We leave it as an exercise to verify that F is, as defined, a functional correspondence and that the domain of this correspondence is **On**.

It now suffices to observe that, for every ordinal α, $F \upharpoonright \alpha = T_\alpha$ (verify). Because T is a E-tower, the equality $T_{\sigma(\alpha)} = E(T_\alpha)$ can be rewritten as:

$$F \upharpoonright \sigma(\alpha) = E(F \upharpoonright \alpha).$$

□

Exercise A.47

(i) Supply the required verification in the proof of Theorem A.46.
(ii) Describe $F \upharpoonright 0$ and $F \upharpoonright \lambda$ when λ is a limit ordinal.

The preceding theorem raises the question of generalising transfinite recursion to functional correspondences other than extending operations. We can in fact reduce general functional correspondences defined on ordinal sequences to extending operations and prove:

Theorem A.48 *Let* G *be a functional correspondence defined on ordinal sequences. There is a unique functional correspondence* F *such that, for any ordinal* α, *the following equality holds:*

$$F(\alpha) = G(F \upharpoonright \alpha).$$

Proof As announced before the statement of the theorem, we reduce G to a representation in terms of a suitable extending operation E. If s_α is an ordinal sequence of length α, we set $E(s_\alpha) = s_\alpha \cup \{(\alpha, G(s_\alpha))\}$. Since $E(s_\alpha)$ is itself an ordinal sequence with α in its domain, follows from the definition that:

$$(E(s_\alpha))(\alpha) = G(s_\alpha)$$

Given the extending operation E, Theorem A.46 implies the existence of F such that $F \upharpoonright \sigma(\alpha) = E(F \upharpoonright \alpha)$. Noting that the ordinal sequence $F \upharpoonright \sigma(\alpha)$ has α in its domain, we obtain the chain of inequalities:

$$F(\alpha) = (F \upharpoonright \sigma(\alpha))(\alpha) = (E(F \upharpoonright \alpha))(\alpha) = G(F \upharpoonright \alpha).$$

The correspondence F is unique by transfinite induction. □

A Set theory

Exercise A.49

(i) Suppose that the formula $\psi(x, y)$ defines G. Suppose $\eta(x, y, z)$ says that x is the ordered pair (y, z). Using η, explicitly write a formula $\varphi(y, a, b)$ saying that y is a function with domain a and codomain b. Use φ to write a formula $\theta(x, a)$ saying that x is an ordinal sequence of length a. Finally, write a formula $\phi(x, y)$ that defines the extending operation E described in the proof of Theorem A.48.

(ii) Prove that the correspondence F from the proof of Theorem A.48 is unique.

We next qualify Theorem A.48 by requiring G to depend on two functional correspondences H, K.

Theorem A.50 *Let a be a set and H, K be functional correspondences defined on sets. There is a unique functional correspondence F defined on the class of ordinals and satisfying the following conditions:*

(1) $F(0) = a$;
(2) $F(\sigma(\alpha)) = H(F(\alpha))$;
(3) $F(\lambda) = K(F[\lambda])$, *where λ is a limit ordinal.*[6]

Proof In order to use Theorem A.48 we must find a functional correspondence G that is defined on ordinal sequences. Given the result we want to achieve, a definition by cases suggests itself. As in Exercise A.49-(i), let $\theta(x, u)$ b a formula saying that x is an ordinal sequence of length u. Moreover, let $h(x, y)$ and $k(x, y)$ be defining formulae for the classes H, K respectively. In order to satisfy condition (1), we use the formula:

$$\theta(x, \emptyset) \wedge y = a, \tag{A.1}$$

which, intuitively, says that $G(0) = a$.

In order to satisfy condition (2), we let $Ord(\alpha)$ abbreviate the formula saying that α is an ordinal and we consider:

$$\exists \beta (\theta(x, \beta) \wedge \exists \alpha (Ord(\alpha) \wedge \beta = \alpha \cup \{\alpha\}) \wedge h(x(\alpha), y)), \tag{A.2}$$

which intuitively says that, whenever x is an ordinal sequence of successor length $\beta = \alpha \cup \{\alpha\}$, $G(x) = H(x(\alpha))$.

In order to satisfy condition (3), we let $Ran(x)$ abbreviate the formula that holds whenever x is the range of a function and we consider:

$$\exists \lambda (\theta(x, \lambda) \wedge \neg \exists \alpha (Ord(\alpha) \wedge \lambda = \alpha \cup \{\alpha\}) \wedge k(Ran(x), y)), \tag{A.3}$$

---

[6] If $\phi(x, y)$ defines F, any element of $F[\lambda]$ is a set a such that it satisfies $\phi(\delta, a)$ for $\delta \in \lambda$.

which intuitively says that, whenever x is an ordinal sequence of limit length λ, $\mathsf{G}(x) = \mathsf{K}(Ran(x))$.

Calling $\varphi_i(x, y)$ the formula (A.i), with $i = 1, 2, 3$, we take G to be the functional correspondence defined by $\varphi_1(x, y) \vee \varphi_2(x, y) \vee \varphi_3(x, y)$. By the previous theorem, there is a unique F defined on the class of ordinals and that satisfies the equality:

$$\mathsf{F}(\alpha) = \mathsf{G}(\mathsf{F} \restriction \alpha).$$

The theorem now follows from the definition of G.

□

Exercise A.51 Use the definition of G to verify the last statement in the proof of Theorem A.50.

We finally come to the form of transfinite recursion that serves our model-theoretic purposes:

Theorem A.52 *Let G be a functional correspondence defined on sets and c be a set. There is a functional correspondence F defined on the class of ordinals whose values, written with ordinal subscripts, satisfy the following conditions:*

(1) $F_0 = c$;
(2) $F_{\sigma(\alpha)} = \mathsf{G}(F_\alpha)$;
(3) $F_\lambda = \bigcup_{\delta \in \lambda} F_\delta$.

Proof We apply Theorem A.50 with the notation $\mathsf{F}(\alpha) = F_\alpha$, by setting $\mathsf{H} = \mathsf{G}$ and K as set-theoretic union.[7]

□

Since Theorem A.43 is a special case of Theorem A.52, we have given a round-robin proof of the equivalence between any two of the theorems proved in this section.

Remark A.53 Because Theorem A.52 ensures the existence of a functional correspondence defined on the class of ordinals, it also guarantees the existence of functions with the same properties defined on any given ordinals. Such functions are simply restrictions of the functional correspondence to specific ordinals. We sometimes only need to work with restrictions of the functional correspondences delivered by transfinite recursion.

A.4.1 An Application of Transfinite Recursion

We use transfinite recursion to show that every vector space (finite-dimensional or infinite-dimensional) has a basis.

---

[7] If x is a set, the formula $\forall y(y \in z \leftrightarrow \exists v(v \in y \wedge y \in x))$ defines $z = \bigcup x$. We take K to be the functional correspondence defined by the last formula.

Theorem A.54 (Hamel) *Every vector space V over a field K has a basis.*

Proof We apply transfinite recursion as given by Theorem A.52 to build larger and larger sets of linearly independent vectors, until we reach one whose span is the whole of V.

Consider the set $S \subseteq \mathcal{P}(V)$, whose elements are sets of vectors that do not generate V. In other words, if $A \in S$, $acl_V(A) \neq V$. Thus, for each $A \in S$, $V - acl_V(A) = A'$ is nonempty. Calling C the set containing A' for each $A \in S$, we assume that there is a choice function f for the nonempty set C. Given A', $f(A')$ selects a vector from this latter set, which is linearly independent of the vectors in A. The existence of f is an axiom of ZFC known as the Axiom of Choice (see Sect. A.6 for more details).

In order to apply transfinite recursion in the form of Theorem A.52, we introduce the functional correspondence G, defined below:

$$\mathsf{G}(X) = \begin{cases} X \cup \{f(X)\} & \text{if } \emptyset \neq X \text{ and } V - acl_V(A) \neq \emptyset \text{ for some } A \in S \\ X & \text{otherwise} \end{cases}$$

Fixing the set $\{v\}$, with $v \in V$ and v distinct from the null vector, Theorem A.52 yields the following **On**-sequence:

(1) $B_0 = \{v\}$;
(2) $B_{\sigma(\alpha)} = \mathsf{G}(B_\alpha)$;
(3) $B_\lambda = \bigcup_{\delta \in \lambda} B_\delta$, with λ limit.

It is not difficult to prove by transfinite induction that, for each ordinal α, B_α is a linearly independent subset of V. We can also use transfinite induction to show that the ordinal sequence we have obtained is an increasing chain relative to inclusion. In particular, we are going to show, by transfinite induction on γ, that:

$$\alpha \subseteq \beta \subseteq \gamma \text{ implies } B_\alpha \subseteq B_\gamma.$$

If $\gamma = 0$, the result vacuously holds. If $\gamma = \sigma(\delta)$, we assume that the result holds for δ. Thus, if $\alpha \subset \beta \subset \gamma$, β must be strictly included in $\delta \cup \{\delta\}$, i.e. $\beta \subseteq \delta$, in which case $B_\alpha \subseteq B_\beta$ by inductive hypothesis. If $\beta = \gamma$, then $\alpha \subseteq \delta \subset \gamma$ implies $B_\alpha \subseteq B_\delta$ by inductive hypothesis. It then suffices to note that $B_\delta \subseteq B_{\sigma(\delta)}$, by the definition of G, and invoke the transitivity of inclusion. Finally, if λ is a limit ordinal and the result holds for any $\delta \in \lambda$, we only need to consider the case $\alpha \subset \lambda = \beta$ (otherwise the inductive hypothesis alone yields the result we seek). Since $\alpha \in \beta$, and $B_\lambda = B_\beta$ is the union of $\{B_\delta\}_{\delta \in \beta}$ we obtain $B_\alpha \subseteq B_\lambda$.

The ordinal sequence we have obtained by transfinite recursion contains a chain of linearly independent subsets of V. Since V has some size κ, with κ an ordinal,[8] we cannot have strict inclusion throughout the chain, or we would obtain, e.g. at

---

[8] We shall see how ordinals can be used as canonical representatives of set sizes in Sect. A.5

stage $\lambda > \kappa$, more linearly independent vectors in V than there are vectors in V, which is impossible. It follows that, at some stage α, we must have $B_\alpha = B_{\sigma(\alpha)}$. We may, in particular, choose α to be the least ordinal at which this happens.

By the definition of **G**, $B_\alpha = B_{\sigma(\alpha)}$ implies that B_α is a nonempty, linearly independent subset $A \subseteq V$ such that $V - acl_V(A)$ is empty. In other words $V = acl_V(A)$ and A is a basis for V.

□

Exercise A.55 Prove by transfinite induction that, for each ordinal α, B_α is a linearly independent subset of V.

Remark A.56 The proof of Hamel's theorem is simple enough to allow a formally explicit definition of **G**. This is not a typical situation. To avoid cumbersome formal details, in the main text we adopt the policy of specifying the correspondences involved in applications of transfinite recursion less formally.

A.5 Cardinal Numbers

Sets are compared relative to size by means of injections. Although we have so far compared ordinals only by similarities, it is clear that we can compare them by means of injections. An ordinal will have larger size than another if it cannot be made similar to it by a suitable reordering. In other words, α is larger in size than β if we cannot use any injection of the set β into α to induce over the whole set α the ordering of β. This remark leads to:

Definition A.57 An ordinal α is said to be **initial** iff there is no $\beta \in \alpha$ such that α admits an ordering induced by β.

There is a formula in the language of **ZFC** that defines the class **Cn** of initial ordinals.[9] We call the elements of this class **cardinal numbers**. It follows from our definition that **Cn** is a subclass of **On**: because it inherits the ordering of **On**, the class **Cn** is a proper, well-ordered class. By Theorem A.32, **Cn** is similar to an ordinal or to the whole class **Cn**. In order to show that the latter is the case, we rely upon:

Theorem A.58 (Hartogs) *For any set A, there is an ordinal α such that α cannot be injected into A.*

Proof We leave it to the reader to verify that an ordinal α can be injected into A iff, for some $M \subseteq A$, there is a well-ordered set $(M, <)$ that is similar to α. Any well-ordered set $(M, <)$, with $M \subseteq A$, is an ordered pair whose first component M is an element of $\mathcal{P}(A)$ and whose second component $<$ is an element of $\mathcal{P}(A \times A)$ (i.e. a set of ordered pairs with components in A). Thus, the class W of all well-ordered

---

[9] It is a tedious exercise, but not one too difficult, to write such a formula explicitly.

A Set theory

sets of the form $(M, <)$, with $M \subseteq A$, is a subset of the set $\mathcal{P}(A) \times \mathcal{P}(A \times A)$ or, equivalently, an element of the set $\mathcal{P}(\mathcal{P}(A) \times \mathcal{P}(A \times A))$.

There is a formula $\phi(x, y, A, W)$ in the language of ZFC with the parameter W that holds between x, y when $x \in W$, y is an ordinal and there is a similarity between them.[10] Since y is uniquely determined,[11] $\phi(x, y, A, W)$ describes a functional correspondence and Replacement applies. It follows that the class of ordinals y such that $\phi(x, y, A, W)$ holds as x ranges over the *set* W is in fact a set. Because **On** is a class but not a set, there is an ordinal α such that $\neg \phi(x, \alpha, A, W))$ is satisfied for every choice of $x \in W$. This is to say that α is not similar to any well-ordered subset of A. In other words, α cannot be injected into A. □

Exercise A.59 All the formulae mentioned in this exercise are first-order $\mathcal{L}_=$-formulae.

  (i) Consider the formula $\eta(y, y_1, y_2)$ from Exercise A.49-(i) and abbreviate it by $y = (y_1, y_2)$. Write a formula $\varphi(x, y, z)$ saying that z is a bijection from x onto y.
 (ii) Write a formula $\zeta(x, x_1, x_2)$ saying that u is the ordered pair (x_1, x_2) where x_2 well-orders x_1.
(iii) Explicitly write $\phi(x, y, A, W)$ from the proof of Theorem A.58, using A, W as parameters.

Exercise A.60 One version of the Axiom of Choice says that, if X, Y are sets, then either $|X| \leq |Y|$ or $|Y| \leq |X|$ (in other words, we can either inject X into Y or Y into X). Use Hartogs' Theorem and the Axiom of Choice to prove that every set can be well-ordered.

By the last exercise, ZFC implies that every set can carry the order type of some ordinal. Taking the least such ordinal, we deduce that every set has the same size as an initial ordinal. We can therefore use initial ordinals, or cardinals, as representatives of set size.

Exercise A.61 Show that **Cn** is not a set. Deduce that **Cn** is similar to **On**.

We have already singled out the set ω of finite ordinals. We are now in a position to show that every element of this set is an initial ordinal i.e. a cardinal number. To this end, we adopt:

Definition A.62 The set A is said to be **finite** iff every injection from A into A is a bijection.

The definition makes intuitive sense: an injection from a n-element set A must send n distinct arguments into n distinct values. It is therefore a permutation of the elements of A. We now define a set A to be infinite iff it is not finite in the sense

---

[10] For an explicit statement of $\phi(x, y, A, W)$ see Exercise A.55.
[11] The well-ordered set x cannot be similar to two distinct ordinals α, β. Otherwise α or β would be similar to one of its initial segments, against Exercise A.30.

of Definition A.62. This definition is in keeping with the natural model-theoretic definition of 'infinite'. We say that A is infinite iff it is a $\mathcal{L}_=$-structure that models the sentence φ_n, stating the existence of at least n distinct objects, for each integer $n \geq 1$. If A is infinite in the latter sense, then $\mathbb{N}$ can be injected into it. In this case, A is also infinite in the sense derived from Definition A.62. Let $f : \mathbb{N} \longrightarrow A$ be an injection: it suffices to consider $f[\mathbb{N}]$ and inject it into a strict subset of itself (we may e.g. send $f(n)$ into $f(n+1)$, for each $n \geq 0$). The function that coincides with the identity over $A - f[A]$, and with the injection just described over $f[A]$, injects A into, but not onto, itself.

If, on the other hand, $g : A \longrightarrow A$ is an injection but not a bijection, then there is some $a \in A - g[A]$. Clearly, $a \neq g(a)$ (because $g(a) \in g[A]$ and $a \notin g[A]$). The injectivity of g implies that $g(a) \neq g(g(a)) = g^2(a)$. Note that, moreover, $g^2(a) \neq a$. A further application of g leads to $g^3(a) \neq g^2(a)$, again by injectivity. We also have $g^3(a) \neq g(a)$ because equality would entail $g^2(a) = a$ (applying g^{-1} to both sides of the original equality). In general (we leave the proof to the reader), $g^n(a)$ is different from $a, \ldots, g^{n-1}(a)$. Thus, we obtain enough objects to guarantee that A models φ_n for each $n \geq 1$.

We now show that every element of ω is finite in the sense of Definition A.62. This is the reason why we have called the elements of ω finite ordinals in Exercise A.35. The same Exercise tells us how to prove that a given first-order condition holds for each element of ω. It suffices to show that the given condition holds for 0 and that it is preserved by the successor operation. Thus, ω admits the following **induction principle**:

if $S \subseteq \omega, 0 \in S$ and, for every $x \in S, x + 1 \in S$, then $S = \omega$.

Exercise A.63 Using the induction principle for ω, show that every element of ω is finite (i.e. any injection from $x \in \omega$ onto itself is already a bijection).

Exercise A.64 Using the induction principle for ω and the previous exercise, show that every element of ω is an initial ordinal.

It follows from the last two exercises that ω is the set of finite cardinal numbers. Note that ω itself is an infinite cardinal. It is infinite because the injection that sends each finite cardinal into its successor is not a bijection (0 is not a value). It is a cardinal because any one of its elements is finite. We call the subclass of **Cn** that only contains ordinals greater or equal to ω **Cn**$_\infty$: it is the class of **infinite cardinal numbers** or **infinite cardinals**.

Exercise A.65 Show that **Cn**$_\infty$ is a class and that, consequently, there is a class similarity between **On** and **Cn**$_\infty$.

Calling $\aleph$ the class similarity from Exercise A.65 and indicating its values by the application of ordinal indices, we can enumerate **Cn**$_\infty$ as follows:

$$\aleph_0, \aleph_1, \aleph_2, \ldots, \aleph_\omega, \ldots, \aleph_\alpha, \ldots,$$

where $\aleph_0 = \omega$ and the elements of the above enumeration are called 'aleph null', 'aleph one' etc. We refer to infinite cardinals collectively as 'alephs'. The class of infinite cardinals is well-ordered by the indexing over **On** and admits proofs by transfinite induction on the ordinal indices of its elements. Moreover, given a cardinal $\aleph_\alpha$, its successor in $\mathbf{Cn}_\infty$ is $\aleph_{\sigma(\alpha)}$ because the class similarity $\aleph$ preserves successors. By the same clue, $\aleph_\lambda$ is a limit point of $\mathbf{Cn}_\infty$ iff λ is.

Exercise A.66 Show that each infinite cardinal is a limit ordinal.

We have seen that ordinals may be used to enumerate any well-ordered set and we already noted that cardinals may be used as representatives of set size. We restate the last remark as:

Theorem A.67 *Every infinite set has the same size as an aleph.*

Proof Let A be an infinite set. It suffices to note that, by Exercise A.60, A can be well-ordered. As a consequence, there is an ordering $<$ of A such that the well-ordered set $(A, <)$ is similar to an ordinal α, by Theorem A.32. If α is an initial ordinal, then the given similarity is also a bijection between A and the infinite cardinal α. If α is not initial, then the set of ordinals β such that $\beta \in \alpha$ and α admits β as an ordering is nonempty. It therefore has a least element γ, which is an initial ordinal. It follows that γ is the cardinal number of A. $\square$

Remark A.68 Without using the Axiom of Choice, we can guarantee that ordinals have cardinal numbers (i.e. that there are bijections from any one ordinal onto some initial ordinal) but we cannot prove Theorem A.67. If, however, we assume that every set can be well-ordered, then every set can be compared with an ordinal and, consequently, with an initial ordinal. In this case Theorem A.67 follows. As we shall see in the next section, the Axiom of Choice is equivalent to the statement that every set can be well-ordered.

By the last theorem, if A, B are infinite and $|A| = |B|$ then there is an aleph of the same size as A, B (if A, B are finite, we reach the same conclusion with respect to finite cardinals). We may therefore think of each cardinal number as a distinguished representative of set size. We can thus conceive of $|\cdot|$ as an operation that assigns to a set A its cardinal number $|A|$ (thus, e.g., $|\mathbb{N}| = \aleph_0$, $|\{1, 2, 3\}| = 3$): we shall therefore write equalities of the form $|A| = \kappa$, where κ designates a cardinal number.

A.5.1 Cardinal Arithmetic

The main reason why we are interested in cardinals, given our model-theoretic focus, is that we can add, multiply and raise them to a cardinal power. These operations are useful because they provide us with a way of counting e.g. the number of formulae in an infinite language or the number of subsets of a specified size that are included in the domain of a given structure.

Conforming to a standard notational convention, we use the Greek letters κ, μ, λ to designate generic cardinals (finite or infinite). We first define cardinal addition, multiplication and exponentiation.

Definition A.69 Let κ, μ, λ be cardinals. Then:

(1) $\kappa + \mu = \lambda$ iff $A \cap B = \emptyset$, $|A| = \kappa$, $|B| = \mu$ and $\lambda = |A \cup B|$;
(2) $\kappa \cdot \mu = \lambda$ iff $|A| = \kappa$, $|B| = \mu$ and $|A \times B| = \lambda$
(3) $\kappa^\mu = \lambda$ iff $(\mu \longrightarrow \kappa)$ is the set of functions with domain μ and codomain κ and $|(\mu \longrightarrow \kappa)| = \lambda$.

As far as finite cardinals are concerned, the operations just defined essentially coincide with the familiar arithmetical operations on $\mathbb{N}$. Exponentiation (item (3)) may seem less obvious than addition or multiplication but an easy check will show that it corresponds to the familiar operation of raising to a power, where the exponent is determined by the common domain of a family of functions.

Exercise A.70 Suppose $|A| = 2$ and $|B| = 3$. Show that there are eight distinct functions from B to A and nine distinct functions from A to B. Thus $|A|^{|B|} = 8$ and $|B|^{|A|} = 9$.

Using Definition A.69, it is possible to deduce many basic equalities and inequalities of cardinal arithmetic. We record a list of useful ones in the statement of the following lemma:

Lemma A.71 Let κ, λ, μ be cardinal numbers. Then:

(1) $(\kappa + \mu) + \lambda = \kappa + (\mu + \lambda)$;
(2) $\kappa + \mu = \mu + \kappa$;
(3) $(\kappa \cdot \mu) \cdot \lambda = \kappa \cdot (\mu \cdot \lambda)$;
(4) $\kappa \cdot \mu = \mu \cdot \kappa$;
(5) $\kappa \cdot (\lambda + \mu) = \kappa \cdot \lambda + \kappa \cdot \mu$;
(6) $\kappa \leq \lambda$ implies $\kappa \cdot \mu \leq \lambda \cdot \mu$;
(7) $\kappa \leq \kappa \cdot \mu$;
(8) $\kappa^{\lambda+\mu} = \kappa^\lambda \cdot \kappa^\mu$;
(9) $(\kappa^\lambda)^\mu = \kappa^{\lambda \cdot \mu}$;
(10) $\kappa \geq 2$ implies $\kappa + \kappa \leq \kappa \cdot \kappa$;
(11) $\kappa \leq \mu$ and $\lambda > 0$ imply $\lambda^\kappa \leq \lambda^\mu$ and $\kappa^\lambda \leq \mu^\lambda$;
(12) $\kappa < 2^\kappa$.

Proof We leave (1) to (7) as exercises and prove (8) to (12). Each item in the statement of the lemma is a direct consequence of Definition A.69 and, in particular, does not depend on the Axiom of Choice. We consider sets A, B, C such that $|A| = \kappa$, $|B| = \mu$ and $|C| = \lambda$.

(8) Let B, C be disjoint.[12] We consider the set $\mathcal{F}$ of functions from the domain $B \cup C$ into A and the set $\mathcal{G}$ of pairs whose first component is a function from C to A and whose second component is a function from B to A. The goal is to show that $\mathcal{F}$ can be injected into $\mathcal{G}$ and, moreover, that $\mathcal{G}$ can be injected into $\mathcal{F}$. Note that a function from $B \cup C$ into A can be split into a unique pair of functions in $\mathcal{G}$ because B, C are disjoint. It follows that $\mathcal{F}$ is injectible into $\mathcal{G}$. Conversely, given a pair of functions $(f, g) \in \mathcal{G}$, we can uniquely define the function $f \frown g$, with domain $B \cup C$. Because B, C are disjoint, f, g do not share any argument and can be naturally concatenated to determine an element of $\mathcal{F}$. This shows that $\mathcal{G}$ can be injected into $\mathcal{F}$.

(9) Let $\mathcal{F}$ be the set of functions with domain B and whose values are functions from C to A. Moreover, let $\mathcal{G}$ be the set of functions from $B \times C$ to A. If $f \in \mathcal{F}$, define a function $g_f \in \mathcal{G}$ by the following condition:

$$g_f(b, c) = (f(b))(c), \text{ with } b \in B, c \in C.$$

where $f(b)$ is a function from C to A that takes a value in A at the argument $c \in C$. Consider the function $H : \mathcal{F} \longrightarrow \mathcal{G}$ defined by:

$$H(f) = g_f.$$

The function H is injective: if f, f' differ at some argument b, the functions $f(b)$ and $f'(b)$ are distinct, i.e. they differ at some argument c. This is to say that g_f is different from $g_{f'}$. Moreover, H is surjective. To see this, consider an arbitrary $g \in \mathcal{G}$ and, for any $b \in B$, the restriction $f_b = g \upharpoonright \{b\} \times C$ of determines a function from C to A, namely $g_b(c) = g(b, c)$, for any $c \in C$. Thus, if we consider $f \in \mathcal{F}$ such that:

$$f(b) = f_b$$

it follows that $g = H(f) = g_f$. We have shown that H is a bijection, i.e. that $|\mathcal{F}| = |\mathcal{G}|$.

(10) Suppose $|A| = \kappa = |A'|$, with A, A' disjoint. Since $|A| \geq 2$, we may suppose $0, 1 \in A'$. It then suffices to note that there is a bijection between $A \cup A'$ and $A \times \{0\} \cup A \times \{1\} \subseteq A \times A'$.

(11) By hypothesis, there is an injection $h : A \longrightarrow B$ and $C \neq \emptyset$. Let c be a fixed element of C. We focus on the first inequality from (11). Given a function f from A to C, we may associate with it the function g_f from B to C by setting:

$$g_f(b) = \begin{cases} h^{-1}(b) & \text{if } b \in h[A] \\ c & \text{otherwise} \end{cases}$$

---

[12] This can always be arranged by taking $B' = B \times \{0\}$ and $C' = C \times \{1\}$ in case B, C are not disjoint.

It is clear that the association that assigns g_f to f is an injective function. The second inequality from (11) can be proved by injecting the set of functions from C to A into the set of functions from C to B. To this end, we assign to a function $f : C \longrightarrow A$ the function $h \circ f : C \longrightarrow B$. The assignment is injective because, if f, f' differ at some $c \in C$, we have $f(c) \neq f'(c)$ and the injectivity of h implies $hf(c) \neq hf'(c)$, i.e. $h \circ f$ is distinct from $h \circ f'$.

(12) By Exercise 4.27-(iii), $|A| < |\mathcal{P}(A)|$. It may be verified that there is a bijection between $\mathcal{P}(A)$ and the set of functions from A to $\{0, 1\}$. By the definition of cardinal exponentiation $|\mathcal{P}(A)| = 2^{|A|}$. We conclude $|A| < 2^{|A|}$. Since $|A| = \kappa$, the desired result follows.

$\square$

However informative, the last lemma does not provide us with the means of carrying out calculations with cardinals. This is because, unless κ, λ are both finite, we do not have, by Lemma A.71 alone, any way of reducing expressions like $\kappa \cdot \kappa$ or $\kappa + \lambda$. We know, however, that the union of two countable sets is countable, i.e. $\aleph_0 + \aleph_0 = \aleph_0$. Since the Cartesian product of n countable sets is countable, we can also deduce:

$$\underbrace{\aleph_0 \cdot \ldots \cdot \aleph_0}_{n \text{ times}} = \aleph_0.$$

Noting that the Cartesian product of n countable sets is a set of functions from a n-element set into a countable set, we obtain the equality:

$$\underbrace{\aleph_0 \cdot \ldots \cdot \aleph_0}_{n \text{ times}} = \aleph_0 = \aleph_0^n.$$

The last results all follow from the equality $\aleph_0 \cdot \aleph_0 = \aleph_0$ and the associativity of cardinal multiplication, i.e. part (3) of Lemma A.71. We are now going to use the last equality as the basis of a transfinite induction on $\mathbf{Cn}_\infty$ and obtain $\kappa \cdot \kappa = \kappa$ for any infinite cardinal. Part (10) of Lemma A.71 then will imply that $\kappa + \kappa = \kappa = \kappa \cdot \kappa$.

Theorem A.72 *If κ is an infinite cardinal, then $\kappa \cdot \kappa = \kappa$.*

Proof We prove the result using transfinite induction in the form of Theorem A.7 adapted to the class $\mathbf{Cn}_\infty$. The inductive hypothesis is that, for any infinite cardinal $\kappa < \lambda$, $\kappa \cdot \kappa = \kappa$. The goal is to use this assumption to show that $\lambda \cdot \lambda = \lambda$. For this transfinite induction to work we rely on the inductive basis $\aleph_0 \cdot \aleph_0 = \aleph_0$, which we take as given.[13]

---

[13] Without an inductive basis, if the implication we wish to prove failed, we would still be able to deduce that there is a least infinite cardinal at which it fails. We could not, however, derive a contradiction from the last fact because, if the infinite cardinal in question was $\aleph_0$, we could not find any infinite cardinals below it—or in fact any cardinals at all—that satisfy the statement of our theorem.

A Set theory

In order to complete the proof, we consider the Cartesian product $\lambda \times \lambda$ and turn it into a conveniently well-ordered set $(\lambda \times \lambda, \prec)$. Under the ordering $\prec$, every initial segment of $\lambda \times \lambda$ has cardinality smaller than λ. By Theorem A.32, $(\lambda \times \lambda, \prec)$ is similar to an ordinal α and $|\alpha| \not< \lambda$. Otherwise, since α, λ are comparable, as ordinals, relative to inclusion, we must have $\lambda \subset \alpha$, in which case $\lambda \in \alpha$ and the well-ordered set $\lambda \times \lambda$ has an initial segment of length λ, against the choice of well-ordering for $\lambda \times \lambda$.

We deduce $\lambda \leq |\alpha| = |\lambda \times \lambda| = \lambda \cdot \lambda$. A direct verification shows that $\lambda \leq \lambda \cdot \lambda$. The last two equalities jointly imply $\lambda \cdot \lambda = \lambda$.

It remains to define $(\lambda \times \lambda, \prec)$ and verify that it has the properties needed for the above argument to work. If $(\alpha, \beta), (\gamma, \delta) \in \lambda \times \lambda$, we set:

$$(\alpha, \beta) \prec (\gamma, \delta) \iff \begin{cases} max\{\alpha, \beta\} \in max\{\gamma, \delta\} \text{ or} \\ max\{\alpha, \beta\} = max\{\gamma, \delta\} \text{ and } \alpha \in \gamma \text{ or} \\ max\{\alpha, \beta\} = max\{\gamma, \delta\} \text{ and } \alpha = \gamma \text{ and } \beta \in \delta \end{cases}$$

where the disjunction is exclusive and *max* is defined in terms of the ordering of **On**. The verification that $\prec$ determines a linear ordering of $\lambda \times \lambda$ is left to the reader.

To show that $\prec$ well-orders $\lambda \times \lambda$, we consider a nonempty set $X \subseteq \lambda \times \lambda$. Because X is nonempty, the set $\{\alpha \in \lambda : \alpha = max\{\alpha, \beta\}$ for some $(\alpha, \beta) \in X\}$ is nonempty and has a least element η. We may now restrict attention to the pairs in X whose maximum is η and consider the nonempty set of their first components. If θ is the least element of the latter set, then (θ, η) is least in X relative to $\prec$. Thus, $(\lambda \times \lambda, \prec)$ is a well-ordered set, as claimed. □

Exercise A.73 Verify that $(\lambda \times \lambda, \prec)$ is a linearly ordered set.

Remark A.74 Theorem A.72 does not require the Axiom of Choice. In presence of Choice, this theorem implies that, for any set A, $|A \times A| = |A|$.

One important consequence of Theorem A.72 is:

Corollary A.75 *For any nonzero cardinals κ, λ, if κ or λ is infinite, then $\kappa + \lambda = \kappa \cdot \lambda = max\{\kappa, \lambda\}$.*

Proof Suppose $\lambda \leq \kappa$. By hypothesis $1 \leq \lambda$. Then κ is infinite and Lemma A.39, part (6), yields the inequalities $\kappa \cdot \lambda \leq \kappa \cdot \kappa$ and $\kappa \leq \kappa \cdot \lambda$. Jointly with Theorem A.67, the last inequalities imply $\kappa \cdot \lambda = \kappa = max\{\kappa, \lambda\}$. The same argument works under the assumption $\kappa \leq \lambda$. By the Axiom of Choice, the cases examined are exhaustive. □

Exercise A.76 The requirement that κ, λ be nonzero depends on the fact that, for any cardinal κ, $0 \cdot \kappa = 0$. Prove the last equality.

Exercise A.77 Suppose $2 \leq \kappa \leq \lambda$, with λ an infinite cardinal. Using parts (11) and (12) of Lemma A.71, as well as Theorem A.67, prove that $2^\lambda = \kappa^\lambda$. Deduce that, in particular, $\aleph_0^{\aleph_0} = 2^{\aleph_0}$.

A.5.2 Regular Cardinals

In the proof of Theorem 9.75 we appeal to the notion of a regular cardinal. Since every cardinal is a limit ordinal, it is the union or, equivalently, the least upper bound of all smaller ordinals. In some cases, it may be possible to choose fewer ordinals to determine the same cardinal: here 'fewer' means 'determining a set of smaller cardinality'. For instance, the cardinal $\aleph_\omega$ is the union of $\aleph_0$ smaller ordinals, namely the cardinals from the sequence $\{\aleph_n\}_{n\in\mathbb{N}}$. When a cardinal κ is the least upper bound of no less than κ smaller ordinals, we call it a **regular cardinal**.

The following definition allows us to formulate the simple results we need concerning regular cardinals.

Definition A.78 If κ is any cardinal and $X \subseteq \kappa$, we say that X is **unbounded** if the least upper bound of its elements (note that X is a set of ordinals) is κ. Otherwise we say that X is bounded.

Any subset X of a regular cardinal κ that has size $< \kappa$ is bounded. If X has a maximum, this is also its least upper bound and it is obviously an ordinal smaller than κ, whose size is $< \kappa$ because κ is initial. If X has no maximum, its least upper bound is the union of the ordinals in X: if this union were κ, then κ would not be regular. A simple consequence of what we have pointed out is:

Lemma A.79 *Let κ be a regular cardinal, $\lambda < \kappa$. Any function from λ into κ has a bounded image.*

Proof Let f be such a function. Since $f[\lambda]$ is of size at most $\lambda < \kappa$, $f[\lambda]$ must be a bounded subset of κ. □

By the Axiom of Choice, every cardinal is an aleph. A **successor cardinal** is an aleph indexed by a successor ordinal. If κ is a cardinal, we denote its successor (the next aleph) by κ^+.

Lemma A.80 *Successor cardinals are regular.*

Proof Let κ^+ be a successor cardinal. We consider a subset $X \subseteq \kappa^+$ such that $\bigcup X = \kappa^+$. Each element of X is an ordinal of size strictly smaller than κ^+, i.e. at most κ. Since $|\bigcup X| \leq |X|\kappa$, we must have $|X| = \kappa^+$. □

A.6 The Axiom of Choice

In the preceding sections we have invoked the Axiom of Choice in a few equivalent forms. In this final section we provide a little information about the Axiom of Choice, provide some of its equivalent formulations and use transfinite recursion to deduce Zorn's lemma from it.

We begin with three possible formulations of Choice.

(1) **(AC1)** Let A be a nonempty set whose elements are nonempty. There is a function $f : A \longrightarrow \bigcup A$, such that, for any $x \in A$, $f(x) \in x$.
(2) **(AC2)** Let A be a nonempty set whose elements are nonempty and pairwise disjoint. There is a set c such that, for any $x \in A$, $c \cap x = \{u\}$ for some $u \in x$.
(3) **(AC3)** Let $A = \{A_i\}_{i \in I}$ be a nonempty set whose elements are nonempty. Then $\Pi_{i \in I} A_i$ is nonempty.

The function whose existence is stated in (AC1) is called a **choice function** and the set whose existence is stated in (AC2) is called a **choice set**.

Exercise A.81

(i) Prove that (AC1) implies (AC2).
(ii) Prove that (AC2) implies (AC3). To this end, consider the nonempty set of mutually disjoint sets of the form $\{i\} \times A_i$, for each $i \in I$.
(iii) Prove that (AC3) implies (AC1). To this end, given the nonempty set A, consider the Cartesian product of its elements.

In view of Exercise A.60, the Axiom of Choice implies the proposition that every set can be well-ordered. It is in fact equivalent to it.

Lemma A.82 *The statement that every set can be well-ordered implies the Axiom of Choice.*

Proof Let A be a nonempty set whose elements are nonempty. By hypothesis, we can well-order A as well as its elements. Suppose that the well-ordered set $(A, <)$ is similar to the ordinal β. Then, we can enumerate A as $\{a_\alpha : \alpha \in \beta\}$. Each a_α can, in turn, be well-ordered and, under a fixed well-ordering, it has a minimal element b_α. The set of ordered pairs of the form (a_α, b_α) is a choice function for A, i.e. (AC1) holds. □

We finally turn to a proof of Zorn's Lemma.

A.6.1 Zorn's Lemma

Let $\mathcal{L}$ be a first-order language whose only non-logical symbol is the binary relation symbol $\leq$. We say that a $\mathcal{L}$-structure $\mathcal{M}$ is a **partially ordered set** or a **partial order** iff it is a model of the following $\mathcal{L}$-sentences:

(a) $\forall x (x \leq x)$ (reflexivity)
(b) $\forall x \forall y ((x \leq y \land y \leq x) \rightarrow x = y)$ (antisymmetry);
(c) $\forall x \forall y \forall x ((x \leq y \land y \leq z) \rightarrow x \leq z)$ (transitivity).

It is easy to see that, if S is a set, then the $\mathcal{L}$-structure $(\mathcal{P}(S), \subseteq)$ is a partial order.
In order to state Zorn's lemma, we need to introduce three order-theoretic notions, namely chain, upper bound and maximal element.

Let $\mathcal{M}$ be a partial order and $A \subseteq M$ a set that is linearly ordered by $\leq^M$. We call A a **chain** of M. A chain is said to have an **upper bound** $b \in M$ iff $a \leq^M b$ for any $a \in A$. An element $a \in M$ is said to be **maximal**, or $\leq^M$-maximal, iff, for any $b \in M$, either b is incomparable with a or $b \leq^M a$.

Theorem A.83 (Zorn's Lemma) *Let $(M, \leq^M)$ be a partial order. If every chain included in M has an upper bound, then M has a $\leq^M$-maximal element.*

Proof See Theorem A.88 below. □

Zorn's lemma can be usefully applied whenever we are able to associate with an object of interest a suitable partial ordering. This is the nontrivial step in typical applications of the lemma: once we know which partially ordered set we are working with, it is often easy enough to verify that chains have upper bounds.

A partial order playing an important role in Sect. 6.5 is described by the next exercise.

Exercise A.84 Let X, Y be sets, with $A \subseteq X$. Consider the set F of pairs (S, f) where $A \subseteq S \subseteq X$ and $f \subseteq X \times Y$. For any $(S, f), (U, g) \in F$, set:

$$(S, f) \leq (U, g) \text{ iff } S \subseteq U \text{ and } f \subseteq g.$$

Show that $(F, \leq)$ is a partially ordered set and that each chain in $(F, \leq)$ has an upper bound.

Exercise A.85 Use induction to show that every finite set can be linearly ordered. Adapt your proof to show that every finite, partially ordered set $(A, <_A)$ can be extended to a linearly ordered set $(A, <)$. This is to say that $<_A$, regarded as a set of ordered pairs, is a subset of $<$ (in other words, if $x, y \in A$ and $x <_A y$, then $x < y$).

Exercise A.86 A typical application of Zorn's lemma proves that every partially ordered set can be linearly ordered. This result can also be arrived at by an application of the compactness theorem.

(i) Let $\mathcal{A} = (A, <_A)$ be $\mathcal{L}_<$-structure that is a partially ordered set. The **positive diagram** of $\mathcal{A}$, denoted by $\mathcal{D}^+(\mathcal{A})$ is the subset of $\mathcal{D}(\mathcal{A})$ whose elements do not contain the symbol $\neg$. Suppose that the sentences in $\mathcal{D}^+(\mathcal{A})$ contain a constant symbol c_a for each $a \in A$. Let φ be a $\mathcal{L}_<$-sentence whose models are linear orders. Show that the set of sentences $S = \{\varphi\} \cup \mathcal{D}^+(\mathcal{A}) \cup \{\neg(c_a = c_b) : a, b \in A\}$ is finitely satisfiable.
(ii) Apply compactness to show that S has a model $\mathcal{M}$. Use Theorem 4.46 to deduce that $\mathcal{A}$ is a substructure of a linear ordering.
(iii) Why is only the positive diagram, but not full diagram of $\mathcal{A}$, needed in the proof just given?

As we know from Chap. 5, the compactness theorem depends on the existence of non-principal ultrafilters, which follows from Zorn's lemma but is a strictly weaker property. Thus, the statement that every partially ordered set can be linearly ordered is not equivalent to the Axiom of Choice. On the other hand, the statement that every linearly ordered set can be well-ordered is equivalent to the Axiom of Choice (for a proof, as well as a thorough discussion of order-theoretic versions of Choice, see [29], pp.111–114).

Exercise A.87 A partially ordered set $(T, \leq)$ is a **tree** if, for any $t \in T$, the set $\{s \in T : s \leq t\}$ is linearly ordered. A **branch** of $(T, \leq)$ is a linearly ordered subset that is not strictly included into any other linearly ordered subset of T (other than, possibly, T, if $(T, \leq)$ is already a linearly ordered set).

(i) Use Zorn's lemma to show that every tree has a branch.
(ii) Suppose that every tree has a branch and show that, in this case, every set can be well-ordered. To this end, consider an arbitrary set A and the class T of injections from an ordinal into A. Show that this class is a set and that, moreover, $(T, \subseteq)$ is a tree. A branch of this tree well-orders A.

We next show that the Axiom of Choice implies Zorn's lemma, using transfinite recursion. For the converse implication see Exercise A.90

Theorem A.88 *(AC1) implies Zorn's lemma.*

Proof Toward a contradiction, we assume the existence of a partially ordered set $(P, \leq)$ that satisfies the following conditions: (i) every chain has an upper bound; (ii) there is no maximal element. We use transfinite recursion to derive a contradiction from (i) and (ii).

First, note that, if $A \subseteq P$ is a chain, there is $x \in P$ such that $a < x$ for every $a \in A$. Otherwise, any upper bound guaranteed by (i) should coincide with an element of A, i.e. the chain A should have a maximum m and no $x \in A$ such that $m < x$, against (ii).

Let S be the set of subsets of P with strict upper bounds. We associate with each $A \in S$ the nonempty set A_s of its strict upper bounds. By (AC1), there is a choice function f that selects one element from each A_s. We extend f to $f^* = f \cup \{(\emptyset, 1)\}$. This extension enables us to introduce a functional correspondence G defined on all ordinal sequences by the following condition, where t_α is an ordinal sequence of length α:

$$\mathsf{G}(t_\alpha) = f^*(\{y \in P : \forall z (z \in P \cap t[\alpha] \to z < y\}).$$

If the image of an ordinal sequence is A_s, for some $A \in S$, G selects a strict upper bound in P for A_s. If the image of an ordinal sequence does not intersect P or is not A_s for any $A \in S$, then G is defined and equals a.

Because G is defined on every ordinal sequence, Theorem A.48 applies. In other words, we can use transfinite recursion to deduce the existence of a functional correspondence F defined on the class of ordinals and such that:

$$\mathsf{F}(\alpha) = \mathsf{G}(\mathsf{F} \restriction \alpha).$$

It can be proved by transfinite induction that, for every ordinal α, $\mathsf{F}(\alpha) \in P$ and $\beta \in \alpha$ imply $\mathsf{F}(\beta) < \mathsf{F}(\alpha)$ (see Exercise A.89). It follows that, if α is any ordinal, there is an injection of α into P (it suffices to restrict attention to the ordinal sequence $F \upharpoonright \alpha$, which is a function). In other words, every ordinal can be injected into the set P, against Hartogs' theorem.

This contradiction shows that, if a partially ordered set $(P, \leq)$ satisfies condition (i), then it must salso satisfy condition (ii). □

Exercise A.89 Suppose that, for every $\beta \in \alpha$, $\mathsf{F}(\beta) \in P$ and that, for every $\gamma \in \beta \in \alpha$, $\mathsf{F}(\gamma) < \mathsf{F}(\beta)$. Show that, in this case, $\mathsf{F}(\alpha) \in P$ and, for any $\beta \in \alpha$, $\mathsf{F}(\beta) < \mathsf{F}(\alpha)$.

Exercise A.90 Assume Zorn's lemma and let A be a nonempty sets whose elements are nonempty and pairwise disjoint. Consider the set $P \subseteq \mathcal{P}(\bigcup A)$, whose elements share with each $x \in A$ at most one element. In other words, If $b \in P$ and $x \in A$, then $b \cap x$ is either $\emptyset$ or an element of x.

  (i) Check that $(P, \subseteq)$ is a partially ordered set.
 (ii) Show that every chain of $(P, \subseteq)$ has an upper bound.
(iii) Apply Zorn's lemma to deduce the existence of a choice set for A. Deduce that Zorn's lemma implies (AC2).

Appendix B
Solutions to Exercises

B.1   Chapter 1

Exercise 1.2

(i) The transitivity of $<$ allows us to deduce $x < x$ from the hypotheses $x < y$ and $y < x$. By the irreflexivity of $<$, we reach a contradiction.

(ii) The minimum of X is any $x \in X$ such that, for every y in X, $y = x$ or $x < y$. If X is finite, we may set $X = \{x_1, \ldots, x_n\}$ for some n in $\mathbb{N}$. We outline a procedure to find the minimum of X: at first we consider x_1 and compare it with $x_2, x_3, \ldots, x_n$ until we find x_i such that $x_i < x_1$ or no x_i of the required kind occurs. In the second case, x_1 is the minimum of X. In the first case, we compare x_i with $x_2, x_3, \ldots, x_{i-1}, x_{i+1}, \ldots, x_n$, proceeding as above. After at most n rounds of comparisons, we determine the minimum of X. A similar procedure searches for the maximum of X. It is clear that the above procedures determine *the* minimum and *the* maximum of X. We may, however, reason by contradiction and suppose that $x \neq y$ are minima of X. Because X is linearly ordered, $x < y$, because x is a minimum, and $y < x$, because y is. We know from part (i) that both relations cannot simultaneously hold. Thus, $x = y$ and the minimum is unique. Similar considerations apply to the maximum of X.

(iii) There exist six distinct linear orders of $\{A, B, C\}$. They are: $A < B < C$, $B < C < A$, $C < A < B$, $C < B < A$, $B < A < C$, $A < C < B$.

Exercise 1.4

(i) Because $\succcurlyeq$ is reflexive, $x \succcurlyeq x$. Consequently, $x \succcurlyeq x$ and $x \succcurlyeq x$. By definition, $x \sim x$.

(ii) If $x \sim y$, then $x \succcurlyeq y$ and $y \succcurlyeq x$. If we read the last relations from right to left, we obtain $y \sim x$.

(iii) Suppose $x \sim y$ and $y \sim x$. Then $x \succcurlyeq y$, $y \succcurlyeq x$, $y \succcurlyeq z$ and $z \succcurlyeq y$. Using the transitivity of $\succcurlyeq$, we obtain $x \succcurlyeq z$ and $z \succcurlyeq x$. We conclude $x \sim z$.

Exercise 1.5

Suppose $[x] > [y]$ and $u \sim x$, $v \sim y$. By the definition of $>$, $x \succ y$. The equivalence $u \sim x$ corresponds to the relations $x \succcurlyeq u$ and $u \succcurlyeq x$. The relation $x \succ y$ holds if $x \succcurlyeq y$ and $x \not\sim y$. From the transitivity of $\succcurlyeq$ we deduce $u \succcurlyeq y$. Moreover, we can rule out $u \sim y$ because the equivalence of u, y would imply $x \sim y$ (by the transitivity of $\sim$). We conclude $u \succ y$. From $u \succ y$ and $y \sim v$ we obtain, by the same argument, $u \succ v$. It now follows from the definition of $>$ that $[u] > [v]$.

Exercise 1.6

(i) The relations $[x] > [y]$ and $[y] > [z]$ imply $x \succ y$ and $y \succ z$. We deduce $x \succ z$ to infer $[x] > [z]$. By the definition of $\succ$, the relations $x \succcurlyeq y$ and $y \succcurlyeq z$ hold, whilst the equivalences $x \sim y$, $y \sim z$ do not.
Moreover, the transitivity of $\succcurlyeq$ implies $x \succcurlyeq z$. We show that $x \sim z$ cannot hold. If it did, by definition we could use $z \succcurlyeq x$ and $x \succcurlyeq z$ to deduce $z \sim y$, a contradiction (this is because $z \succcurlyeq x$ and $x \succcurlyeq y$ imply $z \succcurlyeq y$ and we already have $y \succcurlyeq z$). It follows that $z \not\sim x$ and $x \succ z$.

(ii) Suppose that $x \not\sim y$. Since $\succcurlyeq$ is a complete preorder, the relation $x \succ y$ or the relation $y \succ x$ holds. By the definition of $>$ one of $[x] > [y]$, $[y] > [x]$ must hold.

(iii) Suppose that the application of three ranking criteria to A, B, C gives rise to the respective complete preorders $A \succ B \succ C$, $B \succ C \sim A$ and $C \succ A \succ B$. The corresponding pairwise majority comparisons lead to $A \succ B$, $B \succ C$ e $C \sim A$.

Exercise 1.7

Given the pairwise comparisons:

$$A < B \quad B < C \quad A < C$$
$$B < C \quad C < A \quad B < A$$
$$C < A \quad A < B \quad C < B.$$

interchanging the third element in the first row with the second in the second row and the third element in the thrid row with the first element in the second, we obtain:

$$A < B \quad B < C \quad C < A$$
$$C < B \quad A < C \quad B < A$$
$$C < A \quad A < B \quad B < C.$$

Because no procedure based on pairwise comparisons can discriminate between the above arrangements (they determine the same set of nine pairwise comparisons), no such procedure discriminates between transitive and non-transitive rankings.

Exercise 1.8

For instance:

$$A < B < C < D$$
$$B < C < D < A$$
$$C < D < A < B$$
$$D < A < B < C.$$

Exercise 1.9

(i) We have $b^2(A) = C$, $b^2(B) = A$ and $b^2(C) = B$.
(ii) $b^3 = bb^2$. Since b sends C into A, $bb^2(A) = A$. Similar calculations show that bb^2 fixes B and C.
(iii) The permutation ab interchanges B, C, leaving A fixed. If we apply it twice, A is kept fixed while B, C are interchanged twice, i.e. they are kept fixed. It follows that $abab$ fixes A, B, C, i.e. it coincides with e.
(iv) We note that a fixes only C, ab fixes only A and ab^2 fixes only B. These permutations must therefore be all distinct.
(v) We have $ba(A) = C$, $ba(B) = B$ and $ba(C) = A$. The only permutation of S_3 fixing B is ab^2, thus $ba = ab^2$. By part (iv), $ba = ab^2 \neq ab$.
(vi) If we had $b = b^2$, we could deduce $bb = b^2b = e$, against $b^2 \neq e$. If we had $b^2 = ab$, we could deduce $e = b^2b = ab^2$, against $ab^2 \neq e$ (ab^2 fixes only B and e fixes A, B, C). If we had $b^2 = ab^2$, we could deduce $e = b^2b = ab^2b = a$, contradicting $a \neq e$.

Exercise 1.10

Since $yzw = (yz)w$, we deduce $x(yzw) = x((yz)w)$ and we can apply the associative law to the three permutations x, yz e w, thereby reaching the following chain of equalities:

$$x(yzw) = x((yz)w) = (x(yz))w = (xyz)w = ((xy)z)w = (xy)(zw)$$

Exercise 1.11

(i) Since $b^2 = bb$, associativity yields $b^2abab^2a = b^2((abab)ba)$. By Exercise 1.9-(iii), $abab = e$: using the last equality in the preceding one, we obtain $b^2abab^2a = b^2(ba) = (b^2b)a = b^3a = a$. As for $abab^2abab^2 = abab(babab^2) = babab^2 = ba(bab^2)$, we know from Exercise 1.9-(v) that $ba = ab^2$. In view of the last equality, we obtain $abab^2abab^2 = ba(bab^2) = ba((ba)b^2) = ba(ab^2b^2) = ba((ab)b^3) = ba(ab) = b(aab) = b((aa)b) = b^2$.

(ii) The multiplicative inverses of e, a, ab, ab^2 are e, a, ab, ab^2 respectively. The multiplicative inverse of b is b^2.

Exercise 11.1

Since z has a multiplicative inverse u and $xz = yz$, $(xz)u = (yz)u$. By the associative law, $x = x(zu) = (xz)u = (yz)u = y(zu) = y$.

Exercise 11.3

(i) The nonzero components of the given vector are those of the form $d_{i,j}$ with $i = 2$ or $j = 2$. If none of the indices i, j, k equals 2, then $d_{i,j} + d_{j,k} = 0 + 0 = 0 = d_{i,k}$. If only i equals 2, $d_{i,j} + d_{i,k} = 1 + 0 = 1 = d_{i,k}$. If only k equals 2, $d_{i,j} + d_{j,k} = d_{i,j} - d_{k,i} = 0 - 1 = -1$ and $d_{i,k} = -d_{k,1} = -1$. Since i, j, k are distinct, we do not have to consider the case in which two of them are equal. Strong transitivity holds.
(ii) It suffices to note that $rd_{i,j} + rd_{j,k} = r(d_{i,j} + d_{i,k}) = rd_{ik}$.
(iii) The sum of two vectors has components of the form $d_{i,j} + f_{i,j}$. Then:

$$(d_{i,j} + f_{i,j}) + (d_{j,k} + f_{j,k}) = (d_{i,j} + d_{j,k}) + (f_{i,j} + f_{i,k}) = d_{i,k} + f_{i,k}.$$

(iv) By (ii), rd, sf are strongly transitive. By (iii), their sum is.

Exercise 11.4

If $r_1, r_2, r_3, r_4, r_5, r_6 \in \mathbb{Q}$ are all different from zero, any vector $(q_1, q_2, q_3, q_4, q_5, q_6)$ di $\mathbb{Q}^6$, satisfies the equality:

$$(q_1, q_2, q_3, q_4, q_5, q_6) = \frac{q_1}{r_1}(r_1, 0, 0, 0, 0, 0) + \ldots + \frac{q_6}{r_6}(0, 0, 0, 0, 0, r_6).$$

Consider any five vectors $v_{i_1}, \ldots, v_{i_5} \in \mathbb{Q}^6$, such that the only nonzero component of v_{i_k} is the iki-th ($i_k \in \{1, \ldots, 6\}$). There is $j \in \{1, \ldots, 6\}$ such that the j-th component of each v_{i_k} equals zero. As a result, no vector in $\mathbb{Q}^6$ whose j-th component differs from zero can be represented as a linear combination of $v_{i_1}, \ldots, v_{i_5}$.

Exercise 11.7

(i) $c_\alpha = (1, -1, 0, 1, 0, 0)$, $c_\beta = (1, 0, -1, 0, 1, 0)$, $c_\gamma = (0, 1, -1, 0, 0, 1)$.
(ii) $-c_\alpha = (-1, 1, 0, -1, 0, 0)$ corresponds to the cycle $(1, 3, 2)$. The remaining cases are entirely similari.
(iii) If $r, s \in \mathbb{Q}$, $rc_\alpha + sc_\beta = (r + s, -r, -s, r, s, 0)$. Because the sixth component of c_γ is different from zero, the last vector must differ from every linear combination of the form $rc_\alpha + sc_\beta$. Analogous considerations show that neither of c_α, c_β can be represented as a linear combination of the other two cyclic vectors.
(iv) The cyclic sequence $(1, 2, 4, 3)$ is a concatenation of the cycles $(1, 2, 4) = c_\beta$ and $(4, 3, 1) = (0, -1, 1, 0, 0, -1) = -c_\gamma$. The cyclic vector codified by $(1, 2, 4, 3)$ is $c_\beta - c_\gamma$.
(v) Let r, s, t be rational numbers and $rc_\alpha + sc_\beta + tc_\gamma$ be strongly transitive. The last vector is $(r + s, -r + t, -s - t, r, s, t)$. Strong transitivity implies $3r = s + t$, $3s = -r - t$ and $3t = r - s$, which imply:

$$3(r + s + t) = s + t - r - t + r - s = 0$$

and, consequently, $r + s + t = 0$. From $r = -s - t$ and $3r = s + t$ we deduce $4r = 0$ and, finally, $r = 0$. As a result $s = -t$. Because $3t = r - s = t$, $t = 0 = s$. It now follows that $rc_\alpha + sc_\beta + tc_\gamma$ is the vector $(0, 0, 0, 0, 0, 0)$.

Exercise 11.13

Given the vector $(r + s, -r + t, -s - t, r, s, t)$, we obtain $P(A_1) = \frac{1}{4}(r + s - r + t - s - t) = 0$, $P(A_2) = \frac{1}{4}(-r - s + r + s) = 0$, $P(A_3) = \frac{1}{4}(r - t - r + t) = 0$ and $P(A_4) = \frac{1}{4}(s + t - s - t) = 0$.

Exercise 11.14

(i) The vector determined by the given profile is:

$$(2, -2, -2, -2, -2, 2) = (1, -2, -1, -3, -2, 1) + (1, 0, -1, 1, 0, 1),$$

where the first summand on the right-hand side of the above equalities is strongly transitive, while the second summand is a linear combination of cyclic vectors corresponding to the cycle $(1, 2, 3, 4)$. We may therefore write:

$$(2, -2, -2, -2, -2, 2) = (1, -2, -1, -3, -2, 1) + (c_\alpha + c_\gamma).$$

(ii) The vector determined by the given profile is $(-1, 17, 3) = (4, 12, 8) - 5(1, -1, 1)$, where $(1, -1, 1)$ is the counterpart of c_α for three alternatives.

Exercise 11.19

(i) The set $\mathbb{N}$ equipped with addition is a monoid but not a group (it has the neutral element 0 but no additive inverses).
(ii) The set of positive even numbers equipped with addition is a semigroup but not a monoid since it does not contain the additive neutral element 0.
(iii) If 3 followed from 1 and 2, a structure that satisfies 1 and 2 should be a group. By (i), this is not the case.

Exercise 11.22

When $x = y$, property (b) ensures that for every x there are e, d such that $x + e = x = d + x$. Because the last equalities hold for every x, they must, in particular, hold for $x = e$ and $x = d$. It follows that $e = d + e = d$. The initial equalities can therefore be rewritten as $x + e = x = e + x$. The argument just given also shows that e is unique. When $y = e$, property (b) ensures that for every x there are u, v such that $x + u = e$ and $v + x = e$. Using (a), we obtain $v = v + e = v + (x + u) = (v + x) + u = e + u = u$. The initial equalities can now be rewritten as $u + x = e = x + u$. Every x has a unique inverse u. We conclude that a structure satisfying (a) and (b) must be a group.

Exercise 11.26

If G has three elements, one of them is e and we may call a, b the remaining two. We show that $G = C_3$, which is certainly abelian. Each element of G has a unique multiplicative inverse. Thus, a's inverse can only be a itself or b.

B Solutions to Exercises   383

If $aa = e$, then the inverse of b can only be b itself (if not, a would have two distinct inverses, namely a and b). Moreover, $ab = a$ cannot hold, since it implies $b = (aa)b = a(ab) = aa = e$ and $b \neq e$. We also rule out $ab = b$, which implies $a = e$.

We are forced to conclude $ab = e$, which yields, again, two distinct inverses for a. This contradiction shows that $ab = e$ and that $bb, aa \neq e$. If $aa = b$ then $a = a(ab) = bb = b^2$ and G is $C_3 = \{e, b, b^2\}$.

We are able to rule out $aa = a$ because it implies $aab = e$, i.e. that ab is the inverse of a. Since we had established that the inverse of a is b, the uniqueness of inverses leads to $ab = b$. A contradiction now follows from $ab = e$ and $b \neq e$. Thus $aa = b$ and $G = C_3$. The cyclic group of order 3 is certainly abelian because $bb^2 = bbb = (bb)b = b^2b$.

Exercise 11.28

(i) $F = \{-1\}$ generates $\mathbb{Z}$;
(ii) We reason by contradiction. Suppose that

$$F = \left\{ \frac{p_1}{q_1}, \ldots, \frac{p_n}{q_n} \right\}.$$

is a set of generators. We may assume that all denominators are positive. The set

$$F' = \frac{1}{q_1 \cdots q_n}$$

generates F, since the, iterated sums of $1/(q_1 \cdots q_n)$ or $-1/(q_1 \cdots q_n)$ with itself generate every element of F. For instance, if $p_1 > 0$, $p_1 q_2 \cdots q_n$ iterations of the sum between $\dfrac{1}{q_1 \cdots q_n}$ and itself yield p_1/q_1. It follows that $\mathbb{Q}$ is generated by a single rational number of the form $1/q$, con $q > 0$. Let us consider the smallest prime number k that is not a prime factor of q. Then $1/k$ is not generated by $\{1/q\}$ because $1/q$ only generates numbers whose denominator is q or a divisor of q. This contradiction shows that $\mathbb{Q}$ is not finitely generated.
(iii) An arbitrary, finite subset of $\mathbb{Q} - \{0\}$ has the form:

$$F = \left\{ \frac{p_1}{q_1}, \ldots, \frac{p_n}{q_n} \right\}.$$

Because the integers $p_1, \ldots, p_n, q_1, \ldots, q_n$ can be decomposed into prime factors, let A be the *finite* set of all such prime factors. There is a smallest prime not in A: call it p. If $\mathbb{Q} - \{0\}$ was finitely generated by F, every nonzero rational number could be expressed as a product of (positive or negative) powers of elements of F. Every product of this form has a numerator and denominator that admit decompositions into prime factors in A. It follows that no such product can be equal to p. Thus, F does not generate $\mathbb{Q} - \{0\}$. Because F was chosen to be an arbitrary, finite subset of $\mathbb{Q} - \{0\}$, no such set generates $\mathbb{Q} - \{0\}$. In other words, $\mathbb{Q} - \{0\}$ is not finitely generated.

Exercise 1.24

We only provide the solution to (i), since the remaining parts are equally simple. Let v_i, u_i be the i-th components of u, v, with $i \in \{1, \ldots, 6\}$. The i-th components of su, sv are su_i, sv_i respectively and the i-th component of $su + sv$ is $su_i + sv_i = s(u_i + v_i)$, which is the same as the i-th component of $s(u + v)$. This is true for every i in $\{1, \ldots, 6\}$. It follows that $s(u + v) = su + sv$.

Exercise 12.1

Yes. The action of $\mathbb{Q}$ on the elements of $\mathbb{Q}$, conceived as vectors, amounts to multiplication in $\mathbb{Q}$.

B.2 Chapter 2

Exercise 2.3

By inductive hypothesis, $t_1(x/r, y/s)$ is a τ-term. By definition of τ-term, so is $-t_1(x/r, y/s)$.

Exercise 2.4

(i) The inductive base ends just before the sentence starting with 'So far we have only used the fact ...'. The remaining part of the proof is the inductive step.
(ii) We talk of terms *simpliciter*, as opposed to τ-terms. If t is a variable or constant symbol, t^* is the empty symbol and thus not a term. If t is of the form $+ t_1 t_2$, then t^* is obtained from t by deleting a certain number of consecutive symbols

from some point on. If the deletion begins immediately after t_1 or before t_1, the resulting string is $+ t_1$ or an initial segment thereof.

In the latter case, if we do not obtain the empty string, we obtain $+ t_1^*$, where t^* is not a term by inductive hypothesis. Because $+$ must be followed by two terms, for it to produce a term, and it is followed by none when t_1^* is empty, the string $+ t_1^*$ cannot be a term. If t_1^* is a nonempty string in $+ t_1^*$, the symbol $+$ is followed by one, not two, terms: the resulting string cannot be a term either. We turn to the deletions retaining an initial, nonempty segment of t_2, say t_2^*. By inductive hypothesis, t_2^* is not a term and $+$ is not followed by two terms: the string $+ t_1 t_2^*$ cannot be a term (if we try to import symbols from t_1 into t_2^* to make it a term, the resulting initial segment of t_1 is not a term).

If $t = -t_1$, a nonempty, initial segment has the form $- t_1^*$, which by inductive hypothesis, cannot be a term, since $-$ is not followed by one term (it is followed by none).

Exercise 2.5

(i) $f_1 0 + f_{1/2} 0$ e $f_{-3} 0 + (-f_{-5/4} 0)$ are closed τ''-terms, whereas $f_1 x + f_0 y$ e $0 + x$ are not.

(ii) In this case, the τ-terms coincide with the variables.

Exercise 2.10

The set S of τ'-terms is the smallest set of τ'-strings containing $V \cup \{0, 1\}$ and closed under $+, -$ e $\cdot$. The proof of Lemma 2.2 is easily extended to τ'-terms. We have to prove that, if $t(x, y), r, s$ are τ'-terms, then $t(x/r, y/s)$ is a τ'-term.

Let S be the set of τ'-terms for which the lemma holds. The proof that $V \cup \{0, 1\} \subseteq S$ is as in Lemma 2.2 (0 and 1 are treated in the same way). Closure under $+, -$ is proved as in Lemma 2.2. By inductive hypothesis, if $t_1(x, y), t_2(x, y), r, s \in S$, then $t_1(x/r, y/s), t_2(x/r, y/s) \in S$. It follows that:

$$\cdot t_1 t_1(x/r, y/s) t_2(x/r, y/s) \in S.$$

Closure under $\cdot$ is proved and we can conclude $S = T$.

Exercise 2.16

The formal properties stated in Exercise 11.22 can be formally stated as:

(i) $\forall x \forall y \forall x (x + (y + z) = (x + y) + z)$;
(ii) $\forall x \forall y (\exists u(x + y = y) \land \exists z(z + x = y))$ or, equivalently, $\forall x \forall y \exists u \exists z(x + y = y \land z + x = y)$.

The formal properties stated in Exercise 1.24 can be formally stated as:

(i) $\forall x_1 \forall x_2 (f_s(x_1 + x_2) = f_s x_1 + f_s x_2)$, per ogni $s \in \mathbb{Q}$;
(ii) $\forall x_1 ((f_{r+s} x_1 = f_r x_1 + f_s x_1)$, per ogni $r, s \in \mathbb{Q}$;
(iii) $\forall x_1 (f_{rs} x_1 = f_r(f_s x_1))$, per ogni $r, s \in \mathbb{Q}$;
(iv) $\forall x_1 (f_1 x_1 = x_1)$.

In (i)–(iii) we must take into account an infinitely large set of $\mathcal{L}$-formulae.

Exercise 2.17

(i) The $\mathcal{L}_=$-formula $\exists x_1 \exists x_2 (\neg x_1 = x_2)$ states the existence of two distinct elements. The $\mathcal{L}_=$-formula $\exists x_1 \exists x_2 \exists x_3 (\neg(x_1 = x_2) \land \neg(x_2 = x_3) \land \neg(x_1 = x_3))$ states the existence of three distinct elements.
(ii) $\forall x_1 \exists x_2 (x_2 < x_1) \land \forall x_3 \exists x_4 (x_3 < x_4)$.
(iii) $\forall x_1 \forall x_2 (\neg(x_1 = x_2) \rightarrow \exists x_3 (x_1 < x_3 \land x_3 < x_2))$;
(iv) $\forall x (\forall y (x + y = y + x)) \rightarrow \forall x \forall y (x + y = y + x)$;
(v) $\exists x_1 (\neg(x_1 = 0) \land (x_1 + x_1 + x_1 + x_1 = 0) \land \neg(0 = x_1 + x_1) \land \neg(0 = x_1 + x_1 + x_1))$.

Exercise 2.18

We only consider the sentence $\forall x \forall y (\neg(x = y) \rightarrow (x < y \lor y < x))$. Its subformulae are the original formula itself, $\forall y (\neg(x = y) \rightarrow (x < y \lor y < x))$, $\neg(x = y) \rightarrow (x < y \lor y < x)$ and the subformulae of $\neg(x = y), (x < y \lor y < x)$.

Exercise 2.21

(i)

$$\ell(\forall x(x < y \rightarrow \exists x_2(y < x_2 \land x_2 < x)) \lor x_2 < y) =$$
$$= \ell((\forall x(x < y \rightarrow \exists x_2(y < x_2) \cup \ell(x_2 < y)$$

B Solutions to Exercises

$$
\begin{aligned}
&= \ell((\forall x(x < y \to \exists x_2(y < x_2) \cup \{x_2, y\} \\
&= (\ell(x < y \to \exists x_2(y < x_2)) - \{x\}) \cup \{x_2, y\} \\
&= ((\ell(x < y) \cup \to \ell(\exists x_2(y < x_2))) - \{x\}) \cup \{x_2, y\} \\
&= ((\{x, y\} \cup (\ell(y < x_2) - \{x_2\})) - \{x\}) \cup \{x_2, y\} \\
&= ((\{x, y\} \cup (\{y, x_2\} - \{x_2\})) - \{x\}) \cup \{x_2, y\} \\
&= ((\{x, y\} \cup \{y\}) - \{x\}) \cup \{x_2, y\} \\
&= \{y\} \cup \{x_2, y\} \\
&= \{x_2, y\}.
\end{aligned}
$$

(ii)

$$
\begin{aligned}
\ell(P(y) \to \exists y(P(y) \to (x = z))) &= \ell(P(y)) \cup \ell(\exists y(P(y) \to (x = z)))) \\
&= \{y\} \cup \ell(\exists y(P(y) \to (x = z)))) \\
&= \{y\} \cup (\ell(P(y) \to (x = z)) - \{y\}) \\
&= \{y\} \cup (\ell(P(y)) \cup \ell(x = z)) - \{y\}) \\
&= \{y\} \cup ((\{y\} \cup \{x, z\}) - \{y\}) \\
&= \{y\} \cup \{x, z\} \\
&= \{x, y, z\}.
\end{aligned}
$$

(iii)

$$
\begin{aligned}
\ell(\forall x_1 \exists x_2 \forall x_3(x_1 + x_2 &= x_2) \to \exists y(y + e < e)) \\
&= \ell(\forall x_1 \exists x_2 \forall x_3(x_1 + x_2 = x_2)) \cup \ell(\exists y(y + e < e)) \\
&= (((\ell(x_1 + x_2 = x_2) - \{x_1\}) - \{x_2\}) - \{x_3\}) \cup (\{y\} - \{y\}) \\
&= ((((\{x_1, x_2\} - \{x_1\}) - \{x_2\}) - \{x_3\}) \cup \emptyset \\
&= ((\{x_2\} - \{x_2\}) - \{x_3\}) \\
&= \emptyset - \{x_3\}) \\
&= \emptyset.
\end{aligned}
$$

Exercise 2.30

(i) $\sigma_1(x) = 0, \sigma_1(y) = 1; \sigma_2(x) = 1, \sigma_2(y) = 0; \sigma_3(x) = -2, \sigma_3(y) = 3$.

(ii) Consider the $\mathcal{L}_s$-structure $(\mathbb{N}, s^{\mathbb{N}}, \mathbf{0})$, where $s^{\mathbb{N}}$ is the successor function determined by the equality $s^{\mathbb{N}}(n) = n+1$. The closed $\mathcal{L}_s$-terms $0, s0, ss0, sss0, \ldots$ name $0, 1, 2, 3, \ldots$.

(iii) By hypothesis, $\sigma(x) = \sigma'(x)$ for every variable x. Constant symbols have a fixed interpretation. If f is a k-ary function symbol naming the function $\mathbf{f}$ in a given structure and $t_1, \ldots, t_k$ are terms satisfying the equalities $\sigma(t_i) = \sigma'(t_i)$, $i = 1, \ldots, k$, condition $\star$ ensures:

$$\sigma(ft_1 \ldots t_k) = \mathbf{f}\sigma(t_1) \ldots \sigma(t_k) = \mathbf{f}\sigma'(t_1) \ldots \sigma'(t_k) = \sigma'(ft_1 \ldots t_k).$$

Exercise 2.31

If $t(x_1, \ldots, x_n)$ is a constant symbol, there is nothing to prove. If $t(x_1, \ldots, x_n)$ is a variable, then it is one of $x_1, \ldots, x_n$. If $t(x_1, \ldots, x_n) = x_i$, with $i \in \{x_1, \ldots, x_n\}$, then $t^{M,\sigma}(x_1, \ldots, x_n) = \sigma(x_i) = a_i = s_i^{M,\sigma} = t^{M,\sigma}(x_1/s_1, \ldots, x_n/s_n)$. The inductive step requires that we consider a term $t(x_1, \ldots, x_n)$ obtained by applying the k-ary function symbol f to k terms $t_1(x_1, \ldots, x_n), \ldots t_k(x_1, \ldots, x_n)$. By inductive hypothesis, for $i \in \{1, \ldots, k\}$, $t_1^{M,\sigma}(x_1, \ldots, x_n) = t_1^{M,\sigma}(x_1/s_1, \ldots, x_n/s_n)$. It suffices to note that:

$$(ft_1(x_1, \ldots, x_n) \ldots t_k(x_1, \ldots, x_n))^{M,\sigma}$$
$$= f^M t_1^{M,\sigma}(x_1, \ldots, x_n) \ldots t_k^{M,\sigma}(x_1, \ldots, x_n)$$
$$= f^M t_1^{M,\sigma}(x_1/s_1, \ldots, x_n/s_n) \ldots t_k^{M,\sigma}(x_1/s_1, \ldots, x_n/s_n)$$
$$= (ft_1(x_1/s_1, \ldots, x_n/s_n) \ldots t_k((x_1/s_1, \ldots, x_n/s_n))^{M,\sigma}.$$

The induction is complete.

To illustrate the result just proved, consider the setting of Exercise 2.30-(ii) and the $\mathcal{L}_s$-term $ss(x_1)$. Assigning the number 2 to the variabile x_1 is equivalent to assigning 0 to x_1 and evaluating $ss(x_1)$ in $\mathbb{N}$.

For another illustration, consider S_3, regarded as a $\mathcal{L}_g$-structure. Assigning b^2 to x_1 is equivalent to assigning b to x_1 and evaluating $x_1 + x_1$ in S_3.

Exercise 2.40

(i) We must show two things. First, that $\mathcal{M} \models \forall x \neg \phi[\sigma]$ implies $\mathcal{M} \models \neg \exists x \phi[\sigma]$. Second, that $\mathcal{M} \models \neg \exists x \phi[\sigma]$ implies $\mathcal{M} \models \forall x \neg \phi[\sigma]$. In the first case, the hypothesis $\mathcal{M} \models \forall x \neg \phi[\sigma]$ and Definition 2.32-(6) yield $\mathcal{M} \models \neg \phi[\sigma(x/a)]$

for every $a \in M$. In other words, for every $a \in M$, $\mathcal{M} \not\models \phi[\sigma]$. If we had $\mathcal{M} \models \exists x \phi[\sigma]$, we would reach a contradiction. Thus $\mathcal{M} \not\models \exists x \phi[\sigma]$ or, what amounts to the same thing, $\mathcal{M} \models \neg \exists x \phi[\sigma]$. The argument just given can be run backward.

(ii) By (i), $\mathcal{M} \models \neg \exists x (\neg \phi)[\sigma]$ iff $\mathcal{M} \models \forall x \neg (\neg \phi)[\sigma]$ and it is easy to verify the further equivalence with $\mathcal{M} \models \forall x \phi[\sigma]$.

(iii) By (i) $\mathcal{M} \models \neg \forall x \neg \phi[\sigma]$ iff $\mathcal{M} \models \neg \neg \exists x \phi[\sigma]$ and it is easy to verify the further equivalence with $\mathcal{M} \models \exists x \phi[\sigma]$.

Exercise 2.41

Given $v(\phi) = 1$ and $v(\psi) = 0 = v(\chi)$ we compute (i):

$$v(\psi \vee (\chi \rightarrow \neg \psi)) = max\{v(\psi), v(\chi \rightarrow \neg \psi)\}$$
$$= max\{0, max\{1 - v(\chi), v(\neg \psi)\}\}$$
$$= max\{0, max\{1, 1\}\}$$
$$= max\{0, 1\}$$
$$= 1.$$

(ii) $v(\chi \wedge (\psi \vee \chi)) = 0$; (iii) $v(\chi \rightarrow (\psi \vee (\neg \phi \wedge \chi))) = 1$.

Exercise 2.43

We only give truth-tables for (i) and (v).

ϕ	$\neg \phi$
1	0
0	1

ϕ	ψ	$(\phi \rightarrow \psi)$	$\wedge$	$(\psi \rightarrow \phi)$
1	1	1	**1**	1
0	1	1	**0**	0
1	0	0	**0**	1
0	0	1	**1**	1

Exercise 2.44

Since $\phi \equiv \psi$ amounts to the fact that, in every $\mathcal{L}$-structure, the $\mathcal{L}$-formulae ϕ, ψ have the same evaluation, it suffices to check that the truth-tables of ϕ and ψ coincide. We only check (a) and (e) below.

ϕ	ψ	$\neg$	$(\neg\phi$	$\wedge$	$\neg\psi)$	$\phi \vee \psi)$
1	1	**1**	0	0	0	**1**
0	1	**1**	1	0	0	**1**
1	0	**1**	0	0	1	**1**
0	0	**0**	1	1	1	**0**.

ϕ	ψ	χ	ϕ	$\wedge$	$(\psi \vee \chi)$	$(\phi \wedge \psi)$	$\vee$	$(\phi \wedge \chi)$
1	1	1	1	**1**	1	1	**1**	1
1	1	0	1	**1**	1	1	**1**	0
1	0	1	1	**1**	1	0	**1**	1
1	0	0	1	**0**	0	0	**0**	0
0	1	1	0	**0**	1	0	**0**	0
0	1	0	0	**0**	1	0	**0**	0
0	0	1	0	**0**	1	0	**0**	0
0	0	0	0	**0**	0	0	**0**	0

Exercise 2.45

$v(\phi \leftrightarrow \psi) = 1$ iff $v(\phi) = v(\psi)$ iff $\phi \equiv \psi$.

Exercise 2.46

In order to work with valuations on arbitrary formulae, it is necessary to fix not only a $\mathcal{L}$-structure $\mathcal{M}$ but also an assignment σ. Relative to fixed $\mathcal{M}, \sigma$, the discussion of valuations given in the text applies without changes to formuale.

Exercise 2.47

If ϕ, ψ are $\mathcal{L}$-formulae, $\phi \models \psi$ iff for every $\mathcal{L}$-structure $\mathcal{M}$ and every assignment σ, $\mathcal{M} \models \phi[\sigma]$ iff $\mathcal{M} \models \psi[\sigma]$.

Exercise 2.50

(i) $\neg\phi \vee \neg\psi$; (ii) $(\neg\phi \wedge \neg\psi) \vee (\phi \wedge \psi)$. The formulae ϕ and $\phi \vee \phi$ are distinct disjunctive normal forms of ϕ.

Exercise 2.52

We only discuss (i), (iii), (vii).

In (i), suppose $\mathcal{M} \models \exists x(\phi(x) \vee \psi(x))[\sigma]$. By Definition 2.32-(7), $\mathcal{M} \models \phi(x) \vee \psi(x)[\sigma(x/a)]$, for some $a \in M$. Then $\mathcal{M} \models \phi(x)[\sigma(x/a)]$ or $\mathcal{M} \models \psi(x)[\sigma(x/a)]$. In the first case, by Definition 2.26-(7), $\mathcal{M} \models \exists x\phi(x)[\sigma]$. Definition 2.32-(3) yields $\mathcal{M} \models \exists x\phi(x) \vee \exists x\psi(x)[\sigma]$. An almost identical argument applies in the second case.

In (iii), suppose $\mathcal{M} \models \forall x\phi(x) \vee \forall x\psi(x)$ and fix $a \in M$. If $\mathcal{M} \models \forall x\phi(x)$ then $\mathcal{M} \models \phi(x)[\sigma]$ whenever $\sigma(x) = a$. Consequently, $\mathcal{M} \models \phi(x) \vee \psi(x)[\sigma]$. This is the case for any $a \in M$, so $\mathcal{M} \models \forall x(\phi(x) \vee \psi(x))$. Because the same argument works if $\mathcal{M} \models \forall x\psi(x)$, our conclusion follows in any case.

In (vii), suppose $\mathcal{M} \models \forall x\phi(x) \to \psi[\sigma]$. Our goal is to prove $\mathcal{M} \models \phi(x) \to \psi[\sigma(x/a)]$, for some $a \in M$. Toward a contradiction, let us assume that no $a \in M$ works. Then $\mathcal{M} \models \phi(x) \wedge \neg\psi[\sigma(x/a)]$, for every $a \in M$. In particular, $\mathcal{M} \models \forall x\phi(x)[\sigma]$ and $\mathcal{M} \models \neg\psi[\sigma(x/a)]$. Because $x \notin \ell(\psi)$, $\mathcal{M} \models \neg\psi[\sigma]$ (since the assignments σ and $\sigma(x/a)$ coincide on ψ). We deduce $\mathcal{M} \models \forall x\phi(x) \to \psi[\sigma]$ and $\mathcal{M} \models \forall x\phi(x) \wedge \neg\psi[\sigma]$. The last satisfaction relation is equivalent to $\mathcal{M}\neg(\forall x\phi(x) \to \psi)[\sigma]$. We have reached a contradiction, which implies that $\mathcal{M} \models \phi(x) \to \psi[\sigma(x/a)]$, for some $a \in M$, as desired. By Definition 2.32-(7), $\mathcal{M} \models \exists x(\phi(x) \to \psi)[\sigma]$.

Exercise 2.54

If two closed formulae ϕ, ψ are equivalent, then $\phi \models \psi$ and $\phi \models \psi$. Suppose that, in some $\mathcal{L}$-structure $\mathcal{M}$, $v(\phi) = 1$. This is to say that $\mathcal{M} \models \phi$ and, consequently, $\mathcal{M} \models \psi$. It follows that $v(\psi) = 1$. If, on the other hand, $v(\phi) = 0$, then $\mathcal{M} \models \neg\phi$ and, since $\psi \models \phi$, we can't have $\mathcal{M} \models \psi$ (or $\mathcal{M} \models \phi, \neg\phi$). We must therefore have $\mathcal{M} \models \neg\psi$, i.e. $v(\psi) = 0$. We have shown that $v(\phi) = v(\psi)$ in an arbitrary $\mathcal{L}$-structure $\mathcal{M}$. Thus $\phi \equiv \psi$ in the sense of Sect. 2.3. If, on the other hand $\phi \equiv \psi$, then $v(\phi) = v(\psi)$ in every $\mathcal{L}$-structure $\mathcal{M}$. It follows that $\mathcal{M} \models \phi$ implies $v(\phi) = 1 = v(\psi)$ and $\mathcal{M} \models \psi$: in other words, $\phi \models \psi$. By the same argument, $\psi \models \phi$.

Exercise 2.55

(i) The $\mathcal{L}_=$-formula $\phi(x) = \exists u(\neg(u = x))$ in every domain with at least two elements, whereas $\phi(u)$, i.e. the $\mathcal{L}_=$-formula $\exists u(\neg(u = u))$, is never satisfied. These two formulae cannot therefore be equivalent.
(ii) We suppose $\mathcal{M} \models \phi(x, y)[\sigma]$ and set $\sigma(x) = a$, $\sigma(y) = b$. The formula $\phi(u, v)$ contains free occurrences of u, v that correspond to those of x, y. This

is to say that no free occurrence of x is turned into a bound occurence of u and the same holds for y, v. This arrangement disposes of the problem exemplified by (i). If an assignment σ of $a, b \in M$ to x, y satisfies $\phi(x, y)$ in $\mathcal{M}$, an assignment that takes the same values on u, v is otherwise identical to σ, must satisfy $\phi(u, v)$ in $\mathcal{M}$. An induction on the complexity of ϕ, which we omit, provides a more structured argument.

(iii) Suppose $\mathcal{M} \models \forall x \exists y \phi(x, y)[\sigma]$. Then, for every $a \in M$ there is $b \in M$ such that $\mathcal{M} \models \phi(x, y)[\sigma(x/a, y/b)]$. The last relation holds iff $\mathcal{M} \models \phi(u, v)[\sigma(x/a, y/b, u/a, v/b)]$. Note that the assignments $\sigma(x/a, y/b, u/a, v/b)$ and $\sigma(u/a, v/b)$ coincide on $\phi(u, v)$, in which x, y do not occur free. Definition 2.32-(7) allows us to conclude that, for every $a \in M$, $\mathcal{M} \models \exists v \phi(u, v)[\sigma(u/a)]$. Part (6) of the same definition yields $\mathcal{M} \models \forall u \exists v \phi(u, v)[\sigma]$.

(iv) Suppose $\phi \equiv \psi$ and $\mathcal{M} \models \forall x \phi[\sigma]$, i.e. that, for every $a \in M$, $\mathcal{M} \models \phi[\sigma(x/a)]$. By hypothesis, $\mathcal{M} \models \psi[\sigma(x/a)]$, i.e. $\mathcal{M} \models \forall x \phi[\sigma]$. The same argument run backward leads to $\forall x \phi \equiv \forall x \psi$. The last equivalence holds for any two ϕ, ψ such that $\phi \equiv \psi$. It is easy to verify that $\phi \equiv \psi$ iff $\neg \phi \equiv \neg \psi$, which yields $\forall x \neg \phi \equiv \forall x \neg \psi$ by what we just proved. In this case, we also have $\neg \forall x \neg \phi \equiv \neg \forall x \neg \psi$, i.e. by a previous exercise, $\exists x \phi \equiv \exists x \psi$.

(v) Set $\exists y \phi = \alpha$ e $\exists u \forall v \psi = \beta$. First, we show $\forall x \alpha \vee \beta \equiv \forall x (\alpha \vee \beta)$. To this end, suppose $\mathcal{M} \models \forall x \alpha \vee \beta[\sigma]$. We must consider two cases. In the first case, $\mathcal{M} \models \beta[\sigma]$ and we can deduce $\mathcal{M} \models \alpha \vee \beta[\sigma]$. Since $x \notin \ell(\beta)$, no value of $\sigma(x)$ disrupts the satisfiability of β: it follows that $\mathcal{M} \models \alpha \vee \beta[\sigma(x/a)]$, for every $a \in M$. We obtain $\mathcal{M} \models \forall x (\alpha \vee \beta)[\sigma]$. In the second case, $\mathcal{M} \models \forall x \alpha[\sigma]$, i.e. $\mathcal{M} \models \alpha[\sigma(x/a)]$, for every $a \in M$. Again, no value of $\sigma(x)$ disrupts the satisfiability of β. Using the hypothesis $\mathcal{M} \models \forall x \alpha \vee \beta[\sigma]$ we can therefore conclude $\mathcal{M} \models \forall x \alpha \vee \beta[\sigma(x/a)]$, for every $a \in M$. It now follows that $\forall x \alpha \vee \beta \models \forall x (\alpha \vee \beta)$.

Suppose $\mathcal{M} \models \forall x (\alpha \vee \beta)[\sigma]$: then $\mathcal{M} \models \alpha \vee \beta[\sigma(x/a)]$, for every $a \in M$. If $\mathcal{M} \models \beta[\sigma(x/a)]$, then $\mathcal{M} \models \beta[\sigma]$ and we reach the desired result. If, on the other hand, $\mathcal{M} \models \alpha[\sigma(x/a)]$, for every $a \in A$, we obtain $\mathcal{M} \models \forall x \alpha$, which also yields the desired result.

We now prove $\exists y \phi \vee \beta \equiv \exists y (\phi \vee \beta)$. On the one hand, the hypothesis $\mathcal{M} \models \exists y \phi \vee \beta[\sigma]$ leads to an alternative between $\mathcal{M} \models \exists y \phi[\sigma]$ and $\mathcal{M} \models \beta[\sigma]$. In the second case we run an argument already described, relying upon $y \notin \ell(\phi)$. In the first case, we have $\mathcal{M} \models \phi[\sigma(y/b)]$, for some $b \in M$. Because $\sigma(y)$ does not affect the satisfiability of β, we easily obtain $\mathcal{M} \models \exists y (\phi \vee \beta)[\sigma]$. Analogous considerations enable us to show that $\mathcal{M} \models \exists y (\phi \vee \beta)[\sigma]$ implies $\mathcal{M} \models \exists y \phi \vee \beta[\sigma]$.

Making use of the equivalence $\exists y (\phi \vee \beta) \equiv \exists y \phi \vee \beta$ and of (iv), we arrive at $\forall x \exists y (\phi \vee \beta) \equiv \forall x (\exists y \phi \vee \beta)$. Since we know that $\forall x (\exists y \phi \vee \beta) \equiv \forall x \exists y \phi \vee \beta$, we conclude:

$$\forall x \exists y (\phi \vee \beta) \equiv \forall x \exists y \phi \vee \beta.$$

A similar argument establishes the equivalence $\exists u \forall v (\phi \vee \psi) \equiv \phi \vee \exists u \forall v \psi$.
By (iv) again, we deduce $\exists y \exists u \forall v (\phi \vee \psi) \equiv \exists y (\phi \vee \exists u \forall v \psi) \equiv \exists y \phi \vee \exists u \forall v \psi$.
One further application of (iv) yields:

$$\forall x \exists y \exists u \forall v (\phi \vee \psi) \equiv \exists y (\phi \vee \exists u \forall v \psi) \equiv \forall x (\exists y \phi \vee \exists u \forall v \psi) \equiv \forall x \exists y \phi \vee \exists u \forall v \psi.$$

Exercise 2.57

Let $\mathcal{M} = (\mathbb{N}, R^{\mathbb{N}})$ be a $\mathcal{L}$-structure and let $\sim$ denote the binary relation determined by the condition $(m, n) \in R^{\mathbb{N}}$ iff $m \cdot n = 0$. Then $\sim$ refers to a relation that is symmetric and transitive but not reflexive. Reflexivity fails because $(0, 0) \notin R^{\mathbb{N}}$. If, by contrast, $\mathcal{M} = (\mathbb{N}, \leq^{\mathbb{N}})$ is a $\mathcal{L}$-struttura in which $\sim$ denotes the relation $\leq^{\mathbb{N}}$, i.e. the familiar ordering of $\mathbb{N}$, then $\sim$ names a reflexive and transitive relation that is not symmetric (since, e.g. $2 \leq 3$ but it is not the case that $3 \leq 2$). If, $\mathcal{M} \models \forall x \exists y (x \sim y)$ and $\sim^M$ is symmetric and transitive, consider any $a \in M$. We must have $\mathcal{M} \models \exists y (x \sim y)[\sigma(x/a)]$ for any assignment whose value at x is a. We also have $\mathcal{M} \models x \sim y[\sigma(x/a, y/b)]$. By symmetry $\mathcal{M} \models y \sim x[\sigma(x/a, y/b)]$. Using transitivity, we can infer from $a \sim^M b$ e $b \sim^M a$ the relation $a \sim^M a$. Thus $\mathcal{M} \models x \sim s[\sigma(x/a)]$, which, since a was chosen arbitrarily, allows us to conclude $\mathcal{M} \models \forall x (x \sim x)$.

Exercise 2.58

(i) In $(\mathbb{Z}, \oplus, 0)$, $1 \oplus (1 \oplus 3) = 1 \oplus 4 = 5$ and $(1 \oplus 1) \oplus 3 = 0 \oplus 3 = 3$. The operation $\oplus$ is not associative but admits the neutral element 0. Moreover, every $n \in \mathbb{Z}$ has the inverso $-n$ relative to $\oplus$;

(ii) If $\mathcal{M} = (\mathbb{N}, +, 0)$, the given binary operation is associative and has a neutral element, but admits no additive inverses for numbers different from 0;

(iii) Let $\mathcal{N} = (A, \odot, 1)$, where $A = \{e, a, b\}$. We define the operation $\odot$ by listing its nine values. First, $e = a \odot a = b \odot b = e \odot e$. Then each element of A is its own inverse. Next, we let e be the neutral element relative to $\odot$. Finally, we stipulate $a \odot b = b$ and $b \odot a = a$. It follows that $b \odot (a \odot b) = b \odot b = e \neq b = a \odot b = (b \odot a) \odot b$, i.e. associativity fails.

Exercise 2.59

If $\phi \equiv \psi$, $\mathcal{M} \models \psi$ iff $\mathcal{M} \models \psi$ even as we restrict attention to the models of T. Thus ϕ, ψ are T-equivalent.

Exercise 2.61

Suppose $x \in \ell(\psi)$ and let $\psi(x)$ be T-equivalent to a universal $\mathcal{L}$-formula $\forall y_1 \ldots \forall y_m \alpha(x)$. Toward a contradiction, we assume that $\exists x \psi(x)$ is not T-equivalent to any universal $\mathcal{L}$-formula. There must exist a structure $\mathcal{M} \models T$ such that, for every universal formula χ, there is an assignment $\sigma(x/a)$ with which $\psi(x)$ is satisfied while χ is not, or viceversa. Because $\forall y_1 \ldots \forall y_m \alpha(x)$ is a universal $\mathcal{L}$-formula, there is $\tau(x/a)$ such that $\mathcal{M} \models \psi(x)[\tau(x/a)]$ and $\mathcal{M} \models \neg \forall y_1 \ldots \forall y_m \alpha(x)[\tau(x/a)]$ or, alternatively, $\mathcal{M} \models \neg \psi(x)[\sigma(x/a)]$ and $\mathcal{M} \models \forall y_1 \ldots \forall y_m \alpha(x)[\sigma(x/a)]$. Since $\psi(x)$ and $\forall y_1 \ldots \forall y_m \alpha(x)$, are T-equivalent, either way we reach the contradiction $\mathcal{M} \models \psi(x) \wedge \neg \psi(x)[\sigma(x/a)]$. It follows that $\exists x \psi(x)$ is T-equivalent to a universal $\mathcal{L}$-formula.

Exercise 2.62

Consider an arbitrary $\mathcal{L}$-formula φ, T-equivalent to a universal $\mathcal{L}$-formula χ. It follows that $\neg \chi$ is T-equivalent to an existential $\mathcal{L}$-formula. Since $\neg \chi$ is a $\mathcal{L}$-formula, it is T-equivalent to a universal formula α by hypothesis. This means that $\neg \alpha$ is existential and T-equivalent to $\neg \neg \chi$, in turn equivalent to χ and, as a consequence, T-equivalent to φ.

Exercise 2.64

If $\varphi \equiv \varphi'$ and $\mathcal{M} \neg \varphi[\sigma]$, then $\mathcal{M} \neg \varphi'[\sigma]$, since $\mathcal{M} \models \varphi'[\sigma]$ and $\varphi' \models \varphi$ imply $\mathcal{M} \models \varphi$. Thus $\neg \varphi \models \neg \varphi'$. Interchanging the roles of φ, φ' in the preceding argument, we obtain $\neg \varphi' \models \neg \varphi$. We can now conclude $\neg \varphi \equiv \neg \varphi'$. Having proved the inductive step for $\neg$ and $\vee$, we treat $\wedge$ and $\rightarrow$ using the equivalences $\phi \wedge \psi \equiv \neg(\neg \phi \wedge \neg \psi)$ and $\phi \rightarrow \psi \equiv \neg \phi \vee \psi$. Given the proof of the inductive step for $\exists$, we know that $\psi \equiv \chi$ implies $\neg \exists \neg \phi \equiv \neg \exists \neg \phi'$, which yields $\forall x \phi \equiv \forall x \phi'$.

Exercise 2.65

(i)

$$\forall x \forall y (x = y \rightarrow \forall z (x = z \wedge x = x))$$
$$\equiv \forall x \forall y (x = y \rightarrow \forall z \neg(\neg(x = z) \vee \neg(x = x)))$$
$$\equiv \forall x \forall y (x = y \rightarrow \neg \exists z (\neg(x = z) \vee \neg(x = x)))$$

$$\equiv \forall x \forall y (\neg(x = y) \lor \neg \exists z (\neg(x = z) \lor \neg(x = x)))$$
$$\equiv \neg \exists x \exists y \neg (\neg(x = y) \lor \neg \exists z (\neg(x = z) \lor \neg(x = x))).$$

(ii)

$$\forall x \neg (x = z) \land \forall y (x = y \to \exists z (x = x))$$
$$\equiv \neg \exists x (x = z) \land \forall y (x = y \to \exists z (x = x))$$
$$\equiv \neg \exists x (x = z) \land \forall y (\neg(x = y) \lor \exists z (x = x))$$
$$\equiv \neg \exists x (x = z) \land \neg \exists y \neg (\neg(x = y) \lor \exists z (x = x))$$
$$\equiv \neg (\exists x (x = z) \lor \exists y \neg (\neg(x = y) \lor \exists z (x = x))).$$

Exercise 2.67

(i)

$$\forall x (x = y \to \exists z \forall x (z = x)) \equiv \forall x (\neg(x = y) \lor \exists z \forall x (z = x))$$
$$\equiv \forall x (\neg(x = y) \lor \exists z \forall u (z = u))$$
$$\equiv \forall x \exists z (\neg(x = y) \lor \forall u (z = u))$$
$$\equiv \forall x \exists z \forall u (\neg(x = y) \lor (z = u)).$$

(ii)

$$\forall x \exists y (x = y) \to \exists y \forall x (x = y) \equiv \neg \forall x \exists y (x = y) \lor \exists y \forall x (x = y)$$
$$\equiv \exists x \forall y \neg (x = y) \lor \exists y \forall x (x = y)$$
$$\equiv \exists x \forall y \neg (x = y) \lor \exists u \forall v (v = u)$$
$$\equiv \exists x \forall y \exists u \forall v (\neg(x = y) \lor (v = u)).$$

The same formula may be equivalent to distinct formulae in prenex form. For instance, $\forall x (x = y) \to \exists z (x = z)$ is equivalent to $\exists x \exists z (x = y \to x = z)$ and to $\exists z \exists x (x = y \to x = z)$.

Exercise 2.68

Consider an atomic $\mathcal{L}$-formula $Rt_1 \ldots t_k$, where R is a k-ary relation symbol and the variables occurring in $t_1, \ldots, t_k$ belong to the set $\{x_1, \ldots, x_n\}$. Since σ, σ' coincide on these variables, $t_i^{M,\sigma} = t_i^{M,\sigma'}$ with $i, \ldots, k$. Consequently, $\mathcal{M} \models Rt_1 \ldots t_k[\sigma]$ iff $(t_1^{M,\sigma}, \ldots, t_k^{M,\sigma}) \in R^M$ iff $(t_1^{M,\sigma'}, \ldots, t_k^{M,\sigma'}) \in R^M$ iff $\mathcal{M} \models Rt_1 \ldots t_k[\sigma']$.

As for the inductive step, we restrict attention to $\neg, \vee$ and $\exists$. The inductive hypothesis is that $\mathcal{M} \models \phi[\sigma]$ iff $\mathcal{M} \models \phi[\sigma']$, *for every* pair of assignment whose values coincide on $\{x_1, \ldots, x_n\}$ and the same is true of ψ. Recall that the free variables occurring in ϕ and ψ belong to $\{x_1, \ldots, x_n\}$. The inductive hypotheis implies $\mathcal{M} \models \neg\phi[\sigma]$ iff $\mathcal{M} \models \neg\phi[\sigma']$. Moreover, if $\mathcal{M} \models \phi \vee \psi[\sigma]$, Definition 2.32-(3) implies $\mathcal{M} \models \phi[\sigma]$ or $\mathcal{M} \models \psi[\sigma]$. In the first case, the inductive hypothesis yields $\mathcal{M} \models \phi[\sigma']$, which, in turn, implies $\mathcal{M} \models \phi \vee \psi[\sigma']$ (by Definition 2.32-(3)). The second case is nearly identical. We conclude:

$$\mathcal{M} \models \phi \vee \psi[\sigma] \, implies \, \mathcal{M} \models \phi \vee \psi[\tau].$$

The opposite implication follows from a similar argument.

We finally consider $\exists x_i \phi(x_i)$ (if $x \notin \{x_1, \ldots, x_n\}$, $\exists x \phi \equiv \phi$ and the proof would be complete with one appeal to the inductive hypothesis). If $\mathcal{M} \models \exists x_i \phi(x_i)[\sigma]$, then $\mathcal{M} \models \phi(x_i)[\sigma(x_i/a)]$, for some $a \in M$. Because σ, σ' coincide on $\{x_1, \ldots, x_n\}$, so do $\sigma(x_i/a)$ and $\sigma'(x_1/a)$. Thus $\mathcal{M} \models \phi(x_i)[\tau(x_1/a)]$ sse $\mathcal{M} \models \exists x_i \phi(x_i)[\tau]$. We obtain:

$$\mathcal{M} \models \exists x_i \phi(x_i)[\sigma] \text{ implica } \mathcal{M} \models \exists x_i \phi(x_i)[\tau].$$

Interchanging the roles of σ, σ' in the last argument, we obtain the converse implication.

Exercise 2.74

The $\mathcal{L}_r$-formula $\exists y(x + (y \cdot y) = z)$ works.

Exercise 2.75

The required defining formulae are, respectively, $\phi \wedge \psi, \phi \vee \psi, \neg\phi$ and $(\phi \wedge \neg\psi) \vee (\psi \wedge \neg\phi)$.

Exercise 2.76

Let $\psi(x_1, x_2, \ldots, x_{n+1})$ be the formula $x_1 = x_1 \wedge \phi_X(x_2, \ldots, x_{n+1})$. Then $\psi(x_1, x_2, \ldots, x_{n+1})$ defines $M \times X$.

B Solutions to Exercises

Exercise 2.77

The formula $\exists y \phi_X(x_1, \ldots, x_n, y)$ defines $\pi[X]$.

Exercise 2.82

We stipulate that $\overline{x}$ designates a n-tuple of variables. Let $\eta(\overline{x}, \overline{y}, \overline{z})$ be the conjunction of the following three formulae:

- $\phi_=(\overline{x}, \overline{x})$;
- $\phi_=(\overline{x}, \overline{y}) \to \phi_=(\overline{y}, \overline{x})$;
- $(\phi_=(\overline{x}, \overline{y}) \wedge \phi_=(\overline{y}, \overline{z})) \to \phi_=(\overline{x}, \overline{z})$.

Then the admissibility condition (i) is the sentence:

$$\forall \overline{x} \forall \overline{y} \forall \overline{x} (\partial_\Gamma(\overline{x}) \wedge \partial_\Gamma(\overline{y}) \wedge \partial_\Gamma(\overline{z}) \to \eta(\overline{x}, \overline{y}, \overline{z})).$$

The admissibility condition (ii) is the sentence:

$$\forall \overline{x} \forall \overline{y} ((\partial_\Gamma(\overline{x}) \wedge \partial_\Gamma(\overline{y}) \wedge \phi_=(\overline{x}, \overline{y}) \wedge \phi_P(\overline{x})) \to \phi_P(\overline{y})).$$

The remaining admissibility conditions are easily formalised in an analogous manner. We only verify that (i) and (ii) for R^+ hold in Example 2.79.

Since reflexivity and symmetry are trivial, we consider pairs of integers $(m, n), (p, q), (r, s)$ with $n, q, s \neq 0$ and such that, if $\mathcal{Z}_r \models \phi_=(m, n, p, q) \wedge \phi_=(p, q, r, s)$, then $\mathcal{Z}_r \models \phi_=(m, n, r, s)$.

Our hypothesis amounts to the equalities $mq = pn$ and $ps = rq$, which allow us to deduce $pnr = mrq = mps$, which in turn implies $nr = ms$. It follows that $\mathcal{Z}_r \models \phi_=(m, n, r, s)$, as desired.

As for the admissibility condition (ii), we suppose that $\mathcal{Z}_r \models \phi_=(m, n, m', n') \wedge \phi_=(p, q, p', q') \wedge \phi_=(r, s, r', s')$, where $n, n', q, q', s, s' \neq 0$. From the additional assumption $\mathcal{Z}_r \models \phi_{R_+}(m, n, p, q, r, s)$ we must deduce $\mathcal{Z}_r \models \phi_{R_+}(m', n', p', q', r', s')$.

When we rephrase our assumptions as explicit equalities and recall the definition of R_+, we obtain the equalities $r = nq$ and $s = (mq + pn)$. It thus suffices to show that (r', s') is equivalent to $(m'q' + p'n', n'q')$.

Since $rs' = r's$ by hypothesis, we deduce $(nq)r' = s'(mq + pn)$. Using the equalities $m'n = mn'$ and $p'q = pq'$ is easy to see that $(mq + pn)n'q' = (m'q' + p'n')(nq)$. We have shown that (r', s') is equivalent to $(mq + pn, nq)$ and that the latter pair is equivalent to $(m'q' + p'n', n'q')$. By transitivity, (r', s') is equivalent to $(m'q' + p'n', n'q')$, as desired.

Exercise 2.83

Set $\tau\{P, f, g\}$, where P is a binary relation symbol, f is a unary function symbol and g is a binary function symbol.

The formula $P(f(g(x,y)), g(x,x))$ is an atomic, nested formula in the given signature. It is easy to see that it is equivalent to:

$$\exists u \exists v \exists z (u = g(x,y) \wedge v = f(u) \wedge z = g(x,x) \wedge P(v,z)),$$

in which only unnested atomic formulae occur and the only free variables are the variables occurring free in the original formula. The original, nested formula is also equivalent to:

$$\forall u \forall v \forall z ((z = g(x,x) \wedge v = g(x,y) \wedge u = f(v)) \to P(v,z)),$$

If, more generally, τ contains a binary relation symbol P and a set of function symbols, finitely many of which are used to produce nestings in an atomic formula of the form $P(t_1, t_2)$, we can reduce the last formula to e.g. an equivalent, unnested existential formula by introducing new variables for each level of complexity of t_1, t_2 and writing a list of conjunction that set these variables equal to less complex terms. This could be done iteratively, supposing $t_1 = f(s_1, \ldots, s_k)$, where $s_1, \ldots, s_k$ are terms and writing, in the first instance:

$$\exists u_1 \ldots \exists u_k \exists v (v = f(u_1, \ldots, u_k) \wedge \bigwedge_{i=1}^{k} u_i = s_i \wedge P(v, t_2)).$$

We may then proceed iteratively and add, for each term $s_i = g(r_1, \ldots, r_{n_i})$, existential quantification on $z_1, \ldots, z_{n_i}$ and the equalities $z_{j_k} = r_k$. This procedure reduces $u_i = s_i$ to the unnested formula $u_i = g(z_1, \ldots, z_{n_i})$ but produces the nested atomic subformulae $z_{j_k} = r_k$ ($k = 1, \ldots, n_i$), if r_k is a nested term. The new atomic subformulae, however, involve fewer nestings than the previous ones. Because of this, we may further iterate the procedure described and progressively reduce the number of nestings until we end with atomic subformulae alone. An entirely analogous procedure applies to the construction of a universal, unnested formula equivalent to a given nested atomic formula.

Exercise 2.84

The previous exercise is the inductive base. Given the inductive hypothesis, it is clear that the introduction of a connective or a quantifier does not affect the complexity of the terms occurring in any given formula.

Exercise 2.85

The inductive base follows from the definition of Γ.

Since we have Γ-counterparts of atomic, unnested formulae, we define Γ-counterparts for the connectives in a recursive fashion. Since all binary connectives are treated in the same way, we use $\odot$ as a metavariable ranging on them below.

- $[\neg \phi]_\Gamma = \neg \phi_\Gamma$;
- $[\phi \odot \psi]_\Gamma = \phi_\Gamma \odot \psi_\Gamma$.

Suppose

$$\mathcal{N} \models \phi(f_\Gamma(\bar{a}_1), \ldots, f_\Gamma(\bar{a}_k)) \iff \mathcal{M} \models \phi_\Gamma(\bar{a}_1, \ldots, \bar{a}_k).$$

and

$$\mathcal{N} \models \psi(f_\Gamma(\bar{a}_1), \ldots, f_\Gamma(\bar{a}_k)) \iff \mathcal{M} \models \psi_\Gamma(\bar{a}_1, \ldots, \bar{a}_k).$$

Using the properties of equivalence, it is easy to see that analogous conditions transfer to $\neg \phi$ and $\phi \odot \psi$. the inductive hypothesis for quantifiers is stated relative to a formula $\psi(x_0, x_1, \ldots, x_n)$.

$$\mathcal{N} \models \psi(f_\Gamma(\bar{a}_0), f_\Gamma(\bar{a}_1), \ldots, f_\Gamma(\bar{a}_k)) \iff \mathcal{M} \models \psi_\Gamma(\bar{a}_0, \bar{a}_1, \ldots, \bar{a}_k).$$

If $\mathcal{M} \models \exists \bar{x}_0(\partial \Gamma(\bar{x}_0) \wedge \psi_\Gamma(\bar{x}_0, \bar{a}_1, \ldots, \bar{a}_k)$, then there is a tuple $\bar{a}_0$ such that $f_\Gamma(\bar{a}_0) \in N$ and $\mathcal{N} \models \psi(f_\Gamma(\bar{a}_0), f_\Gamma(\bar{a}_1), \ldots, f_\Gamma(\bar{a}_k))$. We deduce:

$$\mathcal{N} \models \exists x \psi(x, f_\Gamma(\bar{a}_1), \ldots, f_\Gamma(\bar{a}_k)).$$

If, conversely, the last condition holds, we use the surjectivity of f to guarantee that any $b \in N$ such that $\mathcal{N} \models \psi(b, f_\Gamma(\bar{a}_1), \ldots, f_\Gamma(\bar{a}_k))$, is of the form $f_\Gamma(\bar{a}_0)$, in which case it is easy to reconstruct the relevant condition on $\mathcal{M}$.

A trivial modification of the above argument works for the universal quantifier. The induction is complete.

Exercise 2.91

We restrict ourselves to verifying transitivity. If $\mathcal{G} \models \psi(a_1, a_2) \wedge \psi(a_2, a_3)$, then $a_2 - a_1$ and $a_3 - a_2$ are in H. Because $\mathcal{H}$ is a subgroup of $\mathcal{G}$, $(a_3 - a_2) + (a_2 - a_1) = a_3 - a_1 \in H$, i.e. $\mathcal{G} \models \psi(a_1, a_3)$.

We next consider the admissibility condition for $+$. Suppose $\mathcal{G} \models a + b = c \wedge \psi(a, a') \wedge \psi(b, b')$ If $\mathcal{G} \models \psi(c, c')$, it suffices to show that $\mathcal{G} \models \psi(a' + b', c')$. This is easy to see because the assumptions we have yield the equalities $(a + b) - c \in H$

and $c' - c, a' - a, b' - b \in H$. Because $\mathcal{H}$ is a subgroup of $\mathcal{G}$, $a' - a + b' - b = (a'+b')-(a+b) = (a'+b')-c \in H$. Since $c'-c \in H$ and $c'-c-(a'+b')+c \in H$, i.e. $\mathcal{G} \models \psi(a'+b', c')$. The remaining admissibility conditions are satisfied in a similar manner.

Fixing $a \in G$, we finally consider $K = \{c \in G : \mathcal{G} \models \psi(a, c)\}$. It is clear that any element of G of the form $a + c$, with $c \in H$ satisfies $a + c - a \in H$ and is therefore an element of K. If, conversely, $c \in K$, then $d = c - a \in H$ so c is of the form $a + d$ with $d \in H$.

B.3 Chapter 3

Exercise 3.2

If Σ is inconsistent and $M \models \Sigma$, then $M \models \phi \wedge \neg\phi$, which is impossible (a sentence is true in M iff its negation is not). It follows that, if Σ has any models, it must be consistent. Conversely, if Σ has no models, then any sentence (in the language of Σ) would be vacuously a consequence of Σ. For any sentence ϕ, $\phi, \neg\phi \in \Sigma$ and Σ is inconsistent. We have shown that a theory has no models is inconsistent. By contraposition, a consistent theory has at least one model.

Exercise 3.4

If R is a n-ary relation symbol and $t_1, \ldots, t_n$ are terms, then $M \models Rt_1 \ldots t_n$ iff $(t_1^M \ldots t_n^M) \in R^M$. Depending on whether the m-tuple $(t_1^M \ldots t_n^M)$ is or is not in R^M, we have $Rt_1 \ldots t_n \in ThM$ or $\neg Rt_1 \ldots t_n \in ThM$.

If the result we wish to prove holds for ϕ, ψ, it trivially holds for $\neg\phi$ and $\phi \wedge \psi$. Consider $\exists x \phi(x)$. Either there is an assignment $\sigma(x/a)$, with $a \in M$, that satisfies $\phi(x)$ in M or no such assignment exists. In the former case, $M \models \exists x \phi(x)$. In the latter case, $M \models \neg\exists x \phi(x)$. Consequently, $\exists x \phi(x) \in ThM$ or $\neg\exists x \phi(x) \in ThM$.

ThM has at least one model M. By Exercise 3.2, ThM is consistent. If, moreover, $ThM \models \phi$, then $M \models \phi$ and $\phi \in ThM$. This is to say that ThM is deductively closed and thus a theory.

Exercise 3.5

If T is consistent, by Exercise 3.2 it has a model M. The theory ThM is a complete extension of M.

Exercise 3.7

(i) If $M \in s(T)$, then $M \models T$ and, consequently, $M \models S$. Thus $M \in Mod\,S = s(S)$.
(ii) If $\phi \in Th\mathbf{K}$, then $\phi \in Th\mathbf{H}$, since a sentence true in every structure of the class $\mathbf{K}$ is certainly true in every structure of the subclass $\mathbf{H}$.
(iii) If $S = T$, $Mod\,S = Mod\,T$, by substitution. If $S \neq T$, let ϕ be a sentence of S that does not belong to T. Since S is a theory, $\phi \in S$. Since S, T are theories, ϕ is not a consequence of T. There must therefore be a model M of $T \cup \{\neg\phi\}$. Because $M \not\models \phi \in S$, M is not a model of S. We conclude $Mod\,S \neq Mod\,T$. By contraposition, we have shown that $Mod\,S = Mod\,T$ implies $S = T$.

Exercise 3.9

Let G be a finite group. If $0 \neq a \in G$ and $o(G) = n$, we wish to show that $na = 0$. To this end we note that, if no group element from the sequence $a, 2a, 3a, \ldots$ equals zero, the finiteness of G, implies $pa = qa$, with $p \neq q$. We may assume $p > q$. We deduce $pa - qa = (p-q)a = 0$ and set $n = p-q$. If some group element from the sequence $a, 2a, 3a, \ldots$ equals zero, then let it be na. Note that we have considered two cases, one of which in fact never occurs: what matters, for present purposes, is that they both lead to the same conclusion. We conclude $G \models \exists x(x \neq 0 \wedge nx = 0)$, i.e. G is not torsion-free.

Exercise 3.11

Given $p/q + \mathbb{Z}$ and a positive integer $n \in \mathbb{N}$, $p/qn + \mathbb{Z}$ is the coset we want. If σ is an assignment such that $\sigma(y) = p/q + \mathbb{Z}$ and $\sigma(x) = p/qn + \mathbb{Z}$, clearly $\mathbb{Q}/\mathbb{Z} \models nx = y[\sigma]$. The group $\mathbb{Q}/\mathbb{Z}$ is not torsion free because, for any positive integer m, $\sigma(x) = 1/m + \mathbb{Z}$ satisfies $x \neq 0 \wedge mx = 0$ in $\mathbb{Q}/\mathbb{Z}$.

Exercise 3.12

Because G is a linear order, if $0 \neq x$, then $x < 0$ or $0 < x$. If $x < 0$ then $-x + x < -x + 0$, i.e. $0 < -x$. Otherwise $-x < 0$. In any case, 0 is not an endpoint for G. In fact, tor each $y \neq 0$, y is not an endpoint of G. By trichotomy, $0 < y$ or $y < 0$. In the first case $y + 0 < y + y$ and, in the latter case, $y + y < y$. We conclude that G is a linear order without endpoints. Next, we check that G is dense. If $x < y$ we can't have $y - x = 0$, which is equivalent to $y = 0 + x = x$, or $y - x < 0$. To rule out the second possibility, set $y - x = v$ and note that $v < 0$ implies $v + x < x$, i.e. $y < x$. By trichotomy, we must conclude $0 < y - x$. By divisibility, there is $u \in G$ such that $2u = y - x$. Because $0 < 2u = u + u$, we rule out $u = 0$. We can also rule out $u < 0$, which implies $2u < u < 0$. Trichotomy yields $u > 0$, which leads to $u + x > x$ and to $y = 2u + x > u + x$. We have shown that, if $x < y$, then $x < u + x$ and $u + x < y$, i.e. that G is dense. Density ensures that G is infinite. The same conclusion follows from the fact that G has no endpoints.

Exercise 3.14

The term $0x_1 + \ldots + 0x_n$ belongs to T, thus $0 \in L_T(v_1, \ldots, v_n)$. If $v = \lambda_1 v_1 + \ldots + \lambda_n v_n$ and $u = \mu_1 v_1 + \ldots + \mu_n v_n \in L_T(v_1, \ldots, v_n)$, then $(\lambda_1 + \mu_1)x_1 + \ldots + (\lambda_n + \mu_n)x_n \in T$ and $(\lambda_1 + \mu_1)v_1 + \ldots + (\lambda_n + \mu_n)v_n \in L_T(v_1, \ldots, v_n)$. Finally, if $v \in L_T(v_1, \ldots, v_n)$ e $\mu \in K$, then $\mu\lambda_1 x_1 + \ldots + \mu\lambda_n x_n \in T$ and $\mu v \in L_T(v_1, \ldots, v_n)$. The set $L_T(v_1, \ldots, v_n)$ is a subspace V.

Exercise 3.21

(1) Since $(4, 6) = 2(2, 3)$, every vector in $L(v_1, v_2)$ is of the form $\lambda_1 v_1 + \lambda_2 v_2 = (\lambda_1 + 2\lambda_2)v_1$ and, as such, belongs to $L(v_1)$. For any $a, b \in \mathbb{Q}^2$:

$$\frac{3a - 2b}{13}(5, 1) + \frac{5b - a}{26}(4, 6) = (a, b) = \frac{3a - 2b}{13}(5, 1) + \frac{5b - a}{13}(2, 3),$$

$L(v_1, v_3) = \mathbb{Q}^2 = L(v_2, v_3)$.
(2) For instance, $(2, 5, 3) \notin L(v_1, v_2)$ but, if $v_3 = (0, 0, 1)$, then $L(v_1, v_2, v_3) = \mathbb{Q}^3$.
(3) Immediate.
(4) Immediate.

Exercise 3.24

If, for some positive $n \in \mathbb{N}$, $v_1, \ldots, v_n \in V$ are linearly independent, then $dim V \geq n$. If this is the case for each positive integer n, $dim V$ is not an integer. Conversely, if $dim V$ is not an integer $n \in \mathbb{N}$, then V contains a vector $v_1 \neq 0$ and a vector $v_2 \notin L(v_1)$, or $dim V = 1$. The last argument may be iterated to obtain $v_3 \notin L(v_1, v_2)$, $v_4 \notin L(v_1, v_2, v_3)$ and so on. An infinite sequence of linearly independent vector ensues.

Exercise 3.32

By polynomial division, there are $g(x)$ and $r(x)$ such that $f(x) = g(x)(x - a) + r(x)$. We evaluate the last polynomials on $a \in K$, obtaining $0 = f(a) = g(a)(a - $

$a) + r(x)$, which implies $r(x) = 0$. We deduce the existence of a unique polynomial $g(x)$ such that $f(x) = (x - a)g(x)$. If a polynomial of degree n had at least $n + 1$ distinct roots, it could be expressed as a product of at least $n + 1$ distinct factors of degree 1. The resulting product should have, by the previous exercise, degree at least $n + 1 > n$, a contradiction.

B.4 Chapter 4

Exercise 4.3

Let M, N be $\mathcal{L}$-structures and $h : M \longrightarrow N$ a homomorphism. If c is a constant symbol naming $c^M \in M$, then c also names $c^N = h(c^M) \in N$. If x is a variable with value $\sigma(x) = a \in M$, then $h\sigma(x) = h(a)$. Let f be a n-ary function symbols and $t_1(x_1, \ldots, x_n), \ldots, t_n(x_1, \ldots, x_n)$ be terms. By inductive hypothesis, for each $i \in \{1, \ldots, n\}$:

$$h(t_i^M(a_1, \ldots, a_n) = t_i^N(h(a_1), \ldots, h(a_n))$$

As a consequence:

$$h(ft_1^M(a_1, \ldots, a_n) \ldots t_1^M(a_1, \ldots, a_n))$$
$$= f^N(h(t_1^M(a_1, \ldots, a_n), \ldots, h(t_n^M(a_1, \ldots, a_n))$$
$$f^N(t_1^N(h(a_1), \ldots, h(a_n)), \ldots, t_n^N(h(a_1), \ldots, h(a_n)).$$

Exercise 4.7

If a homomorphism $h : M \longrightarrow N$ preserves all quantifier-free formulae, then it strongly preserves them, since the negation of a quantifier-free formula is quantifier-free. As a consequence, if φ is quantifier-free we must have:

- $M \models \varphi(a_1, \ldots, a_m)$ implies $N \models \varphi(f(a_1), \ldots, f(a_n))$;
- $M \models \neg\varphi(a_1, \ldots, a_m)$ implies $N \models \neg\varphi(f(a_1), \ldots, f(a_n))$,

where the second implication is equivalent to:

$$N \models \varphi(f(a_1), \ldots, f(a_n)) \text{ implies } M \models \varphi(a_1, \ldots, a_m).$$

This is to say that h strongly preserves $\varphi(x_1, \ldots, x_n)$. Since this is true of every quantifier-free formula, it is true of the atomic ones. As a consequence h is injective and a strong homomorphism, i.e. it is an embedding.

Conversely, if a homomorphism $h : M \longrightarrow N$ is an embedding, then it strongly preserves every atomic formula. Moreover, if $\phi(x_1, \ldots, x_n) \psi(x_1, \ldots, x_n)$ are strongly preserved by h, then so are $\neg \phi(x_1, \ldots, x_n)$ and $\phi(x_1, \ldots, x_n) \vee \psi(x_1, \ldots, x_n)$. We have just carried out an induction on the complexity of quantifier-free formulae. We conclude that h strongly preserves all of them.

We conclude that a homomorphism preserves every quantifier-free formula iff it is an embedding.

Exercise 4.8

A map that sends one square onto its copy inscribed in the circle is a homomorphism. It is not, however, an embedding. No better choice of homomorphism is possible because the diagonals of one square are not codified by the binary relation on the other square.

Exercise 4.9

Let M, N be linear orders and $h : M \longrightarrow N$ be a homomorphism. If $a, b \in M$ and $a \neq b$, trichotomy implies $a < b$ or $b < a$. Suppose the former. Then, since h is a homomorphism, we can deduce $f(a) < f(b)$, which implies, by trichotomy again (this time in N), $f(a) \neq f(b)$. This is to say that f is injective. It only remains to show that it strongly preserves inequalities. If $f(a) < f(b)$, we cannot have $a = b$, which would imply $f(a) = f(b)$, thus violating irreflexivity in N, since $f(a) < f(b)$ would then imply $f(a) < f(a)$. We cannot have $b < a$ either, which would imply $f(b) < f(a)$, which would violate irreflexivity due to the fact that $f(a) < f(b)$ and the ordering on N is transitive. By trichotomy in M, we can only have $a < b$. This shows that $M \models a < b$ iff $N \models h(a) < h(b)$. It follows that h is an embedding.

Exercise 4.10

The formula $\exists x (x + x = x)$ holds in one structure but fails in the other.

Exercise 4.13

If f is an elementary embedding, it is certainly an embedding. We need the surjectivity of f to prove the converse implication. The proof proceeds by an induction on the complexity of formulae. The inductive base follows immediately from the fact that f is an embedding. The inductive step for the connectives is trivial. Surjectivity plays a role when we assume $\mathcal{N} \models \exists x \phi(x, f(\overline{a}))$. In this case there is

B Solutions to Exercises 405

$b \in N$ such that $\mathcal{N} \models \phi(b, f(\overline{a}))$ and $b = f(a)$ for some $c \in M$. It follows that $\mathcal{N} \models \phi(f(c), f(\overline{a}))$.

Exercise 4.14

Suppose that M has m elements and N has n elements, with $m < n$. This has to be the case because no injection between two finite sets of the same size can fail to be surjective. Then the first-order sentence:

$$\exists x_1 \ldots \exists x_n (x_1 \neq x_2 \wedge \ldots \wedge x_1 \neq x_n \wedge x_2 \neq x_3 \wedge \ldots \wedge x_2 \neq x_n \wedge \ldots \wedge x_{n-1} \wedge x_n)$$

is true in N but not in M. No function from M to N can preserve its negation.

Exercise 4.15

Because the composition of functions is associative and identity maps are trivially elementary, we only have to verify that the composition of two elementary maps is an elementary map.

Suppose $\mathcal{M}_1, \mathcal{M}_2, \mathcal{M}_3$ are in **K** and $f : \mathcal{M}_1 \longrightarrow \mathcal{M}_2, g : \mathcal{M}_2 \longrightarrow \mathcal{M}_3$ are elementary. Then:

$$\mathcal{M}_1 \models \varphi(a_1, \ldots, a_k) \iff \mathcal{M}_2 \models \varphi(f(a_1), \ldots, f(a_k))$$
$$\iff \mathcal{M}_3 \models \varphi(g(f(a_1)), \ldots, g(f(a_k))),$$

i.e. $g \circ f$ is elementary. If we restrict attention to the models of a theory T, we still have a class of structures, so nothing material changes.

Exercise 4.20

(i) Since every integer can be represented as $i + 5k$, with $i \in \{0, 1, 2, 3, 4\}$, $k \in \mathbb{Z}$, we may consider the function $h : \mathbb{Z} \longrightarrow \mathbb{Z}/5\mathbb{Z}$ defined by the equality $h(i+5k) = i+5\mathbb{Z}$. This function is also a homomorphism. Since $h((i+5k) + (j+5k)) = h((i+j)+5(2k))$, if $i+j \leq 4$, then $h((i+j)+5(2k)) = (i+j)+5\mathbb{Z}$ and, if $i + j \geq 5$, then we tale the largest m such that $(i + j) - 5m \geq 0$ and rewrite $(i+j)+5(2k)$ as $(i+j-5m)+5(2k+m)$, with $0 \leq (i+j)-5m < 5$. Then $h((i+5k)+(j+5k)) = ((i+j)-5m)+5\mathbb{Z}$. It is clear that h is surjective and that it is not injective.
In general, since every integer can be represented as $i + nk$, with $n \geq 1, k \in \mathbb{Z}$ and $i \in \{0, \ldots, n-1\}$, we have the homomorphism $h : \mathbb{Z} \longrightarrow \mathbb{Z}/n\mathbb{Z}$ defined by the equality $h(i + nk) = i + n\mathbb{Z}$.
(ii) Given the vector spaces $\mathbb{Q}^2, \mathbb{Q}^3$, show that the function $f : \mathbb{Q}^2 \longrightarrow \mathbb{Q}^3$ defined by the condition $f((p, q)) = (p, q, 0)$ is a $\mathcal{L}_v$-embedding. First, we check

that f is a homomorphism. For vector spaces, this implies that f sends the null vector of $\mathbb{Q}^2$ into the null vector of $\mathbb{Q}^3$, preserves vector addition, and the action of any scalar. Clearly $f((0,0)) = (0,0,0)$ so the null vectors are matched. Moreover:

$$f((p,q) + (r,s)) = f((p+q, r+s)) = (p+q, r+s, 0)0)$$
$$= (p,q,0) + (r,s, = f((p,q)) + f((r,s)).$$

Finally, if $t \in \mathbb{Q}$, $f(t(p,q)) = f((tp, tq)) = (tp, tq, 0) = t(p,q,0) = tf((p,q))$. To show that f is also an embedding, it suffices to show that it is injective. If $(p,q) \neq (r,s)$, then either $p \neq r$ or $q \neq s$, or both. It is easy to see that, in each one of these cases $(p,q,0)$, $(r,s,0)$ will not be equal in each corresponding component. Thus, $f((p,q)) \neq f((r,s))$ and f is injective.

(iii) We consider the function $h : \mathbb{Q}^3 \longrightarrow P_2(\mathbb{Q})$ determined by the equality $f((p,q,r)) = p + qx + rx^2$. Clearly, $f((0,0,0)) = 0$. Moreover:

$$f((p,q,r) + (s,t,u)) = f((p+s, q+t, r+u))$$
$$= (p+s) + (q+t)x + (r+u)x^2 = (p + qx + rx^2)$$
$$+ (s + tx + ux^2) = f((p,q,r)) + f((s,t,u)).$$

Finally, if $t \in \mathbb{Q}$, $f(t(p,q,r)) = f((tp, tq, tr)) = tp + tqx + trx^2 = t(p + qx + rx^2) = tf((p,q,r))$. The function f is clearly surjective. As for its injectivity, suppose $p + qx + rx^2 = s + tx + ux^2$: it follows that $(p-s) + (q-t)x + (r-u)x^2 = 0$, i.e. that $f((p-s, q-t, r-u)) = 0$. It is easy to see that this happens exactly when $p - s = q - t = r - u = 0$, i.e. when $p = s, q = t, r = u$. We have shown that $f((p,q,r)) = f((s,t,u))$ implies $(p,q,r) = (s,t,u)$. This is to say that f is injective. We conclude that f is at once a bijection and a strong homomorphism, i.e. it is an isomorphism.

(iv) If $dim V = k$, then there are $v_1, \ldots, v_k \in V$ such that, for every $w \in V$ there are $q_1, \ldots, q_k \in \mathbb{Q}$ such that $w = q_1 v_1 + \ldots + q_k v_k$. The function $f : V \longrightarrow \mathbb{Q}^k$ defined by the equality $f((q_1, \ldots, q_k)) = q_1 v_1 + \ldots + q_k v_k$ is an isomorphism.

(v) We have a binary relation $\sim$ on $\mathbb{N} \times \mathbb{N}$, defined by the condition:

$$(m,n) \sim (p,q) \text{ iff } m + q = p + n.$$

(a) It is easy to see that $\sim$ is reflexive. By the commutativity of addition in $\mathbb{N}$, $\sim$ is also symmetric. Finally, if $(m,n) \sim (p,q)$ and $(p,q) \sim (r,s)$, then $m + q = p + n$ and $p + s = q + r$ imply:

$$m + q + p + s = p + n + q + r,$$

from which we can cancel p, q to obtain $m + s = n + r$, i.e. $(m, n) \sim (r, s)$.
(b) Let $[m, n]$ be the equivalence class containing (m, n), and $\mathbb{N} \times \mathbb{N}/\sim$ the set of all such equivalence classes. We define:

$$[m, n] +' [p, q] = [m + p, n + q], \quad -'[m, n] = [n, m].$$

We want to check that $+', -'$ are actually functions, i.e. single-valued. To this end, suppose $(m, n) \sim (a, b)$ and $(p, q) \sim (c, d)$. Then $m + b = a + n$ and $p + d = q + c$. As a result:

$$(m + b) + (p + d) = (a + n) + (q + c) \, iff \, (m + p) + (b + d)$$
$$= (a + c) + (n + q) \, iff \, (m + p, n + q)$$
$$\sim (a + c, b + d) \, iff \, [a + c, b + d] = [m + n, p + q].$$

Thus $[m, n] +' [p, q] = [a, b] +' [c, d]$. Moreover, if $(a, b) \sim (m, n)$, then $m + b = a + n$ iff $b + m = n + a$ iff $(b, a) \sim (n, m)$ iff $[b, a] = [n, m]$, i.e. $-'[m, n] = -'[b, a]$.

(c) The $\mathcal{L}_g$-structure $(\mathbb{N} \times \mathbb{N}/\sim, +', -', [0, 0])$ is an abelian group. The associativity of $+'$ can be tediously verified by appealing to the associativity of ordinary addition on $\mathbb{N}$. The operation $+'$ is also clearly commutative. Moreover, $[m, n] +' [0, 0] = [m + 0, n + 0] = [m, n]$ and $[m, n] +' (-'[m, n]) = [m, n] +' [n, m] = [m + n, m + n] = [0, 0]$.

(d) It is easy to check that $h : \mathbb{N} \times \mathbb{N}/\sim \longrightarrow \mathbb{Z}$, defined by the condition $h([m, n]) = m - n$, is an isomorphism. Note, in particular, that $h([m, n]) = h([p, q])$ iff $m - n = p - q$ iff $m + q = n + p$ iff $(m, n) \sim (p, q)$.

Exercise 4.24

Note that, if $M \subseteq N$ and ι_M is an embedding, then:

(1) for any constant symbol c in the signature of $\mathcal{L}$, $\iota(c^M) = c^N$, i.e. $c^M = c^N$;
(2) for any m-ary relation symbol R in the signature of $\mathcal{L}$:

$$R^M(a_1, \ldots, a_m) \iff R^N(\iota_M(a_1), \ldots, \iota_M(a_m)) \iff R^N(a_1, \ldots, a_m),$$

i.e. R^M is a restriction of the relation R^N to M^m;
(3) for any k-ary function symbol f in the signature of $\mathcal{L}$,

$$f^M(a_1, \ldots, a_m) = \iota_M(f^M(a_1, \ldots, a_n))$$
$$= f^N(\iota_M(a_1), \ldots, \iota_M(a_n)) = f^N(a_1, \ldots, a_n),$$

i.e. the operations of $\mathcal{M}$ are just restrictions of the corresponding operations of $\mathcal{N}$.

Exercise 4.27

Under the given restriction, we obtain a surjective embedding $f : M \longrightarrow f[M]$, which, by Exercise 4.13, is an isomorphism.

Exercise 4.28

If $m < n$, with $m, n \in \mathbb{N}$, the same relation continues to hold in $\mathbb{Z}$, which extends the linear ordering of $\mathbb{N}$. Conversely, if $m < n$ in $\mathbb{Z}$ and m, n are natural numbers, they continue to be ordered in the same way when we restrict attention to $\mathbb{N}$, by taking the intersection of the linear ordering on $\mathbb{Z}$ with $\mathbb{N}^2$. As a consequence $\mathbb{N} \sqsubseteq \mathbb{Z}$. However, the $\mathcal{L}_<$-formula $\exists x (x < 0)$ is true in $\mathbb{Z}$ but not in $\mathbb{N}$ (note that this formula has one free variables, to which we have assigned the number 0). Consequently, $\mathbb{N}$ cannot be an elementary substructure of $\mathbb{Z}$.

If we now turn to $\mathbb{Z}, \mathbb{Q}$, we see that the $\mathcal{L}_<$-formula $\exists x (0 < x \wedge x < 1)$ is true in $\mathbb{Q}$ but not in $\mathbb{Z}$. Thus, $\mathbb{Z}$ cannot be an elementary substructure of $\mathbb{Q}$, when both are conceived as $\mathcal{L}_<$-structures.

Exercise 4.30

Let $A \subseteq M$ the subset of M whose elements are named by terms in $Tm(x_1, \ldots, x_n)$. First, we check that A contains the constants of M and that the operations on M can be restricted to operations on A.

Since constant symbols are terms, every constant $c^M \in M$ is also a constant of A. If f^M is a k-ary function and $b_1, \ldots, b_k \in A$, then there are terms $t_1(x_1, \ldots, x_n), \ldots, t_k(x_1, \ldots, x_n)$ such that $b_1 = t_1^M(a_1, \ldots, a_n), \ldots, b_k = t_k^M(a_1, \ldots, a_n)$. Since:

$$f t_1(x_1, \ldots, x_n \ldots t_k(x_1, \ldots, x_n) \in Tm(x_1, \ldots, x_n),$$

it follows that $f^M(t_1^M(a_1, \ldots, a_n), \ldots, t_k^M(x_1, \ldots, x_n)) = f^M(b_1, \ldots, b_k) \in A$. This is to say that every function on M induces a function on A with the same values for any arguments in A.

As for relations on, if R^M is a j-ary relation on M, we simply induce a j-ary relation on A by setting $R^A = R^M \cap A^j$. It is easy to see that, if $b_1, \ldots, b_j \in A$, then:

$$R^M(b_1, \ldots, b_j) \iff R^A(b_1, \ldots, b_j).$$

It follows that, if M is a $\mathcal{L}$-structure, so is A, once the functions and relations of M are restricted to A. By the last equivalence, the inclusion map $\iota : A \longrightarrow M$, which is certainly injective, strongly preserves every atomic formula. It is therefore an embedding of A into M. We conclude that $A \sqsubseteq M$.

Exercise 4.35

It is clear that if A is the domain of a substructure of $\mathcal{N}$ it interprets the symbols in the signature of $\mathcal{N}$. As a consequence, it must contain the constants in N and the values of every k-ary operation on N^k when applied to an argument in A^k.

If the last two conditions hold, then we can interpret the symbols in the signature of $\mathcal{N}$ on the set A and treat it as the domain of a structure $\mathcal{A}$.

Exercise 4.36

(a) Since f is injective, it is automatically a $\mathcal{L}_=$-embedding. Because it is surjective, it is elementary by Exercise 4.13.
(b) Suppose $\mathcal{N} \models \exists x \varphi(x, f(a_1), \ldots, f(a_m))$, with $a_1, \ldots, a_m \in M$. We shall show that $\mathcal{N} \models \exists x \varphi(f(b), f(a_1), \ldots, f(a_m))$, for some $b \in M$. It follows from the last condition and the Tarski-Vaught test that $f[M]$ is the domain of an elementary substructure of $\mathcal{N}$. Since f is a bijection from M to $f[M]$, it is an elementary $\mathcal{L}_=$-embedding. These two facts can be put together to show that $f : M \longrightarrow N$ is $\mathcal{L}_=$-elementary.
In order to apply the Tarski-Vaught test, we note that, if $\mathcal{N} \models \varphi(c, f(a_1), \ldots, f(a_m))$, there is a bijection g from N to N that fixes $f(a_i), i = 1, \ldots, m$, swaps c with an arbitrary, fixed $b \in f[M]$ and keeps every other element of N fixed. By part (i), g is a $\mathcal{L}_=$elementary map, so it preserves every $\mathcal{L}_=$formula. Thus:

$$\mathcal{N} \models \varphi(c, f(a_1), \ldots, f(a_m)) \iff \mathcal{N} \models \varphi(g(c), f(a_1), \ldots, f(a_m)).$$

Exercise 4.37

On the one hand, $(\mathbb{N}, <)$ is certainly an elementary substructure of itself. On the other hand, suppose $(A, <) \preceq (\mathbb{N}, <)$. By definition of structure $A \neq \emptyset$. Suppose $0 \in A$. Since $\mathbb{N} \models \exists x(0 < x \land \forall z(0 < z \rightarrow (z = x \lor x < z)))$, the last formula holds in A and A must, as a result, contain a number $n > 0$ that satisfies it. We can show $n = 1$. If not, then $\mathbb{N}$ would model a formula to the effect that there are numbers between 0 and n: the same formula should hold in A. It could not, if $n \neq 1$ would be a positive natural number that is the immediate successor of 1, which is impossible. Thus $n = 1$. By a similar argument, we can deduce that $2, 3, \ldots, n, \ldots$ are elements of A. It follows that $(A, <)$ is $(\mathbb{N}, <)$. If $m \in A$ and $m \neq 0$, we exploit the fact that $\mathbb{N} \models \exists x(x < m \land \forall z(z < m \rightarrow (z = x \lor z < x)))$ holds to deduce that $m - 1 \in A$. Repeating this argument finitely many times, we conclude that $0 \in A$, in which case we already know that $(A, <) = (\mathbb{N}, <)$.

Exercise 4.43

The set $\mathcal{D}(\mathbb{Z}/3\mathbb{Z})$ contains 24 elements. We have the three equalities $0 = 0, 1 = 1, 2 = 2$ and the three negated equalities $\neg(0 = 1), \neg(0 = 2), \neg(1 = 2)$. The remaining, eighteen atomic sentences, describe the operations of addition and multiplication on the set $\{0, 1, 2\}$. Each operation requires nine atomic sentences, since nine possible distinct choices of arguments (i.e. elements of $\{0, 1, 2\} \times \{0, 1, 2\}$ are available.

If $N' \models \mathcal{D}(\mathbb{Z}/3\mathbb{Z})$, then we consider the objects $a, b, c \in N$ named by the constant symbols 0, 1, 2 respectively and define the injection $f : \mathbb{Z}/3\mathbb{Z} \longrightarrow N$ by the conditions $f(0) = a$, $f(1) = b$ and $f(2) = c$. Since N' models the diagram of $\mathbb{Z}/3\mathbb{Z}$, we have e.g. $a+b = b$. Thus, $f(0+1) = f(1) = b = a+b = f(0)+f(1)$. Similar chains of equalities can be written for addition or multiplication performed on any two elements of $\{a, b, c\}$. It follows that f strongly preserves equalities. As a result, it is a $\mathcal{L}_r$-embedding.

Exercise 4.44

If 1, 2, 3, 4, 5, 6 are constant symbols used to name the elements of M, then the diagram of M contains six equalities, fifteen negated equalities and thirty-six literals of the form $R(i, j)$ or $\neg R(i, j)$, with $i, j \in \{1, \ldots, 6\}$. Overall, the diagram of M contains fifty-seven elements.

If $N \models \mathcal{D}(M)$ and N contains exactly six elements, may may set $N = \{b_1, \ldots, b_6\}$ and $M = \{a_1, \ldots, a_6\}$, where b_i, a_i are named by the constant symbol i in $\mathcal{D}(M)$. The bijection $f(a_i) = b_i$ is an isomorphism because $R^M(a_i, b_i) \iff R(i, j) \in \mathcal{D}(M) \iff R^N(b_i, b_j)$.

If, in $\mathcal{D}(M)$, we replace i by the variable x_i, for each $i \in \{1, \ldots, 6\}$, we obtain a set of 57 literals in six free variables: let $\psi(x_1, \ldots, x_6)$ be their conjunction and $\theta(x_1, \ldots, x_6)$ be the formula $\forall y(x_1 = y \lor \ldots \lor x_6 = y)$. The formula:

$$\exists x_1 \ldots \exists x_6(\psi(x_1, \ldots, x_6) \land \theta(x_1, \ldots, x_6))$$

is modelled by a structure N iff N has exactly six elements that satisfy $\mathcal{D}(M)$. Since M is embeddable in N and any embedding must be surjective, M, N are isomorphic.

Exercise 4.45

Let $|M| = n$. The diagram of M contains only finitely many atomic sentences in which only n distinct constant symbols $c_1, \ldots, c_n$ occur. If we let $\psi(c_1, \ldots, c_n)$ be the conjunction of the elements of $\mathcal{D}(M)$, then the sentence:

$$\exists x_1 \ldots \exists x_n (\psi(x_1, \ldots, x_n) \land \forall y (x_1 = y \lor \ldots \lor x_n = y))$$

identifies M up to isomorphism. The structure M is embeddable into any model N of the last sentence. Because N has exactly n elements, any embedding of M must be surjective, i.e. an isomorphism.

Exercise 4.47

The proof is obtained from that of Theorem 4.49 with minor modifications, e.g. by replacing 'elementary embedding' with 'embedding' or '$\mathcal{ED}(M)$' with '$\mathcal{D}(M)$'.

Exercise 4.50

If R is a k-ary relation symbol, we have:

$$\langle a'_1, \ldots, a'_k \rangle \in R^{M'} \iff \langle h(a'_1), \ldots, h(a'_k) \rangle \in R^N.$$

Thus, all relations are strongly preserved between M' and N. Constants are preserved, since, for any constant $a \in N$, $a' = h^{-1}(a)$ is, by stipulation, a constant of M', i.e. $h(a') = a$. Finally, if f is a m-ary function:

$$f^{M'}(a'_1, \ldots, a'_m) = a' = h^{-1}(f^N(a_1, \ldots, a_n)).$$

Since, by definition $h(a'_i) = a_i$, we obtain:

$$h(f^{M'}(a'_1, \ldots, a'_m)) f^N(h(a'_1), \ldots, h(a'_n)).$$

It follows that h is a strong homomorphism.

B.5 Chapter 5

Exercise 5.3

If C is a decisive coalition relative to f, (a, b) and a given profile in $\Pi A_i \in \mathsf{L}^I$, let $\Pi B_i \in \mathsf{L}^I$ be such that, in π_2, $C = \{i \in I : B_i \models a < b\}$. Because trichotomy holds, if $D = I - C$, then $D = \{i \in I : B_i \models b < a\}$ in π_2. When restricted to $\{a, b\}$ the profiles ΠA_i, ΠB_i are identical. Since $f(\Pi A_i) \models a < b$ and f is pairwise, $f(\Pi B_i) \models a < b$. Thus, C is decisive on any profile.

Exercise 5.7

(i) Let C be a decisive coalition relative to (a, b) and $c \neq a, b$. We consider a profile ΠA_i, in which $C = \{i \in I : A_i \models c < b \wedge a < b\}$ and, for every $i \in I$, $A_i \models c < a$. Since f is unanimous, $f(\Pi A_i) \models c < a$. Since C is decisve over a, b, $f(\Pi A_i) \models a < b$. By transitivity, $f(\Pi A_i) \models c < b$.

(ii) Let C be a decisive coalition relative to (c, b) and $a \neq c, b$. We consider a profile ΠA_i, in which $C = \{i \in I : A_i \models c < a \wedge c < b\}$ and, for every $i \in I$, $A_i \models b < a$. Since f is unanimous, $f(\Pi A_i) \models b < a$. Since C is decisve over c, b, $f(\Pi A_i) \models c < b$. By transitivity, $f(\Pi A_i) \models c < a$.

(iii) Suppose that C is a decisive coalition relative to (a, b) and $c \neq a, b$. Toward a contradiction, we suppose that C is not decisive on (b, a). Then there is a profile ΠA_i such that $C = \{i \in I : A_i \models b < a\}$ and $f(\Pi A_i) \models a < b$, in which case $D = I - C$ is decisive on (a, b). Consider a profile ΠB_i such that $D = \{i \in I : \Pi B_i \models a < b \wedge b < c\}$ and $C = \{i \in I : \Pi B_i \models c < a \wedge b < c\}$. Because f is unanimous, $f(\Pi B_i) \models b < c$. Because D is decisive on (a, b) and C is decisive on (c, a), by parts (i) and (ii), we also have $f(\Pi B_i) \models a < b \wedge c < a$. The resulting cycle in $f(\Pi B_i)$ produces a contradiction. It follows that C is decisive on (b, a) in ΠB_i and, consequently, in every profile.

Exercise 5.12

Clearly, $I \in \mathcal{U}_j$ and $\emptyset \notin \mathcal{U}_j$. If $C, D \in \mathcal{U}_j$, then $j \in C$ and $j \in D$. Consequently $j \in C \cap D$ and $C \cap D \in \mathcal{U}_j$. It follows that $\mathcal{U}_j$ is a proper filter over I. Moreover, if $C \subseteq I$, either $j \in C$, in which case $C \in \mathcal{U}_j$ or $j \notin C$, in which case $j \in I - C$ and $I - C \in \mathcal{U}_j$. We conclude that $\mathcal{U}_j$ is an ultrafilter.

Exercise 5.13

(i) Suppose that $C, D \in \mathcal{G}$, by (iii) iff $C \cap D \in \mathcal{G} \iff I - (C \cap D) \notin \mathcal{G}$. Suppose that $I - (C \cap D) \in \mathcal{G}$: then $C \cap D \cap (I - (C \cap D)) = \emptyset$, against condition (ii). We conclude that $I - (C \cap D) \notin \mathcal{G}$. It follows that $C \cap D \in \mathcal{G}$.

(ii) Let $\mathcal{U}$ be an ultrafilter over a nonempty set I and $C \in \mathcal{U}$. Suppose that $C = D \cup E$ and that $D \cap E = \emptyset$. If $D \in \mathcal{U}$, we are done. If not, $I - D \in \mathcal{U}$ and $E = C \cap I - D \in \mathcal{U}$.

(iii) Set $I = \{i_1, \ldots, i_n\}$. We split I into $\{i_1\}, \{i_2, \ldots, i_n\}$: if $\{i_1\} \in \mathcal{U}$, we are done, otherwise $\{i_2, \ldots, i_n\} \in \mathcal{U}$ and we split the last set into $\{i_2\}, \{i_3, \ldots, i_n\}$. After at most $n - 1$ steps, we identify i_k such that $\{i_k\} \in \mathcal{U}$.

Exercise 5.15

Since f is defined in a pairwise fashion, its values on pairs do not change across profiles. To show that f is an aggregation, we must check that, for any profile ΠA_i, $f(\Pi A_i)$ is a linear order. Suppose $f(\Pi A_i) \models a < b \land b < c$. By definition of f, there are decisive coalitions C, D such that $C = \{i \in I : A_i \models a < b\}$ and $D = \{i \in I : A_i \models b < c\}$. The coalition $C \cap D$ is decisive and, for every $j \in C \cap D$, $A_j \models a < c$. Thus $C \cap D \subseteq E = \{i \in I : A_i \models a < c\} \in \mathcal{U}$. It follows that there is a decisive coalitions whose members rank a below c: the definition of f now implies $f(\Pi A_i) \models a < c$. This is true of any arbitrary alternatives a, b, c. It follows that the f-values are transitive. We next verify that they satisfy trichotomy. If $a \neq b$ are any two distinct alternatives, then, in any profile ΠA_i, I splits into the sets $D = \{i \in I : A_i \models a < b\}$ and $I - D = \{i \in I : A_i \models b < a\}$. Exactly one of these two sets is in $\mathcal{U}$. As a result, either $f(\Pi A_i) \models a < b$ or $f(\Pi A_i) \models b < a$. The irreflexivity of $<$ follows from the fact that, for any profile ΠA_i, $\{i \in I : A_i \models a < a\} = \emptyset \notin \mathcal{U}$.

Exercise 5.17

If $\mathcal{U}$ is not maximal relative to inclusion, then there is a filter $\mathcal{G} \neq \mathcal{P}(I)$ such that $\mathcal{U} \subset \mathcal{G}$, i.e. there is $V \in \mathcal{P}(I)$ such that $V \in \mathcal{G}$ and $V \notin \mathcal{U}$. It follows from the properties of ultrafilters that $I - V \in \mathcal{U}$ and that $\emptyset = V \cap I - V \in \mathcal{G}$, since $\mathcal{G}$ contains the intersections of its elements. As a consequence, $\mathcal{G} = \mathcal{P}(I)$, which contradicts $\mathcal{G} \neq \mathcal{P}(I)$. We must conclude that $\mathcal{U}$ is maximal.

Exercise 5.18

(i) Note that $U = U \cap I \subseteq I$, so $U, I \in \mathcal{F}_U$. If $W \in \mathcal{F}_U$ and $W \subseteq Z$, it is clear that $Z \in \mathcal{F}_U$. It remains to show that $\mathcal{F}_U$ contains the intersections of its elements. If $Z_1, Z_2 \in \mathcal{F}_U$, then there are $Y_1, Y_2 \in \mathcal{F}$ such that $U \cap Y_1 \subseteq Z_1$ and $U \cap Y_2 \subseteq Z_2$. Then $Z_1 \cap Z_2$ contains the common elements of these two sets, which include the common elements of $U \cap Y_1$ and $U \cap Y_2$. It follows that $U \cap Y_1 \cap Y_2 \subseteq Z_1 \cap Z_2$ and $Z_1 \cap Z_2 \in \mathcal{F}_U$. Since $\mathcal{F}$ is a proper filter, if $\mathcal{F}_U$ is not proper, then there is $V \in \mathcal{F}$ such that $U \cap V = \emptyset$. This goes against the choice of U. We conclude that $\mathcal{F}_U$ is a proper filter.

(ii) If for each $V \in \mathcal{F}$, $U \cap V \neq \emptyset$ and $U \notin \mathcal{F}$, part (i) implies that $\mathcal{F}_U$ is a strict extension of $\mathcal{F}$, against its maximality.

(iii) By contraposition, from (ii).

Exercise 5.21

(i) Suppose $R^\sim([f_1], \ldots [f_k])$. Then $V = \{i \in I : R^i(f_1(1), \ldots, f_k(i))\} \in \mathcal{U}$. We set $U_j = \{i : I : f_j(i) = g_j(i)\}$, with $j = 1, \ldots, k$. Thus $i \in U = V \cap U_1 \cap \ldots \cap U_k$ iff $(f_1(i), \ldots, f_k(i)) \in R^i$ and $f_1(i) = g_1(i), \ldots, f_k(i) = g_k(i)$. In other words, $i \in U$ iff $(g_1(i), \ldots, g_k(i)) \in R^i$. Since $U \in \mathcal{U}$, we can conclude $R^\sim([g_1], \ldots [g_k])$. The converse implication is proved in the same way.
(ii) Trivial modifications of the argument from part (i) yield the result.

Exercise 5.24

$$\Pi_{\mathcal{U}} A_i \models t_1([\overline{f}]) = t_2(\overline{f}]) \iff t_1^\sim([\overline{f}]) = t_1^\sim([\overline{f}])$$
$$\iff [(t_1^i([\overline{f(i)}])] = [(t_2^i([\overline{f(i)}])]$$
$$\iff [t_1(\overline{f(i)}) = t_2(\overline{f(i)})] \in \mathcal{U}.$$

Exercise 5.27

(i) Let $\mathcal{U}$ be an ultrafilter over the set I. Suppose $V \cup W \in \mathcal{U}$. Then V or $I - V$ is in $\mathcal{U}$. If V is, we are done. Otherwise $I - V \cap (V \cup W) \in \mathcal{U}$ and $W \supseteq I - V \cap (V \cup W)$.
(ii) If $\Pi_{\mathcal{U}} A_i \models \phi \vee \psi$. Then $E = \{i \in I : A_i \models \phi \vee \psi\} \in \mathcal{U}$. The set E can be rewritten as $C \cup D$, where $C = \{i \in I : A_i \models \phi\}$ and $D = \{i \in I : A_i \models \psi\}$. By part (i) C or D is in $\mathcal{U}$, i.e. $\Pi_{\mathcal{U}} A_i \models \phi$ or $\Pi_{\mathcal{U}} A_i \models \psi$. The converse statement is easily proved.
(iii) It is easy to see that H is an embedding because it is a homomorphism of linear orders. By Theorem 5.23, it is elementary because, if $[f_{a_1}], \ldots, [f_{a_n}] \in H[A]$ (i.e., for every $i \in I$ and $j \in \{1, \ldots, n\}$, $f_{a_j}(i) = a_j \in A$), then $[f_{a_j}] = H(a_j)$ and $\Pi_{\mathcal{U}} A \models \phi(H(a_1), \ldots, H(a_n))$ iff $A \models \phi(a_1, \ldots, a_n)$.

Exercise 5.28

We define the binary relation $<^F$ on F by the condition $a <^F b$ iff $b - a \in P^F$. Since $a - a = 0$, we cannot have $a < a$ for any $a \in F$. If $a <^F b$ and $b <^F c$, then $b - a, c - b \in P^F$. It follows that $(b - a) + (c - b) = c - a \in P^F$ iff $a <^F c$.

Finally, if $a \neq b$, then either $b - a \in P^F$ or $a - b = -(b - a) \in P^F$, i.e. $a <^F b$ or $b <^F a$. It remains to check that (1) and (2) hold. If $a <^F b$, then $b - a \in P^F$ and $b - a = (b + c) - (a + c)$, thus $(b + c) - (a + c) \in P^F$ and $a + c <^F b + c$. If $0 < c^F$, then $(b - a)c = bc - ac \in P^F$ and $ac <^F bc$. Conversely, if F is a linear order that models (1) and (2), then we may define $P^F = \{a \in F : 0 < a\}$. It follows from the definition that $0 \notin P^F$. If $a, b \in P^F$, then $0 < b$, which, by (1), implies $a < a + b$, i.e. $0 < a + b$, by transitivity. Moreover, $0 < b$, implies $a0 < ab$ by (2), i.e. $0 < ab$. We conclude that the sum and product of two positive field elements are positive. Finally, if $0 < a$, then adding $-a$ to both sides of the inequality and using (1), we obtain $-a < 0$: by the same argument, if $0 < -a$, then $a < 0$. In any case, whenever $a \neq 0$, $a \in P^F$ or $-a \in P^F$.

Exercise 5.31

(i) Since $[1/n]$ contains a sequence that is positive over the set of indices $\mathbb{N}$ and smaller than 1 over the same set of indices;
(ii) the sequence 2ϵ is $(1, 2/3, 1/2, 2/5, \ldots)$ and it is smaller than 1 over a cofinite subset of $\mathbb{N}$, namely $\mathbb{N} - \{1\}$. The sequence 3ϵ is smaller than 1 over $\mathbb{N} - \{1, 2\}$;
(iii) in general, the sequence $k\epsilon$ is smaller than 1 over $\mathbb{N} - \{1, 2, \ldots, k\}$;
(iv) fix k and note that $(1/2, 1/3, \ldots, 1/k, 1/k + 1, \ldots)$ is smaller than $(1/k, 1/k, \ldots)$ over the cofinite set $\mathbb{N} - \{1, 2, \ldots, k\}$.

Exercise 5.34

If F is Archimedean ordered and $a \in F$ is an infinitesimal, either $a = 0$ or $a \neq 0$. If $a \neq 0$, then $n < |1/a|$ for each $n \in \mathbb{N}$. This cannot happen, so $a = 0$. Conversely, if the only infinitesimal of F is 0, for any $a \in F$, if $a \neq 0$, $1/a$ is not an infinitesimal. By definition, there is $n \in \mathbb{N}$ such that $1/n < |1/a|$. Because $|1/a| = 1/|a|$, it follows that $|a| < n$. Since a was an arbitrary element of $F - \{0\}$, F is Archimedean.

Exercise 5.36

(i) We consider $-[a] = [-a]$, which is a positive infinitesimal. Since $|[-a]| = [-a]$, the inequality $n[a] < [1]$ holds. Calling $1/[-a]$ the multiplicative inverse of $[-a]$, we deduce $n < 1/[-a]$ for each $n \in \mathbb{N}$. It follows that $1/[-a]$ is not finite. Since $1/[-a] = -1/[a] = |-1/[a]| = 1/|[a]|$, we conclude that $n < 1/|[a]|$.
(ii) If $[a] \in {}^*F$ and $[a]$ is not an infinitesimal, then $[a] \neq 0$. Suppose that $[a]$ is a positive infinitesimal. Then $1/n < [a]$ for some $n \in \mathbb{N}$. It follows that $1/[a] < n$. The multiplicative inverse of $[a]$ is thus finite. A similar argument works for negative infinitesimals.

416　　　　　　　　　　　　　　　　　　　　　　　B Solutions to Exercises

(iii) Clearly [0] is an infinitesimal and, if [a] is an infinitesimal, so is $-[a]$. Because universal properties are preserved under substructures, I is an abelian group. Now suppose that $[a] \in$ I and $[b]$ is a finite element of *F. There is $m \in \mathbb{N}$ such that $|[b]| < m$. Since $|[a]| < 1/mn$ for every $n \in \mathbb{N}$, field arithmetic implies $|[a][b]| = |[a][b]| < 1/n$ for every $n \in \mathbb{N}$. It follows that $[a][b]$ is an infinitesimal. As a result, I is an ideal.
(iv) If $[b]$ is not an infinitesimal, an ideal of *F containing $[b]$ will contain $[b]1/[b] = [1]$ and, as a consequence, the whole of *F.
(v) By part (iv), the only ideal that strictly extends I is *F.

Exercise 5.37

It suffices to note that $[a] +$ I, $[b] +$ I are cosets of an abelian group.

Exercise 5.38

If $[a]$, $[b]$ are distinct infinitesimals one of $[a] < [b]$, $[b] < [a]$ must hold, but $(a) = (b)$.

Exercise 5.40

If $\epsilon < 0$ is any positive quantity, there is $M \in \mathbb{N}$ such that $1/M < \epsilon$. If $kn > 2M+1$, then

$$|1/2M - 1/k| = 1/2M - 1/k > 1/M - 1/2M + 1 = 1/2M(2M + 1) < 1/2M$$

and a similar chain of inequalities holds for n. As a result:

$$|1/n - 1/k| \leq |1/n - 1/2M| + |1/2M - 1/k| < 1/2M + 1/2M = 1/M.$$

Exercise 5.41

(i) Since F is Archimedean, there is $m \in \mathbb{N}$ such that $1/m < p$. With m fixed, $ma < n$ for some positive integer n. We may select n to be the least integer with this property. Then $ma \geq n - 1$.
(ii) Note that $a \geq (n-1)/m$. Thus:

$$0 \leq |a - q| = a - q = a - \frac{n-1}{m} = (a - \frac{n}{m}) + \frac{1}{m} < \frac{1}{m},$$

where $a - n/m < 0$, since $ma < n$.

Exercise 5.42

(i) The hypotheses yield $(0) < (a)$, i.e. $|(a)| = (a)$. Since $[a]$ is also positive, we have $|[a]| = [a]$, i.e. $(|a|) = (a)$;
(ii) we have $-(a) = |(a)|$ and $|a| = -a$. It follows that $(|a|) = (-a) = -(a) = |(a)|$ (recall that $(-a) = -(a)$ is a way of restating the equality $-([a] + \mathbb{I}) = -[a] + \mathbb{I}$;
(iii) since $(a) = (0)$ $|[a]|$ is $[0]$ or a positive infinitesimal. Either way, $(|a|) = (0)$.

The inequality $|(a) - (q)| < (p)$ implies $|(a - q)| < (p)$, which, by the result we have just proved, yields $|(a - q)| < (p)$. By definition of $<$, we can deduce $|[a - q]| < [p]$ and, since $[a - q] = [a] - [q]$, the desired inequality.

Exercise 5.44

(i) Call H the equivalence class containing the sequence $(1, 2, 3, \ldots)$. Then $H \in {}^*\mathbb{N}$ is infinitely large, as well as $H + 1, H + 2, H + 3, \ldots$.
(ii) Since every integer is odd or even, so are H, K. If H, K are both even, then $J = (H + K)/2$ is an integer and it satisfies the requisite inequalities. If H, K are both odd, the same strategy work. Otherwise, we set $J = (H + K + 1)/2$.

Exercise 5.45

(i) if $[s] = n \in \mathbb{N}$, ${}^*z(n)$ contains the sequence $(z(n), z(n), z(n), \ldots)$;
(ii) if $[s] \in {}^*\mathbb{N}$, then there is $U \in \mathcal{U}$ such that $i \in U$ implies $m = s(i) \in \mathbb{N}$ so s is equivalent to a sequence t which coincides with it over U and whose values are all natural numbers. It follows that z is defined on every values of t and, consequently ${}^*z$ is defined at $[t] = [s]$. If a sequence does not takes values in $\mathbb{R} - \mathbb{N}$ over an element of $\mathcal{U}$, then ${}^*z$ is not defined at $[s]$. The domain of ${}^*z$ is therefore exactly ${}^*\mathbb{N}$.

Exercise 5.47

Suppose that z converges to $a, b \in \mathbb{R}$. By Theorem 5.46, for any infinitely large $H \in {}^*\mathbb{N}$, $a \simeq {}^*z(H) \simeq b$. The transitivity of $\simeq$ implies $a \simeq b$. Since $a, b \in \mathbb{R}$, if $a - b$ is an infinitesimal, the only options is $a - b = 0$. Thus $a = b$ and the limit of z is unique.

Exercise 5.48

(i) Show that $z : \mathbb{N} \longrightarrow \mathbb{R}$ is a Cauchy sequence iff .
(ii) If z converges, then for any infinitely large $H, K \in {}^*\mathbb{N}$, ${}^*z(H) \simeq a \simeq {}^*z(K)$, i.e. ${}^*z(H) \simeq {}^*z(K)$. By part (i), z is a Cauchy sequence.
(iii) The sequence z is not convergent. In fact, it takes arbitrarily large values.

Exercise 5.49

(i) The set B of lower bounds for A is nonempty and bounded above. By order-completeness, B has a least upper bound a. If, for some $d \in A$, $a < d$, then d would be an upper bound for B (the elements of B all lie below any element of a) strictly smaller than a. It follows that $d \leq a$ and a is a lower bound for A. Let c be a lower bound for A. Then $c \in B$. Since a is the least upper bound for C, $a \leq c$. Thus, a is not smaller than any lower bound for A. In other words, a is the greatest lower bound for A.
(ii) By part (i), $z(n) \geq a$ for each $n \in \mathbb{N}$. Fix $\epsilon > 0$ and suppose that a is not $lim\, z(n)$. Then, for any $n \in \mathbb{N}$ there is $m > n$ such that $z(n) - a \geq \epsilon$. It follows that, from some n on, $z(n) - a \geq \epsilon$. This is because, if $z(m_1) - a \geq \epsilon$, it is possible to choose $n > m_1$ and determine $m_2 > n > m_1$ such that $z(m_2) - a \geq \epsilon$. Because z is decreasing, whenever $m_1 \leq k \leq m_2$, $z(k) - a \geq \epsilon$. If this is the case, then $a + \epsilon/2$ is a lower bound for z that is also strictly greater than a, against the choice of a. It follows that $a = lim\, z(n)$.

Exercise 5.50

(i) Since c is positive and $1 < c$, $c < c^2$. Iterating this argument, we obtain a decreasing sequence $c, c^2, c^3, \ldots$ Let $a \in \mathbb{R}$ be its limit;
(ii) if H is infinitely large, then $c^H \simeq a$ implies $c^{H+1} \simeq ac$. Since $a \simeq c^{H+1} \simeq ac$ and $\simeq$ is transitive, we deduce $a \simeq ac$.
(iii) Because $a, ac \in \mathbb{R}$, $a \simeq ac$ implies $a - ac = 0$, i.e. $a = ac$. Since $c \neq 1$, we must have $a = 0$.

Exercise 5.51

(i) Because z is a Cauchy sequence, we can find $m \in \mathbb{N}$ such that, $n, k > m$ implies $|z(n) - z(k)| < 1$. Fixing k, we deduce that, for every $n > k > m$, $|z(n) - z(k)| < 1$. Since $|z(n)| - |z(k)| \leq |z(n) - z(k)|| < 1$, we arrive at the inequality $|z(n)| < |z(k)| + 1$, which holds for each $n > k$;

B Solutions to Exercises

(ii) set $m-1 = max\{|z(0)|, \ldots, |z(k-1)|, |z(k)|\}$. Clearly, $m \geq |z(k)|+1$, so $m \geq z(n)$ for each $n > k$. Because m is also an upper bound for $\{|z(0)|, \ldots, |z(k)|\}$, it is an upper bound for the entire sequence of absolute values. This is to say that z is bounded.

Exercise 5.58

Clearly, $I \in \mathcal{F}$. If $V, Z \in \mathcal{F}$, then there are $U_1, \ldots, U_m, W_1, \ldots, W_n$:

$$U_1 \cap \ldots \cap U_m \subseteq V \text{ and } W_1 \cap \ldots \cap W_n \subseteq Z.$$

It follows that:

$$U_1 \cap \ldots \cap U_m \cap W_1 \cap \ldots \cap W_n \subseteq V \cap Z.$$

It is clear that $V \in \mathcal{F}$ and $V \subseteq Z$ implies $Z \in \mathcal{F}$. The filter $\mathcal{F}$ is proper. If $\emptyset \in \mathcal{F}$, there should be $U_1, \ldots, U_m$ such that:

$$U_1 \cap \ldots \cap U_m \subseteq \emptyset,$$

which implies that the finite intersection property fails.

Exercise 5.60

Suppose that each finite subset of S has a model but S has none. Then $S \models \varphi \wedge \neg\varphi$ and so does a finite subset of S, against the hypothesis that all finite subsets have models.

Exercise 5.61

We may associate with S a set of sentences $S(c)$ obtained by replacing x_i with the new constant symbol c_i in an expanded language. If S is finitely satisfiable under suitable assignments, so is $S(c)$. But then $S(c)$ has a model and this model determines an assignment that will satisfy S.

Exercise 5.62

We may or may not expand the language. If we do so, it suffices to add an infinite sequence of new constant symbols c_i, for each $i \in \mathbb{N}$, and consider the set of sentences $T \cup \{c_i \neq c_j : i, j \in \mathbb{N}\}$. A finite subset of the last set is modelled by a sufficiently large, finite model of T. It now follows that the whole set has a model M, which contains an infinite sequence of distinct objects. The model M is therefore infinite.

Exercise 5.64

(i) S_0 is finitely satisfiable by hypothesis. If S_n is finitely satisfiable, then consider S_{n+1}. A finite subset E of S_{n+1} contains finitely many sentences from S_n and finitely many conjunctions of sentences from S_n. Let D be the set of individual conjuncts from the sentences in E but not in S_n. Then $D \cup (E \cap S_n)$ is a finite subset of S_n. By inductive hypothesis, it has a model M. It is easy to see that $M \models E$.

(ii) If S' were not finitely satisfiable, a finite subset of S' would have no models. By construction, such a finite subset must be contained in S_n, for a sufficiently large n, against the finite satisfiability of the latter set. In a similar vein, if $\phi, \psi \in S'$ then there are $m, n \in \mathbb{N}$ such that $\phi S_m, \psi \in S_n$. If $n \geq m$, then $\phi, \psi \in S_n$, which implies $\phi \wedge \psi \in S_{n+1} \subseteq S'$.

(iii) Since S' is unsatisfiable, because S_0 is, S' contains a contradiction and cannot therefore be finitely satisfiable. If there is a finite, unsatisfiable subset E of S', there must be a corresponding, unsatisfiable subset of S_0. It suffices to consider the conjunctions in E that do not already belong to S_0. Their conjuncts may belong to S_0. If not, they belong to S_n, where we can select $n \in \mathbb{N}$ to be the least number with this property. This means that their conjuncts lie in S_{n-1}. If needed, we can break these conjuncts further into constituents that will ultimately lie in S_0. Because only finitely many steps are required, the resulting set of conjuncts is an unsatisfiable, finite subset of S_0.

Exercise 5.66

The ultraproduct $\Pi_{\mathcal{U}} X_k$, where $\mathcal{U}$ extends the filter of cofinite subsets of $\mathbb{N}$, is an infinite $\mathcal{L}_=$-structure.

Exercise 5.67

(1) Let S_1 be an axiomatisation of **K** and S_2 be an axiomatisatoin of its complement. The set $S_1 \cup S_2$ has no models, so there is a finite set $A \subseteq S_1 \cup S_2$ with no models. Note that A must contain sentences from both S_1 and S_2. Let us call α the conjunction of all sentences in $S_1 \cap A$ and β the conjunction of all sentences in $S_2 \cap A$. We claim that the single sentence $\alpha \wedge \neg\beta$ axiomatises **K**. If a structure is in **K**, it clearly models $\alpha \subseteq S_1$. It cannot also model β because $\alpha \wedge \beta$ has no models. Thus, it models $\neg\beta$. Conversely, if a structure M models $\alpha \wedge \neg\beta$, then it models $\neg\beta$ and cannot be in the complement of **K**, whose members are

B Solutions to Exercises

certainly models of $\beta \subseteq S_2$. Since every structure is in **K** or its complement, $M \in \mathbf{K}$.

(2) Let A_n be the set of literals $\{\neg(x_i = x_j) : i, j \leq n\}$ and let $\alpha(x_1, \ldots, x_n)$ the conjunction of all literals in A_n. Let φ_n be the $\mathcal{L}_=$-sentence:

$$\exists x_1 \ldots \exists x_n \alpha(x_1, \ldots, x_n).$$

It is not difficult to see that the set of $\mathcal{L}_=$-sentences $\{\varphi_n : n \geq 1\}$ axiomatises the class of infinite $\mathcal{L}_=$-structures.

(3) If the class of finite $\mathcal{L}_=$-structures was elementary, by part (i) it should be possible to axiomatise it—and thus also the class of infinite $\mathcal{L}_=$-structures—by a single sentence. Let φ be an axiomatisation of the class of infinite $\mathcal{L}$-structures. This axiomatisation must be equivalent to any other axiomatisation for the same class. In particular, for each $n \in \mathbb{N}$ $\{\varphi_n : n \in \mathbb{N}\} \models \varphi$. By Compactness, there is a finite set $A \subseteq \{\varphi_n : n \in \mathbb{N}\}$ such that $A \models \varphi$. Since $\varphi \models \{\varphi_n : n \in \mathbb{N}\}$, we conclude that φ is equivalent to a conjunction of finitely many sentences of the form φ_n. Such a conjunction, however, has finite models. This contradiction shows that the class of finite $\mathcal{L}$-structures is not elementary.

(4) If the class of infinite $\mathcal{L}$-structures had a finite axiomatisation, it would be axiomatised by a sentence φ. Consequently, the class of finite $\mathcal{L}$-structures would be axiomatised by $\neg\varphi$. By part (iii), this is impossible. Thus, the class of infinite $\mathcal{L}$-structures is not finitely axiomatisable. In the light of (ii), it is in fact quasi-finitely axiomatisable.

Exercise 5.68

Let p_i be the i-th prime number. The sentence:

$$\underbrace{1 + \ldots + 1}_{p_i \text{ times}}$$

is true on the set of indices $\{p_j \in P : p_i < p_j\}$, which is cofinite and thus in $\mathcal{U}$. It follows that the characteristic of $\Pi_{\mathcal{U}} F_i$ is not p_i. Since p_i was chosen arbitrarily from P, the characteristic of $\Pi_{\mathcal{U}} F_i$ is not positive. It must therefore be 0. It follows that the class of fields of positive characteristic is not closed under ultraproducts. This class cannot therefore be elementary.

Exercise 5.69

Suppose that T axiomatises the class of finite $\mathcal{L}$-structures. Then T has arbitrarily large, finite models. By Compactness, it must have one infinite model. As for Exercise 5.68, we suppose that T axiomatises the class of fields of positive characteristic and, for each positive $i \in \mathbb{N}$ we call φ_i the sentence:

$$\neg \underbrace{(1 + \ldots + 1)}_{p_i \text{ times}} = 0, \text{ where } p_i \text{ is the i-th prime.}$$

The set of sentences:

$$S = T \cup \{\varphi_i : i \geq 1\}$$

is finitely satisfiable because there are fields of arbitrarily large, positive characteristic (consider $\mathbb{Z}/p\mathbb{Z}$, as p increases). Thus, S has a model, which must have characteristic 0.

B.6 Chapter 6

Exercise 6.2

(i) Because $u \neq 0$, at least one of a, b is not zero. Consequently $a^2 + 2b^2 \neq 0$ and

$$u^{-1} = \frac{a}{a^2 + 2b^2} + \left(-\frac{b}{a^2 + 2b^2}\right)\sqrt{2}$$

(ii) It is easily seen that the sum, difference and product of two numbers of the form $x + y\sqrt{2}$, with $x, y \in \mathbb{Q}$, has the same form. By part (i), quotients behave in the same manner. The field axioms are easily verified to hold.

(iii) An even power of $\sqrt{2}$ is a positive integer and an odd power of $\sqrt{2}$ is a positive, integer multiple of $\sqrt{s}$. Thus, every element of $\mathbb{Q}[\sqrt{2}]$ is of the form $a + b\sqrt{2}$, with $a, b \in \mathbb{Q}$. By the same clue, every element of $\mathbb{Q}(\sqrt{2})$ is of the form:

$$\frac{p(\sqrt{2})}{q(\sqrt{2})} = \frac{a + b\sqrt{2}}{c + d\sqrt{2}}.$$

It suffices to note that:

$$\frac{a + b\sqrt{2}}{c + d\sqrt{2}} \cdot 1 = \frac{a + b\sqrt{2}}{c + d\sqrt{2}} \cdot \frac{c - d\sqrt{2}}{c - d\sqrt{2}}$$

$$= \frac{(a + b\sqrt{2})(c - d\sqrt{2})}{c^2 - 2d^2}$$

$$= \frac{(ac - 2bd) + (bc - ad)\sqrt{2}}{c^2 - 2d^2}$$

$$= \frac{ac - 2bd}{c^2 - 2d^2} + \frac{bc - ad}{c^2 - 2d^2}\sqrt{2} \in \mathbb{Q}[\sqrt{2}].$$

(iv) Since every element $\mathbb{Q}(\sqrt{2})$ is of the form $a + b\sqrt{2} = a \cdot 1 + b \cdot \sqrt{2}$, it can be expressed as a linear combination of $1, \sqrt{2}$ with coefficients in $\mathbb{Q}$. The vectors $1, \sqrt{2}$ are linearly independent over the field $\mathbb{Q}$ (incidentally, they are not linearly independent over the set of real numbers).

Exercise 6.4

$K \neq \emptyset$ because $0/1 = 0 \in K$. Using the field operations, we note that, $p(\alpha)/r(\alpha), q(\alpha)/s(\alpha) \in K$ implies:

$$\frac{p(\alpha)}{r(\alpha)} \pm \frac{q(\alpha)}{s(\alpha)} = \frac{p(\alpha)s(\alpha) \pm q(\alpha)r(\alpha)}{r(\alpha)s(\alpha)} \in S,$$

since K is an integral domain, $r(\alpha)s(\alpha) \neq 0$. If $q(\alpha)/s(\alpha) \neq 0$, then $q(\alpha) \neq 0$ and:

$$\frac{p(\alpha)}{r(\alpha)} \frac{s(\alpha)}{q(\alpha)} \in S.$$

We conclude that $S \subseteq M$.

Exercise 6.5

The argument from Lemma 6.3 allows us to identify $K(\alpha_1, \ldots, \alpha_n)$ with the subset of M whose elements are of the form $p(\alpha_1, \ldots, \alpha_n)/q(\alpha_1, \ldots, \alpha_n)$, where $q(\alpha_1, \ldots, \alpha_n) \neq 0$ and $p(x_1, \ldots, x_n), q(x_1, \ldots, x_n) \in \mathbb{Q}[x_1, \ldots, x_n]$.

Exercise 6.13

If K' was a transcendental extension of K, there would exist $\alpha \in K'$, transcendental over K. The vector space K' over K would then contain an infinite sequence of linearly independent vectors. By Exercise 3.24, K' could not be finite-dimensional, against the hypothesis $[K' : K] = n$. The field K' must therefore be an algebraic extension of K.

Exercise 6.15

Consider any $c \in N$. Since $b_1, \ldots, b_n$ is a basis for N over M, there are scalars $\mu_1, \ldots, \mu_n \in M$ such that:

$$c = \mu_1 b_1 + \ldots + \mu_n b_n.$$

Each scalar is an element of M and, as such, it can be expressed as a linear combination of $a_1, \ldots, a_m$ with coefficients in K. For $i = 1, \ldots, n$, we set:

$$\mu_i = \lambda_{1i} a_1 + \ldots + \lambda_{mi} a_m.$$

Then c:

$$c = \lambda_{11} a_1 b_1 + \lambda_{12} a_1 b_2 + \ldots + \lambda_{1n} a_1 b_n + \ldots + \lambda_{m1} a_m b_1 + \lambda_{m2} a_m b_2 + \ldots + \lambda_{mn} a_m b_n.$$

where $a_i b_j \in N$ and all coefficients are elements of K. This is to say that the mn vectors $a_i b_j$ ($1 \le i \le m, 1 \le j \le n$) generate the vector space N over the field K. It remains to show that the mn vectors just mentioned are linearly independent over K. We begin with the equality:

$$0 = \lambda_{11} a_1 b_1 + \lambda_{12} a_1 b_2 + \ldots + \lambda_{1n} a_1 b_n + \ldots + \lambda_{m1} a_m b_1 + \lambda_{m2} a_m b_2 + \ldots + \lambda_{mn} a_m b_n,$$

which may be rewritten as:

$$c = (\lambda_{11} a_1 + \ldots + \lambda_{m1} a_m) b_1 + \ldots + (\lambda_{1n} a_1 + \ldots + \lambda_{mn} a_m) b_n.$$

Because $b_1, \ldots, b_n$ are linearly independent over M, for $i = 1, \ldots, n$ we can deduce:

$$\lambda_{1i} a_1 + \ldots + \lambda_{mi} a_m = 0.$$

By the linear independence of $a_1, \ldots, a_m$ over K, we can also deduce $\lambda_{1i} = \ldots = \lambda_{1n} = 0$. It now follows that $\{a_i b_j \in M : 1 \le i \le m, 1 \le j \le n\}$ is a basis for N over K and that $[N : K] = mn$.

Exercise 6.17

It suffices to show that every $\beta \in \mathcal{K}(T)$ is algebraic over $\mathcal{K}$. Note that β lies in an extension of $\mathcal{K}$ obtained by adjoining to K only finitely many elements $\alpha_1, \ldots, \alpha_k$ of T. Since $\mathcal{K}(\alpha_1, \ldots, \alpha_k)$ is a finite extension of $\mathcal{K}$, it is algebraic. It follows that β is algebraic over K.

Exercise 6.19

Recall that $K(\alpha, \beta)$ is, by definition, the smallest extension of K containing α and β. We note that, if β is algebraic over K, then β is algebraic over the extension $K(\alpha) = K[\alpha]$. It follows from Theorem 6.11 that $(K(\alpha))(\beta) = (K[\alpha])[\beta]$. Since $(K(\alpha))(\beta)$ is a field obtained from the smallest extension of K containing α by the addition of β, $K \cup \{\alpha, \beta\} \subseteq (K(\alpha))(\beta)$. Consequently, $K(\alpha, \beta) \subseteq (K[\alpha])[\beta]$.

Conversely, any element of the last field is of the form $q_0 + q_1\beta + \ldots + q_n\beta^n$, with $q_i \in K[\alpha], i = 0, \ldots, n$. Thus $K(\alpha, \beta) \supseteq (K[\alpha])[\beta]$ and we can conclude $(K[\alpha])[\beta] = K(\alpha, \beta)$.

Exercise 6.22

If K is a finite field, its elements can be listed as a finite sequence, e.g. $c_1, \ldots, c_n$. The polynomial:

$$((x - c_1)(x - c_2) \cdots (x - c_n)) + 1$$

is an element of $K[x]$ without any roots in K. Consequently, K is not algebraically closed. By contraposition, if K is algebraically closed, it is infinite.

Exercise 6.27

Let K be of 0 characteristic. The field elements $1, 1+1 = 2, 1+1+1 = 3, \ldots$, are all different from 0. Consequently, for each $n \neq 0$, the multiplicative inverse $1/n$ of n is in K. Since K contains m and $1/n$, it also contains m/n and its additive inverse $-m/n$. This is to say that K contains a copy of $\mathbb{Q}$.

If K is of characteristic p, we consider the set $Z_p \subseteq K$ whose elements are $0, 1, 2, \ldots, p-1$. Addition in K, when restricted to Z_p, corresponds to addition on the integers modulo p. This is because $i + j \geq p$ implies $i + j = p + k$ with $k \in \{0, \ldots, p-1\}$ and $p = 0$.

Since, in the integers modulo p, the additive inverse of i is $p - i$, the same happens in K. Thus, if $i, j \in Z_p$ implies $i - j = i + (p - j) \in Z_p$ and Z_p is an additive subgroup of K.

Multiplication in K, when restricted to Z_p, corresponds to multiplication on the integers modulo p. If $i, j \in Z_p$, the product ij is obtained by an iteration of addition on 1, carried out ij times: if k is the largest multiple of p such that $ij \geq pk$, then $ij \in K$ is the sum of 1 and itself iterated $ij - pk$ times (with $ij, pk \in \mathbb{N}$). Thus, $i, j \in Z_p$ implies $ij \in Z_p$. Moreover, if $j \neq 0$, the elements of Z_p that are multiples of j, namely $j, 2j = (1+1)j, 3j = (1+1+1)j, \ldots (p-1)j$ are all nonzero (neither $r \in \{1, \ldots, p-1\}$ nor j is zero and K is an integral domain) and distinct. If two of them, say mj, nj, with $m < n$, were equal, then $mj = nj$ would imply $(n - m)j = 0$, with $0 < n - m < p$. This is impossible because $(m - n)$ is not a multiple of p and we are forced to conclude $j = 0$. If the first $p - 1$ multiples of j are all distinct, they are exactly the nonzero elements of Z_p. One of them is therefore $j^{-1} \in Z_p$. It now follows that Z_p is a subfield of K, which we can identify with $\mathbb{Z}/p\mathbb{Z}$.

Exercise 6.28

The elements of $\mathbb{Q}(X)$, the fraction field associated with $\mathbb{Q}[x]$ are equivalence classes of the form $[p(x), q(x)]$, where $p(x), q(x) \in \mathbb{Q}[x]$ and $q(x) \neq 0$.

Exercise 6.29

We identify $f[A]$ and A. Our next goal is to show that, if K is a field and $f : A \longrightarrow K$ is an embedding, there is an embedding $g : Q(A) \longrightarrow N$ such that $g \upharpoonright A = f$. This result can be proved if we are able to extend f to a suitable embedding whose domain is $Q(A)$. To this end, we change notation slightly and refer to the $\sim$-equivalence class $[m, n]$ of $A \times A_0/\sim$ as m/n. Because $f[A]$, A are identified, we also assume $a/1 = a$. Under these stipulations, the operations of addition and multiplication between $\sim$-equivalence classes can be rewritten as:

$$\frac{m}{n} + \frac{r}{s} = \frac{ms + rn}{ns} \quad \text{and} \quad \frac{m}{n}\frac{r}{s} = \frac{mr}{ns}.$$

If the embedding $f : A \longrightarrow N$ is given, we define $g : Q(A) \longrightarrow N$ by the condition:

$$g(a/b) = f(a)f(b)^{-1},$$

where $f(b)^{-1} \in N$ is the multiplicative inverse of $f(b)$ because $b \neq 0$ and f is injective. Note that $g(a) = g(a/1) = f(a)f(1)^{-1} = f(a)$, i.e. $g \upharpoonright A = f$ (we have used 1 to refer to the multiplicative neutral element in both A and K). It remains to show that g is a $\mathcal{L}_r$-embedding of $Q(D)$ into K. We leave it as an exercise to show that g is injective. Moreover, g is a homomorphism.

Addition is preserved by g because:

$$\begin{aligned}
g\left(\frac{a}{b} + \frac{c}{d}\right) &= g\left(\frac{ad + cb}{bd}\right) \\
&= f(ad + cb)[f(bd)]^{-1} \\
&= f(ad + cb)[f(b)f(d)]^{-1} \\
&= [f(ad) + f(cb)][f(b)^{-1}f(d)^{-1}] \\
&= f(ad)f(b)^{-1}f(d)^{-1} + f(cb)f(b)^{-1}f(d)^{-1} \\
&= f(a)f(d)f(b)^{-1}f(d)^{-1} + f(c)f(b)f(b)^{-1}f(d)^{-1} \\
&= f(a)f(b)^{-1} + f(c)f(d)^{-1} \\
&= g\left(\frac{a}{b}\right) + g\left(\frac{c}{d}\right),
\end{aligned}$$

where, besides the definition of g, we have repeatedly used the fact that f is a homomorphism and the fact that K is a field. Further calculations show that multiplication is preserved, i.e.:

$$g\left(\frac{a}{b}\frac{c}{d}\right) = g\left(\frac{a}{b}\right) g\left(\frac{c}{d}\right).$$

It now follows that g is an injective homomorphism, i.e. an embedding, since no relation symbols belong to $\mathcal{L}_r$.

Exercise 6.30

A field homomorphism $f : K \longrightarrow M$ is in particular an additive group homomorphism: as such, it is injective iff its kernel (i.e. the set $\{a \in K : f(a) = 0\}$) is $\{0\}$, as the reader may verify. If the kernel of f contained $a \neq 0$, we could use the fact that f is a field homomorphism to deduce:

$$1 = f(1) = f(aa^{-1}) = f(a)f(a^{-1}) = 0 f(a^{-1}) = 0,$$

which cannot be the case in a field.

Exercise 6.31

(i) We only check that $\sim$ is transitive. If $(a, n) \sim (b, m)$ and $(b, m) \sim (c, p)$, then $ma = nb$ and $pb = mc$ and, consequently, $n(pb) = n(mc)$. Since $n(pb) = p(nb) = p(ma) = m(ap)$, we obtain $m(ap) = m(nc)$ iff $ap = nc$ iff $(a, n) \sim (c, p)$. Note that we only made use of properties that follow from the definition of mx, with $m > 0$, $x \in H$.
(ii) Let $[a, n]$ be the equivalence class of (a, n). We define $[a, n] + [b, m] = [ma + nb, mn]$ and $-[a, n] = [-a, n]$. It follows that addition in D is associative and commutative with neutral element $[0, m]$, since $[a, n] + [0, m] = [ma + 0, mn] = [ma, mn] = [a, n]$, on account of the fact that $(ma, mn) \sim (a, n)$. Moreover $[a, n] + [-a, n] = [na + (-na), n] = [0, n]$, so additive inverses exist. We conclude that D is a group.
(iii) The group D is divisible. If $0 < m \in N$ is fixed and $[a, n]$ is given, then:

$$\underbrace{[a, mn] + \ldots + [a, mn]}_{m \text{ times}} = [ma, mn] = [a, n].$$

Moreover, D is torsion free. If $n[a, m] = [a, m]$ we could deduce $[na, m] = [a, m]$, i.e. $nma = ma$ iff $(n - 1)ma = 0$, which implies $n = 1$, since $n > 1$ implies that H is not torsion-free, against the hypothesis.

(iv) Define $f : H \longrightarrow D$ by the condition $f(a) = [a, 1]$. It is easy to see that f is a group homomorphism. Moreover, if $f(a) = f(b)$, then $[a, 1] = [b, 1]$ iff $(a, 1) \sim (b, 1)$ iff $a = b$ and f is injective.

Exercise 6.33

We have:

$$\rho(q(x) + \langle p(x)\rangle) + \rho(s(x) + \langle p(x)\rangle) = q(a) + s(a)$$
$$= \rho((q(x) + s(x)) + \langle p(x)\rangle)$$
$$= \rho((q(x) + \langle p(x)\rangle) + (s(x) + \langle p(x)\rangle)).$$

A nearly identical chain of equalities shows that ρ preserves multiplication. As for $'$, we observe that, if $p(x)$ contains a term of the form $a_i x^i$ and $q(x)$ contains a term of the form $b_i x^{k_i}$, then the coefficient of x^{k_i} in the sum $(p(x) + q(x))$ is $h^*(a_i + b_i) = h^*(a_i) + h^*(b_i)$, i.e. the same as the coefficient of x^{k_i} in the sum $p'(x) + q'(x)$. If, in $p(x)q(x)$, $k_i + k_j = k_r + k_s$, let a_i, a_j be the coefficients of x^{k_i}, x^{k_j} in $p(x), q(x)$ respectively and let b_r, b_s be the coefficients of x^{k_r}, x^k_s in the same, respective polynomials. Then the term of $p(x)q(x)$ in $x^{k_i+k_j}$ has a coefficient containing the term $a_i a_j + b_r b_s$. As a result, the corresponding coefficient in $(p(x)q(x))'$ contains the term $h^*(a_i a_j + b_r b_s) = h^*(a_i)h^*(a_j) + h^*(b_r) + h^*(b_s)$, which is clearly a term in the coefficient of $x^{k_i+k_j}$ when the product $p'(x)q'(x)$ is computed.

Exercise 6.34

(i) Since C_1 is an algebraic extension of K and C_2 is algebraically closed, the previous theorem guarantees the existence of an embedding of $h : C_1 \longrightarrow C_2$ such that, if ι_1 is the inclusion of K in C_1 and ι_2 the inclusion of K in C_2, $h\iota_1 = \iota_2$. Thus, $h[C_1]$ is a subfield of C_2 as well as an algebraic extension of K.
(ii) Since C_1 is algebraically closed, so is $h[C_1]$.
(iii) Suppose $a \in C_2$. Then a is algebraic over K. Since $h[C_1]$ is an algebraic extension of K (it is a subset of C_2, which is an algebraic extension), a is algebraic over $h[C_1]$. But $h[C_1]$ is algebraically closed, so $a \in h[C_1]$, In other words, $h[C_1] = C_2$ and h is a surjective embedding.

B.7 Chapter 7

Exercise 7.2

(i) If $\mathcal{N}$ is finite, it may be taken itself to be the sought structure. If $\mathcal{N}$ is infinite, then it has a countable subset X. Thus $max\{|X|, |\mathcal{L}|\} = |\mathcal{L}|$ and the result follows immediately from the Downward Löwenheim-Skolem theorem.

(ii) Let $\mathcal{N}$ be a $\mathcal{L}$-structure of size $\lambda \geq |\mathcal{L}|$. Let μ be an infinite cardinal such that $|\mathcal{L}| \leq \mu \leq \lambda$. Since $\mu \subseteq \lambda$, there is an injection of μ into N, i.e. there is a subset X of N such that $|X| = \mu$. The Downward Löwenheim-Skolem theorem then yields the sought result.

Exercise 7.4

Any expansion by constants of $\mathcal{M}$ that models $\mathcal{ED}(M)$ can also be used as a model of any finite set of literals of the form $\neg(c_\alpha = c_\beta)$ because M is infinite by hypothesis. Thus, the union of $\mathcal{ED}(M)$ and the entire, given set of negated equalities is finitely satisfiable. By Compactness, it has a model $\mathcal{N}$.

Exercise 7.6

(i) Let the $\mathcal{L}$-structure $\mathcal{M}$ be a finite with $|M| = n$. There is a $\mathcal{L}$-sentence φ_n whose models are isomorphic to $\mathcal{M}$. Because any elementary substructure of $\mathcal{M}$ must be a model of φ_n, the only candidate is $\mathcal{M}$, since no substructure of $\mathcal{M}$ with fewer than n elements can be isomorphic to it.

(ii) Let T be a $\mathcal{L}$-theory with an infinite model $\mathcal{M}$. Setting $|M| = \lambda$, we see that, by the upward Löwenheim-Skolem theorem, $\mathcal{M}$ has an elementary extension $\mathcal{N}$ of cardinality strictly greater than λ. Because $\mathcal{M} \equiv \mathcal{N}$, $\mathcal{N}$ is an infinite model of T and we can iterate the argument just given. If, moreover, $\kappa = |M| \geq |\mathcal{L}|$, fixing $\mu > \kappa$, we may add to $\mathcal{L}$ μ new constant symbols and apply the upward Löwenheim-Skolem theorem to determine an elementary extension $\mathcal{N} \succeq \mathcal{M}$ whose size is at least κ. If it is exactly κ, we are done. Otherwise N has a subset of size $\kappa \geq |\mathcal{L}|$ and we can apply the downward Löwenheim-Skolem theorem. The argument we have relied upon proves Theorem 7.5.

(iii) Suppose that T is a $\mathcal{L}$-theory with an infinite model $\mathcal{M}$. If $|M| = |\mathcal{L}|$ there is nothing to prove. If $|M| < |\mathcal{L}|$, we use part (ii) to obtain an elementary extension of $\mathcal{M}$ of the desired size. If $|M| > |\mathcal{L}|$, we invoke the downward Löwenheim-Skolem instead.

Exercise 7.9

(i) Let $\mathcal{G}$ be $\mathcal{L}_g$-structure that models the group axioms. If $\mathcal{G}$ is infinite, $|G| \geq \aleph_0 = |\mathcal{L}_g|$. By the downward Löwenheim-Skolem theorem, $\mathcal{G}$ has a countably infinite, elementary substructure (which may coincide with $\mathcal{G}$), since it has a countably infinite subset X. This substructure is obviously a subgroup.

(ii) Let $\lambda \leq |G|$.

- By the Downward Löwenheim-Skolem theorem, $\mathcal{G}$ has an elementary subgroup $\mathcal{H}$ of cardinality λ.
- Consider the set:

$$N_b = \{g_1 b^{\epsilon_1} g_1^{-1} \cdots g_k b^{\epsilon_k} g_k^{-1} : \epsilon_i \in \{-1, 1\}, i = 1, \ldots, k, g_i \in H\}$$

We first show that $\mathcal{N}_b \sqsubseteq \mathcal{H}$. It is clear that every element of N_b is in H, by the definition of N_b. Writing two elements of N_b clearly determines an element of N_b so we have closure under group multiplication. Finally, the group elements:

$$g_1 b^{\epsilon_1} g_1^{-1} \cdots g_k b^{\epsilon_k} g_k^{-1} \, and \, g_k b^{\epsilon_k} g_k^{-1} \cdots g_1 b^{\epsilon_1} g_1^{-1}$$

are group inverses and both belong to N_b. We conclude that $\mathcal{N}_b$ is a subgroup of $\mathcal{H}$. As for normality, if $h \in H$, we have:

$$h(g_1 b^{\epsilon_1} g_1^{-1} \cdots g_k b^{\epsilon_k} g_k^{-1})h^{-1} = (hg_1)b^{\epsilon_1}(g_1^{-1} h^{-1})h \cdots (hg_k)b^{\epsilon_k}(g_k^{-1} h^{-1})$$
$$= (hg_1)b^{\epsilon_1}(hg_1)^{-1} \cdots (hg_k)b^{\epsilon_k}(hg_k)^{-1}$$
$$\in N_b.$$

- Note that, if $\mathcal{H}$ is not simple, then it must have a proper, normal subgroup $\mathcal{N}$. Suppose $b \in \mathcal{N}$. Then $\mathcal{N}_b$ is a normal subgroup of $\mathcal{H}$ by (ii) and it is proper because it is a subgroup of $\mathcal{N}$, as one may easily check. It follows that, if $\mathcal{H}$ is not simple, then there is $b \in H$ such that $\mathcal{N}_b$, is a proper subgroup of $\mathcal{H}$. To rule this out, we introduce a model-theoretic consideration.

If $a \in H$, we note that $a \in \mathcal{G} = \mathcal{N}'_b = \{g_1 a^{\epsilon_1} g_1^{-1} \cdots g_k a^{\epsilon_k} g_k^{-1} : \epsilon_i \in \{-1, 1\}, i = 1, \ldots, k, g_i \in G\} = \mathcal{G}$, by the simplicity of $\mathcal{G}$. It follows that:

$$\mathcal{G} \models \exists x_1 \ldots \exists x_k (a = (x_1 + \epsilon_1 b + (-x_1)) + \cdots + (g_k + \epsilon_k b + (-x_k^{-1})))$$

where we recall that $\mathcal{L}_g$ forces us to adopt additive notation for formulae. Because $\mathcal{H}$ is an elementary substructure of $\mathcal{G}$ and $a, b \in H$, the last formula transfers to $\mathcal{H}$ with the same parameters and we conclude that $\mathcal{N}_b, \mathcal{H}$ coincide. This is true of any b, so $\mathcal{H}$ is simple.

Exercise 7.11

Suppose that, if $\mathcal{M}, \mathcal{N} \models T$ implies $\mathcal{M} \equiv \mathcal{N}$. If T were incomplete, there would be a sentence φ such that $T \not\models \varphi$ and $T \not\models \neg\varphi$. In this case, we could find $\mathcal{M}, \mathcal{N} \models T$ such that $\mathcal{M} \models \varphi$ and $\mathcal{M} \models \neg\varphi$, which contradicts our assumption.

Exercise 7.13

A complete theory T with finite models has only one finite model up to isomorphism. Thus, if a theory has finite models and it is also categorical in some infinite cardinal, it cannot be complete.

Exercise 7.16

Since IS has only infinite models and it is e.g. $\aleph_0$-categorical, it is complete. Alternatively, we may note that any injection between infinite sets is elementary (recall Chap. 3). Thus, given two infinite sets A, B, we may elementarily embed both into the larger of them (using the identity map in one case). It follows that IS admits joint elementary embeddings: consequently, it is complete. We reach the same conclusion by noting that, given $A, B \models$ IS, since the Axiom of Choice implies $|A| \leq |B|$ or $|B| \leq |A|$, we have either an elementary embedding from A to B or one from B to A. In any case $A \equiv B$ and the completeness of IS follows again.

Exercise 7.17

Let $\mathcal{L}_E$ be the language $\mathcal{L}_= \cup \{E\}$, where E is a binary relation symbol.

(i) Let $\varphi_n(x_1, \ldots, x_n)$ be the $\mathcal{L}_=$-formula asserting that $x_1, \ldots, x_n$ are all distinct. Let S_0 be the conjunction of the three sentences stating the reflexivity, symmetry and transitivity of E. We next consider the infinite sequence of sentences $S_1 = \{\psi_n\}_{n \geq 1}$, where ψ_n is the sentence:

$$\exists x_1 \ldots \exists x_n \left(\varphi(x_1, \ldots, x_n \wedge \bigwedge_{1 \leq i, j \leq n} \neg E x_i x_j) \right).$$

The models of $S_0 \cup S_1$ have infinitely many distinct equivalence classes. We finally turn to the sequence $S_2 = \{\theta_n\}_{n \geq 1}$, containing sentences of the form:

$$\forall x_1 \ldots \forall x_n \left((\varphi(x_1, \ldots, x_n) \wedge \bigwedge_{1 \leq i,j \leq n} Ex_i x_j)) \to \exists x_{n+1} (\bigwedge_{i=1}^{n} \neg Ex_{n+1} x_i) \right).$$

Models of $S_0 \cup S_1 \cup S_2$ have infinitely many equivalence classes, none of which may be finite. We may set $\mathsf{EQ}^\infty = S_0 \cup S_1 \cup S_2$.

(ii) Clearly, EQ^∞ has only infinite models, so Vaught's test applies. It is easy to see that two countable models of EQ^∞ must have equivalence classes that are countably infinite (if they had at least one uncountable equivalence class, they would not be countable models). Moreover, countable models must have countably many distinct equivalence classes since, again, if they had uncountably many such classes, they would be of uncountable size. It follows that EQ^∞ is $\aleph_0$-categorical and, consequently, complete.

Exercise 7.20

Let VS_K^∞ be the theory of infinite vector spaces over K.

(i) By the Löwenhiem-Skolem theorems, there is a model $\mathcal{V}$ of VS_K^∞ such that $\kappa = |V| > |K|$. By Theorem 7.18, if $\mathcal{V}, \mathcal{W} \models \mathsf{VS}_K^\infty$ and $|V| = |W| = \kappa$, then $\mathcal{V}, \mathcal{W}$ are isomorphic.

(ii) If K is a countable field, then K is at most countably infinite, i.e. $K \leq \aleph_0 \leq |V|$, where V is the domain of an arbitrary model of VS_K^∞. Theorem 7.18 applies for any cardinal $\kappa > \aleph_0$.

(iii) Parts (i) and (ii) license applications of Vaught's test, which entail that, for any field K, VS_K^∞ is a complete theory.

(iv) Suppose K is a field of characteristic p. Then VS_K has models of cardinality $p, p^2, p^3, \ldots p^n, \ldots$. Since the sizes of its finite models are unbounded, VS_K has infinite models by Compactness. The VS_K-sentence saying that there are exactly p distinct vectors is true in a model of VS_K and false in larger finite or infinite models.

(v) If K is a finite field, a complete extension of VS_K may be obtained by fixing the number of vectors. Thus, it suffices to add a VS_K-sentence saying that there are e.g. p vectors. Note that there are infinitely many distinct complete extensions of VS_K, since there are infinitely many distinct ways of fixing the number of vectors.

Exercise 7.22

(i) Any two reals s, r are linearly dependent over $\mathbb{R}$ because s/r is a scalar and $s/r \cdot r = s$. On the other hand, $1, \sqrt{2}$ are linearly independent over the field

$\mathbb{Q}$. If q was a rational such that $q \cdot 1 = \sqrt{2}$, $\sqrt{2}$ would be a rational number, a contradiction. Thus, $\mathbb{R}$ is not one-dimensional as a vector space over $\mathbb{Q}$.

(ii) Any finite dimensional vector space over $\mathbb{Q}$ is isomorphic to the vector space $\mathbb{Q}^n$, for some natural number n. But each such vector space is countably infinite, whereas $\mathbb{R}$ is uncountable.

(iii) Note that $\mathcal{P}(\mathbb{Q})$ contains every lower cut of the rationals, i.e. $\mathbb{R}$ is injectible into $\mathcal{P}(\mathbb{Q})$. It is easy to verify that $|\mathbb{Q}| = |\mathbb{N}|$ implies $|\mathcal{P}(\mathbb{Q})| = |\mathcal{P}(\mathbb{N})|$. By cardinal arithmetic, the set X of all functions from $\mathbb{N}$ to the set $\{0, 1\}$ has cardinality $2^{\aleph_0}$. Thus, the set of infinite binary sequences X has the same size as $|\mathcal{P}(\mathbb{N})|$. It now follows that there is an injection from $\mathcal{P}(\mathbb{Q})$ to X. As a result $|\mathbb{R}| \leq |X|$. The converse inequality was proved in Sect. 4.3. The Cantor-Bernstein theorem now implies $|\mathbb{R}| = 2^{\aleph_0}$.

Exercise 7.23

(i) For ease of notation, we focus on the case $n = 2$. Fix $(s_1, s_2) \in R^2$. As an element of the vector space $\mathbb{R}$ over $\mathbb{Q}$, s_1 can be written as a linear combination of the form $\lambda_1 x_1 + \ldots + \lambda_k x_k$, where $x_1, \ldots, x_k$ is a subset of a fixed uncountable basis of $\mathbb{R}$ over $\mathbb{Q}$. Similarly, $s_2 = \mu_1 y_1 + \ldots + \mu_h y_h$. We deduce:

$$(s_1, s_2) = \lambda_1(x_1, 0) + \ldots + \lambda_k(x_k, 0) + \mu_1(0, y_1) + \ldots + \mu_h(0, y_h).$$

This shows that the set of vectors of the form $(x, 0)$ or $(0, x)$, with $x \in B$, is indeed a basis B' for $\mathbb{R}^2$ over $\mathbb{Q}$.

(ii) Since $B' = B \times \{0\} \cup \{0\} \times B$, where the last union is disjoint, cardinal arithmetic yields $|B| = |B'|$. Any bijection between B and B' may easily be extended to a vector space isomorphism between $\mathcal{R}^n$ and $\mathcal{R}$. The reducts of the last two structures to $\mathcal{L}_g$ are therefore isomorphic under the same isomorphism.

Exercise 7.25

We only verify only one of the vector space axioms, namely $\forall x((f_r + f_s)(x) = f_{r+s}(x))$, for the sake of illustration. The remaining verifications proceed along similar lines. Consider $r, s \in Q$ and set $r = m/n$, $s = p/q$, with $n, q > 0$. Then $r + s = (mq + np)/nq$ and f_{r+s} sends any fixed vector $a \in M$ into a unique vector $b \in M$ such that $(mq+np)a = nqb$. On the other hand, $f_r(a) = b_1$ and $f_s(a) = b_2$, with $ma = nb_1$ and $pa = qb_2$. Thus:

$$nqb = (mq + np)a = qnb_1 + qnb_2 = qn(b_1 + b_2)$$

from which we deduce that $b = b_1 + b_2$. Thus, $f_{r+s}(a) = f_r(a) + f_s(a)$.

Exercise 7.28

It suffices to show that, for any $q, s \in \mathbb{Q}$, $q < s$ implies $f(q) < f(s)$. The converse implication and injectivity follow, using trichotomy. If $q < s$, there is a rational q_n such that the initial segment $\{q_1, \ldots, q_n\}$ in the enumeration of $\mathbb{Q}$ used in the proof contains q, s. By construction $f_n(q) < f_n(s)$ and f agrees with f_n on $\{q_1, \ldots, q_n\}$.

Exercise 7.29

Every model $\mathcal{V} \models \mathsf{VS}_\mathbb{Q}^\infty$ contains at least one vector $v \neq 0$, where 0 is the null vector of $\mathcal{V}$. Thus, it contains $L(v)$, the subspace generated by v, which is easily seen to be isomorphic to $\mathbb{Q}$. Set $f(q) = qv$, for each $q \in \mathbb{Q}$. Then f is an injective linear function from $\mathbb{Q}$ into $\mathcal{V}$, i.e. an embedding. Because $\mathcal{V}$ is an arbitrary model of $\mathsf{VS}_\mathbb{Q}^\infty$, the $\mathcal{L}_v$-structure $\mathbb{Q}$ is algebraically prime.

Exercise 7.32

(i) The theory $\mathsf{DLO}_{+,+}$ is $\aleph_0$-categorical. If the endpoints of any two countable models $\mathcal{M}, \mathcal{N}$ of $\mathsf{DLO}_{+,+}$ are deleted, the resulting linear orders without endpoints are isomorphic. Any isomorphism between them can be extended to an isomorphism between $\mathcal{M}, \mathcal{N}$ by matching the respective endpoints. Vaught's test thus applies and $\mathsf{DLO}_{+,+}$ is complete.

(ii) The argument from part (i) continues to apply when we only have one lower endpoint or one upper endpoint to delete. In the first case, we obtain the completeness of $\mathsf{DLO}_{+,-}$, the theory of dense linear orders with one lower endpoint. In the second case we obtain the completeness of $\mathsf{DLO}_{-,+}$, the theory of dense linear orders with one upper endpoint.

(iii) If an extension of DLO is to be complete, it must in particular model the sentence $\forall x \exists y (y < x)$ or its negation and the sentence $\forall x \exists y (x < y)$ or its negation. Thus, there can only be four complete extensions of DLO and they are precisely $\mathsf{DLO}_{-,-}$, $\mathsf{DLO}_{-,+}$, $\mathsf{DLO}_{+,-}$ and $\mathsf{DLO}_{+,+}$. Given one of the last four theories, any two of its models are elementarily equivalent. We may say that the models of DLO are partitioned into four equivalence classes relative to the relation $\equiv$ (note that these equivalence classes are not sets).

Exercise 7.33

(i) We proceed by induction on the complexity of formulae, restricting attention to the propositional connectives only. The inductive base is clear: $x = x$ holds for any assignment of $a \in M$ to x and the formula $x < x$ fails for any assignment of $a \in M$ to x. If $\varphi(x)$ always fails, then $\neg\varphi(x)$ always holds, and conversely. If at least one of $\varphi(x)\psi(x)$ always holds, so does their disjunction and if they both always fail, then so does their disjunction. This completes the induction.

B Solutions to Exercises   435

(ii) If, for any model of $\mathsf{DLO}_{+,-}$, the formula $\exists y(y < x)$ was satisfied iff the quantifier-free formula $\phi(x)$ was, then $\phi(x)$ should hold under certain assignments and fail under others (e.g. when the minimum of a model is assigned to x), which is impossible by part (i).

(iii) Given a model $\mathcal{N}$ of $\mathsf{DLO}_{+,-}$, let a be its minimum and let $\mathcal{M}$ be the substructure of $\mathcal{N}$ with domain $\{x \in N : b \leq x\}$, with $b \in N$ and $b \neq a$. Clearly, $\mathcal{M}$ is a model $\mathsf{DLO}_{+,-}$ but it is not an elementary substructure of $\mathcal{N}$ because $\exists y(y < x)$ holds in $\mathcal{N}$ when b is assigned to x but fails in $\mathcal{M}$ under the same assignment.

Exercise 7.35

Let $\lambda > \aleph_0$ be an infinite cardinal and L a set of cardinality λ such that $(L, <^L) \models \mathsf{DLO}_{-,-}$.

(i) It is easy to see that $\mathcal{M}$ is a linear order. For any $a \in M$, a belongs to some copy of $(L, <^L)$, say the copy indexed by $m \in \mathbb{N}$. Since L has no endpoints, there is $b < a$ (note that this is the case for $m = 0$, in particular). A similar argument shows that $\mathcal{M}$ has no maximum. If $a < b \in M$ and a, b belong to the same copy of of $(L, <^L)$, its density implies the existence of $c \in M$ such that $a < c$ and $c < b$. If a, b belong to distinct copies of of $(L, <^L)$, then there is $c \in b$, with c, b in the same copy of $(L, <^L)$, such that $c < b$: consequently $a < c$. Thus, density holds.

(ii) Essentially the same argument used in part (i) applies.

(iii) The common cardinality of $\mathcal{M}, \mathcal{N}$ is λ. We show that no isomorphism can exist between these models of $\mathsf{DLO}_{-,-}$. We proceed by contradiction and assume that an isomorphism f does in fact exist. For any $n \in \mathbb{N}$, let L_n ($\mathbb{Q}_n$), be the n-th copy of $(L, <^L)$ (($\mathbb{Q}, <$)) occurring in $\mathcal{N}$.

We prove by transfinite induction that, for each $\kappa < \lambda$, $f[\mathbb{Q}_\kappa] \subseteq L_0$. This result shows that e.g. L_1 is not in the range of f and that f cannot be an isomorphism, against the assumption.

We suppose that, for any $\gamma < \beta < \lambda$, $f[\mathbb{Q}_\gamma] \subseteq L_0$. If $f[\mathbb{Q}_\beta]$ is not included in L_0, then there is $a \in Q_\beta$ such that $f(a) \in L_m$ with $m \neq 0$. Because $|\bigcup_{\gamma \leq \beta} Q_\beta| < \lambda$, there is $d \in L_0 - \bigcup_{\gamma \leq \beta} Q_\beta f[\mathbb{Q}_0]$ such that, for some $b \in M$, $f(b) = d$. Note that, by construction, b cannot be in $\mathbb{Q}_\gamma$ with $\gamma \leq \beta$. As a consequence, $b \in Q_\delta$ for some $\delta > \beta$. It follows that $a < b$ but $f(b) < f(a)$, against the assumption that f is an isomorphism.

Exercise 7.36

Let $\lambda > \aleph_0$ be an infinite cardinal.

(i) Irreflexivity is clear. Transitivity follows from a proof by cases. If $(\alpha, p) <_L (\beta, q)$ and $(\beta, q) <_L (\gamma, r)$, suppose $\alpha \in \beta$ and $\beta = \gamma$. Then $\alpha \in \gamma$ and $(\alpha, p) <_L (\gamma, r)$. The remaining three cases are very similar. Trichotomy follows from the fact that ordinals and rational numbers are linearly ordered by $\in$ and $<$ respectively. There is no $<_L$-least element because, even though there is a least ordinal, namely 0, there isn't a least pair $(0, p)$, with $p \in \mathbb{Q}$. Since λ is limit, for any pair (α, p), L contains the pair $(\alpha + 1, p)$, which is strictly greater. It follows that there is no $<_L$-greatest element. If $(\alpha, p) <_L (\beta, q)$, then, whether or not $\alpha = \beta$, the pair (α, r), with $p < r$ and $r < q$, lies between the given pairs. We conclude that $(L, <_L)$ is a dense linear order without endpoints of size λ.
(ii) The arguments given in part (i) may be replicated with minor changes.
(iii) Consider the subset of L whose elements are of the form $(0, p)$, with $p \in \mathbb{Q}$. It is easy to see that this set is the least copy of $\mathbb{Q}$ in $(L, <_L)$ and, due to the reversal of ordinal comparisons in $<^L$, it i the greatest copy of $\mathbb{Q}$ in $(L, <^L)$.
(iv) Let f be the function from ω_1 into L defined by the condition $f(\alpha) = (\alpha, p)$. Since $\beta, \gamma \in \omega_1$ implies $\beta \in \gamma$ iff $(\beta, p) <_L (\gamma, p)$, the function f is an embedding of ω_1 into $(L, <^L)$.
(v) If g was an isomorphism from $(L, <^L)$ to $(L, <_L)$, the composition gf would embed ω_1 into $(L, <_L)$. We must have $gf(0) = (\beta, p)$, with $\beta \neq 0$. Otherwise ω_1 would be injectible into the countable set of pairs below $(0, p)$ relative to $<_L$.
(vi) The function $gf[\omega_1]$ cannot be embedded into the elements of finitely many copies of $\mathbb{Q}$, whose union is countable. Thus, the elements of $gf[\omega_1]$ occur in an infinite sequence of distinct copies of $\mathbb{Q}$ within $(L, <^L)$. Their first components thus determine an infinite, descending chain of ordinals below β, against the fact that β is a well-ordered set. This contradiction is arrived at under the assumption that g is an isomorphism between $(L, <^L)$ and $(L, <_L)$. We conclude that no such isomorphism exists.

Exercise 7.39

(i) If $q_1\sqrt{2} + q_2\sqrt{3} = 0$, with $q_1, q_2 \in \mathbb{Q}$, squaring both sides we obtain $2q_1^2 + 3q_2^2 = -2q_1q_2\sqrt{6}$, which is an equality between a rational and an irrational number (the irrationality of $\sqrt{6}$ may be proved along the same line as the irrationality of $\sqrt{2}$). It follows that $\sqrt{2}, \sqrt{3}$ are linearly independent over $\mathbb{Q}$. Since, on the other hand, they determine a root of the polynomial $x_1^2 + x_2^2 - 5$, they are algebraically dependent over $\mathbb{Q}$.
(ii) We know from Chap. 3 that the elements of $\mathbb{Q}(r)$ are of the form $f(r)/g(r)$, with f, g polynomials in $\mathbb{Q}[x]$ evaluated on r and $g \neq 0$. Since $\mathbb{Q}[x]$ is countable, $\mathbb{Q}[x]^2$ is. Moreover, the map sending $f(r)/g(r)$ into $f(x)/g(x)$ is injective. Thus $\mathbb{Q}(r)$ is countable. Because $\mathbb{R}$ is uncountably large, $\mathbb{Q}(r)$ leaves out uncountably many reals by cardinal arithmetic. Since each real that

is algebraic over $\mathbb{Q}(r)$ has a minimum polynomial with coefficients in the countable field $\mathbb{Q}(r)$, there are only countably many such reals. We conclude that $\mathbb{R}/Q(r)$ must contain real numbers that are not algebraic over $\mathbb{Q}(r)$, i.e. that are transcendental over $\mathbb{Q}(r)$.

(iii) Set $s \in S$. By hypothesis, there is no $f(x) \in K[x]$ such that $f(x)$ is not the constant polynomial 0 and $f(s) = 0$. It follows that s does not have a minimum polynomial over K. Thus, s is transcendental over K. Because, in particular, s is the root of no polynomial over K of degree 1, $s \notin K$ and $S \cap K = \emptyset$.

(iv) Consider the set of ordered pairs (f, g), with $f, g \in K[x_1, \ldots, x_n]$, $g \neq 0$. We introduce the equivalence relation $(f, g) \sim (h, k)$ iff $fk = gh$. The domain of $K(x_1, \ldots, x_n)$ is the quotient $K[x_1, \ldots, x_n]^2 / \sim$. Addition between pairs is defined by the stipulation:

$$(f, g) \oplus (h, k) = (fk + hg, gk)$$

where $gk \neq 0$ because $g, k \neq 0$ and $K[x_1, \ldots, x_n]$ is an integral domain. Multiplication between pairs is defined by the stipulation:

$$(f, g) \otimes (h, k) = (fh, gk)$$

Note that the equivalence class of $(0, 1)$, with 1 the unit of K, is a neutral element relative to addition, while the equivalence clsass of $(1, 1)$ is a multiplicative unit. It is not difficult to verify that $K(x_1, \ldots, x_n)$ is a ring. Note that $(f, g) \not\sim (0, 1)$ iff $f \neq 0$: whenever this is the case (g, f) belongs to an equivalence class in the domain of $K(x_1, \ldots, x_n)$ and, since $(g, f) \otimes (f, g) = (1, 1)$, (f, g) has a multiplicative inverse. It follows that $K(x_1, \ldots, x_n)$ is a field.

There is no problem using the construction just defined on an arbitrary set X of variables. We still define $K(X)$ as a fraction field whose elements are pairs (f, g), where $f, g \in K[x_1, \ldots, x_n]$, for sufficiently large n and $x_1, \ldots, x_n \in X$, and $g \neq 0$. Since any ring $K[x_1, \ldots, x_n]$ based on a finite subset of X is an integral domain, we may define addition of pairs as above. Our construction results again in a field extension of K.

(v) First, we prove that (a) implies (b). Since S is algebraically independent over K, $s_1, \ldots, s_n$ are transcendental over K: in particular, s_1 is. Consider the field of fractions $K(s_1, \ldots, s_{j-1})$. If $s : j$ was not transcendental over it, there is a minimum polynomial $m(x_j)$ over $K(s_1, \ldots, s_{j-1})$ such that $m(s_j) = 0$. But, if we replace $s_1, \ldots, s_{j-1}$ in $m(x_j)$ with $x_1, \ldots, x_{j-1}$, we obtain a nonzero polynomial $m(x_1, \ldots, x_j)$ with coefficients in K and such that $m(s_1, \ldots, s_j) = 0$, against the fact that S is algebraically independent over K.

Next, we assume (b) and prove (c). By part (iv) we know that we can write the elements of $K(s_1, \ldots, s_n)$ as 'ratios' $f(s_1, \ldots, s_n)/g(s_1), \ldots, s_n)$, with $f, g \in K[x_1, \ldots, x_n]$ and $g \neq 0$. For convenience, we use $\overline{x}, \overline{x}$ as abbreviations of the sequences $x_1, \ldots, x_n$, $s_1, \ldots, s_n$ respectively. We identify equivalent ratios (i.e. ratios where $f/g \sim h/k$ in the sense that $fk = gh$). Thus:

$$f(\overline{x})/g(\overline{x}) \sim h(\overline{x})/k(\overline{x}) \text{ iff } f(\overline{x})k(\overline{x}) = h(\overline{x})g(\overline{x}) \text{ iff } f(\overline{s})k(\overline{s})$$
$$= g(\overline{s})h(\overline{s}) \text{ iff } f(\overline{s})/g(\overline{s}) = h(\overline{s})/k(\overline{s}).$$

Then the function θ defined by the condition:

$$\theta\left(\frac{f(\overline{s})}{g(\overline{s})}\right) = \frac{f(\overline{x})}{g(\overline{x})}$$

is injective and, by construction, also a bijection. Denoting sum in $K(s_1, \ldots, s_n)$ by $\oplus$ and sum in $K(x_1, \ldots, x_n)$ by $+$, we obtain:

$$\theta\left(\frac{f(\overline{s})}{g(\overline{s})} \oplus \frac{h(\overline{s})}{k(\overline{s})}\right) = \theta\left(\frac{f(\overline{s})k(\overline{s}) + h(\overline{s})g(\overline{s})}{g(\overline{x})k(\overline{x})}\right)$$
$$= \frac{f(\overline{x})k(\overline{x}) + h(\overline{x})g(\overline{x})}{g(\overline{x})k(\overline{x})}$$
$$= \frac{f(\overline{x})}{g(\overline{x})} + \frac{h(\overline{x})}{k(\overline{x})}$$

It is easy to see that multiplication, additive inverses and multiplicative inverses are preserved. We conclude that θ is a field isomorphism.

It now remains to show that (c) implies (a). To this end, we suppose that $f(\overline{s}) = 0$, with $f(\overline{x}) \in K[\overline{x}]$. Our goal is to show that f is 0. Since there is an isomorphism θ from $K(\overline{s})$ onto $K(\overline{x})$, our hypothesis implies $\theta(f(\overline{s})) = \theta(0))$, i.e. $f(\overline{x}) = 0$.

Exercise 7.41

(i) The argument from part (i) of the previous exercise may be replicated here, relative to $\mathbb{Q}$, rather than $\mathbb{Q}(r)$.
(ii) Note that α is of the form $f(r)/g(r)$ for some $f, g \in \mathbb{Q}[x]$ with $g \neq 0$. If x_1 is evaluated on r and x_2 on $f(r)/g(r)$, then the polynomial $g(x_1)x_2 - f(x_1)$ is evaluated on $g(r)(f(r)/g(r)) - f(r)) = 0$.
(iii) Since r is transcendental over $\mathbb{Q}$, $\{r\}$ is an algebraically independent subset of $\mathbb{R}$ over $\mathbb{Q}$ by part (v)–(a) of the previous exercise. If this subset could be extended to a larger, algebraically independent subset, it should be possible to add to it at least one more element $\alpha \in \mathbb{R}$. Part (ii) and part (v)–(b) of the previous exercise imply that $\{r, \alpha\}$ is not an algebraically independent set. It now follows that $\{r\}$ is a maximal, algebraic independent set, i.e. it is a transcendence base of $\mathbb{R}$ over $\mathbb{Q}$. It is most definitely not a basis of the infinite-dimensional vector space $\mathbb{R}$ over the field $\mathbb{Q}$.

Exercise 7.43

Suppose that S is a transcendence base of $\mathcal{M}$ over K and consider $a \in \mathcal{M} - K(S)$. If $a \in L$ were transcendental over $K(S)$, Theorem 7.42 would imply the algebraic independence of $S \cup \{a\}$ over K, against the maximality of S. It follows that a is algebraic over $K(S)$ and, consequently, that $\mathcal{M}$ is an algebraic extension of $K(S)$. Conversely, if $\mathcal{M}$ is an algebraic extension of $K(S)$, S must be maximal because, if it wasn't, it should be possible to find $a \in L$ and obtain the algebraically independent extension of $S \cup \{a\}$ over K. By Theorem 7.42, this can be done only if $a \in M$ is transcendental over $K(S)$, a possibility that we have ruled out.

Exercise 7.44

The only polynomial over K that is 0 when evaluated vacuously on the empty set is 0 itself. Thus $\emptyset$ is algebraically independent over K. We may thus consider the family $\mathcal{T}$ of algebraically independent subset of $\mathcal{M}$ over K and rest assured that $\mathcal{T}$ is not empty. Any chain in $\mathcal{T}$ has an upper bound. If $\{S_i\}_{i \in I}$ is such that, for any $j, k \in I$, either $S_j \subseteq S_k$ or $S_k \subseteq S_j$, then $\bigcup \{S_i\}_{i \in I}$ is algebraically independent because any one of its finite subsets is contained in some algebraically independent set from $\mathcal{T}$. Zorn's lemma applies and guarantees the existence of a maximal subset of L that is algebraically independent over K. This is a transcendence base of $\mathcal{M}$ over K.

Exercise 7.45

(i) First, note that $T \neq \emptyset$. This is because $\emptyset \subset S$ and T would not be a maximal independent set over K. If no element of T is transcendental over $\mathcal{K}(S_1)$, then the field extension $\mathcal{K}'(T)$, where $\mathcal{K}' = \mathcal{K}(S_1)$, is an algebraic extension of $\mathcal{K}(S_1)$. Because T is a transcendence base of $\mathcal{M}$ over K, $\mathcal{M}$ is an algebraic extension of $\mathcal{K}(T) \subseteq \mathcal{K}'(T)$ and, consequently, an algebraic extension of $\mathcal{K}'(T)$. By Theorem 6.18, $\mathcal{L}$ is an algebraic extension of $\mathcal{K}(S_1)$. Exercise 7.43 now implies that S_1 is a transcendence base of $\mathcal{M}$ over K. Since, however, S is algebraically independent, S_1 is not maximal, against the definition of transcendence base. This contradiction implies that some $t_1 \in T$ is transcendental over $\mathcal{K}(S_1)$.
(ii) If s_1 were transcendental over $\mathcal{K}(T_1)$, we could deduce that $\{s_1\} \cup T_1$ is algebraically independent over K, against the maximality of S. It follows that s_1 is algebraic over $\mathcal{K}(T_1)$.
(iii) In view of part (ii), the extension field $\mathcal{K}'(T_1)$ generated by $\mathcal{K}(T_1)$ and $\{s_1\}$ is an algebraic extension of $\mathcal{K}(T_1)$. Because $\mathcal{K}'(T_1)$ contains a transcendence base of $\mathcal{M}$, every element of $\mathcal{M}$ not in $\mathcal{K}'(T_1)$ must be algebraic over $\mathcal{K}'(T_1)$ (since it is already algebraic over a subfield of the latter field). It follows from

Theorem 6.11 that $\mathcal{M}$ is an algebraic extension of $\mathcal{K}(T_1)$. By Exercise 7.43, T_1 is a transcendence base of $\mathcal{M}$ over $\mathcal{K}$.
(iv) Note that $T \supset \{t_1\}$. Otherwise, $\{t_1, s_2, \ldots, s_n\}$ would properly extend the transcendence base $\{t_1\}$. Thus, T is not a singleton. Now note that S_2 is algebraically independent over $\mathcal{K}$ and that it is obtained by deleting one element from the transcendence base T_1. Since $T - \{t_1\} \neq \emptyset$, we can reason as in part (i) to show that there is $t_2 \in T$ transcendental over $\mathcal{K}(S_2)$. Iterating parts (ii) and (iii), we obtain the transcendence base T_2 and can carry out another iteration of the same argument, if any s_i are left in T_2. We stop when none of them is left, i.e. after n steps.
(v) By construction $T_n \subseteq T$. If it missed out any elements of T, it would have a proper extension that is algebraically independent over $\mathcal{K}$, against the maximality of T. It follows that $T = T_n$.

Exercise 7.47

On the other hand, an element of $\mathcal{K}(S)$ has the form $h(\overline{s})/k(\overline{s})$, with $h(\overline{x}), k(\overline{x}) \in \mathcal{K}[X]$. Since $\mathcal{K} \subseteq \mathcal{K}(\bigcup_{s \in S} T_s)$, $h(\overline{s}), k(\overline{s})$ may also be regarded as polynomials with coefficients in $\mathcal{K}(\bigcup_{s \in S} T_s)[X]$. As a result, $h(\overline{s})/k(\overline{s})$ may be regarded as an element of $\mathcal{K}(\bigcup_{s \in S} T_s)(S)$.

Exercise 7.50

If the transcendence degree of $\mathcal{L}$ over $\mathcal{K}$ equals 0, then $\emptyset$ is a transcendence base of $\mathcal{M}$ over $\mathcal{K}$. In this case, by Exercise 7.43, $\mathcal{M}$ is an algebraic extension of the field generated by $\mathcal{K} \cup \emptyset$, i.e. $\mathcal{K}$ itself.

Conversely, if $\mathcal{M}$ is an algebraic extension of $\mathcal{K}$, its transcendence base over $\mathcal{K}$ extends $\emptyset$. But, for any $a \in \mathcal{M}$, $\{a\}$ is algebraic over $\mathcal{K}$, so $\emptyset$ is maximal, i.e. a transcendence base of $\mathcal{M}$ over $\mathcal{K}$.

Exercise 7.51

Set $S_b = S - \{b\}$. The hypothesis that a is algebraic over $\mathcal{K}(S)$ implies the existence of a minimum polynomial of a, i.e. $m(x) \in \mathcal{K}(S)[x]$ such that $m(a) = 0$. The hypothesis that a is not algebraic over $\mathcal{K}(S_b)$ implies that b must occur among the coefficients of $m(x)$. We may thus write $m(x) = m(x, b)$, in order to highlight the occurrence of a parameter, and note that $m(a, b) = 0$. This is to say that $m(a, y) \in K(S_b\{a\})$ has b as a root, i.e. b is algebraic over $\mathcal{K}(S_b \cup \{a\})$.

B Solutions to Exercises   441

Exercise 7.54

It is easy to see that, if $a, b \in M$, there are m, n such that $a \in K_m, b \in K_n$. Since $K_n \supseteq K_m$ and K_n is the domain of a field, $a \pm b$, ab and a/b are elements of K_n and thus of M.

Exercise 7.56

$K_0 = K$ is countable by assumption. We replicate the construction of Lemma 7.52 to obtain a field extension $\mathcal{K}_1'$. By the Downward Löwenheim-Skolem theorem, there is a countable, elementary substructure $\mathcal{K}_1 \preceq \mathcal{K}_1'$. In $\mathcal{K}_1$, every polynomial in $\mathbb{Q}[x]$ has a root, since in particular existential sentences transfer from $\mathcal{K}_1'$ to $\mathcal{K}_1$. This argument shows that, if $\mathcal{K}_n$ is countable, it is possible to choose $\mathcal{K}_{n+1}$ to be a countable extension. As a result, $\bigcup_{n \geq 1} \mathcal{K}_n$ is countable, being a countable union of countable sets.

Exercise 7.58

(i) By definition of $\mathbb{Q}(A)$, A is a subset of $\mathbb{Q}(A)$ and, *a fortiori*, of $\overline{\mathbb{Q}(A)}$;

(ii) If $a \in \overline{\mathbb{Q}(A)}$ then a is the root of a polynomial $p(x)$ with coefficients in $\mathbb{Q}(A)$. Let $a_0, \ldots, a_n$ be the coefficients of $p(x)$. Then $p(x)$ is a polynomial with coefficients in $\mathbb{Q}(A_0)$, with $A_0 = \{a_0, \ldots, a_n\}$. Since $p(x)$ can be written as a product of linear factors with coefficients in $\overline{\mathbb{Q}(A_0)}$, $a \in \overline{\mathbb{Q}(A_0)}$.

(iii) $A \subseteq B$ implies $\mathbb{Q}(A) \subseteq \mathbb{Q}(B) \subseteq \overline{\mathbb{Q}(B)}$. Thus, $\mathbb{Q}(A)$ is included in the algebraically closed field $\overline{\mathbb{Q}(B)}$. It follows that $\overline{\mathbb{Q}(A)} \subseteq \overline{\mathbb{Q}(B)}$.

(iv) Let $\mathcal{K}$ be an algebraically closed field containing $\overline{\mathbb{Q}(A)}$. The relative algebraic closure of $\overline{\mathbb{Q}(A)}$ in $\mathcal{K}$ is, up to isomorphism, $\overline{\mathbb{Q}(A)}$. But, if $a \in \mathcal{K}$ is algebraic over $\overline{\mathbb{Q}(A)}$, then a is the root of a polynomial with coefficients in $\overline{\mathbb{Q}(A)}$. An such polynomial has its roots already in $\overline{\mathbb{Q}(A)}$. Thus $\overline{\overline{\mathbb{Q}(A)}} = \overline{\mathbb{Q}(A)}$.

(v) Suppose $A \subseteq \overline{\mathbb{Q}(B)}$ and $B \subseteq \overline{\mathbb{Q}(C)}$. By part (iii), the second inclusion implies $\overline{\mathbb{Q}(B)} \subseteq \overline{\mathbb{Q}(\overline{\mathbb{Q}(C)})}$. Note that $\mathbb{Q}(\overline{\mathbb{Q}(C)})$ is the smallest subfield of $\mathcal{L}$ generated by $\mathbb{Q} \cup \overline{\mathbb{Q}(C)} = \overline{\mathbb{Q}(C)}$. Because $\overline{\mathbb{Q}(C)}$ is already a subfield of $\mathcal{L}$, it coincides with $\mathbb{Q}(\overline{\mathbb{Q}(C)})$. We have shown $\overline{\mathbb{Q}(B)} \subseteq \overline{\mathbb{Q}(C)}$, which, together with the first inclusion from our hypothesis, implies $A \subseteq \overline{\mathbb{Q}(C)}$.

Exercise 7.61

If $p(x_1, \ldots, x_n) \in F[X]$, with $|X| = |S_1| = |S_2|$, then, for any $s_1, \ldots, s_n \in S_1$:

$$f_1(p(s_1, \ldots, s_n)) = p(f(s_1), \ldots, f(s_n)).$$

Moreover, if $p(x_1, \ldots, x_n), q(x_1, \ldots, x_m) \in F[X]$, then, for any $s_1, \ldots, s_n, t_1, \ldots, t_m \in S_1$:

$$f_2\left(\frac{p(s_1, \ldots, s_n)}{q(t_1, \ldots, t_m)}\right) = \frac{f_1(p(s_1, \ldots, s_n))}{f_1(q(t_1, \ldots, t_m))} = \frac{p(f(s_1), \ldots, f(s_n))}{q(f(t_1), \ldots, f(t_m))}.$$

Exercise 7.68

(i) If $\mathcal{M} \preceq \mathcal{N}$, and $\varphi(x)$ is algebraic over A in $\mathcal{M}$, suppose $\varphi(x)$ has exactly n solutions in M. Then $\mathcal{M} \models \exists^{=n} x \varphi(x)$ and, since $\mathcal{M} \equiv \mathcal{N}$, this is the case iff $\mathcal{N} \models \exists^{=n} x \varphi(x)$. Thus $\varphi(x)$ is algebraic over A in $\mathcal{N}$ and its solutions are all in M.
(ii) Immediate from the definition of algebraic closure.
(iii) If $a \in A \subseteq M$, the formula $x = a$ is algebraic over A in $\mathcal{M}$. Thus $a \in \subseteq acl_{\mathcal{M}}(A)$, for each $a \in A$.
(iv) A formula $\varphi(x)$ that is algebraic over A in $\mathcal{M}$ is certainly also algebraic over B in $\mathcal{M}$.

Exercise 7.69

(i) By way of illustration, consider $1/2 x^3 + 2/3 x + 5 \in \mathbb{Q}[x]$. The $\mathcal{L}_r$-formula defining its roots in L is:

$$\exists y_1 \exists y_2 ((y_1 + y_1 = 1) \wedge (y_2 + y_2 + y_2 = 1 + 1)$$
$$\wedge\; y_1 x^3 + y_2 x + (1 + 1 + 1 + 1 + 1+) = 0).$$

(ii) By part (i), $acl_{\mathcal{M}}(\emptyset)$ contains every element of M that is algebraic over the prime subfield $\mathbb{Q}$, i.e. $acl_{\mathcal{M}}(\emptyset)$ contains the relative algebraic closure of $\mathbf{Q}$ in $\mathcal{M}$. Because $\mathcal{M}$ is algebraically closed, $\overline{\mathbf{Q}} \subseteq acl_{\mathcal{M}}(\emptyset)$.
(iii) By hypothesis, any algebraic $\mathcal{L}_r$-formula $\psi(x, \overline{a})$ with parameters in $A \subseteq L$ defines the same finite set as a quantifier-free formula $\varphi(x, \overline{a})$.
By the remarks preceding Exercise 2.47 in chapter two, $\varphi(x, \overline{a})$ may be written as a disjunctive normal form, i.e. a disjunction of conjunctions in which the conjunct are either atomic formulae or negated atomic formulae, i.e. literals. We may therefore express $\varphi(x, \overline{a})$ as:

$$\theta_1(x, \overline{a}) \vee \ldots \vee \theta_m(x, \overline{a}),$$

where $\theta_i(x, \overline{a})$ is a conjunction of literals. If no positive literals occur in $\theta_i(x, \overline{a})$, then this formula contains a finite list of **polynomial inequations**, i.e. atomic formulae of the form $\neg(f(x, \overline{a}) = 0)$ where $f(x)$ is a polynomial with coefficients in the substructure of $\mathcal{M}$ generated by A. Because a polynomial with coefficients in M has only finitely many roots and M is infinite, any negated atomic formula, or polynomial inequation, defines an infinite subset of M. This infinite set is in particular *cofinite* (recall the definition given in Sect. 5.4), because it only excludes from M the finitely many roots of a fixed polynomial. Moreover, any finite conjunction of polynomial inequations defines a cofinite set because the intersection of finitely many cofinite sets can only exclude finitely many elements of M.

By contrast, a single polynomial equation or a conjunction of polynomial equations defines a finite set. It follows that the algebraic formulae among $\theta_1, \ldots, \theta_m$ are exactly those containing at least one positive literal.

(iv) It follows from part (ii) that, if the disjunction $\theta_1(x, \bar{a}) \vee \ldots \vee \theta_m(x, \bar{a})$, is algebraic over A, then it defines a finite union of finite sets of polynomial roots. This is true of any quantifier-free formula and thus, by our assumption, of every formula. We conclude that $acl_\mathcal{M}(A)$ only contains polynomial roots. It must therefore coincide with $\overline{\mathbb{Q}(A)}$.

(v) This is a special case of part (iii) with $A = \emptyset$.

Exercise 7.72

The set $acl_\mathcal{V}(A)$ is, in this case, the linear closure of A in $\mathcal{V}$. To see why, note that any algebraic formula must be equivalent to a quantifier-free formula over A. Such a formula may be assumed in disjunctive normal form. Each disjunct is a conjunction of equalities and negated equalities. Since any equality in one free variable x describes x as a linear combination of elements of A, equalities are algebraic and define single vectors. Negated equalities define, by contrast, infinite sets of vectors. Thus, any algebraic formula must contain one equality in each disjunct and, as a consequence, must define a finite set of vectors generated by A. The totality of algebraic formulae thus adds to A the linear combinations of its elements, thus generating the linear closure of A. Since, in a vector space, algebraic closure coincides with linear closure, independence straightforwardly reduce to linear independence. If $A, \mathcal{V}$ are as in the previous exercise, B is independent over A iff for any $b \in B$, b does not lie in the linear closure of A, i.e. iff b cannot be expressed as a linear combination of vectors from A. When e.g. $B = \{v_n\}$ and $A = \{v_1, \ldots, v_{n-1}\}$, independence reduces to the fact that v_n is linearly independent of $\{v_1, \ldots, v_{n-1}\}$.

Exercise 7.74

If $\mathcal{M} \models \mathsf{ACF}_0$, a basis B for $A \subseteq M$ is an algebraically independent subset of A such that $\overline{\mathbb{Q}(B)} = acl_\mathcal{M} B = acl_\mathcal{M} A = \overline{\mathbb{Q}(A)}$. Thus $\mathbb{Q}(B)$ is a field extension of $\mathbb{Q}$ such that $\overline{\mathbb{Q}(A)}$ is algebraic. It follows that B is a transcendence base for $\overline{\mathbb{Q}(A)}$ over $\mathbb{Q}$.

B.8 Chapter 8

N.B.: for notational convenience, we use $\bar{x}$ as an abbreviation of the list of variables $x_1, \ldots, x_n$ and $\exists \bar{x}$ as an abbreviation of the string of quantifiers $\exists x_1, \ldots, \exists x_n$. We adopt similar abbreviations for lists of elements of set M and strings of universal quantifiers.

Exercise 8.3

For any $\mathcal{L}$-formula $\phi(\bar{x})$ and any $\mathcal{L}$-theory T, $T \models \forall \bar{x}(\phi(\bar{x}) \leftrightarrow \phi(\bar{x}))$. Thus, T-equivalence is reflexive. Its symmetry is trivial. To conclude the proof, we suppose that, given a $\mathcal{L}$-theory T and $\mathcal{L}$-formulae ϕ, ψ, χ:

$$T \models \forall \bar{x}(\phi(\bar{x}) \leftrightarrow \psi(\bar{x})) \text{ and } T \models \forall \bar{x}(\phi(\bar{x}) \leftrightarrow \chi(\bar{x})).$$

Fix an arbitrary model $\mathcal{M}$ of T and $\bar{a} \in M$. If $\mathcal{M} \models \phi(\bar{a})$, then by hypothesis $\mathcal{M} \models \psi(\bar{a})$. The last relation in turn implies $\mathcal{M} \models \chi(\bar{a})$. If $\mathcal{M} \models \chi(\bar{a})$, we may reverse the argument. Because $\mathcal{M}$ and $\bar{a}$ have been selected arbitrarily, what we have shown applies to any model of T and any n-tuple of elements in the domain of that model. Thus:

$$T \models \forall \bar{x}(\phi(\bar{x}) \leftrightarrow \chi(\bar{x})).$$

We conclude that T-equivalence is transitive and, thus, an equivalence relation.

Exercise 8.5

Note that, relative to any $\mathcal{L}$-theory T, the $\mathcal{L}$-formula $\exists \bar{x} \phi(\bar{x}, \bar{z})$, with $\bar{z}$ occurring free in ϕ, is T-equivalent to $\exists \bar{y} \phi(\bar{y}, \bar{z})$, where $\bar{y}$ is a list of new variables, disjoint from either $\bar{x}$ or $\bar{z}$. We may also arrange for the lists of free and bound variables to be disjoint (i.e. no variable is free in som occurrences and bound in others).

Using this fact, we note that a disjunction of existential $\mathcal{L}$-formulae is T-equivalent to a disjunction of the form:

$$\exists \bar{x}_1 \phi(\bar{x}_1, \bar{z}_1) \vee \ldots \vee \exists \bar{x}_m \phi(\bar{x}_m, \bar{z}_m)$$

where the lists of variables involved are pairwise disjoint. For an arbitrary $\mathcal{L}$-theory T, the last formula is T-equivalent to:

$$\exists \bar{x}_1, \ldots, \exists \bar{x}_m (\phi(\bar{x}_1, \bar{z}_1) \vee \ldots \vee \phi(\bar{x}_m, \bar{z}_m))$$

which is an existential $\mathcal{L}$-formula. The result follows from the fact that equivalence is T-equivalence relative to any $\mathcal{L}$-theory.

Exercise 8.9

Let $\mathcal{M}$ be an existentially closed field. This is to say that, if $\mathcal{N}$ is a field extending $\mathcal{M}$ and $\mathcal{N} \models \exists \overline{x} \phi(\overline{x}, \overline{a})$, with $\overline{a} \in M$, then $\mathcal{M} \models \exists \overline{x} \phi(\overline{x}, \overline{a})$. Whenever $\phi(x, \overline{a})$ is an atomic formula in one free variable, the condition $\mathcal{N} \models \exists x \phi(x, \overline{a})$ asserts that a polynomial in the variable x with coefficients in M, depending on $\overline{a}$, has a root in N. Existential closure thus implies that any polynomial with coefficients in M has a root in M. This is to say that $\mathcal{M}$ is algebraically closed.

Exercise 8.8

(i) On the one hand, if condition (2) holds for all existential formulae, it certainly holds for primitive ones. If, on the other hand, it holds for primitive formulae, we note that, given an existential formula $\exists \overline{x} \phi(\overline{a}, \overline{x})$, the matrix $\phi(\overline{a}, \overline{x})$ may be assumed in disjunctive normal form. In this case, the whole formula may be written as:

$$\exists \overline{x} (\phi_1(\overline{x}) \vee \ldots \vee \phi_m(\overline{x})),$$

where each $\phi_i(\overline{x})$ is a conjunction of literals. It is not difficult to check that the last formula is equivalent to:

$$\exists \overline{x} (\phi_1(\overline{a}, \overline{x}) \vee \ldots \vee \exists \overline{x} \phi_m(\overline{a}, \overline{x})).$$

A $\mathcal{L}$-structure $\mathcal{M}$ with $\overline{a} \in M$ thus models an existential formula iff it models a disjunction of primitive formulae iff it models at least one primitive formula from the disjunction. Thus, condition (2) restricted to primitive formulae implies condition (2) generalised to existential formulae.

(ii) Suppose $\mathcal{M}, \mathcal{N} \models \mathsf{EQ}^\infty$ and $\mathcal{M} \sqsubseteq \mathcal{N}$. Let $\varphi(\overline{a}, \overline{x})$ be a primitive formula in the language of EQ^∞ with parameters $\overline{a} \in M$. The literals occurring in the matrix of φ have one of the forms $Ex_i x_j$, $Ex_i a$, $\neg Ex_i y_j \neg Ex_i a$, $\neg (x_i = a)$, $\neg (x_i = y_j)$, $x_i = y_j$, $x_i = a$. We may disregard equalities by carrying out substitutions.

Suppose $\mathcal{N} \models \exists \overline{x} \varphi(\overline{a}, \overline{x})$. The negative literals in φ tell us across how many distinct equivalence classes the elements of N that satisfy the given primitive formulae are to be distributed and how many of them are distinct. The positive literals tell us how many of these elements lie within the same equivalence class.

Thus, in order to satisfy $\exists \overline{x} \varphi(\overline{a}, \overline{x})$ in $\mathcal{N}$, we have to consider only finitely many distinct equivalence classes and only finitely many distinct objects within each equivalence class. Some of these equivalence classes contain elements of M so they include corresponding equivalence classes in M. Other equivalence classes are not fixed by a parameter. Since, however, M contains infinitely many equivalence classes, and they are all infinite, M contains enough, suitably distributed objects to play the role of $\bar{b}$ in N. It follows that $\mathcal{M} \models \exists \bar{x} \varphi(\bar{a}, \bar{x})$.

Exercise 8.12

If $\mathcal{M} \models \mathsf{DLO}_{+,-}$, let a_0 be least in M. We assign the symbol P in the expanded language of $\mathsf{DLO}^*_{+,-}$ to the singleton $\{a_0\}$. It is clear that, for any $a \in M$, $\mathcal{M} \models P(a)$ iff $a = a_0$ iff $\mathcal{M} \models \forall y(\neg(y < a))$.

If $\mathsf{DLO}^*_{+,-}$ is complete, any two of its models satisfy the same sentences in the language of $\mathsf{DLO}^*_{+,-}$. Thus, they satisfy the same sentences in the language of $\mathsf{DLO}_{+,-}$ and the last fact must therefore be true of their reducts to this language. It follows that $\mathsf{DLO}_{+,-}$ must be complete.

If $\mathcal{N} \models \mathsf{DLO}^*_{+,-}$, let a_0 be least in N. If $\mathcal{M}$ is a submodel of $\mathcal{N}$, suppose $a \neq a_0$ is least in M. Note that, in this case $\mathcal{M} \models P(a)$ and, since atomic formulae are preserved by substructures, $\mathcal{N} \models P(a)$. But now P^N is at least a pair and it is easy to see that the interpretation of P in $\mathcal{N}$ must be a singleton (otherwise we can deduce $\mathcal{N} \models \forall y(\neg(y < a)) \wedge \forall y(\neg(u < a_0)))$.

It follows that P^M, which must be a singleton, cannot be any other singleton than $\{a_0\}$.

Exercise 8.14

The ordered set of nonnegative rational numbers $(Q_0^+, <)$ can be embedded into any dense linear order with a least element $\mathcal{M}$. The proof imitates that of Lemma 7.27. We only have to extend the embedding obtained in that proof by mapping 0 into the least element of M. Because $\mathsf{DLO}^*_{+,-}$ is model-complete, $(Q_0^+, <)$ is a prime model.

It follows that any two models of $\mathsf{DLO}^*_{+,-}$ include the same (up to isomorphism) elementary substructure $(Q_0^+, <)$. Thus, any two models of $\mathsf{DLO}^*_{+,-}$ must be elementarily equivalent. It follows that $\mathsf{DLO}^*_{+,-}$ is complete. By the previous exercise, $\mathsf{DLO}_{+,-}$ is complete.

Exercise 8.15

We may simply extend the approach used for $\mathsf{DLO}^*_{+,-}$ and expand the language of the latter theory by the addition of a new unary predicate Q. We obtain $\mathsf{DLO}^*_{+,+}$ by adding to $\mathsf{DLO}_{+,-}$ the axiom (in the expanded language):

$$\forall x(Q(x) \leftrightarrow \forall y(\neg(x < y))).$$

B Solutions to Exercises   447

The proof of model-completeness for $\mathsf{DLO}^*_{+,+}$ is not essentially different from the proof of Lemma 8.13 and only requires remarks about literals with parameters of the form $Q(a_0), \neg Q(b_k), \neg Q(a_i)$ entirely similar to the remarks made for the corresponding literals involving P. By model-completeness, the set of rationals $(0, 1) = \{q \in \mathbb{Q} : 0 \leq q \wedge q \leq 1\}$, regarded as a linear order, is a prime model of $\mathsf{DLO}^*_{+,+}$ (modify Lemma 7.27 again, matching least and greatest elements). By Exercise 8.14, the latter theory is complete and so is $\mathsf{DLO}_{+,+}$.

Exercise 8.16

(i) The first axioms says that R is a symmetric relations. The next two axioms say, respectively, there there is exactly one point bearing R to nothing else, not even itself, and that there are exactly three irreflexive points relative to R. The last axiom specifies that two of these points are related to each other. The penultimate axiom says that, apart from the three special points just described, any other point is R-related only to itself.

(ii) Suppose that $\mathcal{M} \sqsubseteq \mathcal{N} \models T$ and $\mathcal{N} \models \exists y \varphi(y, \overline{a})$, with $\overline{a} \in M$. Then there is $b \in N$ such that $N \models \varphi(b, \overline{a})$. If b is one of the three special, irreflexive points of N, then, since these points are shared with M and $\mathcal{M} \models \exists y \varphi(y, \overline{a})$. Otherwise b is a point of N that is only related to itself. It is clear that any self-map of $\mathcal{N}$ that fixes the three special points and permutes the remaining points is an automorphism. Thus, if b is not a special point of N, there is an automorphism that preserves $\overline{a}$ and sends b into a non-special point $a_0 \in M$. Because automorphisms preserves first-order information $\mathcal{N} \models \varphi(a_0, \overline{a})$. The Tarski-Vaught criterion is satisfied so T is model-complete.

(iii) If $\mathcal{M} \models T$, let $a, b \in M$ be two the special objects of M, with a unrelated to any other point of M. Since any subset of M determines a substructure of $\mathcal{M}$, this is the case with both $\{a\}$ and $\{b\}$. In both cases, R is restricted to the empty set, so the resulting substructures are isomorphic, because they both contain exactly one element. Their common diagram is $\mathcal{D}(\mathcal{A}) = \{c = c, \neg R(c, c)\}$.

If we expand $\mathcal{M}$ to $\mathcal{M}_a$ and $\mathcal{M}_b$, which interpret the constant symbol c occurring in $\mathcal{D}(\mathcal{A})$ on a, b respectively, then it is clear that $\mathcal{M}_a, \mathcal{M}_b \models T \cup \mathcal{D}(\mathcal{A})$.

Nonetheless, these two expansions of $\mathcal{M}$ are not elementarily equivalent in the expanded language because one of them satisfies the sentence $\exists y(R(c, y))$ and the other does not. It follows that $(T \cup \mathcal{D}(\mathcal{A})) \models$ is an incomplete theory.

Exercise 8.18

(i) The literals we must consider contain the symbols '$=, <, S$'. Because we are working with linear orders, negated equalities or inequalities can be reduced to inequalities on account of trichotomy (e.g. $\mathcal{N} \models \neg(b < a)$ iff $\mathcal{N} \models b = a$

or $\mathcal{N} \models a < b$). Equalities may be eliminated by substitution. The remaining literals are of the form $S(a, b)$ or $\neg S(a, b)$.

By the definition of S, $\mathcal{N} \models \neg S(a, b)$ iff $\mathcal{N} \models \neg(a < b)$ or $\mathcal{N} \models \exists z (a < z \wedge z < b)$. It follows that, when we encounter $\neg S(a, b)$, we may be able to replace it with $\neg(a < b)$ or with the pair of conditions $a < z, z < b$ containing the new variable z, which we can always satisfy by assigning to z the immediate successor of a.

In short, we can satisfy $\varphi(\overline{y}, \overline{a})$ in $\mathcal{N}$ iff we can satisfy a system of literals containing only inequalities and atomic formulae of the form $S(y_i, a_j)$ or $S(a_j, y_i)$.

(ii) If b_j lies between a_i and a_{i+1} and there are only finitely many elements of N between the last two objects, then the chain with endpoints a_i, a_{i+1} is included in M because M contains the immediate successors of its elements and it takes finitely many successors to reach a_{i+1} from a_i. Thus, any b_j occurring in a finite chain between two parameters of $\varphi(\overline{y}, \overline{a})$ is already an element of M.

If, on the other hand, a_i, a_{i+1} contains an infinite chain, then it contains infinitely many elements of M, because M is closed under successors. Thus, any inequality of the form $a_i < x < a_{i+1}$ that is satisfied in N is also satisfied in M.

The last remarks tell us that the system of simultaneous inequalities occurring in $\varphi(\overline{y}, \overline{a})$ can always be satisfied in M. Literals of the form $S(a_j, y_i)$ can only be satisfied in M because M has no greatest element and it contains the successors of its elements by axiom (b). Literals of the form $S(y_i, a_j)$ are also satisfied in M because M has no least element and axiom (c) holds.

It follows that $\mathcal{M} \models \exists \overline{y} \varphi(\overline{y}, \overline{a})$.

(iii) By Robinson's test, dLO$_{-,-}$ is model-complete. The ordered set $(\mathbb{Z}, <)$ is a prime model of the theory, which is therefore also complete.

Exercise 8.19

We can introduce a unary predicate symbol P in the same way we did when working with DLO$_{+,-}$. Robinson's test applies by an argument nearly identical to that in the previous exercise. We only have to consider literals of the form $P(x), \neg P(x)$. Positive ones can only be satisfied by an object that must be in any submodel of a model. Negative literals are satisfied by all but one object in any model so they are certainly satisfied in any submodel of a model.

Exercise 8.20

One of the models of dLO$_{+,+}$ has exactly three objects ordered as $a < b < c$. Call this model $\mathcal{M}$. Then $\mathcal{M}$ models the first-order sentence φ saying that there are exactly three distinct objects, but φ is clearly false in any infinite model of

B   Solutions to Exercises   449

$\mathcal{M}$. An infinite discrete linear order with endpoints may be obtained from $\mathbb{N} = \{0, 1, 2, 3, \ldots\}$, endowed with its natural ordering, by adjoining its reverse ordering $\mathbb{N}^* = \{\ldots, 3, 2, 1, 0\}$. The set $\mathbb{N} \cup \mathbb{N}^*$ may then be linearly ordered as follows:

$$0 < 1 < 2 < 3 \ldots < 3 < 2 < 1 < 0.$$

Since any infinite discrete, linear order must either have two endpoints, or one, or none, and the theories of infinite discrete linear orders with or without endpoints are all complete, it follows that any two elementarily equivalent infinite, discrete linear orders must either both have a least element, or both have a greatest element, or have both endpoints, or none.

Exercise 8.21

(i) A model of Rel^∞ contains exactly exactly two distinct points that are related. Every other point is related only to itself. Moreover, the two distinct, related points display one of two mutually exclusive configurations (either each point is also related to itself, or none of the two is). It follows that the theory cannot be complete because each configuration can be described by a sentence in the language of Rel^∞ that holds in one model and fails in another.
(ii) Consider $\mathcal{M}, \mathcal{N} \models \mathsf{Rel}^\infty$ such that $\mathcal{M} \sqsubseteq \mathcal{N}$ and $\mathcal{N} \models \exists y \varphi(y, \bar{a})$, with $\bar{a} \in \mathcal{M}$. If $\mathcal{N}$ contains exactly two related, irreflexive points, so does the submodel $\mathcal{M}$. Thus, if $\mathcal{N} \models \varphi(b, \bar{a})$ and b is one of the related, irreflexive points, $b \in M$. If b is a reflexive and isolated point not in M, there is an automorphism of $\mathcal{M}$ that sends it into such a point, while keeping the parameters fixed. Nothing changes if $\mathcal{N}$ contains exactly two related and reflexive points. The Tarski-Vaught criterion applies and $\mathcal{M} \preceq \mathcal{N}$. Model-completeness follows.

Exercise 8.22

If T is a complete and $\mathcal{M}$ if a finite model of T, suppose $|M| = n$. By completeness, every model of T has a domain of n elements. Thus, in particular, the only submodel of any $\mathcal{M} \models T$ is $\mathcal{M}$, which is certainly an elementary substructure of itself. It follows that T is model-complete.

Exercise 8.26

(i) Suppose that $\mathcal{M} \models T$. Then an arbitrary expansion $\mathcal{M}_a$, with $a \in M$ singled out as a distinguished element, continues to be a model of T. Beczuse $T \models$

$\phi(c)$, $\mathcal{M}_a \models \phi(c)$, where c names a. Because this result holds for an aribtrary expansion of $\mathcal{M}$, it is clear that $\mathcal{M} \models \phi(a)$, for each $a \in M$, i.e. $\mathcal{M} \models \forall x \phi(x)$. Since $\mathcal{M}$ is an arbitrary model of T, we conclude $T \models \forall x \phi(x)$.

(ii) It suffices to consider the conjunction $\forall \overline{x} \phi(\overline{x}) \wedge \forall \overline{y} \psi(\overline{y})$, where it is possible to assume that $\overline{x}$ and $\overline{y}$ are disjoint (if not, we consider the equivalent form of the original formula in which the relevant universal quantification affect disjoint sets of variables). If $\mathcal{M}$ models the last conjunction, then for any independent choices of $\overline{a}, \overline{b}$, $\mathcal{M} \models \phi(\overline{a}) \wedge \psi(\overline{b})$. Fix $\overline{a}$: then, however the objects $\overline{b} \in M$ are chosen, $\phi(\overline{a}) \wedge \psi(\overline{y})$ is satisfied by $\overline{b}$. This is to say that:

$$\mathcal{M} \models \forall \overline{y}(\phi(\overline{x}) \wedge \forall \overline{y} \psi(\overline{y})).$$

The same argument leads to:

$$\mathcal{M} \models \forall \overline{x} \forall \overline{y}(\phi(\overline{x}) \wedge \psi(\overline{y})).$$

If we think of $\overline{x}, \overline{y}$ as a single string $\overline{z}$, we arrive at the form:

$$\mathcal{M} \models \forall \overline{z}(\phi(\overline{z}) \wedge \psi(\overline{z})),$$

since we require, as usual, that the fee variables within ϕ, ψ occur among (without necessarily coinciding with) $\overline{z}$.

If, conversely:

$$\mathcal{M} \models \forall \overline{z}(\phi(\overline{z}) \wedge \psi(\overline{z})),$$

then it is easy to see that universal quantification can be distributed over the conjuncts.

Exercise 8.28

Suppose $\mathcal{M} \sqsubseteq \mathcal{N} \models T$. Then $\mathcal{N} \models T_\forall$ and, since universal sentences are hereditary, we deduce $\mathcal{M} \models T_\forall$. If, on the other hand, $\mathcal{M} \models T_\forall$, we only have to show that $T \cup \mathcal{D}(\mathcal{M})$ is finitely consistent to find $\mathcal{N} \models T$ that extends $\mathcal{M}$. But finite consistent can be established exactly as in the proof of Theorem 8.29.

Exercise 8.31

(i) If T has an existential axiomatisation, embeddings preserve each of the axioms. Thus, extensions of models of T are themselves models of T.

B Solutions to Exercises   451

(ii) Our goal is to show that any model of $T_\exists$ is already a model of T. Thus, we fix
$\mathcal{N} \models T_\exists$. If $\mathcal{M} \models T$, suppose f embeds $\mathcal{M}$ into an elementary extension $\mathcal{N}_1$
of $\mathcal{N}$. By hypothesis, $\mathcal{N}_1$ models T. Since $\mathcal{N}_1 \equiv \mathcal{N}$, it follows that $\mathcal{N} \models T$.
(iii) Toward a contradiction, we assume that $T \cup T_\neg$ has no models. By Compactness, this is to say that a finite subset A of $T \cup T_\neg$ has no models. Since $T, T_\neg$, individually taken, have models, A contains finitely many sentences from T and finitely many sentences, say $\delta_1, \ldots, \delta_n$, from $T_\neg$. Since these sentences cannot hold simultaneously, the conjunction of the finite subset from T implies the negation of the conjunction $\delta_1 \wedge \ldots \wedge \delta_n$, i.e. the sentence:

$$\neg\delta_1 \vee \ldots \vee \neg\delta_n.$$

Note that the last formula is a equivalent to a disjunction of existential sentences (since $\delta_i \in T_\neg$ implies that δ_i is of the form $\neg\gamma_i$, with γ existential) and thus, by an earlier exercise, to an existential sentence φ.
It follows that $\mathcal{N} \models \varphi$ and, consequently, $\mathcal{N} \models \neg\delta_i$ for some $i \in \{1, \ldots, n\}$. Since, however, $\delta_i \in T_\neg$, it also follows that $\mathcal{N} \models \delta_i$, a contradiction.
(iv) Let $\mathcal{M}$ be a model of $T \cup T_\neg$. Consider $\mathcal{ED}(\mathcal{N}) \cup \mathcal{D}(\mathcal{M})$ where new constant symbols in $\mathcal{ED}(\mathcal{N}), \mathcal{D}(\mathcal{M})$ respectively are disjoint.
If the last set is inconsistent, then $\mathcal{ED}(\mathcal{N})$ and some finite subset $\{\gamma_1, \ldots, \gamma_n\} \subseteq \mathcal{D}(\mathcal{M})$ cannot be simultaneously satisfied. In other words:

$$\mathcal{ED}(\mathcal{N}) \models \neg\gamma_1 \vee \neg\gamma_n.$$

In particular $\mathcal{ED}(\mathcal{N}) \models \neg\gamma_i(c_1, \ldots, c_k)$ for some $i \in \{1, \ldots, n\}$, where $c_1, \ldots, c_k$ are constant symbols occurring in the diagram of $\mathcal{M}$ and not belonging to the common language of $\mathcal{M}, \mathcal{N}$.
By an easy generalisation of Exercise 8.26-(i), $\mathcal{ED}(\mathcal{N}) \models \forall x_1 \ldots \forall x_k \neg\gamma_i(x_1, \ldots, x_k)$, where the last sentence is equivalent to the negated existential sentence $\neg\exists x_1 \ldots \exists x_k \gamma(x_1, \ldots, x_k)$.
By the definition of $T_\neg$, the last sentence is in $T_\neg$. It follows that, for a suitable expansion by constants $\mathcal{M}'$ of $\mathcal{M}$, $\mathcal{M}' \models \neg\exists x_1 \ldots \exists x_n \gamma(x_1, \ldots, x_k)$ and $\mathcal{M} \models \gamma(c_1, \ldots, c_k)$.
This contradiction shows that $\mathcal{ED}(\mathcal{N}) \cup \mathcal{D}(\mathcal{M})$ has a model $\mathcal{N}'_1$. Let $\mathcal{N}_1$ be its reduct to the language of $\mathcal{N}$.
(v) Now we have $\mathcal{N} \preceq \mathcal{N}_1$ and an embedding from $\mathcal{M}$ to $\mathcal{N}_1$. The argument from part (i) applies and we conclude that $T, T_\exists$ have the same consequences, i.e. T has an existential axiomatisation.

Exercise 8.34

A model-complete theory with finite models has only finite models. Conversely, if, in the proof of the theorem, we happen to select a finite $\mathcal{M}$, we can still carry out the given construction of an extension $\mathcal{N}$, but the hypothesis of the theorem forces $\mathcal{M}, \mathcal{N}$ to be isomorphic since $\mathcal{N}$ models the diagram of $\mathcal{M}$ and is of the same size as $\mathcal{M}$ (if larger, then it would contain at least $n + 1$ distinct objects, if $|M| = 1$, and we could reflect the sentence saying that there are at least $n + 1$ distinct objects into $\mathcal{M}$). By the construction of $\mathcal{N}$ and the isomorphism with $\mathcal{M}$, it is clear that an expansion of the last structure must model $\varphi(\overline{c})$. The proof then proceeds as shown.

Exercise 8.36

Consider an arbitrary $\mathcal{M}_\beta$ such that $\overline{a} \in M_\beta$. If $\overline{a} \in M_\gamma$ and $\gamma \neq \beta$, either $\gamma \in \beta$ or $\beta \in \gamma$. In the former case, since M_γ is the domain of a substructure of $\mathcal{M}_\beta$, it is closed under the operations named by the symbols in $\mathcal{L}$. In particular $f^{M_\gamma}(\overline{a}) = b \in M_\gamma \subseteq M_\beta$. Since the inclusion map is an embedding, the parametric equality $f(\overline{a}) = b$ is preserved in $\mathcal{M}_\beta$ and $f^{M_\gamma}(\overline{a}) = b = f^{M_\gamma}(\overline{a})$. A similar argument holds when $\beta \in \gamma$. It follows that the interpretation of f is uniquely determined on each k-tuple from M^k.

Now let us consider $\mathcal{M}_\beta$ and the union $\mathcal{M}$. Clearly $\mathcal{M}_\beta$ and M contain the same constants since they are both $\mathcal{L}$-structures. Moreover, for any m-ary relation symbol R, the definition of R^M implies:

$$\mathcal{M}_\beta \models R^{M_\beta}(\overline{a}) \iff \mathcal{M} \models R^M(\overline{a}).$$

By the way functions are defined, we see that, for any k-ary function symbol R:

$$\mathcal{M}_\beta \models f^{M_\beta}(\overline{a}) = b \iff \mathcal{M} \models f^M(\overline{a}) = b.$$

It is clear that the inclusion map is an embedding of $\mathcal{M}_\beta$ into $\mathcal{M}$, i.e. $\mathcal{M}_\beta \sqsubseteq \mathcal{M}$.

Exercise 8.37

We already know that $\mathcal{M}_\beta \sqsubseteq \mathcal{M}$. The inductive base therefore is proved. The inductive step for $\neg, \vee$ does not pose any problems other than stating the inductive hypothesis correctly (i.e. relative to every $\beta \in \alpha$). The inductive hypothesis for $\exists$ is that $\psi(y, \overline{x})$ is a formula such that, for each $\gamma \in \alpha$:

$$\mathcal{M}_\gamma \models \psi(b, \overline{a}) \iff \mathcal{M} \models \psi(b, \overline{a}).$$

We next suppose $\mathcal{M} \models \exists y \psi(y, \overline{a})$ and aim to prove that $\mathcal{M}_\beta \models \exists y \psi(y, \overline{a})$. It follows that, for some $b \in M$, $\mathcal{M} \models \psi(b, \overline{a})$. By construction, $b \in M_\gamma$, for some $\gamma \in \alpha$. Since ordinals are comparable relative to $\in$, we have $\gamma \in \beta$, $\gamma = \beta$ or $\beta \in \gamma$. The first two possibilities pose no problems because they yield the sought result in view of the inductive hypothesis.

The third possibility requires an appeal to the inductive hypothesis, which entails $\mathcal{M}_\gamma \models \exists y \varphi(y, \overline{a})$. Because $\mathcal{M}_\beta \preceq \mathcal{M}_\gamma$, we conclude $\mathcal{M}_\beta \models \exists y \varphi(y, \overline{a})$.

On the other hand, we suppose that $\mathcal{M}_\beta \models \exists y \psi(y, \bar{a})$ and aim to prove that $\mathcal{M} \models \exists y \psi(y, \bar{a})$. If not, $\mathcal{M} \models \forall y \neg \psi(y, \bar{a})$. In particular, for any $c \in \mathcal{M}_\beta$ such that $\mathcal{M}_\beta \models \psi(c, \bar{a})$, we have $\mathcal{M} \models \neg \psi(c, \bar{a})$. Since, however, the proof of induction carried out so far already implies that every formula with parameters in $\mathcal{M}_\beta$ is reflected by $\mathcal{M}$ into $\mathcal{M}_\beta$, we arrive at the contradiction $\mathcal{M}_\beta \models \varphi(c, \bar{a}) \wedge \neg \psi(c, \bar{a})$.

Exercise 8.39

(i) Let $\forall \bar{x} \psi(\bar{x})$ be a universal sentence and $\mathcal{M}$ be the union of the chain $\{\mathcal{M}_\beta\}_{\beta \in \alpha}$. Let $\bar{a} \in M$ be a tuple of the same length as $\bar{x}$. This tuple is included in $\mathcal{M}_\gamma$ for some $\gamma \in \alpha$. By hypothesis, $\mathcal{M}_\gamma \models \psi(\bar{a})$. Since $\mathcal{M}_\gamma \sqsubseteq \mathcal{M}$ and $\psi(\bar{x})$ is a quantifier-free formula, we conclude $\mathcal{M} \models \psi(\bar{a})$. This argument applies to any tuple in M, so $\mathcal{M} \models \forall \bar{x} \psi(\bar{x})$.

(ii) We reason almost exactly as in part (i) and, given $\forall x \exists y \psi(x, y)$, we suppose $\mathcal{M} \models \exists y \psi(a, y)$. Then $a \in \mathcal{M}_\gamma$ for some $\gamma \in \alpha$ and, by hypothesis, $\mathcal{M}_\gamma \models \exists y \psi(a, y)$. Because existential formulae are preserved by extensions and $\mathcal{M}$ is an extension of $\mathcal{M}_\gamma$, we deduce $\mathcal{M} \models \exists y \psi(a, y)$ for each $a \in M$. In other words $\mathcal{M} \models \forall x \exists y \psi(a, y)$.

The same argument would have worked with strings of quantifiers, i.e. with a formula of the form $\forall \bar{x} \exists \bar{y} \psi(\bar{x}, \bar{y})$, with ψ quantifier-free.

Exercise 8.42

(i) Since we are first building a chain whose starting point is already of size κ, we may modifiy our recursion to control size increase. In particular, we can ensure that, if $\beta = \alpha + 1$ and $|\mathcal{M}_\beta|$ is of size κ, we apply the Downward Löwenheim-Skolem theorem to move to an elementary substructure $\mathcal{M}_\alpha^*$ of $\mathcal{M}_\alpha$, as given in the recursion. The substructure in question is generated by the set $M_\beta \cup \{\bar{a}\}$, where $\bar{a}$ is a finite sequence that satisfies φ_α in $\mathcal{M}_\alpha$. Then $|M_\alpha^*| = \kappa$.

It can now be proved by induction that every $\mathcal{M}_\alpha$, with $\alpha \in \kappa$, is of size κ. The base and successor step are already taken care of. If, at a limit ordinal $\lambda < \kappa$, any element of a chain is of size κ, the union of the chain is itself of size κ.

This induction ensures that $\mathcal{N}_0$ is of size κ. Because we obtained each $\mathcal{N}_m$ in the proof by repeating the recursion used to obtain $\mathcal{N}_0$, and our starting point is always a structure of size κ, $|\mathcal{N}_m|$ can be chosen to have size κ, for each $m \in \mathbb{N}$.

(ii) Suppose that, for some $\mathcal{N} \models T$, $\mathcal{N} \sqsupseteq \mathcal{N}_0$ and $\mathcal{N} \models \exists \bar{x} \varphi(\bar{x}, \bar{a})$, with $\bar{a} \in \mathcal{N}_0$. Since $\exists \bar{x} \varphi(\bar{x}, \bar{a})$ is φ_δ in our enumeration of existential sentences, and $\mathcal{N}$ is an extension of $\mathcal{M}_\delta \sqsubseteq \mathcal{N}_0$. This is to say that $\mathcal{M}_{\delta+1}$ contains $\bar{b}$ such that $\mathcal{M}_{\delta+1} \models \varphi(\bar{b}, \bar{a})$, i.e. $\mathcal{M}_{\delta+1} \models \exists \bar{x} \varphi(\bar{x}, \bar{a})$. Since existential sentences are preserved by extensions, we can conclude $\mathcal{N}_0 \models \exists \bar{x} \varphi(\bar{x}, \bar{a})$.

(iii) If $\mathcal{N} \sqsubseteq \mathcal{N}^*$ and $\mathcal{N}^* \models \exists \bar{x} \varphi(\bar{x}, \bar{a})$, the constant symbols $\bar{a}$ name elements of N, i.e. elements of some $\mathcal{N}_m$. By construction, $\mathcal{N}_{m+1} \models \exists \bar{x} \varphi(\bar{x}, \bar{a})$ and, since extensions preserve existential formulae $\mathcal{N} \models \exists \bar{x} \varphi(\bar{x}, \bar{a})$.

Exercise 8.45

For any $\mathcal{M}, \mathcal{N} \models T$, let f, g be embeddings of $\mathcal{M}, \mathcal{N}$ respectively into a joint extension $\mathcal{N}^*$. Then $\mathcal{M}, \mathcal{N}$ may be identified with substructures of $\mathcal{N}^*$. Because T is model-complete, they are in fact elementary substructures of $\mathcal{N}^*$. Thus, they are elementarily equivalent to the same extension and, as a result, to each other. Since $\mathcal{M}, \mathcal{N}$ were chosen arbitrarily, T is complete.

Exercise 8.48

Let $\delta_1(\overline{c}), \ldots, \delta_m(\overline{c})$ be quantifier-free sentences from $\mathcal{D}(\mathcal{N})$, where $\overline{c}$ is a (possibly empty) list that names elements of $N - M$, such that:

$$\mathcal{ED}(\mathcal{M}) \models \neg(\delta_1(\overline{c}) \wedge \ldots \wedge \delta_n(\overline{c})).$$

Then $\mathcal{ED}(\mathcal{M}) \models \neg \exists \overline{x}(\delta_1(\overline{x}) \wedge \ldots \wedge \delta_n(\overline{x}))$. Because $\mathcal{M} \subseteq \mathcal{N}$ and an expansion $\mathcal{M}^+$ of $\mathcal{M}$ models $\mathcal{ED}(\mathcal{M})$, it follows that $\mathcal{N} \models \neg \exists \overline{x}(\delta_1(\overline{x}) \wedge \ldots \wedge \delta_n(\overline{x}))$, contradicting the fact that $\delta_i(\overline{c}) \in \mathcal{D}(\mathcal{N})$ for each $i = 1, \ldots, n$. It follows that $\mathcal{ED}(\mathcal{M}) \cup \mathcal{D}(\mathcal{N})$ has a model.

On the other hand, supposed that f elementarily embeds $\mathcal{M}$ in $\mathcal{B}$ and g embeds $\mathcal{N}$ in $\mathcal{B}$. Consider an existential formula $\exists \overline{x} \psi(\overline{x}, \overline{a})$, with parameters in M, such that $\mathcal{N} \models \exists \overline{x} \psi(\overline{x}, \overline{a})$. Because embeddings preserve existential formulae, $\mathcal{B} \models \exists \overline{x} \psi(\overline{x}, f(\overline{a}))$. Because f, g agree on $\overline{a}$, $\mathcal{B} \models \exists \overline{x} \psi(\overline{x}, g(\overline{a}))$. Since g is elementary, we conclude $\mathcal{M} \models \exists \overline{x} \psi(\overline{x}, \overline{a})$.

Exercise 8.50

Suppose that $T \models \neg(\varphi_1 \wedge \ldots \wedge \varphi_m)$. Since the conjunction of universal sentences $\varphi_1 \wedge \ldots \wedge \varphi_m$ is equivalent to a single universal sentence, its negation is equivalent to an existential sentence, which must be in $T_{\forall\exists}$ because it follows from T. Thus, $\mathcal{M}_0 \models \neg(\varphi_1 \wedge \ldots \wedge \varphi_m)$, but $\mathcal{M}_0$ also models $\varphi_1 \wedge \ldots \wedge \varphi_m$ by the definition of $T^\forall$. We have reached the desired contradiction. It follows that $T \cup T^\forall$ is consistent.

Exercise 8.51

(i) We suppose that $T \cup T_{\varphi,\forall\exists}$ is inconsistent. If so, there is a finite subset $\{\psi_1(\overline{c}), \ldots, \psi_m(\overline{c})\}$ of $T_{\varphi,\forall\exists}$ such that $T \models \neg(\psi_1(\overline{c}), \ldots, \psi_m(\overline{c}))$. Using the definition of $T_{\varphi,\forall\exists}$, we infer:

$$T \models \varphi(\overline{c}) \to \psi_1(\overline{c}), \ldots, \psi_m(\overline{c})$$

and

$$T \models \neg(\psi_1(\overline{c}), \ldots, \psi_m(\overline{c})) \to \neg\varphi(\overline{c}).$$

Thus, we may deduce:

$$T \models \neg\varphi(\overline{c}), which in turn implies T \models \neg\exists \overline{x}\varphi(\overline{x}).$$

It now follows that $\varphi(\overline{x})$ is T-equivalent to $\neg(x = x)$, where x is any variable from the list $\overline{x}$.

(ii) It suffices to ensure that all tuples of quantified variables involved are disjoint.

(iii) We suppose $\mathcal{M}_0 \models T \cup \{\neg\varphi(\overline{c})\} \cup T_{\varphi,\forall\exists}$ and work with an expanded language $\mathcal{L}_0$ in which every element of $\mathcal{M}_0$ is named by a constant symbol. Define $T^\forall$ as the set of $\mathcal{L}_0$-sentences true in $\mathcal{M}_0$ and note that $\mathcal{D}(\mathcal{M}_0) \subseteq T^\forall$. Then the set $T \cup T^\forall \cup \{\varphi(\overline{c})\}$ has a model $\mathcal{N}$. If not, $T \cup \{\varphi(\overline{c})\} \models \neg(\psi_1(\overline{c}) \wedge \ldots \wedge \psi_m(\overline{c}))$, where $\psi_i(\overline{c}) \in T^\forall$. Since the last set is closed under conjunctions, $\psi_1(\overline{c}) \wedge \ldots \wedge \psi_m(\overline{c}) \in T^\forall$ and, as a result, $\mathcal{N} \models \exists\overline{x}(\psi_1(\overline{x}) \wedge \ldots \wedge \psi_m(\overline{x}))$. However, the fact that $\mathcal{N} \models T \cup \{\varphi(\overline{c})\}$ implies $\mathcal{N} \models \neg\exists\overline{x}(\psi_1(\overline{x}) \wedge \ldots \wedge \psi_m(\overline{x}))$. This contradiction shows that there is a model $\mathcal{N}_0$ of $T \cup T^\forall \cup \{\varphi(\overline{c})\}$. Clearly, $\mathcal{M}_0$ can be embedded in to $\mathcal{N}_0$. Moreover, we cannot have an existential sentence $\exists\overline{x}\theta(\overline{x}, \overline{a})$ hold in $\mathcal{N}_0$ and not hold in $\mathcal{M}_0$. If this were the case, we would have $\mathcal{M}_0 \models \forall\overline{x}\neg\theta(\overline{x}, \overline{a})$ and the last sentence should belong to $T^\forall$. Since $\mathcal{N}_0 \models T^\alpha$, we reach a contradiction.

We now know that there is $\mathcal{M}_1$ into which $\mathcal{M}_0$ can be elementarily embedded and $\mathcal{N}_0$ can be embedded. Because $\mathcal{M}_0 \preceq \mathcal{M}_1$, $\mathcal{M}_1 \models T \cup \{\neg\varphi(\overline{c})\} \cup T_{\varphi,\forall\exists}$ and we can iterate the construction just described.

It is clear that each $\mathcal{N}_1$ will be a model of $\varphi(\overline{c})$, which is preserved by their union, while the union of the elementary chain determined by the $\mathcal{M}_i$ is a model of $\varphi(\overline{c})$. We thus arrive at two chains with the same union, which models a sentence and its negation. This contradiction implies $T \cup T_{\varphi,\forall\exists} \models \{\neg\varphi(\overline{c})\}$.

(iv) By Compactness, the consequence relation $T \cup T_{\varphi,\forall\exists} \models \{\neg\varphi(\overline{c})\}$ implies that, for a finite set $A \subseteq T_{\varphi,\forall\exists}$, $T \cup A \models \varphi(\overline{c})$. Because the conjunction of the elements of A is, up to equivalence, an element $\psi(\overline{c})$ of $T_{\varphi,\forall\exists}$, we can conclude that:

$$T \models \psi(\overline{c}) \to \varphi(\overline{c}).$$

The opposite implication follows from the definition of $T_{\varphi,\forall\exists}$. Because $\overline{c}$ does not occur in T:

$$T \models \forall x(\psi(\overline{x}) \leftrightarrow \varphi(\overline{x})).$$

Exercise 8.56

(i) If ψ is a $\mathcal{L}^m$-formula, it may contain relation symbols from the expanded language. Note that, for each one of these symbols T^m contains an axiom stating its T^m-equivalence to a $\mathcal{L}$-formula. Thus, by the substitution theorem specialised to T^+, the formula ψ^* obtained by replacing any occurrence of a new relation symbol in ψ by its T^m-equivalent, is T^m-equivalent to ψ. But ψ^* is a $\mathcal{L}$-formula, so it must be equivalent, by construction, to some $\mathcal{L}^m$-atomic formula. Quantifier elimination now follows.

(ii) Suppose $\mathcal{N} \models \varphi(f(\overline{a}))$, with $\overline{a} \in M$. Since $\mathcal{N} \models T^m$, $\mathcal{N} \models R_\varphi(f(\overline{a}))$. Because f is an embedding, the last condition holds iff $\mathcal{M} \models R_\varphi(\overline{a})$ iff $\mathcal{M} \models \varphi(\overline{a})$. It follows that f is elementary.

Exercise 8.57

(i) The definition of Θ is deliberately given to support a proof by induction on the complexity of formulae. Condition (a) gives the inductive hypothesis and condition (b) imply the inductive step for connectives. Condition (c) is designed to yield the inductive step for the existential quantifier.

(ii) If $\psi \in \Sigma$ is strongly preserved by models of T and $\overline{a} \in M$, we have $\mathcal{M} \models \psi(\overline{a}) \iff \mathcal{N} \models \psi(f(\overline{a}))$. For any $\mathcal{L}$ formula $\phi(\overline{x})$, there is a T-equivalent formula $\psi(\overline{x})$ in Σ such that $\mathcal{M}, \mathcal{N} \models \forall \overline{x}(\phi(\overline{x}) \leftrightarrow \psi(\overline{x}))$. We can use the last equivalence to deduce:

$$\mathcal{M} \models \phi(\overline{a}) \iff \mathcal{N} \models \phi(f(\overline{a})).$$

It follows that f is elementary.

Exercise 8.59

(i) Suppose $\mathcal{M}, \mathcal{N} \models T$ and $\mathcal{M} \subseteq \mathcal{N}$. If $\mathcal{N} \models \exists y \varphi(y, \overline{a})$ then $\mathcal{N} \models \exists y \varphi(\overline{a}, y) \to \varphi(t(\overline{a}), \overline{a})$ as well. It follows that $\mathcal{N} \models \varphi(f_{\exists \varphi}(\overline{a}), \overline{a})$ and, since $\mathcal{M}$ is a substructure of $\mathcal{N}$, $f_{\exists \varphi}(\overline{a}) \in M$. The Tarski-Vaught criterion applies and $\mathcal{M} \preceq \mathcal{N}$.

(ii) Suppose $\exists y \varphi(\overline{x}, y)$ is a primitive formula. Because T has Skolem functions:

$$T \models \forall \overline{x}(\exists y(\varphi(\overline{x}, y) \leftrightarrow \varphi(\overline{x}, t(\overline{x})).$$

Because we can eliminate existential quantifiers from primitive formulae, Theorem 8.55 applies and quantifier-elimination follows.

Exercise 8.61

Consider the formula $\exists y \varphi(\overline{x}, y)$. The function symbol $F_{\exists \varphi(\overline{x}, y)}$ is an extralogical symbol of T_s. If $\mathcal{N} \models T$, then let $\mathcal{N}'$ be expanded to the augmented language obtained by adding $F_{\exists \varphi(\overline{x}, y)}$ to $\mathcal{L}$.

By the definition of $F_{\exists \varphi(\overline{x}, y)}^{\mathcal{N}}$ devised in the proof, it is clear that, for any tuple $\overline{a} \in N$, either $\mathcal{N} \models \exists y \varphi(\overline{a}, y)$ or $\mathcal{N} \not\models \exists y \varphi(\overline{a}, y)$.

In the first case, $\mathcal{N} \models \varphi(\overline{a}, F^{\exists \varphi(\overline{x}, y)}(\overline{a}))$ so the conditional $\mathcal{N} \models \exists y \varphi(\overline{a}, y) \to \varphi(\overline{a}, F_{\exists \varphi(\overline{x}, y)}(\overline{a}))$ holds. In the second case, the same conditional holds vacuously. Because $\overline{a}$ is arbitrary, condition (1) is satisfied relative to $\exists y \varphi(\overline{x}), y)$. Because the last formula was also chosen arbitrarily (1) holds.

To verify that the theory T_s has Skolem functions, we fix a model $\mathcal{M} \models T_s$, a tuple $\overline{a} \in M$ and a $\mathcal{L}^s$-formula $\exists y \varphi(\overline{a}, y)$. By construction, $\exists y \varphi(\overline{a}, y)$ is a $\mathcal{L}_n$-formula, for some $n \in \mathbb{N}$. It follows that condition (1) relative to $\exists y \varphi(\overline{x}, y)$ is a $\mathcal{L}_{n+1}$-sentence in T_{n+1}, i.e. that it is in T^s. It follows that T^s satisfies condition (2).

Exercise 8.62

Expanding the language $\mathcal{L}$ to $\mathcal{L}_m$ involves adding only $|\mathcal{L}|$ new symbols to $\mathcal{L}$. Thus, the total number of symbols of $\mathcal{L}_m$ is the infinite cardinal $|\mathcal{L}|$ and $|\mathcal{L}_m| = |\mathcal{L}|$. Since T has infinite models, there is one of size $|\mathcal{L}|$.

As for the addition of Skolem functions to $\mathcal{L}$, there are at most $|\mathcal{L}|$ of them, so the above argument can be replicated.

Exercise 8.64

It is easy to prove by induction on the complexity of formulae that, if $\overline{a}, \overline{b} \in M$ satisfy exactly the same atomic formulae, then they satisfy exactly the same quantifier-free formulae. Thus $\mathcal{M} \models \theta(\overline{a}) \iff \mathcal{M} \models \theta(\overline{b})$. Since $\varphi(\overline{x})$ is T-equivalent to $\theta(\overline{x})$, we obtain:

$$\mathcal{M} \models \varphi(\overline{a}) \leftrightarrow \theta(\overline{a}) \iff \mathcal{M} \models \varphi(\overline{b}) \leftrightarrow \theta(\overline{b}).$$

It is easy to see that:

$$\mathcal{M} \models \varphi(\overline{a}) \leftrightarrow \varphi(\overline{b}),$$

as desired.

Exercise 8.66

(i) Suppose that $\bar{a}, \bar{b} \in M$ satisfy exactly the same literals in the language of IS. This is to say that $\bar{a}, \bar{b}$ are lists of distinct objects of the same length. If $a_{n+1} \neq a_1, \ldots, a_n$ is used to extend $\bar{a}$, the only literals satisfied by a_{n+1} with parameters in $\bar{a}$ are $x = x$ and $\neg(x = a_j)$, for each a_j in the list $\bar{a}$. Because M is infinite, it is clear that some $b_{n+1} \in M$ can be found, which satisfies exactly the same literals as a_{n+1} with respect to the parameters $\bar{b}$. Since M is an arbitrary model of IS, the latter theory has quantifier elimination.

(ii) If $M \models \text{DLO}_{-,-}$, suppose that $\bar{a}, \bar{b} \in M$ satisfy exactly the same literals in the language of $\text{DLO}_{-,-}$. This is to say that the system of inequalities determined by $\bar{a}$ is replicated by $\bar{b}$. If a_{n+1} is adjoined to $\bar{a}$, it satisfies a fixed finite set of inequalities in the parameters $\bar{a}$. If we switch to the parameters $\bar{b}$ (so that a_i is sent into b_i, for any $i \in \{1, \ldots, n\}$, if the tuples considered are n-tuples), we can easily use the density of M, as well as the absence of a maximal or minimal element, to ensure the existence of $b_{n+1} \in M$ that satisfies the given conditions with parameters $\bar{b}$. It follows that $\text{DLO}_{-,-}$ has quantifier elimination. Thus, $\text{DLO}_{-,-}$ is model-complete: since $\mathbb{Q} \sqsubseteq \mathbb{R}$ are two models of the theory, model-completeness entails $\mathbb{Q} \preceq \mathbb{R}$. Finally, by quantifier elimination the definable subsets of an arbitrary model of $\text{DLO}_{-,-}$ are finite unions of intervals and singletons (defined by disjunctive normal forms of quantifier-free formulae).

(iii) Suppose that $\bar{a}, \bar{b} \in M$ satisfy exactly the same literals in the language of EQ_∞. Then we may partition each tuple into the subsets belonging to distinct equivalence classes. If we extend $\bar{a}$ by adding a_{n+1}, then we may pick an object that is equivalent to some of the $\bar{a}$ or to none. Since there are infinitely many distinct equivalence classes and they are all infinite, we can replicate the selection relative to $\bar{b}$.

(iv) Since $\text{dLO}_{+,-}$ has a finite relational vocabulary, it suffices to show that (2) from the statement of Theorem 8.65 fails in a model of $\text{dLO}_{+,-}$. We may pick $\mathcal{N}$, the ordered set $\mathbb{N}$, as the model in question. If $\bar{a}$ is the list $1, 2, 4$ and $\bar{b}$ is the list $0, 1, 4$, it is easy to see that these lists satisfy exactly the same literals. Extend the first list by adding $a_{n+1} = 3$. We must be able to extend $\bar{b}$ by an element b_{n+1} in such a way that $\mathcal{N} \models S(1, b_{n+1})$. The only possibility is $b_{n+1} = 2$. But now we have $\mathcal{N} \models S(3, 4) \land \neg S(2, 4)$. Condition (2) of Theorem 8.65 fails. Adding P to the language does not affect the example just constructed.

Exercise 8.68

If $\mathcal{A}$ is an integral domain and $f : \mathcal{A} \longrightarrow M$, with $M \models \text{ACF}$, then we know from Chap. 4 that we can extend f to an embedding $\overline{f}$ of the fraction field $\mathcal{Q}(A)$ of $\mathcal{A}$ into M. By Theorem 6.32 any algebraic extension of $\mathcal{Q}(A)$, $\overline{f}$ can be extended to an embedding g. Since $\overline{\mathcal{Q}(A)}$ is an algebraic extension of $\mathcal{Q}(A)$, we are done.

Exercise 8.69

(i) Let A be a finite set. A IS-closure of A is any countably infinite superset of A. Since there is a bijection between any two countably infinite sets, IS-closures of finite sets are isomorphic.
(ii) Suppose that K is an infinite field. The substructure of an infinite vector space $\mathcal{V}$ over K are its linear subspaces. Only one of these subspaces is finite, i.e. the subspace containing the null vector alone. The other subspaces are infinite and, as models of the theory, they are each its own T-closure. As for 0, any 1-dimensional linear subspace constitutes a T-closure.
If, on the other hand, K is a finite field, its substructures are vector spaces but they may be finite. If $\mathcal{A}$ is a finite-dimensional subspace of $\mathcal{V}$, we note that $\mathcal{V}$ must be infinite-dimensional in order to be an infinite vector space. It is therefore possible to extend a basis of $\mathcal{A}$ to a countably infinite, linearly independent subset of V, and use the subspace generated by this infinite basis as a T-closure of $\mathcal{A}$.

Exercise 8.73

It is clear that quantifier elimination implies model-completeness. Model-completeness implies that embeddings are elementary and thus simple as well. We conclude the proof by assuming that, if $\mathcal{A}, \mathcal{B} \models T$ and $f : \mathcal{A} \longrightarrow \mathcal{B}$ is an embedding, then f is simple. We prove condition (2) from Lemma 8.70 for every formula, using an induction on the complexity of formulae. This is trivial for atomic formulae. The inductive step for $\neg, \vee$ is easy. We are left with the inductive hypothesis that, given $\exists y \varphi(\overline{x}, y)$, $\varphi(\overline{x}, y)$ is T-equivalent to a quantifier-free formula $\theta(\overline{x}, y)$. It follows from the inductive hypothesis that:

$$T \models \forall \overline{x} (\varphi(\overline{x}) \leftrightarrow \theta(\overline{x}, y)).$$

Because f is simple and existential formulae are preserved by extensions, we deduce that, for any $\mathcal{A}, \mathcal{B} \models T$:

$$\mathcal{A} \models \theta(\overline{x}, y)) \iff \mathcal{B} \models \theta(\overline{x}, y)).$$

By T-equivalence, the last result carries over to $\exists y \varphi(\overline{x}, y)$. The induction is complete.

Exercise 8.46

The only quantifier-free $\mathcal{L}_r$-sentences are of the form $m = 0$ or $\neg (m = 0)$. Let T be a complete extension of ACF. Either $T \models m = 0$ for some $m \in \mathbb{N}$, or $\neg (m = 0)$ for each $m \geq 1$. Because ACF has quantifier elimination, every $\mathcal{L}_r$-sentence is ACF-equivalent to a literal of the form $m = 0$ or $\neg(m = 0)$. Thus, no other possibilities can arise than those described. It follows that ACF has exactly two completions.

If $T \models \neg(m = 0)$ for each $m \geq 1$, then $T = \mathsf{ACF}_0$. If $T \models m = 0$ for some $m \in \mathbb{N}$, there is a prime p such that $T \models p = 0$. To see this, it suffices to consider the decomposition of m into prime factors and note that the models of T are, in particular, integral domains.

Thus, if $m = 0$, there is a prime factor p of m such that $p = 0$. This p is also least. If not, suppose $\mathcal{M} \models T$ and that the least number q such that $\mathcal{M} \models q = 0$ is smaller than p. In $\mathbb{N}$, there is an integer m such that $mq < p < (m+1)q$. It follows that $k = p - mq < q$. Thus, $\mathcal{M} \models \neg(k = 0)$ but then $\mathcal{M} \models (k + mq) = p$, i.e. $\mathcal{M} \models \neg(p = 0)$.

We conclude that there is a least prime such that $T \models p = 0$. In this case, $T = \mathsf{ACF}_p$, which we already know to be complete.

Exercise 8.75

(i) If $\mathcal{M} \models T_{0,s}$, then $\mathcal{M}$ contains a distinguished object $0^{\mathcal{M}}$. Because M is closed under the operation $s^{\mathcal{M}}$, it must contain the whole sequence $0, s^{\mathcal{M}}(0), s^{\mathcal{M}} s^{\mathcal{M}}(0), \ldots$, which determines a substructure of $\mathcal{M}$ isomorphic to $\mathcal{N}$ under the bijection $f : \mathbb{N} \longrightarrow M$ given by the condition $f(n) = \underbrace{s^{\mathcal{M}} \ldots s^{\mathcal{M}}}_{n \, times}(0^{\mathcal{M}})$.

(ii) By part (i), we know that $\mathcal{M} \models T_{0,s}$ contains one isomorphic copy of $\mathcal{N}$. Let A be the domain of this isomorphic copy: if $a \in M - A$, then $s^{\mathcal{M}}(a), s^{\mathcal{M}} s^{\mathcal{M}}(a), \ldots$ are elements of $M - A$ (if any one of them was in A, then we could deduce that $a \in A$). Moreover, $a \neq 0^{\mathcal{M}}$ implies that there is b_1 such that $s^{\mathcal{M}}(b_1) = a$ and, since $b \notin A$, we can iterate this argument to obtain $b_2, b_3, \ldots$. It follows that $M - A$ contains an isomorphic copy B of $(\mathbb{Z}, s)$. If $M - (A \cup B) \neq \emptyset)$, suppose $c \in M - (A \cup B)$. Then c determines an isomorphic copy of $(\mathbb{Z}, s)$ that is disjoint from B. In general, as long as we have not exhausted M, we may identify new isomorphic copies of $(\mathbb{Z}, s)$. Since each of them is countably infinite, if $|M| = \kappa > \aleph_0$, there must be precisely κ of them. It follows that, if $\mathcal{A}, \mathcal{B} \models T_{0,s}$ with $|A| = \kappa = |B|$, then each of $\mathcal{A}, \mathcal{B}$ contains exactly one copy of $\mathcal{N}$ and κ copies of $(\mathbb{Z}, s)$. It follows that $\mathcal{A}, \mathcal{B}$ are isomorphic. Since $T_{0,s}$ only has infinite models (due to the axioms in (4), which prevent loops), Vaught's test applies and $T_{0,s}$ is complete.

B Solutions to Exercises   461

(iii) By the definition of completeness, if T, S are complete $\mathcal{L}$-theories and $T \subseteq S$, then $T = S$. Since $T_{0,s} \subseteq Th(\mathcal{N})$, part (ii) implies equality. It follows that $T_{0,s}$ is an axiomatisation of $Th(\mathcal{N})$.
(iv) A finite subset T' of $T_{0,s}$ must omit infinitely many sentences of the form (4). This omission allows sufficiently long, finite loops as models of T'. Since T', $T_{0,s}$ do not have the same models, T' cannot be an axiomatisation of $T_{0,s}$. Moreover, any finite set of sentences T' in the language of $T_{s,0}$ is a set of consequence of $T_{0,s}$. By Compactness, it follows from a finite part of $T_{0,s}$. Since any finite part of $T_{0,s}$ allows finite loops, T' also does, and cannot be a finite axiomatisation of $T_{s,0}$. We conclude that $T_{0,s}$ is not finitely axiomatisable.
(v) If $\mathcal{A} \sqsubseteq \mathcal{M} \models T_{0,s}$, $\mathcal{A}$ may be a submodel of $\mathcal{M}$ or not. In the latter case, it is possible to obtain a submodel $\mathcal{A}^+ \sqsupseteq \mathcal{A}$ by adding, for any $a \in A$ other than 0^M and lacking predecessors, the elements of M that are predecessors of A (i.e. b such $s^M(b) = a$, c such that $s^M(c) = b$, and so on). It is easy to see that $\mathcal{A}^+$ is a $T_{0,s}$-closure of $\mathcal{A}$.
It therefore suffices to show that $T_{0,s}$ is model-complete. We apply Robinson's Test and consider $\mathcal{A} \sqsubseteq \mathcal{B} \models T_{0,s}$. Suppose $\mathcal{B} \models \exists \bar{y} \phi(\bar{a}, \bar{y})$, with $\bar{a} \in A$ and $\phi(\bar{x}, \bar{y})$ is a conjunction of literals. Given the parameters from A, the literals we encounter must be of one of the following form (where a is a generic element from the list $\bar{a}$):

$$\neg(y = a), \neg(y = y'), s^n(y) = y', \neg(s^n(y) = y'), s^n(y)$$
$$= as^n(a) = y, \neg(s^n(a) = y), \neg(s^n(y) = a).$$

We rule out equalities of the form $s^n(y) = s^m(a)$ or $s^n(y) = s^m(y')$ because the injectivity of s always allows us to reduce to a $T_{0,s}$-equivalent form belonging to the above list.
Our goal is to show that the literals in ϕ have solutions already in A. Because of this, we do not have to consider $s^n(y) = a s^n(a) = y$ which can have solutions *only* in A. As for the negated equalities, $\neg(y = a), \neg(y = y')$, if their variables already occur in an equality of the form $s^n(y) = a s^n(a) = y$, they automatically hold in A because there cannot be any loops. If y, y' do not occur in any of the above equalities, then the inequalities $\neg(y = a), \neg(y = y')$ can be satisfied in $\mathcal{A}$ in virtue of the fact that A is infinite. This is also true of $s^n(y) = y'$: if e.g. y occurs in a literal of the form $s^m(y) = a s^m(a) = y$, then $y \in A$ and, consequently, $y' \in A$.
We turn to negative literals of the form $\neg(s^n(a) = y), \neg(s^n(y) = a)$. If y already occurs in a positive literal with parameters within ϕ, then it can only be chosen within A. If it does not, then there are infinitely many choices from A for y. We may reduce $\neg(s^n(y) = y')$ to the above analysis by noting that, if one of y, y' must be chosen in A, then the fact that A is infinite allows us to choose it from A as well. Otherwise y, y' may be freely chosen within A.

Robinson Test applies and we conclude that $T_{0,s}$ is model-complete. Because the substructures of tis models have $T_{0,s}$-closures, it follows that $T_{0,s}$ has quantifier elimination.

(vi) We show that $Th(\mathcal{N}_s)$ is not model complete. Consider the substructure of $\mathcal{N}$ obtained by deleting 0 from $\mathbb{N}$, which we call $(\mathbb{N}^0, s)$. Since $\mathcal{N}_s$ and $(\mathbb{N}^0, s)$ are isomorphic (under the isomorphism $n \mapsto n+1$), they are elementarily equivalent. Thus $(\mathbb{N}^0, s)$ is a submodel of $\mathcal{N}_s$. It is not difficult to see that, despite being a submodel $(\mathbb{N}^0, s)$ is not an elementary substructure of $\mathcal{N}_s$. In fact, $\mathcal{N}_s \models \exists y(1 = s(y))$ but $(\mathbb{N}^0, s) \not\models \exists y(1 = s(y))$. It follows that model-completeness fails for $Th(\mathcal{N}_s)$: we conclude that quantifier elimination must also fail.

Exercise 8.77

Toward a contradiction, we suppose $\mathcal{M}_{\bar{c}} \models T \cup \Gamma_{\bar{c}} \cup \{\neg\varphi(\bar{c})\}$. Let $\bar{a} \in M$ be the list named by $\bar{c}$. We call $\langle \bar{a} \rangle$ the substructure of $\mathcal{M}$ generated by the tuple $\bar{a}$. It is important to note that any element of $\langle \bar{a} \rangle$ is named by a closed $\mathcal{L}_{\bar{c}}$-term. This is because, by definition, $\langle \bar{a} \rangle$, the elements of $\langle \bar{a} \rangle$ are precisely the interpretations of the $\mathcal{L}$-terms in the free variables $\bar{x}$ when these variables are evaluated on $\bar{a} \in M$.

The remark is important because it allows us to assume that $\mathcal{D}(\langle \bar{a} \rangle)$ is a set of $\mathcal{L}_{\bar{c}}$-sentences. We may then easily prove by contradiction that $T \cup \mathcal{D}(\langle \bar{a} \rangle) \cup \{\varphi(\bar{c})\}$ has a model $\mathcal{N}_{\bar{c}}$. If not, then we could find a finite subset $\{\delta_1, \ldots, \delta_m\} \subseteq \mathcal{D}(\langle \bar{a} \rangle)$ such that:

$$T \models \bigwedge i = 1^m \delta_i \to \neg\varphi(\bar{c}) \iff T \models \varphi(\bar{c}) \to \neg\bigwedge_{i=1}^m \delta_i.$$

It now follows that $\neg\bigwedge_{i=1}^m \delta_i \in \Gamma_c$ and $\mathcal{M} \models \neg\bigwedge_{i=1}^m \delta_i \wedge \bigwedge_{i=1}^m \delta_i$, a contradiction. Thus, we obtain $\mathcal{N}_{\bar{c}} \models T \cup \mathcal{D}(\langle \bar{a} \rangle) \cup \{\varphi(\bar{c})\}$.

The reducts $\mathcal{M}, \mathcal{N}$ of $\mathcal{M}_c, \mathcal{N}_c$ have thus a common substructure $\langle \bar{a} \rangle$ and $\mathcal{M} \models \neg\varphi(\bar{a})$ while $\mathcal{N} \models \varphi(\bar{a})$, against (2).

Exercise 8.79

Note that any two models of $T \cup \mathcal{D}(\mathcal{A})$ may be regarded as models of T with a common substructure $\mathcal{A}$. Substructure-completeness is thus a restatement of condition (2). If $\mathcal{M}, \mathcal{N} \models T$ have a common substructure $\mathcal{A}$, then these models have expansions $\mathcal{M}_A, \mathcal{N}_A \models T \cup \mathcal{D}(\mathcal{A})$. Condition (2) simply says that $\mathcal{M}_A \equiv \mathcal{N}_A$, i,e, that $T \cup \mathcal{D}(\mathcal{A})$ is complete.

Exercise 8.78

If conditions (2) and (3) are restricted to primitive formulae, then, by condition (1), primitive formulae are T-equivalent to quantifier free formulae. This is enough to have quantifier elimination by Theorem 8.55.

Exercise 8.84

Let ϕ be DAG-sentence ϕ that is DAG-equivalent to a quantifier-free DAG-sentence. For any $\mathcal{M} \models$ DAG we must have $\mathcal{M} \models \phi \leftrightarrow \theta$.

We consider two models of DAG, namely the structure $Q = (\mathbb{Q}, +^{\mathbb{Q}}, -^{\mathbb{Q}}, 0^{\mathbb{Q}})$ and its substructure Z whose domain is $\{0\}$. If ϕ is $\exists x \neg (x = 0)$, since $Q \models$ DAG, we must have $Q \models \phi \leftrightarrow \theta$. Since $Q \models \phi$, it follows that $Q \models \theta$. Because $Z \sqsubseteq Q$ and θ is quantifier-free, $Z \models \theta$. It is however clear that $Z \models \neg \exists x \neg (x = 0)$. Thus we have a model of DAG that fails to satisfy the biconditional $\phi \leftrightarrow \theta$.

We conclude that $\exists x \neg (x = 0)$ cannot have a quantifier-free DAG-equivalent. It follows that DAG cannot have quantifier elimination.

Exercise 8.85

Note that the theory of $(\mathbb{Q}, +^{\mathbb{Q}}, 0^{\mathbb{Q}})$ is a subtheory of the theory of $(\mathbb{Q}, +^{\mathbb{Q}}, -^{\mathbb{Q}} 0^{\mathbb{Q}})$ $\models$ DAG$^+$. This subtheory contains in particular all of the axioms of DAG$^+$ with the exception of $\forall x (x + (-x) = 0)$. However, it contains the sentence $\forall x \exists y (x + y = 0)$. Let T_{DAG} be the theory axiomatised by the axioms of DAG in which the function symbol $-$ does not occur together with the sentence $\forall x \exists y (x + y = 0)$. It is clear that any model of T_{DAG} can be expanded to a model of DAG. Thus, any two uncountable models of T_{DAG} must be isomorphic and T_{DAG} is a complete theory. It follows that it coincides with the theory of $(\mathbb{Q}, +^{\mathbb{Q}}, 0^{\mathbb{Q}})$.

By Lindström's test, T_{DAG} is model-complete. Now suppose $\mathcal{A} \sqsubseteq \mathcal{M} \models T_{DAG}$. Then $\mathcal{A}$ is a monoid and the divisible hull of the subgroup generated by $\mathcal{A}$ in $\mathcal{M}$ is its T_{DAG}-closure. It now follows that T_{DAG} has quantifier elimination.

Exercise 8.87

If $b \in V_1 - V_2$, $b \notin V$, since $V \subseteq V_1 \cap V_2$. Note, however, that if $\varphi(\overline{a}, y)$ contained at least one positive literal, b would be forced to lie in V. We conclude that $\varphi(\overline{a}, y)$ is a conjunction of negative literals, each saying that y is not a specified vector in the linear closure of $\overline{a}$. Since V_2 is a proper extension of V, it is possible to find $c \in V_2 - V$, which automatically satisfies the negative literals in $\varphi(\overline{a}, y)$.

Exercise 8.88

In view of the previous exercise, Theorem 8.76 applies.

Exercise 8.89

The theory ODAG^+ has algebraically prime models, namely ordered divisible hulls, i.e. the ordered analogue of the divisible hulls we know how to construct from substructures of models of DAG. In order to prove that ODAG^+ has quantifier elimination, it suffices to consider models $\mathcal{M} \sqsubseteq \mathcal{N}, \overline{a} \in M$ and a primitive formula $\exists x \varphi(x, \overline{a})$ modelled by $\mathcal{N}$. We have to prove that $\mathcal{M} \models \exists x \varphi(x, \overline{a})$.

Because $\mathcal{N}$ is linearly ordered, we may assume that no negative literals occur in $\varphi(x, \overline{a})$. The positive literals may then be partitioned into equalities and inequalities. If any equalities occur, because they are of the form $m_1 a_1 + \ldots + m_k a_k = nx$, with $a_1, \ldots, a_k$ in $\overline{a}$, they have unique solutions in M. The result is therefore immediate if at least one literal in the conjunctio $\varphi(x, \overline{a})$ is an equality.

If not, we only have inequalities of the form $m_1 a_1 + \ldots + m_k a_k < qx$ or $qx < m_1 a_1 + \ldots + m_k a_k$. Suppose there are n such inequalities. By divisibility, we can reduce each one of them to the form $x < b_i$ or $b_i < x$, with $i = 1, \ldots, n$ and $b_i \in M$ the d such that:

$$q_i d = m_1 a_1 + \ldots + m_k a_{k_i} -$$

If $c \in M$ satisfies the n inequalities determine by b_i, we can rely on the fact that $\mathcal{M}, \mathcal{N}$ are both discrete linear orders without endpoints and find $a \in M$ with the same relative position as c. It follows that $\mathcal{M} \models \exists x \varphi(x, \overline{a})$. We conclude that ODAG^+ has quantifier elimination.

Exercise 8.93

Both DAG^+ and ODAG^+ have quantifier elimination. Thus, any formula $\phi(x)$ in the language of one or the other theory is equivalent to a disjunctive normal form $\theta(\overline{x})$, which defines the union of the sets defined by its disjuncts. Each disjunct defines, in turn, the intersection of the sets defined by its literals.

In the case of DAG^+, literals define either singletons or complements of singletons. Their finite intersections, if nonempty, are either singletons or cofinite sets. Finite unions of singletons are finite sets and finite unions of singletons and cofinite sets are cofinite. It follows that DAG^+ is a strongly minimal theory.

The case of ODAG^+ is not very different but we are able to restrict attention to positive literals only because of trichotomy. Literals may be singletons or intervals with or without endpoints. Finite intersections of intervals and singletons are either points or intervals. Finite unions of such intersections are therefore unions of points and intervals. We conclude that ODAG^+ is o-minimal.

B   Solutions to Exercises   465

Exercise 8.94

(i) A definable subspace of $\mathcal{V}$ other than $\mathcal{V}$ itself and the subspace consisting of the null vector alone must contain at least one vector v other than the null vector. If $\mathcal{V}$ is one-dimensional, this subspace coincides with the set defined by the formula $x = x$. If $\mathcal{V}$ is at least two-dimensional, then any strict subspace $\mathcal{W}$ of $\mathcal{V}$ must be infinite and have dimension strictly lower than $dim V$. This is to say that a basis of W leaves out some linearly independent vectors in $V - W$: but any such vector generates an infinite subspace of $\mathcal{V}$. It follows that W is an infinite set with an infinite complement. No such set can be defined by a parametric L_v-formula $\varphi(x)$.
(ii) If a L_v-formula $\varphi(x_1, x_2)$ defined pairs x_1, x_2 of linearly independent vectors in V, then $\neg \varphi(x_1, x_2)$ would define linearly dependent pairs in V and, fixing a parameter $v \in V$, the formula $\neg \varphi(x_1, v)$ would define the infinite, one-dimensional subspace generated by v, which is impossible by part (i).

Exercise 8.95

If T is strongly minimal, for any model of $\mathcal{M}$ let $\varphi(M)$ be the subset of M defined by $\varphi(x)$. Strong minimality coincides with the fact that $M \cap \varphi(M) = \varphi(M)$ is finite or cofinite, for any $\mathcal{M} \models T$.

Exercise 8.96

Since ACF has quantifier-elimination, its definable subsets are defined by quantifier-free formulae. As such, they define finite unions of sets that result form intersecting (finite) sets of roots of specific polynomials with (cofinite) sets of non-roots of specific polynomials. Thus, the definable subsets of any model of ACF are either finite or cofinite and ACF is a strongly minimal theory.

Exercise 8.102

Let B_M be a basis for $\mathcal{M}$. If the map $f : M \longrightarrow N$ is an isomorphism, $f[B_M] = B_N$ is a subset of N of the same size as B_M. It suffices to show that B_N is actually a basis for $\mathcal{N}$. Consider any $b \in N$: because f is surjective, there is $c \in M$ such that $f(c) = b$. Since, moreover, $c \in acl_\mathcal{M}(B_M)$, there is an algebraic formula $\phi(x)$ such that $\mathcal{M} \models \phi(c)$. Consequently, $\mathcal{N} \models \phi(f(c))$ and ϕ (or its f-image, if it is a parametric formula) is algebraic in $\mathcal{N}$ since the property that it has exactly m solutions in M can be stated as a first-order sentence that transfers to $\mathcal{N}$. It follows that $b \in acl_\mathcal{N}(B_N)$. Since b is an arbitrary element of N, $N = acl_\mathcal{N}(B_N)$. It only remains to show that B_N is an independent set. To this end, fix $b^* \in B_N$.

There is $c^* \in M$ such that $f(c^*) = b^*$. If b^* were dependent on $B_N - \{b^*\}$, $b^* \in acl_\mathcal{N}(B_N - \{b^*\})$, i.e. there is an algebraic formula ϕ with parameters in $B_N - \{b^*\}$ such that $\mathcal{N} \models \phi(b^*)$. Because f is an isomorphism, $\mathcal{M} \models \phi(c^*)$, where ϕ has parameters in $B_M - \{c^*\}$, against the independence of c^* from $B_M - c^*$. We conclude that B_N is a basis for $\mathcal{N}$. Thus $\mathcal{M}, \mathcal{N}$ have the same dimension, since $|B_M| = |f[B_M]| = |B_N|$.

Exercise 8.105

The theory ThM is model-complete. We prove it using Robinson's test. Consider an existential formula $\exists x_1 \ldots \exists x_n \varphi(x_1, \ldots, x_n)$. Because existential quantifiers distribute over disjunction, we may assume that the matrix of the above formula is a conjunction of literals. Suppose $N \models ThM$ and M is a substructure of N. What an existential formula in the language $\{E\}$ can say is that $x_1, \ldots, x_n$ belong to certain distinct equivalence classes or to the same equivalence classes (and also that they are all distinct or not). Because only finitely many equivalence classes are involved and only finitely many distinct objects are involved, the existential formula we are considering is certainly modelled by $\mathcal{M}$ because M is partitioned into infinitely many, arbitrarily large equivalence classes.

It follows that any formula in the language of ThM is equivalent to an existential formula *modulo* ThM. We may now focus on $\mathcal{M}$ and consider its definable subsets. If we consider a formula $\varphi(x)$ without parameters, it defines $\emptyset$, the whole of M or a cofinite subset of M. To see this, consider $\exists y(Exy)$. Because E^M is reflexive, the last formula defines the whole of M. Similarly, the formula $\exists y \exists z(Exy \wedge \neg Exz)$ defines M because any $x \in M$ lies in some equivalence class and there are at least two distinct equivalence classes. If, in general, we have a conjunction of literals without parameters, this conjunction does not define the whole of M when x is required not to be equivalent to n distinct elements.

In this case, x cannot be in any of the equivalence classes with $1, 2, \ldots, n$ elements, but it can be in any other equivalence class. Only negated equalities and negated equivalences enable us to restrict the subset of M defined by an existential formula, unless we use parameters.

If we do, then a single positive literal with a parameter locates x within an equivalence class. In this case the set we define is a singleton or a distinguished equivalence class: the defining formula is therefore algebraic.

If we use parameters and negative literals, we rule x out of finitely many equivalence classes but x is otherwise unconstrained, so we define a cofinite subset of M. Since a union of finite and cofinite subsets is finite or cofinite, it follows that the definable subsets of M are finite or cofinite. Thus, M is a minimal structure.

By the upward Löwenheim-Skolem theorem, ThM has an elementary extension $\mathcal{N}$ of cardinality $\kappa > \aleph_0$. Note that ThM contains sentences to the effect that, for each $n \geq 1$, there is exactly one equivalence class of size n (the following statement can be formalised in $\{E\}$: for any n-tuples $\overline{x}, \overline{y}$, if the elements of each tuple are all distinct and all equivalent to each other, then each element of $\overline{x}$ is identical with some element of $\overline{y}$). It follows that $\mathcal{N}$ has one infinite equivalence class C, since the equivalence classes it shares with $\mathcal{M}$ only determine a countably infinite domain. For any $b \in C$, the formula Exb defines C, i.e. a subset of N that is infinite and has an infinite complement (including at least M). It follows that $\mathcal{N}$ is not minimal and, consequently, that $\mathcal{M}$ is not strongly minimal.

B.9 Chapter 9

Exercise 9.2

Let $\mathcal{M}$ be a $\mathcal{L}$-structure and $a \in M$. Show that the set $\{\varphi(x) : \mathcal{M} \models \varphi(a)\}$, where $\varphi(x)$ is a $\mathcal{L}$-formula in which at most x occurs free, is a complete 1-type.

Exercise 9.3

Show that, for any $n \geq 1$, every n-type can be extended to a complete n-type.

Exercise 9.7

Let $\mathcal{M}$ be a $\mathcal{L}$-structure $\mathcal{M}$ with $|M| \geq \aleph_0$. Then $q = \{\neg(x = a) : a \in M\}$ is a 1-type over M because any finite subset of q is satisfied in $\mathcal{M}$. Trivially, q cannot be realised by $\mathcal{M}$.

Exercise 9.9

We may e.g. take the 1-type $q = \{(0 < x)\} \cup \{x < 1/n : n \in \mathbb{N}\}$, which is finitely satisfied in $\mathbb{Q}$ but can only be realised in the elementary extensions of $(\mathbb{Q}, <)$ that do not contain positive infinitesimals. We know that $(\mathbb{R}, <)$ is not one of them.

Exercise 9.10

Let $\mathcal{M} \models \mathsf{DLO}_{-,-}$ and $p(x)$ be a complete 1-type in the language $\mathcal{L}_<$ that $\mathcal{M}$ realises. Because $\mathsf{DLO}_{-,-}$ admits quantifier elimination, each formula $\psi(x) \in p(x)$ is equivalent to a formula in disjunctive normal form, say $\varphi_1(x) \vee \ldots \vee \varphi_n(x)$, where $\varphi_i(x)$ is a conjunction of literals.

Knowing which inequalities $p(x)$ contains allows us to determine which literals it must contain because, if $p(x)$ does not contain an inequality, it must contain its negation. If the literals in $p(x)$ did not include the conjuncts of any $\varphi_i(x)$, with $i = 1, \ldots, n$, then $p(x)$ would be finitely inconsistent because it would contain

$\psi(x)$ and a finite set of literals that refute each disjunct of $\psi(x)$. Conversely, if $p(x)$ was finitely satisfiable, the literals in $p(x)$ would have to include the conjuncts of some $\varphi_j(x)$, by the completeness of $p(x)$.

We finally note that, for any $\mathcal{L}_<$-formula $\theta(x)$, either $\theta(x)$ or $\neg\theta(x)$ is in $p(x)$. But $\theta(x) \in p(x)$ iff the literals in a conjunct of its disjunctive normal form are in $p(x)$. It follows that, if a complete 1-type $q(x)$ contains exactly the same literals as $p(x)$, then it must be $p(x)$. It is now obvious that $q(x)$ must be realised in $\mathcal{M}$.

Exercise 9.11

Let $\Gamma = \{p_\beta : \beta \in \alpha\}$ be an enumeration of the complete n-types over A. Consider p_0 and let $\mathcal{M}_A$ be an expansion of $\mathcal{M}$ in which new constant symbols name the parameter set A. If $\mathcal{M}_A$ realises p_0, we set $\mathcal{N}_0 = \mathcal{M}_A$. Otherwise, because $\mathcal{ED}(\mathcal{M}) \cup \{p_0\}$ is finitely satisfiable, there is $\mathcal{N}_0 \succeq \mathcal{M}_A$ such that $\mathcal{N}_0$ realises p_0.

Now suppose $\mathcal{N}_\beta$ realises p_γ for each $\gamma \in \beta$. We let $\mathcal{N}_{\beta+1}$ be an elementary extension of $\mathcal{N}_\beta$ that realises p_γ, for each $\gamma \in \beta$, as well as p_β. After α iterations of this process we obtain an elementary chain $\{\mathcal{N}_\beta : \beta \in \alpha\}$. Let $\mathcal{N}$ be the union of this chain. By construction, $\mathcal{N}$ realises each type in Γ.

Exercise 9.12

(i) Clearly $[\neg(x = x =)] = \emptyset$ and $[x = x] = S_n(M/A)$.
(ii) If $[\varphi]$, $[\psi]$ are basic open sets, then $[\varphi] \cap [\psi]$ is the set of complete n-types over A in which both φ and ψ occur. Such complete types contain $\varphi \wedge \psi$ as well (finite consistency would fail if they contained its negation). It follows that $[\varphi] \cap [\psi] \subseteq [\varphi \wedge \psi]$. Using completeness, it is ease to see that the last inclusion is actually an equality. It now follows that the intersection of two basic open sets is itself a basic open set. A similar argument shows that $[\varphi] \cup [\psi] = [\varphi \vee \psi]$.
(iii) Suppose U, V are open sets and let p be a complete n-type over A in $U \cap V$. Because each of U, V is a union of basic open sets, $p \in [\varphi] \subseteq U$ and $p \in [\psi] \subseteq V$ for some φ, ψ. It follows that $p \in [\varphi \wedge \psi]$. It is easy to see that, if the last condition holds, then $p \in U \cap V$. Thus, $p \in U \cap V$ iff $p \in [\varphi \wedge \psi]$, i.e. any element of $U \cap V$ is contained in a basic open set, which is itself included in the intersection. The intersection $U \cap V$ may thus be viewed as the union of open sets of the form $[\varphi \wedge \psi]$. As such, it must itself be open.
Next, suppose $p \in \bigcup_{i \in I} U_i = U$, where each U_i is an open set. Then there is $j \in I$ such that $p \in U_j$ and, since U_j is a union of basic open sets, $p \in [\varphi] \subseteq U_j$ for some formula φ. Since $[\varphi] \subseteq U_j \subseteq U$, the set U is open.
(iv) A basic open set $[\varphi]$ is trivially open, as the union $[\varphi] \cup [\varphi]$. It is easy to see that the complement of of $[\varphi]$ is the basic open set $[\neg\varphi]$. It follows that $[\varphi]$ is closed as well.

(v) Let B be the collection of all basic open sets contained in some U_i from the given family of open sets. If $S_n(M/A)$ can be expressed as a union of finitely many open sets from B, it can certainly be expressed as a finite union of open sets of the form U_i.

Toward a contradiction, we suppose that $S_n(M/A)$ cannot be expressed as a union of finitely many open sets from B and use our assumption to deduce that $S_n(M/A) \neq \bigcup_{i \in I} U_i$. Because the last inequality contradicts our hypothesis, we conclude that $S_n(M/A)$ can in fact be expressed as a union of finitely many open sets from B, from which our desired conclusion follows.

Now wee note that, if finite subset of B, say $\{[\varphi]_1, \ldots, [\varphi]_n\}$ leaves out some element of $S_n(M/A)$, the intersection $[\neg\varphi]_1 \cap \ldots \cap [\neg\varphi]_n = [\neg\varphi_1 \wedge \ldots \wedge \neg\varphi_n]$ is nonempty, i.e. it contains a complete n-type p over A. Because types are finitely consistent, $\{\neg\varphi_1, \ldots, \neg\varphi_n\}$ is satisfiable (a suitable expansion of the language may be used to treat each φ_i as a sentence).

It follows that the set $\Gamma = \{\neg\varphi : [\varphi] \in B\}$ is a type. Let q be a complete type extending Γ. Clearly, $q \in S_n(M/A)$ but, by construction, $q \notin [\varphi]$ for each $[\varphi] \in B$. As a consequence, the union of B does not coincide with $q \in S_n(M/A)$. It follows that $S_n(M/A) \neq \bigcup_{i \in I} U_i$, against our hypothesis.

Exercise 9.13

(i) Let $q(x)$ be the minimum polynomial of b over K. Then $q(x)$ divides $m(x)$. Since, however, $m(x)$ is irreducible, $q(x), m(x)$ must be associates, i.e. they have the same degree and $m(x) = aq(x)$ with $a \in K$. Because $m(x)$ is monic, $a = 1$ and $m(x) = q(x)$.

(ii) If $f(x) = 0$ is a polynomial equation in $tp_{\mathcal{F}}(a/K)$, then a is a root of $f(x)$. It follows that $f(x) = q(x)m(x)$ for some polynomial $q(x)$. Thus $\mathcal{F} \models \forall x(m(x) = 0 \to f(x) = 0)$, as desired.

(iii) If $f(x) = 0$ is a polynomial equation not in $tp_{\mathcal{F}}(a/K)$, then $\neg(f(x) = 0) \in tp_{\mathcal{F}}(a/K)$. Now note that, if b is a root of $m(x)$, then b cannot also be a root of $f(x)$. This is because $f(x)$ would be divisible by $m(x)$, in which case $f(x) = 0$ must be in $tp_{\mathcal{F}}(a/K)$. We conclude that no root of $m(x)$, is also a root of. This is enough to conclude that $\mathcal{F} \models \forall x(m(x) = 0 \to \neg(f(x) = 0))$.

(iv) Any quantifier-free formula $\psi(x) \in tp_{\mathcal{F}}(a/K)$ can be written as a disjunction of conjunctions. Each conjunction is of the form $A_1(x) \wedge \ldots \wedge A_n(x)$, where $A_i(x)$ is a literal, i.e. e polynomial equation of the form $f(x) = 0$ or the negation of such an equation. Since a realises $\psi(x)$ in $\mathcal{F}$, a realises every literal in at least one disjunct of $\psi(x)$. If the disjunct in question contains at least one positive literal, then this literal is $f(x) = 0$ with $f(x) = q(x)m(x)$. If the disjunct in question is $\neg(f(x) = 0)$, by part (iii) no root of $m(x)$ can be a root of $f(x)$. It follows that any root of $m(x)$ satisfies $\psi(x)$. Consequently $\mathcal{F} \models \forall x(m(x) = 0 \to \psi(x))$.

(v) Note that $m(x)$ is an algebraic formula over K because it defines a finite subset of F. Consider an arbitrary formula $\theta(x, \bar{b})$, in which a tuple $\bar{c}$ of parameters from K occurs. By quantifier elimination, the formula $\theta(x, \bar{y})$, in which no parameters occur, is ACF-equivalent to a quantifier-free formula $\eta(x, \bar{y})$. Thus:

$$\mathcal{F} \models \theta(a, \bar{c}) \leftrightarrow \eta(a, \bar{c}).$$

Suppose $\mathcal{F} \models m(a) \rightarrow \eta(a, \bar{c})$. Part (iv) implies that, for any root b of $m(x)$, $\mathcal{F} \models m(b) \rightarrow \eta(b, \bar{c})$. We conclude $\mathcal{F} \models m(b) \rightarrow \theta(b, \bar{c})$. This suffices to deduce: $\mathcal{F} \models \forall x (m(x) \rightarrow \theta(x, \bar{c}))$.

Since, for any formula $\theta(x) \in tp_\mathcal{F}(a/K)$, we trivially have $\mathcal{F} \models m(a) \rightarrow \eta(a)$, the above argument shows that $m(x)$ implies every formula in the 1-type $tp_\mathcal{F}(a/K)$.

Exercise 9.16

Consider the algebraic formula in $p(\bar{x})$, say $\phi(\bar{x})$, with the smallest number of solutions in M and let $\theta(\bar{x})$ be any formula in $p(\bar{x})$. If $\mathcal{M} \models \exists \bar{x}(\phi(\bar{x}) \wedge \neg \theta(\bar{x}))$, then the formula $\phi(\bar{x}) \wedge \neg \psi(\bar{x})$ is algebraic a defines a strictly smaller, finite set than the set defined by $\phi(\bar{x})$.

Exercise 9.17

If $p(\bar{x}) \neq q(\bar{x})$, there is a formula $\psi(\bar{x})$ such that $\psi(\bar{x}) \in p(\bar{x})$ and $\neg \psi(\bar{x}) \in q(\bar{x})$. Suppose both $p(\bar{x}), q(\bar{x})$ are in $[\varphi(\bar{x})]$. Then $\mathcal{M} \models \forall \bar{x}(\varphi(\bar{x}) \rightarrow \psi(\bar{x}))$ and $\mathcal{M} \models \forall \bar{x}(\varphi(\bar{x}) \rightarrow \neg \psi(\bar{x}))$, which jointly imply $\mathcal{M} \models \forall \bar{x}(\varphi(\bar{x}) \rightarrow \psi(\bar{x}) \wedge \neg \psi(\bar{x}))$. But $\varphi(\bar{x})$ is in $p(\bar{x})$ and thus it is satisfied in $\mathcal{M}$ and $\psi(\bar{x}) \wedge \neg \psi(\bar{x})$ cannot evidently be satisfied in any structure. We conclude that $p(\bar{x}) = q(\bar{x})$.

Next, suppose that $p(\bar{x})$ is an isolated point of $S_n(M/A)$. This is to say that, for some formula $\varphi(\bar{x})$, $[\varphi(\bar{x})] = \{\bar{x}\}$. If $\psi(\bar{x}) \in p(\bar{x})$, then $\mathcal{M} \models \exists \bar{x}(\varphi(\bar{x}) \wedge \psi(\bar{x}))$. We cannot also have $\mathcal{M} \models \exists \bar{x}(\varphi(\bar{x}) \wedge \neg \psi(\bar{x}))$, otherwise $[\varphi(\bar{x})]$ would consist of at least two distinct types. We must therefore have $\mathcal{M} \models \neg \exists \bar{x}(\varphi(\bar{x}) \wedge \neg \psi(\bar{x}))$, i.e.:

$$\mathcal{M} \models \forall \bar{x}(\varphi(\bar{x}) \rightarrow \psi(\bar{x})).$$

Exercise 9.19

The implication from (i) to (ii) is the previous exercise. In order to establish the implication from (ii) to (iii), we only have to prove one side of the biconditional. If $\mathcal{M} \models \forall \bar{x}(\varphi(\bar{x}) \rightarrow \psi(\bar{x}))$, but $\psi(\bar{x}) \notin p(\bar{x})$, the completeness of $p(\bar{x})$ implies $\neg \psi(\bar{x}) \in p(\bar{x})$. By (ii), we deduce $\mathcal{M} \models \forall \bar{x}(\varphi(\bar{x}) \rightarrow \neg \psi(\bar{x}))$, against the satisfiability of $\varphi(\bar{x})$. To prove the implication from (iii) to (i), note that $\varphi(\bar{x}) \in p(\bar{x})$. If $[\varphi(\bar{x})] \neq \{p(\bar{x})\}$, there are $q(\bar{x}) \in [\varphi(\bar{x})]$ and $\theta(\bar{x}) \in p(\bar{x})$ such that $\neg\theta(\bar{x}) \in q(\bar{x})$. Because $q(\bar{x})$ is finitely satisfiable, we deduce:

$$\mathcal{M} \models \exists\bar{x}(\varphi(\bar{x}) \wedge \neg\theta(\bar{x})) \iff \mathcal{M} \models \exists\bar{x}\neg(\varphi(\bar{x}) \to \theta(\bar{x}))$$
$$\iff \mathcal{M} \models \neg\forall\bar{x}(\varphi(\bar{x}) \to \theta(\bar{x})),$$

which contradicts the hypothesis, since $\theta(\bar{x}) \in p(\bar{x})$. It follows from this contradiction that $[\varphi(\bar{x})]$ has exactly one element, namely $p(\bar{x})$.

Exercise 9.20

Let $p(\bar{x})$ be an isolated type of $S_n(M/A)$. By the previous exercise, there is $\varphi(\bar{x}) \in p(\bar{x})$ such that, for each $\psi(\bar{x}) \in p(\bar{x})$, $\mathcal{M} \models \forall\bar{x}(\varphi(\bar{x}) \to \psi(\bar{x}))$.

Because $p(\bar{x})$ is finitely satisfiable, there is a tuple $\bar{a}$ of elements of M such that $\mathcal{M} \models \varphi(\bar{a})$. It now follows that $\mathcal{M} \models \psi(\bar{a})$ for each $\psi(\bar{a}) \in p(\bar{x})$. We conclude that $p(\bar{x}) \subseteq tp_{\mathcal{M}}(\bar{a}/A))$. Equality follows from the fact that $p(\bar{x})$ is complete.

Because algebraic types are isolated, the above argument applies to them, too.

Exercise 9.22

(i) Let $\mathcal{M}$ be a $\mathcal{L}$-structure and $A \subseteq B \subseteq M$. If $\varphi(\bar{x})$ implies every formula in $tp_{\mathcal{M}}(\bar{a}/B)$, then it implies every formula in $tp_{\mathcal{M}}(\bar{a}/A) \subseteq tp_{\mathcal{M}}(\bar{a}/B)$.
(ii) It suffices to note that $tp_{\mathcal{M}}(\bar{a}/A) \subseteq tp_{\mathcal{M}}(\bar{a}, \bar{b}/A)$.
(iii) Let $\psi(\bar{x}, \bar{y})$ isolate $tp_{\mathcal{M}}(\bar{a}, \bar{b}/A)$. We first consider the formula $\psi(\bar{x}, \bar{b})$, which is realised by $\bar{a}$. This formula belongs to $tp_{\mathcal{M}}(\bar{a}, /A \cup \{\bar{b}\})$. As a consequence:

$$\theta(\bar{x}, \bar{b}) \in tp_{\mathcal{M}}(\bar{a}/A \cup \{\bar{b}\}) \implies \theta(\bar{x}, \bar{y}) \in tp_{\mathcal{M}}(\bar{a}, \bar{b}/A)$$
$$\iff \mathcal{M} \models \forall\bar{x}\forall\bar{y}(\psi(\bar{x}, \bar{y}) \to \theta(\bar{x}, \bar{y}))$$
$$\implies \mathcal{M} \models \forall\bar{x}(\psi(\bar{x}, \bar{b}) \to \theta(\bar{x}, \bar{b})).$$

It follows that $\psi(\bar{x}, \bar{b})$ isolates $tp_{\mathcal{M}}(\bar{a}/A \cup \{\bar{b}\})$.

Next we consider the formula $\exists\bar{x}\psi(\bar{x}, \bar{y}) \in tp_{\mathcal{M}}(\bar{b}/A)$. Because $p_{\mathcal{M}}(\bar{b}/A) \subseteq tp_{\mathcal{M}}(\bar{a}, \bar{b}/A)$, we deduce:

$$\theta(\bar{y}) \in tp_{\mathcal{M}}(\bar{b}/A) \implies \theta(\bar{y}) \in tp_{\mathcal{M}}(\bar{a}, \bar{b}/A)$$
$$\iff \mathcal{M} \models \forall\bar{x}\forall\bar{y}(\psi(\bar{x}, \bar{y}) \to \theta(\bar{x}, \bar{y}))$$
$$\iff \mathcal{M} \models \forall\bar{y}\forall\bar{x}(\psi(\bar{x}, \bar{y}) \to \theta(\bar{y}))$$
$$\implies \mathcal{M} \models \forall\bar{y}(\exists\bar{x}\psi(\bar{x}, \bar{y}) \to \theta(\bar{y})).$$

It follows that $\exists\bar{x}\psi(\bar{x}, \bar{y})$ isolates $tp_{\mathcal{M}}(\bar{b}/A)$.

(iv) If $a \in acl_{\mathcal{M}}(A)$, then a satisfies an algebraic formula over A. It follows that $tp_{\mathcal{M}}(a/A)$ is algebraic and thus isolated.

Exercise 9.24

We provide a round-robin proof, starting with the implication from (1) to (2). Let $\varphi(\overline{x})$ be a formula such that $[\varphi(\overline{x})] = \{p(\overline{x})\}$. If condition (2) fails, there is a model of T, say $\mathcal{M}$, such that $\mathcal{M} \models \exists \overline{x}(\varphi(\overline{x}) \wedge \neg \psi(\overline{x}))$, with $\psi(\overline{x}) \in p(\overline{x})$. This is to say that $\{\varphi(\overline{x}), \neg\psi(\overline{x})\}$ is consistent with T. It follows that the last set is included in a complete n-type $q(\overline{x}) \neq p(\overline{x})$ such that $q(\overline{x}) \in [\varphi(\overline{x})]$, against (1).

Next, we assume (2) and prove (3). To this end, suppose $T \models \forall \overline{x}(\varphi(\overline{x}) \to \psi(\overline{x}))$. If $\psi(\overline{x}) \notin p(\overline{x})$, the completeness of $p(\overline{x})$ implies $\neg\psi(\overline{x}) \in p(\overline{x})$ and we can easily use the last condition with the hypothesis (2) to obtain a contradiction. It follows that $\psi(\overline{x}) \in p(\overline{x})$.

Finally, we assume (3) and prove (1). We proceed almost exactly as in the final part of Exercise 9.19. Proceeding by contradiction, we may assume $[\varphi(\overline{x})] \supseteq \{p(\overline{x}), q(\overline{x})\}$ with $\theta(\overline{x}) \in p(\overline{x})$ and $\neg\theta(\overline{x}) \in q(\overline{x})$. Because $q(\overline{x}) \in S_n(T)$, there is a model of T, say $\mathcal{M}$ such that: $\mathcal{M} \models \exists \overline{x}(\varphi(\overline{x}) \wedge \psi(\overline{x}))$. This contradicts, as in Exercise 9.19, the fact that, by (3), $T \models \forall \overline{x}(\varphi(\overline{x}) \to \psi(\overline{x}))$.

Exercise 9.25

If every point $p(\overline{x}) \in S_n(M/A)$ is isolated, then $\{p(\overline{x})\} = [\varphi(\overline{x})]$ for a suitable choice of $\varphi(\overline{x})$. We obtain an open cover of $S_n(M/A)$ by taking the union of all the basic open sets determined by the corresponding isolated points. By Compactness, this open cover has a finite subcover, i.e. $S_n(M/A)$ is the union of finitely many singletons and thus a finite set.

Conversely, if $|S_n(M/A)| < \aleph_0$, $|S_n(M/A)| = \{p_1(\overline{x}), \ldots, p_k(\overline{x})\}$ for some $k \in \mathbb{N}$. Fix $i \in \{1, \ldots, k\}$. Because the types $p_1(\overline{x}), \ldots, p_k(\overline{x})$ are all distinct, for each $j \neq i$ there is $\theta_{ij}(\overline{x})$ such that $\neg\theta_{ij}(\overline{x}) \in p_i(\overline{x})$ and $\theta_{ij}(\overline{x}) \in p_j(\overline{x})$. As a consequence:

$$\left[\bigwedge_{i \neq j} \neg\theta_{ij}(\overline{x}) \right] = \{p_i(\overline{x})\}$$

for any $i \in \{1, \ldots, k\}$. In other words, every type in $S_n(M/A)$ is isolated.

Exercise 9.26

(i) Let p be a non-algebraic n-type in $S_n(M/A)$. Consider the set Δ whose elements are formulae of the form $\neg\psi(\overline{x})$, where $\psi(\overline{x})$ algebraic in M over B. It suffices to show that $p \cup \Delta$ is finitely consistent.
To this end, consider, on the one hand, $\{\theta_1(\overline{x}), \ldots, \theta_k(\overline{x})\} \subseteq p$. Because p is complete, $\theta(\overline{x}) = \theta_1(\overline{x}) \wedge \ldots \wedge \theta_k(\overline{x})$ is a non-algebraic formula in p with infinitely many realisations in M. A finite subset of Δ, on the other hand, is of the form $\{\neg\psi_1(\overline{x}), \ldots, \neg\psi_m(\overline{x})\}$ and the conjunction of its elements, call it $\psi(\overline{x})$ defines a cofinite subset of M. It follows that the subsets of M defined by $\theta(\overline{x})$ and $\psi(\overline{x})$ have a nonempty intersection, i.e. that $p \cup \Delta$ is finitely consistent. The latter set is clearly a type q' over B and it does not contain any algebraic formula. Let q be a complete extension of q' over B: then $q \in S_n(M/B)$ cannot be algebraic. If it were, it should contain an algebraic formula $\psi(\overline{x})$ over B, but, by construction, $\neg\psi(\overline{x})$ is already in q'.

(ii) Suppose $p(\overline{x}) \in S_n(M/A)$ is algebraic and let $\varphi(\overline{x})$ be an algebraic formula that isolates $p(\overline{x})$. Note that $\varphi(\overline{x}) \in p(\overline{x})$ and the realisations of $\varphi(\overline{x})$ in M coincide with those of $p(\overline{x})$. If $\varphi(\overline{x})$ has exactly k distinct realisations, $\exists^{=k}\overline{x}\varphi(\overline{x}) \in p(\overline{x})$. We have already shown in Exercise 9.20 that M realises its isolated types and thus, in particular, $p(\overline{x})$. No elementary extension of M can add further realisations of $p(\overline{x})$ because any such elementary extension must be a model of the sentence $\exists^{=k}\overline{x}\varphi(\overline{x}) \in p(\overline{x})$.
Conversely, suppose that $p(\overline{x}) \in S_n(M/A)$ contains only non-algebraic formulae. Let $\mathcal{M}^*$ be an expansion of $\mathcal{M}$ that models the elementary diagram of $\mathcal{M}$. Let $\mathcal{L}^*$ be the expanded language associated with $\mathcal{M}^*$. We add n new constant symbols $d_1, \ldots, d_n$ and consider the set of sentences $p(\overline{d})$, obtained from $p(\overline{x})$ by replacing each occurrence of x_i in a formula of $p(\overline{x})$ with an occurrence of d_i. For each n-tuple $(c_1, \ldots, c_n) = \overline{c}$ of constant symbols in $\mathcal{L}^*$ designating a realisation of $p(\overline{x})$, consider the formula:

$$\neg(d_1 = c_1) \wedge \ldots \wedge \neg(d_n = c_n).$$

Let Δ be the set of all conjunctions of the above form. We claim that $\mathcal{ED} \cup p(\overline{d}) \cup \Delta$ is finitely satisfiable. This is because, if $\psi_1(\overline{d}), \ldots, \psi_k(\overline{d}) \in p(\overline{d})$, set $\theta(\overline{d}) = \psi_1(\overline{d}) \wedge \ldots \wedge \psi_k(\overline{d}) \in p(\overline{d})$. By assumption, $\theta(\overline{x})$ is not algebraic, i.e. it defines an infinite subset of M. Because of this, for any finite subset F of Δ, it is possible to find a tuple that satisfies $\theta(\overline{x})$ and that differs from the finitely many tuples referred to in F. In short $\{\theta(\overline{d})\} \cup F$ is modelled by $\mathcal{M}^*$. Compactness now implies that $\mathcal{ED} \cup p(\overline{d}) \cup \Delta$ has a model $\mathcal{N} \succeq \mathcal{M}$ with at least one realisation of $p(\overline{x})$ that is not in $\mathcal{M}$, a contradiction. We conclude that $p(\overline{x})$ contains one algebraic formula, i.e. that it is an algebraic type.

Exercise 9.27

By quantifier elimination, a complete type $p \in S_1(M/A)$ is uniquely determined by the literals it contains (these literals determine the disjunctive normal forms allowed in p and then we can use the fact that every other formula is IS-equivalent to a disjunctive normal form as in Exercise 9.13). If a 1-type contains any positive literals, they must be of the form $x = a$, with $a \in A$. Clearly, each such type is algebraic. We are thus left with the single 1-type q containing $\neg(x = a)$ for each $a \in A$. Because no quantifier-free formula in q is algebraic, q is the only non-algebraic type in $S_1(M/A)$. A similar argument shows that, if $\mathcal{M}$ is an infinite vector space over K, the algebraic 1-types over A are realised by vectors in the linear closure of A, while the single non-algebraic 1-type over A is realised by any vector independent of that algebraic closure.

Exercise 9.30

(2) $\implies$ (1) is immediate because we can take $\mathcal{M}$ to be $\mathcal{N}$ and A to be C. $\mathcal{E}$-separability follows from the fact that the elements of N are distinct types over C. To prove the converse implication, we note that, by (1), $\mathcal{M}$ has a scaling structure $\mathcal{N}$ that is isomorphic to it. Let τ be an isomorphism from $\mathcal{M}$ onto $\mathcal{N}$. Then, for any element $s(x)$ of N there is a such that $\tau(a) = s(x) = tp_\mathcal{M}(a/a_1, \ldots, a_n)$. Because $\mathcal{N}$ and $\mathcal{M}$ are isomorphic, types are preserved by τ and we obtain:

$$\begin{aligned} s(x) &= \tau(a) \\ &= tp_\mathcal{M}(a/a_1, \ldots, a_n) \\ &= tp_\mathcal{N}(\tau(a)/\tau(a_1), \ldots, \tau(a_n)) \\ &= tp_\mathcal{N}(s(x)/\tau(a_1), \ldots, \tau(a_n)). \end{aligned}$$

It now suffices to set $\tau(a_i) = c_i$.

Exercise 9.34

Let $\{\psi_1(x, f(\overline{c})), \ldots, \psi_n(x, f(\overline{c}))\}$ be a finite subset of $p(x)$. By the definition of $p(x)$ we have $\mathcal{M} \models \exists x \bigwedge_{i=1}^n \psi(x, \overline{c})$. Because f is partial elementary, we deduce $\mathcal{N} \models \exists x \bigwedge_{i=1}^n \psi(x, f(\overline{c}))$.

Exercise 9.36

We are given a chain of elementary maps $\{f_\beta : \beta \in \alpha\}$ from $\mathcal{M}$ to $\mathcal{N}$ and we have to show their union f is an elementary map with domain M and codomain N. First we note that the union of a chain of functions is itself a function. Moreover, the domain of f is M by construction.

Finally, suppose $\mathcal{M} \models \varphi(\bar{b}, \bar{a})$, with the elements of $\bar{a}$ in A and the elements of $\bar{b}$ in $M - A$. Because each element of the last tuple occurs at some point along the enumeration of $M - A$, these elements must all belong to the domain of f_δ for some ordinal $\delta \in \alpha$. Because f_δ is elementary, we have:

$$\mathcal{M} \models \varphi(\bar{b}, \bar{a}) \implies \mathcal{N} \models \varphi(f_\delta(\bar{b}), f_\delta(\bar{a})).$$

But f coincides with f_δ on the latter's domain. The argument continues to work if we begin with the assumption $\mathcal{N} \models \varphi(f(\bar{b}), f(\bar{a}))$, precisely because f will coincide with some f_δ on the finite subset of M determined by the tuple $\bar{b}, \bar{a}$.

Exercise 9.38

As a union of elementary maps, f is certainly elementary by the previous exercise. We only have to show that the domain of f coincides with N. First, we note that, for every $b \in N$, there is a least $i \in \mathbb{N}$ such that $b \in N_i$. Because, by construction, $N_i \subseteq M_{i+1}$, b is in the domain of f_{k+1} and thus of $f \subseteq f_{k+1}$. Since b was chosen arbitrarily in N, the domain of f is N.

Exercise 9.39

Consider a finite subset of $q(x)$, e.g. $\{\psi_1(x, f(\bar{b}_1)), \ldots, \psi_k(x, f(\bar{b}_k))\}$. Because $tp_\mathcal{M}(a/A)$ is a type, we have $\mathcal{M} \models \exists x(\psi_1(x, \bar{b}_1) \wedge \ldots \wedge \psi_k(x, \bar{b}_k))$. Because f is elementary, we also have:

$$\mathcal{M} \models \exists x(\psi_1(x, \bar{b}_1) \wedge \ldots \wedge \psi_k(x, \bar{b}_k)) \iff \mathcal{N} \models$$
$$\exists x(\psi_1(x, f(\bar{b}_1)) \wedge \ldots \wedge \psi_k(x, f(\bar{b}_k))).$$

It follows that $\mathcal{N} \models \exists x(\psi_1(x, f(\bar{b}_1)) \wedge \ldots \wedge \psi_k(x, f(\bar{b}_k)))$, i.e. $q(x)$ is a type over $f[A]$. By construction, it is also clear that $q(x)$ is a complete 1-type over $f[A]$.

We next suppose that $\varphi(x, \bar{b})$ isolates $tp_\mathcal{M}(a/A)$. Because f is elementary, for any $\psi(x, \bar{d}) \in tp_\mathcal{M}(a/A)$ we have:

$$\mathcal{M} \models \forall x(\varphi(x, \bar{b}) \to \psi(x, \bar{d})) \iff \mathcal{N} \models \forall x(\varphi(x, f(\bar{b})) \to \psi(x, f(\bar{d}))).$$

Because every formula in $q(x)$ is obtained from a formula in $tp_\mathcal{M}(a/A)$ by replacing its parameters with their f-images, it follows that $q(x)$ is isolated. The last argument is reversible, i.e. if, knowing that $q(x) \in S_1(N/f[A])$, we are provided with the added information that $q(x)$ is isolated, it follows that $tp_\mathcal{M}(a/A)$ is isolated.

Exercise 9.44

If a were algebraic over K, it should have a minimum polynomial with coefficients in K, against the fact that K is algebraically closed. This remark concludes the first verification. As for the second verification, we note that b satisfies every (negative) literal in $q(x)$ and, consequently, every quantifier-free formula in $q(x)$. By quantifier elimination for ACF, b realises every formula in $q(x)$.

Exercise 9.45

Let $\mathcal{V}$ be an infinite vector space over the field K. It suffices to show that $\mathcal{V}$ is homogeneous. To this end we consider $A \subseteq V$ such that $|A| < |V|$ and an elementary map $f : A \longrightarrow V$. Note that f can be extended to an embedding from the linear closure of A into V (it suffices to consider a maximal, linearly independent subset of A). This embedding is elementary by quantifier elimination. We may therefore directly assume that A is the domain of a subspace $\mathcal{A}$ of $\mathcal{V}$. Let us next fix $b \in V - A$ and consider $tp_\mathcal{V}(b/A)$.

This 1-type contains all literals of the form $\neg(x = a)$, with $a \in A$. It cannot contain any positive literal or b would be a vector in A. The literals satisfied by b uniquely determine the set of quantifier-free formulae satisfied by b and, by quantifier elimination, the whole type $tp_\mathcal{V}(b/A)$. Let us now consider the set of formulae $q(x)$ obtained from $tp_\mathcal{V}(b/A)$ by replacing each occurrence of a in a formula of $tp_\mathcal{V}(b/A)$ with an occurrence of $f(a)$.

The set $q(x)$ contains all literals of the form $\neg(x = f(a))$, with $a \in A$. Because $V - f[A] \neq \emptyset$ ($|A| = |f[A]|$), there is a vector $d \in V$ that simultaneously satisfies the literals in $q(x)$. Now note that every quantifier-free formula over $f[A]$ must, in disjunctive normal form, contain a disjunct whose conjuncts are negative literals. If not, at least one conjunct must have been obtained by a positive literal in $tp_\mathcal{V}(b/A)$, which is impossible. It follows that d satisfies every quantifier-free formula in $q(x)$. By quantifier elimination, d satisfies every formula in $q(x)$, which is therefore a type. In fact, because $tp_\mathcal{V}(A)$ is a complete type, so is $q(x) = tp_\mathcal{V}(d/f[A])$.

It is easy to see that $f \cup (b, d)$ extends f and that it is a partial elementary function.

Exercise 9.52

Let $\mathcal{M}$ be a countable structure with $A \subseteq M$ and $|A| < |M|$: then A is finite. Suppose $f : A \longrightarrow M$ is partial elementary and fix $b \in M - A$. By atomicity, $tp_\mathcal{M}(b/A)$ is isolated. In view of Exercise 9.39, the set $q(x)$, obtained from $tp_\mathcal{M}(b/A)$ by replacing each formula $\varphi(x, \overline{a})$ in the latter type with $\varphi(x, \overline{b})$, is an isolated type over the finite set $f[A]$. Atomicity implies that some $c \in M$ realises $q(x)$. Because, for each $a \in A$, $x \neq a$ is in $tp_\mathcal{M}(b/A)$, c differs from each element of $f[A]$ and $f \cup \{(b, c)\}$ is an elementary one-point extension of f.

Exercise 9.53

By the previous exercise, any countable atomic structure is homogeneous. If $\mathcal{M}, \mathcal{N}$ are countable and atomic, Theorem 9.47 implies that we only have to verify the equality $S_k(M) = S_k(N)$ for each $k \geq 1$. To this end, we note that, for any fixed k, $p \in S_k(M)$ implies that p is isolated, by atomicity. Because p does not contain any parameters, it is realised by $\mathcal{N}$, which, as any structure, realises all of its isolated types: thus $p \in S_k(N)$. The last argument can be reversed and it establishes the equality we need. Since k was chosen arbitrarily, theorem 9.47 implies that $\mathcal{M}, \mathcal{N}$ are isomorphic.

Exercise 9.54

If $\mathcal{M}$ is finite, let $|M| = n$. There is an existential sentence $\exists x_1 \ldots \exists x_n \varphi(x_1, \ldots, x_n)$ in the language of T that fixes $\mathcal{M}$ up to isomorphism (the matrix of the last sentence essentially describes the diagram of $\mathcal{M}$). Fix $a_1, \ldots, a_k \in M$ and consider $tp_{\mathcal{M}}(a_1, \ldots, a_k)$. By the choice of $a_1, \ldots, a_k$, $\mathcal{M} \models \exists x_{k+1} \ldots \exists x_n \varphi(b_1, \ldots, b_k, x_{k+1}, \ldots, x_n)$. It follows that, calling $c_{k+1}, \ldots, c_n$ the elements of $M - \{a_1, \ldots, a_k\}$:

$$\mathcal{M} \models \varphi(b_1, \ldots, b_k, c_{k+1}, \ldots, c_n).$$

If $\mathcal{M} \models \exists x_{k+1} \ldots \exists x_n \varphi(b_1, \ldots, b_k, x_{k+1}, \ldots, x_n)$, there are $d_{k+1}, \ldots, d_n$ such that:

$$\mathcal{M} \models \varphi(b_1, \ldots, b_k, d_{k+1}, \ldots, d_n).$$

It follows that the function sending $a_1, \ldots, a_k, c_{k+1}, \ldots, c_n$ into $b_1, \ldots, b_k, d_{k+1}, \ldots, d_n$ is an automorphism of $\mathcal{M}$. As a consequence, for any $\theta(x_1, \ldots, x_k) \in tp_{\mathcal{M}}(a_1, \ldots, a_k)$, $\mathcal{M} \models \theta(b_1, \ldots, b_k)$. Thus:

$$\mathcal{M} \models \exists x_{k+1} \ldots \exists x_n \varphi(b_1, \ldots, b_k) \to \theta(b_1, \ldots, b_k).$$

Because $b_1, \ldots, b_k$ is an arbitrary element of M^k, we conclude:

$$\mathcal{M} \models \forall x_1 \ldots \forall x_k (\exists x_{k+1} \ldots \exists x_n \varphi(x_1, \ldots, x_k) \to \theta(b_1, \ldots, b_k)).$$

In other words, $\exists x_{k+1} \ldots \exists x_n \varphi(x_1, \ldots, x_k)$ isolates $tp_{\mathcal{M}}(a_1, \ldots, a_k)$. It follows that $\mathcal{M}$ is atomic.

Exercise 9.56

(i) For any closed term $t \in M$, the sentence $\exists x(t = x)$ is valid so it is added to a suitable stage in our construction of T^*, together with $t = c$, for c a suitable

constant symbol. It follows that any closed term is equivalent to a constant symbol.
(ii) For any m-ary function symbol f, the value of $f^{M^+}(c_1, \ldots, c_m)$ is a constant equivalent to the closed term $f(c_1, \ldots, c_m)$. Constant symbols name themselves as constants.
(iii) The inductive base is immediate by construction. Connectives are easily dealt with. Suppose $\exists x \varphi(x) \in T^+$. By construction, there is a constant symbol c such that $\varphi(c) \in T^+$. The inductive hypothesis (which can be restricted to sentences alone) implies that $\mathcal{M}^+ \models \varphi(c)$ and, consequently, $\mathcal{M}^+ \models \varphi(c)$. The last argument can be reversed because T^+ is complete.

Exercise 9.57

It suffices to partition the even-numbered steps of our recursion into two sequences (e.g. the sequence of multiples of four and its complement in the set of even numbers) in order to ensure that, at each step of one sequence, a formula of p_1 fails to be satisfied by the relevant tuple of constant symbols while, at each step of the other sequence, a formula of p_2 fails to be satisfied by the same tuple.

Exercise 9.59

To show that T is not λ-categorical, let $\mathcal{D}$ be the model whose domain is the set D (any constant symbol in C may be interpreted on a fixed element of D) and $\mathcal{C}$ be an extension of $\mathcal{D}$ that contains the constant symbols in C. Any elementary function from D to C must fix every element of D: as a consequence, it cannot be surjective. Thus, $\mathcal{D}, \mathcal{C}$ cannot be isomorphic.

Quantifier elimination follows from the fact that $\mathcal{D}$ is an algebraically prime model of T. If $\mathcal{M} \sqsubseteq \mathcal{N}$ and $\mathcal{M}, \mathcal{N} \models T$, let $\varphi(x, \overline{y})$ be a conjunction of literals, i.e. equalities or negated equalities. If $\mathcal{N} \models \exists x \varphi(x, \overline{a})$, with $\overline{a}$ a tuple of elements of M, it is clear that, if a positive literal occurs in φ, $\mathcal{M} \models \exists x \varphi(x, \overline{a})$ and that, if no positive literals occurs in φ, the fact that M is infinite implies the same conclusion. By the Shoenfield-Robinson criterion for quantifier elimination, it follows that T has quantifier elimination.

As a result, $\mathcal{D}$ is a prime model of T and, if $\mathcal{M}, \mathcal{N} \models T$, $\mathcal{M} \equiv \mathcal{D}$ and $\mathcal{D} \equiv \mathcal{N}$. It follows that T is a complete theory.

Exercise 9.61

If $\mathcal{M}$ is finite, Exercise 9.54 implies that $\mathcal{M}$ is atomic. Moreover, T can only have finite models because it is a complete theory. As a result, $\mathcal{M}$ is trivially a prime model of T and it is countable.

If $\mathcal{M}$ is infinite, then T has only infinite models. We first suppose that $\mathcal{M}$ is a prime model of T. Because T is countable, it has a countably infinite model. Since a prime model must be elementarily embeddable into any model of T, it has to be countable. If $\mathcal{M}$ was not atomic, we could omit at least one of its non-isolated types, thus obtaining a countable model into which $\mathcal{M}$ could not be elementarily embedded. It follows that $\mathcal{M}$ is atomic. Conversely, if $\mathcal{M}$ is countable and atomic, Theorem 9.51 implies that $\mathcal{M}$ is a prime model of T.

The algebraic closure of the field $\mathbb{Q}$ is a prime model of ACF_0, thus countable and atomic. The vector space $\mathbb{Q}$ over itself is also countable and atomic, being a prime model of $\mathsf{VS}_\mathbb{Q}^\infty$.

Exercise 9.62

The linear order $(\mathbb{R}, <)$ is atomic. This can be shown by relying on the fact that two finite, ordered tuples of reals are related by a piecewise linear function that is an automorphism of the ordered reals. The ordered, divisible abelian group $(\mathbb{Q}, <, +, -, 0)$ is, by quantifier elimination, a prime model of ODAG^+. By Lemma 9.60, it is atomic.

Exercise 9.64

In this case $\zeta(\overline{x})$ is T-equivalent to $\bot$.

Exercise 9.65

(1) Clearly, $\overline{a} \sim \overline{a}$ by the identify automorphism. If α is an automorphism of $\mathcal{M}$, then α^{-1} is, too. Thus, if $\overline{a} \sim \overline{b}$ (via α), $\overline{b} \sim \overline{a}$ (via α^{-1}). Finally, if $\overline{a} \sim \overline{b}$ and $\overline{b} \sim \overline{c}$, calling α, β the automorphisms sending $\overline{a}$ into $\overline{b}$ and $\overline{b}$ into $\overline{c}$ respectively, we see that $\beta \cdot \alpha$ sends $\overline{a}$ into $\overline{c}$: because the composition of two automorphisms of $\mathcal{M}$ is an automorphism of $\mathcal{M}$, $\overline{a} \sim \overline{c}$.

(2) First, note that, if $\overline{a} \sim \overline{b}$, then $\overline{b}$ realises $tp_\mathcal{M}(\overline{a})$. In other words, the tuples belonging to the same orbit of M^n all realise the same (isolated) type. By $\aleph_0$-categoricity, there are only finitely many n-types to be realised. It suffices to rule out that distinct orbits do not realise the same complete n-type. To this end, we observe that, if $tp_\mathcal{M}(\overline{a}) = tp_\mathcal{M}(\overline{b})$, there is a partial, elementary map from $\overline{a}$ to $\overline{b}$. Because $\mathcal{M}$ is atomic (condition (2) of the Ryll-Nardzewski theorem), it is homogeneous and thus strongly homogeneous. It follows that f can be extended to an automorphism of $\mathcal{M}$, i.e. $\overline{a} \sim \overline{b}$. It is therefore impossible that there should be distinct orbits whose elements realise the same n-type. It follows that there are as many distinct orbits as there are distinct types in $S_n(T)$. Because $S_n(T)$ is finite, $\mathcal{A}$ is oligomorphic.

(3) The dense linear order $(\mathbb{Q}, <)$ and the structure $(E, \sim)$ consisting of countably many, countably infinite equivalence classes (with $\sim$ the relevant equivalence relation) have oligomorphic automorphism groups.

Exercise 9.67

(i) Suppose $p \in S_n(M/A)$ with $|A| < \kappa$. Replace each formula $\phi(x_1, \ldots, x_n) \in p$ with the formula $\exists x_2 \ldots \exists x_n \phi(x_1, \ldots, x_n)$. The replacements determine a finitely consistent set of formulae in which only x_1 occurs free. This set can be extended to a type $q_1 \in S_1(M/A)$, which is realised, say by a_1, in M. We next consider the set of parameters $|A \cup a_1| < \kappa$ and the type p_1, obtained from p by replacing $\phi(x_1, \ldots, x_n)$ with $\phi(a_1, x_2, \ldots, x_n)$. We obtain q_2 from p_1 by the introduction of existential quantifiers binding $x_3, \ldots, x_n$ in each formula from p_1. Then q_2 is a 1-type realised by $a_2 \in M$. We deduce that p_2, obtained from p by replacing $\phi(x_1, \ldots, x_n)$ with $\phi(a_1, a_2, \ldots, x_n)$ is a type. Iterating the procedure described so far, we arrive at p_{n-1} and can use κ-saturation directly over $A \cup \{a_1, \ldots, a_{n-1}\}$ to conclude that there is a realisation $a_n \in M$ of p_{n-1}. But this is to say that the n-tuple $a_1, \ldots, a_n$ realises p.

(ii) Let A be a subset of M: then A has cardinality smaller than κ for any infinite cardinal κ. Any type $p \in S_1(M/A)$ contains the set of sentences true in $\mathcal{M}$, which includes an existential sentence φ characterising this structure up to isomorphism. By definition of type, p is realised in an elementary extension of $\mathcal{M}$ that is also a model of φ. There is only one such elementary extension, namely $\mathcal{M}$ itself.

Exercise 9.69

To show that $p(x)$ is a complete 1-type over $f[A]$ we reason exactly as in Exercise 9.39. To show that g is elementary, we only have to note that, for any formula $\varphi(x, \overline{a})$, we have:

$$\mathcal{M} \models \varphi(a, \overline{a}) \iff \mathcal{M} \models \varphi(b, f(\overline{a})).$$

Exercise 9.73

Let $\{a_0, a_1, a_2, \ldots\}$, $\{b_0, b_1, b_2, \ldots\}$ be enumerations of M, N respectively. We build an isomorphism by means of a back-and-forth process. In the initial stage, we consider $tp_{\mathcal{M}}(a_0)$, which is a type p in $S_1(N)$, by elementary equivalence. Let b_i be the element of least index in the enumeration of N to realise p. The function f_0 sending a_0 into b_i is partial elementary.

In the second stage of the procedure we consider $b_j \in N$ with the least index not in the image of f_0 (this is b_0 unless $f_0(a_0) = b_0$). The type $tp_\mathcal{N}(b_j/b_i)$ determines a complete 1-type $q \in S(M/\{a_0\})$, which is realised in $\mathcal{M}$ by saturation. We choose the realisation a_i of least index in the given enumeration of M and set $f_1 = f_0 \cup \{(a_i, b_j)\}$.

We proceed to define f_2 by fixing a_k of least index such that a_k is not in the domain of f_1 and considering $tp_\mathcal{M}(a_k/\{a_0, a_i\})$. By the construction described so far, we ensure that a_n certainly belongs to the domain of f_{2n}, while b_n certainly is a value of f_{2n+1}.

The function $f = \bigcup_{n \in \mathbb{N}} f_n$ is therefore a bijection between M and N. To prove that it is elementary, consider the formula $\varphi(x_0, \ldots, x_n)$ and suppose that $\mathcal{M} \models \varphi(a_0, \ldots, a_n)$. It follows that $\varphi(a_0, \ldots, a_{n-1}, x_n) \in tp_\mathcal{M}(a_n/\{a_0, \ldots, a_{n-1}\})$. By the definition of f, $f(a_n)$ realises the type obtained from $tp_\mathcal{M}(a_n/\{a_0, \ldots, a_{n-1}\})$ by replacing the parameters in each formula with $f(a_0), \ldots, f(a_{n-1})$. It follows that $\mathcal{N} \models \varphi(f(a_0), \ldots, f(a_n))$.

The argument can be reversed because f is surjective. We conclude that f is elementary. In particular, $\mathcal{M}, \mathcal{N}$ are isomorphic.

Exercise 9.76

(i) By hypothesis, $S_1(M/A) \leq \kappa$. Moreover, there are only $|M|$ finite subsets of M. Because $|M| = |\mathcal{L}|$, the set S of all 1-types over a finite subset of A is bounded above by $|\mathcal{L}| \cdot \kappa = |\mathcal{L}| + \kappa$. By an elementary chain construction based on an enumeration of S combined with the Downward Löwenheim-Skolem theorem we can realise each type of S in an elementary extension of $\mathcal{M}$ whose size is $\leq |\mathcal{L}| + \kappa$. The union of the chain so obtained has size $\leq |\mathcal{L}| + \kappa$ and realises every type in S.

(ii) Given $\mathcal{N}_0$ from the previous exercise, we only have to iterate the same construction to obtain an elementary chain $\{\mathcal{N}_i : i \in \omega\}$. If $\mathcal{N}$ is the union of this chain and $A \subseteq N$ is finite, there is $\mathcal{N}_i$ such that $A \subseteq N_i$. Thus, any 1-type over A is realised in $\mathcal{N}_{i+1}$ and, consequently, in $\mathcal{N}$.

Exercise 9.81

(i) Suppose $\mathcal{M}$ is atomic and consider $p \in S_m(M/A)$. Since $p = p(\overline{x}, \overline{a})$, we consider the $(m + n)$-type $p(\overline{x}, \overline{y}) \in S_{n+m}(M)$, which is isolated. By Exercise 9.39, $p(\overline{x}, \overline{a})$ is also isolated.

(ii) Suppose $\mathcal{M}$ is atomic and let $A \subseteq M$ be the set $\{a_1, \ldots, a_n\}$. If $p \in S_1(M/A)$, then $p \in S_1(M_A)$ and, since $\mathcal{M}_A$ is atomic, p is isolated. It follows that p is realised in $\mathcal{M}_A$ and, consequently, p is realised in $\mathcal{M}$ over the finite set of parameters A. As a result, $\mathcal{M}$ is $\aleph_0$-saturated.

(iii) Let $\overline{\mathbb{Q}} \models \mathsf{ACF}_0$ be the algebraic closure of the field $\mathbb{Q}$. Because $\overline{\mathbb{Q}}$ is a prime model of ACF_0, it is atomic and thus $\aleph_0$-saturated by part (iii). By Theorem 9.78, ACF_0 is small. The same argument works for ACF_p, with p prime, taking the algebraic closure of the prime subfield $\mathbb{Z}_p$.

B.10 Chapter 10

Exercise 10.3

(i) The verification is routine. We only show the existence of multiplicative inverse. Note that $(0,0)$ is the additive neutral element of $(\mathbb{R}^2, \oplus, \odot)$ and that $(1,0)$ is its multiplicative unit. Clearly, $(0,0) \neq (1,0)$. Moreover, if $(a,b) \neq (0,0)$, then at least one of a, b is nonzero and $\gamma = a^2 + b^2 \neq 0$. Thus, the fractions a/γ and $-b/\gamma$ are defined. The multiplicative inverse of (a,b) is $(a/\gamma, -b/\gamma)$.

(ii) Since the ideal generated by x^2+1 is maximal, the quotient $\mathbb{R}[x]/(x^2+1)$ is a field. The elements of $\mathbb{R}[x]/(x^2+1)$ are of the form $a + bx + (x^2+1)$. Since $x + (x^2+1)$ is a root of the polynomial $Y^2 + 1 = 0$, $(x + (x^2+1))(x + (x^2+1)) = -1 + (x^2+1)$. It follows that

$$(a + bx + (x^2+1))(c + dx + (x^2+1)) = (ab - cd) + (bc + ad)x + (x^2+1).$$

The function $\theta : R[x]/(x^2+1) \longrightarrow \mathbb{R}^2$ defined by the condition $\theta(a + bx + (x^2+1)) = (a,b)$ is easily seen to be an isomorphism between the quotient $\mathcal{L}_r$-structure with domain $R[x]/(x^2+1)$ and $(\mathbb{R}^2, \oplus, \odot)$.

Exercise 10.8

Let $f(x) \in K[x]$ a non-constant polynomial. We may choose $f(x)$ to be monic and irreducible since, if it is not, we only need to consider one of its monic, irreducible factors. Since $\mathcal{K}$ is perfect, $f(x)$ is separable. There is a simple, algebraic field extension of $\mathcal{K}$ in which $f(x)$ has a root α. Since $f(x)$ is the minimum polynomial of α, the fact that $\mathcal{K}$ is separably closed now implies $\alpha \in K$. It follows that $\mathcal{K}$ is algebraically closed.

Exercise 10.15

(a) Consider an arbitrary ideal I including S. Then I must contain every finite sum of the form $f_1 s_1 + \ldots + f_k s_k$, with $s_1, \ldots, s_k \in S$, $f_1, \ldots, f_k \in K[x_1, \ldots, x_n]$. It is easy to see that the sums and differences of the finite sums just described are finite sums of the same form. In fact, such sums determine an ideal included in any I extending S. It follows that the elements of J are finite sums of the given form. It follows that any point $\bar{a}$ at which every element of S vanishes is a root of every finite sum of the form $f_1 s_1 + \ldots + f_k s_k$, i.e. a point at which every element of J vanishes. The converse statement is clear.

(b) By Hilbert's basis theorem, there is $n > 0$ such that $V = (f_1, \ldots, f_n)$, with $f_1, \ldots, f_n \in K[x_1, \ldots, x_n]$. As a result, the parametric $\mathcal{L}_r$-formula $f_1(\bar{x}) = 0 \wedge \ldots \wedge f_n(\bar{x}) = 0$ defines V.

Exercise 10.17

(i) If every $f \in J$ vanishes at $\bar{a}$, then, in particular, every $f \in I \subseteq J$ does. In symbols, $\mathcal{Z}(J) \subseteq \mathcal{Z}(I)$.
(ii) If f vanishes over V, it clearly vanishes over $U \subseteq V$. Thus, $\mathcal{I}(V) \subseteq \mathcal{I}(U)$.
(iii) Any n-tuple of $\mathbb{A}^n$ is vacuously a zero of the constant polynomial 0.
(iv) The constant polynomial 1 vanishes nowhere.
(v) Let J be an ideal of $K[x_1, \ldots, x_n]$. If $f \in J$, then f vanishes over $\mathcal{Z}(J)$. Consequently, f belongs to the ideal of polynomials vanishing over $\mathcal{Z}(J)$, which is $\mathcal{I}(\mathcal{Z}(J))$.
(vi) If $\bar{a} \in V$ then every polynomial in $\mathcal{I}(V)$ vanishes in particular at $\bar{a}$, so $\bar{a} \mathcal{Z}(\mathcal{I}(V))$.
(vii) Suppose $\mathcal{I}(V) = \mathcal{I}(W)$. If $\bar{a} \in V$, then any $g \in \mathcal{I}(W) = \mathcal{I}(V)$ must vanish at $\bar{a}$. Because W is a variety, let $g_1 = 0, \ldots, g_n = 0$ be defining literals for it. We have just verified that $\bar{a}$ satisfies these literals, so $\bar{a} \in W$. It follows that $V \subseteq W$. The converse inclusion is verified in a similar manner. It follows that $V = W$, i.e. $\mathcal{I}$ is injective.
As for $\mathcal{Z}$, if V is a variety and $f_1 = 0, \ldots, f_m = 0$ are defining literals for it, then $V = \mathcal{Z}(J)$, with $J = (f_1, \ldots f_k)$.

Exercise 10.23

(i) If Q is primary and $ab \in Q$, then $ab \in \sqrt{Q} \supseteq Q$. Because $\sqrt{Q}$ is prime, $a \in \sqrt{Q}$ or $b \in \sqrt{Q}$. Suppose $a \notin Q$. Then there is $m \geq 1$ such that $b^m \in Q$. Conversely, suppose $ab \in Q$ implies $a \in Q$ or $b^m \in Q$ for some integer $m \geq 1$. If $cd \in \sqrt{Q}$, then $c^k d^k \in Q$, for some $k \geq 1$. If $c^k \in Q$, then $c \in \sqrt{Q}$. Otherwise $c^k \notin Q$ implies $d^{kn} \in Q$, for some $n \geq 1$. In this case we conclude $d \in \sqrt{Q}$ and $\sqrt{Q}$ is a prime ideal.
(ii) If $ab \in Q$ and Q is prime, then $a \notin Q$ implies $b^1 \in Q$. By part (i), Q is primary.
(iii) Suppose that every zero divisor in R/Q is nilpotent. Since Q is the zero element of R/Q, a zero divisor $b + Q$ is a nonzero element such that $(a+Q)(b+Q) = Q$, for some nonzero a. Because $b+Q$ is nilpotent, $b^m \in Q$

for some $m \geq 1$. Now suppose $ab \in Q$, i.e. $(a+Q)(b+Q) = Q$. If $a \notin Q$, then $b+Q = Q$, i.e. $b \in Q$, or $b+Q$ is a zero divisor. In the latter event, $b^m \in Q$. It follows that Q is primary.

If, on the other hand, Q is primary and $b+Q$ is a zero divisor in R/Q, then there is $a+Q$ such that $a \notin Q$ and $ab+Q = Q$. Since $b^m \in Q$, $b+Q$ is nilpotent.

(iv) Clearly, $p \in \sqrt{(p^m)}$: it follows that $(p) \subseteq \sqrt{(p^m)}$. Conversely, suppose $q \in \sqrt{(p^m)}$. Then there is $k \geq 1$ sch that $q^k \in (p^m)$, i.e. $q^k = np^m = (np^{m-1})p \in (p)$. It follows that p divides q^k and thus q. Consequently, $q \in (p)$ and $\sqrt{(p^m)} = (p)$. Since (p) is a maximal ideal of $(\mathbb{Z}, +, -, \cdot, 0, 1)$, it is prime. By part (ii), the ideal (p^m) is primary.

Exercise 10.27

(i) If $g \in \sqrt{Q_1 \cap Q_2}$, then there are $m, n > 0$ such that $g^n \in Q_1$ and $g^m \in Q_2$. Consequently, $g^{mn} \in Q_1 \cap Q_2$ and $g \in \sqrt{Q_1 \cap Q_2}$. The converse inclusion is verified in a similar way.
(ii) By the minimality of $\{Q_1, \ldots, Q_m\}$, $\sqrt{Q_i} \neq \sqrt{Q_j}$, i.e. $P_i \neq P_j$.
(iii) Suppose that, for some $i \in \{1, \ldots, m\}$, $P_i \supseteq \bigcap_{j \neq i} P_j$. Suppose $a \in \bigcap_{j \neq i} Q_j \subseteq \bigcap_{j \neq i} P_j \subseteq P_i$. Because $P_i = \sqrt{Q_i}$, $a^k \in Q_i$ for some positive k. Since a is in each Q_j, so is a^k: we conclude $a \in \bigcap_{i=1}^m Q_i$ because this intersection is a radical ideal containing a^k. It follows that $a \in Q_i$, i.e. $\bigcap_{j \neq i} Q_j \subseteq Q_i$, which is impossible.
(iv) The set $\{P_1, \ldots, P_m\}$ determines a primary decomposition of I that consists of m distinct elements by (ii). This decomposition is minimal by (iii).

Exercise 10.29

(i) Let J be any ideal of $K[x_1, \ldots, x_n]$. Then it is easy to see that $\sqrt{J}$ is a radical ideal and that $\mathcal{Z}(J) = \mathcal{Z}(\sqrt{J})$, since $K[x_1, \ldots, x_n]$ is an integral domain. Consequently, $\mathcal{I}(\mathcal{Z}(J)) = \mathcal{I}(\mathcal{Z}(\sqrt{J})) = \sqrt{J}$, where the last equality depends on the *Nullstellensatz*.
(ii) If I, J are radical ideals and $\mathcal{Z}(I) = \mathcal{Z}(J)$, then $\mathcal{I}(\mathcal{Z}(I)) = \mathcal{I}(\mathcal{Z}(J))$ and the *Nullstellensatz* implies $I = J$.
(iii) Let J be a radical ideal. By the *Nullstellensatz*, $\mathcal{I}(\mathcal{Z}(J)) = J$. Since $\mathcal{Z}(J)$ is a variety V, $J = \mathcal{I}(V)$.
(iv) Relative to $M[x_1, \ldots, x_n]$, we have the inclusion $\mathcal{Z}(f_1, \ldots, f_n) \subseteq \mathcal{Z}(g)$, where we $(f_1, \ldots, f_n$ and (g) are the ideals generated by $f_1, \ldots, f_n$ and g respectively. Consequently, $\mathcal{I}(\mathcal{Z}(g)) \subseteq \mathcal{I}(\mathcal{Z}(f_1, \ldots, f_n))$ and we can deduce:

$$g \in \sqrt{(g)} \subseteq \sqrt{(f_1, \ldots, f_n)}.$$

By definition of radical ideal, there is $k > 0$ such that $g^k \in (f_1, \ldots, f_n)$.

Exercise 10.31

(i) Suppose that G is not surjective. Then there is $\bar{a}$ such that $\bar{a}$ is not in the image of G. Consider the field $\mathsf{E} \subseteq \overline{\mathbb{F}_p}$ generated by $\bar{a}$ and the coefficients of $f_1, \ldots, f_n$. Then E a finite subfield of $\overline{\mathbb{F}_p}$ on which G is defined. Since G is injective by hypothesis and E is a finite set, $G \to E^n$ is injective and, consequently, surjective. But $\bar{a} \in E^n$ is not in the image of G. This contradiction shows that G is in fact a bijection.

(ii) Since each f_i is a $\mathcal{L}_r$-term, the following are $\mathcal{L}_r$-sentences.

   (a) $\theta_1 := \forall x_1 \ldots \forall x_n \forall y_1 \ldots \forall y_n (\bigvee_{i=1}^{n} \neg(x_i = y_i) \to \bigvee_{i=1}^{n} \neg(f_i(\bar{x}) = f_i(\bar{y})))$.
   (b) $\theta_2 := \forall x_1 \ldots \forall x_n \exists y_1 \ldots \exists y_n (\bigwedge_{i=1}^{n} f_i(y_1, \ldots, y_n) = x_i)$.

   We may set $\varphi := \theta_1 \to \theta_2$.

(iii) If $\mathcal{K}$ is an algebraically closed field of positive characteristic p, it is elementarily equivalent to $\overline{\mathbb{F}_p}$. By part (i), $\overline{\mathbb{F}_p} \models \varphi$. Elementary equivalence now implies $\mathcal{K} \models \varphi$.

(iv) If $\mathsf{ACF}_0 \not\models \varphi$ then $\mathsf{ACF}_0 \models \neg\varphi$. By compactness, a finite subset of ACF_0 implies φ. This finite subset excludes only finitely many positive characteristics, so it is also a subset of ACF_p, for sufficiently large p. It now follows $\mathsf{ACF}_p \models \neg\varphi$, which is impossible. It follows that $\mathsf{ACF}_0 \models \varphi$.

Exercise 10.34

Suppose $MR^M \varphi \geq 1$. Then there is an infinite sequence of disjoint subsets of $\varphi(M)$ and the latter must be infinite. By contrast, if $\varphi(M)$ is infinite, then it contains a sequence $\{a_i\}_{i \in \mathbb{N}}$, which can be associated with the sequence of formulae $\{x_i = a_i\}_{i \in \mathbb{N}}$, each of which defines a singleton included in $\varphi(M)$. By definition $MR^M(x_i = a_i) \geq 0$, so we can conclude that $MR^M \varphi \geq 1$.

It follows from the above equivalence that, if $MR^M \varphi = 0$, $\varphi(M)$ must be nonempty and finite (if not, $MR^M \geq 1$). If $\varphi(M) \neq \emptyset$ is finite, then it is impossible to find an infinite sequence of pairwise disjoint, definable subsets of $\varphi(M)$. Thus $MR^M \varphi \geq 0$ but $MR^M \varphi \not\geq 1$, i.e. $MR^M \varphi = 0$.

Suppose $\varphi(M)$ is minimal in M. Then $\varphi(M)$ is infinite, i.e. $MR^M \varphi \geq 1$. We cannot have an infinite sequence of definable, pairwise disjoint subsets of $\varphi(M)$, each of Morley rank 1 because, if $A_1, A_2, \ldots$ was such a sequence, with defining formulae $\psi_1, \psi_2, \ldots$, each of the $\psi_i(M)$ would be a cofinite subset of $\varphi(M)$. But any two cofinite subsets have a cofinite intersection, so they cannot be disjoint. It follows that $MR^M \varphi \geq 1$ but $MR^M \varphi \not\geq 2$. In other words $MR^M \varphi = 1$.

Let $\mathcal{M} \models \mathsf{ACF}$. Since M is defined by $x = x$ and ACF is a strongly minimal theory, M is minimal and $MR^M(M) = 1$.

Exercise 10.36

We proceed by transfinite induction. If $\alpha = 0$, the result holds vacuously.

If $\alpha = \beta+1$ and $MR^M(\varphi) \geq \beta+1$, then $MR^M(\varphi) \geq \beta+1$ and there is an infinite sequence of mutually inconsistent formula ψ_i such that $MR^M(\psi_i) \geq \beta$ and $\mathcal{M} \models \forall \overline{x}(\psi_i \to \varphi)$. It follows that $MR^M(\psi_i) \leq MR^M(\varphi)$, i.e. $MR^M(\psi_i) \in \{\beta, \beta+1\}$. If, for each $i \in \mathbb{N}$, $MR^M(\psi_i) = \beta+1$, we could deduce $MR^M(\varphi) = \beta+2$, which is impossible. Thus, there is at least one $j \in \mathbb{N}$ such that $MR^M(\psi_j) = \beta$ and ψ_j implies φ in $\mathcal{M}$. We can use ψ_j and the inductive hypothesis to conclude that, for any $\gamma < \beta$ there is η such that $MR^M(\eta) = \gamma$ and η implies ψ_j. Thus η implies φ in $\mathcal{M}$. This completes the inductive step for a successor ordinal.

If $\alpha = \lambda$ and λ is a limit ordinal, we consider $\beta < \lambda$. Then $\beta+2 < \lambda$. Since $MR^M(\varphi) \geq \lambda$, it follows that $MR^M(\varphi) \geq \beta+2, \beta$. Thus, there are $\theta_1, \theta_2, \ldots$, each of Morley rank at least $\beta+1$ and implying φ in $\mathcal{M}$. If all of the θ_i were of Morley rank λ, the contradiction $MR^M(\varphi) \geq \lambda+1$ would follow. As a consequence, there is $j \in \mathbb{N}$ such that $MR^M(\theta_j) = \delta < \lambda$. We have $\beta+1 \leq \delta$. Thus, by inductive hypothesis, there is η such that $MR^M(\eta) = \beta$ and η implies θ_j, which in turn implies φ in $\mathcal{M}$.

Exercise 10.38

It suffices to prove by transfinite induction that $MR^M(\varphi) \geq \alpha$ implies $MR^M(\psi) \geq \alpha$. The only mildly nontrivial case involves the successor step. If $MR^M(\varphi) \geq \alpha$ and α is a successor ordinal, then there is an ordinal β such that $\alpha = \beta+1$. There is an infinite sequence of formulae $\{\theta_i(\overline{x})\}_{i\in\mathbb{N}}$ defining a sequence of pairwise disjoint subsets of $\varphi(M)$, each of Morley rank $\geq \beta$. Since the free variables in each $\theta_i(\overline{x})$ occur among $x_1, \ldots, x_n$, they also occur among $x_1, \ldots, x_n, y_1, \ldots, y_m$. As a consequence, we obtain a sequence $\{\chi_i(\overline{x}, \overline{y})\}_{i\in\mathbb{N}}$ of formulae that define pairwise disjoint subsets of $\psi(M)$, each of Morley rank $\geq \beta$. It follows that $MR^M(\psi) \geq \beta+1$.

Exercise 10.41

Let $\mathcal{M}$ be a $\mathcal{L}$-structure, $A \subseteq M$.

(i) Suppose X is minimal and let ψ be a $\mathcal{L}$-formula defining X. Then $MR^M(\psi) = 1$ and, for any $\mathcal{L}$-formula φ, either $X \cap \varphi(M)$ or $X \cap \neg\varphi(M)$ is finite, i.e. either $\psi \wedge \varphi$ has Morley rank 0, or $\psi \wedge \neg\varphi$ does. It follows that $MR^M(\varphi) = 1$ or $MR^M(\neg\varphi) = 1$. If, conversely X is 1-strongly minimal, then $MR^M(\psi) = 1$ and X is infinite. Moreover, for any $\mathcal{L}$-formula φ, $1 = MR^M(\psi) = max\{MR^M(\varphi \wedge \psi), MR^M(\neg\varphi \wedge \psi)\}$. By 1-strong minimality, either $MR^M(\psi \wedge \varphi) < 1$ or $MR^M(\psi \wedge \neg\varphi) < 1$. It follows that one of $\psi \wedge \varphi, \psi \wedge \neg\varphi$ defines a set that is empty or finite. Consequently X is minimal.

B Solutions to Exercises

(ii) Define $X \sim_\alpha Y$ by the condition $MR^M((X - Y) \cup (Y - X)) < \alpha$. Since $X - X = \emptyset$ and $MR^M(X - X) = -\infty < \alpha$, $X \sim_\alpha X$ follows, for any definable X. It is easy to see that $\sim_\alpha$ is symmetric. In order to verify transitivity, we suppose $MR^M((X - Y) \cup (Y - X)) < \alpha$ and $MR^M((Y - Z) \cup (Z - Y)) < \alpha$. ADD DETAILS FROM TEXTBOOK

(iii) Let $\psi_1(\overline{x}), \ldots \psi_n(\overline{x})$ be $\mathcal{L}_A$-formulae such that $MR^M(\varphi \wedge \neg \psi_i) < \alpha$. Set $\Psi = \psi_1 \wedge \ldots \wedge \psi_n$. We have:

$$MR^M(\Psi \wedge \varphi) = MR^M((\neg\psi_1 \wedge \varphi) \vee \ldots \vee (\neg\psi_n \wedge \varphi))$$
$$= \max\{MR^M(\neg\psi_i \wedge \varphi)\}$$
$$< \alpha.$$

As a consequence, $MR^M(\varphi \wedge \Psi) = \alpha \geq 0$ and $MR^M(\Psi) \geq 0$, i.e. Ψ defines a nonempty subset of M.

(iv) Since $MR^M(\varphi) = \alpha$, the set defined by the given condition is a type. It is a complete type by the definition of α-strong minimality.

Exercise 10.45

(i) If $MR^M(\varphi) = \infty$, the definition of q implies $\varphi \in q$ since $MR^M(\varphi \wedge \neg\varphi) = -\infty < \infty$. A fortiori $\varphi \in p$. If $\theta \in p$, either $\theta \in q$ or $\theta \notin q$. In the first case $MR^M(\varphi \wedge \neg\theta) < \infty$ and, since $\infty = MR^M(\varphi) = MR^M((\varphi \wedge \theta) \vee (\varphi \wedge \neg\theta)) = \max\{MR^M(\varphi \wedge \theta), MR^M(\varphi \wedge \neg\theta)\}$, we must have $MR^M(\varphi \wedge \theta) = \infty$. Because $\mathcal{M} \models \forall \overline{x}(\varphi \wedge \theta \to \theta)$, we conclude $MR^M(\theta) = \infty$.
If, on the other hand, $\theta \notin q$, we cannot have $\neg\theta \in q$, i.e. $MR^M(\varphi \wedge \theta) = \infty$, and $MR^M(\varphi) = \infty$, as above. Because θ is an arbitrary $\mathcal{L}_A$-formula in p, the Morley rank of p is ∞.

(ii) When $MR^M(\varphi) = \alpha$, we set $\psi \in q$ iff $MR^M(\varphi \wedge \neg\psi) < \alpha$. We can then repeat the argument in (i) almost to the letter and deduce $MR^M(\theta) \geq \alpha$ for any $\theta \in p$. It follows that the Morley rank of p is α.

Exercise 10.54

(i) Suppose that $f, g \in \mathfrak{P}(p)$. Then $f = 0, g = 0 \in p$. Because p is a complete n-type the literals $f \pm g = 0$ are in p. Moreover, if $f \in K[x_1, \ldots, x_n]$ and $g = 0$ is in p, $fg = 0$ is in p and, by definition $fg \in \mathfrak{P}(p)$. It follows that $\mathfrak{P}(p)$ is an ideal. If $fg\mathfrak{P}(p)$ then $fg = 0$ is in p and, since $K[x_1, \ldots, x_n]$ is an integral domain, $f = 0$ or $g = 0$, i.e. $f \in \mathfrak{P}(p)$ or $g \in \mathfrak{P}(p)$. Thus $\mathfrak{P}(p)$ is prime.

(ii) Suppose that the literals $f_1(\bar{x}) = 0, \ldots, f_m(\bar{x}) = 0$, $\neg(g_1(\bar{x}) = 0), \ldots, \neg(g_n(\bar{x}) = 0)$ are in $\mathfrak{T}(J)$. Because J is a prime ideal, $K[x_1, \ldots, x_n]/J$ is an integral domain and can be extended to its fraction field Q_J. Set $y_i = x_i + J$, with $i = 1, \ldots, m$. Then it is clear that $f_i(\bar{y}) = 0$ modulo J, i.e. that the literals $f_1(\bar{x}) = 0, \ldots, f_m(\bar{x}) = 0$ are all satisfied in the integral domain $K[x_1, \ldots, x_n]/J$ and, via inclusion, in Q_J. This is true of the negative literals. The same literals continue to be modelled in the algebraic closure of Q_J.

(iii) Since the positive and negative literals of the n-type q discussed in part (ii) are fixed, the quantifier-free formulae in an extension of q are uniquely determined. More explicitly, a disjunctive normal form D is finitely consistent with q iff the literals in this type include the conjuncts in at least one disjunct of D. By quantifier elimination, this is enough to fix every $\mathcal{L}_r$-formula in a complete extension of q.

(iv) Suppose $f \in J$. Then $f = 0$ is in $\mathfrak{T}(J)$ and, consequently, $f \in \mathfrak{P}(\mathfrak{T})(J)$. Thus $J \subseteq \mathfrak{P}(\mathfrak{T})(J))$. The converse inclusion is established in a similar way. Now suppose p is a complete n-type. Since it is uniquely determined by its literals, we may restrict attention to them alone. Thus, if $f = 0$ is in p, then $f \in \mathfrak{P}(p)$ by definition and $f = 0$ is in $\mathfrak{T}(\mathfrak{P})(p)$. If, on the other hand, $\neg(g = 0) \in p$, then $g \notin \mathfrak{P}(p)$, and, by definition of $\mathfrak{T}$, $\mathfrak{T}(\mathfrak{P}(p))$ containes $\neg(g = 0)$. This shows that $p \subseteq \mathfrak{T}(\mathfrak{P})(p)$. Since both types involved are complete, equality follows.

Exercise 10.59

If W is not irreducible, then it has a unique representation of the form $W_1 \cup \ldots \cup W_k$, where $W_1, \ldots, W_k$ are irreducible. Since each W_i is definable, so is their union. By Lemma 10.37, $MR^K(W) = max\{MR^K(W_1), \ldots, MR^K(W_k)\} = dim(W)$.

Exercise 10.61

Consider $\mathcal{M} \models \mathsf{ACF}$ and a countable set $A \subseteq M$. We wish the evaluate the size of $S_1(A)$. By Lemma 10.1, it suffices to determine the size of the set of algebraic 1-types over A, since we already know that there is exactly one non-algebraic type in $S_1(A)$. Each algebraic 1-type is isolated by a polynomial equation with parameters in A. The set of polynomials with coefficients in the subfield of $\mathcal{M}$ generated by A is countable, because the generated subfield is. It follows that $S_1(A) \leq \aleph_0$.

B.11 Chapter 11

Exercise 11.1

Let $\mathcal{F} = (F, P^F, +^F, -^F, \cdot^F, 0^F, 1^F)$ be an ordered field with P^F the associated set of positive elements. For notational ease, we drop F-superscripts in what follows. If $-1 \in P$, then $1 = (-1) \cdot (-1) \in P$. Consequently $1 + (-1) = 0 \in P$, which is impossible. It follows that $-1 \notin P$.

Next suppose $a \in F - \{0\}$ is such that, for some $b \in F$, $a = b \cdot b$. Note that $b \neq 0$. If $b \in P$, then $b \cdot b \in P$ so $a \in P$ i.e. $a - 0 \in P$ and $0 < a$. If $b \notin P$, then $-b \in P$ and $a = (-b) \cdot (-b) \in P$. We conclude $\mathcal{F} \models \forall x((\neg(x = 0) \land \exists y(x = y \cdot y)) \to 0 < x)$.

Suppose $a, b, c, d \in F$ and $a < b$, $ac = 1 = bd$. If $c = d$, $ac = 1 = bd = bc$, i.e. $a = b$, against our hypothesis. Thus $c \neq d$. Consider the product $c(b-a)d = c - d$. Either $P(c-d)$ or $P(d-c)$. Suppose $P(d-c)$. Since $P(b-a)$ by hypothesis, c, d cannot both be positive, or $c - d$ would be. If $P(c)$ then $P(0 - d)$. Since $P(c)$ and $P(ac)$ both hold, a must be positive (otherwise $P(-a)$ and $P(-ac)$ iff $P(-1)$, which we have ruled out). It follows that b is positive and $P(-bd)$ iff $P(-1)$. The hypotheses $P(-c)$ and $P(d)$ lead to contradiction in a similar manner. We conclude that $P(c-d)$.

Exercise 11.3

Suppose that $\mathcal{F}$ is formally real and $0 = a_1^2 + \ldots + a_k^2$, with $a_i \in F$, $(i = 1, \ldots, k)$. If $a_i \neq 0$ for some $i \in \{1, \ldots, k\}$, then a_i has a multiplicative inverse $1/a_i$ and the given equality reduces to:

$$-1 = \frac{a_1^2}{a_i^2} + \ldots + \frac{a_{i-1}^2}{a_i^2} + \frac{a_{i+1}^2}{a_i^2} + \ldots + \frac{a_k^2}{a_i^2},$$

which is impossible. We conclude $a_1 = \ldots = a_k = 0$. The converse implication is obvious, since 1 is a square.

Exercise 11.4

Let $\mathcal{F}$ be a formally real field and call $\Sigma(F^2)$ the set of sums of squares of nonzero elements of F. This is to say that $a \in \Sigma(F^2)$ implies the existence of a positive $n \in \mathbb{N}$ and of $a_1, \ldots, a_n \in F - \{0\}$ such that $a = a_1^2 + \ldots + a_n^2$. Show that:

(i) clear by the definition of 'formally real';
(ii) if $a, b \in \Sigma(F^2)$ the expression $a + b$ is certainly a sum of nonzero squares. The product ab is of the form:

$$(a_1^2 + \ldots + a_n^2)(b_1^2 + \ldots + b_m^2) = \sum_{i \le n, j \le m} (a_i b_j)^2.$$

Since one of the terms in the last summation is the product of two nonzero squares and F is an integral domain, ab is a sum of nonzero squares.

(iii) Let $a = a_1^2 + \ldots + a_n^2$. Then $1/a = a_1^2/a^2 + \ldots + a_n^2/a^2$.

Exercise 11.6

(i) $b \in T$ because $b = 0 + b \cdot 1$;

(ii) suppose $u, v \in T$ and set $u = c + bd$, $v = c' + bd'$. Then $u + v = (c + c') + b(d + d') \in T$ because $c + c', d + d' \in P_0^F$. We also have $uv = cc' + b(cd' + c'd + dd') \in T$;

(iii) Since T includes P_0^F, it must include $\Sigma(F^2) \cup \{0\}$;

(iv) If $-1 \in T$ then $c + bd = -1$ with $d \ne 0$, since $-1 \notin P_0^F$. It follows that $1/d \in P_0^F$. Consequently $(1+c)/d = -b$ and $-b \in P_0^F$, against our choice of b.

Exercise 11.7

(i) Because P^K is closed under sums $b_1 a_1^2 + \ldots + b_m a_m^2$ is in P^K if each term of the last sum is in P^K. This is the case because $b_i \in P^F \subseteq P^K$, $a_i^2 \in P^K$ (e.g. by Exercise 11.1 and P^K is closed under multiplication. Because, in any ordered field, -1 is not positive, -1 cannot in particular be of the form $b_1 a_1^2 + \ldots + b_m a_m^2$.

(ii) By part (i), if $\mathcal{K}$ could be expanded to an ordered extension of $\mathcal{F}$, we should have $\Sigma_p \subseteq P^K$. Thus, in particular, $-1 \notin \Sigma_p$. If, conversely, $-1 \notin \Sigma_p$ we modify the definition of S in the proof of Theorem 11.5 by replacing $\Sigma(F^2) \cup \{0\}$ with $\Sigma_p \supseteq \Sigma(F^2) \cup \{0\}$. A maximal element of S determines an ordering of K in which Σ_p is a set of positive elements. Because $P^F \subseteq \Sigma_p$, $\mathcal{K}$ can be expanded to an ordered extension of $\mathcal{F}$.

Exercise 11.13

(i) Note that $c > 1$. Thus $d > 1$ and $d^n > d^k$ for any $k \in \{0, \ldots, n-1\}$. We evaluate:

$$|a_0 + a_1 d + \ldots + a_{n-1} d^{n-1}|$$
$$\leq |a_0| + |a_1| d + \ldots + |a_{n-1}| d^{n-1}$$
$$\leq b(1 + d + \ldots + d^{n-1})$$
$$\leq b(d^{n-1} + \ldots + d^{n-1})$$
$$= nbd^{n-1}.$$

Since $d > c$, $d > nb$. Consequently, $d^n = dd^{n-1} > nbd^{n-1}$. It follows that $p(d)$ must be positive.

(ii) The argument from part (i) for $-d$ leads to the inequality:

$$|\sum_{i=0}^{n-1} a_i(-d)^i| < d^n \text{ iff } -d^n + |\sum_{i=0}^{n-1} a_i(-d)^i| < 0.$$

Since n is odd, $-d^n = (-d)^n$ and the last inequality says that, even when the sum $\sum_{i=0}^{n-1} a_i(-d)^i$ is nonnegative, adding $-d^n$ to it determines a negative value. It follows that $p(-d) < 0$.

(iii) Lemma 11.8 guarantees a root of $p(x)$ in an ordered, algebraic extension of $\mathcal{F}$. Since, however, $\mathcal{F}$ is real closed, it does not have proper, ordered and algebraic extensions. It follows that $p(x)$ has a root in F. If $p(x)$ were not monic, with leading coefficient $a_n \neq 0$ we could simply divide its coefficients by a_n throughout. The resulting monic polynomial has a root in F and, consequently, $p(x)$ must have one.

Exercise 11.14

Suppose $\mathcal{F}$ is real closed, $0 \neq a \in F$. Since there is a unique ordered expansion of $\mathcal{F}$, one of a, $-a$ is a positive element in this expansion. If a is positive but there is no b such that $a = b^2$, we can reason as in the proof of Lemma 11.12 to arrive at a contradiction. If $-a$ is positive, we argue in the same way. Thus one of a, $-a$ must be a square. Clearly, exactly one of them can be a square because $a = b^2$ and $-a = c^2$ imply $0 = b^2 + c^2$, against the fact that $\mathcal{F}$ is formally real.

Exercise 11.19

If $\mathcal{A} \sqsubseteq \mathcal{M} \models \mathsf{RCF}_<$, then 0^M, $1^M \in A$ and, since $\mathcal{A}$ must be closed under $+^M$ and $-^M$, it contains a copy of $\mathbb{Z}$.

Exercise 11.22

If $\mathcal{M} \models \mathsf{RCF}_<$ and $\mathcal{A} \sqsubseteq \mathcal{M}$, we consider the ordered fraction field $\mathcal{Q}_< \sqsubseteq \mathcal{M}$. We next take $\mathcal{Q}_{alg}$, the relative algebraic closure of the field $\mathcal{Q}$ in $\mathcal{M}$. The field $\mathcal{Q}_{alg}$ is formally real because the $\mathcal{L}_r$-reduct of $\mathcal{M}$ is. If, moreover, $p(x) \in Q[x]$ is a polynomial of odd degree, then $p(x) \in M[x]$ and, consequently, a root α of $p(x)$ lies in M. By definition of relative algebraic closure, $\alpha \in Q_{alg}$. Finally, the ordering of $\mathcal{M}$ induces an ordering of $\mathcal{Q}_{alg}$ that extends the ordering of $\mathcal{Q}_<$.

We conclude that $\mathcal{Q}_{alg}$ is a formally real field with the intermediate value property for polynomials, i.e. $\mathcal{Q}_{alg} \models \mathsf{RCF}$. With its induced order, this field is a real closed, ordered extension of $\mathcal{Q}_<$. By definition, $\mathcal{Q}_{alg}$ is a real closure of $\mathcal{Q}$ and its $\mathcal{L}_{or}$-expansion induced by $\mathcal{M}$ is an order closure of $\mathcal{Q}_<$, which is in particular a model of $\mathsf{RCF}_<$.

Exercise 11.26

Suppose σ, τ are extensions of the isomorphism between $\mathcal{F}_1^<, \mathcal{F}_2^<$ that satisfy the statement of Theorem 11.25. We may therefore consider them to be isomorphisms between the ordered real closures $\mathcal{R}_1^<, \mathcal{R}_2^<$ of the corresponding fields.

Suppose $\alpha \in R_1^<$. We shall show that $\sigma(c) = \tau(c)$. Because $\mathcal{R}_1^<$ is, as a field, an algebraic extension of $\mathcal{F}_1$, α is algebraic over F_1. Let $p(x) \in F_1[x]$ be its minimum polynomial. Then $R_1^<$ contains m roots of $p_1(x)$ for some positive integer m.

The field isomorphism σ extends to a ring isomorphism from $F_1[x]$ to $F_2[x]$. Under the latter isomorphism, let $p_1(x)$ take the value $p_2(x)$. By Lemma 11.23, $p_2(x)$ has exactly m roots in $R_2^<$.

Because σ is an order-preserving function, we have:

$$a_1 < \ldots < a_m \iff \sigma(a_1) < \ldots < \sigma(a_m).$$

If we apply the same argument with τ, we obtain the polynomial $p_2(x)$ as the image of $p_1(x)$, because σ, τ agree on F_1. It follows that $\tau(a_1) < \ldots < \tau(a_m)$ are the m roots of $p_2(x)$. Because ordered is preserved, we must have $\tau(a_i) = \sigma(a_i)$ for each $i \in \{1, \ldots, m\}$. To see this in a model-theoretic fashion, note that:

$$\mathcal{R}_1^< \models (p_1(a_1) = 0 \wedge \forall z(p_1(z) = 0 \to a \leq z)$$

and that σ, τ strongly preserve all first-order $\mathcal{L}_{or}$-formulae with parameters. It follows that $\sigma(a_1), \tau(a_1)$ are both the least root of $p_2(x)$, i.e. $\sigma(a_1) = \tau(a_1)$. Because we can write first-order conditions that describe a_i as the least root of $p_1(x)$ that is distinct from $a_1, \ldots, a_{i-1}$, we obtain $\sigma(a_i) = \tau(a_i)$ for each $i \in \{1, \ldots, m\}$. Because, for a specific value j, $a_j = \alpha$, we have in particular shown that $\sigma(\alpha) = \tau(\alpha)$. Since no particular hypothesis was made about α, σ, τ are the same isomorphism between $\mathcal{R}_1^<$ and $\mathcal{R}_2^<$.

B Solutions to Exercises

Exercise 11.28

(i) A negative literal of the form $\neg(t_1 = t_2)$ is equivalent to the quantifier-free formula $t_1 < t_2 \vee t_2 < t_1$ *modulo* a linear order. A similar equivalence is available for a negative literal of the form $\neg(t_1 < t_2)$. Using the substitution theorem, it is possible to reduce any quantifier-free formula to a negation-free formula in which no quantifiers occur.

(ii) Let $\psi(\bar{a}, y)$ be a conjunction of literals and suppose $\mathcal{N} \models \exists y \psi(\bar{a}, y)$. By part (i), we may replace $\psi(\bar{a}, y)$ by an equivalent disjunctive normal $\theta(\bar{a}, y)$ form in which no negations occur. If we make the disjuncts of $\theta(\bar{a}, y)$ explicit, we may rewrite it as:

$$\theta_1(\bar{a}, y) \vee \ldots \vee \theta_m(\bar{a}, y),$$

where each $\theta_i(\bar{a}, y)$ is a conjunction of positive literals. Since the existential quantifier distributes over disjunction, our initial condition becomes:

$$\mathcal{N} \models \exists y \theta_1(\bar{a}, y) \vee \ldots \vee \exists y \theta_m(\bar{a}, y).$$

The last condition holds iff at least one disjunct is satisfied in $\mathcal{N}$. Any such disjunct is satisfied in $\mathcal{M}$ by hypothesis, so $\mathcal{M} \models \exists y \varphi(\bar{a}, y)$.

Exercise 11.29

We may consider the real field R and the formula $\exists y(x = y \cdot y)$, which defines the set of nonnegative reals.

Exercise 11.31

By quantifier elimination the real closure $\mathcal{R}_<$ of $\mathsf{Q}_<$, the ordered field of rational numbers, is a prime model of $\mathsf{RCF}_<$. Thus, if σ is a $\mathcal{L}_{or}$-sentence and $\mathcal{M}, \mathcal{N} \models \mathsf{RCF}_<$, we have:

$$\mathcal{M} \models \sigma \iff \mathcal{R}_< \models \sigma \iff \mathcal{N} \models \sigma.$$

It follows that $\mathcal{M} \equiv \mathcal{N}$. Since $\mathcal{M}, \mathcal{N}$ are arbitrary, real closed ordered fields, $\mathsf{RCF}_<$ is complete. The argument just given is not disrupted by taking $\mathcal{L}_r$-reducts, so RCF is complete, too.

Exercise 11.32

Let R be the real field, which is a real closed field. Because RCF is complete, $\mathcal{M} \models$ RCF implies $\mathcal{M} \equiv$ R. Conversely, if $\mathcal{M} \equiv$ R, $\mathcal{M} \models$ RCF since RCF $\subseteq Th(\mathbb{R})$.

Exercise 11.35

We may proceed in two ways. $\mathcal{N}$ and $\mathcal{M}$ can be ordered in a unique way and the ordering of $\mathcal{N}$ induces an ordering of $\mathcal{M}$. Thus, the only expansion of $\mathcal{M}$ to an ordered field is as an ordered subfield of $\mathcal{N}$. An alternative argument is based on the fact that $\mathcal{M}_< \models a < b$ is equivalent to $\mathcal{M}_< \models \exists y((b - a) = y^2)$. Since the last formula is modelled by $\mathcal{M}$ and preserved by extensions, it is modelled by $\mathcal{N}$ and, consequently, by $\mathcal{N}_<$. Since $\mathcal{N}_< \models a < b$, we conclude that the ordering of M is extended by the ordering of N.

Exercise 11.36

Let $b \in M - \{a\}$. Since $\mathcal{M}$ is, as a $\mathcal{L}_<$-structure, a dense linear order without endpoints, there are $c, d \in M$ such that $\mathcal{M} \models c < b$ and $\mathcal{M} \models b < d \wedge d < a$. It follows that $b \in (c, d)$. Since b is an arbitrary element of $M - \{a\}$, it follows that the latter set is a union of basic open sets, i.e. an open set. Its complement $\{a\}$ is therefore closed.

If C, D are closed subsets of M, it is easy to see that $C \cup D$ is the complement of $M - C \cap M - D$. The latter set is open because the intersection of two open intervals is open. It follows by an inductive argument that any finite union of closed sets is closed. Thus, in particular, any finite subset of M is closed.

Exercise 11.37

If $\{a\} \cup (b, c)$ was open, it would be a union of basic open sets. But any basic open set containing a intersects the complement of $\{a\} \cup (b, c)$ so this set cannot be open. If it were closed, its complement should be open and thus a union of basic open sets. It is clear, however, that the subset of M defined by $x < a \vee (a < x \wedge x < b) \vee x = b \vee x = c \vee c < x$ cannot be such a union. It follows that $\{a\} \cup (b, c)$ is neither open nor closed. This is also true of the set $\{b\} \cup (b, c)$.

Exercise 11.38

A definable subset of M is of the form $\varphi(M)$, for some quantifier-free $\mathcal{L}_{r<}$-formula $\varphi(x)$ in disjunctive normal form with parameters in M. We may assume that the connective $\neg$ does not occur in $\varphi(x)$. Each disjunct of $\varphi(x)$ containing a positive

B Solutions to Exercises 495

literal of the form $x = a$ is equivalent to this single literal (we may only consider satisfiable disjuncts, since the unsatisfiable ones, if any, may just be deleted). Disjuncts that are not positive literals define intersections of open intervals, which are themselves open intervals.

Thus $\varphi(x)$ defines a finite union of open intervals and points. If at least one of its points does not belong to any of the given open intervals, $\varphi(M)$ cannot be open by the argument given in the previous exercise. Otherwise $\varphi(M)$ is a finite union of open intervals. If these intervals are disjoint, we are done. If not, we note that the union of two overlapping, open intervals is an open interval. Thus, once we replace unions of overlapping intervals with single open intervals, we end up with finite unions of disjoint intervals.

Exercise 11.40

If, for any open interval (a, b), $f^{-1}[(a, b)]$ is a finite union of open intervals, then it is open in M and f is continuous. If, conversely, f is continuous, then let $\psi(x, y)$ be the formula that defines f. Then the formula $\theta(x) := \exists y(\psi(x, y) \wedge a < y \wedge y < b)$ defines $f^{-1}[(a, b)]$. By the previous exercise, $\theta(x)$ is equivalent to a quantifier-free formula that defines a finite union of open intervals.

Exercise 11.41

If $f[U]$ is not definably connected, then there are definable, nonempty open sets V_1, V_2 such that $V_1 \cup V_2 = f[U]$. Note that $U = f^{-1}[V_1 \cup V_2] = f^{-1}[V_1] \cup f^{-1}[V_2]$. Because f is continuous and definable $f^{-1}[V_1]$ and $f^{-1}[V_2]$ are finite unions of open intervals. Each of these unions is thus a nonempty, definable open set. It follows that U is not definably connected.

Exercise 11.44

Suppose $V \subseteq M$ is open, $v \in V$ and $v = f(a, b)$. This is to say that $(a, b) \in f^{-1}[V]$. By hypothesis there are open intervals $I = (a_1, a_2)$, $J = (b_1, b_2)$ such that $a \in I$, $b \in J$ and $f^{-1}[V] \supseteq I \times J$. Since $I \times J$ is a basic open set and we can find suitable I, J for each $v \in V$ that is a value of f, it follows that $f^{-1}[V]$ is a union of basic open sets and thus an open set. We conclude that f is continuous.

Exercise 11.45

Consider an open set $V \subseteq M$. Since g is continuous $g^{-1}[V]$ is an open subset U of $M \times M$. Let $(a, b) \times (c, d)$ be a basic open subset of U. Because f is continuous and the identity function is continuous $f^{-1}[(a, b)] \times (c, d)$ is open in $M \times M$. It is clear that, if $(u, v, w) \in f^{-1}[(a, b)] \times (c, d)$, then $(f(u, v), w)) \in (a, b) \times (c, d) \subseteq U$ and $h(u, v, w) = g(f(u, v), w) \in V$. We conclude that h is continuous.

Exercise 11.47

If $\mathcal{M}$ has quantifier elimination, it must be o-minimal. It follows that $\mathcal{M} \models \mathsf{RCF}_<$. If, on the other hand, $\mathcal{M} \models \mathsf{RCF}_<$, it follows from Theorem 11.27 that it has quantifier elimination.

Exercise 11.48

The axioms of real closed ordered fields, can be stated in the language of signature $\{<, +, \cdot, 0, 1\}$. Let them determine the set RCF^-. Any model $\mathcal{R}$ of $\mathsf{RCF}^-_<$ can be expanded to a model of $\mathsf{RCF}_<$ because the formula $x + y = 0$ defines the unary function $-^R$, on which the unary function symbol $-$ can be interpreted.

Using the latter fact, we can in easily see that $\mathsf{RCF}^-_<$ has algebraically prime models and that it satisfies the criterion for quantifier elimination used for $\mathsf{RCF}_<$. To see this, suppose $\mathcal{M}^-, \mathcal{N}^- \models \mathsf{RCF}^-_<$ and $\mathcal{N}^- \models \exists y \varphi(\bar{a}, y)$, where $\varphi(\bar{a}, y)$ is a conjunction of positive literals in $\mathcal{L}^-_{<r}$. By expanding $\mathcal{M}^-, \mathcal{N}^-$ to $\mathcal{M}, \mathcal{N} \models \mathsf{RCF}_<$ respectively, we can use an earlier argument to show that $\mathcal{M} \models \exists y \varphi(\bar{a}, y)$. Because the last condition does not depend on the interpretation of $-$, we conclude $\mathcal{M}^- \models \exists y \varphi(\bar{a}, y)$.

It now follows that $\mathsf{RCF}^-_<$ has quantifier elimination. Consequently, any definable subset of $\mathsf{RCF}^-_<$ is defined by a quantifier-free formula. Such a formula lies among the quantifier-free $\mathcal{L}_{r<}$-formulae, which we have shown to define finite unions of points and intervals. As a result, $\mathsf{RCF}^-_<$ is o-minimal.

Exercise 11.51

Let $\bar{a} \in \mathbb{N}^m$ and $\varphi(\bar{x}) \in \mathfrak{F}$. In view of the previous exercise, because $\mathcal{R}^- \models \mathsf{RCF}^-_<$, there is a quantifier-free formula $\psi(\bar{x})$ such that:

$$\mathcal{R}^- \models \varphi(\bar{a}) \leftrightarrow \psi(\bar{a}).$$

Because quantifier-free formulae are in $\mathfrak{F}$, the biconditional $\varphi(\bar{x}) \leftrightarrow \psi(\bar{x})$ is in $\mathfrak{F}$. We can now use the fact that $\mathcal{R}^-$ is a $\mathfrak{F}$-elementary extension of $\mathcal{N}$ to deduce:

$$\mathcal{N} \models \varphi(\bar{a}) \leftrightarrow \psi(\bar{a}).$$

The last condition is true of every m-tuple in $\mathcal{N}$. We can therefore conclude:

$$\mathcal{N} \models \forall \bar{x}(\varphi(\bar{x}) \leftrightarrow \psi(\bar{x})).$$

B Solutions to Exercises 497

Exercise 11.52

By Morleysation, we can guarantee that an expansion of $(\mathbb{N}, +^{\mathbb{N}}, \cdot^{\mathbb{N}}, 0^{\mathbb{N}}, 1^{\mathbb{N}})$ has quantifier elimination. If this expansion were decidable $Th(\mathbb{N})$ would be.

Exercise 11.54

Consider a primitive $\mathcal{L}_{r<}$-formula of the form $\exists y \varphi(x_1, \ldots, x_n, y)$. If $\mathcal{R} \models \mathsf{RCF}_<$, $\varphi(x_1, \ldots, x_n, y)$ defines a semi-algebraic set $A \subseteq R^{n+1}$. The given formula defines, in turn, a projection of A on the first n coordinates. It follows from the Tarski-Seidenberg theorem that it defines a semi-algebraic set. This is to say that there is a quantifier-free formula $\psi(x_1, \ldots, x_n)$ such that $\mathcal{R} \models \forall \overline{x}(\exists x \varphi(\overline{x}, y) \leftrightarrow \psi(\overline{x}))$. The Tarski-Seidenberg theorem on its own does not tell us whether the last sentence transfers to other real closed fields. By contrast, quantifier elimination guarantees that this sentence simultaneously holds in every ordered real closed field. We could reach this conclusion from the Tarski-Seidenberg theorem if we knew that $\mathsf{RCF}_<$ is a complete theory.

Exercise 11.55

Consider the formula:

$$\varphi(\overline{y}) := \exists x (p_1(x, \overline{y}) \lessgtr 0 \land \ldots \land p_s(x, \overline{y}) \lessgtr 0),$$

where each conjunct is a literal determined by the function f. Quantifier elimination implies the existence of a quantifier-free formula $\mathsf{B}(\overline{y})$ that is equivalent to the above existential formula *modulo* $\mathsf{RCF}_<$. By definition, a quantifier-free formula is a Boolean combination of polynomial equalities and inequalities. It is clear that, for any $\mathcal{R} \models \mathsf{RCF}_<$, $\mathcal{R} \models \mathsf{B}(\overline{y})$ iff $\mathcal{R} \models \varphi(\overline{b})$ iff there is $a \in R$ such that a is a simultaneous solution of the inequalities $p_1(x, \overline{b}) \lessgtr 0, \ldots, p_1(x, \overline{b}) \lessgtr 0$.

Exercise 11.56

(i) By hypothesis $\mathcal{R} \models \forall \overline{x}(\psi(\overline{x}) \leftrightarrow \varphi(\overline{x}))$. It follows that $\varphi(R^m) = \psi(R^m)$. Because RCF is complete, $\mathcal{R}_1 \models \forall \overline{x}(\psi(\overline{x}) \leftrightarrow \varphi(\overline{x}))$. Consequently, $S_1 = \psi(R_1^m)$.
(ii) We have $\theta(R^m) = T \supseteq S = \varphi(R^m)$. This conditions is equivalent to $\mathcal{R} \models \forall \overline{x}(\varphi(\overline{x}) \rightarrow \theta(\overline{x}))$. By the completeness of RCF, the last sentence transfers to $\mathcal{R}_1$. Consequently, $T_1 \supseteq S_1$.
(iii) Suppose $\mathcal{R} \models \varphi(\overline{a})$. Because RCF is model complete, $\mathcal{R}_1 \models \varphi(\overline{a})$. In other words $\overline{a} \in \varphi(R_1^m)$ for each $\overline{a} \in \varphi(R^m)$. In short, $S \subseteq S_1$. This exercise shows why we have referred to S_1 as an extension of S. If S is finite, there is $n \in \mathbb{N}$

such that $|S| = n$. A sentence to the effect that S has exactly n elements holds in $\mathcal{R}$. By the completeness of RCF, this sentence transfers to $\mathcal{R}_1$. It follows that $\varphi(R_1^m)$ has exactly n elements and, since it extends S, it must coincide with it.

(iv) The set $R^m - S$ is defined by $\neg\varphi(\overline{x})$. In $\mathcal{R}_1^m$. the same formula defines $R_1^m - S_1$, which therefore is the extension $(R^m - S)_1$. The set $S \cup T$ is defined by the formula $\varphi(\overline{x}) \vee \theta(\overline{x})$, which in turn defines $S_1 \cup T_1$ in R_1^m.

Exercise 11.57

Let C be a set of new constant symbols of size κ and C^m an enumeration of all the m-tuples of constant symbols from C (thus $|C^m| = \kappa$). Consider the set $\mathcal{ED} \cup \Gamma$, where $\Gamma = \{\neg(c_i = c_j) : i, j \in \lambda\} \cup \{\varphi(\overline{c}) : c \in C^m\}$.

It is easy to see that $\mathcal{ED} \cup \Gamma$ is finitely satisfiable. A model $\mathcal{R}_1$ of this set of sentences is an elementary extension of $\mathcal{R}$ in which the extension S_1 has size at least κ. If its size is exactly κ, we may set $\mathcal{R}_1 = \mathcal{R}_\kappa$. Otherwise, since $R \subseteq R_1$, we may apply the downward Löwenheim-Skolem theorem and consider the elementary substructure of $\mathcal{R}_1$ generated by $R \cup C$. The size of this structure is exactly κ, so the extension of S in this structure, which we may call $\mathcal{R}_\kappa$, has exactly size κ.

Exercise 11.58

The extension $(\Pi[S])_1$ is defined by the formula:

$$\exists \overline{x} \zeta(\overline{x}, \overline{y}) := \exists x_1 \ldots \exists x_{n+1}(\eta(x_1, \ldots, x_{n+1}) \wedge \bigwedge_{i=1}^{n} x_i = y_i).$$

The function Π_1 is a set of $(2n + 1)$-tuples in R_1^{2n+1} defined by the formula:

$$\eta(x_1, \ldots, x_{n+1}) \wedge \bigwedge_{i=1}^{n} x_i = y_i).$$

The image of Π_1 on S_1, which is defined by $\eta(x_1, \ldots, x_{n+1})$ is defined by the formula:

$$\exists \overline{x} \zeta(\overline{x}, \overline{y})$$

as above. Thus $\Pi_1[S_1] = (\Pi[S])_1$.

Exercise 11.59

Suppose $\varphi(x_1, \ldots, x_m, y)$ defines f. The sentence:

$$\forall x_1 \ldots \forall x_m \forall y \forall z \varphi(x_1, \ldots, x_m, y) \wedge \varphi(x_1, \ldots, x_m, z) \rightarrow y = z$$

says that the function defined by φ is injective. By the completeness of RCF (or RCF$_<$), this formula holds in $\mathcal{R}$, where φ defines f, iff it holds in $\mathcal{R}_1$, where φ defines f_1.

An entirely similar argument relative to the sentence:

$$\forall y \exists x_1 \ldots \exists x_m \varphi(x_1, \ldots, x_m, y)$$

establishes the analogous equivalence for surjectivity.

Exercise 11.60

The formula $\psi(x, y) := (x = 0 \wedge y = 0) \vee (0 < x \wedge y = x) \vee (x < 0 \wedge y = -x)$ works.

Exercise 11.65

We reason by contraposition and suppose that $R(x_1)$ is not formally real. Then we can find $f_1/g_1 + \ldots + f_k/g_k$ such that:

$$-1 = \frac{f_1^2}{g_1^2} + \ldots + \frac{f_k^2}{g_k^2}.$$

With $g_1, \ldots g_k \neq 0$ and where not all f_i ($i \in \{1, \ldots, k\}$) can be identically zero. Let $\gamma = g_1 \cdots g_k$ and $\gamma_i = g_1 \cdots g_{i-1} \cdot g_{i+1} \cdots g_k$. Then the above equality can be rewritten as:

$$-\gamma(x)^2 = (f_1(x)\gamma_1(x))^2 + \ldots + (f_k(x)\gamma_k(x))^2.$$

Because R is infinite, there is $a \in R$ that differs from the roots of each f_i and each g_i. When we evaluate the latter polynomial at a, we obtain:

$$-\gamma(a)^2 = (f_1(a)\gamma_1(a))^2 + \ldots + (f_k(a)\gamma_k(a))^2.$$

Since $\gamma(a)^2 \neq 0$ and the right-hand side of the last equation contains nonzero summands, we conclude that R is not formally real. This shows that, if $\mathcal{R}$ is formally real, then $\mathcal{R}(x_1) = \mathcal{K}$ is formally real. By the same argument $K(x_2) = R(x_1, x_2)$ is formally real. Further iterations of this argument lead to the conclusion that $R(x_1, \ldots, x_n)$ is formally real.

Exercise 11.68

Suppose $\theta(x, y_1, \ldots, y_n)$ defines f. Then $\exists y_1 \ldots \exists y_{i-1} \exists y_{i+1} \ldots \exists y_n \theta(x, y_i)$ defines the i-th projection f_i of f, with $i = 1, \ldots, n$. Clearly, for each i, Lemma 11.67 holds for f_i. We thus obtain, for each $i = 1, \ldots, n$, a partition $I_{i1} \cup I_{in_i} \cup D_i$. Since each partition is R, their intersection is R. Moreover, since this same intersection is obtained as an intersection of unions of disjoint intervals and finite sets of points, it is itself such a union. Since, moreover, each f_i is continuous on the intervals of the respective partition, every projection is continuous on the intersections of these intervals. This is enough to ensure the continuity of f.

B.12 Chapter 12

Exercise 12.1

Consider the finitely generated structure:

$$\mathcal{D} = \langle a_0, \ldots, a_{m-1}, b_0, \ldots, b_{n-1} \rangle \sqsubseteq \mathcal{M}.$$

The elements of D are designated by evaluating terms of the form:

$$t(x_0, \ldots, x_{m-1}, y_0, \ldots, y_{n-1})$$

on the generators $a_0, \ldots, a_{m-1}, y_0, \ldots, y_{n-1}$.

When we consider terms whose variables occur among $x_0, \ldots, x_{m-1}$ only, and evaluate them on $a_0, \ldots, a_{m-1}$, we obtain the elements of $\mathcal{A}$. When we consider terms whose variables occur among $y_0, \ldots, x_{n-1}$ only, and evaluate them on $b_0, \ldots, b_{n-1}$, we obtain the elements of $\mathcal{B}$. Because atomic formulae are preserved by substructures and extensions, if R is a k-ary relation symbol and $c_0, c_{k-1} \in \mathcal{A}$, we have $\mathcal{A} \models R(c_0, \ldots, c_k)$ iff $\mathcal{M} \models R(c_0, \ldots, c_k)$ iff $\mathcal{D} \models R(c_0, \ldots, c_k)$. It follows that the inclusion of $\mathcal{A}$ into $\mathcal{D}$ is an embedding and the same can be said of the inclusion of $\mathcal{B}$ into $\mathcal{D}$.

Exercise 12.3

Suppose the signature contains the relation symbols $R_1, \ldots, R_k$, where the i-th relation symbol is n_i-ary, for $i = 1, \ldots, k$. A m element domain A is turned into a

$\mathcal{L}$-structure once the k relation symbols in the signature are interpreted on subsets of A^{n_i}, $i = 1, \ldots k$. Because $|A^{n_i}| = m^{n_i}$, there are:

$$2^{m^{n_1}} \cdot \ldots \cdot 2^{m^{n_k}}$$

ways of interpreting the given relation symbols and thus only finitely many non-isomorphic structures based on a n-element set. This is true of each $n \in \mathbb{N}$ to there are only countably many isomorphism types.

Exercise 12.7

- (1) $\implies$ (2) Suppose $\mathcal{M}$ is algebraically ω-homogeneous in the sense of Definition 12.6. We shall prove that this definition implies (2). Let $\mathcal{A}, \mathcal{B} \subseteq \mathcal{M}$ be finitely generated substructures and $f_0 : \mathcal{A} \longrightarrow \mathcal{M}$, $f_1 : \mathcal{A} \longrightarrow \mathcal{B}$ be embeddings. We may suppose $\mathcal{A} = \langle \overline{a} \rangle$ and $\mathcal{B} = \langle \overline{b} \rangle$.
  Because f_1 is an embedding, $\mathcal{B}$ models the diagram of $\mathcal{A}$ and there is $\mathcal{C} \sqsupseteq \mathcal{A}$ such that $\mathcal{C}$ is isomorphic to $\mathcal{B}$.
  f_0 and f_1 determine two distinct isomorphisms, from $\mathcal{A}$ to $f_0[\mathcal{A}]$ and to $f_1[\mathcal{A}]$ respectively. We use algebraic ω-homogeneity repeatedly to obtain the isomorphisms $\hat{f}_0, \hat{f}_1^{-1}$, which will extend the corresponding functions.
  Since $\mathcal{C}$ is finitely generated (as an isomorphic copy of $\mathcal{B}$), let $\overline{d}$ be a set of generators. There is a tuple $\overline{e}$ of elements of $\mathcal{M}$ such that f_0 extends to an isomorphism $\hat{f}_0$ from $\langle \overline{a}, \overline{d} \rangle$ to $\langle \hat{f}_0(\overline{a}), \overline{e} \rangle$, where $\hat{f}_0(\overline{d}) = \overline{e}$.
  In a similar manner we can extend $f_1^{-1} : f_1[\mathcal{A}] \longrightarrow \mathcal{A}$ to an isomorphism $\hat{f}_1^{-1} : \mathcal{B} \longrightarrow \mathcal{C}$, such that, if $f_1(a) \in f_1[\mathcal{A}]$, then $\hat{f}^{-1}(f_1(a)) = a$.
  It now suffices to set $g = \hat{f}_0 \circ \hat{f}_1^{-1}$ to obtain, for any $a \in \mathcal{A}$:

$$g \circ f_1(a) = \hat{f}_0 \circ \hat{f}_1^{-1}(f_1(a)) = \hat{f}_0(a) = f_0(a).$$

- (2) $\implies$ (3). Condition (3) is only a special case of (2), in which f_0 is an inclusion.
- (3) $\implies$ (1). If $\mathcal{A} = \langle \overline{a} \rangle$, $c \in M$, and f_0 is an isomorphism between $\mathcal{A}$ and $\mathcal{B}$ condition (3) yields the diagram:

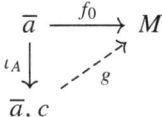

Because g extends f_0, when restricted to its image it sends $\langle \overline{a}, c \rangle$ into a finitely generated substructure $\mathcal{C}$ that extends $\mathcal{B}$. Thus $\mathcal{C}$ contains some $d = g(c)$ and it is isomorphic to $\langle \overline{a}, c \rangle$.

Exercise 12.10

To build the desired chain of isomorphism, we start from an arbitrary finitely generated structure $\mathcal{A}$ in the common age of $\mathcal{M}, \mathcal{N}$. Let $\mathcal{M}_0, \mathcal{N}_0$ its isomorphic copies in $\mathcal{M}, \mathcal{N}$ respectively. The partial isomorphism $f_0 : \mathcal{M}_0 \longrightarrow \mathcal{N}_0$ is the starting point of our construction.

Let $\{m_i\}_{i \in \mathbb{N}}, \{n_i\}_{i \in \mathbb{N}}$ be fixed enumerations of $M - M_0, N - N_0$ respectively. We associate the chains $\{\mathcal{M}_i\}_{i \in \mathbb{N}}, \{\mathcal{N}_i\}_{i \geq 1}$ of finitely generated substructures with the initial segments of the given enumerations.

To complete our recursive construction we assume that a partial isomorphism $f_i : \mathcal{M}_i \longrightarrow \mathcal{N}_i$ is given, such that its domain contains $\{m_j\}_{j < i}$ and its image contains $\{n_j\}_{j < i}$. Our goal is to extend f_i to a partial isomorphism $f_{i+1} : \mathcal{M}_{i+1} \longrightarrow \mathcal{N}_{i+1}$ with the same property extended to $i + 1$. In particular, m_i will belong to the domain of f_{i+1} and n_i to its image.

Because $\mathcal{M}, \mathcal{N}$ have the same age, the finitely generated substructure $\mathcal{M}_i$ is embedded in $\mathcal{N}$ by e.g. g_0 and it is included into $\mathcal{A}_{i+1}$, generated by $M_i \cup \{m_i\}$. Thus, we can extend g_0 to $g : A_{i+1} \longrightarrow N$ using the algebraic ω-homogeneity of $\mathcal{N}$.

The next step is to consider the finitely generated substructure with domain $g[A_{i+1}]$, which is included in the finitely generated substructure $g[A_{i+1}] \cup \{n_i\}$ and embedded into M by g^{-1}. An application of algebraic ω-homogeneity in $\mathcal{N}$ yields an extension of g^{-1}. We call M_{i+1} the image of this extension and N_{i+1} its domain. Because, restricted to these substructures, the embedding is a partial isomorphism, we define its inverse to be f_{i+1}. It is clear that M_{i+1}, as defined, contains $A_{i+1} \supseteq M_i \cup \{m_i\}$.

Since $A_{i+1} \supseteq N_i$ by construction, it is also clear that $f_{i+1}[M_{i+1}]$ contains $N_i \cup \{n_i\}$. This completes our recursive construction.

Exercise 12.12

Suppose $\mathcal{M}$ is ultrahomogeneous and that $\mathcal{A} = \langle \bar{a} \rangle, \mathcal{B} = \langle \bar{b} \rangle$ are finitely generated substructures of $\mathcal{M}$. Suppose, furthermore, that $f : A \longrightarrow B$ is an isomorphism.

Let us fix $c \in M$ and consider $\mathcal{A}_1 = \langle \bar{a} c \rangle$. By ultrahomogeneity, f can be extended to an automorphism of $\mathcal{M}$, which we continue to call f by an abuse of notation. Because f is defined on M, there is $d \in M$ such that $f(c) = d$.

It suffices to show that $\mathcal{B}_1 = \langle \bar{b} d \rangle$ is isomorphic to $\mathcal{A}_1$ under f. First, we can prove by an induction on the complexity of terms (left to the reader) that, if $t(x_1, \ldots, x_k)$ is a k-ary term and $\sigma_1, \ldots, \sigma_k$ are terms whose free variables occur among $\bar{x}, y$:

$$f(t^{\mathcal{A}_1}(\sigma^{\mathcal{A}_1}(\bar{a}, c), \ldots \sigma_k(\bar{a}, c))) = t^{\mathcal{B}_1}(\sigma^{\mathcal{B}_1}(\bar{b}, d), \ldots \sigma_k^{\mathcal{B}_1}(\bar{b}, d)).$$

In other words, when restricted to $\mathcal{A}_1$, f has $\mathcal{B}_1$ as its image. Because f is an automorphism of $\mathcal{M}$, this restriction is certainly an isomorphism between finitely generated substructures. Thus, algebraic ω-homogeneity holds.

Exercise 12.15

A substructure of a finite linear order is a finite linear order: this is enough to show that L satisfies **HP**. Given two finite linear orders $\mathcal{A}_1 = (L_1, <_1)$, $\mathcal{A}_2 = (L_2, <_2)$, we may arrange for them to be disjoint by choosing isomorphic copies. Assuming them disjoint, we consider the finite set $L_1 \cup L_2$ and note that $\mathcal{A}_3 = (L_1 \cup L_2, <_3)$ is a linear order when $<_3 = <_1 \cup <_2 \cup L_1 \times L_2$. Because $\mathcal{A}_1, \mathcal{A}_2$ can be naturally embedded into $\mathcal{A}_3$, **JEP** holds.

Exercise 12.20

Let $\mathcal{U}, \mathcal{W}$ finitely generated, and thus finite, vector subspaces of $\mathcal{M}$. Let $f : \mathcal{U} \longrightarrow \mathcal{W}$ be an isomorphism. Then $\mathcal{U}, \mathcal{W}$ have bases of the same size $k \in \mathbb{N}$, say $B_U = \{u_0, \ldots, u_{k-1}\}$ and $B_W = \{w_0, \ldots, w_{k-1}\}$. Since isomorphisms send linearly independent vectors into linearly independent vectors, we may assume that f sends B_U into B_W and, in particular, that $f(u_i) = w_i, i = 1, \ldots, k$.

We can extend B_U to a basis $\{u_i\}_{i \in \mathbb{N}}$ of $\mathcal{M}$ and B_W to another basis $\{u_i\}_{i \in \mathbb{N}}$ of $\mathcal{M}$. We next extend f to B_W by stipulating $f(u_i) = w_i$ for each $i \in \mathbb{N}$ (by an abuse of notation, we adopt f to describe the extension of the original f). Finally, we extend f to the whole of $\mathcal{M}$ by the condition:

$$f_1(\lambda_1 u_{i_1} + \ldots + \lambda_n u_{i_n}) = \lambda_i w_{i_1} + \ldots + \lambda_n w_{i_n},$$

for any finite subset of $u_{i_1}, \ldots, u_{i_n}$ of $\{u_i\}_{i \in \mathbb{N}}$. Because every vector of $\mathcal{M}$ is a linear combination of finitely many w_i, the above stipulation ensures that f is surjective. The fact that the sequence $\{w_i\}_{i \in \mathbb{N}}$ is independent further ensures that $f(a) = 0$ implies $a = 0$, i.e. that f is also injective. Because f is linear by construction, it is an automorphism of $\mathcal{M}$ that extends an automorphism between finitely generated substructures of $\mathcal{M}$. We conclude that $\mathcal{M}$ is ultrahomogeneous.

Exercise 12.24

If it is possible to carry any linear order $a_1 < \ldots < a_m$ into any other linear order $b_1 < \ldots < b_m$ by means of an automorphism, then it is possible to carry any linear order of k elements into any other such linear order, when $k \leq m$, simply by extending the given linear order arbitrarily to the length m and finding an automorphism that links the extensions.

As for n-point uniqueness, it is clear that, if an automorphism fixing n points is the identity, an automorphism fixing $k > n$ must also be the identity. Finally, $m \leq n$ because $n < m$ leads to an immediate inconsistency. If $n < m$ then two linear orders of length m that coincide in their first n elements and differ otherwise are related by an automorphism, but this automorphism can only be the identity, which is impossible.

Exercise 12.30

Suppose $\mathcal{L}$-structure $\mathcal{M}$ is a saturated $\mathcal{L}$-structure and $A \subseteq M$ is a set of cardinality $< |M|$. Let p be a 1-type over $A \subseteq M$ relative to the $\mathcal{L}^-$-reduct $\mathcal{M}^-$ of $\mathcal{M}$. Note that p is finitely consistent in $\mathcal{M}$, i.e. it is a type in $\mathcal{M}$. As such, it is contained in some element q of $S_1^{\mathcal{M}}(A)$. Because $\mathcal{M}$ is saturated, there is $c \in M$ such that c realises q. Because $p \subseteq q$, c realises p. It follows that $\mathcal{M}^-$ is saturated.

Exercise 12.31

Since $\mathcal{M}_n$ is an expansion of the ordered rationals, any one of its automorphisms is also an automorphism of the ordered rationals. Since the set of automorphisms of a fixed structure is a group relative to functional composition, $Aut(\mathcal{M}_n)$ is a subgroup of $Aut(\mathbb{Q})$.

B.13 Appendix A

Exercise A.2

For any $\{a, b\} \subseteq M$, we must have aRb or bRa. Moreover, for any $\{a, b, c\} \subseteq M$, if aRb and bRc, we cannot have cRa. If we did, we could not have aRb, aRc. But, by hypothesis, we cannot have bRa, bRc either, or cRa, cRb, against the definition of (M, R). It follows that aRc must hold (since one of aRc, cRa has to hold) and R is transitive. Trichotomy now follows from irreflexivity, since aRb, bRa cannot simultaneously hold. Asymmetry also follows and we conclude that (M, R) is a linear order.

Exercise A.3

Let $\mathcal{M} = (M, <)$ be an infinite, linearly ordered set. If $\mathcal{M}$ is not well-ordered, then there is $S \subseteq M$ such that $S \neq \emptyset$ and S has no least element. Since S is nonempty, we may fix $a_0 \in S$. Because a_0 cannot be least in S, there is $a_1 < a_0$. The process can be repeated because a_1 cannot be least in S either. By dint of iteration we obtain an infinite sequence $\{a_i\}_{i \in \mathbb{N}}$ such that, for each $i \in \mathbb{N}$, $a_i > a_{i+1}$. Conversely, if such infinite sequence occurs in $\mathcal{M}$, then it is itself a nonempty subset of M without a least element: M cannot therefore be well-ordered by $<$.

Exercise A.4

Suppose that Σ was a set of $\mathcal{L}$-sentences isolating the class of well-ordered sets. We add to $\mathcal{L}$ an infinite sequence of constant symbols $\{c_i\}_{i \in \mathbb{N}}$ and consider the following set of sentences in the language expanded by constants:

$$\Sigma' = \Sigma \cup \{c_i < c_j : j < i\}.$$

Any finite subset of Σ' consists in general of a finite number of inequalities involving constant symbols and a finite number of sentences from Σ. The sentences from Σ hold in every well-ordered sets: in particular, they hold in the linear order $(\mathbb{N}, <)$. Given finitely many inequalities involving the constant symbols $c_{i_1}, \ldots, c_{i_k}$, it is easy to expand $(\mathbb{N}, <)$ to a $\mathcal{L}'$-structure, with $\mathcal{L}' = \{<, c_{i_1}, \ldots, c_{i_k}\}$, in which the constant symbols are interpreted on integers $n_1, \ldots, n_k$ ordered as the given inequalities dictate. It follows that any finite subset of Σ' is satisfied in an expansion of $(N, <)$. By Compactness, Σ' has a model $\mathcal{M}$, which must contain the infinite descending chain determined by the interpretations of the constant symbols $\{c_i : i \in \mathbb{N}\}$. The previous Exercise implies that $\mathcal{M}$ is not a well-ordered set. It follows that Σ does not single out the class of well-ordered set.

Exercise A.6

Let $(M, <)$ be a well-ordered set and $a \in M$ be a limit point relative to $<$. Consider the nonempty set $S \subseteq M$ defined by the condition $x < a$. The set S is nonempty because it contains the least element of M, which we may call a_0. Moreover, if $S = \{a_0\}$ we could deduce that a is the successor of a_0, against the choice of a. Thus, S contains at least two elements. Since a_0 is least in M, we have $a_0 < a_1$, with $a_1 \in S$. By the same argument we rule out the possibility that no elements of S lie strictly between a_1 and a. Proceeding in this manner, we identify an infinite sequence included in S (the countable version of the Axiom of Choice is implicitly being appealed to).

Exercise A.8

Toward a contradiction, suppose that $S \neq M$. It follows that $M - S \neq \emptyset$. Let a be the least element of $M - S$. Clearly a is not the least element of M by (i). Moreover, a is not a successor point of M, otherwise it should follow its successor $b \in S$ and, by (ii), we could deduce $a \in S$. Finally, a cannot be a limit point of M or, for any

$b < a$, $b \in S$, which implies $a \in S$ by (iii). Since every point of M can only be its least element, a successor or a limit point, it follows that $M - S$ is empty, i.e. $M = S$.

Exercise A.11

If g was another similarity between $(M, <^M)$ and the same initial segment of $(N, <^N)$, we could use it to determine the automorphism $g^{-1} \circ f$ of $(M, <^M)$ onto itself. By Lemma A.10, $g^{-1} \circ f = \iota$, where ι is the identity on M. Applying g to both sides of the last equality we obtain $f = g$.

Exercise A.12

(i) If $f(c)$ was not least in N, then there would be $b \in N$ with $b < f(c)$. Because f is surjective, $b = f(a)$ for some $a \in M$. But c is least in M, so $c < a$ implies $f(c) < f(a)$, i.e. $f(c) < b$, a contradiction. It follows that $f(c)$ is least in N.
(ii) Suppose b is the successor of a in M. Clearly $f(a) < f(b)$ and, if $f(b)$ was not the successor of $f(a)$, we could find $d \in N$ such that $f(a) < d$ and $d < f(b)$. Because f is surjective, there is $c \in M$ such that $f(c) = d$. Moreover, because f is order-preserving both ways, we can deduce $a < c$ and $c < b$, against the hypothesis that b is the successor of a.
(iii) If a is a limit point of M, then it is not the least element of M or a successor point of M. Consequently, $f(a)$ is not the least element of N (or we could reach a contradiction using the argument from (i) relative to f^{-1}) and it is not a successor point of N (or we could use the argument from (ii) relative to f^{-1} to obtain a contradiction). The only possibility left is that $f(a)$ should be a limit point of N.

Exercise A.15

(i) Suppose X is transitive and consider $Y \subseteq X$. If $z \in \bigcup Y$, then there is y such that $z \in y$ and $y \in Y$. But $Y \subseteq X$ implies $y \in X$ and the transitivity of X implies $y \subseteq X$. Thus $z \in X$ and we conclude $\bigcup Y \subseteq X$. If, conversely, the last inclusion holds for any subset of X, we fix $x \in X$ and note that $\{x\} \subseteq X$. It follows that $\bigcup\{x\} \subseteq X$ but $\bigcup\{x\} = x$. Thus, $x \subseteq X$ and X is transitive.
(ii) For any of the given sets it is easy to see that each element (if any) is also a subset.

Exercise A.17

(i) Each of the sets from Exercise A.15-(ii) is linearly ordered by membership. Its elements are listed in such a way that each of them contains the preceding sets as its own elements. Finite linear orders are well-ordered sets. Transitivity has already been verified.
(ii) Let β be an initial segment of α. Then $\beta \subseteq \alpha$ is linearly ordered by $\in$. Thus, if $\gamma \in \beta$, then $\gamma \in \alpha$ and $\gamma \subseteq \alpha$. But, for each $\delta \in \gamma$, we must have $\delta \in \beta$ by the transitivity of $\in$: this means that $\gamma \subseteq \beta$, i.e. that β is transitive.
(iii) If $\beta \in \alpha$, then $\beta \subseteq \alpha$. Now consider the initial segment of α $S = \{\gamma \in \alpha : \gamma \in \beta\}$. Because β is a subset of α, S is the set of all elements of β, i.e. β itself. By part (ii), S is an ordinal and so is β.
(iv) Let α be an ordinal and suppose $\beta, \delta \in \alpha$. By part (iii), β, δ are ordinals, so $\beta \in \delta$ or $\delta \in \beta$ implies $\beta \subseteq \delta$ or $\delta \subseteq \beta$. It follows that inclusion linearly orders α. Moreover, if S is a nonempty subset of α and γ is its least element relative to $\in$, then, because each element of S is transitive, it is clear that γ is a subset of each, i.e. the least element of S relative to inclusion. It follows that α is well-ordered by inclusion.

Exercise A.19

(i) If α is an ordinal and $\alpha \in \alpha$, then Lemma A.18 implies $\alpha \subset \alpha$, which is impossible.
(ii) For any set α, $\emptyset \subseteq \alpha$. If $\alpha \neq \emptyset$ is an ordinal, then $\emptyset \subset \alpha$ and Lemma A.18 implies $\emptyset \in \alpha$. Because $\emptyset$ has no elements, it must be the least element of α.

Exercise A.20

Since α is an ordinal, it is well-ordered by $\in$ and it is transitive. The set $\alpha \cup \{\alpha\}$ is a linear order relative to $\in$ with maximum α. If $S \neq \emptyset$ is a subset of $\alpha \cup \{\alpha\}$, either it is a subset of α, in which case it has a least element relative to $\in$, because α is an ordinal, or it contains α. In the latter case, unless $S = \{\alpha\}$, the least element of S is the least element of $S \cap \alpha$. What we have observed shows that $\alpha \cup \{\alpha\}$ is a well-ordered set relative to $\in$. If $\beta \in \alpha \cup \{\alpha\}$, then $\beta \in \alpha$ or $\beta = \alpha$. In the former case, the transitivity of α implies $\beta \subseteq \alpha$. In the latter case, it suffices to note that $\alpha \subseteq \alpha \cup \{\alpha\}$. Thus, $\alpha \cup \{\alpha\}$ is an ordinal.

Exercise A.22

$\gamma_1, \gamma_2 \in \bigcup S$. There are ordinals $\beta_1, \beta_2 \in \alpha$ such that $\gamma_i \in \beta_i$, with $i = 1, 2$. Since $\beta_i \subseteq \alpha$, $\bigcup S \subseteq \alpha$. Thus, $\bigcup S$ is a set of ordinals that inherits the well-order over α. Moreover, if $\gamma \in \bigcup S$, we note that there must be $\delta \in S$ such that $\gamma \in \delta$. Now, for any $\gamma_1 \in \gamma$, since δ is an ordinal, $\gamma \subseteq \delta$ and $\gamma_1 \in \delta$. But $\gamma_1 \in \delta$ and $\delta \in S$ imply $\gamma_1 \in \bigcup S$. It follows that $\gamma \subseteq \bigcup S$, i.e. that the last set is transitive.

Exercise A.25

Let ψ_1 be the formula $\forall y(\neg(y \in y))$, ψ_2 the formula $\forall y \forall w(y \in w \vee w \in y \vee y = w)$ and ψ_3 the formula $\forall y \forall w \forall z((y \in w \wedge w \in z) \rightarrow y \in z)$ $\varphi_2(x)$ is the formula $\forall y \forall w \forall z((y \in x \wedge w \in x \wedge z \in x) \rightarrow \psi_1 \wedge \psi_2 \wedge \psi_3)$. $\varphi_3(x)$ is the formula $\forall y((\neg(y = \emptyset) \wedge y \subseteq x) \rightarrow \exists z(z \in y \wedge \forall w(w \in y \rightarrow z = w \vee z \in w)))$.

Exercise A.27

Let S be the set supposedly defined by $x = x$. The formula $x \in S \wedge \neg(x = x)$ is equivalent to the formula $\neg(x = x)$ and, by Replacement, it defines a set $T \subseteq S$. Note that, by its definition, T satisfies the condition $T \in T$ iff $T \notin T$, which is equivalent to the contradiction $T \in T$ and $T \notin T$. It follows from this contradiction that $x = x$ does not define a set but only a class.

Exercise A.30

Let $(M, <)$ be a well-ordered set and $(N, <)$ be a proper initial segment of $(M, <)$ (we do not change symbols do indicate the restriction of $<$ to $N \subseteq M$). Then $M - N \neq \emptyset$. Fix $d \in M - N$: then, in M, for any $a \in N$, $a < d$. if there existed a similarity f between $(M, <)$ and $(N, <)$, $f(d) < d$, since $f(d)$ should necessarily be an element of N. Because f is an order-preserving function and $d, f(d) \in M$, we are forced to deduce $ff(d) < f(d)$. Further applications of f determine an infinite, descending chain, whose existence contradicts the hypothesis that $(M, <)$ is a well-ordered set. It follows that f cannot exist, i.e. that $(M, <), (N, <)$ cannot be similar.

Exercise A.34

We derive a contradiction from the assumption that there are ordinals not in P. This is to say that the collection of ordinals not in P is nonempty and, consequently, has a least element. The latter cannot be $\emptyset$, a successor or a limit ordinal, by hypothesis. But any ordinal falls under one of these three categories.

Exercise A.35

(i) If N satisfies (a), (b), then $\omega \subseteq N$ by definition of ω.
(ii) By (i), if $S \subseteq \omega$ satisfies (a) and (b), then $\omega \subseteq S$, because ω is a subset of any set satisfying (a), (b). Consequently $S = \omega$.
(iii) We consider the set $S \subseteq \omega$ whose elements are ordinals. Clearly, $\emptyset \in S$. Moreover, if $x \in S$, then $x \cup \{x\}$ is an ordinal, by a previous Exercise, and $x \cup \{x\} \in S$. It follows that S satisfies (a), (b). By part (ii), $S = \omega$. The first four elements of ω have been listed in the statement of Exercise .
(iv) Let $S \subset \omega$ consist of the elements of ω that are not limit ordinals. Clearly $\emptyset \in S$. Moreover, if $x \in S$, then $x \cup \{x\}$ is a successor ordinal. Since S satisfies conditions (a), (b), it follows that $S = \omega$, i.e. ω does not contain any limit ordinals. As for ω, it cannot be a successor ordinal: if it had the form $x \cup \{x\}$, then $x \in \omega$ would imply $\omega \in \omega$, which is impossible. It follows that ω is a limit ordinal and, in particular, the smallest such ordinal.

Exercise A.37

The transfinite induction principle we have in the case of F-towers may be stated as follows: Let T be a F-tower. If S is a subclass of T such that:

(i) 0 is in S;
(ii) if x is in S, then F(x) is in S;
(iii) if C is a chain whose elements are in S, then $\bigcup C$ is in S;

in this case, S and T coincide. Now consider F-towers T, S and let S' be the class of elements of S that are also in T. It is easy to see that $0 \in S'$ and that, if x is in S', then x is in T, in which case F(x) must also be in T (because T is a F-tower). Since F(x) is also in S (because S is a F-tower), we conclude that F(x) is in S'. Finally, let C be a chain whose elements are in S'. Because its elements are in T by inductive hypothesis $\bigcup C$ is. The same statement holds relative to S. It now follows that S' coincides with S: in other words, S is a subclass of T. The argument we have provided continues to work after we interchange S and T: thus, S and T coincide.

Exercise A.38

(i) Call x in T right-normal iff, for any y in T, R(y, x). Analogously, x in T left-normal iff, for any y in T, R(x, y).
(ii) Suppose that x is right-normal. We shall show that it must also be left-normal. By condition (1) R($x, 0$) holds. Next, we suppose that R(x, y) and note that, by right-normality, R(y, x) also holds. By condition (2), R(x, F(y)) must hold. Finally, if C is a chain and R(x, y) for each $y \in C$, then by condition (3) we deduce R($x, \bigcup C$). Transfinite induction now implies that x is left-normal.

(iii) Condition (1) says that 0 is right-normal. Now suppose that x is right-normal and consider any $y \in \mathsf{T}$. Certainly $\mathsf{R}(y, x)$ but, because any right-normal element of T is left-normal by part (ii), we also have $\mathsf{R}(x, y)$ and condition (2) now yields $\mathsf{R}(y, \mathsf{F}(x))$. This is to say that $\mathsf{F}(x)$ is right-normal. Finally, if C is a chain and every $x \in C$ is right-normal, then we have $\mathsf{R}(y, x)$ for any $y \in \mathsf{T}$. Condition (3) yields $\mathsf{R}(y, \bigcup C)$. Thus $\bigcup C$ is right-normal. We conclude that every element of T is right-normal.

(iv) Fix x, y in T. Because y is right-normal, $\mathsf{R}(x, y)$ must hold.

Exercise A.40

It suffices to note that, whatever y is, $0 \subseteq x$.

Exercise A.41

If a F-tower T was a set Lemma A.39 would imply that any two of its elements are comparable relative to inclusion. Thus, T would be a chain and $\bigcup \mathsf{T} \in \mathsf{T}$, since T contains the unions of its chains. Moreover $\bigcup \mathsf{T} \subset \mathsf{F}(\bigcup \mathsf{T}) \in \mathsf{T}$. But $\mathsf{F}(\bigcup \mathsf{T}) \in \mathsf{T}$ implies $\mathsf{F}(\bigcup \mathsf{T}) \subseteq \bigcup \mathsf{T}$. We have arrived at a contradiction.

Exercise A.47

(i) Note that $\mathsf{F} \restriction \alpha = T_\alpha$ is a set of ordered pairs that belong to the ordinal sequences in T whose length is at most α. Such ordinal sequences determine a chain whose maximum is T_α. It follows that $\mathsf{F} \restriction \alpha \subseteq T_\alpha$. Equality holds by the definition of F.

(ii) $\mathsf{F} \restriction 0$ is $\emptyset$ and $\mathsf{F} \restriction \lambda = T_\lambda = \bigcup_{\beta \in \lambda} T_\beta = \bigcup_{\beta \in \lambda} (\mathsf{F} \restriction \beta)$.

Exercise A.49

(i) First, $\varphi(y, a, b)$ is the formula $\forall y_1 (y_1 \in y \leftrightarrow \exists v_1 \exists v_2) \eta(y_1, v_1, v_2) \wedge v_1 \in a \wedge v_2 \in b \wedge \forall x(x \in a \rightarrow \exists z \exists w(z \in b \wedge \eta(w, x, z) \wedge w \in y) \wedge \forall x \forall z \forall x_1 \forall x_2 \forall z_2 ((\eta(x, x_1, x_2) \wedge \eta(z, x_1, z_2) \wedge x \in y \wedge z \in y) \rightarrow z_2 = x_2))$.
Second, $\theta(x, a)$ is the formula $\exists y \varphi(x, a, y)$. Finally, the extending operation E is defined by the formula $\exists u \exists v (\theta(x, u) \wedge \varphi(y, u, v) \wedge \forall z(z \in y \leftrightarrow (z \in x \vee \psi(u, z))))$.

(ii) Suppose that F and H were distinct functional correspondences satisfying the statement of the theorem. There must be a least ordinal α at which they differ. This is to say that $\mathsf{F} \restriction \alpha = \mathsf{H} \restriction \alpha$. But, in this case:

$$F(\alpha) = G(F \restriction \alpha) = G(H \restriction \alpha) = H(\alpha),$$

a contradiction. It follows that F, H are the same class.

Exercise A.51

The verification is straightforward and simply requires spelling out conditions (6.2) and (6.3), since condition (6.1) was already spelled out in the proof of the theorem.

Exercise A.55

The set B_0 is clearly linearly independent. If B_α is a linearly independent set of vectors, the definition of G guarantees that $B_{\sigma(\alpha)}$ is obtained by adding to B_α one vector that does not belong to its linear closure. The resulting extension of B_α is therefore linearly independent. Finally, if λ is a limit ordinal and, for each $\delta \in \lambda$, B_δ is linearly independent, then the union of these sets is B_λ. If B_λ were not linearly independent, it would contain a linearly dependent set $\{v_1, \ldots, v_k\}$. Because the sets B_α are linearly ordered by inclusion, there is some $\delta \in \lambda$ such that $\{v_1, \ldots, v_k\} \subseteq B_\delta$, i.e. B_δ is not linearly independent, a contradiction. Thus, B_λ must be linearly independent.

Exercise A.60

By the Axiom of Choice, any two sets are comparable relative to their cardinality. Given an arbitrary set X, Hartogs' theorem guarantees the existence of an ordinal α such that α cannot be injected in X. The Axiom of Choice then implies that X can be injected in α. Let f be an injection of X into α. For any $x, y \in X$, we set:

$$x < y \text{ iff } f(x) \in f(y).$$

The above definition of $<$ well-orders X.

Exercise A.61

If **Cn** was a set A, by a previous exercise $\bigcup A$ would be an ordinal. By Hartogs' theorem, there is an ordinal not injectible in $\bigcup A$. Consider the least such ordinal α. Because there is no injection from α to $\bigcup A$, there is no injection from α to an element of A. In other words, α is not injectible in any element of **Cn**: in particular, it cannot be an element of **Cn**. We have identified an initial ordinal that does not belong to **Cn**, against the definition of **Cn**. This contradiction shows that **Cn** is a subclass of **On** that is not a set. By Theorem A.32 there is a class similarity between **Cn** and **On**.

Exercise A.63

Clearly, 0 is finite because the only injection from 0 into itself is the empty function, which is vacuously a bijection. Suppose that every injection of x into itself is surjective and consider an injection f from $x \cup \{x\}$ into itself. Either $f(x) = x$ or $f(x) \neq x$. In the former case, $f \restriction x$ is a bijection by inductive hypothesis and, consequently, f is surjective, too. In the latter case, $b = f(x) \in x$. If x is not a value of f, then f injects $x \cup \{x\}$ into x, which is not possible because $f \restriction x$ already exhausts x by inductive hypothesis. Thus, x is a value of f and there is $a \in x$ such that $f(a) = x$. As a consequence, $f \restriction (x \cup \{x\}) - \{a\}$ is an injection and, if we delete from it the pair (x, b) and replace it with (a, b), we obtain an injection of x onto itself by the inductive hypothesis. It follows that $f \restriction (x \cup \{x\}) - \{a\}$ injects its domain onto x, i.e. that f is, again, a bijection.

Exercise A.64

Clearly, 0 is initial. If some nonempty $b \in \omega$ were not initial, there would be $a \in b$ such that there exists a bijection f between a, b. Because a, b are ordinals $a \in b$ implies $a \subset b$ and, since f is an injection, the previous exercise implies that it is already a bijection from a onto itself. In other words, it cannot be surjective because $b - a \neq \emptyset$. This contradiction shows that b must be initial. Since b was chosen arbitrarily among the nonzero ordinals in ω, we conclude that every element of ω is an initial ordinal.

Exercise A.65

If $\mathbf{Cn}_\infty$ was a set, since ω is a set, we could conclude that **Cn** was a set, against Exercise A.61.

Exercise A.66

If α is an infinite, successor ordinal, then $\alpha = \beta \cup \{\beta\}$ and adding one element to the infinite set β does not affect its cardinality. Thus, α cannot be initial. It follows that, if α is initial, it must be a limit ordinal.

Exercise A.70

Set $A = \{a, b\}$ and $B = \{0, 1, 2\}$. A function from A to B must independently assign one of three values to a and then one of three values to b. There are $3 \times 3 = 9$ ways of doing it. On the other hand, a function from B to A must independently assign one of two values to each of three elements: there are $2 \times 2 \times 2 = 8$ ways of doing it.

Exercise A.73

Irreflexivity is clear, because none of the conditions defining $\prec$ can hold between a pair (α, β) and itself. Trichotomy reduces to the trichotomy of ordinal comparisons. Transitivity requires the tedious verification of many different cases. We consider only one, since the remaining ones are disposed of in essentially the sam way. Suppose $(\alpha, \beta) \prec (\delta, \gamma)$ and $(\delta, \gamma) \prec (\eta, \theta)$. If $max\{\alpha, \beta\} \in max\{\delta, \gamma\}$ and $max\{\delta, \gamma\} = max\{\eta, \theta\}$ but $\delta \in \eta$, then $max\{\alpha, \beta\} \in max\{\eta, \theta\}$ and $(\alpha, \beta) \prec (\eta, \theta)$.

Exercise A.76

$0 \cdot \kappa$ is the cardinal number associated with the Cartesian product $\emptyset \times \kappa$, which cannot contain any ordered pairs, because no first components are available. Thus $\emptyset \times \kappa = \emptyset$ and $0 \cdot \kappa = 0$, as desired.

Exercise A.77

By part (11) of Theorem A.67, we deduce $2^\lambda \leq \kappa^\lambda$. By part (12) of the same theorem, $\kappa < 2^\kappa$, which implies $\kappa^\lambda \leq 2^{\kappa \cdot \lambda} = 2^\lambda$ since, by hypothesis, $max\{\kappa, \lambda\} = \lambda$. The inequalities $\kappa^\lambda \leq 2^{\kappa \cdot \lambda}$ and $2^\lambda \leq \kappa^\lambda$ jointly imply $2^\lambda = \kappa^\lambda$.

Exercise A.81

(i) Let f be a choice function for A, by (AC1). Set $c = f[A]$.
(ii) A choice set for the family $\{\{i\} \times A_i\}_{i \in I}$ is a set of ordered pairs of the form (i, a_i), where $a_i \in A_i$. Such a set is a function with domain I and such that $f(i) = a_i \in A_i$, i.e., by definition, an element of the Cartesian product $\Pi_{i \in I} A_i$.
(iii) Let ΠA be the Cartesian product of the elements of A, which are nonempty. By (AC3), this Cartesian product is nonempty. An element of the latter Cartesian product is a function that assigns to each element of A one of its elements, i.e. a choice function for A.

Exercise A.84

The relation $\leq$ is reflexive and antisymmetric, because inclusion is. It is transitive for the same reason. Consider a chain $\{(S_i, f_i)\}_{i \in I}$ of elements of F and let $S = \bigcup_{i \in I} S_i$, $f = \bigcup_{i \in I} f_i$. Since $A \subseteq S_i$, $A \subseteq S$. Moreover, if $x \in S$, then there is $i \in I$ such that $x \in S_i$. In this case $f_i(x)$ is defined and this value coincides with $f_j(x)$, for each $j \in I$ such that $x \in S_j$ (e.g. if $(S_j, f_j) \leq (S_i, f_i)$, then $f_j \subseteq f_i$ amd $f_j(x) = f_i(x)$). Thus, $f(x) = f_i(x)$ is uniquely determined. It follows that $(S, f) \in F$ and it is easy to see that (S, f) is an upper bound for the chain $\{(S_i, f_i)\}_{i \in I}$.

Exercise A.85

Let $(A, \leq)$ be a partially ordered set, with A finite. If $|A| = 1$, then it is clear that A is already a linear order. Next, suppose that any n-element partial order can be extended to a linear order. Consider $|A| = n + 1$: because A is finite, it has a maximal element a, i.e. an element that is not smaller than any other element of A. By deleting a from A, we obtain a n-element partial order that can be extended to a linear order. By adding a as the maximum of the newly obtained linear order, we arrive at a linear order that extends the partial ordering over A. This completes the induction.

Exercise A.86

(i) By the previous exercise, a finite partial order modelling a finite subset of $\mathcal{D}^+(\mathcal{A})$ can be extended to a model of φ. Thus $\{\varphi\} \cup \mathcal{D}^+(\mathcal{A}) \cup \{\neg(c_a = c_b) : a, b \in A\}$ is finitely satisfiable.
(ii) By Compactness, $\{\varphi\} \cup \mathcal{D}^+(\mathcal{A})\{\neg(c_a = c_b) : a, b \in A\}$ has a model $\mathcal{M}$. Clearly $\mathcal{M}$ is a linear order. Consider the function $f : A \longrightarrow M$ defined by the condition $f(a) = c_a^M$. If $a <^A b$, a sentence of the form $c_a < c_b$ belongs to $\mathcal{D}^+(A)$ and, consequently, $c_a^M < c_b^M$. The function that sends $a \in A$ into $c_a^M \in M$ is injective and order-preserving. Thus, it is possible to identify $f[A]$, a substructure of $\mathcal{M}$, with a linear order extending the partial ordering of A.
(iii) It was unnecessary to use the full diagram of $\mathcal{A}$ in the proof just given because we cannot require the order of $\mathcal{M}$ to be reflected by $\mathcal{A}$ since, in particular, $\mathcal{A}$ is partially ordered and not linearly ordered.

B Solutions to Exercises   515

Exercise A.87

(i) The linearly ordered subsets of T are partially ordered by inclusion. Given a chain of linearly ordered subsets of T, its union is easily checked to be a linearly ordered subset of T. By Zorn's lemma, there is a maximal, linearly ordered subset of T, that is a branch.
(ii) Given a set A, the class T of injections from an ordinal into A is a set. To see this, note that, by Hartogs' theorem, there is an ordinal α that is not injectible into A. If α is the least ordinal for which this happens, each $\beta \in \alpha$ is injectible into A. For each β there is a set of possible injections, which is a subset of the set of functions from β to A. The union of the relevant sets of injections yields the set T. Consider the partial order $(T, \subseteq)$. If $t \in T$, then any $s \subseteq t$ is an injection from an ordinal β to A, with $\beta \in \alpha$. More precisely, s is the restriction of t to $\beta \in \alpha$. If $u \subseteq t$, then u is the restriction of t to some ordinal $\gamma \in \alpha$. Because α is linearly ordered by $\in$, $\gamma \in \beta$ or $\beta \in \gamma$. Thus, either $s \subseteq u$ or $u \subseteq s$. It follows that the set $\{s \in T : s \subseteq t\}$ is linearly ordered by $\subseteq$. In other words $(T, \subseteq)$ is a tree. By hypothesis, T has a branch t. By construction t is an injection from some ordinal α into A that cannot be further extended. If t were not surjective and $a \in A$ was not a value of t, then $t \cup (\sigma(\alpha), a)$ would be a proper extension of t, against the fact that t is a branch. Thus, t is a bijection and A can be ordered like α via t.

Exercise A.89

Suppose that, for each $\beta \in \alpha$, $\mathsf{F}(\beta) \in P$ and that, for every $\gamma \in \beta \in \alpha$, $\mathsf{F}(\gamma) < \mathsf{F}(\beta)$. Then $\mathsf{F} \upharpoonright \alpha$ is a chain in P and, by definition of F, $\mathsf{F}(\alpha) = \mathsf{G}(\mathsf{F} \upharpoonright \alpha)$ is a strict upper bound for that chain. In other words, $\mathsf{F}(\alpha) \in P$ and, for any $\beta \in \alpha$, $\mathsf{F}(\beta) < \mathsf{F}(\alpha)$.

Exercise A.90

(i) The powerset $\mathcal{P}(A)$ is partially ordered by inclusion and $(P, \subseteq)$ is a substructure.
(ii) Let C be a chain of $(P, \subseteq)$ and consider $\bigcup C$. Clearly, $\bigcup C$ is an upper bound for C in the partial order over $\mathcal{P}(A)$. We have to check that $\bigcup C \in P$. An arbitrary element a of $\bigcup C$ is an element of some $c \in C$. But $c \in P$, so a shares with each element of A at most one element. Since a was arbitrary, $\bigcup C \in P$.
(iii) By Zorn's lemma, P contains a maximal element m. Suppose that, for some $a \in A$, $a \cap m \neq \emptyset$. Since $a \neq \emptyset$, there is $b \in a$. The set $\{b\}$ is certainly in P and not in m. Thus, $m \cup \{\{b\}\}$ is a proper extension of m, against its maximality. It follows that m is a choice set for A, i.e. (AC2) holds.

Bibliography

1. Abbot, S.: Understanding Analysis. Springer, New York (2015)
2. Arrow, K.: Social Choice and Individual Values. Wiley, New York (1951)
3. Artin, E., Schreier O.: Algebraische Konstruktion reeller Körper'. Hamburgische Abhandlungen **5**, 85–99 (1927)
4. Artin, E.: 'Eine Kennzeichnung der reell abgeschlossenen Körper'. Hamburgische Abhandlungen **5**, 225–231 (1927)
5. Axler, S.: Linear Algebra Done Right. Springer, New York (2015)
6. Baldwin, J.: Model Theory and the Philosophy of Mathematical Practice: Formalization Without Foundationalism. Cambridge University Press, Cambridge (2018)
7. Blyth, T.S., Robertson, E.F.: Basic Linear Algebra, 2nd edn. Springer, London (2005)
8. Bochnak, J., Coste, M., Roy, M.-F.: Real Algebraic Geometry. Springer, Berlin (1998)
9. Buechler, S.: Essential Stability Theory. Cambridge University Press, Cambridge (2017)
10. Button, T., Walsh, S.: Philosophy and Model Theory. Oxford University Press, New York (2018)
11. Cameron, P.J.: Groups of order-automorphisms of the rationals with prescribed scale type. J. Math. Psychol. **33**, 163–171 (1989)
12. Cox, D.A., Little, J., O'Shea, D.: Ideals, Varieties, and Algorithms, 5th edn. Springer, New York (2015)
13. Corry, L.: Modern Algebra and the rise of Mathematlcal Structures. Birkhäuser, Basel (2004)
14. Dummit, D.S., Foote, R.M.: Abstract Algebra, 3rd edn. Wiley, New York (2004)
15. Eckert, D., Herzberg, F.S.: The birth of social choice theory from the spirit of mathematical logic: Arrow's theorem in the framework of model theory. Studia Logica **106**, 893–911 (2018)
16. Enderton, H.B.: Elements of Set Theory. Academic Press, New York (1977)
17. Enderton, H.B.: A Mathematical Introduction to Logic. Academic Press, New York (2001)
18. Enderton, H.B.: Computability Theory: An Introduction to Recursion Theory. Academic Press, New York (2011)
19. Goldblatt, R.: Lectures on the Hyperreals. Springer, New York (1998)
20. Gouvea, S.: p-adic Numbers. Springer, New York (2010)
21. Gray, J.: A History of Abstract Algebra. Springer, Berlin (2018)
22. Hedman, S.: A First Course in Logic. Oxford University Press, New York (2004)
23. Herstein, N.I.: Topics in Algebra. Wiley, New York (1975)
24. Hewitt, E., Stromberg, K.: Real and Abstract Analysis, 1st edn. Springer, New York (1965)
25. Hodges, W.: Model Theory. Cambridge University Press, Cambridge (1993)
26. Hodges, W.: A Shorter Model Theory. Cambridge University Press, Cambridge (2002)

27. Hodges, W.: A Visit to Tarski's seminar on elimination of quantifiers. In: van Benthem, J., Gupta, A., Parikh, R. (eds,) Proof, Computation and Agency, Synthese Library 352, pp. 53–66. Springer, New York (2011)
28. Hölder, O.: Die Axiome der Quantität und die Lehre vom Mass. Berichte über die Verhandlungen der Königlich Sächsischen Gesellschaft der Wissenschaften zu Leipzig, Mathematisch-Physikalike Classe **53**, 1–64 (1901)
29. Howard, P., Rubin, J.: The axiom of choice and linearly ordered sets. Fundamenta Mathematicae **97**, 111–122 (1977)
30. Howie, J.M.: Fields and Galois Theory. Springer, London (2000)
31. Hrbacek, K., Jech, T.: Introduction to Set Theory, 3rd edn. Marcel Dekker, Basel (1999)
32. Hrbacek, K., Lessmann, O., O'Donovan, R.: Analysis with Ultrasmall Numbers. CRC Press, Boca Raton (2015)
33. Humphreys, J.F.: A Course in Group Theory. Oxford University Press, Oxford (1996)
34. Hungerford, T.W.: Algebra. Springer, New York (2000)
35. Kirby, J.: An Invitation to Model Theory. Cambridge University Press, Cambridge (2019)
36. Krantz, D.H., Luce, R.D., Suppes, P., Tversky, A.: Foundations of Measurement Volume I: Additive and Polynomial Representations. Wiley, New York (1971)
37. Lascar, D.: Perspectives historiques sur les rapports entre la théorie des modèles et l'algèbre. Revue d'histoire des Mathématiques **4**, 237–260 (1998)
38. Lauwers, L., Van Liedekerke, L.: Ultraproducts and aggregation. J. Math. Econ. **24**, 217–237 (1995)
39. Lewis, C.I.: Facts, systems, and the unity of the world. J. Phil. **20**, 141–151 (1923)
40. Luce, R.D., Krantz, D.H., Suppes, P., Tversky, A.: Foundations of Measurement Volume III: Representation, Axiomatization, and Invariance. Academic Press, New York (1990)
41. Luxemburg, W. A.: What is nonstandard analysis? Am. Math. Month. **80**, 38–67 (1973)
42. Manzano, M.: Model Theory. Oxford University Press, New York (1999)
43. Mal'tsev, A.I.: The Metamathematics of Algebraic Systems. North-Holland, Amsterdam (1971)
44. Marcja, A., Toffalori, C.: A Guide to Classical and Modern Model Theory. Springer, Dordrecht (2003)
45. Marker, D.: Model Theory: An Introduction. Springer, New York (2002)
46. Marker, D., Slaman T.A.: Decidability of the Natural Numbers with the Almost-All Quantifier. arXiv:math/0602415 (2006)
47. Mcintyre, A.: Model completeness. Stud. Logic Found. Math. **90**, 139–180 (1977)
48. Mcintyre, A., McKenna, K., van den Dries, L.: Elimination of quantifiers in algebraic structures. Adv. Math. **47**, 74–87 (1983)
49. McPherson D.: Sharply multiply homogeneous permutation groups, and rational scale types. Forum Mathematicum **8**, 501–508 (1996)
50. Meisters, G.H., Monk, J.D.: Construction of the reals via ultrapowers. Rocky Mountain J. Math. **3**, 141–158 (1973)
51. Morley, M.D.: Categoricity in power. Trans. Am. Math. Soc. **114**, 514–538 (1965)
52. Moschovakis, Y.: Notes on Set Theory. Springer, New York (2002)
53. Niederée, R.: On the reference to real numbers in fundamental measurement: A model-theoretic approach. In: Roskam, E., Suck, R. (eds.), Progress in Mathematical Psychology, pp. 3–23. North-Holland, Amsterdam (1987)
54. Niederée, R.: What do numbers measure? A new approach to fundamental measurement. Math. Soc. Sci. **24**, 237–276 (1992a)
55. Niederée, R.: Maßund Zahl: Logisch-modelltheoretische Untersuchungen zur Theorie der fundamentalen Messung. Peter Lang, Frankfurt a.m. (1992b)
56. Neumann, P.M.: The structure of finitary permutation groups. Archiv der Mathematik **27**, 3–17 (1976)
57. Prest, M.: Model Theory and Modules. Cambridge University Press, Cambridge (1988)
58. Prestel, A., Delzell, C.: Mathematical Logic and Model Theory: A Brief Introduction. Springer, London (2011)

59. Robinson, A.: Complete Theories. North-Holland, Amsterdam (1956)
60. Robinson, A.: Nonstandard Analysis. North-Holland, Amsterdam (1966)
61. Robinson, A.: Formalism 64. In: Proceedings of the 1964 International Congress for Logic, Methodology and Philosophy of Science, pp. 228–246. North-Holland, Amsterdam (1964)
62. Robinson, A.: From a formalist's point of view. Dialectica **23**, 45–49 (1969)
63. Robinson, A., Zakon, E.: Elementary properties of ordered abelian groups. Trans. Am. Math. Soc. **96**, 222–236 (1960)
64. Rothmaler, P.: Introduction to Model Theory. Taylor & Francis, New York (1999)
65. Saari, D.G.: Geometry of Voting. Springer, New York (1994)
66. Saari, D.G.: A new way to analyze paired comparison rules. Math. Oper. Res. **39**, 647–655 (2014)
67. Saari, D.G.: Basis for binary comparisons and non-standard probabilities. Phil. Trans. Roy. Soc. A **374**, 20150103 (2015)
68. Saari, D.G.: Arrow, and unexpected consequences of his theorem. Public Choice **179**, 133–144 (2019)
69. Saari, D.G.: Seeking consistency with paired comparisons: A systems approach. Theory Decis. **91**, 377–402 (2021)
70. Sarbadhikari, H., Srivastava, S.M.: A Course on Basic Model Theory. Springer, Singapore (2017)
71. Sinaceur, H.: Corps et modéles. Essai sur l'histoire de l'algébre rèelle. Vrin, Paris (1991)
72. Slinko, A.: Algebra for Applications, 2nd edn. Springer Nature, Cham (2017)
73. Smullyan, R.M., Fitting, M.: Set Theory and the Continuum Problem. Dover, New York (2010)
74. Tarski, A.: Contributions to the theory of models I. Indagationes Mathematicae **16**, 572–581 (1954)
75. Tarski, A.: Logic, Semantics, Metamathematics, 2nd edition. Translated by J.H. Woodger and edited by J. Corcoran. Hackett, New York (1983)
76. Van den Dries, L.: Tame Topologies and O-minimal Structures. Cambridge University Press, Cambridge (1998)
77. Van der Waerden, B.L.: A History of Algebra: From Al-Khwarizmi to Emmy Noether. Springer, Berlin, Heidelberg (1985)
78. Velleman, D.J.: How to Prove It: A Structured Approach. Cambridge University Press, Cambridge (2006)

Index

A
Absolute value, 111
Affine n-space, 261
Aggregation, 99
Algebraic, v
   closure (model-theoretic), 170
   closure of a field, 138
   element, 133
   formula, 171
   geometry, vi
Amalgamation property, 325
Arity, 24
Arrow's theorem, 102
Artin-Lang theorem, 316
Artin-Schreier theorem, 295
Ascending Chain Condition, 75
Assignment, 35
Atomic, vii
   formula, 28
Automorphism, 82
Axiomatisation, 64
Axiom of Choice, 372

B
Back-and-forth construction, 235
Bolzano-Weierstrass Theorem, 121
Boolean
   algebra, 127
   combination, 311

C
Cantor-Bernstein Theorem, xvi
Cardinal
   finite, 366
   infinite, 366
   number, xvi, 364
   regular, 372
   successor, 372
Cauchy sequence, 114
Chain, 190, 374
   elementary, 191
   union of, 190
Chevalley's theorem, 262
Commutative ring, 70
Compactness Theorem, 122
Completeness
   of ACF_0, 170
   of ACF_p, 170
   of DAG^+, 155
   of $\mathsf{DLO}_{-,-}$, 158
   of VS_K^∞, 154
   of $\mathsf{RCF}_<$, 304
   sequential, 114
Condorcet's paradox, 3
Connective, 29
Consequence, 43
Contraposition, xviii
Convergent sequence, 114
Coset, 60

D
Degree
   of a field extension, 132
   of a polynomial, 72
   of transcendence, 165
Diagram, 88
   elementary, 89
   positive, 374

Dimension
    of an irreducible variety, 280
    of a vector space, 69
    of models of a strongly minimal theory, 211
    model-theoretic, 173
Divisible hull, 142

E
Elementary
    bijection, 211
    class, 64
    embedding, 80
    equivalence, 81
    extension, 82
    partial map, 80
    substructure, 82
Embedding, 79
Equivalence relation, 4
Equivalent sentences, 41
Exchange
    lemma, 210
    property, 173

F
Field, 70
    algebraically closed, 139
    algebraic extension of, 134
    Archimedean ordered, 111
    of characteristic 0, 140
    finite extension of, 136
    formally real, 288
    of fractions, 141
    o-minimal, 305
    order closed, 293
    perfect, 256
    of positive characteristic, 140
    real closed, 293
    separably closed, 256
    simple extension of, 131
    transcendental extension of, 134
Filter, 102
    maximal, 105
Finite intersection property, 122
Formula
    closed, 32
    in disjunctive normal form, 41
    existential, 47
    open, 32
    in prenex form, 47
    primitive, 180
    universal, 47
Fraïssé limit, 325

Function
    injective, xiv
    rational, 141
    surjective, xiv

G
Galois connection, 63
Group, 19
    abelian, 19
    of automorphisms, 82
    divisible, 66
    oligomorphic, 243
    o-minimal, ordered, 215
    ordered, 67
    torsion-free, 66

H
Hereditary property, 322
Hilbert's basis theorem, 75
Hilbert's Nullstellensatz, 264
Hilbert's 17th problem, 316
Homomorphism, 78

I
Ideal, 71
    maximal, 71
    primary, 264
    prime, 74
    radical, 263
Inclusion map, 82
Induction
    arithmetical, xvi
    on the complexity of formulae, 30
    on the complexity of terms, 25
    transfinite, 348
Integral domain, 70
Interpretation, 55
Isomorphism, 80

J
Joint embedding property, 322

L
Language
    first-order, 29
    signature of, 28
    size of, 61
Lattice, 127
Limit
    point of a well-ordered set, 349

Index 523

Limit (of a sequence of real numbers), 114
Lindenbaum algebra, 127
Lindström's test, 192
Linear closure, 68
Linear independence, 68
Linear order, 3, 63
    dense, 156
    discrete, 183
Literal, 87
Łos' theorem, 107
Löwenheim-Skolem Theorem
    downward, 147
    upward, 149

M

Measurement
    scale, 334
    theory, 226
Milnor's Curve Selection theorem, 319
Minimal
    formula, 209
    structure, 52, 209
Model
    algebraically prime, 182, 205
    κ-saturated, 244
    κ-universal, 244
    monster, 249
    prime, 208
Morley degree, 272
Morley rank
    of a formula, 269
    of a type, 274
Morleysation, 197
m-point homogeneity, 334

N

Normal subgroup, 338
n-point uniqueness, 334

O

Omitting Types theorem, 239
Ordinal, 350
    finite, 356

P

Pairwise majority, 2
Polynomial
    elementary symmetric, 256
    irreducible, 134
    minimal, 136

    monic, 134
    symmetric, 256
Projection of an algebraic set, 262

Q

Quantifier elimination, 195
Quotient group, 60
Quotient ring, 71

R

Ring
    nilradical of, 263
    Noetherian, 73
    of polynomials, 72
Ryll-Nardzewski theorem, 242

S

Sentence, 32
Set
    algebraic, 261
    constructible, 261
    definable, 52
    open, 221
    partially ordered, 347
    semi-algebraic, 311
    well-ordered, 347
Skolem functions, 198
Stone space, 222
Strongly minimal
    formula, 209
    structure, 209
    theory, 209
Structure
    algebraically ω-homogeneous, 324
    atomic, 237
    existentially closed, 180
    homogeneous, 232
    κ-homogeneous, 232
    strongly κ-homogeneous, 232
    ultrahomogeneous, 325
Subformula, 31
Successor
    point of a well-ordered set, 349

T

Tarski-Seidenberg theorem, 311
Tarski-Vaught test, 85
Term, 25
    closed, 27

Theory, 63
  of algebraically closed fields, 139
    of characteristic 0, 141
    of characteristic p, 141
  of a class of structures, 64
  complete, 64
  decidable, 308
  of dense linear orders without endpoints, 156
  finitely axiomatisable, 64
  of infinite sets, 65
  κ-categorical, 151
  κ-stable, 285
  model-complete, 178
  o-minimal, 215
  of ordered, divisible abelian groups, 67
  of real closed fields, 296
  of real closed ordered fields, 297
  small, 247
  substructure complete, 204
  of torsion-free, abelian groups, 66
  of torsion-free, divisible abelian groups, 66
  of vector spaces over a field $\mathcal{K}$, 68
Topological compactness, 222
Topological space, 221
  totally disconnected, 222
Transfinite recursion, 358
Type
  algebraic, 223
  complete, 220
  isolated, 223

U
Ultrafilter, 102
  principal, 102
Ultrapower, 109
Ultraproduct, 106
Unnested
  atomic formula, 29
  term, 29

V
Variable, 24
  bound, 31
  free, 31
Variety
  algebraic, 261
  generic point of, 282
  irreducible, 279
Vaught's test, 151
Vector
  cyclic, 13
  space, 68
  strongly transitive, 10

Z
Zorn's lemma, 374

GPSR Compliance

The European Union's (EU) General Product Safety Regulation (GPSR) is a set of rules that requires consumer products to be safe and our obligations to ensure this.

If you have any concerns about our products, you can contact us on

ProductSafety@springernature.com

In case Publisher is established outside the EU, the EU authorized representative is:

Springer Nature Customer Service Center GmbH
Europaplatz 3
69115 Heidelberg, Germany

www.ingramcontent.com/pod-product-compliance
Ingram Content Group UK Ltd.
Pitfield, Milton Keynes, MK11 3LW, UK
UKHW020111240426

470311UK00006B/34